MYOCARDIOLOGY

Proceedings of the Third Annual Meeting of the
INTERNATIONAL STUDY GROUP FOR RESEARCH IN CARDIAC METABOLISM
and of the Conference on "Preventive Myocardiology" of the
AMERICAN COLLEGE OF CARDIOLOGY

Third Annual Meeting of the

INTERNATIONAL STUDY GROUP FOR RESEARCH IN CARDIAC METABOLISM

29 June – 1 July, 1970,
Stowe, Vermont, U.S.A.

Chairman of Organization Committee

G. Rona, *M.D., Ph.D., F.R.C.P. (C)*
Director of Laboratories, The Lakeshore General Hospital; Professor of Pathology, McGill University, Montreal, Canada

Program Chairman

E. Bajusz, *M.D., Ph.D., F.A.C.C.*
Director, Division III, Bio-Research Institute, Cambridge and Boston, Massachusetts; Professor (Invited), Department of Pathology, University of Montreal, Montreal, Canada

Organization Committee

J. B. Caulfield, *M.D.*
Pathologist, Department of Pathology, Massachusetts General Hospital, Boston, Massachusetts

H. A. Heggtveit, *M.D.*
Associate Professor of Pathology, Faculty of Medicine, University of Ottawa, Ottawa, Canada

K. I. Melville, *M.Sc., M.D.C.M.*
Professor, Department of Pharmacology and Therapeutics, McGill University, Montreal, Canada

A. Schwartz, *B.S., M.S., Ph.D.*
Professor and Head, Division of Myocardial Biology, Baylor College of Medicine, Texas Medical Center, Houston, Texas

Local Committee

N. R. Alpert, *Ph.D.*
Professor and Chairman, Department of Physiology and Biophysics, University of Vermont, College of Medicine, Burlington, Vermont

H. C. Herrlich, *Ph.D.*
Cardiovascular Research Unit, Medical Center Hospital of Vermont, DeGoesbriand Unit, University of Vermont, Burlington, Vermont

Conference on "Preventive Myocardiology" of the

AMERICAN COLLEGE OF CARDIOLOGY

26-28 June, 1970, Stowe, Vermont

Program Director

W. Raab, *M.D., F.A.C.C.*
Emeritus Professor of Experimental Medicine, Department of Medicine, University of Vermont, College of Medicine, Burlington, Vermont

Co-Directors

E. Bajusz, *M.D., Ph.D., F.A.C.C.*
Director, Division III, Bio-Research Institute, Cambridge and Boston Massachusetts; Professor (Invited), Department of Pathology, University of Montreal, Montreal, Canada

E. Lepeschkin, *M.D., F.A.C.C.*
Professor of Medicine, Director, Cardiovascular Research Unit, University of Vermont, College of Medicine, Burlington, Vermont

Recent Advances in Studies on Cardiac Structure and Metabolism, Volume 1

MYOCARDIOLOGY

edited by

E. Bajusz, M.D., Ph.D.
Boston

G. Rona, M.D., Ph.D.
Montreal

UNIVERSITY PARK PRESS
Baltimore • London • Tokyo

Library of Congress Catalog Card No. 73-171251
ISBN 0-8391-0095-7

Printed in the United States of America
by Port City Press, Inc., Baltimore, Maryland

Library of Congress Cataloging in Publication Data
Main entry under title:

Myocardiology.

(Recent advances in studies on cardiac structure and metabolism, v. 1)
"Proceedings of the third annual meeting of the International Study Group for Research in Cardiac Metabolism and of the Conference on 'Preventive Myocardiology' of the American College of Cardiology."
Includes bibliographical references.
1. Heart–Muscle–Congresses. 2. Heart–Muscle–Diseases–Congresses. I. Bajusz, Eörs, ed. II. Rona, George, 1924– ed. III. International Study Group for Research in Cardiac Metabolism. IV. American College of Cardiology. V. Conference on "Preventive Myocardiology," Stowe, Vt., 1970. VI. Series.
RC667.R4 vol. 1 RC685.M9 616.1′2 73-171251
ISBN 0-8391-0095-7

Contents

Dedication

His colleagues profoundly regret that the Stowe meetings, for the success of which Dr. Wilhelm Raab so wholeheartedly worked, have proved to be the last conferences he ever attended. The ISRCM suffered the grievous loss of one of its most distinguished founding members in the untimely death, on September 21, 1970, of Dr. Raab. His widely known scientific contributions to the development of modern cardiology are summarized by Dr. David Lehr in the present volume. Dr. Raab is remembered by countless friends not only as a brilliant investigator and stimulating teacher but also as a modest man of culture with a ready sense of humor and broad interests beyond his area of research. It is a fitting tribute to his memory that these Proceedings, which cover a number of subjects to which Dr. Raab made such major contributions, should be dedicated to his honor.

Contributing Authors

Abelmann, W. H., *Boston, Mass.* (USA)
Agostoni, A., *Milan* (Italy)
Albin, R., *Chicago, Ill.* (USA)
Allen, J. C., *Houston, Tex.* (USA)
Alousi, A. A., *Ardsley, N.Y.* (USA)
Angelakos, E. T., *Philadelphia, Pa.* (USA)
Apstein, C. S., *New York, N.Y.* (USA)
Aschenbrenner, V., *Chicago, Ill.* (USA)
Bajusz, E., *Boston, Mass.* (USA)
Balazs, T., *Philadelphia, Pa.* (USA)
Baroldi, G., *Milan* (Italy)
Bednarik, B., *Brno* (Czechoslovakia)
Beards, J. A., *Ardsley, N.Y.* (USA)
Bencosme, S. A., *Kingston, Ont.* (Canada)
Berger, J. M., *Montreal, Que.* (Canada)
Beznak, M., *Ottawa, Ont.* (Canada)
Binaghi, P. C., *Milan* (Italy)
Bing, R. J., *Pasadena, Calif.* (USA)
Blanchaer, M. C., *Winnipeg, Man.* (Canada)
Blascheck, J. A., *Montreal, Que.* (Canada)
Brachfeld, N., *New York, N.Y.* (USA)
Brusis, O. A., *Burlington, Vt.* (USA)
Buja, L. M., *Bethesda, Md.* (USA)
Cantin, M., *Montreal, Que.* (Canada)
Carballo, L. C., *Philadelphia, Pa.* (USA)
Case, R. B., *New York, N.Y.* (USA)
Chau, R., *New York, N.Y.* (USA)
Cipriano, P. R., *New Orleans, La.* (USA)
Coleman, H. N., *Rochester, Minn.* (USA)
Connor, W. E., *Iowa City, Iowa* (USA)
Crass, M. F., III, *Omaha, Neb.* (USA)
Cronin, R. F. P., *Montreal, Que.* (Canada)
Crow, C. A., *Houston, Tex.* (USA)
Daniels, J. B., *Philadelphia, Pa.* (USA)
De la Iglesia, F. A., *Sheridan Park, Ont.* (Canada)
Detweiler, D. K., *Philadelphia, Pa.* (USA)
Dhalla, N. S., *Winnipeg, Man.* (Canada)
Dusek, J., *Montreal, Que.* (Canada)
Ettinger, P. O., *Newark, N.J.* (USA)
Evans, C., *New York, N.Y.* (USA)
Evans, M. E., *London* (England)
Fedelesova, M., *Bratislava* (Czechoslovakia)
Ferrans, V. J., *Bethesda, Md.* (USA)
Fizel, A., *Bratislava* (Czechoslovakia)
Fizelova, A., *Bratislava* (Czechoslovakia)
Friedman, I., *Jerusalem* (Israel)
Frollo, I., *Bratislava* (Czechoslovakia)
Gardell, C., *Montreal, Que.* (Canada)
Garvey, H. L., *Ottawa, Ont.* (Canada)
Gertz, E. W., *Boston, Mass.* (USA)
Gery, I., *Jerusalem* (Israel)
Gmeiner, R., *New York, N.Y.* (USA)
Goodman, T., *Ottawa, Ont.* (Canada)
Grice, H. C., *Ottawa, Ont.* (Canada)
Gudbjarnason, S., *Detroit, Mich.* (USA)
Gueyikian, I., *New York, N.Y.* (USA)
Gunning, J. F., *Rochester, Minn.* (USA)
Hames, C. G., *Claxton, Ga.* (USA)
Harigaya, S., *Houston, Tex.* (USA)
Hasnain, S., *Ottawa, Ont.* (Canada)
Heggtveit, H. A., *Ottawa, Ont.* (Canada)
Heier, G., *Berlin* (German Democratic Republic)
Henderson, A. H., *Boston, Mass.* (USA)
Hibbs, R. G., *Bethesda, Md.* (USA)
Hoak, J. C., *Iowa City, Iowa* (USA)
Holec, V., *Bratislava* (Czechoslovakia)
Hollenberg, C. H., *Toronto, Ont.* (Canada)
Horvath, E., *Montreal, Que.* (Canada)
Hunt, B., *Ottawa, Ont.* (Canada)
Huttner, I., *Montreal, Que.* (Canada)
Jacobson, B. E., *Winnipeg, Man.* (Canada)
Jeffrey, F. E., *Boston, Mass.* (USA)
Jennings, R. B., *Chicago, Ill.* (USA)
Jesrani, M. U., *Newark, N.J.* (USA)
Judd, J. T., *Cincinnati, Ohio* (USA)
Kako, K. J., *Ottawa, Ont.* (Canada)
Kaplan, J., *New York, N.Y.* (USA)
Khan, M. I., *Newark, N.J.* (USA)
Kikuchi, T., *Ottawa, Ont.* (Canada)
King, M. P., *Philadelphia, Pa.* (USA)
Klassen, G. A., *Montreal, Que.* (Canada)
Klein, I., *New York, N.Y.* (USA)
Korecky, B., *Ottawa, Ont.* (Canada)
Kovacs, K., *Montreal, Que.* (Canada)
Krajny, M., *Montreal, Que.* (Canada)
Krause, E. G., *Berlin* (German Democratic Republic)
Laufer, A., *Jerusalem* (Israel)
Lehr, D., *New York, N.Y.* (USA)
Leone, A., *Pisa* (Italy)
Lepeschkin, E., *Burlington, Vt.* (USA)
Levitsky, S., *Bethesda, Md.* (USA)
Luginbuhl, H., *Bern* (Switzerland)
Lumb, G., *Toronto, Ont.* (Canada)
Lutmer, R. F., *Cincinnati, Ohio* (USA)
Maca, R. D., *Lincoln, Neb.* (USA)
Markov, A., *Newark, N.J.* (USA)
Mathes, P., *Detroit, Mich.* (USA)
McCollum, W. B., *Houston, Tex.* (USA)
McIntosh, C. L., *Bethesda, Md.* (USA)

Meier, R., *Claxton, Ga.* (USA)
Melville, K. I., *Montreal, Que.* (Canada)
Minelli, R., *Pavia* (Italy)
More, R. H., *Montreal, Que.* (Canada)
Munro, I. C., *Ottawa, Ont.* (Canada)
Nadkarni, B. B., *Ottawa, Ont.* (Canada)
Nair, K. G., *Chicago, Ill.* (USA)
Ojiambo, H. P., *Nairobi* (Kenya)
Oldewurtel, H. A., *Newark, N.J.* (USA)
Olewine, D. A., *Statesboro, Ga.* (USA)
Olson, R. E., *St. Louis, Mo.* (USA)
Onnen, K., *Berlin* (German Democratic Republic)
Opie, L. H., *London* (England)
Oratz, M., *New York, N.Y.* (USA)
Oster, K. A., *Bridgeport, Conn.* (USA)
Patterson, D. F., *Philadelphia, Pa.* (USA)
Petrie, R., *Montreal, Que.* (Canada)
Poupa, O., *Goteborg* (Sweden)
Raab, W., *Burlington, Vt.* (USA)
Rabinowitz, M., *Chicago, Ill.* (USA)
Radice, F., *Milan* (Italy)
Ramsey, F. H., *Statesboro, Ga.* (USA)
Ravens, K. G., *Detroit, Mich.* (USA)
Regan T. J., *Newark, N.J.* (USA)
Roberts, W. C., *Bethesda, Md.* (USA)
Rona, G., *Montreal, Que.* (Canada)
Rose, K. D., *Lincoln, Neb.* (USA)
Rothschild, M. A., *New York, N.Y.* (USA)
Russek, H. I., *New York, N.Y.* (USA)
Russek, L. G., *New York, N.Y.* (USA)
Safer, B., *Philadelphia, Pa.* (USA)
Salem, F. A., *Ottawa, Ont.* (Canada)
Saroff, J., *Cincinnati, Ohio* (USA)
Scheuer, J., *Pittsburgh, Pa.* (USA)
Schreiber, S. S., *New York, N.Y.* (USA)
Schwartz, A., *Houston, Tex.* (USA)
Seelig, M. S., *Bloomfield, N.J.* (USA)
Selye, H., *Montreal, Que.* (Canada)
Shen, A. C., *Chicago, Ill.* (USA)
Shipp, J. C., *Gainesville, Fla.* (USA)
Simpson, M. T., *Augusta, Ga.* (USA)
Siska, K., *Bratislava* (Czechoslovakia)
Skosey, J. L., *Chicago, Ill.* (USA)
Sobel, H., *Denver, Colo.* (USA)
Solymoss, B., *Montreal, Que.* (Canada)
Sonnenblick, E. H., *Boston, Mass.* (USA)
Sordahl, L. A., *Houston, Tex.* (USA)
Sosa-Lucero, J. C., *Sheridan Park, Ont.* (Canada)
Stabilini, R., *Milan* (Italy)
Stam, A., Jr., *Boston, Mass.* (USA)
Taichman, G. C., *Ottawa, Ont.* (Canada)
Theodosis, D., *Montreal, Que.* (Canada)
Thomas, M., *London* (England)
Toth, S., *Montreal, Que.* (Canada)
Urbanek, E., *Brno* (Czechoslovakia)
Ursick, J. A., *Lincoln, Neb.* (USA)
Varga, S., *Montreal, Que.* (Canada)
Vasku, J., *Brno* (Czechoslovakia)
Wagner, R., *Boston, Mass.* (USA)
Warner, E. D., *Iowa City, Iowa* (USA)
Wexler, B. C., *Cincinnati, Ohio* (USA)
Williams, W. H., *Bethesda, Md.* (USA)
Williamson, J. R., *Philadelphia, Pa.* (USA)
Wollenberger, A., *Berlin* (German Democratic Republic)
Wrogemann, K., *Winnipeg, Man.* (Canada)
Yates, J. C., *Winnipeg, Man.* (Canada)
Young, R. C., *London* (England)
Zak, R., *Chicago, Ill.* (USA)
Ziegelhoffer, A., *Bratislava* (Czechoslovakia)
Zsoter, T., *Toronto, Ont.* (Canada)

Editorial Foreword

Since 1967 the International Study Group for Research in Cardiac Metabolism (ISRCM) has sponsored annual meetings and symposia to coordinate world-wide research in myocardiology (i.e., studies in cardiac metabolism and in structural-metabolic-functional interrelationships). The participants at the Third Annual Meeting, which was the first international congress of this young organization on the American continent, were greeted by the following message received from the President of the United States of America:

THE WHITE HOUSE

WASHINGTON

June 17, 1970

It gives me great pleasure to welcome and greet the participants in the Third Annual -- and first American -- Meeting of the International Study Group for Research in Cardiac Metabolism.

In recent years we have witnessed unparalleled progress against heart disease. This progress has not come easily. It represents a culmination of the work of many scientists from many laboratories and many nations. And your presence here emphasizes that the frontiers of knowledge are being further extended through cooperative international research efforts. We look to your organization, and others like it, to foster and stimulate the useful exchange of ideas and information that is so basic to our continuing success in meeting a challenge that faces all of us.

My best wishes to each of you for a successful and productive meeting.

Richard Nixon

Medical research is indeed not confined by frontiers. International collaboration is most essential in achieving further progress.

Such progress, however, brings with it new challenges. For example, the current information explosion has created a gap between the knowledge that investigative myocardiology has accumulated, and the delivery and application of this knowledge by the clinical cardiologist at the practical level.

Our challenge is thus to translate the experimental findings into practical and clinically applicable methods. This is most obvious and urgent with respect to the prevention of myocardial disease. Therefore, upon the suggestion of the late Dr. Wilhelm Raab, the Third Annual Meeting of the ISRCM was held in Stowe, Vermont, conjointly with the Conference on "Preventive Myocardiology" of the American College of Cardiology. The present volume is a carefully chosen selection of the important information presented at these two interrelated meetings.

On behalf of the Organization Committee and the Officers of the ISRCM, we wish to express our gratitude to the following pharmaceutical companies and research institutions for their generous financial support, which helped to defray the costs of the Third Annual Meeting:

Ayerst Laboratories, Montreal, Canada
Ayerst, International, New York, N.Y.
Eli Lilly & Co., Indianapolis, Indiana
Geigy Pharmaceuticals, Ardsley, N.Y.
ICI Pharmaceutical Division, Alderley Park, Macclesfield, Cheshire, England
Knoll Pharmaceutical Co., Orange, N.J.
Mason Research Institute, Worcester, Mass.
Merck Sharp & Dohme of Canada Ltd., Montreal, Canada
Merck Sharp & Dohme Co., West Point, Pa.
Schering Corporation, Bloomfield, N.J.
Smith Kline & French Laboratories of Canada, Montreal, Canada
Smith Kline & French Laboratories, Philadelphia, Pa.
Spezialchemie GmbH & Co., Munich, West Germany
Warner-Lambert Research Institute, Morris Plains, N.J.

We would also like to acknowledge the assistance of those who served on the Organization Committee and on the Local Committee. We are extremely grateful to Mrs. G. Henderson for her superb secretarial assistance and to Mrs. B. Sacra for her indispensable aid in editing this volume. Our sincere thanks are due also to the editorial staff of University Park Press for much help and tolerance.

Eörs Bajusz and George Rona

REMARKS BY DR. DAVID LEHR AT THE CLOSING SESSION OF THE SEMINAR ON PREVENTIVE MYOCARDIOLOGY OF THE AMERICAN COLLEGE OF CARDIOLOGY ON JUNE 28, 1970, IN STOWE, VERMONT

Ladies and Gentlemen:

I stand before you as just one member of the audience and as one participant in this exciting and richly rewarding meeting, to express, in your name, our deepest thanks to the man who is the spirit behind this conference, who made it all possible, and who worked untiringly to ensure its success.

Permit me to impose upon your time for some additional minutes for the purpose of saying a few words about this man, a fellow Austrian and scientist whom I deeply admire and respect.

Dr. Wilhelm Raab is now 75 years old. He has thus continued to work untiringly far beyond his obligatory retirement age. Dr. Raab is a very modest, selfless, self-effacing man. Like Baron Trapp of Austria, who was one of his close friends, he is fiercely independent and antiauthoritarian. He loves freedom and liberty more than his personal safety and security. He is a man who has been uncompromising all his life in his adherence to the highest principles of ethics and morality. Some of you may have seen the motto in the Trapp Family Lodge here in Stowe. It says, "Do unto others as though you were the others." This is indeed Dr. Raab's principle. Throughout his life, he has unwaveringly followed the precept that Shakespeare so beautifully expressed in Hamlet, "This above all, to thine own self be true."

He left Austria, the country of his birth, which he deeply loved, voluntarily in 1939, because he did not want to live in a Nazi country and because he wanted to identify with the persecuted and take a personal stand against tyranny.

Only a few weeks ago, I was privileged to listen to an autobiographical speech that Dr. Raab gave before the Austrian Medical Society at a testimonial dinner held in his honor. His deeply honest and beautiful thoughts still reverberate in my mind.

It must be a source of deep satisfaction for Dr. Raab to find, in a conference such as this one, that the scientific concepts which he stood for almost alone for so many years have borne fruit and have finally come to be recognized all over the world. Surely, his preoccupation with the metabolism of the myocardium may not have been the most spectacular and dramatic subject, and his tenacious preaching about the cardiotoxic effect of the catecholamines, in particular, did take a very long time to gain acceptance. I heard him many years ago speak of himself as a catecholamaniac. There is no doubt in my mind that this stubborn perseverance and belief in the correctness of his ideas contributed much to their ultimate acceptance.

Yet, Dr. Raab's pioneering scientific contributions reflect only one part of his multifaceted personality. Many of you are probably unaware that Dr. Raab is an excellent violinist, accomplished painter, and philosopher in his own right; in other words, a humanist in the finest sense of the word.

As you probably know, he is leaving us soon, since he has recently remarried and intends to make his new home in Munich. But I am sure that he leaves us only physically; he will remain in close contact. It is also comforting to know that his two sons are following in his footsteps; they are both accomplished scientists and on their way toward emulating their great father.

Some day, I hope that Dr. Raab can be persuaded to write his autobiography. It would make fascinating reading. Many men in highest places have been his close friends. Some of you may know that he was appointed to the White House Conference on Physical Fitness by President Kennedy.

Finally, I would like to ask Dr. Raab to forgive these ramblings. If they sound disjointed, I may offer in apology that these are off-the-cuff remarks. I did not even have an inkling that this honor would come my way when I entered this hall one hour ago. But I am sure that he can feel that my words come from the heart, and the heart of the matter is a matter of heart. Is there a better way to end one's remarks at a meeting on the heart?

Ladies and Gentlemen, may I ask you to honor Dr. Raab by a standing ovation.

Mailing address: David Lehr, M.D. Dept. of Pharmacology New York Medical College, Flower and Fifth Avenue Hospitals, Fifth Avenue at 106th St., New York, N.Y. 10029 (USA).

University of Vermont, Department of Medicine,
Burlington, Vermont, USA

WHY "MYOCARDIOLOGY?"

W. RAAB†

A conspicuous paradox in today's Western medicine is the perennial disproportion between the alarmingly high incidence of premature disability and death from degenerative heart disease, on the one hand, and the lack of large-scale, organized educational and practical preventive initiative and counteraction on the other.

The United States of America, in particular, occupies a leading position, not only in curative achievements and efforts, including the development of artificial hearts, but also in the mass production of civilization-induced "artificial" heart disease.

As far as active preventive measures are concerned, only very little has been done here, up to the present, except for the Surgeon General's and the American Heart Association's anti-smoking campaigns, and for some rather casual public admonitions with regard to diet and physical activity, (7).

By contrast, massive practical prophylactic programs are being carried out abroad in countries much less afflicted with heart disease, and with less sophisticated populations. In such places, highly effective prevention-oriented professional and lay educational systems, thousands of rural reconditioning centers (5), clinical prophylactoria, vast popular sports organizations, and the prevalence of less body-softening living habits have contributed to the maintenance of cardiac health conditions that are far superior to those in the environment of affluent Western supercivilization.

The reasons given by responsible agencies and individuals for their reluctance to take determined remedial steps in our own permanent national cardiac emergency are routinely summarized in the statement that "more research is needed" before systematic nationwide action can be recommended or legislated here.

No one will deny the desirability of more research and, in fairness to existing though limited initiative in this respect, it should be pointed out that the National Heart Institute did commit more than 50 million dollars to the National Diet-Heart Study and the Coronary Drug Project, two monumental undertakings planned for periods of five to nine years and aimed entirely at an assessment of the presumably antiatherogenic effectiveness of certain dietary measures and drugs. Unfortunately, the National Diet-Heart Study was recently abandoned for budgetary reasons.

† Deceased September 21, 1970

Both of the projects just named illustrate a decisive short-coming in the current "official" approach to the problem of degenerative heart disease. They represent a perpetual adherence to traditional, but outdated concepts that imply a purely, or almost purely, coronary vascular origin of fatal myocardial lesions. But until recently, only minimal if any attention has been focused on those at least equally important, nonvascular cardiometabolic factors that participate fundamentally in the broad pathogenic spectrum of pluricausal myocardial, so-called coronary heart disease.

Despite the long recognized inadequacy of the exclusively mechanistic vascular interpretation of degenerative heart disease (4, 5, 7), the apparent appeal of its comparative simplicity as well as a virtually ineradicable ambiguity and confusion in clinical semantics have created a firm psychological block among most cardiologists and other physicians in dealing with the complex nonvascular aspects of the heart muscle's degenerative pathology.

Regardless of the variously reported nonexistence of thrombi and vascular occlusions in up to more than 50% of myocardial so-called infarctions (2, 3, 10), and regardless of frequent gross discrepancies between the incidence, degree and location of coronary vascular *vs.* myocardial structural lesions, such terms as "coronary occlusion," "coronary thrombosis." "coronary atherosclerosis," "coronary heart disease," "coronary artery disease" or plainly "a coronary" are indiscriminately and interchangeably used in clinical practice and textbooks and, especially, in epidemiologic reports. Unfortunately, some of the latter, in referring to nonautopsied patient material, glibly employ the terms "athero-sclerotic heart disease" or "occlusive coronary disease" or "atherosclerosis," without any clear proof of the presence or absence of arterial occlusion, or of the degree and actual pathogenic involvement and significance of existing coronary lesions in the individual instances used for statistical evaluation.

Much of this chaotic state of current terminology is caused by the power of a prejudicial and often misleading semantic tradition, and by the difficulty of overcoming it by coping with the total multidisciplinary problem-complex in question, rather than with incoherent detailed observations and nonintegrated splinter facts. These unfavorable psychological conditions are reinforced by illogical, fashionable expressions, such as for example, "coronary care unit" (referring to highly useful organizations where much is being done for the care of the heart muscle but practically nothing whatsoever for the care of the coronary arteries). The confusion is maintained and aggravated by the unfamiliarity of many specialized fraction experts and policymakers with masses of essential, but widely scattered and disjointed, data from the extensive world literature. Thus, numerous important products of the current information explosion remain frustratingly unutilized.

It hardly seems necessary to point out that a rational approach to preventive action against any disease requires, first of all, an at least rudimentary insight into those basic pathogenic factors at which preventive measures are specifically to be aimed. For this reason, our Seminar is intended to serve as an interim guide toward identification and integration of the manifold cardiometabolic pathogenic elements of degenerative heart disease, and toward logically deducible broad principles of practical primary and secondary preventability. The task was made difficult by the volume and diversity of the pertinent international literature, and of the huge body of significant experimental and clinical observations that have accumulated in the last decades. Hope for a solution of many still unanswered important questions is being dimmed rather than brightened by the increasingly bewildering mass of more and more specified, and less and less coordinated, details. Full preoccupation with the perfection of highly sophisticated research techniques inevitably diverts the vision of some of the most capable minds from the scope and essence of those great problem-complexes, into the texture of which countless disorganized details need to be fitted in order to acquire their real usefulness and the credit they deserve. The present lull in government-supported research may offer a fruitful opportunity for a creative reviewal and digestion of much information that has been produced in the past without receiving sufficient attention.

We believe that, at this juncture, the human, social, and economic need of scientifically rational preventive action against rampant degenerative heart disease justifies an attempt to clamber out of a paralyzing maze of "micro-facts" into the wider, open spaces of integrative "macro-concepts," where the specific targets for immediate and future practical action should become more clearly discernible.

Because of the essential differences that separate the problems concerning prevention of vascular, i.e., coronary arterial disease, from those concerning the heart itself, i.e., the myocardium, the distinctive use of the terms "preventive angiology" *vs.* "preventive cardiology," or, even more specifically, "preventive myocardiology," is being proposed for the sake of precision in reasoning and communication. The papers presented at our Seminar will concentrate primarily on preventive myocardiology in the logical sense of the term.

Profound, potentially pathogenic, and mutually aggravating influences of sympathoadrenal catecholamines and of adrenal corticoids upon the oxygen economy and electrolyte balance of the heart muscle have been intensively investigated over many years (1, 6–9). They will be considered as specific targets for preventive action against so-called CHD. This popular, but somewhat worn-out cliché for "coronary heart disease" may more appropriately be read as meaning "cardiac hypoxic dysionism," in keeping with present-day knowledge and concepts (7).

REFERENCES

1. BAJUSZ, E.: Electrolytes and Cardiovascular Diseases. (Williams & Wilkins, Baltimore, Vol. I, II 1965/6).

2. BAROLDI, G. and SCOMAZZONI, G.: Coronary Circulation in the Normal and Pathologic Heart (Dept. of the Army, Office of the Surgeon General, U.S. Goverment Printing Office 1967).
3. EHRLICH, J. C. and SHINOHARA, Y.: Low incidence of coronary thrombosis in myocardial infarction. Arch. Pathol. *87:* 432–437 (1964).
4. MYASNIKOV, A. L.: Myocardial necroses of coronary and noncoronary genesis. Amer. J. Cardiol. *13:* 35–48 (1964).
5. RAAB, W. (Ed.): Prevention of Ischemic Heart Disease (C. C Thomas, Springfield, Ill. 1966).
6. RAAB, W.: Myocardial electrolyte derangement–crucial feature of pluricausal–so-called "coronary" heart disease (dysionic cardiopathy). Ann. N. Y. Acad. Sci. 147, Art. *17,* 627–686 (1969).
7. RAAB, W.: Preventive Myocardiology–Fundamentals and Targets (C. C Thomas, Springfield, Ill. 1970).
8. SELYE, H.: The Chemical Prevention of Cardiac Necroses (Ronald Press, New York 1958).
9. SELYE, H.: The Pluricausal Cardiopathies (C. C Thomas, Springfield, Ill. 1961).
10. SPAIN, D. M. and BRADESS, V. A.: The relationship of coronary thrombosis to coronary atherosclerosis and ischemic heart disease. Amer. J. Med. Sci. *240:* 701–715 (1960).

Myocardial Metabolism, the Development of a New Science

INTRODUCTORY REMARKS

RICHARD J. BING

Professor of Medicine,
University of Southern California Medical School;
President of the International Study Group
for Research in Cardiac Metabolism,
Huntington Memorial Hospital,
Pasadena, California, USA

Studies on cardiac metabolism are based on the broad studies of Lavoisier, Helmholtz, Voit, and Rubner. In 1927, Tigersted published in his *Physiology of Circulation* the first comprehensive chapter on cardiac metabolism. From these beginnings, work on cardiac metabolism progressed to the heart-lung preparation, and in the chapter of Evans in 1939, in *Recent Advances of Physiology,* data can be found on the oxygen usage of the isolated heart and the heart-lung preparation. In 1946, catheterization of the coronary sinus permitted studies on the human heart beating in its normal environment *in situ* and on the measurement of coronary blood flow. Within recent years, nutritional coronary blood flow could be directly determined without intubation of the coronary sinus. Studies utilizing coronary sinus catheterization have revealed the ability of the heart to extract foodstuffs from the blood. This work has progressed to an investigation of enzymatic pathways in heart muscle, and of synthesis and turnover of materials in cardiac muscle. These studies have become closely connected with cardiac contractility and its relation to metabolism.

The field of cardiac metabolism, therefore, has progressed from *in vitro* to *in vivo* preparation and, finally, to the study of intermediary metabolism in heart muscle. The future will see further application of biochemical and biophysical methods to the study of cardiac metabolism. For the clinician, a study of this growing field has become important, primarily because of a more thorough understanding of the mechanisms underlying heart failure, myocardial infarction, and cardiomyopathies.

Division of Myocardial Biology, Baylor College of Medicine,
and Fondren-Brown Cardiovascular Research and Training Center,
The Methodist Hospital, Houston, Texas, USA

STUDIES ON MITOCHONDRIA AND THE MUSCLE RELAXING SYSTEM [1]

A. SCHWARTZ,[2] L. A. SORDAHL, and
W. B. McCOLLUM

The cardiac muscle cell of most mammalian species, including human, is characterized by numerous mitochondria, sometimes up to 40% of dry weight. A number of different isolation procedures are described, along with the ultrastructural characteristics and biochemical functions of each. It is concluded that when mitochondria are isolated in a medium containing KCl and EDTA, both morphological and biochemical integrity are maintained for relatively long periods of time. A portable perfusion chamber is described, which maintains both the heart and lungs in an isolated condition for protracted periods of time. Insulin, KCl, and glucose are added to the perfusing medium (blood) in order to maintain viability. Mitochondria isolated from both human and animal hearts thus maintained exhibited a remarkable retention of respiratory control, oxygen consumption, and ADP : O ratios for up to 1 week. Even after significant depression of respiratory control, addition of $MgCl_2$ restored normal activity.

The relationship between ultrastructural configuration and respiratory state in heart mitochondria was studied in detail, and it was concluded that alterations of inner membrane configurations are not directly associated with either respiratory state or electron transport in heart mitochondria. Similar studies carried out in liver mitochondria, however, did show a direct correlation between the morphological appearance of isolated mitochondria and the respiratory state. It is probable that the ultrastructural appearance of isolated mitochondria depends more on ion transport or osmotic conditions, or both, than on anything else. The ultrastructural configuration of isolated mitochondria from humans, Syrian hamsters, and other animal species will be described. Isolated heart mitochondria can employ energy derived from either electron transport or from ATP for transporting ions, particularly potassium and calcium. Valinomycin-induced uptake and histone-induced extrusion of potassium are most active in heart mitochondrial preparations. It is of interest that numerous local anesthetics that are cationic in nature can successfully compete with histones for certain anionic sites on the membrane and reduce the potency of

[1] Supported by United States Public Health Service Grants HE 05435 and HE 07906.
[2] Recipient of a Career Development Award (K_3 HE 11, 875).

the histone for potassium transport. These anesthetic agents can also inhibit valinomycin-induced potassium uptake, but much more of the agent is required than that needed to prevent histone effect.

Sucrose-density gradient centrifugation of isolated mitochondria from normal and from hypertrophied and failing rabbits was studied. Normal mitochondria exhibited one distinct band and "clean" electron micrographs; aged normal mitochondria exhibited a second gradient band. It is of interest that freshly isolated normal mitochondria treated with calcium also exhibited multiple gradient bands. Mitochondria isolated from chronically-stressed rabbits, described above, also exhibited two or more gradient bands.

Developing chronic hypertrophy with and without failure was studied in detail by placing an ameroid clip on the proximal portion of the ascending aorta of rabbits; this surgical procedure slowly constricts the blood vessel over a period of weeks and months. After 12 days to 2 weeks, the heart weight to body weight ratio increased by 30–40%; after 4 weeks, in some animals, the ratio doubled, with no or minimal clinical signs of congestive heart failure (table I). After this period, significant congestive heart failure accompanied by massive hypertrophy was evident. Heart mitochondria isolated during the hypertrophic

Table I

Mitochondria from Hypertrophied and Failing Rabbit Hearts

Animal (No. days post op.)	$\frac{\text{Heart wt.}}{\text{Body wt.}} \times 10^{-3}$ g	RCI	QO_2	Ca^{++} Content nmoles/mg	No. of gradient peaks
Controls (–)	1.7–2.0	9.2	230	5.7–6.1	1
Hypertrophy (16) (no failure)	2.76	8.9	322	8.6	1
Hypertrophy (29) (no failure)	2.64	9.4	315	5.4	2
Failing (31) (slight)	2.34	6.5	188	4.9	2 +
Failing (24) (marked)	4.2	7.0	98		

Ameroid clips were placed on the ascending aorta, causing increasing stenosis of the vessel. Significant increases in heart weight to body weight ratios were noted quite early. Control data are from 12 animals. Mitochondria isolated from these hearts exhibited a lower respiratory control index (RCI; ratio of rate of resting respiration to active phosphorylating respiration) as frank signs of failure appeared. The rate of respiration during active phosphorylation (QO_2; expressed as nanoatoms O_2/min/mg) markedly increased during hypertrophy and then dropped below control levels as failure progressed. Calcium content of the isolated mitochondria appeared to be elevated in early hypertrophy and subsequently declined. Mitochondria from normal and early hypertrophic hearts exhibited single discreet peaks on sucrose density gradients; however, as hypertrophy progressed, two distinct peaks were observed and, in the animal with early (slight) failure signs, two peaks with a prominent shoulder were seen. Hypertrophy (no failure) indicates no gross pathological signs of failure were observed. Failing (slight) indicates some pleural fluid was present, with increases in lung weight and liver weight to body weight ratios. Failure (marked) indicates the animal was in obvious respiratory difficulty with large amounts of pleural fluid and ascites and significant increases in lung and liver ("nutmeg") weight to body weight ratios.

phase (with no failure) revealed significantly higher rates of oxygen consumption and respiratory control. Ultrastructural studies strengthen our suggestion that mitochondria increase in number and, in essence, become "super" in function. During the frank failing stage, the usual depression of energy production was observed, confirming previous studies using other animal models. These data are consistent with Meerson's three-stage hypothesis of heart failure. It is suggested that calcium may play a role in the failing process.

Sarcoplasmic reticulum fragments isolated from normal and failing animals (Syrian hamster, rabbit, and man) exhibit a significant diminution in the rate of calcium binding and calcium uptake. A stopped-flow spectrophotometric procedure for recording rapid calcium kinetics is described. According to this system, the rate of calcium binding of dog CRS equals 11–12 nmoles/mg protein within 100 msec; maximum binding = 90; maximum uptake is in excess of 5,000. Isolated CRS fulfills the suggested *in vivo* role of relaxation. Both binding and uptake were unaffected by ouabain, catecholamines, or by beta-blocking agents. However, La^{3+} appeared to inhibit the rate of binding and rate of release of calcium.

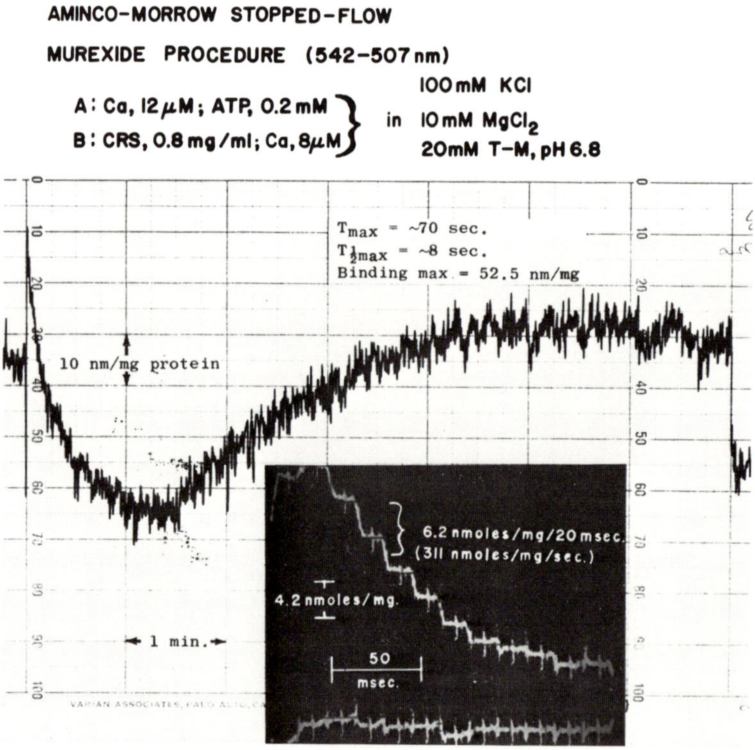

Fig. 1 Aminco-Morrow stopped-flow murexide procedure.

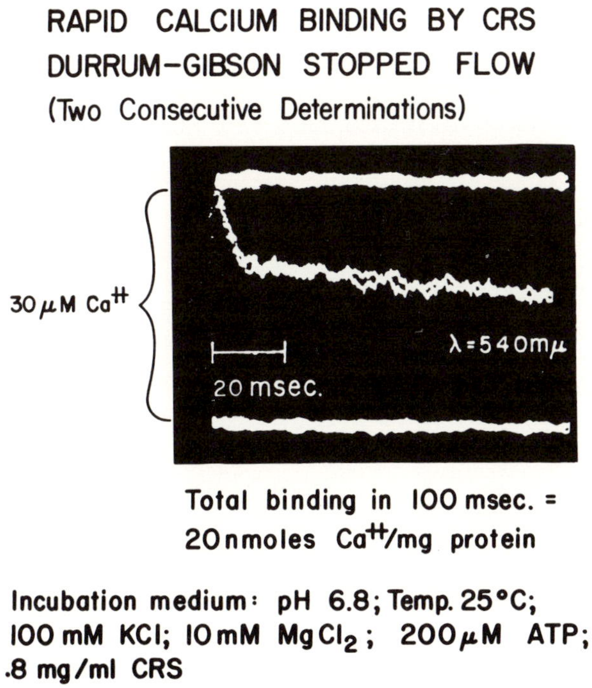

Fig. 2 Rapid calcium binding by CRS Durrum-Gibson stopped flow.

Recently, we have employed, in addition to Aminco-Morrow stopped-flow spectrophotometer, a Durrum single-beam stopped-flow device. The latter has an extremely rapid time constant and the mixing rate is less than 2 msec. The instrument can be conveniently calibrated, and turbidity does not interfere to any significant extent. Using the active dog CRS, we have been able to demonstrate even higher rates of initial calcium binding than that indicated above. Figures 1 and 2 compare traces obtained using both stopped-flow spectrophotometers. We have calculated that the binding rate is clearly in excess of that required for relaxation of mammalian heart muscle. Again, we have not been able to find specific effects of cardiac glycosides, catecholamines, or any of the currently employed antiarrhythmic agents, on either binding or release of calcium by CRS.

These studies indicate the importance of both energy-producing and energy-utilizing processes in mitochondria, particularly as related to heart disease. An important aberration of intracellular calcium metabolism not associated with mitochondria in a number of failing preparations was found. A depression of calcium binding in the relaxing system (sarcoplasmic reticulum) of hypertrophied and failing rabbit hearts was observed, suggesting that physiological dynamics of the heart may be altered even prior to severe congestive heart failure.

Mailing address: Dr. Arnold Schwartz, Division of Myocardial Biology, Baylor College of Medicine, Texas Medical Center, Houston, Texas 77025 (USA).

Wayne State University, School of Medicine,
Detroit, Michigan, U.S.A., and
Science Institute, University of
Iceland, Reykjavik, Iceland

INHIBITION OF ENERGY TRANSFER IN ISCHEMIC HEART MUSCLE

S. GUDBJARNASON

The kinetic heterogeneity of ATP and creatine phosphate (CP) breakdown observed in tetanized skeletal muscle by Hohorst *et al.* (4) led these authors to conclude to a functional compartmentation of ATP, where 85% of the adenine nucleotides do not react with creatine kinase (CK) at a rate comparable to the rate of CP breakdown.

Studies by Harary *et al.* (3) indicate that minor changes in ATP content affect the spontaneous contraction of cultured cells. The authors concluded that a small drop in total ATP may represent a large drop in a compartmentalized area of the cell, functionally organized for supplying ATP for the spontaneous beating.

Our studies on acutely ischemic heart muscle suggest the existence of functional compartmentation of ATP and CP, and indicate that energy-rich phosphate transfer from the site of synthesis in mitochondria to the site of utilization by contractile proteins may be of critical importance in maintenance of muscle contraction in heart muscle.

This study examines the relationship between ATP and CP in heart muscle in two experimental models: a) in the intact noninfarcted area of the infarcted heart at various levels of energy rich phosphates; and b) in acutely ischemic heart muscle, where an equilibrium between ATP and CP does not appear to exist.

MATERIALS AND METHODS

Experimental myocardial ischemia and infarction was produced in mongrel dogs under pentobarbital anesthesia and artificial respiration, by ligation of several brances of the anterior descending and circumflex coronary arteries (1, 2). The early changes in energy-rich phosphates and lactate following acute coronary artery occlusion were studied in nine dogs. Biopsy samples were obtained before and at various intervals (15, 30, and 60 sec, and 3, 10 and 30 min) after the ligations, as described previously (1). Biopsy samples were also obtained from the infarcted and noninfarcted muscle during the postinfarction period (1, 2, 4, and 10 days after infarction).

RESULTS AND DISCUSSION

Figure 1 illustrates the well-known reduction in CP levels of infarcted heart muscle following infarction, as well as the unexpected diminution in the CP levels of the noninfarcted and apparently nonischemic heart muscle during the postinfarction period. Figure 2 illustrates similar changes in the ATP levels of infarcted and noninfarcted muscle. The reasons for the marked reduction in CP and ATP levels of noninfarcted muscle are not known, but the mechanisms may be related: a) to an increase in energy utilization per unit of surviving muscle, which may be out of proportion to the rate of synthesis of high-energy phosphates, and b) to an increase in irreversible adenine nucleotide degradation accompanying the increase in workload and increased energy utilization. An imbalance between energy utilization and energy production, induced for example by hypoxia (5), by epinephrine infusion (5), or by uncoupling of oxidative phosphorylation (9), leads to a markedly increased rate of adenine nucleotide degradation and to a reduction in the total nucleotide pool. A similar condition may also prevail in the overloaded, noninfarcted muscle, where the energy imbalance may persist for a considerable period of time until the compensatory processes have developed a new steady state.

Figure 3 illustrates the relationship between the tissue content of ATP and the tissue content of CP in normal muscle, in noninfarcted muscle, and in

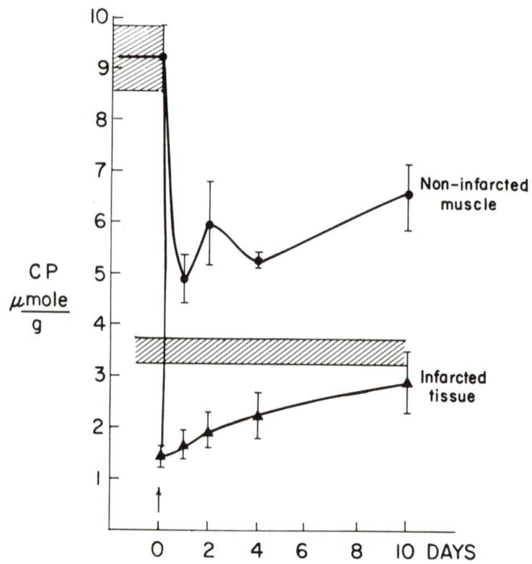

Fig. 1 Changes in tissue content of creatine phosphate (CP) in infarcted and noninfarcted heart muscle during the postinfarction period. The vertical lines represent ± standard error of the mean; N = 13 for the control group (before infarction), N = 8 one day after infarction, N = 18 two days after, N = 5 four days after, and N = 10 ten days after infarction. The lower shaded area represents the limit of survival; below this level of CP the muscle cells do not appear to survive or maintain contraction.

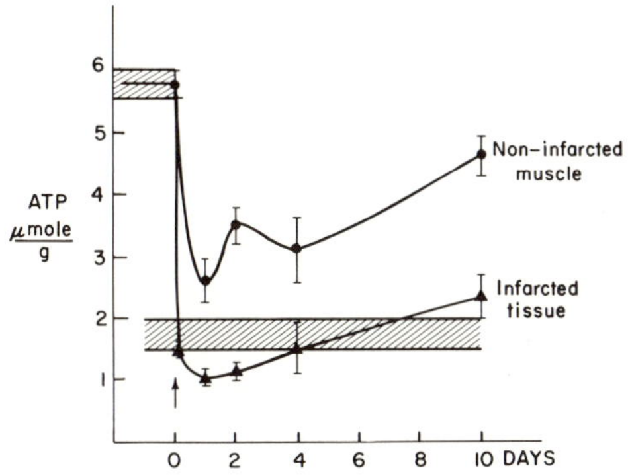

Fig. 2 Changes in tissue content of ATP in infarcted and noninfarcted heart muscle during the postinfarction period. The description is the same as for figure 1.

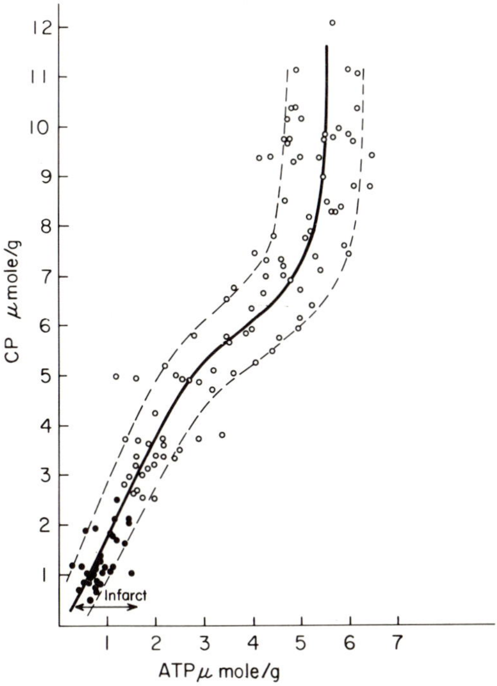

Fig. 3 The relationship between the tissue content of creatine phosphate (CP) and ATP in noninfarcted heart muscle (open circles) and infarcted heart muscle (filled circles) during the postinfarction period.

infarcted tissue (filled circles) during the postinfarction period. The figure demonstrates that the CP stores can be reduced to a level of 7 μmole/g before there is a significant reduction in the ATP level. When the level of CP diminishes below a level of 7 μmole/g, there is a marked diminution in the ATP levels. The lowest levels of energy-rich phosphates at which the heart muscle is able to contract *in vivo* and survive is 1.5–2.0 μmole of ATP/g of tissue and 2.5–4.0 μmole of CP/g, provided the muscle is not ischemic.

In acutely ischemic heart muscle, the relationship between ATP and CP is quite different from that seen in the noninfarcted muscle (fig. 4). In the ischemic muscle, we observe a rapid diminution in CP level, from 9.2 to 2.5 μmole/g, before a marked reduction in ATP levels takes place. The cessation of contractile activity in the ischemic area occurs about 1–2 min after coronary occlusion, at a time when the ATP levels are still relatively high. The ischemic muscle stops contracting at an ATP level of 4.5–4.9 μmole/g, when only about 20% of the ATP has been utilized, whereas the nonischemic muscle can maintain contraction and survive at ATP levels as low as 1.5–2.0 μmole/g. This observation indicates that it is not the lack of ATP in the muscle that is responsible for cessation of muscle contraction.

One of the earliest metabolic changes observed in ischemic heart muscle is the rapid rise in anaerobic glycolysis and lactate accumulation. Katz and Hecht

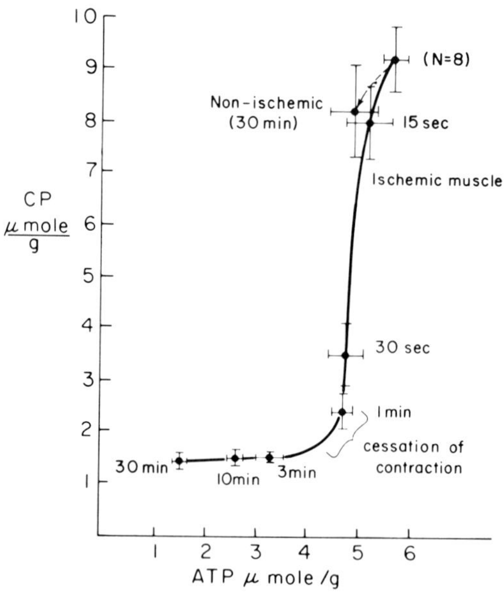

Fig. 4 The relationship between the tissue content of creatine phosphate (CP) and ATP in ischemic heart muscle, following acute coronary artery occlusion. The vertical and horizontal lines represent ± standard error of the mean of CP and ATP determinations.

have proposed (6) that the cessation of muscle contraction may be due to depressed responsiveness of the contractile proteins to calcium ions caused by intracellular acidosis. According to this hypothesis, the increase in hydrogen-ion concentration in ischemic cells causes a displacement of calcium ions bound to troponin, resulting in inhibition of actin-myosin interaction and reduced contractility. The increased inhibition of the actin-myosin interaction should then result in a reduction in the rate of energy utilization, i.e., utilization of ATP and CP. Our experimental results indicate a more complex mechanism and illustrate a rapid utilization of CP, but utilization of only a portion of the ATP stores during the first 60 sec of ischemia.

Figure 5 illustrates the relationship between the lactate level and ATP and CP levels in acutely ischemic muscle. The lactate accumulation is paralleled by a rapid and continuous depletion of CP stores. The ATP content shows an early decrease that levels off with an increase in lactate content. After 1 min of ischemia and cessation of contractile activity in the ischemic muscle, the ATP depletion increases again and the ATP level decreases progressively with increasing lactate content.

Figure 6 illustrates the rate of lactate accumulation and the rates of decrease of ATP and CP stores at different times after coronary artery occlusion. It may be seen from this figure that the rate of ATP depletion diminishes already during the second period of observation, *i.e.*, 15–30 sec after coronary occlusion,

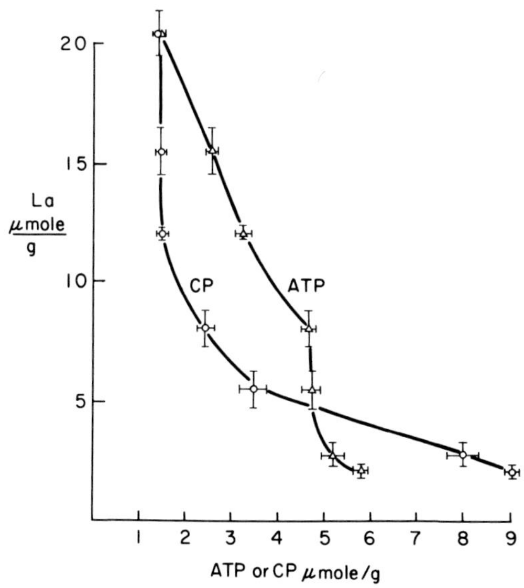

Fig. 5 The relationship between the increase in lactate (La) content of ischemic heart muscle and the decrease in CP and ATP content, following acute coronary artery occlusion. The points correspond to the points in figure 4.

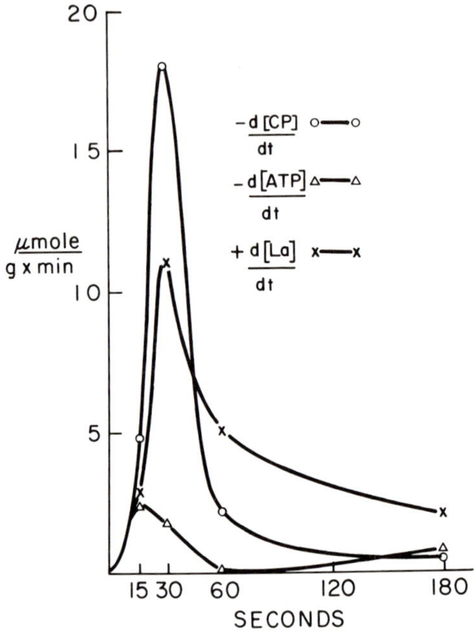

Fig. 6 The rate of change in tissue content of creatine phosphate (CP), ATP, and lactate (La), following acute coronary artery occlusion. –dCP/dt = rate of decrease of CP stores; –dATP/dt = rate of decrease of ATP stores; +dLa/dt = rate of lactate accumulation.

whereas the rates of CP breakdown and lactate accumulation increase further during this period. The rate of glycolysis reaches a maximum after 30 sec and decreases rapidly after that, although the tissue level of lactate continues to increase. The difference in rates of ATP and CP diminution becomes even more evident when they are related to the rate of lactate accumulation (fig. 7). The figure shows a correlation between the rates of CP breakdown and lactate accumulation, whereas the rate of decrease of ATP stores diminishes with an increase in the rate of lactate accumulation.

From the rate of decrease in ATP and CP stores and from the rate of lactate accumulation, we can compute the approximate amount of energy utilized per beat and per gram of muscle. Table I illustrates the results of such a calculation. The sum of Δ ATP, Δ CP, and Δ La represents the total amount of high energy phosphates derived from tissue stores and anaerobic glycolysis (Δ La equals the amount of ATP synthesized by the glycolytic pathway). During the first period of observation, 0–15 sec after occlusion, the total value is low owing to contribution from residual oxygen and oxidative phosphorylation. The amount utilized during the second period of observation, 15–30 sec after occlusion, is 0.393 µmole of ~P per gram and per beat, and is similar to the values reported by Mommaerts *et al.* (7, 8) on tetanized muscle. The energy utilization

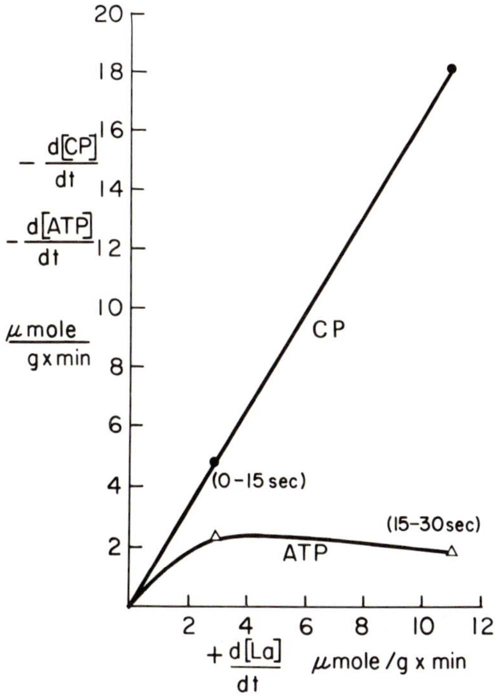

Fig. 7 The relationship between the rate of lactate accumulation (+dLa/dt) and the rates of depletion of CP (–dCP/dt) and ATP (–dATP/dt) stores following acute coronary artery occlusion.

diminishes subsequently and is only 0.096 μmole/g X beat during the third period of observation, *i.e.*, 30–60 sec after occlusion. During this period the rate of glycolysis diminishes by 54% and the rate of decrease in ATP stores is very small. This drop in anaerobic energy production and utilization of ATP stores is followed by cessation of contractile activity in the ischemic muscle 1–2 min after coronary occlusion.

The rapid depletion of CP in ischemic heart muscle is not accompanied by a corresponding decrease in ATP levels, although such a reduction would be expected because of the activity of CK and the concomitant decrease in ATP and CP in noninfarcted muscle, when CP falls below 7 μmole/g (fig. 3).

These observations suggest that a deficient transfer of energy-rich phosphate from the site of synthesis in mitochondria to the site of utilization for muscle contraction may contribute to the early cessation of contractile activity.

Very little is known about the intracellular energy transfer in heart muscle cells. Actively contracting cells produce and utilize energy at a very rapid rate and must have an efficient system for the transfer of this energy from the site of synthesis to the site of utilization by contractile proteins.

Table 1

Rate of Change in CP, ATP and Lactate (La) Levels in Acutely Ischemic Heart Muscle (N = 9)

Period of Ischemia	– dCP/dt (μmole/g × min)	– dATP/dt (μmole/g × min)	+ dLa/dt (μmole/g × min)	Heart rate (beats/min)	Σ Δ CP + Δ ATP + Δ La† / g × beat
0–15 sec	4.80	2.36	2.92	76.8 ± 4.6	0.131
15–30 sec	18.04	1.84	11.08	78.8 ± 4.8	0.393
30–60 sec	2.16	0.10	5.08	76.3 ± 4.8	0.096
1–3 min	0.45	0.72	1.98	73.3 ± 3.5	0.043
3–10 min	0.0	0.09	0.49	73.8* ± 4.0	(0.008)*

*No muscle contraction in the ischemic area.

† Σ Δ CP + Δ ATP + Δ La = Sum of CP and ATP decrease and La increase per gram of tissue and per beat.

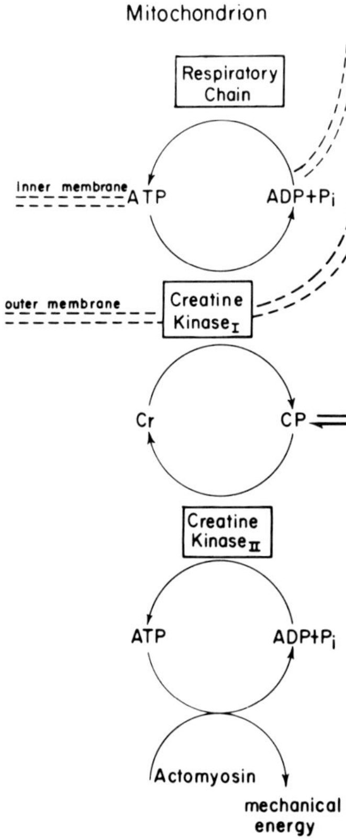

Fig. 8 Compartmentation of ATP and CP in heart muscle and possible paths of high-energy phosphate transfer from the site of synthesis in mitochondria to the site of utilization by contractile proteins.

A possible reaction in the control of transfer of intramitochondrial high energy phosphate to the extramitochondrial site of energy utilization is the mitochondrial CK (fig. 8). Heart muscle possesses three isoenzymes of CK: two cytoplasmic and one mitochondrial CK. The function of the mitochondrial CK has not been clearly established, but this enzyme constitutes 35–45% of total heart muscle CK and is located in the outer mitochondrial membrane.

It is suggested that at the time of acute ischemia and hypoxia, the earliest effect may be seen in mitochondrial redox-systems shifting to the reduced state, accompanied by a proton release. An early inhibition of the pH-sensitive mitochondrial CK may subsequently prevent the essential energy transfer required to maintain muscle contraction.

The rapid rise in the lactate level and hydrogen-ion concentration may not only displace Ca^{2+} from troponin binding sites and in this manner reduce contractility, but may also disturb intracellular energy transfer and lead to a rapid depletion of extramitochondrial energy stores. The early inhibition of anaerobic glycolysis observed 30–60 sec after coronary occlusion illustrates the early inhibition of extramitochondrial ATP production. Diminution in the rate of regeneration of a small pool of extramitochondrial ATP of high turnover rate, immediately available for muscle contraction, may reduce the availability of ATP for muscle contraction as well as impair other ATP dependent processes such as ion transfer.

SUMMARY

The relationship between the rate of decrease in high energy phosphate stores and the rate of lactate accumulation in acute myocardial ischemia was examined in dogs following coronary artery ligation. The ischemic muscle stops contracting at a relatively high ATP level of 4.5 μmole/g, when only about 20% of the ATP has been utilized, whereas the nonischemic muscle survives and maintains contraction at ATP levels as low as 1.5–2.0 μmole/g. The rapid increase in anaerobic glycolysis coincides with an early inhibition of utilization of ATP stores without an interference with the utilization of CP stores. The kinetic heterogeneity of the ATP and CP breakdown suggests an inhibition of transfer of intramitochondrial high-energy phosphate to the extramitochondrial compartment. The anaerobic, extramitochondrial ATP production reaches a maximum 15–30 sec after coronary occlusion and diminishes rapidly thereafter. The energy utilization of ischemic muscle falls from 0.393 μmole ~P/g × beat during the period 15–30 sec after coronary occlusion to 0.096 μmole ~P/g × beat during the period 30–60 sec after ischemia. Cessation of contractile activity appears to be the result of a reduction in rate of regeneration of a small pool of extramitochondrial ATP of high turnover rate available for muscle contraction and other ATP requiring processes such as ion transport.

REFERENCES

1. BRAASCH, W.; GUDBJARNASON, S.; PURI, P. S.; RAVENS, K. G. and BING, R. J.; Early changes in energy metabolism in the myocardium following acute coronary artery occlusion in anesthetized dogs. Circulat. Res. *23:* 429–438 (1968).
2. GUDBJARNASON, S.; DESCHRYVER, C.; CHIBA, C.; YAMANAKA, J. and BING, R. J.: Protein and nucleic acid synthesis during the reparative processes following myocardial infarction. Circulat. Res. *15:* 320–326 (1964).

3. HARARY, J.; SERAYCLARIAN, M. and GOSCHENSON, L. E.: Effects of lipids on contractibility of cultured cells, in TANZ, R. D.; KAVALER, F. and ROBERTS, J. Factors Influencing Myocardial Contractility. (Academic Press, New York 1967). pp. 231–243.

4. HOHORST, H. J.; REIM, M. and BARTELS, H.: Studies on the creatine kinase equilibrium in muscle and the significance of ATP and ADP levels. Biochem. Biophys. Res. Commum. *7:* 142–146 (1962).

5. KATORI, M. and BERNE, R.: Release of adenosine from anoxic hearts, relationship to coronary flow. Circulat. Res. *19:* 420–425 (1966).

6. KATZ, A. M. and HECHT, H. H.: The early "pump" failure of the ischemic heart. Amer. J. Med. *47:* 497–502 (1969).

7. MOMMAERTS, W. F. H. M.: Energetics of muscular contraction. Physiological Reviews, *49:* 427–508 (1969).

8. MOMMAERTS, W. F. H. M.; SORAYDORIAN, K. and WALLNER, A.: Demonstration of phosphocreatine splitting as an early reaction in contracting frog sartorius muscle. Biochem. Biophys. Acta *63:* 75–81 (1962).

9. RICHMAN, H. and WYBONNY, L.: Adenine nucleotide degradation in the rabbit heart. Amer. J. Physiol. *207:* 1139–1145 (1964).

Mailing address: Sigmundur Gudbjarnason, Ph.D., Science Institute, University of Iceland, Dunhaga 3, Reykjavik (Iceland).

Institute of Circulation Research, German Academy of Sciences,
Berlin-Buch, German Democratic Republic

CYCLIC AMP LEVELS IN THE ACUTELY ISCHEMIC MYOCARDIUM OF THE DOG AND GUINEA PIG

A. WOLLENBERGER, E.-G. KRAUSE, and G. HEIER

At the First Annual Meeting of the International Study Group, we presented data showing that arrest of the coronary flow leads to a rapid mobilization of endogenous cardic norepinephrine, to an activation of the myocardial phosphorylase system, and to a strong acceleration of glycolysis, in the myocardium of the dog, rabbit, and rat (2; *cf.* also 4). These changes were either absent or they were delayed and much less prominent in the guinea pig. Since the enzymatic activations and the stimulation of glycolysis in the ischemic heart of the former species were largely prevented by treatment with the β-adrenergic receptor blocker, pronethalol, they were attributed to an adrenergic mechanism presumably operating at the level of adenyl cyclase. In accordance with accepted concepts regarding the pathway of action of catecholamines on muscle glycogenolysis, it was postulated that an increase in the level of cardiac adenosine 3′, 5′-monophosphate (cyclic AMP) was the event linking endogenous norepinephrine liberation and the activation of the glycogen phosphorylase system in the acutely ischemic myocardium.

In the meantime, evidence has been obtained in support of this hypothesis (3). The present report adds to this evidence and shows that, following the onset of myocardial ischemia, cardiac cyclic AMP levels do not rise as promptly in the guinea pig as they do in the dog.

METHODS

Thorax and pericardium of adult dogs were opened under pentobarbital anesthesia and positive pressure breathing. A first sample of the left ventricular wall, weighing 50–100 mg, was quickly frozen *in situ* with a freezing clamp at the time designated as "0 sec"; simultaneously, the ascending aorta was cut. Other ventricular muscle samples of the same heart were taken in succession at various intervals by the same technique. Dogs that had received an intravenous injection of 7 mg/kg of pronethalol 20 min prior to 0 sec were treated in the same manner. Thorax and pericardium of adult guinea pigs were opened under diallyl barbiturate-urethane anesthesia. The entire heart was frozen *in situ* before and at various times after sectioning of the ascending aorta. Cyclic AMP in the ventricular muscle was determined by a modification (3) of the method of Breckenridge (1).

RESULTS

Figure 1 shows that cyclic AMP rises in the heart of the dog immediately after arrest of the blood flow. The rise was not quite as rapid as in earlier experiments (3), in which maximum values were reached after 5 sec; nevertheless, it was rapid enough to attain maximum values after 10 to 15 sec. The elevated level was then maintained for at least 10 min. As reported earlier (3), cyclic AMP did not rise in the heart of pronethalol-treated dogs during the first minute following arrest of blood flow, but it increased during the second minute (fig. 1). In the guinea pig heart, the ischemia-induced rise in cyclic AMP occurred with a delay of at least 5 sec (table I).

DISCUSSION

In our previous experiments (3), measurements of cardiac cyclic AMP levels were limited to the first minute of ischemia. We were not certain, at that time, how long the levels of the nucleotide would stay elevated in the ischemic myocardium of the dog and thought that they might perhaps return to the aerobic value after about 3 min. The present results show that such is not the case and that the elevated level of cyclic AMP is maintained for 10 min or longer.As pointed out in our earlier study (3), the rapid initial rise in cardiac cyclic AMP immediately following the arrest of blood flow is, in all likelihood, due to increased cyclic AMP formation by adenyl cyclase. The prolonged maintainance of the elevated cyclic AMP level in the ischemic tissue may conceivably, and at least in part, have been the consequence of an acumulation

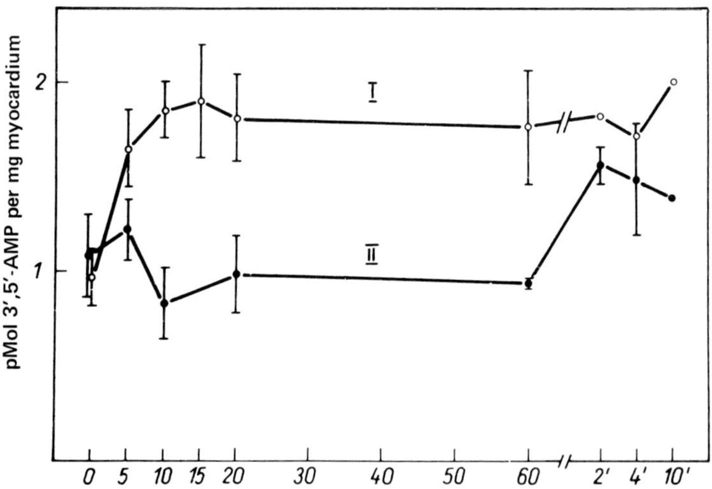

Fig. 1 Levels of cyclic AMP in dog myocardium following arrest of blood flow. I, Untreated controls; II, dogs treated with pronethalol.

Table 1

Cyclic AMP Levels in the Ventricular Muscle of the Guinea Pig before and after Arrest of Blood Flow

Time after arrest of blood flow (sec)	No. of animals	Cyclic AMP (nmoles/g)
0	6	1.22 ± 0.57*
5	5	1.22 ± 0.39
15	6	2.80 ± 0.73
60	5	3.15 ± 0.43

*Means ± standard deviations.

of acid products of glycolysis, causing a decrease in the activity of cyclic nucleotide phosphodiesterase, the enzyme hydrolyzing cyclic AMP to 5′-AMP. The same effect may also have been responsible for the increase in cyclic AMP observed in the heart of the pronethalol-treated dogs after the first minute of ischemia.

In the guinea pig heart, cyclic AMP does not begin to rise as early after the onset of ischemia as it does in the dog heart. However, the delay is relatively short (5–10 sec) and can hardly be attributed to an inhibition of phosphodiesterase by accumulation of acid glycolytic products. Perhaps the very small amounts of norepinephrine that are liberated in the heart muscle of the guinea pig in acute myocardial ischemia (4) may suffice to cause a stimulation of adenyl cyclase activity in the tissue.

SUMMARY

In the dog, arrest of the blood flow causes a rapid rise in cardiac cyclic AMP. The elevated level of this nucleotide is maintained for at least 10 min. In the guinea pig, the ischemia-induced rise in cardiac cyclic AMP occurs with a delay of 5 to 10 sec.

REFERENCES

1. BRECKENRIDGE, B.M.: The measurement of cyclic adenylate in tissues. Proc. Nat. Acad. Sci. *52:* 1580–1586 (1964).
2. WOLLENBERGER, A. and KRAUSE, E.-G.: Humoral and non-humoral regulation of carbohydrate breakdown in the ischemic myocardium. Abstr. 1st Ann. Meet. Internat. Study Group Res. Cardiac Metab., Dubrovnik 1968.
3. WOLLENBERGER, A.; KRAUSE, E.-G. and HEIER, G.: Stimulation of 3′, 5′-cyclic AMP formation in dog myocardium following arrest of blood flow. Biochem. Biophys. Res. Commun. *36:* 664–670 (1969).
4. WOLLENBERGER, A,; KRAUSE, E.-G. und SHAHAB, L.: Die Regulation der anaeroben Energiegewinnung im ischämischen Myokard; in BRÜSCHKE, G. (Ed.), Forschung von heute Gesundheit von morgen, pp. 151–168, VEB Verlag Volk und Gesundheit, Berlin 1969.

Mailing address: Prof. A. Wollenberger, Institut für Kreislaufforschung, Deutsche Akademie der Wissenschaften zu Berlin, Linderberger Weg 70, 1115 Berlin-Buch (Germany/DDR).

Division of Myocardial Biology, Baylor College
of Medicine, Houston, Texas, U.S.A.

PHYSIOLOGIC AND PHARMACOLOGIC ROLES FOR THE CARDIAC Na^+, K^+-ATPase TRANSPORT SYSTEM [1]

J. C. ALLEN[2] and A. SCHWARTZ[3]

The Na^+, K^+-ATPase enzyme system has been shown to be directly involved in the maintenance of Na^+ + K^+ gradients across cell membranes. Since Skou first demonstrated inhibition of the Na^+, K^+-ATPase by ouabain in 1960, many investigators (see Skou [9] and references cited) have shown this to be a specific and reproducible effect. These data have led to the suggestion that this enzyme system is the pharmacologic receptor for cardiac glycosides (1–5, 7–9). This hypothesis suggests that an interaction of the drug with the Na^+, K^+-ATPase is the first step in a sequence of reactions resulting ultimately in increased cardiac contractility.[4] The kinetics of the two temporal extremes of the sequence, i.e., drug-receptor interaction and muscle contraction, must be dealt with separately, since the intermediate steps are unknown. In this paper, we would like to present information relating some biochemical parameters of the initial step in the sequence, i.e., drug-receptor interaction, to the *in vivo* sensitivity of the various species studied.

The sensitivity *in vitro* of the Na^+, K^+-ATPase to cardiac glycosides is as follows: human~beef~dog>rabbit>guinea pig> rat. These coincide with the suggested *in vivo* species sensitivity to the drug.

Figure 1 shows that the stability at 0°C of the complex formed between cardiac Na^+, K^+-ATPase isolated from various species and H^3-ouabain reflects the sensitivity of the species from which the enzyme was isolated. The H^3-ouabain-enzyme complex (O-E) from various species was also formed in the presence of either Mg^{++} + P_i or ATP + Mg^{++} + Na^+. The ease of reversibility of (O-E) at 23°C, 37°C, and 45°C was inversely related to species sensitivity to digitalis. However, in each species studied, (O-E) formed with Mg^{++} + P_i was more stable than that formed with ATP + Mg^{++} + Na^+. We have also shown that the inhibition of the enzyme by ouabain is time-dependent and that the magnitude of this time-dependency again depends on the species from which the enzyme was isolated

[1] Supported by United States Public Health Service grants HE 05435 and HE 07906 and by the National Science Foundation Grant GB 6895.

[2] Postdoctoral Fellow of the USPHS (1 FO2 HE 43042–01).

[3] Recipient of a Career Development Award (K_3 HE 11,875).

[4] The reader is referred to the work of Besch and Schwartz (5) for a possible explanation of the drug's action.

Fig. 1 Stability of ouabain-enzyme complex. H^3-ouabain was bound to cardiac Na^+, K^+-ATPase preparations at 37°C, under optimal conditions (2). The pellets were resuspended in EDTA at 0–4°C, and washed by resuspension and recentrifuged.

(fig. 2). The horizontal line for inhibition of the rat enzyme indicates that no time-dependency was discernible under these conditions.

Earlier, we demonstrated time-dependent binding of H^3-ouabain to beef heart Na^+, K^+-ATPase (3). Figure 3 shows the retarding effect of K^+ on this binding capacity with enzyme from different species. Again note that K^+ is more effective in less sensitive species. An interpretation of these data is that the ouabain-enzyme complex forms more readily with enzyme from sensitive species, and K^+ is less effective in retarding this effect. This also indicates that K^+ depresses the *rate* of H^3-ouabain binding in sensitive species rather than the capacity for binding. If the reaction is allowed to proceed, however, the binding eventually reaches its maximum value.

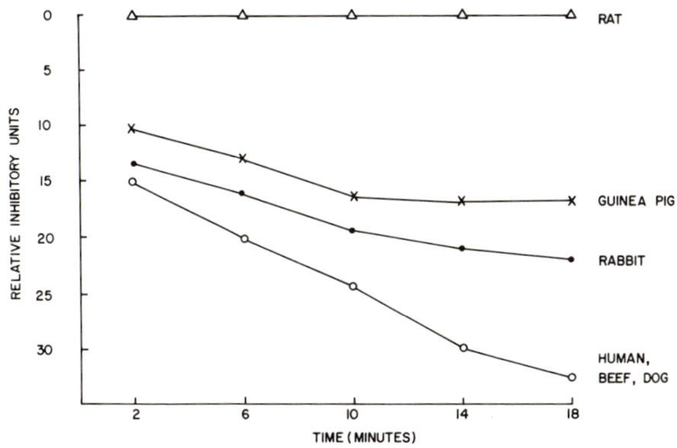

Fig. 2 Relative time-dependencies of ouabain-induced inhibition of cardiac Na^+, K^+-ATPase isolated from different species. The inhibitory effect is expressed in relative units, with rat enzyme showing no time-dependency, a straight line. See Schwartz *et al.* (8) for assay details.

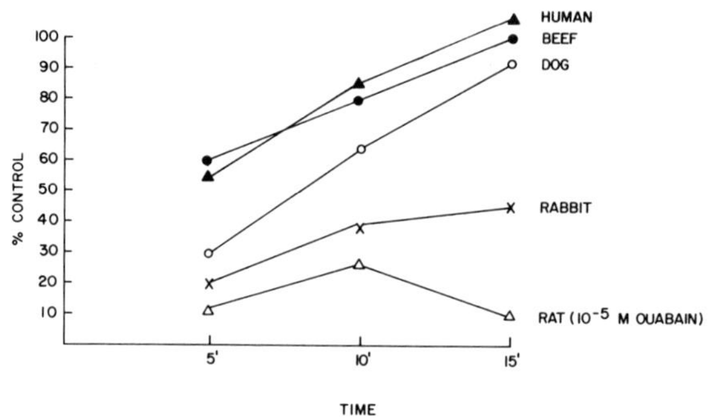

Fig. 3 The retarding effect of K^+ on the ouabain binding to Na^+, K^+-ATPase from various species. All conditions were optimal for binding (8).

Recently, we obtained control post-mortem (2 h) human cardiac tissue. The appearance of all the hearts was normal, and in each case the isolated Na^+, K^+-ATPase compared favorably with values obtained from control hearts of experimental animals (20–25 μmoles P_i/mg protein/h). Figure 4 shows an ouabain-inhibition curve on human cardiac Na^+, K^+-ATPase. Note the increase in sensitivity of the preparation with time.

Experiments performed with canine hearts *in situ* have shown a direct correlation between positive inotropic effect and inhibition of Na^+, K^+-ATPase (4). Furthermore, we have now shown with these perfused dog hearts (3) that, *in vivo*, the drug binds specifically to the Na^+, K^+-ATPase, and that drug binding to other subcellular fractions is due to the presence in the fraction of fragments of membrane bound Na^+, K^+-ATPase.

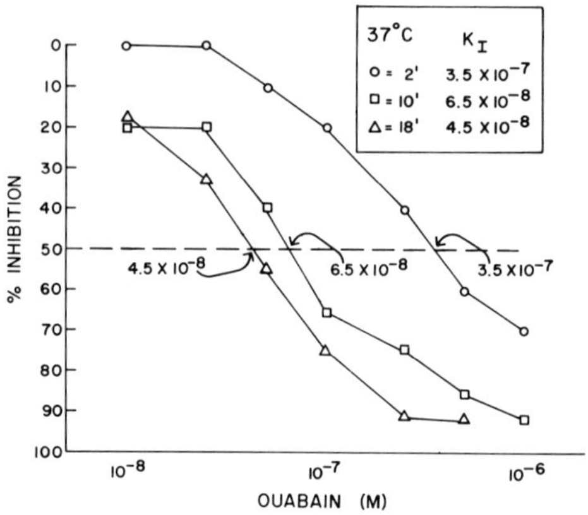

Fig. 4 Ouabain inhibition curves on Na^+, K^+-ATPase from control human cardiac tissue. K_I = drug concentration giving half-maximal inhibition. Spectrophotometric assay was performed as previously described (8).

Na^+, K^+-ATPase isolated from recipient hearts during the transplantation procedure showed very low activity when compared with controls (control = 20.5, diseased = 3 (n = 10)). In the light of the animal experiments mentioned above, a possible explanation for this low activity is that the drug remained bound throughout enzyme isolation; the patients were all digitalized prior to and during the operative procedure.

These new data strengthen our contention that the Na^+, K^+-ATPase is the pharmacologic receptor for cardiac glycosides, and these results are now verified by enzyme data from control and diseased human hearts.

SUMMARY

The interaction of cardiac glycosides with Na^+, K^+-ATPase from both human and animal hearts manifests itself by time-dependent enzyme inhibition and concomitant time-dependent specific drug binding to the enzyme. The stability of the drug-enzyme complex reflects the sensitivity of the species from which the enzyme is isolated, *i.e.*, the more sensitive the species, the more stable the complex. These data support the contention that Na^+, K^+-ATPase is the pharmacologic receptor for cardiac glycosides.

REFERENCES

1. ALLEN, J. C.; BESCH, H. R., Jr.; GLICK, G. and SCHWARTZ, A.: H^3-ouabain binding to Na^+, K^+-ATPase and cardiac relaxing system of perfused dog heart. Mol. Pharmacol., *6:* 441–443 (1970).
2. ALLEN, J. C. and SCHWARTZ, A.: A possible biochemical explanation for the insensitivity of the rat to cardiac glycosides. J. Pharmacol. Exp. Therap. *168:* 42–47 (1969).
3. ALLEN, J. C. and SCHWARTZ, A.: Effects of K^+, temperature and time on ouabain interaction with the cardiac Na^+,K^+-ATPase: Further evidence supporting an allosteric site. J. Mol. Cell. Cardiol. *1:* 39–45 (1970).
4. BESCH, H. R., Jr.; ALLEN, J. C.; GLICK, G. and SCHWARTZ, A.: Correlation between the inotropic action of ouabain and its effects on subcellular enzyme systems from canine myocardium. J. Pharmacol. Exp. Therap. *171:* 1–12 (1970).
5. BESCH, H. R., Jr. and SCHWARTZ, A.: On a mechanism of action of digitalis, J. Mol. Cell. Cardiol., *1:* 195–199 (1970).
6. GLYNN, I. M.: The action of cardiac glycosides on ion movements. Pharmacol. Rev. *16:* 381–407 (1964).
7. REPKE, K.: Effects of digitalis on membrane adenosine triphosphatase of cardiac muscle. In Proceedings of the Second International Pharmacological Meeting, Vol. 4, p. 65–88. Pergamon Press, London (1965).
8. SCHWARTZ, A.; ALLEN, J. C. and HARIGAYA, S.: Possible involvement of cardiac Na^+,K^+-ATPase in the mechanism of action of cardiac glycosides. J. Pharmacol. Exp. Therap. *168:* 31–41 (1969).
9. SKOU, J. C.: Enzymatic basis for active transport of Na^+ and K^+ across cell membrane. Physiol. Rev. *45:* 596–617 (1965).

Mailing address: Dr. Julius C. Allen, Division of Myocardial Biology, Baylor College of Medicine, Texas Medical Center, Houston, Texas 77025 (USA).

Johnson Research Foundation, University of Pennsylvania,
Philadelphia, Pennsylvania, USA

FUNCTIONAL SIGNIFICANCE OF THE MALATE-ASPARTATE SHUTTLE FOR THE OXIDATION OF CYTOPLASMIC REDUCING EQUIVALENTS IN RAT HEART [1]

B. SAFER[2] and J. R. WILLIAMSON[3]

Metabolism of glucose and lactate in the cytosolic compartment of the myocardial cell is necessarily linked to an equivalent oxidation of reduced nicotinamide adenine dinucleotide (NADH) by the mitochondrial respiratory chain for every pyruvate molecule entering the citric acid cycle. Since the early work of Lehninger (8), however, it has been conclusively demonstrated that oxidation of NADH generated in the cytosolic compartment is not accomplished by the direct transfer of reduced pyridine nucleotides across the permeability barrier of the inner mitochondrial membrane (9, 12). Two alternative hydrogen shuttle mechanisms, first proposed by Bücher and Klingenberg (1), will now be considered.

An essential feature of any hydrogen shuttle is that an oxidized member of a redox couple is reduced by cytosolic NADH, and in turn reduces the prosthetic group of an appropriate mitochondrial dehydrogenase linked to the electron transport chain. As shown in figure 1, in the malate shuttle, oxaloacetate (OAA) oxidizes NADH via malic dehydrogenase in the cytosolic compartment; malate then enters the mitochondrial compartment where OAA and NADH are regenerated. Thus, reducing equivalents from cytosolic NADH are transferred into the mitochondrial compartment without the necessity of direct transfer of NADH across the permeability barrier of the inner mitochondrial membrane. To complete this cycle, OAA must then re-enter the cytosolic compartment. Due to the relative impermeability of the inner mitochondrial membrane for OAA (6), it must first be transaminated with glutamate via aspartate aminotransferase to form α-ketoglutarate (α-KG) and aspartate; these can then enter the cytosolic compartment and reform OAA and glutamate. In the α-glycerophosphate (α-GP) shuttle, dihydroxyacetone phosphate (DHAP) oxidizes cytosolic NADH. α-Glycerophosphate thus formed can then be oxidized by mitochondrial flavoprotein-linked α-glycerophosphate oxidase to reform dihydroxyacetone-P

[1] Supported by grants from the United States Public Health Service, GM 12202, and by a grant-in-aid from the American Heart Association.

[2] National Institutes of Health Postdoctoral Fellow.

[3] Established Investigator of the American Heart Association.

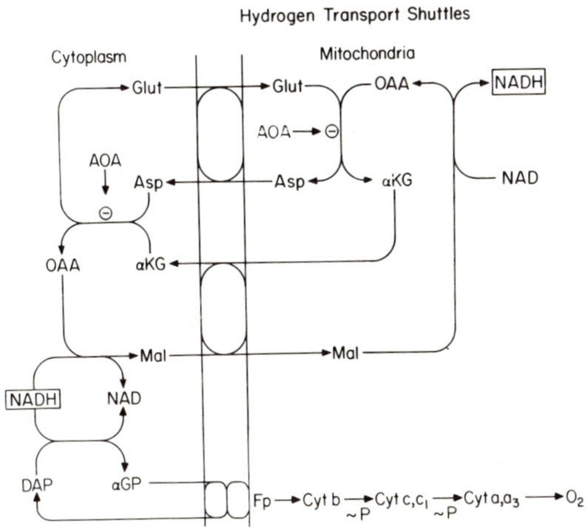

Fig. 1 Hydrogen transport shuttles in rat heart. Metabolic intermediates of the malate-aspartate shuttle and the α-glycerophosphate-dihydroxyacetone phosphate shuttle are shown. In the malate-aspartate shuttle, cytosolic NADH is oxidized via malate dehydrogenase, and malate is oxidized via malate dehydrogenase in the mitochondrial compartment. The sites of inhibition of aspartate aminotransferase by aminooxyacetic acid are indicated. Intramitochondrial NADH, generated by the malate-aspartate shuttle, is oxidized by NADH dehydrogenase of the electron transport chain. In the α-glycerophosphate shuttle, cytosolic NADH is oxidized via α-glycerophosphate dehydrogenase. α-Glycerophosphate then donates the electrons directly into the flavoprotein region of the respiratory chain via α-glycerophosphate oxidase.

and complete the cycle, the reduced flavin being reoxidized by the mitochondrial respiratory carriers.

In order to determine the relative functional significance of these proposed hydrogen shuttle mechanisms in the intact perfused heart, two distinct approaches were utilized. First, aminooxyacetate (AOA), which was first shown by Hopper and Segal (4) to be a competitive inhibitor of the amino acid substrates of transaminases, was used to limit flux of hydrogen specifically through the malate shuttle by inhibiting aspartate aminotransferase (10). If the malate shuttle were the major hydrogen transport mechanism in the myocardial cell, aminooxyacetic acid would be expected to decrease the rate of aerobic glycolysis and lactate oxidation, and at the same time produce a more reduced state of the cytosolic NAD-system but a more oxidized state of the mitochondrial NAD-system. Secondly, the hormone triiodothyronine, which was first shown by Lee and Lardy (7) to induce an increase in total tissue activity of α-glycerophosphate oxidase, was used to stimulate possible reducing equivalent transfer through the α-glycerophosphate shuttle, since this enzyme is believed to be rate limiting under normal conditions. If the α-glycerophosphate shuttle were the major hydrogen transport mechanism, one might expect that any effect of aminooxyacetic acid on the rate of glucose or lactate utilization would be small.

METHODS

Hearts from male Holtzman rats (200–220 g body wt) were perfused according to the method of Williamson and Krebs (14). Specific activity of α-glycerophosphate oxidase was increased by daily subcutaneous injections of triiodothyronine over 96 h. For metabolic studies, hearts were preperfused with substrate-free Krebs-Ringer bicarbonate buffer for 15 min and were then transferred to a recirculating perfusion system with various substrates added to a total volume of 20 ml. At the end of 45 min, during which rates of substrate utilization were obtained by assay of aliquots of perfusate, hearts were quickly frozen with Wollenberger clamps cooled with liquid nitrogen. Powdered tissue was then extracted with 6% PCA and metabolic intermediates assayed according to published methods of Williamson (13). Direct simultaneous readout of flavin and pyridine nucleotides was obtained using a compensated double fluorometer recently developed in this laboratory (11).

RESULTS AND DISCUSSION

A direct method for distinguishing between oxidation-reduction changes of pyridine nucleotides in the cytosolic and mitochondrial compartments of perfused heart is presented in figure 2. Simultaneous readout of flavin and pyridine nucleotide fluorescence changes were measured by surface fluorometry (11). Reduction of pyridine nucleotides and flavoproteins are indicated by an upward deflection of the traces. Since it has been shown by Scholz *et al.* (11) that fluorescent flavoproteins are exclusively located within the mitochondrial compartment and are in rapid equilibrium with the mitochondrial pool of pyridine nucleotides, any fluorescent response of pyridine nucleotides in the absence of similar changes of flavoprotein fluorescence can be interpreted as

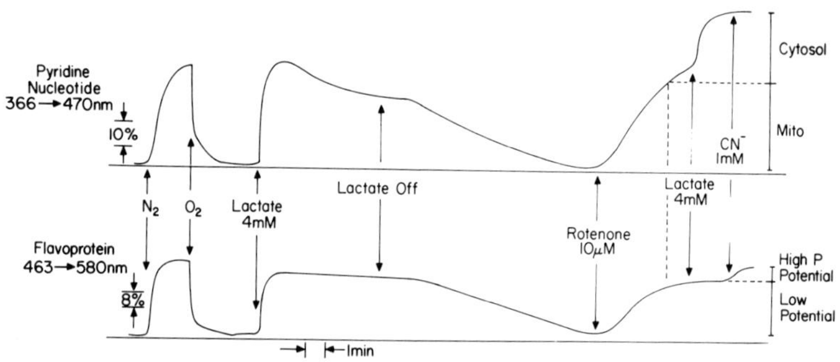

Fig. 2 Oxidation-reduction state of flavin and pyridine nucleotides in rat heart. Rat hearts were perfused in a modified Langendorff flow-through system. Initially, no exogenous substrate was supplied. A compensated double fluorometer was used to simultaneously excite and monitor pyridine nucleotide and flavoprotein fluorescence at the indicated wave lengths. Reduction of both pyridine nucleotide and flavoprotein is indicated by an upward deflection of both traces.

redox changes of pyridine nucleotides occurring exclusively within the cytosolic compartment. Production of a cycle of anoxia or infusion of lactate induces a rapid reduction of both flavin and pyridine nucleotides. It should be noted that while infusion of lactate produced a rapid monotonic reduction of flavoproteins to a constant steady-state level, pyridine nucleotide reduction was characterized by a transient overshoot. These changes can be interpreted as an initial reduction of the cytosolic pool of pyridine nucleotides produced by the infusion of lactate and associated with very high lactate/pyruvate ratios. Generation of pyruvate in the cytosol via lactate dehydrogenase rapidly restores a normal lactate/pyruvate ratio, and the redox state of pyridine nucleotide in the cytosolic compartment then falls to a new stady state level. Following re-establishment of the initial steady state redox baseline characteristic of respiration on endogenous substrates, addition of rotenone, which blocks electron transport at the first phosphorylation site between the high and low potential pools of flavoproteins, again produced a reduction of both flavin and pyridine nucleotides. As indicated by the broken line in figure 2, however, it can be seen that at a time when flavoprotein reduction is essentially complete, further pyridine nucleotide reduction is still occurring. Subsequent addition of lactate then produced a further reduction only of pyridine nucleotides, indicating that the slow response was of cytosolic origin. Thus, of the maximum pyridine nucleotide fluorescence response obtained in the rat heart, it can be seen that approximately 40% can be attributed to a distinct cytosolic pool, the remainder being mitochondrial. Subsequent addition of CN^- produced no further pyridine nucleotide reduction, but is merely shown to illustrate separation of the flavoprotein fluorescence response into a low potential pool located on the substrate side of the rotenone inhibitory site (80% of the total) and a high potential pool located on the oxygen side of the rotenone inhibitory site. A similar separation is seen in isolated heart mitochondria (2).

The effect of aminooxyacetic acid on the redox state of the cytosolic and mitochondrial pools of pyridine nucleotide is shown in figure 3. A control cycle of flavin and pyridine nucleotide reduction produced by infusion of lactate is shown as the first addition. Addition of 0.4 mM aminooxyacetic acid, followed by a second cycle of lactate infusion, clearly illustrates that when intracellular hydrogen transfer by the malate shuttle is blocked, only the cytosolic pyridine nucleotide pool becomes reduced. Since generation of pyruvate via lactate dehydrogenase and its entry into the citric acid cycle is prevented by the inability of the cell to dispose of the cytosolic NADH, no reduction of mitochondrial pyridine nucleotides or flavin pools is observed. As a control, it can be seen that subsequent infusion of pyruvate produced a completely normal cycle of reduction of both flavin and pyridine nucleotides. Since NADH oxidation by the α-glycerophosphate shuttle would be associated with an equivalent of pyruvate oxidation in the citric acid cycle, it may be tentatively

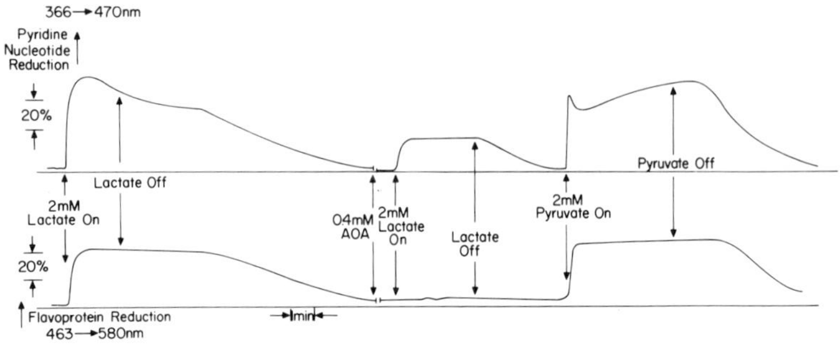

Fig. 3 Effect of aminooxyacetic acid on the oxidation-reduction state of flavin and pyridine nucleotides in rat heart. The conditions were similar to those of figure 2.

concluded that in the normal rat heart, the malate shuttle is the primary mechanism for transfer of reducing equivalents from the cytosolic to the mitochondrial compartment.

The effect of aminooxyacetic acid on several parameters relevant to the hydrogen shuttle function in hearts from normal and hyperthyroid rats is shown in table I. Daily subcutaneous injection of 100 μg/100 g body weight of triiodothyronine induced a 6-fold increase of α-glycerophosphate oxidase activity. However, in both normal and hyperthyroid rat hearts, the measured tissue concentration of α-glycerophosphate remained far below the *Km* of the

Table I

Effects of Aminooxyacetic Acid (AOA) on Perfused Hearts from Normal and Hyperthyroid Rats

α-Glycerophosphate activity was measured in rat heart mitochondria isolated by the method of Chance and Hagihara (3). The reaction mixture contained 4 mg mitochondrial protein in 2.5 ml of buffer containing 100 mM potassium phosphate, 1 mM KCN, 10 mM $MgCl_2$, 0.01 mM cytochrome *c*. Reduction of cytochrome *c* was measured in a Hitachi split beam spectrophotometer. The reaction was started by the addition of α-L-glycerophosphate to the measure cuvette to an initial concentration of 25 mM. Due to very low rates of α-glycerophosphate oxidation by mitochodria from normal rats, the *Km* obtained is only approximate.

Treatment	AOA conc.	α-GP oxidase Activity	α-GP oxidase *Km*	α-GP conc.	Tissue ratios of: α-GP/DHAP	Tissue ratios of: Lactate/Pyruvate
	mM	*μmole/mg mito/min*	*mM*			
Normal	0	7.9	0.5–1.6	0.06	13 ± 2	55 ± 4
Normal	0.4			3.53	62 ± 8	415 ± 62
Hyperthyroid	0	47.7	1.6	0.04	8 ± 1	22 ± 2
Hyperthyroid	0.4			0.56	21 ± 4	423 ± 91

enzyme, as measured by the procedure of Pette (personal communication) and by oxygen polarigraphic techniques. Thus, little functional significance of the α-glycerophosphate shuttle might be expected even in the hyperthyroid state. The tissue ratios of α-glycerophosphate/dihydroxyacetone phosphate and lactate/pyruvate, from which the redox state of cytosol pyridine nucleotides can be calculated, also show that when lactate is provided as the major substrate for the isolated perfused rat heart, aminooxyacetic acid produces a very large reduction of the cytosolic pyridine nucleotide pool in both normal and hyperthyroid rat hearts. Only in this highly unphysiological state does the tissue concentration of α-glycerophosphate rise to levels where significant rates of hydrogen flux through the α-glycerophosphate shuttle might be anticipated.

As shown in table II, the rate of lactate utilization by the normal and hyperthyroid rat heart is inhibited to approximately the same extent by aminooxyacetic acid. In agreement with the fluorometric studies presented previously, the state of reduction of the cytosolic pyridine nucleotide pool, as determined by the perfusate lactate/pyruvate ratio, increased dramatically. In contrast, the rate of pyruvate utilization and the perfusate lactate/pyruvate ratio were not appreciably affected by aminooxyacetic acid in hearts from either normal or hyperthyroid rats. This finding is expected, since metabolism of pyruvate does not involve generation of NADH in the cytosolic compartment of the cell, while metabolism of lactate must be accompanied by an equivalent transfer of hydrogen from the cytosolic into the mitochondrial compartment.

Since aminooxyacetic acid only produced a 60–70% inhibition of lactate utilization, it was not known at this point whether this represents complete inhibition of the malate-aspartate shuttle with the remaining 30% of the hydrogen transfer by other hydrogen shuttle mechanisms, or whether hydrogen transfer by the malate-aspartate shuttle was only partially inhibited. In other

Table II

Effect of Aminooxyacetic Acid (AOA) on Substrate Utilization in Perfused Rat Hearts

The rate of lactate and pyruvate utilization was measured by removing 1 ml aliquots of perfusate at 15 and 45 min. Initial studies demonstrated that the utilization was essentially linear during this interval. Lactate/pyruvate ratios shown were measured in the perfusate.

Treatment	AOA conc.	Lactate uptake	Percentage of change	Lactate/Pyruvate	Pyruvate uptake	Lactate/Pyruvate
	mM	*μmole/g dry wt/h*			*μmole/g dry wt/h*	
Normal	0	264 ± 7		29	767 ± 10	0.31
Normal	0.4	78 ± 8	– 70	1007	714 ± 26	0.26
Hyperthyroid	0	496 ± 10		13	1018 ± 15	0.94
Hyperthyroid	0.4	189 ± 9	– 62	295	959 ± 30	0.55

studies with a flow-through, rather than a recirculating perfusion system, aminooxyacetic acid was found to inhibit lactate utilization by more than 90% with hearts from both normal and hyperthyroid rats. A possible explanation for the lack of complete inhibition found with the recirculating system can be deduced from the data given in table III. Tissue contents of various metabolic intermediates in rat hearts perfused with lactate are shown. All citric acid cycle intermediates were lowered by aminooxyacetic acid in both normal and hyperthyroid rat hearts. Thus, not only is hydrogen flux through the malate-aspartate shuttle diminished, but the limitation of pyruvate entry into the citric acid cycle also leads to loss of cycle intermediates. Decreased flux through the citric acid cycle can also explain the lower levels of creatine phosphate and ATP observed in the presence of aminooxyacetic acid. The increase of tissue aspartate could explain the lack of complete inhibition of hydrogen transfer by aminooxyacetic acid, since this is a competitive inhibitor with respect to aspartate (10). The possibility of a limited flux through the α-glycerophosphate shuttle in the presence of aminooxyacetic acid, caused by the greatly increased concentrations of α-glycerophosphate, cannot be entirely eliminated. However, if it operates, it does not appear to be more active in the hyperthyroid than in normal rat hearts. It may be noted that this conclusion is different from that of Isaacs *et al.* (5).

The rate of oxygen utilization by the isolated rat heart perfused with lactate was found to be 3,100 μatoms/g dry wt/h. Since the rate of lactate utilization was only 264 μmole/g dry wt/h (corresponding to 1,580 μatoms oxygen equivalents/g dry wt/h), oxidation of endogenous fatty acids was assumed to account for the remainder of the oxygen consumption. It was possible,

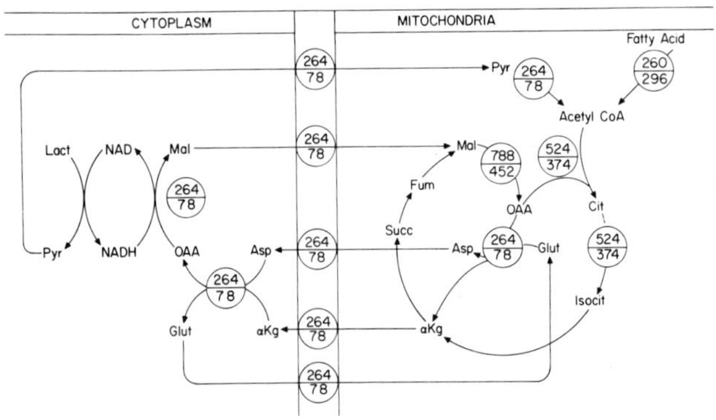

Fig. 4 Flux through the malate-aspartate shuttle and citric acid cycle in rat heart perfused with lactate. Flux rates in μmole/g dry wt/h through the various enzymatic pathways indicated were calculated from the rate of lactate utilization and oxygen consumption. Values in the lower half of the circles are rates in the presence of 0.4 mM aminooxyacetic acid. Oxygen utilization was obtained by measuring oxygen tension in the venous effluent with a Clark electrode and recording the volume of perfusate over a 10 min interval.

Table III

Effects of Aminooxyacetic Acid (AOA) on Hearts from Normal and Hyperthyroid Rats Perfused with 5 mM L(+)lactate

Hearts were quickly frozen with Wollenberger clamps cooled with liquid nitrogen at the end of a 45 min period of perfusion. Metabolic changes in the perfusion fluid were measured over the time interval from 15 to 45 min. Metabolites were extracted with 2.5 ml 6% PCA from the frozen powder and assayed according to Williamson (13). The tissue oxaloacetate (OAA) was calculated from the relationship $\frac{\text{malate}}{\text{OAA}} \times \frac{\text{lactate}}{\text{pyruvate}} = 4$, and the total tissue malate content (13).

Treatment	AOA conc.	Malate	OAA (calc.)	Citrate	α-Keto glutarate	Aspartate	Glutamate	Creatine-P	ATP/ADP
	mM		*mμmole/g dry wt*				*μmole/g dry wt*		
Normal	0	794	6.6	953	404	1.90	16.1	21	3.9
Normal	0.4	235	0.06	80	125	6.44	4.7	11	2.6
Hyperthyroid	0	472	7.4	507	358	3.20	15.8	9	3.2
Hyperthyroid	0.4	270	0.23	89	57	9.71	9.0	4	2.1

therefore, to calculate the relative flow rates of acetyl-CoA into the citric acid cycle from lactate and fatty acid metabolism. The transport rates of intermediates of the malate-aspartate shuttle across the mitochondrial membrane can also be calculated from the rate of lactate uptake. Flux rates of intermediates through these pathways expressed in μmole/g dry wt/h are indicated in the upper half of the circles shown in figure 4. Aminooxyacetic acid decreased oxygen consumption by approximately 30%, while lactate utilization was inhibited by 70%. New flux rates through these enzymatic steps in the presence of aminooxyacetic acid are shown in the lower half of the circles. In addition to the inhibition of hydrogen flux through the malate-aspartate shuttle and of pyruvate entry into the citric acid cycle produced by aminooxyacetic acid, metabolism of fatty acids is increased, but cannot maintain normal citric acid cycle flux. Thus, decreased energy production by the citric acid cycle probably explains the decreased tissue ATP and creatine-P values observed after treatment of hearts with aminooxyacetic acid with lactate as sole exogenous substrate.

SUMMARY

The malate-aspartate shuttle appears to be the primary mechanism for transfer of reducing equivalents from the cytosol to the mitochondrial electron transport chain in rat heart. Not only does the rate of hydrogen transfer by this shuttle regulate the redox state of NAD in the cytosolic compartment, but it also controls pyruvate entry into the citric acid cycle and, therfore, the energy balance of the heart. In addition, the lactate/pyruvate ratio, widely used as an indicator of tissue hypoxia, is not necessarily related to the state of tissue oxygenation, but must always be considered in terms of the rate of reducing equivalent production in the cytosol, and the rate of hydrogen transfer by the malate-aspartate shuttle to the respiratory chain. It remains to be determined whether an impairment of the malate-aspartate H-transport shuttle is involved in heart failure associated with certain disease states.

REFERENCES

1. BÜCHER, T. und KLINGENBERG, M.: Wege des Wasserstoffs in der lebendigen Organisation. Ang. Chem. *70:* 552–570 (1958).
2. CHANCE, B.; ERECINSKA, M.; LEE, C.-P. OSHINO, R.; OHNISHI, T. and PRING, M.: Flavoproteins of the respiratory chain. In: H. KAMIN, Flavins and Flavoproteins (University Park Press, Baltimore 1971).
3. CHANCE, B. and HAGIHARA, B.: Initiation of succinate oxidation in aged pigeon heart mitochondria. Biochem. Biophys. Res. Commun. *3:* 1–5 (1960).
4. HOPPER, S. and SEGAL, H. L.: Kinetic studies of rat liver glutamic-alanine transaminase. J. Biol. Chem. *237:* 3189–3195 (1962).
5. ISAACS, G. H.; SACKTOR, B. and MURPHY, T. A.: The role of the α-glycerophosphate cycle in the control of carbohydrate oxidation in heart and in the mechanism of action of thyroid hormone. Biochem. Biophys. Acta *177:* 196–203 (1969).
6. LARDY, H. A.; PAETKAU, V. and WALTER, P.: Paths of carbon in gluconeogenesis and lipogenesis: the role of mitochondria in supplying precursors of phosphoenolpyruvate. Proc. Nat. Acad. Sci. U. S. A. *53:* 1410–1415 (1965).
7. LEE, Y. and LARDY, H. A.: Influence of thyroid hormones on L-α-glycerophosphate

dehydrogenase and other dehydrogenases in various organs of the rat. J. Biol. Chem. *240:* 1427–1436 (1965).

8. LEHNINGER, A. L.: Phosphorylation coupled to oxidation of dihydrodiphosphopyridine nucleotide. J. Biol. Chem. *190:* 345–359 (1951).
9. PURVIS, J. L. and LOWENSTEIN, J. M.: The relation between intra- and extramitochondrial pyridine nucleotides. J. Biol. Chem. *236:* 2794–2803 (1961).
10. ROGNSTAD, R. and KATZ, J.: Gluconeogenesis in the kidney cortex. Effects of D-malate and aminooxyacetate. Biochem. J. *116:* 483–491 (1970).
11. SCHOLZ, R.; THURMAN, R. G.; WILLIAMSON, J. R.; CHANCE, B. and BÜCHER, Th.: Flavin and pyridine nucleotide oxidation-reduction changes in perfused rat liver. J. Biol. Chem. *244:* 2317–2324 (1969).
12. VON JAGOW, G. and KLINGENBERG, M.: Pathways of hydrogen in mitochondria of *Saccharomyces carlsbergensis.* European J. Biochem. *12:* 583–592 (1970).
13. WILLIAMSON, J. R.: Glycolytic control mechanisms. I. Inhibition of glycolysis by acetate and pyruvate in the isolated, perfused rat heart. J. Biol. Chem. *240:* 2308–2321 (1965).
14. WILLIAMSON, J. R. and KREBS, H. A.: Acetoacetate as fuel of respiration in the perfused rat heart. Biochem. J. *80:* 540–547 (1961).

Mailing address: Brian Safer, M.D., Johnson Research Foundation, University of Pennsylvania, Philadelphia, Pennsylvania 19104 (USA).

Radioisotope Service, New York Veterans Administration Hospital, the Department of Medicine, New York University School of Medicine, and the Department of Biochemistry, New York University School of Dentistry, New York, New York, USA

EARLY CHANGES IN ACUTE CARDIAC OVERLOAD: STUDIES ON ADENYL CYCLASE ACTIVITY, CYCLIC 3′5′-AMP, AND MYOSIN SYNTHESIS [1]

S. S. SCHREIBER, I. KLEIN, M. ORATZ,
C. EVANS, I. GUEYIKIAN, and M. A. ROTHSCHILD

INTRODUCTION

Prolonged increase in cardiac contractile action is associated with myocardial hypertrophy and increased protein synthesis (10, 19). With the use of *in vitro* perfusion, the early changes in protein synthesis following the onset of acute overload (increased aortic pressure) have been traced in the young guinea pig heart unaffected by the endocrine and neural homeostatic mechanisms of the body. Measurement of total left ventricle protein showed that there was an increased incorporation of labeled ^{14}C-lysine into the protein as early as 2–3 h after onset of the *in vitro* stress (27), and the augmented incorporation was not due to alterations in lysine transport since measurements of ^{14}C-lysine specific activity within the cell were found to be the same in overloaded or control ventricles (30). Furthermore, microsomes isolated from ventricles subjected to the stress of increased aortic pressure also showed greater capacity to incorporate ^{14}C-lysine or -leucine into protein than did microsomes from hearts perfused under control conditions (28). The increased microsome incorporation of labeled amino acid was seen in the overloaded microsomes whether incubated in their own cell sap or in cell sap from controls, and the change was noted as early as 1 h after the onset of increased aortic pressure (28). Studies with the artificial messenger polyuridylic acid (polyU) and ^{14}C-phenlalanine showed that polyU increased the capacity of microsomes from control ventricles to incorporate ^{14}C-phenylalanine, indicating that free functioning ribosomes were available in the control microsomes. However, although the microsomes from overloaded ventricles incorporated more ^{14}C-phenylalanine than the controls, the addition of polyU did not further stimulate incorporation (26). The latter data suggested that the overloaded microsomes were saturated with mRNA and, hence, polyU had no effect. Preinjection and perfusion with actinomycin D, an inhibitor of RNA synthesis, prevented the increased protein incorporation of labeled amino acids by microsomes from overloaded ventricles but had no effect

[1] Supported in part by a grant from the National Institutes of Health, No. 09562.

on control microsomes. Microsomes from actinomycin-treated and overloaded ventricles could now be stimulated by polyU. The polyU data and the results with the actinomycin experiments (26) supported the thesis that new mRNA was formed and released to the ribosomes shortly after onset of overload and the aggregation with the mRNA resulted, at least initially, in increased protein synthesis. This was supported by the finding that ^{3}H-uridine incorporation was not increased in ribosomal RNA but was increased in the RNA of heavier aggregates, and secondly by the obliteration of the latter by actinomycin D (26). Although nuclei obtained from overloaded left ventricles did not appear to have more messenger activity when incubated with liver ribosomes in an *in vitro* system, nuclear RNA polymerase studies (29) demonstrated that manganese-stimulated polymerase activity forming DNA-like RNA or mRNA (22) was increased as early as 20 min after onset of acute overload (increased aortic pressure) while the magnesium-stimulated polymerase activity forming ribosomes like RNA (22) was active after 1 h or more of stress (29).

Thus, the data suggested that acute onset of increased aortic pressure in the intact perfused heart was followed within 20 min by increased mRNA associated with increased microsome incorporation of labeled amino acids as early as 1 h after onset of overload and, finally, by an evident increase in total protein incorporation of labeled amino acid after 2–3 h of the stress. As stress continued, both mRNA and ribosomal RNA increased to allow continued increases in myocardial protein synthesis (22, 24).

The present report presents data concerned with two questions raised in the studies of acute overload and protein synthesis. First, what type of protein is synthesized early in acute cardiac overload due to increased aortic pressure; and second, can the trigger initiating the increased synthesis be identified? To answer the first question, the synthesis of myosin, myoglobin, and collagen was studied in perfused hearts subjected to acute increase in aortic pressure after 3 h of stress. The possible trigger mechanism was then studied with evaluation of the adenyl cyclase activity in ventricles subjected to the acute stress during the period of onset of stress to the first evidence of increased messenger-like RNA 20 min later. A time of 10 min after onset of perfusion was selected as the period of study. In addition, direct measurements of the cyclic 3′5′-AMP in the ventricles were made in hearts perfused with a ^{3}H-adenosine precursor.

METHODS

Young guinea pigs were used in this study (average weight 250–300 g) and hearts prepared for perfusion as described previously (27, 28, 30). The heart is doubly cannulated with an inflow cannula secured in the left ventricle and the outflow cannula secured in the aorta. Perfusion was directed into the left ventricle, which then pumped the fluid load out of the aorta. (The right atrium and ventricle were not perfused, except for the coronary circulation, which was pumped out the opened pulmonary artery against no resistance.) Both the aortic flow and the coronary flow (emptying out the opened pulmonary artery) were directed back to 500 ml of perfusate, which was recirculated. The aortic pressure was controlled by a variable aperture flow meter and was monitored continuously by a pressure transducer connected to the aorta by a T-tube. Ventricular flow was controlled by a polystaltic pump and monitored by a flow meter. Control perfusion consisted of a ventricular load of 7 ml/min and aortic pressure of 30–35 mm Hg. Under these conditions, the hearts beat spontaneously with no evidence of failure during the 3-h study period. Overload was accomplished by narrowing the aortic outflow so that the aortic pressure was increased to 60–70 mm Hg while the ventricular inflow was maintained at control levels. The perfusate, a modified Krebs-Henseleit amino acid solution (27, 28, 30), was maintained at 37°C and oxygenated with 95% O_2-5% CO_2 with the pH at 7.45 ± 0.05.

In this study, we have defined "synthesis" of protein as incorporation of labeled amino acid ^{14}C-lysine into total protein, myosin, or myoglobin, and synthesis of collagen as the conversion of ^{14}C-proline into ^{14}C-hydroxyproline.

Myosin

After completion of perfusion with ^{14}C-lysine, the auricles and right ventricular wall were removed from the preparation. The septum was considered part of the left ventricle. Myosin extraction and purification was carried out as described by Davis *et al.* (5) at 0–4°C, the isolated myosin tested for purity by a gel electrophoresis method described by Heywood *et al.* (13) and Small *et al.* (32). Possible RNA contamination of myosin was not found in the isolated myosin preparation, since analysis by an orcinol method (26) was negative. The isolated myosin was analyzed for protein content and radioactivity due to ^{14}C-lysine (27) and the cpm incorporated/mg myosin determined. Intracellular ^{14}C-lysine specific activity was also determined by amino acid analysis of the cytoplasmic supernatant fluid, following protein precipitation. Myosin synthesis was expressed as μmoles lysine incorporated/g myosin nitrogen (N) based on the intracellular lysine specific activity (^{14}C-lysine in myosin/g myosin N divided by intracellular ^{14}C-lysine/μmole lysine).

Myoglobin

After completion of perfusion with ^{14}C-lysine, myoglobin was extracted from the left ventricle by the methods described by Theorell and De Duve (35) and Miyoshi *et al.* (20) with the modification that 10 mg of carrier myoglobin obtained from sperm whale muscle was added to the homogenizing solution. The addition of the carrier myoglobin allowed the isolation and purification of small amounts of myoglobin found in the left ventricles. Since there was no significant difference in weights of control and overloaded ventricles, the contribution of endogenous myoglobin did not produce a significant error in analysis. Identification of the isolated myoglobin and the purity were tested by determining the absorption spectrum between 260 and 620 mμ. The spectra of the preparations followed the patterns described for pure myoglobin (actually converted to metmyoglobin in the preparation) (20, 35). In addition, the purity was tested by acrylamide gel electrophoresis, and four distinct bands were seen as described (20). Contamination with hemoglobin was not evident. After isolation of the pure myoglobin, the protein was analyzed for radioactivity as described earlier (27) and the ^{14}C-lysine incorporated/g myoglobin determined based on intracellular ^{14}C-lysine specific activity. Synthesis was expressed as μmole lysine incorporated/g myoglobin N.

Connective tissue

The conversion of ^{14}C-proline into ^{14}C-hydroxyproline was taken as a measure of collagen synthesis. The hearts were perfused for 3 h with ^{14}C-proline in the standard amino acid Krebs-Henseleit solution (excluding hydroxyproline) and the left ventricles isolated as described above. An aliquot of the left ventricle was analyzed for incorporation of ^{14}C-proline into total ventricular protein. The remainder was homogenized in 0.9% NaCl and then hydrolyzed in 6 N HCl for 48 h, dried *in vacuo,* and the residue washed to remove the HCl. The resulting redried hydrolysate was analyzed for hydroxyproline with a split stream amino acid analyzer (Dr. Louis Fishman, Department of Biochemistry, New York University School of Dentistry) and aliquots counted in a scintillation counter. Intracellular ^{14}C-proline specific activity could not be accurately determined because of inadequate separation of the proline on our columns. Therefore, the synthesis of collagen, conversion of ^{14}C-proline to ^{14}C-hydroxyproline, was compared with simultaneously measured incorporation of labeled proline into total left ventricular protein.

Adenyl cyclase activity in perfused control and overloaded left ventricles

Control and overloaded left ventricular muscle was tested for adenyl cyclase activity after 10 min of perfusion with standard Krebs-Henseleit solution. Following completion of the perfusion, the ventricle was isolated, and a portion, 250–300 mg, cut finely and homogenized in 0.25 M sucrose with a Teflon pestle glass tube homogenizer at 1°C. Centrifugation and determination of protein was as described by Levey and Epstein (18). Adenyl cyclase activity was assayed as recently described by Krishna *et al.* (16) and Levey and Epstein (18). The particulate matter obtained from the cardiac homogenate was incubated at 37°C for 3 min with ^{32}P-α labeled ATP in a volume of 0.15 ml. ATP-^{32}P activity was $7 \times 10^5 - 7 \times 10^6$ counts/min/μmole, and the ATP concentration used in all studies was 5 mM. Paired experiments, control and overload, were run on the same day, and the particulate matter isolated from each, incubated simultaneously with solutions containing identical ATP-^{32}P-specific activites. After 3 min, the reaction was stopped with stopping solution (18) containing ^{14}C-cyclic 3′5′-AMP, (1,500–2,500 counts/min per incubation tube) and cyclic 3′5′-AMP, 1 μmole as carrier, and the mixture boiled for 3 min. The ^{14}C-cyclic 3′5′-AMP acted as an index of recovery (30–35%) during the procedure (16, 18). After boiling, the incubate was fractionated on a Dower-50 column as described by Levey and Epstein (18) and cyclic 3′5′-AMP isolated. The ^{32}P- and ^{14}C-cyclic 3′5′ AMP were then counted in a liquid scintillation counter in Bray's solution. The purity of the cyclic 3′5′-AMP isolated by the column method was verified according to the method of Krishna *et al.* (16) and supported by our studies using the same techniques. ATP-^{32}P, in the activity used, did not appear in the isolated fraction from the Dowex column after treatment with $ZnSO_4$ and $Ba(OH)_2$ (16, 18). The ^{14}C-cyclic 3′5′-AMP allowed the calculation of the total ^{32}P cyclic 3′5′ AMP formed in the incubation mixture (*i.e.*, counts due to ^{32}P-cyclic AMP/counts due to ^{14}C-cyclic AMP multiplied by the counts due to ^{14}C-cyclic AMP in original mixture). The ^{32}P-cyclic 3′5′-AMP divided by the ATP-^{32}P specific activity in the incubate corrected for the protein concentration (mg) allowed expression of the results as cyclic 3′5′-AMP pmole/mg left ventricle protein.

Direct measurement of cyclic 3′5′-AMP in the perfused and overloaded left ventricle

Guinea pig hearts were prepared for perfusion as described above, with cannulation of the left ventricle as well as the aorta. However, the aorta itself was back perfused (Langendorff procedure) with 30–35 ml of amino acid fortified Krebs-Henseleit solution (27) containing 1 mc of ^{3}H-adenosine (sp.act. 26 mc/μmole), which gave approximately 6–10 × 10^6 cpm/ml in a liquid scintillation counter. The perfusate containing the labeled adenosine was recirculated by a varistaltic pump that lifted the perfusate to a vented syringe (10 ml capacity) 60–62 cm above the heart. The fluid then entered the aorta via a no. 18 needle and catheter and the coronary perfusion pressure was maintained at 60–62 cm of H_2O throughout the loading period. In addition, temperature was maintained at 37°C, by suspending the heart in a doubly jacketed chamber with circulating bath in the outer jacket. The perfusate leaving the cardiac coronary circulation was allowed to drop into a small double jacketed reservoir and was then recirculated by the varistaltic pump to the elevated syringe 60 cm above the heart. After 30 min of perfusion with the ^{3}H-adenosine, the heart and its ventricular and aortic cannulae was rinsed with nonradioactive Krebs-Henseleit solution and attached to the perfusate apparatus, described in the methods above, used for control and overload perfusion (27). After 2 min of equilibration at control levels, the left ventricles were perfused as described in "Methods" above (27) with nonradioactive amino-acid Krebs-Henseleit solution for exactly 10 min. Two hearts were run on the same day with identical ^{3}H-adenosine concentrations and loading periods; one was then perfused for 10 min with the nonradioactive perfusate under control perfusion conditions (ventricular perfusion flow 7 ml/min and aortic pressure 30–35 mm Hg) and the second perfused for 10 min under left ventricular overload with the same perfusion flow but with increased aortic pressure (60–70 mm Hg). At the end of the control or overload perfusion, the cannulae, the auricles, and right ventricle were immediately cut away and the left ventricle dropped into liquid nitrogen. The heart in liquid N_2 was pulverized and the powder transferred to a glass homogenizer with 1.5 ml of ice-cold perchloric acid (5%), 0.25 ml of 0.05 M cyclic 3′5′-AMP and 1,500 counts/min of ^{14}C-cyclic 3′5′-AMP as carriers, according to the techniques described by Aurbach and Houston (2) and the methods used to isolate cyclic AMP in adenyl cyclase studies (16, 18). After homogenization and wash with 5% perchloric acid, the homogenate and wash were pooled and centrifuged at 6,000 × *g* and 0°C for 10 min and the precipitate saved for determination of protein (27). The supernatant fluid was neutralized (pH 7.6) with 4 M KOH and, after standing 1 h at 1°C, the mixture was again centrifuged at 6,000 × *g* to remove the $KClO_4$ formed. The supernatant obtained from the latter is designated F.

The validity of the determination of the amount of cyclic 3′5′-AMP present in the left ventricle was directly dependent on the available ^{3}H-ATP formed from the ^{3}H-adenosine. Measurements of the ^{3}H-ATP could not be made until the end of perfusion. Since control or overloaded hearts were loaded with ^{3}H-adenosine in identical fashions, it may be assumed that the ATP activity was the same in both at the onset of the control or overload perfusions. Furthermore, since ATP levels in control and overloaded hearts were found to be the same (26), the specific activity of the cardiac ^{3}H-ATP determined at the end of the 10-min perfusion was used in the calculations of the cyclic AMP present at that time.

Therefore, an aliquot of the cardiac supernatant fluid F above was analyzed for ATP concentration (25) and the rest treated to obtain the specific activity of the ^{3}H-ATP as well as the radioactive ^{3}H-labeled cyclic 3′5′-AMP. Two 1 ml aliquots of F were used for the initial column separation and the collections pooled. Each 1 ml of supernatant fluid F was placed on a Dowex column (50 × 8 H^+, mesh 100–200, 0.4 × 8.5 cm (2, 7)). The column was eluted with distilled water. The 2nd and 3rd ml of effluent (called F-1) was saved for determination of ATP specific activity. The 7th–12th ml of effluent (called F-3) was also saved for the determination of cyclic 3′5′-AMP labeled with ^{3}H (2,7).

Fraction F-3 was treated twice with $Ba(OH)_2$ and $ZnSO_4$ to precipitate nucleotides according to the methods of Krishna *et al.* (16). The purity of the cyclic 3′5′-AMP remaining after precipitation has been verified (7, 8, 16). Nevertheless, when the column was tested with ^{3}H-adenosine, it was found that a small amount of ^{3}H-adenosine contaminated the F-3 fraction (approximately 100 counts/min appearing from the 1–2 $\times$ 10^6 counts/min total activity placed on the column). To further purify the cyclic 3′5′-AMP, the F-3 fraction obtained after column elution and $Ba(OH)_2$ and $ZnSO_4$ treatment was pooled with an additional F-3 fraction from the same ventricle and similarly treated. The pooled fractions were dried, dissolved in an ethanol-ammonia mixture, and run on silica gel chromatography according to methods described by Dighe *et al.* (7). The cyclic 3′5′-AMP was sharply separated by this method, and no ^{3}H-adenosine activity was found to contaminate the final cyclic AMP fraction. This final fraction contained ^{3}H-cyclic AMP as well as ^{14}C-cyclic AMP and nonradioactive cyclic AMP, which were used as carriers. Both ^{3}H- and ^{14}C-cyclic AMP were counted in a liquid scintillation counter, ratio of ^{3}H cpm/^{14}C cpm determined, and the total ^{3}H-cyclic 3′5′-AMP in the original 1 ml of ventricular homogenate was calculated by multiplying this ratio by the amount of ^{14}C activity added to the homogenate as carrier.

Fraction F-1 above contained most of the ATP, ADP, and 5′-AMP, while most of the adenosine was retained by the column (16). Some ^{3}H-labeled inosine and hypoxanthine were also presumed to be present (3). An aliquot was used to determine the ATP concentration of F-1 and then 9 μmole of nonradioactive ATP were added as carrier to each ml of F-1. The F-1 fraction was then chromatographed on a formate column according to the method of Herbert *et al.* (12). The eluate, consisting of increasing formate concentrations, was run through an Isco UV analyzer. The ATP peak was easily discernible with an eluate concentration of 1–1.2 M ammonium formate, and ATP separated from most contaminating nucleotides, ADP, and hypoxanthine. The fraction equivalent to the central region of the ATP peak was saved. The latter was treated with activated charcoal (33) to remove the ATP and possible remaining contaminants from the formate solution, the charcoal eluted with ethanol-ammonia water (33), dried, and rechromatographed on paper prepared as described by Eggleston and Hems (8). The latter chromatography eliminated any small fraction of contaminating nucleotide, especially ITP and its derivatives (8,15), that may have slipped through the formate separation. The ATP isolated on the paper was eluted, the ATP concentration determined by the firefly method used previously (25), and an aliquot counted for ^{3}H radioactivity in a scintillation counter. The specific activity of the isolated purified ATP (cpm due to ^{3}H-ATP per μmole total ATP, radioactive plus carrier) was determined. Since the original F-1 ATP concentration, before addition of carrier, was known, the cpm/ml due to ^{3}H-ATP in the original F-1 fraction, and hence in the ventricle, could be determined. The carrier dilution of the original ATP concentration was 50:1 and, therefore, the original F-1 ATP content produced no more than a 2% error in the specific activity determination.

With the ^{3}H-cyclic 3′5′-AMP and the ^{3}H-ATP specific activities of F-1 fraction (identical to that of the left ventricle *per se*) known, the pmoles cyclic AMP formed from tritiated ATP could be calculated and were corrected for mg protein in the heart.

RESULTS

Myosin, myoglobin, and collagen synthesis

Intracellular lysine specific activity was determined in order to evaluate any difference in incorporation in the labeled lysine into control or overloaded total left ventricle protein, myosin, and myoglobin. After 3 h of perfusion with perfusate containing identical amounts of ^{14}C-lysine and stable lysine (0.2 mM lysine containing 15 μc ^{14}C-lysine) there was no significant difference in intracellular ^{14}C-lysine specific activity (70,000 cpm/μmole lysine ± 8,000 S.E. for the controls and 68,000 ± 7,000 for the overloaded).

Incorporation of labeled lysine into total protein of overloaded left ventricles was significantly increased after 3 h of perfusion (table I). Some of these data have been reported previously (25, 30). At the same time, measurements of incorporation of ^{14}C-lysine into myoglobin were not significantly different between control and overloaded ventricles (table I). On the other hand, incorporation of labeled lysine into isolated myosin was significantly increased in overloaded left ventricles after the 3 h of increased aortic pressure (68 μmoles ^{14}C-lysine/g protein N compared with 51 in the control, $P<0.005$) (31).

Left ventricular hydroxyproline concentration was determined in 5 overloaded and 5 control hearts after 3 h of perfusion, by the methods described by Buccino *et al.*(4), and no significant difference was seen (table II). In 8 control and 8 overloaded hearts, ^{14}C-proline incorporation into left ventricle protein was measured after 3 h of overload or control perfusion. In 7 of the 8 paired studies, the overloaded left ventricular protein incorporated more ^{14}C-proline than the control (+ 15 to + 108%), with a mean increase of 42%. In the same hearts, the conversion of ^{14}C-proline to hydroxyproline was also measured

Table I

Incorporation of ^{14}C-Lysine into Left Ventricle Protein, Myoglobin, and Myosin

	Control	Overload	
Total LV protein (55)*	34	56	$P<0.001$
Myoglobin (9)	85	84	NS†
Myosin (12)	51	68	$P<0.005$

*The values are expressed as μmole lysine/g protein N. Some of the studies were reported previously (25,30). The numbers in parentheses indicate the numbers of hearts used in each group.

†NS, not significant.

Table II

Hydroxyproline in Left Ventricles from Hearts Perfused for 3 h

Control	Overload
μmole/g wet ventricle weight	
2.8	3.6
3.4	3.9
3.5	3.9
4.8	3.5
3.1	4.4
Mean 3.5	3.8

(table III). No consistent change in formation of hydroxyproline was seen, with 3 increasing in overload, 4 decreasing, and 1 not changed (31).

Adenyl cyclase activity in left ventricles from hearts perfused for 10 min under control or overload conditions

Paired perfusion studies, one control and one overload, were run on the same day, and the adenyl cyclase activity of the left ventricles of each determined simultaneously. Preliminary data obtained with 5 pairs of hearts (table IV) showed that the adenyl cyclase activity of the overloaded ventricles was 11–97% greater than that of the controls.

Direct measurement of cyclic 3′5′-AMP in control and overloaded left ventricles

Measurements of ATP concentration (μmole/mg protein) in left ventricles perfused for 10 min under control or overloaded conditions showed no

Table III

^{14}C-Hydroxyproline in Left Ventricles

^{14}C-Proline in perfusate	^{14}C-Hydroxyproline specific activity in left ventricles		
	Control	Overload	Overload/control change
cpm/ml × *10*$^{-3}$	*cpm/μmole*		%
47	66	86	+ 30
47	52	34	– 34
47	45	45	0
47	69	65	– 6
280	262	391	+ 49
280	144	535	+ 271
280	344	219	– 36
280	282	186	– 35

These studies were reported previously (31).

Table IV

Adenyl Cyclase Activity of Perfused Left Ventricles

	Control	Overloaded	Overload/Control Change %
1	132	150	+11
2	74	136	+85
3	167	202	+22
4	95	188	+97
5	300	392	+31
			Mean +49

Control and overloaded hearts in each experiment were perfused for 10 min on the same day and adenyl-cyclase activity determined in the left ventricle of each at the same time with identical solutions. See Methods. The results are expressed as pmole 3′5′-cyclic AMP/mg protein.

consistent trend. These data are similar to those reported previously (26) and there appeared to be no change in the ATP pool. There was also no consistent difference noted in the specific activity of the ^{3}H-ATP found in the heart after 10 min of perfusion (table V). Preliminary studies showed no consistent difference in the ^{3}H-cyclic 3′5′-AMP concentration found at the end of the 10-min perfusion period, between overload and control.

DISCUSSION

The sequence of events leading to increased protein synthesis in the overloaded heart has been traced in an *in vitro* system, with evidence suggesting that one of the earliest responses to the stress of increased aortic pressure is the formation of mRNA (29) with polysome aggregates programmed to increase incorporation of labeled amino acids into protein, followed by increased ribosomal RNA synthesis (22) and continued increase in protein synthesis (10, 19). The present study indicates that the protein synthesized early in stress does not include all of the cardiac protein. Of the three parameters studied, myosin synthesis, as measured by incorporation of labeled lysine, was increased, whereas collagen and myoglobin synthesis were not significantly altered after 3 h of overload. (31).

Although increases in cardiac myoglobin have been reported after 2 weeks of anoxia (23) and after 150 days of high altitude (1), no significant change from controls was seen in the present study 3 h after onset of the increased aortic pressure. On the other hand, increased ^{14}C-lysine incorporation into isolated myosin afforded strong evidence of increased myosin synthesis. Earlier studies of myosin in stressed hearts *in vivo* have shown that alterations in myosin synthesis were not seen in chronic heart failure in animals subjected to experimental aortic stenosis (6); and myosin obtained from failing dog hearts was chemically

Table V

^{3}H Cyclic 3′5′-AMP in Overloaded and in Control Left Ventricles

Experiment	Loading perfusate ^{3}H-adenosine activity	Left ventricle ATP concentration	Left ventricle ^{3}H-ATP activity	Left ventricle ^{3}H cyclic 3′5′ AMP
	cpm/ml	*μmole/mg protein*	*cpm/μmole*	*pmole/mg protein*
Overload I	4.6×10^{6}	0.020	6.25×10^{6}	26.5
Control I	4.6×10^{6}	0.012	6.47×10^{6}	42.8
Overload II	6.0×10^{6}	0.016	9.17×10^{6}	19.6
Control II	6.0×10^{6}	0.018	10.6×10^{6}	21.2
Overload III	6.1×10^{6}	0.026	1.3×10^{7}	27.1
Control III	6.1×10^{6}	0.031	1.9×10^{7}	19.6

identical to that from controls (5). Electron microscope studies by Hatt *et al.* (11) have shown that, after experimental aortic stenosis, there is a myofibrillar breakdown within a few hours followed by new fibrillar formation. The authors interpreted their findings as a process of increased breakdown followed by increased synthesis of the contractile protein. The data of the present study showing an increased incorporation of lysine into isolated myosin supports the conclusions reached with the electron microscope, since breakdown and synthesis of myofibrils would require formation of myosin. An early increase in myosin synthesis is also supported by the findings of Meerson *et al.*(19) that myofibrillar protein synthesis increases 2–7 days after *in vivo* stenosis is produced.

One of the later responses to stress of the overload type has been an increased connective tissue. Buccino *et al.*(4) found that myocardial collagen concentration, measured by hydroxyproline concentration, was increased in the hypertrophied stressed right ventricle after 20–50 days in most animals subjected to experimental pulmonary artery stricture. Earlier responses with collagen are not reported, but increased incorporation of thymidine into nuclei of endothelial cells and presumably fibroblasts indicated interstitial cell proliferation at 1 day after onset of stress (21). The latter would be expected to precede supportive tissue synthesis as the stress continued.

In view of the role played by the internal messenger cyclic 3′5′-AMP in responses to tested endocrine factors, the role of this messenger was tested in the *in vitro* system. Direct measurements of cyclic AMP formation were carried out in the present study, and the preliminary data obtained from paired control and overloaded ventricles perfused on the same day showed no consistent difference in isolated labeled and purified cyclic 3′5′-AMP. It is understood that the cyclic AMP is rapidly destroyed and even the immediate freezing of the heart in liquid nitrogen at the termination of perfusion might not be rapid enough to evaluate the level of labeled cyclic AMP at any one time. It is also possible that the cyclic AMP is increased at an intracellular site, and measurements of over-all cellular cyclic 3′5′–AMP would not show such local increases.

On the other hand, preliminary measurements of adenyl cyclase activity in ventricles from control and overloaded hearts did show a significant trend, with the activity of the particulate matter obtained from overloaded ventricles consistently higher than that of simultaneously run controls.

The actual significance of increased adenyl cyclase activity in the activity in the acutely overloaded perfused heart remains to be clarified. The relationship between the positive inotropic effect of catecholamines and cyclic 3′5′-AMP has been well documented (9, 34). In the perfused heart, the only available catecholamines are those stored in the myocardium, and although it is possible that these stores are used in overload to initiate adenyl cyclase activation with increased formation of cyclic 3′5′-AMP, there are no data as yet to support this contention. Other possibilities may include removal of an inhibitor during overload perfusion, as well as activation of adenyl cyclase by mechanical overload *per se.* Further studies including localization of the adenyl cyclase activity on nuclei or the sarcotubular system (9) during acute overload are in progress and should permit evaluation of this system in an endocrine-free situation.

It has been shown that histones may inhibit RNA synthesis and may possibly act as DNA repressors (14). The observation that cyclic 3′5′-AMP stimulates histone phosphorylation and that the phosphorylation may modify histone DNA interactions suggests that cyclic AMP may induce RNA synthesis (17). It is interesting to speculate that the adenyl cyclase activity increase in acute overload, seen in the present study, increases cyclic AMP at the nuclear site, resulting in the increased mRNA synthesis reported previously (26).

Thus, the present data support the suggestion that increased mechanical stress in overload leads to wear and tear of the contractile elements, which requires new myosin formation fairly early in the stress. As the stress continues, continued synthesis of the contractile proteins and cytoplasmic proteins and connective tissues leads to hypertrophy. Preliminary studies of adenyl cyclase activity of ventricles, 10 min after onset of stress, suggests that cyclic 3′5′-AMP may be involved in the trigger that initiates the progression of mRNA, rRNA, and protein synthesis.

SUMMARY

Cardiac overload is followed by increased protein synthesis leading to hypertrophy. Studies of the mechanisms involved in the increased protein synthesis have been carried out *in vitro* in the acutely overloaded perfused guinea-pig heart. About 20 min after the onset of the stress, nuclear RNA polymerase activity in Mn^{++}, forming a DNA-like RNA, is increased; and as stress continues, formation of Mg^{++} stimulated ribosome-like RNA is increased. Microsomes from overloaded ventricles are programmed as early as 1 h to increase incorporation of labeled amino acids into protein, and total left ventricle protein synthesis is increased after 2–3 h. Studies of synthesis of myosin, myoglobin, and collagen show that myosin synthesis is increased after 3 h of overload, while that of myoglobin and collagen are not consistently different from controls. Evaluation of cyclic 3′5′-AMP as the possible trigger initiating the sequence of events leading to increased protein synthesis was carried out with direct measurements of cyclic AMP as well as adenyl cyclase activity in overload. Although total cyclic AMP did not increase, adenyl cyclase activity was increased after 10 min of overload. It is interesting to speculate that acute overload somehow increases adenyl cyclase activity, which may increase cyclic 3′5′-AMP at or near a nuclear site. The latter may, in turn, act on histone repressors of DNA and permit the formation of mRNA and, later, ribosomal RNA, leading to increased protein synthesis.

ACKNOWLEDGMENT

The authors would like to acknowledge the secretarial assistance of Miss Judith Rosenblum

REFERENCES

1. ANTHONY, A.; ACKERMAN, E. and STROTHER, G.: Affects of altitude acclimatization on rat myoglobin. Changes in myoglobin content of skeletal and cardiac muscle. Amer. J. Physiol. *196:* 512–516 (1959).

2. AURBACH, G. D. and HOUSTON, B. A.: Determination of 3′5′ AMP with a method based on a radioactive phosphate exchange reaction. J. Biol. Chem. *243:* 5935–5940 (1968).

3. BAER, H. P. and DRUMMOND, G. E.: Catabolism of adenine nucleotides by the isolated perfused rat heart. Proc. Soc. Exp. Biol. Med. *127:* 33–36 (1968).

4. BUCCINO, R. A.; HARRIS, E.; SPANN, J. F., Jr. and SONNENBLICK, E. H.: Response of myocardial connective tissue to development of experimental hypertrophy. Amer. J. Physiol. *216:* 425–428 (1969).

5. DAVIS, J. O.; CARROLL, W. R.; TRAPASSO, M. and YANKOPOULOS, N. A.: Chemical characterization of cardiac myosin from normal dogs and from dogs with chronic congestive heart failure. J. Clin. Invest. *39:* 1463–1471 (1960).

6. DESCHRYVER, C. and GUDBJARNASON, S.: Myosin synthesis in chronic heart failure. Proc. Soc. Exp. Biol. Med. *119:* 961–963 (1965).

7. DIGHE, P. K.; PAHUJA, D. N. and SHAH, D. H.: Separation of cyclic 3′5′ AMP from other adenine nucleotides. J. Chromatog. *40:* 449–552 (1969).

8. EGGLESTON, L. V. and HEMS, R.: Separation of adenosine phosphates by paper chromatography and the equilibrium constant of the myokinase system. Biochem. J. *52:* 156–160 (1952).

9. ENTMAN, M. L.; LEVEY, G. S. and EPSTEIN, S. E.: Mechanism of action of epinephrine and glucagon on the canine heart. Circulat. Res. *25:* 429–438 (1969).

10. GUDBJARNASON, S.; TELERMAN, M. and BING, R. J.: Protein metabolism in cardiac hypertrophy and heart failure. Amer. J. Physiol. *206:* 294–298 (1964).

11. HATT, P. Y.; LEDOUX, C.; BONVALET, J. P. et GUILLEMOT, H.: Lyse et synthese des proteins myocardiques au cours de l'insuffisance cardiaque experimentale. Arch. des. mal. du coeur. *12:* 1703–1721 (1965).

12. HERBERT, E.; POTTER, V. R. and TAKAGI, Y.: Nucleotide metabolism. IV. The phosphorylation of 5′uridine nucleotides by cell fractions from the rat liver. J. Biol. Chem. *213:* 923–940 (1955).

13. HEYWOOD, S. M.; DOWBEN, R. M. and RICH, A.: The identification of polyribosomes synthesizing myosin. Proc. Nat. Acad. Sci. *57:* 1002–1099 (1967).

14. HUANG, R. C. and BONNER, J.: Histone, suppressor of chromosomal RNA synthesis. Proc. Nat. Acad. Sci. *48:* 1216–1222 (1962).

15. KREBS, H. A. and HEMS, R.: Some reactions of adenosine and uridine phosphates in animal tissues. Biochim. Biophys. Acta *12:* 172–180 (1953).

16. KRISHNA, G.; WEISS, B. and BRODIE, B. B.: A simple sensitive method for the assay of adenyl cyclase. J. Pharm. Exp. Therap. *163:* 379–385 (1968).

17. LANGAN, T.: Histone phosphorylation: stimulation by adenosine 3′5′ monophosphate. Science *164:* 579–580 (1968).

18. LEVEY, G. S. and EPSTEIN, S. E.: Activation of adenyl cyclase by glucagon in cat and human heart. Circulat. Res. *24:* 151–156 (1969).

19. MEERSON, F. Z.; ALEKHINA, E. M.; ALEKSANDROV, P. N. and BAZARDJAN, A. E.: Dynamics of nucleic acid and protein synthesis of the myocardium in compensatory hyperfunction and hypertrophy of the heart. Amer. J. Cardiol. *22:* 337–348 (1968).

20. MIYOSHI, K.; SAIJO, K.; KURYU, Y. and OSHIMA, Y.: Abnormal ultraviolet spectrum in Duchenne type of progressive muscular dystrophy. Science *142:* 490–491 (1963).

21. MORKIN, E. and ASHFORD, T. P.: Myocardial DNA synthesis in experimental cardiac hypertrophy. Amer. J. Physiol. *215:* 1409–1413 (1968).

22. NAIR, K. G.; CUTILETTA, A. F.; ZAK, R.; KOIDE, T. and RABINOWITZ, M.: Biochemical correlates of cardiac hypertrophy. Circulat. Res. *23:* 451–462 (1968).

23. POEL, W. E.: Effect of anoxic anoxia on myoglobin concentration in striated muscle. Amer. J. Physiol. *156:* 44–51 (1949).

24. POSNER, B. E. and FANBURG, B. L.: Ribonucleic acid synthesis in experimental and cardiac hypertrophy in rats. Circulat. Res. *23:* 127–145 (1968).
25. SCHREIBER, S. S.; EVANS, C.; ORATZ, M. and ROTHSCHILD, M. A.: Effects of hyperbaric oxygen on protein synthesis in the mammalian heart. Amer. J. Physiol. *212:* 35–38 (1967).
26. SCHREIBER, S. S., ORATZ, M., EVANS, C., SILVER, E. and ROTHSCHILD, M. A.: Effect of acute overload on cardiac muscle mRNA. Amer. J. Physiol. *215:* 1250–1259 (1968).
27. SCHREIBER, S. S., ORATZ, M. and ROTHSCHILD, M. A.: Protein synthesis in the overloaded mammalian heart. Amer. J. Physiol. *211:* 314–318 (1966).
28. SCHREIBER, S. S.; ORATZ, M. and ROTHSCHILD, M. A.: Effect of acute overload on protein synthesis in cardiac muscle microsomes. Amer. J. Physiol. *213:* 1552–1555 (1967).
29. SCHREIBER, S. S.; ORATZ, M. and ROTHSCHILD, M. A.: Nuclear RNA polymerase activity in acute hemodynamic overload in the perfused heart. Amer. J. Physiol. *217:* 1305–1309 (1969).
30. SCHREIBER, S. S.; ORATZ, M. and ROTHSCHILD, M. A.: The initiation of protein synthesis in the acutely overloaded perfused heart. Proceedings of the Cardiac Workshop, University of Vermont School of Medicine, August 1969 (Academic Press, New York), (in press).
31. SCHREIBER S. S.; ORATZ, M.; ROTHSCHILD, M. A.; EVANS, D. C. and GUEYIKIAN, I.: Myosin myoglobin and collagen synthesis in acute cardiac overload. Amer. J. Physiol. (in press).
32. SMALL, P. A.; HARRINGTON, W. F. and KIELLEY, W. W.: The electrophoretic homogeneity of myosin subunits. Biochim. Biophys. Acta *49:* 462–470 (1961).
33. SMITH, I.: Chromatographic and Electrophoretic Techniques, p. 242 (Interscience Publisher, Inc., New York 1961).
34. SUTHERLAND, E. W. and ROBISON, G. A.: Metabolic effects of catecholamines. Pharmacol. Rev. *18:* 145–161 (1966).
35. THEORELL, H. and deDUVE, C.: Crystalline human myoglobin from heart muscle and urine. Arch. Biochem. *12:* 113–124 (1947).

Mailing address: Dr. Sidney S. Schreiber, Radioisotope Service,
Veterans Administration Hospital, New York, N. Y. 10010 (USA).

Myocardial Metabolism Laboratory, Department of Medicine, Division of Cardiology, Cornell University Medical College, and Institute for Muscle Disease, Inc., New York, USA

THE PATHOGENESIS OF MYOCARDIAL CELL DEATH [1]

N. BRACHFELD[2]

We might well consider the pathogenesis of the process of myocardial cell death by first examining those factors that sustain cellular viability. A variety of histological, histochemical, biochemical, and electrical phenomena are initiated by the onset of cellular ischemia. The clinical manifestations of these changes are determined in part by the fact that such shifts in normal homeostasis do not appear to bear a close temporal relationship.

The clinician's chief concern is whether or not the injury induced by coronary occlusion will cause temporary and reversible changes or permanent damage requiring *de novo* reparative tissue synthesis, *i.e.,* whether or not he is dealing with ischemia or cell necrosis (cell death). It is self-evident that an early distinction between these two states is crucial in effectively designing a program for treating coronary artery disease, and will also help to define the permissible limits of nonperfusion of the arrested heart at surgery as well as the maximal allowable transport time of a cardiac transplant from donor to recipient.

Ultimate vulnerability to myocardial ischemia is not a fixed phenomenon, but is influenced by multiple factors. Models of myocardial ischemia must be precise in describing all parameters of production as well as the steady state environment of the tissue so traumatized. Investigators continue to be handicapped by the lack of adequate animal models of coronary artery disease, and extrapolation of information obtained from the experimental laboratory animal may be misleading. When the entire heart is subjected to surgical anoxic arrest, the presence or absence of concomitant perfusion, the pH, ionic composition, and substrate composition of perfusing fluids (if used), the environmental temperature, and the state of the coronary circulation before arrest, all bear directly on its recovery of function when reperfused with well-oxygenated blood. The severe ischemic insult rarely produces a homogeneous, localized myocardial infarction. More frequently, it produces a heterogeneous mixture (5) of progressively changing proportions of necrotic and living tissue

[1] Supported by research grants from the National Institutes of Health (PH-43-67-1439 and HE-06216), the New York Heart Association, the American Medical Association Education and Research Foundation, and the Muscular Dystrophy Associations of America, Inc.

[2] Recipient of a Career Development Award from the National Advisory Heart Council of the National Institutes of Health, Bethesda, Maryland.

determined by the location of the occlusion, the state of the collateral circulation, and the location of the infarct.

Initially, the ischemic injury causes potentially reversible cellular changes that gradually become fixed and permanent when unrelieved ischemia progresses to necrosis and cell death. The severity of the response is determined by the state of the cell before the injury, the level of endogenous glycogen stores, and the condition of the arterial supply of the heart. There is a significant difference in cellular mortality, as well as in biochemical and histological changes, when the large infarct caused by occlusion of a component artery of a diffusely diseased system with a poor collateral circulation is compared with other areas of ischemic tissue that are supplied with a poor but potentially good collateral supply. In the former, the intrinsic glycogen store of the cell is the only possible substrate available; in the latter, a restricted circulation provides opportunity for significant anaerobic metabolism, changes in pH, and variations in protein and lipid metabolism. Viability also may be regulated in part by the rapidity with which blood supply to the ischemic area can be re-established. Reversibly injured cells in an area distal to a coronary occlusion recover rapidly when circulation has been restored. Stainable glycogen is resynthesized, and contractile activity ensues. The need for a high level of efficient aerobic metabolic activity to meet the hemodynamic demands made upon the injured tissue seriously compromises its potential for recovery and explains the differences in the rate of survival between anoxic arrest induced at surgery and that caused by coronary artery disease. The demands of muscle contraction help seal the fate of borderline ischemic cells. The actual disintegration of muscle fibers is dependent in part upon the activity of the muscle during the period of ischemia (5).

Bing *et al.* (3) have stressed the fact that survival of the major physiological and biochemical functions of the heart need not follow the same "time table." The factors that determine normal cardiac excitability are most vulnerable to anoxia or severe ischemia; those concerned with energy production and utilization survive for much longer periods. Thus, loss of electrical potential may doom the cell at a time when other essential functions are viable and cell injury is potentially reversible.

HISTOLOGY AND HISTOCHEMISTRY OF MYOCARDIAL ISCHEMIA.

The use of the phenothiazine dye, methylene blue, and other redox sensitive color indicators permits gross identification of an ischemic zone within 10 to 15 sec of inception. Although diagnostic changes are not apparent by light microscopy for from 1–2 h after the onset of ischemia, histochemical changes become manifest in 20–30 min. Fortunately, the electron microscope shortens this time considerably and can provide visual evidence of initial diagnostic alterations with 5–15 min. Examination of routine histological sections to determine the extent and severity of ischemic tissue damage is both difficult and misleading. Changes may be delayed or due to secondary causes. Conversely, recent evidence indicates that drastic visual changes may be induced in cell constituents without causing or signifying permanent damage (8). Loss of contractile activity and relaxation of myofibers seen after induction of ischemia are not necessarily permanent and tell little about ultimate disposition. Other reversible alterations include swelling of endothelia, hyperemia of capillaries and venules, margination of leukocytes, an influx of fluid into the cell, and marked reduction in ATP concentration. Glycogen depletion begins almost immediately, and is accompanied by a decrease in phosphorylase activity. Small neutral lipid droplets stainable with oil red O appear opposite the A bands of sarcomeres, at the poles of sarcolemmal nuclei, and between myofibrils adjacent to mitochondria (12, 14). Mitochondrial swelling is accompanied by swelling of the endoplasmic reticulum, and may be due to uncoupling of oxidative phosphorylation or to intracellular ionic changes, particularly those involving calcium transport (10).

Studies on hepatic parenchymal cells after induction of hypoxia have demonstrated cellular protein denaturation with 30 min of interruption of the blood supply and at a stage when damage is still reversible (9). Glinsmann and Ericsson (6) have shown that organelles, portions of cytoplasm, and mitochondria damaged by hypoxia are segregated within the ischemic cell and subsequently are digested within intracellular cytosegresomes through the action of lysosomal hydrolases. These pathological changes in protein structure are probably induced by the intracellular acidosis that accompanies anaerobic glycolysis. Although the alterations noted above are severe, they are nevertheless potentially reversible. If blood flow to the ischemic tissue is re-established before irreversible changes are induced, recovery is rapid and cells return to full functional activity.

IRREVERSIBLE ISCHEMIC CELL DEATH

There is rarely a sharp line of demarcation between reversible and irreversible ischemic damage. More frequently, there is a gradual transition as more and more cells pass beyond the point of recovery. Death in the ischemic traumatized zone appears to follow an exponential curve. At 22 min post-occlusion, the first irreversibly damaged cells appear. Even after 45 min of total hypoxia, 35 to 66% of the cells remain viable (7).

The point of irreversible injury undoubtedly is controlled by the combined effects of metabolic breakdown, autolysis and protein denaturation of advanced degree, and an increased permeability of the sarcolemma that permits excessive loss of coenzymes, enzyme molecules, and electrolytes. Eventually, the biochemical processes of energy production cannot maintain cellular integrity. This point in time is probably reached between 20 and 60 min post-occlusion (7, 9, 14).

It has been noted that the irreversibly injured cells that lie toward the center of the infarcted zone die almost unaltered, whereas those closest to normally perfused areas, usually at the periphery of the infarct, develop marked changes characteristic of infarction. This has also been seen when transient occlusion of a coronary artery is followed by restoration of normal circulation (13). Restoration of flow or persistent inadequate flow permits exchange of substrate, intermediary products of metabolism, fluid, and electrolytes between dead or injured cells and the general circulation, and accentuates histologically apparent changes.

Light and electron microscopy indicate that the endocardium is most resistant to ischemic damage. The anatomical integrity of the endocardial zone often is maintained despite surrounding areas of massive infarction, its Purkinje fibers retain viability, and glycogen stores are abundant. Survival of this zone has been attributed to nourishment via direct diffusion of nutrients from the cardiac chambers or via thebesian veins.

BIOCHEMICAL ASPECTS OF MYOCARDIAL ISCHEMIA

Significant changes in carbohydrate, lipid, and protein metabolism, and alterations in cytoplasmic, mitochondrial, and lysosomal enzyme systems have been described in reversibly injured ischemic cells.

Glycolysis is normally an inefficient and quantitatively insignificant pathway for production of energy by the heart. During severe ischemia, areas of irreversibly injured myocardium are incapable of maintaining adequate stores of ATP solely by glycolytic means. Nevertheless, anaerobic energy production may make a critical contribution to the maintenance of cell viability, to the survival of excitability, and to the function of nodal and conducting tissue until a competent collateral circulation is established (physiologically or surgically) or vasodilator tonus is induced. It may play an important role in promoting the recovery of the acutely anoxic heart.

Anoxic dependence on glycolysis implies an increase in glucose consumption for even marginal cellular function. We have shown that myocardial glucose extraction increases 504% and glucose consumption, 90%, when moderate, localized degrees of ischemia (flow reduced to 63% of control) are induced in the dog heart *in vivo* (4). Oxygen consumption and glucose extraction coefficient showed a striking, fixed, inverse relationship (fig. 1). Biochemical evidence for enhanced glycolysis also is indicated by changes in lactate metabolism, and by an accumulation of glycolytic intermediates in the reduced redox state.

Increased concentrations of lactate may be found in coronary sinus samples within 6 sec after ligation of a coronary artery. Biopsy of the resultant area of ischemia demonstrates a marked increase in tissue lactate and α-glycerophos-

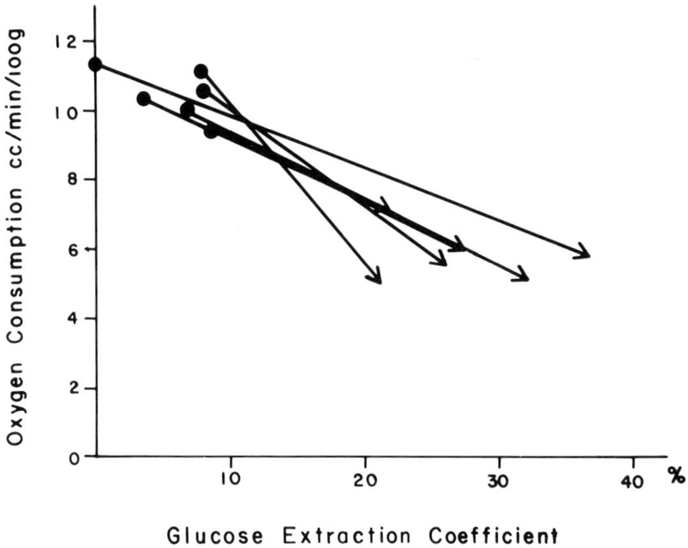

Fig. 1 Relationship of myocardial oxygen consumption to percentage of glucose extracted (glucose extraction coefficient). Control determinations compared with a reduction in flow of 40%

phate concentration. Coronary sinus pH may fall more than 0.4 pH units within 2 min.

Other studies in our laboratory and that of Weissler *et al.* (15) concerning severe myocardial anoxia in the isolated perfused rat heart supported these findings and demonstrated that the inclusion of glucose in circulating perfusate not only enhanced the electrical and mechanical performance of the heart during anoxia, but also permitted more rapid recovery.

EFFECTS OF MYOCARDIAL TISSUE ACIDOSIS

The fall in intracellular pH induced by glycolysis may be critical in determining the eventual fate of ischemic tissue. The heart is extremely susceptible to the accumulation of hydrogen ions. The pH at which activity becomes impossible is significantly higher in the heart than in skeletal muscle. Myocardium also demonstrates a much poorer intrinsic buffering capacity, which enhances the degree of change in pH caused by production of lactic acid. Myocardial refractoriness to vasoactive and inotropic agents and an enhanced susceptibility to the onset and persistence of potentially lethal dysrhythmias during acidosis are well known and may relate to a marked decrement in cardiac pacemaker activity. Acidosis can also cause a sharp fall in left ventricular contractility, in the index of isovolumic work, and in coronary flow.

Most enzyme systems responsible for the maintenance of all intracellular synthetic and energy-producing systems have pH optima within the normal physiological range and are quite sensitive to changes in ionic concentration. Bate-Smith and Bendall (2) have shown that ATP disappears most rapidly at a pH of 6.3, a level at which soluble ATPase shows maximal activity. The proteolytic enzymes normally confined to lysosomes all show acid pH optima approached only by an ischemic cellular environment conducive to intraorganelle physiological or generalized pathological autolysis.

FUNCTION OF MYOCARDIAL ACID HYDROLASES (LYSOSOMES) IN PATHOLOGICAL ISCHEMIC AUTOLYSIS

DeDuve (5a, b) hypothesizes that the primary role of the subcellular particle known as a lysosome or cytosegresome is autophagic digestion. Such autophagic vacuoles appear and soon acquire a complement of acid hydrolases whenever the cell must sacrifice a portion of its own cytoplasm in response to fasting, anoxia, or active catabolism. The products of digestion diffuse through the lysosomal membrane into the cell sap. Indigestible material accumulates within the vacuole until it is finally extruded through the cell membrane. The group of enzymes contained within these particles are capable of digesting the most important constituents of the cell. Destruction of the lysosomal membrane permits enzyme release into the cell sap and may be the primary mechanism for pathological autolysis. Anoxia may stimulate formation of autophagic vacuoles, or the lipoprotein membrane of pre-existing lysosomes may be labilized with release of lytic enzymes into the cytoplasm and digestion of susceptible substrate; in this case, the cell contents themselves. Release *in vitro* follows incubation at an acid pH.

The myocardium undoubtedly is more sensitive to the effects of ischemia than any other tissue in the body and, therefore, presents a favorable environment for hydrolase activity. Anaerobic production of excessive amount of lactic acid may overcome the buffering capacity of the sarcoplasm and decrease cellular pH to a point within the range of enzyme activation.

Solubilization and activation of lysosomal enzymes may occur in response to the acidosis that accompanies a fall in pO_2 rather than to the anoxia itself. They suggest that autolysis of anoxic cells may start with rupture of the lysosomal membrane. The process that renders the particles fragile is rapid and occurs early. As much as 80% of the lysosomal enzymes may be released at a time when morphological and chemical cellular changes are still very discrete (fig. 2).

The full significance of these particle-bound enzymes in the myocardial response to ischemia is uncertain. The mechanism of their release and activation in the myocardium, and techniques for neutralizing or inhibiting release are intriguing subjects of current research.

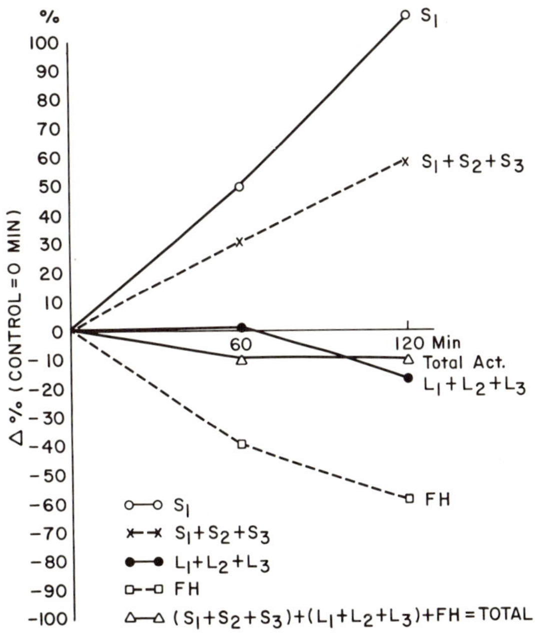

Fig. 2 Redistribution of cathepsin enzyme activity from inactive sites within the lysosomal particle and bound to the final homogenate fraction. Assay was performed following cellular fractionation by preparative ultracentrifugation. Note progressive increase in "free" supernatant activity derived from the cell sap as a function of time, post-mortem. S_1 = first supernatant extraction; $S_1 + S_2 + S_3$ = total supernatant activity (3 extractions);$L_1 + L_2 + L_3$ = total large granule activity (bound); FH = final homogenate activity. Total cathepsin activity is represented by the sum of supernatant, large granule, and final homogenate assay, and did not show a significant change.

EFFECTS OF ISCHEMIA ON MYOCARDIAL LIPID METABOLISM

Myocardial lipid storage has most commonly been associated with acute infectious processes, poisoning, and malignancy, but there is significant histological evidence of lipid storage induced by myocardial ischemia in the experimental animal and man. In animals exposed to low atmospheric pressures or after ligation of the descending ramus of the left coronary artery, the heart shows cytoplasmic accumulation of fat droplets. Storage is temporary and lipids disappear completely after a competent collateral circulation to the ischemic area has been established (14). Droplets are found only in living, actively metabolizing but ischemic tissue, which indicates that they are not an expression of autolytic degeneration. Cells that died rapidly did not accumulate neutral lipids. We have demonstrated that the availability of glucose markedly enhances incorporation of palmitate into myocardial neutral lipids while simultaneously suppressing oxidation of free fatty acid (FFA) (11) (fig. 3). In the anoxic state,

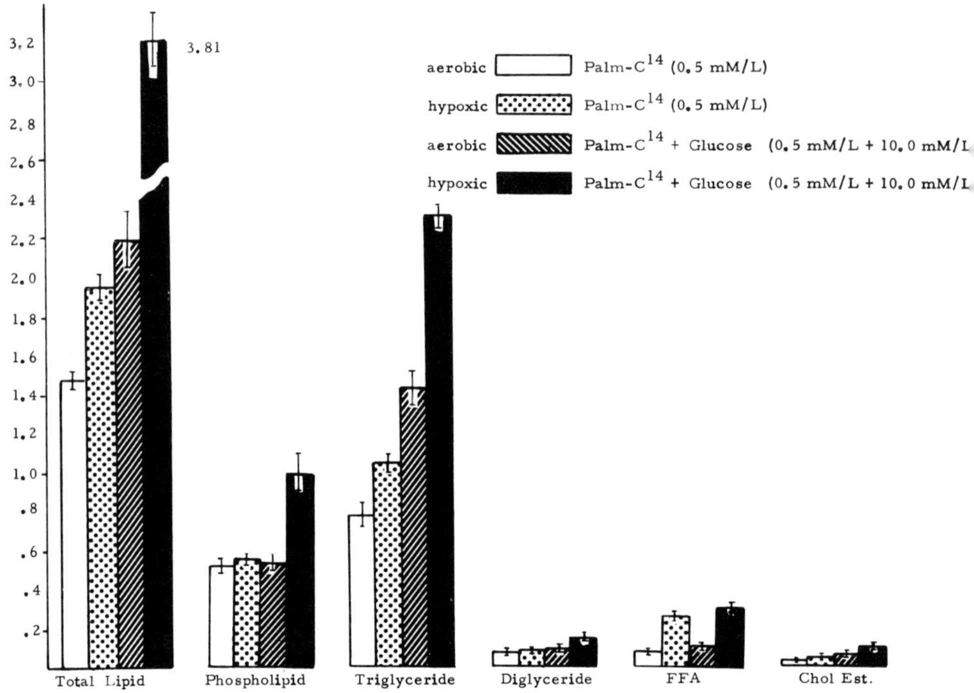

Fig. 3 The effects of glucose on the distribution of recoverable palmitate-C^{14} activity, following aerobic and hypoxic perfusion of the isolated rat heart preparation. Note that the increase in total lipid activity is expressed primarily in the phospholipid and triglyceride fractions and that this enhanced storage as neutral lipid is dependent upon the presence of glucose in the perfusate.

oxidation of extracted lipid decreases by more than 80%, this fraction being recoverable in storage form as neutral lipid. In the absence of glucose, anoxia induced a similar decrease in oxidation of FFA, but such changes are associated with a decreased uptake of FFA and are not accompanied by the previously noted reciprocal increase in tissue neutral lipid. Thus, in contrast to glucose, cellular extraction of FFA appears to bear no fixed relation to the ability of the cell to oxidize this substrate. Uptake and storage occurs under hypoxic conditions as long as sufficient glucose is available to support an actively glycolyzing system and despite the markedly reduced production of energy occurring under these conditions.

In the poorly perfused anoxic myocardium unable to oxidize extracted and esterified FFA's, lipid stores may continue to accumulate and may eventually contribute to the decreased contractility associated with the ischemic myocardium. These accumulations may be toxic and may persist for 2 weeks after coronary artery ligation. It is clinically significant that the severity and duration of fatty changes parallel the severity and duration of the period of myocardial hyperirritability, which suggests that these changes may trigger fatal dysrhythmias.

α-Glycerophosphate is a most important precursor in triglyceride synthesis, since it supplies the glycerol skeleton for esterification of FFA. It is an intermediary product of glycolysis and, when present in increased concentration, as noted during anaerobiosis, may serve as a stimulus for FFA esterification. It is not unreasonable to speculate that during all but the most severely ischemic states, anaerobic glycolysis may produce sufficient ATP for intracellular transport, and activation of extracted FFA, and that it concomitantly may enhance esterification of intracellular FFA by supplying high concentrations of the α-glycerophosphate required for glyceride synthesis. The increase in circulating arterial free fatty acid concentration reported to exist in some patients post-myocardial infarction, may enhance myocardial uptake and esterification, and may well relate to subsequent dysrhythmias.

The foregoing observations permit us to hypothesize a sequence of events ultimately leading to the death of the myocardial cell unless the process is halted before changes have become irreversible. It is impossible to establish their exact temporary relationships, since many occur simultaneously. I make no attempt to ascribe a cause and effect relationship to any but the major events in the outline presented below.

The onset of cellular anoxia causes an immediate fall in creatine phosphate and ATP concentration and a rise in Pi, ADP, and AMP. There is an increase in anaerobic glycolysis, supported by an increase in glucose consumption, aided by enhanced cellular transport and phosphorylation. Glycogenolysis, also, is stimulated by an increase in phosphorylase activity. The activity of phosphofructokinase increases many-fold and there is a shift of glycolytic intermediates to a reduced redox state. Lactate consumption falls and its production is augmented, as is that of α-glycerophosphate. Esterification of FFA is enhanced by the increased concentration of α-glycerophosphate, and droplets of neutral lipid begin to accumulate in the perimitochondrial zone. Intracellular pH begins to fall as the capacity of intracellular buffer systems is gradually exceeded. The decrease in available energy permits ionic shifts, with loss of intracellular potassium and diffusion into the cell of sodium ions and extracellular fluid. Osmotic pressure within the cell gradually increases because of autolytic breakdown of proteins, accumulation of free molecules, and shifts in electrolyte balance. Mitochondrial swelling and changes in endoplasmic reticulum now are visible histologically, as is the loss of glycogen and the appearance of clear fluid-filled intracellular spaces. Cellular synthetic processes cease. Abnormal deflections in the ECG are induced by changes in membrane potential, which reflect the failing function of the ionic pump. There is undoubtedly some protein denaturation at this stage and autophagic ingestion by increased numbers of lysosomal particles digesting portions of the living cell. The cell nucleus becomes slightly distorted.

Unless the circulation is restored, these *reversible* changes progress and become irremediable. The pH continues to fall, cathepsin and proteinase activators appear, the membrane of the lysosomal particle ruptures and releases acid hydrolases, which are rapidly activated and may now function at their pH optima. The combined damage of metabolic breakdown, autolysis, and protein denaturation destroys any potential for energy production, additional enzyme systems fail, the sarcolemma permits leakage of larger molecules, and the nucleus ceases to function. The cell is dead, it has lost its integrity as a functional unit, and a distinction between the internal and external environment ceases to exist. At this stage, actual histological appearance is determined by the relative approximation of the dead tissue to an active circulation (as discussed).

When the area of totally infarcted myocardium is so extensive as to rupture or to interfere mechanically with maintenance of perfusion pressure or cardiac output, death of the entire organism results. Often, however, the outcome depends on the metabolic behavior of the reversibly injured cell. Clinical and experimental evidence indicates that severe, localized ischemia is an acute but temporary phenomenon that is soon resolved. It is induced by permanent coronary occlusion, and also by periods of temporary occlusion both before and after resumption of flow. The increase in circulating catecholamines accompanying coronary thrombosis indicates that adequate perfusion is prevented, not only by the decrease in perfusion pressure and flow induced by obstruction of the large vessels, but also by a temporary increase in peripheral myocardial vascular resistance caused by arteriolar vasoconstriction. Under optimal circumstances, collateral vascular channels open, blood is shunted from adequately perfused areas, and satisfactory aerobic metabolic activity can be reactivated. Treatment of coronary occlusion should be based on rapid revascularization of the myocardium or its equivalent before persistent or progressive increase in ischemia leads to permanent and irreversible damage. Other approaches may well include the development of techniques that will permit a chronic reduction or cessation in cardiac work load without interfering with normal activity (hypotensive pharmacologic agents, beta blockade, pump substitution, *etc.*), and the utilization of biochemical techniques that will reduce metabolic demand, enhance anaerobic enhanced glycolytic metabolic activity and preserve the integrity of lysosomal particles so as to prolong the period of reversibility and encourage the growth or function of collateral vessels.

SUMMARY

An early distinction between the early and reversible changes of cellular ischemia and the irreversible changes of cell necrosis is essential to a rational program for the treatment of coronary artery disease, to a definition of the permissible limits of nonperfusion of the anoxic arrested heart at open-heart surgery, and to a determination of the maximal allowable transport time for cardiac transplantation. The factors in the myocardial cellular environment that support cellular viability and help to maintain homeostasis are reviewed. Reversible, surgically induced anoxic arrest is contrasted with the irreversible arrest and cell necrosis that often attends the myocardial anoxia of coronary occlusion. The enhancement of glycolysis and the other changes in carbohydrate metabolism and the depressed oxidation and lipid storage associated with alterations in lipid metabolism are outlined and their interrelationships are discussed. The effect of the myocardial tissue acidosis, which accompanies ischemic metabolism, is an important factor in determining the reversibility of the effects of ischemia. The probable role of myocardial acid hydrolases in pathological ischemic autolysis is described. A sequence of cellular events induced by ischemia and leading to ultimate cell death is outlined.

REFERENCES

1. BALSAVER, A. M.; MORALES, A. R. and WHITEHOUSE, F. W.: Fat infilitration of myocardium as a cause of cardiac conduction defect. Amer. J. Cardiol. *19:* 261–265 (1967).

2. BATE-SMITH, E. C. and BENDALL, J. R.: Changes in muscle after death. Brit. Med. Bull. *12:* 230–235 (1956).

3. BING, R. F.; KARDESCH, M.; HOGANCAMP, C. E. and MICHAL, G.: The survival of excitability, energy production and energy utlization of the heart. Trans. Assoc. Amer. Physician *71:* 152–161 (1958).

4. BRACHFELD, N. and SCHEUER, J.: Metabolism of glucose by the ischemic dog heart. Amer. J. Physiol. *212:* 603–606 (1967).

5. CAULFIELD, J. and KLIONSKY, B.: Myocardial ischemia and early infarction. An electron microscopic study. Amer. J. Path. *35:* 489–524 (1959).

5a. DE DUVE, C.: Lysosomes: A new group of cytoplasmic particles. *In* Subcellular Particles, edited by T. Hayashi, pp. 128–159 (Ronald Press, New York 1959).

5b. DE DUVE, C., and BEAUFAY, H.: Tissue fractionation studies: Influence of ischemia on the state of some bound enzymes in rat liver. Biochem. J. *73:* 610–616 (1959).

6. GLINSMANN, W. H. and ERICSSON, J. L.: Observations on the subcellular organization of hepatic parenchymal cells. Evolution of reversible alterations induced by hypoxia. Lab. Invest. *15:* 762–778 (1966).

7. JENNINGS, R. B.; KALTENBACH, J. P.; SOMMERS, H. M.; BAHR, G. R. and WORTMAN, W. R.: Studies of the dying myocardial cell; In: T. N. JAMES and J. W. KAYES, The Etiology of Myocardial Infarction, Chap. 12. pp. 189–205 (Little, Brown, Boston 1963).

8. JUDAH, J. D.; AHMED, K., and McLEAN, A. E.: Pathogenesis of cell necrosis. Fed. Proc. *24:* 1217–1221 (1965).

9. MAJNO, G.; LA GATTUTA, M. and THOMPSON, T. E.: Cellular death and necrosis: chemical, physical and morphologic changes in rat liver. Virchows Arch. Path. *333:* 421–465 (1960).

10. MEESSEN, H.: Ultrastructure of the myocardium. Its significance in myocardial disease. Amer. J. Cardiol. *22:* 319–327 (1968).

11. SCHEUER, J. and BRACHFELD, N.: Myocardial uptake and fractional distribution of palmitate-1-C^{14} by the ischemic dog heart. Metabolism *15:* 945–954 (1966).

12. SHNITKA, T. K. and NACHLAS, M. M.: Histochemical alterations in ischemic heart muscle and early myocardial infarction. Amer. J. Pathol. *42:* 507–521 (1963).

13. SOMMERS, H. and JENNINGS, R. B.: Experimental acute myocardial infarction. Lab. Invest. *13:* 1491–1503 (1964).
14. WARTMAN, W. B.; JENNINGS, R. B.; YOKOYAMA, H. O. and CLABAUGH, G. F.: Fatty change of the myocardium in early experimental infarction. Arch. Path. *62:* 318–323 (1956).
15. WEISSLER, A. M.; KRUGER, F. A.; BABA, N.; SCARPELLI, D. G.; LEIGHTON, R. F. and GALLIMORE, J. K.: Role of anaerobic metabolism in the preservation of functional capacity and structure of anoxic myocardium. J. Clin. Invest. *47:* 403–416 (1968).

Mailing address: Norman Brachfeld, M.D., Myocardial Metabolism Laboratory, Department of Medicine, Division of Cardiology, Cornell University Medical College, 525 East 68th Street, New York, N.Y. 10021 (U.S.A.).

Department of Physiology, University of Ottawa, Ottawa, Canada

THE HORMONAL RESTORATION OF THE INTRINSIC CONTRACTILE FUNCTION OF THE ATROPHIC HEARTS OF HYPOPHYSECTOMIZED RATS [1]

B. KORECKY, G. C. TAICHMAN, R. MINELLI, and M. BEZNAK

INTRODUCTION

One of the consequences of hypophysectomy (Hx) in mammals is the cessation of growth concurrent with a decrease in the heart weight to body weight ratio (2). The latter observation indicated a gradual reduction in the existing mass of the heart, amounting to 30–40% of the myocardial tissue after several weeks. This atrophy can be normalized partially by the administration of thyroxine (T_4) and totally by growth hormone (GH) plus T_4 as compared with normal unaffected age controls. It was shown previously (3) that the severely depressed hemodynamic function of Hx animals can be also restored to normal by subsequent treatment with either T_4 alone or T_4 in conjunction with GH. This hemodynamic impairment and its hormonal restoration are believed to be primarily due to a direct intervention upon the myocardium. This conclusion is supported by observations from the isolated heart-lung preparation where both the venous return and the peripheral resistance were controlled (9).

In the aforementioned experiments, the heart was viewed as a compression pump whose competency was assessed by its reserve capacity. With such an experimental arrangement, the contractile state of these hearts could only be expressed in hemodynamic terms, *i.e.*, cardiac work. It now seemed appropriate to view the heart as a muscle as well, thereby judging its contractility in terms of the intrinsic mechanical properties expressed as variations in tension, length, rates of changes in tension and length, and their time relationships (6, 20).

Three types of experiments were designed for this purpose. The cardiac muscle mechanics of rats after Hx and subsequent treatment were studied: a) from isolated papillary muscles; b) from ejecting and imposed isovolumic beats measured, using whole hearts in open-chest animals; c) from the isovolumic phase of normal ejecting beats, using whole hearts in closed-chest animals. Results of these experiments have previously been reported in abstract form (10, 22, 23).

---

[1] Supported by grants from the Medical Research Council of Canada and from the Ontario Heart Foundation.

METHODS

Male rats (Charles River CD 200–230 g) were hypophysectomized and maintained on the standard laboratory diet for 5–6 weeks. The hormonal treatment was started in animals showing no growth and consisted of increasing doses of T_4 (Sodium-l-thyroxine; Sigma) for 3–6 weeks, reaching the total dose of 2–5 mg per rat. GH (NIH-GH-S9) was administered in daily doses of 0.5 mg for a period of 3–6 weeks.

Isolated papillary muscle

Anterior papillary muscles were isolated from left ventricles of rats and mounted between a pair of plexiglass clamps in a muscle bath. The lower clamp was rigidly attached to a force transducer, while the upper one was hooked to an isotonic-isometric lever with an attached LVDT transducer. Isotonic contractions at different preloads and afterloads could be monitored on an oscilloscope and recorded on a direct writing recorder. The bathing solution was bubbled with 95% O_2 and 5% CO_2 and the muscles were stimulated by a pair of platinum electrodes by supra threshold square wave stimuli once per 40 sec (14).

Whole heart, open-chest animal

The procedure used here has been reported by others (7,11). Rats were anesthetized with nembutal (40 mg/kg i.p.) and placed on a heated operating table. Their body temperatures were monitored throughout the experiment and kept at a constant level (±0.5°C). Positive pressure breathing (97% O_2 plus 3% CO_2) was initiated, the thorax was opened by a midline incision, and the heart and aorta were gently exposed. A loop of surgical thread was placed around the root of the ascending aorta, while an aneurism needle was fixed over the vessel. Intracavitary pressures were monitored through a short bevelled hypodermic needle (#18) introduced into the left ventricle through the apex, connected to a pressure transducer and displayed on an oscilloscope, as well as recorded on a rectilinear recorder. The pressures were simultaneously differentiated with respect to time (dp/dt) and continuously recorded. A series of normal ejecting beats were first recorded, after which the aorta was gradually constricted until the maximum intracavitary pressure was achieved (25).

In this study, a series of left ventricular pressure-volume curves were determined as well. The heart was quickly excised and the aorta was cannulated. After the ventricle was sealed by tying the atrium, the chamber was cleared of air bubbles and the heart was immersed in a 4°C saline solution enriched with glucose. Pressure-volume curves were determined by a stepwise introduction of known volumes while noting the increments of pressure change. Following this, the heart was frozen in liquid nitrogen at zero transmural pressure and sectioned transversely between the apex and base. The chamber dimensions (*i.e.*, internal diameter and wall thickness) were then determined (15).

Whole heart, closed-chest animal

After anesthesia (nembutal 40 mg/kg i.p.) a catheter (thin-walled silastic tubing) was passed retrograde down the right carotid artery, the aorta, and eventually into the left ventricle. Intracavitary pressures (P) were recorded during a series of normal ejecting beats and simultaneously differentiated with respect to time (dp/dt). Phase-plane plots of pressure signals with corresponding derivatives were recorded on a storage oscilloscope and photographed (25). Contractile element shortening velocity was calculated by the simplified formula $\frac{dp/dt}{KP}$ during the isovolumic phase of the normal beat. Extrapolation to zero load allowed estimation of the maximal intrinsic velocity of the contractile element (Vmax) in muscle lengths/sec (13, 19, 24). The K in this equation refers to the coefficient of the modulus of the series elastic element (16), being distinct for each group.

RESULTS

Isolated Papillary Muscle

The results of experiments on isolated papillary muscle (fig. 1) indicate that after Hx the maximum velocity of shortening (Vmax), the maximum developed tension (MT), and the mean rate of its development (mean dp/dt) are decreased, while the time to maximum tension (TMT) is prolonged. Administration of GH had no effect; while T_4 alone or in conjunction with GH normalized Vmax, the MT remained somewhat lower, due to the shortened TMT.

The maximum dp/dt, which reflects the changes of velocity of the contractile element, was less depressed after Hx than the Vmax (14). The only explanation for this discrepancy would be a change in the stiffness of the series elastic component. When the modulus of elasticity was calculated from isotonic after-loaded contractions, it was found to be increased after Hx and normalized after subsequent treatment with T_4 or T_4 + GH (fig. 2). These results show the direct effect of Hx and subsequent hormonal effect of T_4 upon the intrinsic contractile properties of the cardiac muscle. This finding is rather unexpected, because experimentally induced hyperthyroidism (the administration of T_4 to normal cats) produced no change in series elasticity (17).

Whole heart, open-chest animal

In the experiments on whole hearts in open-chest animals, an attempt was made to assess the inherent changes in myocardial contractility from normal

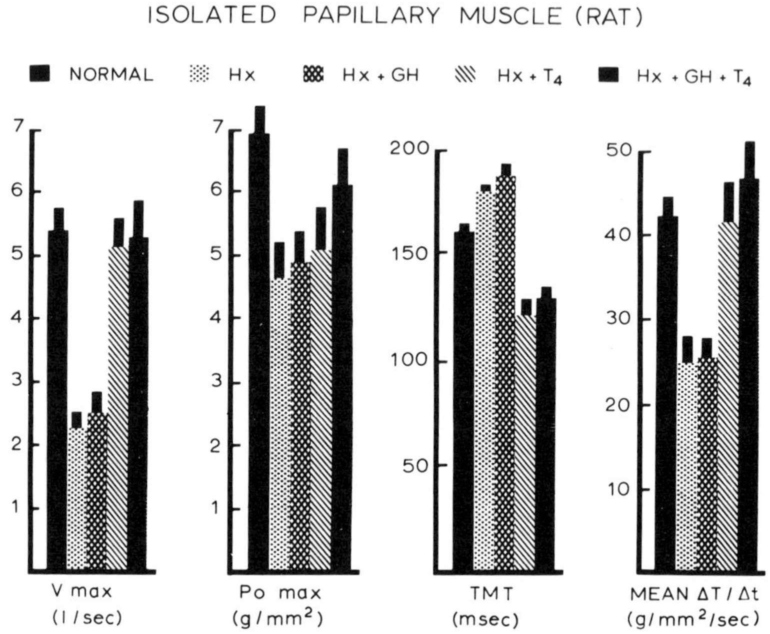

Fig. 1 Changes in maximum velocity of shortening (Vmax in muscle lengths per sec), maximum developed tension (Po in g/mm^2 of cross-sectional area), time to maximum tension (TMT in msec) and mean rate of tension development ($\Delta T/\Delta t$) after hypophysectomy (Hx). Mean values ± SEM are given. Number of animals in individual groups varies from 11–15.

ejecting beats and maximum isovolumic beats. During this imposed isovolumic beat, the geometry of the left ventricle undergoes less variation than during the normal auxotonic beat. The larger than normal end-diastolic volume extends the majority of fibers higher on their length-active tension curves (8), and an infinite afterload (completely obstructed aorta) is imposed on the myocardial fibers (21). With the constancy of these two factors (18), the maximum peak systolic pressure and its maximum rate of pressure development should reflect inherent changes in the inotropic state of these hearts (7). It was therefore felt that the isovolumic left ventricle is the closest *in situ* approximation to the isometrically beating papillary muscle (5).

As can be seen from figure 3, the peak isovolumic systolic pressure decreases after Hx, remains so with GH treatment, while T_4 alone or in combination with GH normalized the value. Because pressure within a spherical chamber, in this case the left ventricle, is a function of stress, midwall radius, and wall thickness (1), it was felt that forces developed within the myocardial wall should be expressed in terms of mid-circumferential wall stress where variations in size due to hormonal intervention would be taken into consideration. Figure 4 shows

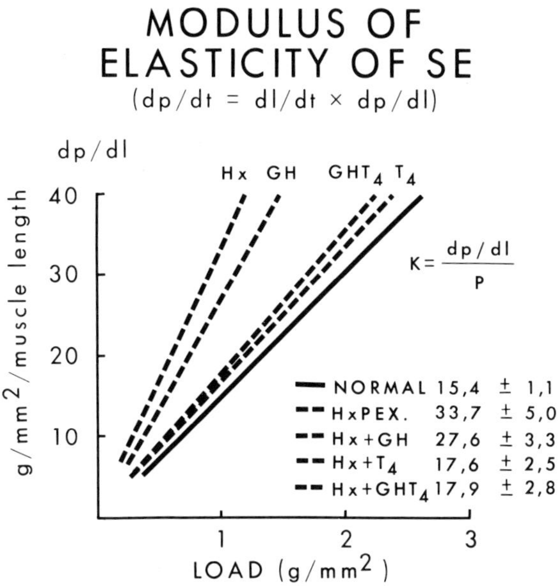

Fig. 2 Coefficient of the modulus of elasticity (per muscle length) determined from isotonic afterloaded contractions in rat papillary muscles. Full line represents normal values, broken lines represent the values after hypophysectomy, and after hypophysectomy and subsequent treatment with thyroxine (T_4) and growth hormone (GH).

that the decrease in peak wall-stress after Hx was significantly improved by T_4 but remained lower when compared with normal animals. A similar result was seen with papillary muscle (fig. 1). The maximum rate of pressure development (max dp/dt) during the isovolumic beat served as a measure of the maximum intensity of mechanical reactions occurring within the beating heart (12, 25). The results indicate that the changes previously reported *in vitro* (fig. 1) are concurrent with the changes observed in the whole heart (fig. 5).

Whole heart, closed-chest animal

Our final goal was to measure the changes in the velocity of shortening of the contractile element during the isovolumic phase of an ejecting beat in closed-chest animals and extrapolate it to zero afterload (Vmax). The results of this experiment, summarized in figure 6, indicate that the Vmax is indeed depressed after Hx and that T_4 is the principal hormone that restores it to normal limits. Since, in this case, there was a minimal intervention with the circulation of the animal and the thorax was left intact, these results should have the highest biological validity. The general pattern is identical with the Vmax measured in isolated papillary muscles; and when the changes in Vmax determined in both experiments were compared on a percentage basis (fig. 7),

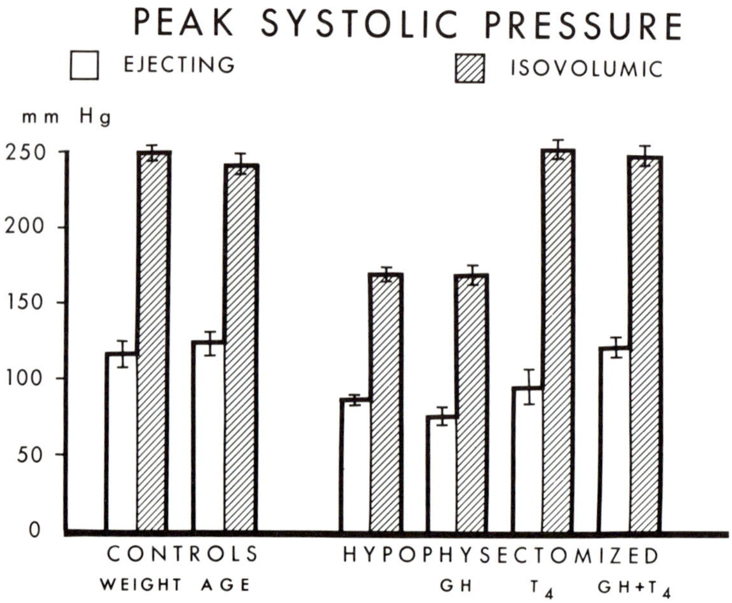

Fig. 3 Peak systolic pressures of the left ventricle recorded from open-chest rats. Values were taken from normal ejecting beats and isovolumic beats after total aortic obstruction. Mean ± SEM are shown for each group.

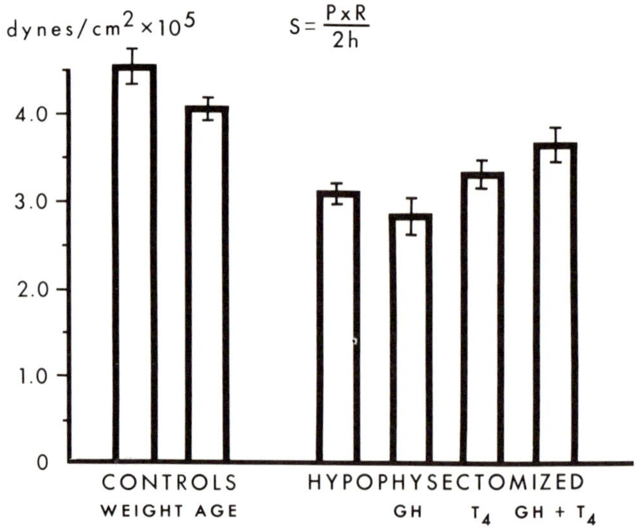

Fig. 4 Peak equatorial mid-wall stress calculated from a modified Laplace Formula, where P, R, and h refer to maximum systolic pressure, midwall radius, and wall thickness, respectively, from maximum isovolumic beats. Values are expressed as mean ± SEM for each group.

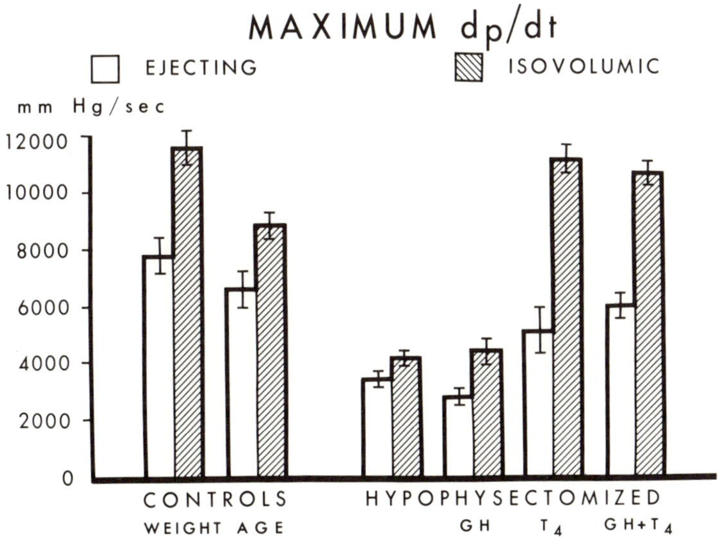

Fig. 5 Maximal rate of pressure development (max dp/dt) measured from open-chest rats. Values were taken from normal ejecting beats and isovolumic beats after a total aortic obstruction. Mean ± SEM are shown for each group.

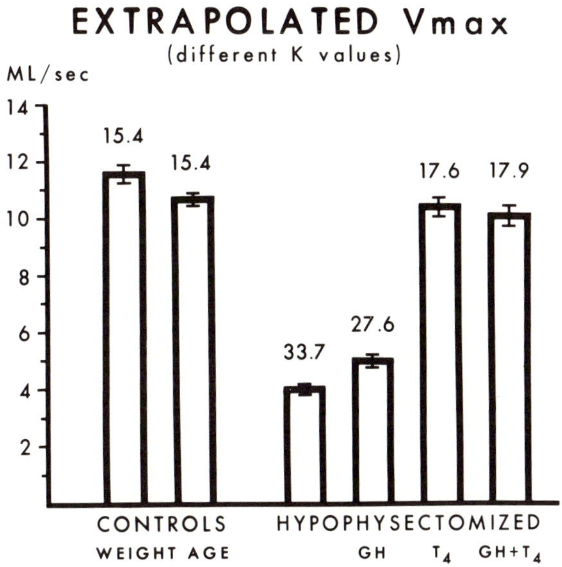

Fig. 6 Maximal velocity of the contractile element shortening (Vmax) extrapolated from the isovolumic portion of the ejecting beat measured from closed-chest rats. Values that appear above the bars for each group refer to K values used in calculating Vmax.

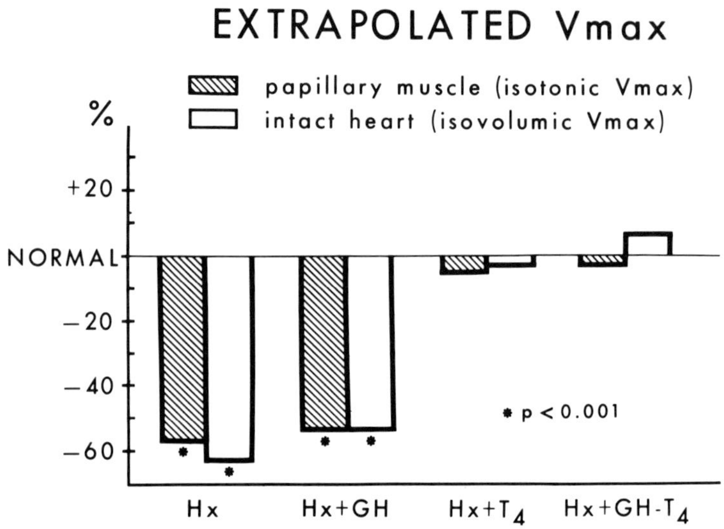

Fig. 7 A comparison between Vmax measured *in vitro* (isotonic contractions of papillary muscles) and *in vivo* (ejecting beats using closed-chest animals). Values are expressed as percentage deviations from mean values of normals.

more than adequate corroboration was achieved and the validity of the results substantially strengthened.

CONCLUSIONS

The results of our experiments obtained in rats at three different levels, *i.e.*, in isolated papillary muscle, in whole heart of open-chest animals, and in whole heart of closed-chest-anesthetized animals, indicate the fundamental importance of thyroxine on the intrinsic contractile properties of the cardiac muscle. This conclusion is in agreement with that of others (4, 7), who have reported the direct effects of thyroxine on the normal myocardium. At all three levels, administration of this hormone after hypophysectomy restored the severely depressed maximum velocity of shortening to normal values. This is reflected by the restoration of decreased maximum dp/dt values as well; but because of the changes in series elasticity after hypophysectomy, this index does not quantitatively follow the changes in velocity of shortening. The maximum developed tension, pressure, and stress, all of which are products of the rate of force development and its duration, are also depressed after hypophysectomy but to a lesser degree than the velocity, since the duration of tension development is prolonged. The administration of thyroxine restores the maximum developed pressure to normal, but the maximum developed tension and stress, normalized per cross-sectional area, remain lower.

SUMMARY

An analysis of the intrinsic cardiac muscle mechanics of the atrophic heart of hypophysectomized rats has been conducted on the isolated papillary muscle, on the whole heart, using open-chest animals, and on the whole heart, using closed-chest animals. The hemodynamic impairment associated with hypophysectomy can now be explained in terms of its primary cardiac muscle dysfunction, *i.e.*, a 55–65% reduction in Vmax and a 30–35% reduction in P_0max. The lack of thyroxine appears to be the principal cause in decreasing contractility, since its administration to hypophysectomized rats re-establishes the normal myocardial contractile state.

ACKNOWLEDGMENTS

The authors wish to thank Mrs. A. M. McKerral and Miss J. M. Hibberd for their valuable technical assistance. Growth hormone was generously supplied by the Endocrinology Study Section of the National Institutes of Health, Bethesda, Maryland.

REFERENCES

1. BADEER, H. S.: Contractile tension in the heart. Amer. Heart J. *66:* 432–433 (1963).
2. BEZNAK, M.: The behaviour of the weight of the heart and the blood pressure of albino rats under different conditions. J. Physiol. (London) *124:* 44–63 (1954).
3. BEZNAK, M.: Hormonal influences in regulation of cardiac performance. Circulat. Res.*15*(Suppl 11)*:* 141–150 (1964).
4. BUCCINO, R. A.; SPANN, J. R. Fr.; POOL, P. E.; SONNENBLICK, E. H. and BRAUNDWALD, E.: Influence of the thyroid state on the intrinsic contractile properties and energy stores of the myocardium. J. Clin. Invest. *46:* 1669–1682 (1967).
5. FISHER, V. J.; LEE, R. J.; MARLON, A. M. and KAVALER, F.: Paired electrical stimulation and the maximal contractile response of the ventricle. Circulat. Res.*20:* 520–553 (1967).
6. FRY, S. L.; GRIGGS, D. M., Jr., and GREENFIELD, J. C., Jr.: Myocardial mechanics: Tension-velocity-length relationships of heart muscle. Circulat. Res.*14:* 73–85 (1964).
7. GOODKIND, M. J.: Left ventricular myocardial contractile response to aortic constriction in the hyperthyroid guinea pig. Circulat. Res.*22:* 605–614 (1968).
8. GRIMM, A. F. and WHITEHORN, W. V.: Myocardial length-tension sarcomere relationships. Amer. J. Physiol. *214:* 1378–1387 (1968).
9. KORECKY, B.; BEZNAK, M. and KORECKA, M.: Effect of hypophysectomy and hormone treatment on the heart-lung preparation of rats. Canad. J. Physiol. Pharmacol.*44:* 13–20 (1966).
10. KORECKY, B. and MINELLI, R.: Hormonal effects on the contractile and series elastic components of rat myocardium. Physiologist *12:* 273, 1969.
11. KRAMES, B. B.; NORTHUP, D. W. and VAN LIERE, E. J.: Pressure development in the hypertrophied heart. Proc. Soc. Exp. Biol. Med. *124:* 150–152 (1967).
12. MASON, D. T.: Usefulness and limitations of the rate of rise of intraventricular pressure (dp/dt) in evaluation of myocardial contractility in man. Amer. J. Cardiol. *23:* 516–527 (1969).
13. MASON, D. T.; SPANN, J. F., Jr. and ZELIS, R.: The maximum intrinsic velocity of the myocardium (Vmax) in man: Estimation from the rate of pressure rise and intraventricular pressure throughout isovolumic left ventricular contraction. Circulation *38*(Suppl VI)*:* (1968).
14. MINELLI, R. and KORECKY, B.: Effect of growth hormone and thyroxine on isometric contractile mechanics of cardiac muscle. Canad. J. Physiol. Pharmacol. *47:* 545–552 (1969).

15. MONROE, R. G., GAMBLE, W. J.; LAFARGE, C. G.; KUMAR, A. E. and MANASEK, F. J.: Left ventricular performance at high end-diastolic pressures in isolated, perfused dog hearts. Circulat. Res. *26:* 85–99 (1970).
16. PARMLEY, W. W. and SONNENBLICK, E. H.: Series elasticity in heart muscle. Its relation to contractile element velocity and proposed muscle models. Circulat. Res. *20:* 112–123 (1967).
17. PARMLEY, W. W.; SPANN, J. F., Jr.; TAYLOR, R. R. and SONNENBLICK, E. H.: The series elasticity of cardiac muscle in hyperthyroidism, ventricular hypertrophy and heart failure. Proc. Soc. Exp. Biol. Med. *127:* 606–609 (1968).
18. REEVES, T. J. and HEFNER, L. L.: Isometric contraction and contractility in the intact mammalian ventricle. Amer. Heart. J. *64:* 525–538 (1962).
19. SHIMOSATO, S.: Isovolemic intraventricular pressure change: An index of myocardial contractility during anesthesia. Anesthesiology, *31:* 327–333 (1969).
20. SONNENBLICK, E. H.: Implications of muscle mechanics in the heart. Fed. Proc. *21:* 975–990 (1962).
21. SONNENBLICK, E. H. and DOWNING, S. E.: Afterload as a primary determinant of ventricular performance. Amer. J. Physiol. *204:* 604–610 (1963).
22. TAICHMAN, G. C. and KORECKY, B.: Changes of myocardial contractility after hormonal intervention. Canad. Physiol. *1:* 59 (1970).
23. TAICHMAN, G. C. and KORECKY, B.: Contractile state of the whole heart after hormonal intervention. Fed. Proc. *29:* 321 (1970).
24. TAICHMAN, G. C. and KORECKY, B.: Determination of myocardial Vmax in closed chest anesthetized rats. Proc. Canad. Fed. Biol. Soc. *13:* 597 (1970).
25. TAICHMAN, G. C. and KORECKY, B.: (Unpublished data).
26. WALLACE, A. G.; SKINNER, N. S., JR' and MITCHELL, J. H.: Hemodynamic detrimants of the maximal rate of rise of left ventricular pressure. Amer. J. Physiol. *205:* 30–36 (1963).

Mailing address: Dr. B. Korecky, Department of Physiology, University of Ottawa, Faculty of Medicine, Ottawa (Canada).

Department of Physiology, University of Manitoba, Canada, and
Department of Biochemistry, St. Louis University,
St. Louis, Missouri, USA

ENERGY STATE AND ULTRASTRUCTURE OF THE SUBSTRATE-DEPLETED HEART [1]

N. S. DHALLA, J. C. YATES, and R. E. OLSON

Abnormalities in myocardial ultrastructure and high energy phosphate stores have been reported in various experimental models of heart failure (1, 2, 6, 7, 9, 15-17, 20); however, the cause-effect relationship of these events in the heart is far from clear. In this study, therefore, we wish to investigate a time-course of changes in the endogenous fuel supply, energy state, ultrastructure, and myocardial contractility of the substrate-depleted heart. It has already been reported that an isolated rat heart perfused with substrate-free medium forms an excellent model of heart failure (14). When the heart perfused with substrate-free medium failed to generate contractility, glycogen, triglycerides, ATP, and creatine phosphate (CP) were observed to decrease appreciably, whereas both ADP and AMP levels increased without changes in the phosphatide content of the heart (3). Likewise, the levels of glycolytic intermediates, citric acid cycle intermediates, and amino acids, which are readily transaminated to citric acid cycle intermediates, were also reduced in this preparation (12, 14). Furthermore, a decline in catecholamine stores (5) and a defective calcium pump in the sarcoplasmic reticulum and mitochondria (4, 13) of the substrate-depleted heart were observed.

MATERIALS AND METHODS

Isolated rat hearts were perfused by the conventional Langendorff technique at 37°C. The perfusion medium, substrate-free Krebs-Henseleit solution containing 1.25 mM Ca^{++} was gassed with 95% O_2 and 5% CO_2. A resting tension of 1 g was applied to the heart on starting the perfusion. The contractile force and heart rate were recorded on the Grass polygraph via a force displacement transducer (FT-03). At different intervals of perfusion, the hearts were frozen with a Wollenberger clamp precooled in liquid N_2 for biochemical determinations. Glycogen was extracted in hot 30% KOH solution and estimated by the glucose oxidase method (11). Triglycerides were extracted by the method of Folch *et al.* (10) and estimated according to the method of Wells and Dittmer (19). Adenine nucleotides and CP were determined enzymatically

[1]Supported by the Medical Research Council of Canada, Manitoba Heart Foundation, Missouri Heart Association, and National Institutes of Health Grant HE 09772.

using fluorometric methods (8), whereas inorganic phosphate (Pi) was determined according to the methods published elsewhere (18). The hearts were also fixed at different intervals of perfusion with substrate-free medium by perfusing with a sodium cacodylate (0.1 M) buffer, pH 7.4, containing 5% glutaraldehyde, for electron microscopic studies. Left ventricle tissues were further fixed with 1% osmium tetroxide in 0.1 M phosphate buffer, dehydrated with acetone, and embedded in araldite 502 resin. The sections were cut using glass knives, stained with uranyl acetate, and examined with a Philips EM-100 electron microscope.

RESULTS

On perfusing the rat heart with aerobic substrate-free medium, contractile force was maintained for 30 min and thereafter declined toward zero in a subsequent period of 90 min. A slight decrease in heart rate was also observed during the course of a 2 h perfusion (fig. 1). The changes in the concentration of glycogen and triglycerides in the substrate-depleted heart are shown in figure 2. After 30 min of perfusion with substrate-free medium, the levels of triglycerides in the heart did not change, while they declined to about 25% of the control value after 2 h of perfusion. On the other hand, glycogen decreased to about 35% of the control value in 30 min and was 25% of the control in 2 h.

The levels of high energy phosphate stores (CP, ATP, ADP, and AMP) were also determined during the course of perfusion of the heart with substrate-free medium for 2 h. It was observed that the initial rate of decline of CP concentration was faster than that of ATP (fig. 3). It was interesting to find, however, that ATP and CP were present to the extent of about 45 and 25% of the control values, respectively, after 2 h of perfusion, at which time the heart failed to generate contractility. From figure 4, it can be seen that the ATP/AMP

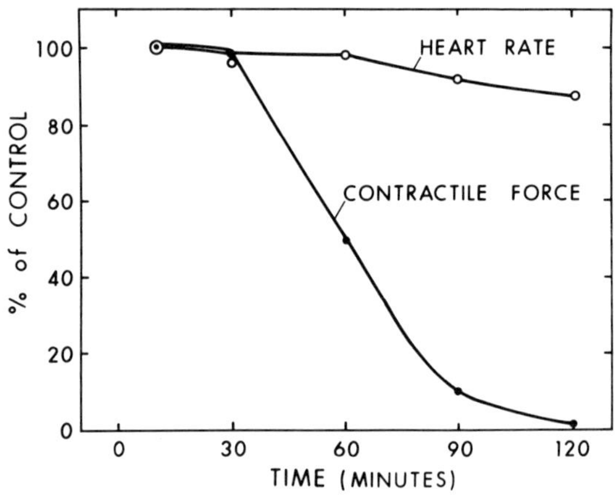

Fig. 1 Changes in contractile force and heart rate of the isolated rat heart perfused with substrate-free medium for 2 h.

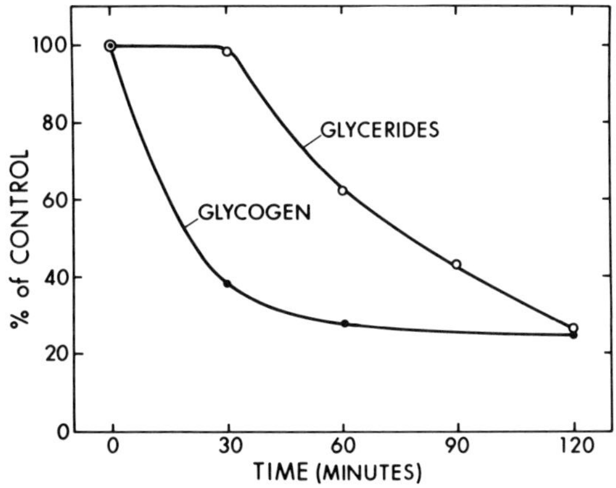

Fig. 2 Changes in glycogen and triglyceride levels of the isolated rat heart perfused with substrate-free medium for 2 h.

ratio did not change until 30 min and thereafter reached a minimal level at 60 min of perfusion with substrate-free medium. On the other hand, the ATP/ADP ratio declined throughout the perfusion period of 2 h. Both the phosphate potential (ATP $\times$ 10^3/ADP $\times$ Pi) and CP/Pi ratio showed a sharp decline on perfusing the heart with substrate-free medium for 2 h (fig. 5).

Sections of the left ventricle were examined with the electron microscope after perfusing the hearts for 10, 30, 60, and 120 min with substrate-free medium. No change in the ultrastructure of the heart perfused for 30 or 60 min was observed except that a moderate degree of swelling of vesicles underneath the sarcolemma was apparent (fig. 6, middle panel). In comparison to the

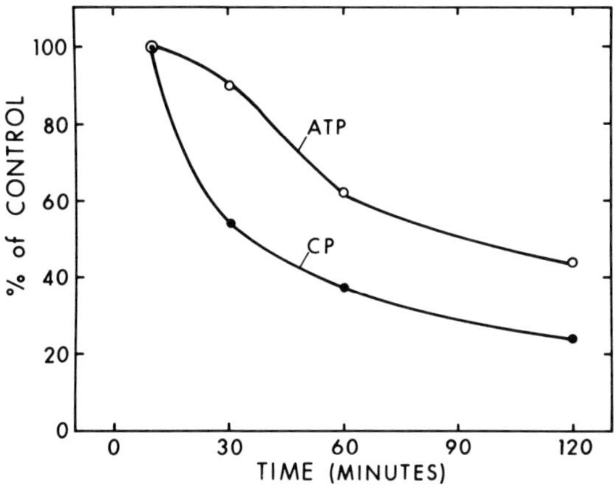

Fig. 3 Changes in creatine phosphate (CP) and ATP levels of the isolated rat heart perfused with substrate-free medium for 2 h.

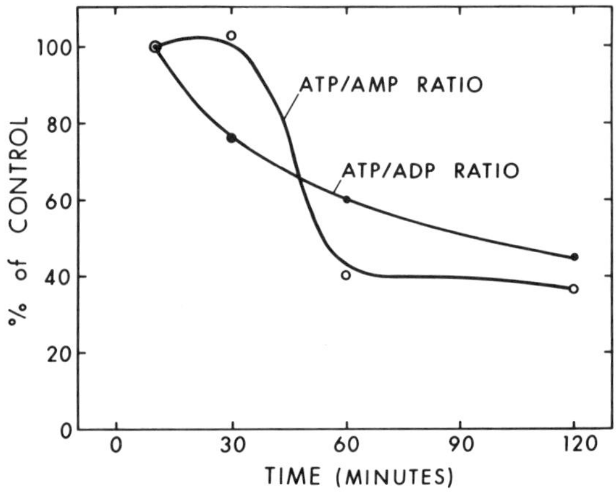

Fig. 4 Changes in ATP/ADP ratio and ATP/AMP ratio in the isolated rat heart perfused with substrate-free medium for 2 h.

control hearts perfused for 10 min (fig. 6, upper panel), the hearts perfused for 2 h with substrate-free medium showed the presence of enlarged vesicles underneath the sarcolemma (fig. 6, lower panel). The changes at the intercalated disk in the hearts perfused for 2 h with substrate are depicted in figure 7. In comparison with the control heart (fig. 7, upper panel), the intercalated disk in the substrate-depleted heart was found to be associated with enlarged elliptical spaces (fig. 7, lower panel). The ultrastructure of the sarcotubular system of the control heart is shown in figure 8 (upper panel). On perfusing the heart for 2 h with substrate-free medium, swelling of T-tubules (fig. 8, middle panel) and of the sarcoplasmic reticulum (fig. 8, lower panel) was observed.

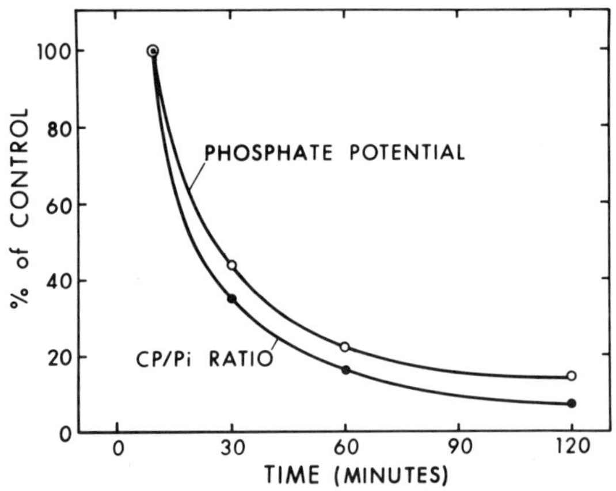

Fig. 5 Changes in phosphate potential (ATP × 10^3/ADP × Pi) and CP/Pi ratio in the isolated rat heart perfused with substrate-free medium for 2 h.

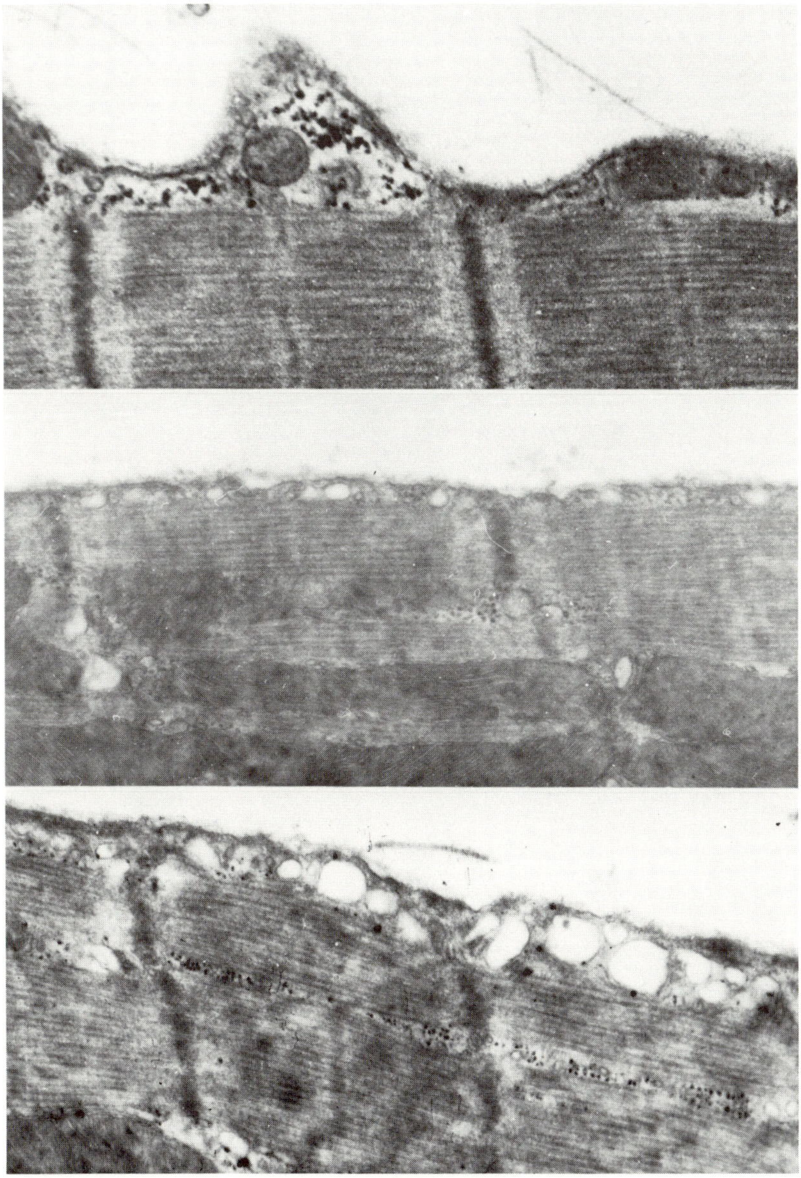

Fig. 6 Changes under the cell membrane of the isolated rat heart perfused with substrate-free medium. Upper panel: control heart; middle panel: heart perfused for 60 min; lower panel: heart perfused for 2 h. × 28,100.

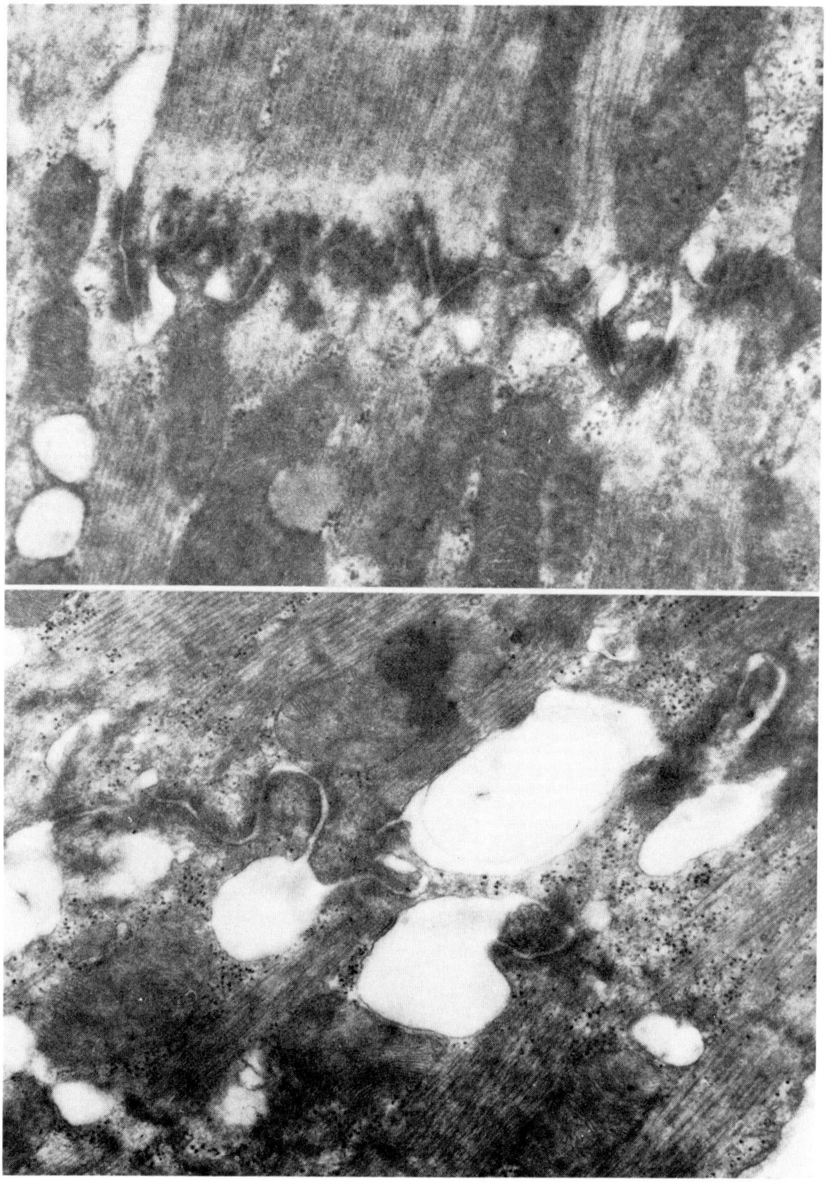

Fig. 7 Changes at the intercalated disk of the isolated rat heart perfused with substrate-free medium for 2 h. Upper panel: control heart; lower panel: heart perfused for 2 h with substrate-free medium. × 23,700.

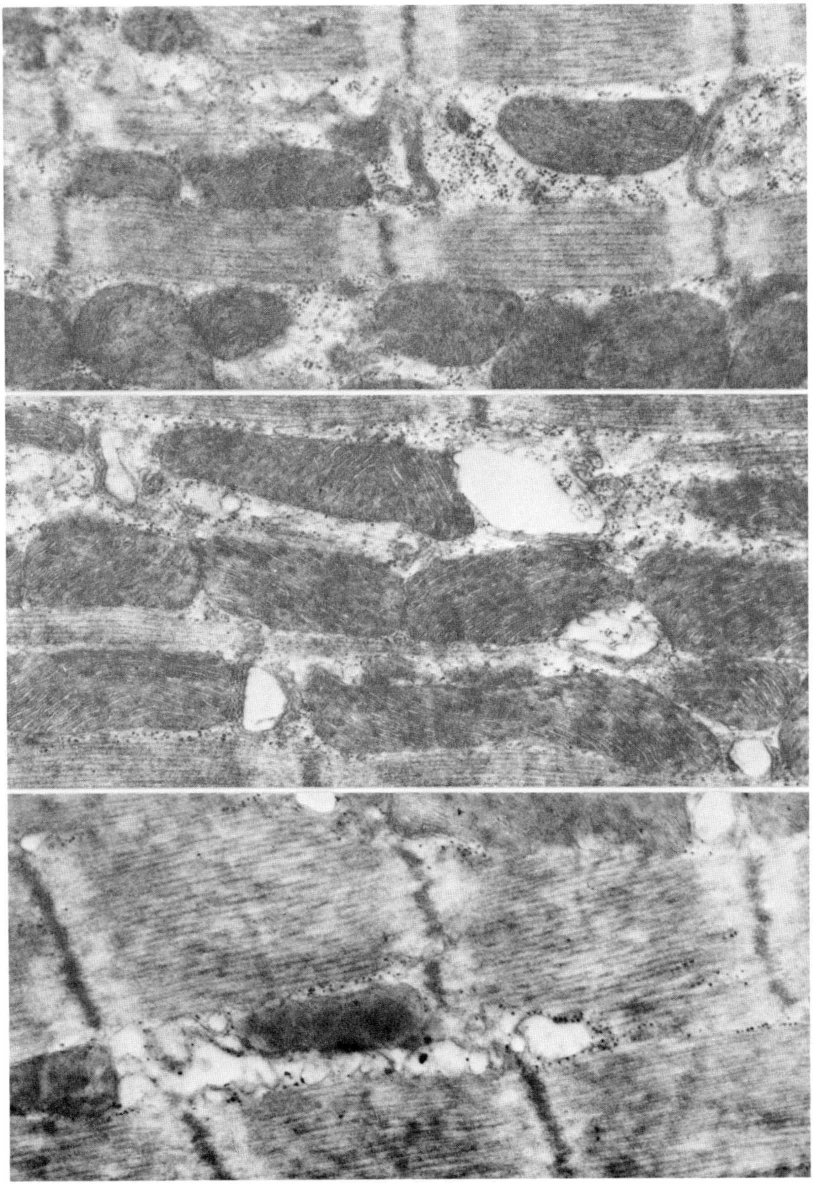

Fig. 8 Changes at the sarcotubular system of the isolated rat heart perfused with substrate-free medium for 2 h. Upper panel: control heart; middle panel: substrate-depleted heart indicating swelling of transverse tubules; lower panel: substrate-depleted heart indicating swelling of sarcoplasmic reticulum. × 22,300.

Figure 9 reveals the swelling of mitochondria in hearts perfused for 2 h with a substrate-free medium. Disruption of mitochondria and the formation of laminar membranous bodies in the substrate-depleted heart is illustrated in figure 10. Damage to the ultrastructure of cristae of some swollen mitochondria in the substrate-depleted heart was also observed (fig. 11). In certain areas of the substrate-depleted heart, sarcomeres were highly contracted (fig. 9, lower panel) and their adjacent sarcomeres were stretched with distorted Z-band and disrupted myofibrils (fig. 12).

DISCUSSION

In this study, we have shown that glycogen, CP, the phosphate potential, CP/Pi ratio, and ATP/ADP ratio decreased considerably before the contractile force began to decline on perfusing the heart with substrate-free medium. At the onset of decrease in contractility, the ATP/AMP ratio and triglyceride level started to decline, while ATP was reduced by about 15% of the control value. These findings suggest that heart failure in this experimental model may have ensued owing to an insufficiency of the energy generation process. This observation that the fuel supply is not mobilized at an adequate rate to maintain contractility is more clearly apparent at the time of the 50% decline in contractile force, when glycogen, the ATP/AMP ratio, phosphate potential, and CP/Pi ratio reached their minimal levels, while ATP, CP, the ATP/ADP ratio and triglycerides decreased markedly. The assumption is further supported by the fact that the addition of glucose as a source of exogenous fuel supply into the perfusion medium was found to fully restore the contractile force in the isolated heart perfused with substrate-free medium for 1 h (14).

The most interesting observation of this study is that the heart failed to generate contractility when a sufficient amount of endogenous fuel supply, glycogen, and triglyceride, and about 40% of ATP, and 25% of CP were still present. We consider that the inability of the substrate-depleted heart to generate contractile force may not be entirely due to an insufficiency of the rate of energy production, but may also be partly due to a defect in the process of energy utilization. Such reasoning is based on our observation that the failure in the substrate-depleted heart is associated with an impaired calcium pump in both sarcoplasmic reticulum and mitochondria (4, 13). Swelling of mitochondria and sarcoplasmic reticulum in the substrate-depleted heart also supports this contention. Furthermore, damage to contractile proteins, which utilize the main portion of cellular energy in the heart, and the appearance of enlarged vesicles under the myocardial cell membrane, where energy is used for ion transport, were observed in this study. A defect in the process of energy utilization in the late phases of failure of the substrate-depleted heart is also realized from the reduced ability of exogenous substrate such as glucose to restore contractility (14).

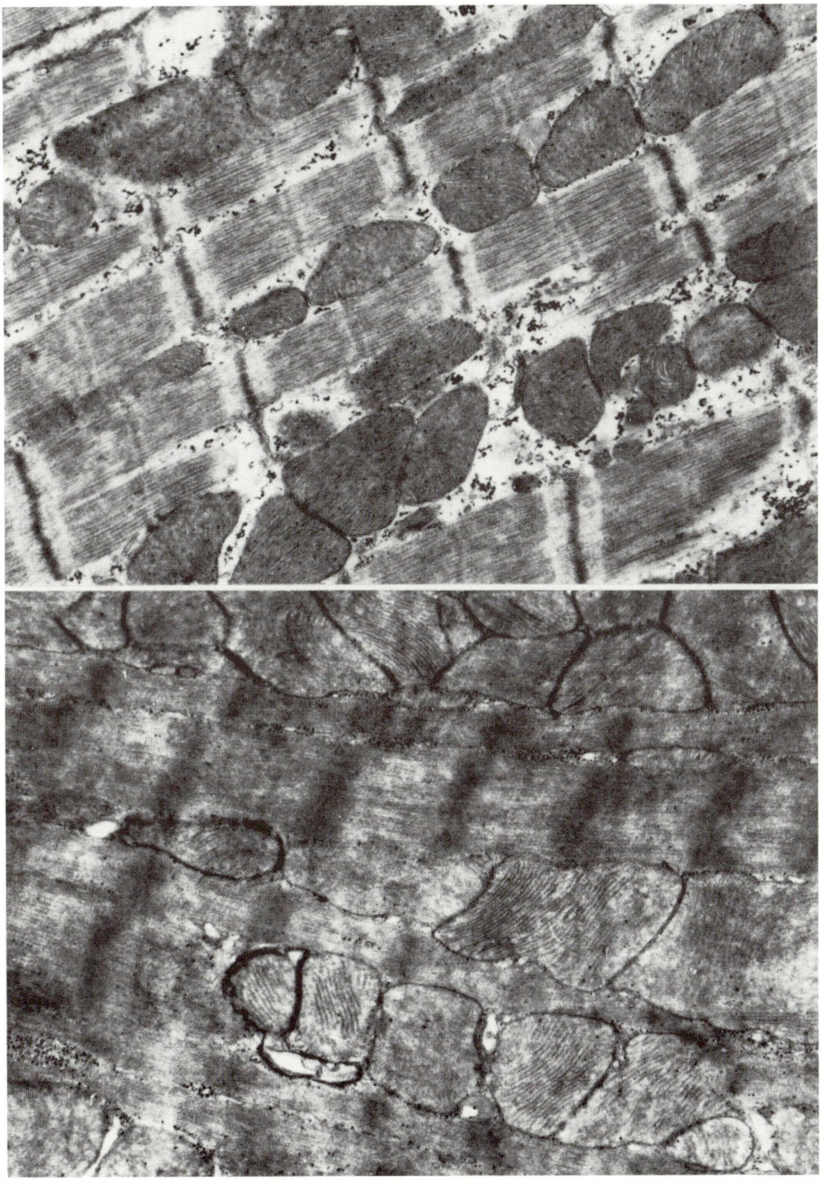

Fig. 9 Ultrastructural changes in the isolated rat heart perfused for 2 h with substrate-free medium. Upper panel: control heart; lower panel: substrate-depleted heart indicating swollen mitochondria and supercontracted sarcomeres. × 18,100.

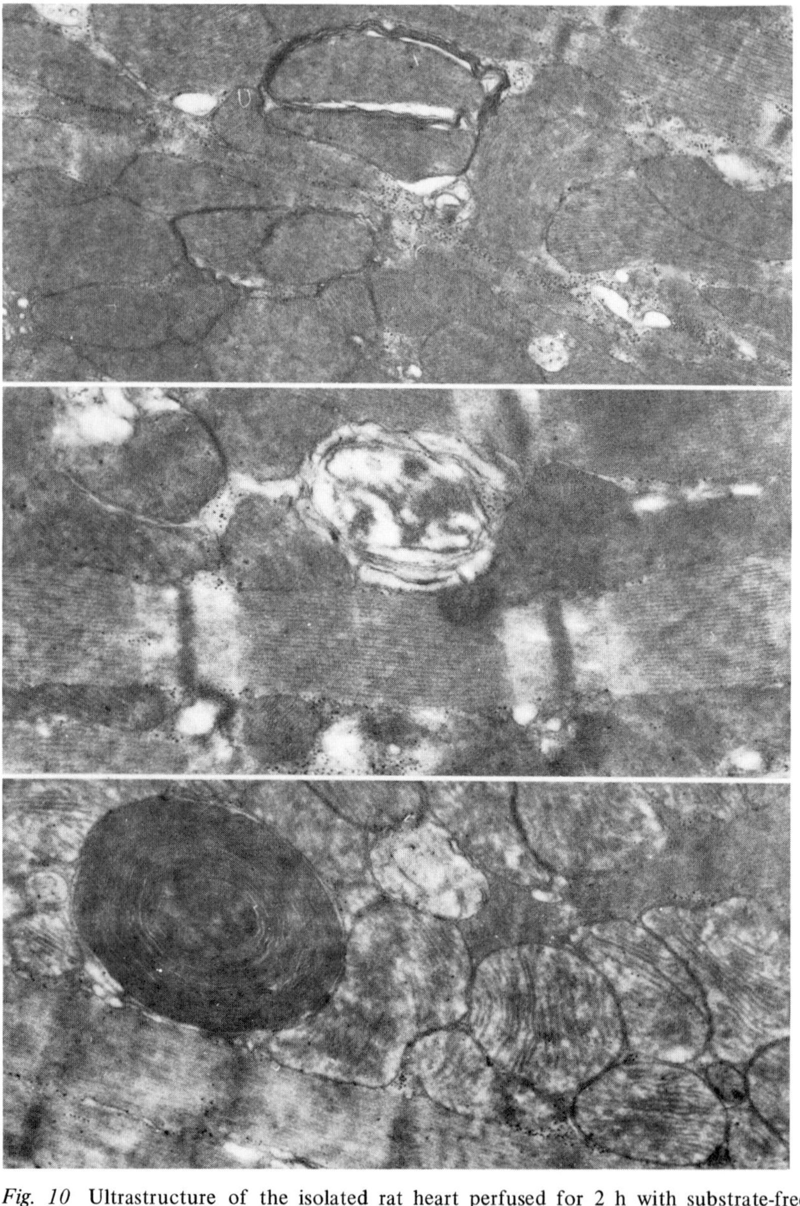

Fig. 10 Ultrastructure of the isolated rat heart perfused for 2 h with substrate-free medium, indicating the formation of coiled membranes around mitochondria (upper panel), disruption of mitochondria (middle panel), and the presence of a laminar body with concentrically arranged layers of membranes (lower panel). × 18,250.

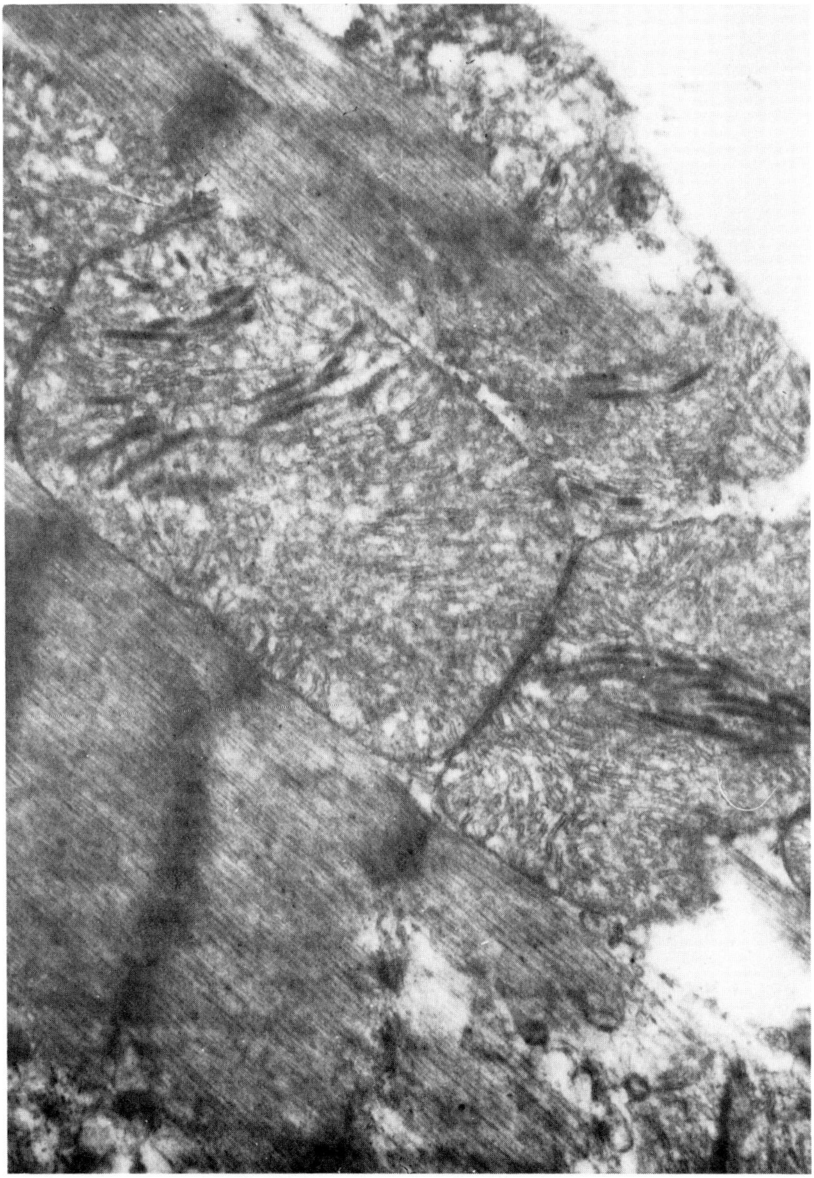

Fig. 11 Ultrastructure of the isolated rat heart perfused for 2 h with substrate-free medium, indicating some damage to the mitochondrial cristae. × 24,640.

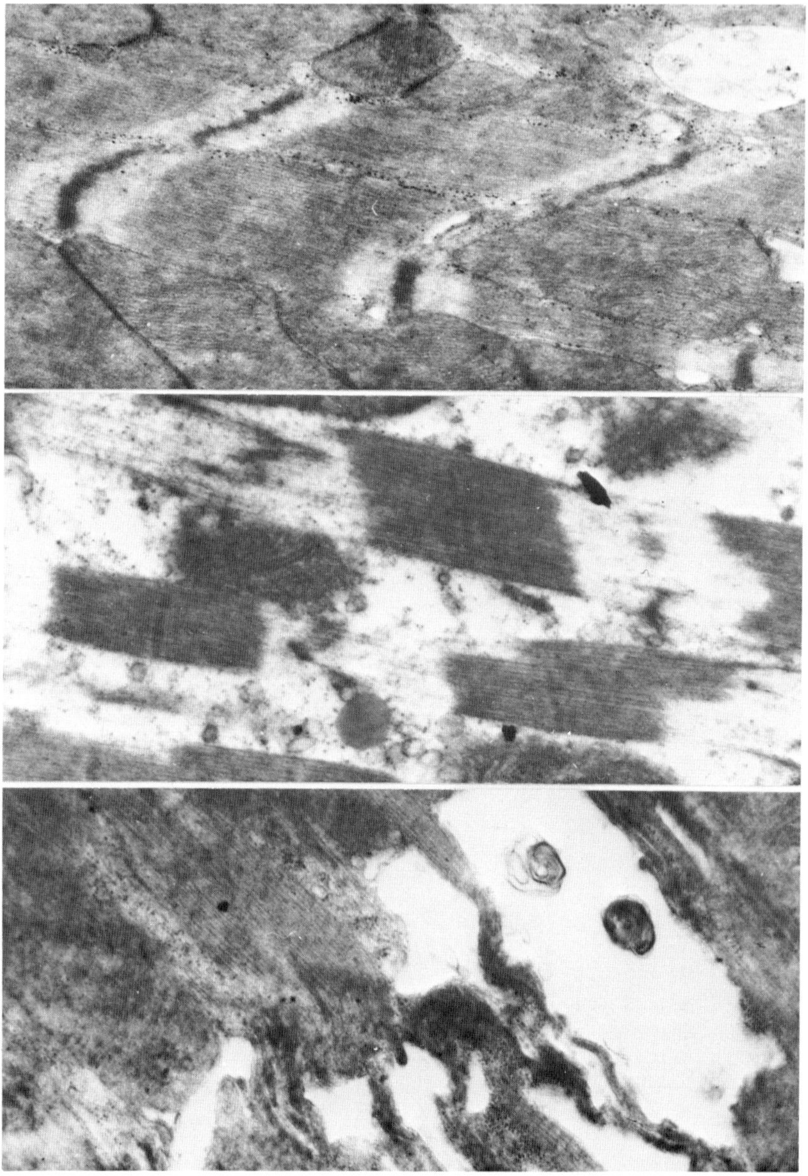

Fig. 12 Ultrastructure of the isolated rat heart perfused for 2 h with substrate-free medium, indicating distorted Z-line (upper panel), and disruption of myofibrils (lower panel). × 18,550.

In this experimental model of heart failure, we have clearly shown a dissociation between the changes in the high energy phosphate stores and the ultrastructure of the heart. Since the changes, particularly those in the phosphate potential, CP/Pi ratio and ATP/AMP ratio, preceded the changes in the ultrastructure of the substrate-depleted heart, it is likely that the ultrastructural changes in this condition may be a result of the low energy state of the heart. It therefore appears that an insufficient operation of the process of energy production in this model induces a decline in cardiac function and leads to a reduction in the high energy phosphate stores. This low energy state of the heart may produce damage to the myocardial ultrastructure, thus resulting in a defect in the process of energy utilization and, subsequently, in the inability of the heart perfused with substrate-free medium to generate contractility.

SUMMARY

In an isolated rat heart perfused with substrate-free medium, we have observed that glycogen, creatine phosphate, the phosphate potential, and the creatine phosphate to phosphate potential ratio decline considerably, prior to a fall in myocardial contractility. The initial decline in contractility was associated with a decrease in the ATP/AMP ratio or triglyceride levels; however, the mobilization of endogenous fuel supply appears to be inadequate to maintain contractility. The inability of the heart to generate contractile force was apparent at the time when a considerable amount of endogenous fuel supply and high energy phosphate stores were still present in the heart. No change in myocardial ultrastructure was observed until the energy state of the heart reached a very low level. At the time of complete heart failure, marked swelling of the sarcotubular system, enlarged vesicles under the sarcolemma, and enlarged elliptical spaces at the intercalated disk were noted. In addition to the formation of coiled membranous bodies in association with mitochondria, scattered swelling and disruption of mitochondria was observed. In some areas, sarcomeres were highly contracted and the adjacent sarcomeres were stretched with distorted Z-band and disrupted myofibrils. These results suggest that ultrastructural changes are secondary to the low energy state of the myocardium.

REFERENCES

1. BAHR, G. F. and JENNINGS, R. B.: Ultrastructure of normal and asphyxic myocardium of the dog. Lab. Invest. *10:* 548–571 (1961).
2. BRYANT, R. E.; THOMAS, W. A. and O'NEAL, R. M.: An electron microscopic study of myocardial ischemia in the rat. Circulat. Res. *6:* 699–709 (1958).
3. DHALLA, N. S. and OLSON, R. E.: Correlation of the endogenous fuel supply and contractile force in the isolated perfused rat heart. Fed. Proc. *26:* 352 (1967).
4. DHALLA, N. S.; RANGEL, T. and OLSON, R. E.: Calcium uptake and ATPase activity of sarcoplasmic reticulum and mitochondria during the course of spontaneous heart failure. Pharmacologist *10:* 208 (1968).
5. DHALLA, N. S.; SULAKHE, P. V.; BHAGAT, B. and OLSON, R. E.: Inotropic responses and catecholamine stores during heart failure by substrate-lack. Fed. Proc. *28:* 416 (1969).
6. DHALLA, N. S.; SUN, C. N. and OLSON, R. E.: Effect of substrate-lack and hypoxia on the function and ultrastructure of the heart. Circulation *36 (Suppl).:* 67 (1967).
7. DHALLA, N. S.; WALZ, D. A.; MCDONALD, V. A. and OLSON, R. E.: Metabolic and functional changes due to hypoxia in the isolated rat heart. Pharmacologist *9:* 222 (1967).

8. ESTABROOK, R. W. and MAITRA, P. K.: A fluorometric method for the quantitative microanalysis of adenine and pyridine nucleotides. Anal. Biochem. *3:* 369–382 (1962).

9. FEINSTEIN, M. B.: Effects of experimental congestive heart failure, ouabain, and asphyxia on the high energy phosphate and creatine content of the guinea pig heart. Circul. Res. *10:* 333–346 (1962).

10. FOLCH, J.; LEES, M. and SLOANE-STANLEY, G. H.: A simple method for the isolation and purification of total lipides from animal tissues. J. Biol. Chem. *226:* 497-509 (1957).

11. HUGGETT, A. ST. G. and NIXON, D. A.: Use of glucose oxidase, peroxidase and O-dianisidine in determination of blood and urinary glucose, Lancet ii: 368–370 (1957).

12. MATOUSHEK, R. A.; DHALLA, N. S. and OLSON, R. E.: Effect of perfusion with substrate-free medium on glycolytic and critic acid cycle intermediates in rat heart. Fed. Proc. *27:* 303 (1968).

13. MUIR, J. R.; DHALLA, N. S.; ORTEZA, J. M. and OLSON, R. E.: Energy linked calcium transport in subcellular fractions of the failing rat heart. Circul. Res. *26:* 429–438 (1970).

14. OLSON, R. E. and DHALLA, N. S.: Biochemical events in the isolated perfused rat heart during the onset of cardiac failure. Int. Symp. on Herzinsuffizienz, Georg Thieme Verlag, Stuttgart, p. 325–333 (1968).

15. POOL, P. E.; COVELL, R. W.; CHIDSEY, C. A. and BRAUNWALD, E.: Myocardial high energy phosphate stores in acutely induced hypoxic heart failure. Circul. Res. *19:* 221-229 (1966).

16. POOL, P. E.; SPANN, J. F.; BUCCINO, R. A.; SONNENBLICK, E. H. and BRAUNWALD, E.: Myocardial high energy phosphate stores in cardiac hypertrophy and heart failure. Circul. Res. *21:* 365–373 (1967).

17. SUN, C. N.; DHALLA, N. S. and OLSON, R. E.: Formation of gigantic mitochondria in hypoxic isolated perfused rat hearts. Experientia *25:* 763–764 (1969).

18. WAHLER, B. E. und Wollenberger, A.: Zur Bestimmung des orthophosphates neben saure-molybdat-labilen Phosphorsaureverbindungen. Biochem. Z. *329:* 508–520 (1958).

19. WELLS, M. A., and DITTMER, J. C.: The quantitative extraction and analysis of brain polyphosphoinositides. Biochemistry *4:* 2459–2467 (1965).

20. WOLLENBERGER, A. and SCHULZE, W.: Mitochondrial alteration in the myocardium of dogs with aortic stenosis. J. Biophys. Biochem. Cytol. *10:* 285–288 (1961).

Mailing address: , N. S. Dhalla, Department of Physiology, Faculty of Medicine, University of Manitoba, Winnipeg 3, Manitoba (Canada).

McGill University Medical Clinic,
Montreal General Hospital, Montreal, Quebec, Canada

MYOCARDIAL METABOLISM IN EXPERIMENTAL CARDIOGENIC SHOCK

R. F. P. CRONIN, T. ZSOTER, and C. H. HOLLENBERG

The syndrome of protracted hypotension that develops in some patients shortly after acute myocardial infarction is a well recognized clinical entity that has been termed cardiogenic shock. The fact that approximately 20% of patients survive this complication with or without the aid of intravenous pressor therapy has led to speculation that cardiogenic shock results from a potentially reversible abnormality of myocardial function in the noninfarcted areas of the heart (4). The most likely abnormality one might suspect is inadequacy in supply or utilization of oxygen; but studies by Bing *et al.* (1) in experimental cardiac shock, which we have been able to confirm in our laboratory, demonstrate a marked fall in coronary vascular resistance, normal or near normal O_2 saturation in coronary mixed venous blood, and a normal relationship between left ventricular work and left ventricular oxygen uptake. These findings prompted the present investigation into the changes in myocardial utilization of a wider range of substrates in experimental cardiogenic shock.

METHODS

A total of 24 dogs weighing between 14 and 16 kg were studied. They were fasted for not less than 16 h, anesthetized with intravenous pentobarbital sodium (25 mg/kg), and allowed to breathe air spontaneously. Catheters were manipulated from the neck vessels into the aorta and midcoronary sinus and their lumen kept free of clot by a slow infusion of saline solution; no heparin was used. In 12 animals, after a 20 min control period, cardiogenic shock was produced by embolizing the coronary tree with 3 mg/kg of a suspension of plastic microspheres, according to methods previously described (5). A fall in mean arterial blood pressure of 35% or more below the control level was required as a criterion for an adequate shock preparation. The remaining 12 animals, composing the control group, were prepared in the same way but coronary embolization was not performed. Simultaneous arterial and coronary venous blood samples were drawn anaerobically from both groups every 15 min. Blood samples were analyzed spectrophotometrically for oxygen saturation and for oxygen capacity by the manometric method. Glucose was determined by the ferricyanide technique. The concentrations of lactate and pyruvate were measured enzymatically. Free fatty acid concentrations were determined according to the method of Dole and Meinertz (6) and triglycerides by the method of Van Handel and Zilversmit (9). In a separate group of 8 shocked animals similarly prepared but heparinized, serial measurements of coronary blood flow were made at similar time intervals, using the I^{131}-iodo-aminopyrine technique of Krasnow *et al.* (8).

RESULTS AND DISCUSSION

The data from the control group of animals and from the animals subjected to experimental cardiogenic shock are summarized in table I. The shocked group showed a marked fall in blood pressure and this degree of hypotension was maintained throughout the experimental period. Previous experience with this shock preparation has indicated that the fall in arterial pressure is accompanied by a disproportionately greater fall in cardiac output, so that total peripheral resistance is increased. On the average, left ventricular work as reflected by the left ventricular time-tension index falls 52% with the induction of shock and myocardial O_2 uptake is reduced 28% (3). The arterial blood of all animals in both groups remained fully saturated. The arteriovenous (A-V) O_2 difference across the heart remained relatively constant in the control group but showed a significant increase in the shocked group. Myocardial O_2 extraction increased from 54% in the control group to 74% in the shocked group 45 min after the induction of cardiogenic shock.

Arterial lactate concentration, A-V lactate difference, and myocardial lactate extraction remained constant in the control group. In the shocked group, arterial lactate began to rise 15 min after coronary embolization, and a significant reduction in myocardial lactate extraction occurred. Myocardial lactate extraction had virtually ceased 45 min after shock, with 5 out of 11 animals showing a net production of lactate across the myocardium. Arterial "excess" lactate calculated from the change in lactate pyruvate ratio showed a significant rise in the shocked animals and rose progressively with time.

A rise in arterial glucose was noted immediately after coronary embolization. Glucose extraction, instead of diminishing progressively with time as in the control group, showed a slight increase but was not significantly greater following shock. Rise in arterial glucose may be a response to endogenous sympathetic activity evoked by the fall in blood pressure, since a similar rise in arterial glucose without a change in myocardial glucose extraction has been reported by Edwards *et al.* (7) in animals during hemorrhagic shock.

These data are quite similar to those of Brachfeld and Scheuer (2), who studied the metabolic consequences of severe myocardial ischemia produced in the open-chest dog by reducing coronary flow. The decrease in myocardial work evoked by these workers corresponds to that in the present study and the changes observed in uptake of oxygen, glucose, lactate, and pyruvate were similar. These authors did not study changes in myocardial lipid

Table I

Hemodynamic and Metabolic Data in Experimental Cardiogenic Shock [1]

Time Period	No. in series	Art. B.P. (mm Hg)	Art. O_2 (vol %)	Cor. sinus O_2 (vol %)	Myocard. O_2 extraction %	Art. lactate (mM/l)	Cor. sinus lactate (mM/l)	Myocard. lactate extraction %	Excess lactate (mM/l)	Art. glucose (mg %)	Cor. sinus glucose (mg %)	Myocard. glucose extraction %	Art. FFA (μEq/l)	Cor. sinus FFA (μEq/l)	Myocard. FFA extraction %	Art. TG (mg %)	Cor. sinus TG (mg %)	Myocard. TG extraction %
Before coronary embolization																		
15 min	12 C	134 ± 6	19.8 ± 0.6	9.9 ± 1.0	50	0.68 ± 0.08	0.43 ± 0.06	39		85 ± 1.2	82 ± 1.0	3.5	908 ± 63	861 ± 57	5.2	57 ± 9.1	54 ± 8.2	5.2
	12 S	123 ± 3	20.0 ± 0.7	10.7 ± 1.4	47	0.52 ± 0.08	0.35 ± 0.08	33		80 ± 1.1	76 ± 1.3	5.0	724 ± 83	724 ± 72	8.6	48 ± 5.2	46 ± 4.8	4.2
5 min	12 C	130 ± 7	19.9 ± 0.7	8.4 ± 0.8	58					86 ± 1.3	85 ± 0.9	1.2						
	12 S	127 ± 4	20.2 ± 0.6	8.3 ± 0.7	59					81 ± 1.0	79 ± 1.4	2.5						
After coronary embolization																		
15 min	12 C	134 ± 5	20.1 ± 0.8	9.7 ± 0.9	52	0.70 ± 0.11	0.46 ± 0.09	34	0.02 ± 0.07	88 ± 1.3	87 ± 1.2	1.1	984 ± 64	840 ± 58	14.6	50 ± 7.3	48 ± 6.7	4.0
	12 S	68 ± 5	19.7 ± 0.7	7.0 ± 0.7	65	1.27 ± 0.23	1.06 ± 0.19	17	0.32 ± 0.21	93 ± 2.5	88 ± 1.7	5.4	750 ± 76	738 ± 65	1.6	50 ± 5.5	44 ± 4.3	12.0
30 min	12 C	129 ± 11	20.0 ± 0.6	8.9 ± 0.8	56					89 ± 1.4	88 ± 1.2	1.1						
	11 S	70 ± 6	19.5 ± 0.9	7.3 ± 0.9	63					91 ± 1.8	87 ± 1.7	4.4						
45 min	12 C	138 ± 7	20.8 ± 0.8	9.2 ± 1.2	56	0.76 ± 0.10	0.53 ± 0.09	32	−0.06 ± 0.08	88 ± 1.4	87 ± 1.8	1.1	964 ± 61	809 ± 61	16.1	48 ± 7.0	46 ± 6.1	4.2
	11 S	70 ± 9	19.9 ± 0.7	5.2 ± 0.9	74	1.48 ± 0.27	1.43 ± 0.22	5	0.76 ± 0.42	91 ± 2.0	84 ± 1.9	7.7	809 ± 95	702 ± 86	13.2	42 ± 4.2	45 ± 4.5	7.1

[1] Mean values ± S.E.; C, control series; S, shock series; FFA, non-esterified free fatty acids; TG, triglycerides.

metabolism. Early in the shock period, we observed markedly diminished free fatty acid extraction with increased triglyceride extraction, while this situation was seen to be reversed during the late shock phase. The interpretation of extraction values can often lead to erroneous conclusions and it would be of interest to know the changes in coronary blood flow that occurred following coronary embolization. Because of technical problems that impede the measurement of coronary flow in the unheparinized closed-chest dog, an estimation of coronary flow was made in separate groups of 8 control and 8 shocked animals given heparin; these determinations are summarized in table II. It can be seen that the degree of hypotension produced in this group of shocked animals was similar to that recorded in the shocked group in which the metabolic studies were carried out; thus, these coronary flow values probably accurately reflect the changes in coronary flow that occurred in the latter group.

If the mean coronary flow values obtained from the 8 shocked animals studied separately are applied to the mean extraction values for all substrates in the group of shocked animals in whom metabolic studies were performed, uptake values can be calculated. The results are expressed as a percentage change in uptake from values obtained in the control group. As illustrated in figure 1, the percentage uptake of oxygen falls to an average of 32% below control levels early in shock and recovers slightly at the 45 min mark. This reduction in aerobic metabolism reflects the reduction in left ventricular work performed at these periods. Reduction in glucose uptake parallels the reduction in O_2 uptake at both periods. Lactate uptake is proportionately reduced early in shock but, at the 45-min mark, lactate extraction has virtually ceased. Lactate may be produced by the heart as well as utilized by that organ and the methods

Table II
Coronary Flow in Experimental Cardiogenic Shock

Time Period	No. of subjects	Arterial B.P. (mm Hg)	Coronary flow (cc/100g/min)
Before coronary embolization			
15 min	8 C	130 ± 5	135 ± 15
	8 S	133 ± 7	149 ± 22
5 min	8 C	132 ± 6	140 ± 16
	8 S	128 ± 5	141 ± 17
After coronary embolization			
15 min	8 C	128 ± 4	145 ± 12
	8 S	71 ± 4	92 ± 4
30 min	8 C	132 ± 6	130 ± 13
	7 S	68 ± 6	88 ± 14
45 min	8 C	136 ± 7	141 ± 10
	7 S	66 ± 8	86 ± 14

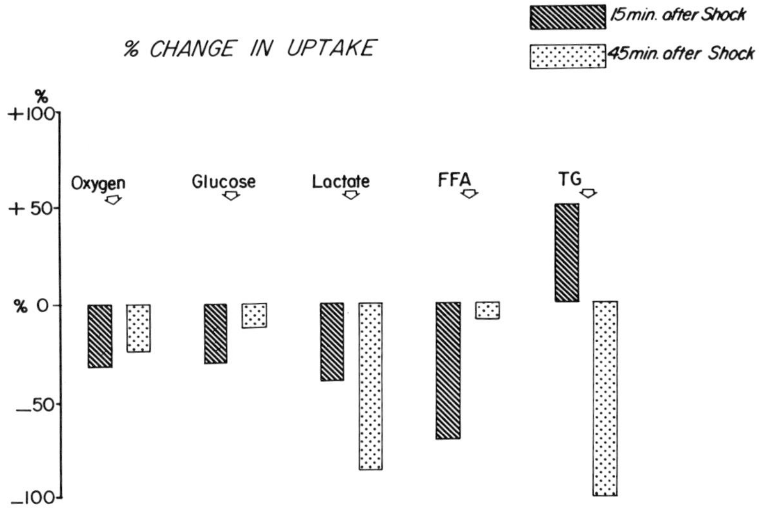

Fig. 1 Change in uptake of various substrates by the heart, following induction of shock by coronary embolization. Mean changes in a group of 12 shocked animals as compared with 12 control animals.

employed permit only an assessment of the net uptake of lactate across the myocardium; nevertheless, the apparent cessation of lactate uptake that was observed is most probably due to lactate production by ischemic muscle, while the normally perfused areas of myocardium continued to extract lactate.

In addition to plasma triglycerides, there exists a considerable lipid pool in the heart, not only in the subepicardial adipose tissue but also in intracellular fat droplets. Free fatty acids and triglycerides can be converted one to the other in perfused heart muscle under the influence of cardiac lipoprotein lipase without entering into energy-producing metabolic pathways. The apparent cessation of free fatty acid uptake early in shock with an extraordinary rise in triglyceride utilization that reverses itself 30 min later suggests such a mechanism. Activation of lipoprotein lipase with a sudden rapid increase in lipolytic conversion of triglycerides to free fatty acids is, therefore, a possible and perhaps likely explanation of the wide fluctuation in the apparent uptake of these substrates. Although the work of Bing *et al.* (1) with a similar shock preparation has shown several hours delay in the liberation of myocardial enzymes, lipoprotein lipase was not assayed and the location of this enzyme close to the cell membrane and in proximity to the capillaries suggests that its liberation might well occur much earlier.

SUMMARY

Myocardial metabolism was studied in 24 dogs. Under sodium pentobarbital anesthesia, catheters were placed in the aorta and coronary sinus; the animals did not receive heparin.

Blood pressure was recorded and blood samples withdrawn from each site every 15 min for 75 min. Samples were analyzed for O_2 content and capacity, glucose, lactate, pyruvate, free fatty acids, and triglycerides. In 12 dogs, cardiogenic shock was produced by closed-chest coronary embolization with microspheres; 12 animals served as controls. No significant metabolic change was observed in the control animals throughout the experimental period. In the shocked animals, the arterial blood remained fully saturated, but a mean increase in myocardial O_2 extraction was observed. Arterial glucose levels did not change and myocardial glucose extraction showed a slight increase that was also not significant. Mean arterial lactate levels rose significantly after induction of cardiogenic shock and myocardial lactate extraction became markedly diminished or ceased entirely, whereas no change in pyruvate levels was noted. The most striking metabolic change observed in the shocked group was a fall in free fatty acid extraction with the onset of cardiogenic shock; this change was accompanied by increased triglyceride extraction. Free fatty acid extraction had returned to control levels 45 min after the onset of shock, and net triglyceride production across the myocardium was observed. It is suggested that coronary embolization may cause an activation of myocardial lipoprotein lipase.

REFERENCES

1. BING, R. J.; CASTELLANOS, A.: GRADEL, E.: LUPTON, C. and SIEGEL, A: Experimental myocardial infarction: Circulatory, biochemical and pathologic changes. Amer. J. Med. Sci. *232:* 533–554 (1956).
2. BRACHFELD, N. and SCHEUER, J.: Metabolism of glucose by the ischemic dog heart. Amer. J. Physiol. *212:* 603–606 (1967).
3. CRONIN, R. F. P.; Effect of isoproterenol and norepinephrine on myocardial function in experimental cardiogenic shock. Amer. Heart J. *74:* 387–395 (1967).
4. CRONIN, R. F. P.; MOORE, S. and MARPOLE, D. G.: Shock following myocardial infarction. A clinical survey of 140 cases. Canad. M.A.J. *93:* 57–63 (1965).
5. CRONIN, R. F. P. and ZSOTER, T.: Hemodynamic effects of rapid digitalization in experimental cardiogenic shock. Am. Heart J. *69:* 233–240 (1965).
6. DOLE, V. P. and MEINERTZ, H.: Microdetermination of long-chain fatty acids in plasma and tissues. J. Biol. Chem. *235:* 2595–2599 (1960).
7. EDWARDS, W. S.; SIEGEL, A., and BING, R. J.: Studies on myocardial metabolism III. Coronary blood flow, myocardial oxygen consumption and carbohydrate metabolism in experimental hemorrhagic shock. J. Clin. Invest. *33:* 1646–1661 (1954).
8. KRASNOW, N.; LEVINE, H. J.; WAGMAN, R. M. and GORLIN, R.: Coronary blood flow measured by 1–131 iodo-antipyrine. Circulat. Res. *12:* 58–62 (1963).
9. VAN HANDEL, E. and ZILVERSMIT, D. B.: Micromethod for direct determination of serum triglycerides. J. Lab. Clin. Med. *50:* 152–157 (1957).

Mailing address: R. F. P. Cronin, M.D., Department of Medicine, The Montreal General Hospital, Montreal 109, Quebec (Canada).

Department of Biochemistry, Colorado State University,
Fort Collins, Colorado, USA

EFFECT OF AGE ON CARDIAC METABOLISM

H. SOBEL

A casual perusal through the cardiovascular literature will reveal the fact that investigators rarely select experimental animals with due regard for age. In fact, it is safe to say that most studies have been carried out with animals that have not yet reached maximum growth. Yet, with the exception of congenital heart diseases or rheumatic conditions, most of the clinical problems that involve the heart occur in persons in their post-maturity state.

In the human, coincident with or soon after cessation of growth, a great variety of physiological decrements begin to appear and they increase with advancing age (1). The progression of the aging syndrome culminates in overt clinical senesence.

Experimental gerontologists consider the problems of three types of cells: replicating cells, cells that replicate under certain stimuli, and the fixed post mitotics. The last include nonreplaceable cells that are supposed to exist with the organism until the death of the latter unless they should die prematurely. The heart cells, among this category, are incapable of replication and must accrue changes over a lifetime that result from the temporal operation of the genetic program and those introduced by the environment. That true muscle cell nuclei do not divide was recently reconfirmed by Moss and Leblond (2) for growing rats. These authors reported, however, that satellite cells associated with muscle fibers are capable of undergoing mitosis and that one or both daughter nuclei may incorporate into the associate fiber. It would be interesting to learn whether this phenomenon occurs in adult animals.

Aging theories that relate to nonreplaceable cells are concerned with: a) problems of transcription of genetic information and protein synthesis, b) problems involving structural components of the cells, and c) problems of delivery of necessary molecules at an adequate rate.

PROBLEMS OF TRANSCRIPTION OF GENETIC INFORMATION AND PROTEIN SYNTHESIS

Damage to nuclear material not manifest in DNA→ DNA replication has been inferred from histological evidence of change in staining characteristics, chromatin content, and variations and aberrations in nuclear structure including invagination of the nuclear membrane and inclusions (3). Munnell and Getty (4) reported nuclear lobulation, invagination of the nuclear membrane, appearance of binucleate cells, and evidence of amitotic division in the aging canine myocardium. Such changes have also been noted in cells of hypertrophic hearts. It was suggested that the nuclear changes might represent an attempt on the part of the cell to restore the nucleo-cytoplasmic ratio, since myocardial cell size increased continuously with age. Mazia (5) had developed the concept that mitotic cells exhibit growth of the cytoplasm until a critical ratio is reached. This serves as the signal for DNA synthesis. There is some evidence that the number of cells in the adult human heart is constant without regard to the size of that organ (6). Sobel *et al.* (7) observed a relatively constant DNA content in dog hearts with age, even though heart weight increased. Munnell and Getty found that myocardial cell size increased continually with age (4). The implication is that larger individuals will have a larger cardiac cytoplasmic mass than smaller individuals. The significance of this is not clear.

Damage to nuclear material is subject to repair; and renewal of DNA occurs in nondividing cells (8). While it has been suggested that the renewal process itself may be subject to age-associated damage, Pelc (9) reported no change in the rate of renewal with age.

The possibility that DNA → RNA transcription may be affected with aging has been suggested by various workers. Decrease in template potential as a result of masking of segments of the genome that had been previously available as sites for RNA synthesis has been proposed by Von Hahn and Fritz (10) and by Samis (11). Phytilä and Sherman (12), working with thymus DNA and nucleoprotein, reported age-related reduction in template activity of partially purified nucleoprotein preparations but found no age-related differences with DNA or purified nucleoprotein preparation. Falzone *et al.* (13), working with native salt extracted and heat denatured chromatin from liver, found no differences in template activity. Devi *et al.* (14) observed that DNA dependent RNA synthesis decreased with age in the rat liver. It did not appear that RNA polymerase activity is affected as much as the unwinding of the two strands of the DNA double helix. Samis and Wulff (15) reported an absence of age-related differences in template activity of rat liver chromatin. Experiments of this nature have not been carried out with heart DNA.

Errors in the translation of the genetic message of DNA into protein synthesis have been implicated during aging by a number of investigators (16, 17). This process could involve problems with any of the components of the protein synthesizing complex. Although protein synthesis has been carried out with ribosomes from heart muscle (18), the effect of age on this system has not been tested. Judging from studies on the hearts of dogs (7), there is no evidence that age affected the quantity of RNA, the contractile or cytoplasmic proteins including myoglobin. Indeed, they increased with the age-associated hypertrophy. It is, however, still possible that subtle problems of protein synthesis do occur in aging hearts. For example, Z-disc alterations have been reported in some aged animals (19, 20).

PROBLEMS INVOLVING STRUCTURAL COMPONENTS OF THE CELLS

Accumulation of pigment in certain non-replicating cells has been known for more than a century. Lipofuscin represents a conglomerate of substances that contain fluorescing lipoprotein. In the human heart, Strehler *et al.* (21) observed a linear relationship of accumulation of lipofuscin with age. However, in the oldest ages, the dispersion of data was very large; so while some individuals had a considerable portion of their myocardial cells occupied with this material, others had no more than much younger persons. Munnell and Getty (22) observed a similar accrual in the myocardia of dogs, but the rate of accrual of lipofuscin was five times as great with marked differences in the older animals. However, numerous observations suggest that a variety of factors can induce the formation of lipofuscin in various sites. Goldfischer and Bernstein (23) found this material to be present in the livers of human fetuses and in newborn, and they concluded that the accumulation of lipofuscin is a function of metabolites sequestered in lysosomes and is not a result of aging. Goldfischer *et al.* (24) suggested the evolution of certain lysosomes into residual bodies that contain oxidized lipids, melanin, heme, metals, and assorted debris into what is then designated "lipofuscin". The composite of data suggest that lipofuscin accumulation in some way reflects the history of the organism. The possible effect of the occupation of a large volume of myocardial cell on the function of the heart has been the subject of much attention, but the matter has not been resolved. Dock (25) pointed out that the force of ejection (I–J of the ballistocardiogram) decreases and the pigment increases by the same amount each decade.

It is now generally agreed that the vitamin-E deficiency pigment is a stage in the evolution of lipofuscin. This would suggest that lipofuscin results from lack of protection against hyperoxia and peroxidized lipids; however, this material has also been observed following hypoxia (26). Glinsmann and Ericsson (27) have found that organelles, portions of cytoplasm and mitochondria in cells damaged by hypoxia, are segregated within the ischemic cell and are subsequently digested within cytosegresomes.

One may inquire into the integrity of the transcription–translation sequence in nonreplicating cells with aging, but there is also the matter of the integrity of nonreplaceable components of the cell aside from DNA. The possibility that membrane systems and organelles may fail with age as a result of peroxidation of lipoprotein constituents has been considered by Packer *et al.* (28). In a study with growing mice, Strehler (29) concluded that various lipids, particularly membrane constituents and as yet uncharacterized proteins of several tissues, are not synthetized at appreciable rates. Thus, membrane constituents must be included as major potential sites of deterioration during aging. If the membranes are indeed longlived, there is the potential of damage to the plasma membranes, endoplasmic reticulum, Golgi apparatus, and mitochondria (30).

Although a great deal is known about the structure and function of the mitochondria (31), the literature concerning the effect of age on the mitochondria on nonreplicating cells is confusing. Changes have been reported in tinctorial properties, size, shape, and fragmentation (32), as well as in various biochemical characteristics of the mitochondria in old cells. Tauchi and Sato (33) stated that in human hepatic cells mitochondria decrease in number and increase in size with age. Tauchi *et al.* (34) observed changes in the cristae and noted that individual mitochondria were encircled by elements of endoplasmic reticulum. Bizarre shapes were induced in the mitochondria of the livers of senile rats treated with thyroxine. Bogatskaya (35) reported that the number of mitochondria in rat myocardial cells decreases with age. Burch *et al.* (36) observed enlarged mitochondria that contained large glycogen particles, in the hearts of aging drosophila. However, Gold *et al.* (37), in studying parameters of oxidative phosphorylation in liver, heart, and kidney of rats 12 to 27 months of age, concluded that there is no decrement in the functional integrity of mitochondria with aging. Mitochondria are self-replicating even in nonreplicating cells (38) and there is no difference with age in their turnover rate in the rat liver (39), at least. The data lead to a tentative suggestion that individual mitochondria function normally but that there may be relatively reduced numbers for the total cardiac mass. However, in the reviewer's experience, mitochondrial mass can increase under hypoxic stimulus in old animals. The effect of age on the replication of mitochondrial DNA requires investigation.

It is well known that mitochondria undergo rapid alterations in structure and function as a result of hypoxia or a great variety of toxic substances. It would be highly significant to compare the changes observed as the result of hypoxia with changes reported to occur in the mitochondria of aging cells. Ou and Tenney (40) observed an increase in number but not in size of mitochondria in hearts of cattle acclimatized to high altitude. Individual mitochondria exhibited increase in electron transport, O_2 uptake, and cytochrome a_3 oxidase. Reznik and Hansen (41) reported enlargement and alteration in patterns of the mitochondria in myofibers that had been exposed to acute ischemia. They concluded that the unusual mitochondria were more indicative of a compensatory process than of a primary abnormality. Jennings and co-workers observed fine structural abnormalities in the mitochondria of ischemic, irreversibly injured dog myocardial cells (42). Such mitochondria cannot perform integrated functions of the Krebs cycle (43). However, reversibly injured cells may be maintained in an adverse state for over 30 min. Mitochondrial swelling with enlargement, and shrinkage of cristae occur probably as a result of intracellular acidosis (44) due to a marked increase in lactate production. These damaged mitochondria eventually participate in the formation of altered lysosomes. It is known that myocardial cells are almost completely dependent upon aerobic, mitochondrial metabolism for their energy requirements. Fatty acids provide the major substrate for energy production. With hypoxia, this source of energy is rapidly curtailed and glycolysis increases; but this compensating mechanism cannot provide for more than a small part of energy needed to maintain ATP production (45). The reduced ability to metabolize fatty acid as an early manifestation of hypoxia is consistent with a need for CoA. This has been established by the stimulatory effect of this substance on the oxidation of pyruvate in damaged mitochondria (43). The possibility that, with aging, there is some compromising of cardiac energetics perhaps consistent with that seen in an early stage in reversibly injured cells is to be considered.

PROBLEMS OF DELIVERY OF NECESSARY MOLECULES AT AN ADEQUATE RATE

This leads to the question whether oxygen and nutrient molecules reach the myocardial cells at an adequate rate. The reviewer (46) proposed that age-changes in the connective tissues and interstitial space would result in reduced rate of transport of essential molecules to the cells, which in turn would lead to the progression of aging phenomena. In the heart, this would be an obvious result of impaired coronary flow that results from problems of atherosclerosis and their sequelae. But even in their absence, diffusion problems exist. It has been argued that diffusion gradients in the extracellular fluid can hardly be a limiting factor in oxygen transport to cells as long as the average oxygen concentration in the fluid is higher than a few micromols per liter (47). Zierler (48) states that it has been assumed that the acting force for the diffusion of oxygen is the gradient in partial pressure of O_2, that the capillary wall is a relatively trivial barrier to diffusion, that the tissues are isotropic, and that diffusion through any tissue under any given set of circumstances can be characterized by a single diffusion coefficient. There are experimental results that appear to challenge some of these assumptions. Since intercapillary distance is the most potent determinant of mean tissue Po_2, the latter depends critically on the number and distribution of functioning capillaries. Myers and Honig (49) found that, in the heart, functional intercapillary distance is much greater than morphological intercapillary distance except under maximal blood flow. Hort (50) points out that muscle fiber thickness and local capillary distance show a considerable variability. Bassingthwaighte's data (51) suggest that concentration gradients are not the only factors in driving small molecules.

The Po_2 at the level of the mitochondria is very nearly zero (52), and the cells barely receive enough oxygen to accommodate their needs (53). McFarland (54) believes that many of the age-changes that occur in the nervous system can be simulated by hypoxia. Luebbers (55) demonstrated that almost all available oxygen pressure in the capillaries is required for oxygen transport. All oxygen molecules arriving at the mitochondrial membrane are immediately metabolized. Fabel (56) reported that oxygen supply to the heart is at a critical level. There are no capillaries between the endocardium and the innermost layer of the myocardium. In some areas, oxygen tension was even lower than had previously been believed to be the critical oxygen tension for mitochondria. Thus, small increases in demand (as by uncoupling of oxidative phosphorylation) or reduction in rate of delivery of oxygen could have serious consequences. The well-known decrease in glucose tolerance with age (57) has been ascribed to problems of diffusion of both glucose and insulin (58). Another critical factor involving diffusion may relate to the rate of transport of free fatty acids. Since they are albumin-bound, let us consider the transvascular passage of labeled albumin into the interstitium (table I). The clearance of albumin (and presumably all molecules carried by albumin, including hormones and very likely larger molecules) was seen to decrease with age (59); transport into the mycardial interstitium also decreased. It was later shown that this phenomenon could have been related in some way to the reduction of cardiac hyaluronic acid with increasing age (60). This substance decreases with age wherever it is found (61). Because of the ability of hyaluronic acid to serve as a molecular sieve (62), it could affect the metabolism of extravascular plasma proteins. It is of more than passing interest that estradiol at the proper dose level increased hyaluronic acid in the skin (63). The effect on hyaluronic acid in the heart is unknown. Furthermore, there is a tendency for circulating albumin to decrease with age (64) and this could affect the transport of free fatty acids. It is known that chylomicra may be used a a source of nutrients by the myocardium (65). While the mechanism of this is not clear, it may relate to the rich supply of myocardial lipoprotein lipase that probably participates in the hydrolysis of triglycerides in the interstitial space (66). The clearance of chylomicrons from the plasma decreases with age (67, 68), an observation that may be related to the diffusional problem and to the fact that, with age, the distribution of chylomicron size may be affected (69). This may result from age-associated changes in the formation of chylomicrons at the intestinal-lacteal site (70). The apparent thickening of basement membranes with age (61) could affect transport of protein-bound free fatty acids and hormones. Finally, according to Schwartz (71), amyloidosis is a concomitant of the aging heart, and the quantity increases with age. This may be the result of immune phenomena. It has been shown that transvascular passage of labeled albumin into the interstitial space in hyperimmunized mice is decreased (72). This observation was thought to be related to alteration in the ground substance architecture as a result of antigen-antibody precipitation. Such processes would be expected to affect rate of transport of free fatty acids and hormones to the myocardial cells. Whereas myocardial infarction occurred in dogs 6–8 years of age that were fed an atherogenic diet, this had not been observed previously in several hundred dogs that were presumably much younger (73) although the degree of hypercholesterolemia and atherosclerosis was similar in both groups. The finding was related to the diffusion problem discussed above.

Table I

A. Retention Volumes of Dog Hearts following Intravenous Injection of Human I^{131}-albumin. Values expressed as Ml Equivalent of Plasma/g of Heart; Individual Values Are Given, Average in Parentheses

Circulation time	Age: <1 Yr.	1–2 Yr.	3–5 Yr.	7–9 Yr.	10 Yr. or older
6 min	0.0083, 0.0068 0.0076, 0.0093 0.0065, 0.0086 (0.0078)	0.0086, 0.0071 0.0076 (0.0077)		0.0057, 0.0100 0.0090, 0.0079 0.0047 (0.0074)	0.0126, 0.0071 0.0078, 0.0078 0.0093, 0.0108, 0.0127 (0.0097)
60 min	0.0226, 0.0222 0.0239 (0.0229)	0.0291, 0.0267 0.0214, 0.0245 0.0195 (0.0242)		0.0210, 0.0126 0.0122, 0.0149 0.0128 (0.0147)	0.0209, 0.0208 0.0134, 0.0143, 0.0177 (0.0174)
120 min	0.0282, 0.0239 0.0334, 0.0575 0.0588 (0.0403)	0.0393, 0.0293 0.0275, 0.0255 (0.0304)	0.0304, 0.0331 0.0312, 0.0262 0.0339 (0.0309)	0.0273, 0.0302 0.0163 (0.0246)	0.0153, 0.0267 0.0148, 0.0203 0.0217 (0.0197)

B. Clearance of Human I^{131}-albumin from Dog Plasma (C_p in 60 or 120-min Specimen/C_p in 6-min Specimen); Individual Values Are Given, Average in Parentheses

Circulation time	<1 Yr.	1–2 Yr.	3–5 Yr.	7–9 Yr.	10 Yr. or older
60 min	0.782, 0.813, 0.885 0.750, 0.927 (0.831)	0.925, 0.893, 0.882 (0.900)	0.881, 0.910, 0.930 0.979, 0.912 (0.922)	0.922, 0.912, 0.952 (0.928)	0.934, 1.000 0.923, 0.920 (0.944)
120 min	0.710, 0759, 0.764 0.646, 0.769 (0.729)	0.863, 0.779, 0.730 0.769 (0.785)	0.786, 0.795, 0.831 0.823, 0.785 (0.804)	0.800, 0.804, 0.836 (0.813)	0.872, 0.895 0.824, 0.820 0.790 (0.840)

Taken from Sobel *et al.* (59).

Are the well-known age-associated physiological decrements, including decreased cardiac output (74), decline in force of contraction (75), decrease in the amplitude of the ventricular electrocardiographic complexes, and reduction (74) with maximal effort of maximal heart rate, a result of the above? Do they speak for problems of energetics or the patency of the conduction and pumping system? Are they the result of autoregulation in the face of reduced demand? Or are they simply a consequence of atherosclerotic coronary vascular disease? And what of the age-associated increase in predisposition to auricular fibrillation, premature contraction and heart failure, on exposure to injury or overload?

A vast ocean of information has still to be acquired in order to understand the metabolism of the aging heart and its implication for clinical problems. Yet, investigators have paid too little attention to the age of their experimental animals. The greatest problem in experimental gerontology is the lack of suitable, aged animals. The experimental cardiologist must work with the gerontologist to resolve this matter.

SUMMARY

Investigators concerned with cardiovascular metabolism do not ordinarily consider the effect of age as a variable. The problems of aging myocardial cells fall under at least three categories: a) transcription of genetic information and protein synthesis; b) structural components of the cells; and c) delivery of necessary molecules at an adequate rate. Pertinent information relative to these matters has been reviewed. These problems and others may affect the over-all metabolism of the aging heart. The need for experimenters in cardiac metabolism to give attention to these matters is considered essential.

REFERENCES

1. SOBEL, H.: When does human aging start? Gerontologist. *6:* 17–22 (1966).
2. MOSS, F. P. and LEBLOND, C. P.: Nature of dividing nuclei in skeletal muscle of growing rats. J. Cell Biol. *44:* 459–462 (1970).
3. ANDREW, W.: Changes in the nucleus with advancing age of the organism; in B. L. STREHLER, Adv. Geront. Res., *1:* 87–108 (Academic Press, New York, 1964).
4. MUNNELL, J. F. and GETTY, R.: Nuclear lobulation and amitotic division associated with increasing cell size in the aging canine myocardium. J. Geront. *23:* 363–369 (1968).
5. MAZIA, D.: Mitosis and the physiology of cell division; in J. BRACHET and A. E. MIRSKY, The Cell, *3:* 77–412 (Academic Press, New York, 1961).
6. LINZBACH, A. J.: Heart failure from the point of view of quantitative anatomy. Amer. J. Cardiol. *5:* 370–382 (1960).
7. SOBEL, H.; THOMAS, H., and MASSERMAN, R.: Myocardial proteins in dogs of various ages. Proc. Soc. Exp. Biol. Med. *116:* 918–921 (1964).
8. PELC, S. R.: Turnover of DNA and function. Nature *219:* 162–163 (1968).
9. PELC, S. R.: Renewal of DNA in non-dividing cells and aging. Exp. Geront. *1:* 215–222 (1965).
10. VON HAHN, H. P. and FRITZ, E.: Age-related alterations in the structure of DNA. III. Thermal stability of rat liver DNA, related to age, histone content and ionic strength. Gerontologia *12:* 237–250 (1966).
11. SAMIS, H. V., JR.: A concept of biological aging: the role of compensatory processes. J. Theoret. Biol. *13:* 236–250 (1966).

12. PYHTILÄ, M. J. and SHERMAN, F. G.: Age-associated studies on thermal stability and template effectiveness of DNA and nucleoproteins from beef thymus. Biochem. Biophys. Res. Comm. *31:* 340–344 (1968).

13. FALZONE, J. A.; SAMIS, H. V. and WULFF, V. J.: Cellular compensations and controls in the aging process. J. Geront. *22:* No. 4, Part II, 42–52 (1967).

14. DEVI, A.; LINDSAY, P.; RAINA, P. L. and SARKAR, N. K.: Effect of age on some aspects of the synthesis of ribonucleic acid. Nature. *212:* 474–475 (1966).

15. SAMIS, H. V., JR. and WULFF, V. J.: The template activity of rat liver chromatin. Exp. Geront. *4:* 111–117 (1969).

16. MEDVEDEV, ZH. A.: Aging at the molecular level and some speculations concerning maintaining the functioning of systems for replicating specific macromolecules; in N. W. SHOCK, Biological Aspects of Aging, *3:* 255–266 (Columbia Univ. Press, New York, 1962).

17. WULFF, V. J.; QUASTER, H. and SHERMAN, F. G.: An hypothesis concerning RNA metabolism and aging. Proc. Nat. Acad. Sci. Wash. *48:* 1373–1375 (1962).

18. RAMPERSAD, O. R. and WOOL, I. G.: Protein synthesis by ribosomes from heart muscle: effect of insulin and diabetes. Science. *149:* 1102–1103 (1965).

19. MUNNELL, J. F. and GETTY, R.: Canine myocardial Z-disc alterations resembling those of nemaline myopathy. Lab. Invest. *19:* 303–308 (1968).

20. FAWCETT, D. W.: The sporadic occurrence in cardiac muscle of anomalous Z bands exhibiting a periodic structure suggestive of tropomyosin. J. Cell Biol. *36:* 266–270 (1968).

21. STREHLER, B. L.; MARK, D. D.; MILDVAN, A. S. and GEE, M. V.: Rate and magnitude of age pigment accumulation in the human myocardium. J. Geront. *14:* 430–439 (1959).

22. MUNNELL, J. F. and GETTY, R.: Rate of accumulation of cardiac lipofuscin in the aging canine. J. Geront. *23:* 154–158 (1968).

23. GOLDFISCHER, S. and BERNSTEIN, J.: Lipofuscin (aging) pigment granules of the newborn human liver. J. Cell Biol. *42:* 253–261 (1969).

24. GOLDFISCHER, S.; VILLAVERDE, H. and FORSCHIRM, R.: The demonstration of acid hydrolase, thermostable reduced diphosphopyridine nucleotide tetrazolium reductase and peroxidase activities in human lipofuscin pigment granules. J. Histochem. Cytochem. *14:* 641–652 (1966).

25. DOCK, W.: How some hearts age. J. Amer. Med. Assoc. *195:* 148–150 (1966).

26. SULKIN, N. M. and SULKIN, D. F.: Age differences in response to chronic hypoxia on the fine structure of cardiac muscle and autonomic ganglion cells. J. Geront. *22:* 485–501 (1967).

27. GLINSMANN, W. H. and ERICSSON, J. L.: Observations on the subcellular organization of hepatic parenchymal cells. II. Evolution of reversible alterations induced by hypoxia. Lab. Invest. *15:* 762–777 (1966).

28. PACKER, L.; DEAMER, D. W., and HEATH, R. L.: Regulation and deterioration of structures in membranes; in B. L. STREHLER, Adv. Geront. Res., *2:* 77–120 (Academic Press, New York, 1967).

29. STREHLER, B. L.: Molecular biology of aging. Naturwissenschaften. *56:* 57–61 (1969).

30. BARBER, A. A. and BERNHEIM, F.: Lipid peroxidation: its measurement, occurrence and significance in animal tissues; in B. L. STREHLER, Adv. Geront. Res., *2:* 355–403 (Academic Press, New York, 1967).

31. LEHNINGER, A. L.: The Mitochondrion, p 263 (W.A. Benjamin, New York, 1964).

32. BERTOLINI, A. M.: Gerontologic Metabolism. 697 pp. (C. C Thomas, Springfield, Ill., 1969).

33. TAUCHI, H. and SATO, T.: Age changes in size and number of human hepatic cells. J. Geront. *23:* 454–461 (1968).
34. TAUCHI, H.; ADACHI, F. and SHAMATO, M.: Effect of age on the cellular changes experimentally produced in the rat liver. Proc. 7th Internat. Cong. Geront., Vienna, Austria. *1:* 127–135 (1966).
35. BOGATSKAYA, L. N.: Age-determined characteristics of oxidation processes in the cardiac muscle. Z. Alterforsch. *21:* 377–381 (1969).
36. BURCH, G. E.; SOHAL, R. S. and FAIRBANKS, L. D.: Senescent changes in the heart of *Drosophila replita* Wollaston. Nature *225:* 286–288 (1970).
37. GOLD, P. H.; GEE, M. V. and STREHLER, B. L.: Effect of age on oxidative phosphorylation in the rat. J. Geront. *23:* 509–512 (1968).
38. CUZNER, M. L.; DAVISON, A. N. and GREGSON, N. A.: Turnover of brain mitochondrial lipids. Biochem. J. *101:* 618–626 (1966).
39. FLETCHER, M. J. and SANADI, D. R.: Turnover of rat liver mitochondrial components in adult and senescent rats. J. Geront. *16:* 255–257 (1961).
40. OU, L. C. and TENNEY, S. M.: Properties of mitochondria from hearts of cattle acclimatized to high altitude. Resp. Physiol. *8:* 151–159 (1970).
41. REZNIK, M. and HANSEN, J. L.: Mitochondria in degenerating and regenerating skeletal muscle. Arch. Path. *87:* 601–608 (1969).
42. HERDSON, P. B.; SOMMERS, H. M. and JENNINGS, R. B.: A comparative study of the fine structure of normal and ischemic dog myocardium with special reference to early changes following temporary occlusion of a coronary artery. Amer. J. Path. *46:* 367–377 (1965).
43. JENNINGS, R. B.; HERDSON, P. B. and SOMMERS, H. M.: Structural and functional abnormalities in mitochondria isolated from ischemic dog mitochondria. Lab. Inv. *20:* 548–557 (1969).
44. BRACHFELD, N.: Maintenance of cell vitality. Circulation 39–40 (Suppl. IV): 202–219 (1969).
45. SCHEUER, J.: Myocardial metabolism in cardiac hypoxia. Amer. J. Cardiol. *19:* 385–391 (1967).
46. SOBEL, H.: Aging of connective tissue and molecular transport. Gerontologia *14:* 235–254 (1968).
47. BOAG, J. W.: Oxygen diffusion and oxygen depletion problems in radiobiology; in EBERT and HOWARD, Current Topics in Radiation Research. *5:* 141–195 (North-Holland Publishing Company, Amsterdam, 1969).
48. ZIERLER, K. L.: Diffusion of gases in peripheral tissue; in Physiol. in the Space Environment. *2:* 79–86 (Publication 1485B, Nat. Acad. of Sci., Nat. Res. Council, 1967).
49. MYERS, W. W. and HONIG, C. R.: Number and distribution of capillaries as determinants of myocardial O_2 tension. Amer. J. Physiol. *207:* 653–660 (1964).
50. HORT, W.: Capillarization of the myocardium under normal and pathological conditions; in D. W. LUEBBERS; V. C. LUFT; G. THEROS and E. WITZLEB, Oxygen Transport in Blood and Tissues. 150-158 (Thieme, Stuttgart, 1968).
51. BASSINGTHWAIGHTE, J. B.: Blood flow and diffusion through mammalian organs. Science. *167:* 1347–1353 (1970).
52. CHANCE, B. P.; COHEN, F.; JÖBSIS, F. and SCHOENER, B.: Intracellular oxidation-reduction states *in vivo.* Science *137:* 499–508 (1962).
53. MILLER, J. A.; ZAKHARY, R. and MILLER, F. S.: Hypothermia, asphyxia and cardiac glycogen in guinea pigs. Science *144:* 1226–1227 (1964).
54. MCFARLAND, R. A.: Experimental evidence of the relationship between aging and oxygen want: in search of a theory of aging. Ergonomics *6:* 339–366 (1963).

55. LUEBBERS, D. W.: Intercapillary oxygen transport and intracellular oxygen concentration. Biochem. Sauerst. Colloq. Ges. Biol. Chem. 19th. 67–92 (1968).

56. FABEL, H.: Normal and critical O_2-supply of the heart; in D. W. LUEBBERS; V. C. LUFT; G. THEROS and E. WITZLEB, Oxygen Transport in Blood and Tissues. 159–171 (Thieme, Stuttgart, 1968).

57. ANDRES, R.: Aging, glucose tolerance, and diabetes–a proposal; in L. GITMAN, Endocrines and Aging, 200–207 (C. C Thomas, Springfield, Ill., 1967).

58. SOBEL, H.: Glucose tolerance, age, and insulin. Lancet. ii: 515 (1967).

59. SOBEL, H.; MASSERMAN, R. and PARSA, K.: Effect of age on the transvascular passage of I^{131} labeled albumin in hearts of dogs. J. Geront. *19:* 501–504 (1964).

60. SOBEL, H. and HEWLETT, M. J.: Effect of age on hyaluronic acid in hearts of dogs. J. Geront. *22:* 196–198 (1967).

61. SOBEL, H.: Aging of the ground substance in connective tissue; in B. L. STREHLER, Adv. Geront. Res. *2:* 205–283 (Academic Press, New York, 1967).

62. OGSTON, A. G. and PHELPS, C. F.: The partition of solutes between buffer solutions and solutions containing hyaluronic acid. Biochem. J. *78:* 827–831 (1961).

63. SOBEL, H. and COHEN, R. A.: Effect of estradiol on hyaluronic acid in the skin of aging mice. Steroids. *16:* 1 (1970).

64. BIRREN, J. E.; BUTLER, R. N.; GREENHOUSE, S. W.; SOKOLOFF, L. and TARROW, M. R.: Human aging: A Biological and Behavioral Study. *328:* (1963). U. S. Government Printing Office.

65. SCHEUER, J. and OLSON, J. E.: Metabolism of exogenous triglyceride by the isolated perfused rat heart. Amer. J. Physiol. *212:* 301–307 (1967).

66. MUIR, J. R.: The regional production of lipoprotein lipase in man. Clin. Sci. *34:* 261–270 (1968).

67. SOBEL, H. and THOMAS, H. V.: Effect of age and restraint on postprandial lipemia clearance in dogs. Proc. Soc. Exp. Biol. Med. *112:* 206–209 (1963).

68. BECKER, G. H.; MEYER, J. and NECHELES, H.: Fat absorptions in young and old. Gastroenterology *14:* 80–92 (1950).

69. SOBEL, H.: Influence of age on events leading to atherosclerotic heart disease in dogs. Sixth International Conference of Gerontology, Copenhagen, Denmark (1963).

70. SOBEL, H: Unpublished observations.

71. SCHWARTZ, P.: Cardiovascular amyloidosis in the aged. Geriatrics *24:* 81–97 (1969).

72. SOBEL, H. and MODABBER, F. Z.: Transvascular passage of albumin–I^{131} into skin of immunized mice. Experientia *24:* 953 (1968).

73. SOBEL, H.; MONDON, C. E. and STRAUS, R.: Spontaneous and stress-induced myocardial infarction in aged atherosclerotic dogs. Circulat. Res. *11:* 971–981 (1962).

74. SHOCK, N. W.; ANDRES, R.; LANDOWNE, M.; NORRIS, A. H.; SIMONSON, E. and SWARTZ, F. C.: Aging of the cardiovascular system. The Heart and Circulation. *1:* 558–584 (1964). Ed. Second National Conference on Cardiovascular Diseases.

75. HARRISON, T. R.; DIXON, K.; RUSSELL, R. O., JR.; BIDWAI, P. S. and COLEMAN, H. N.: The relation of age to the duration of contraction, ejection, and relaxation of the normal human heart. Amer. Heart J. *67:* 189–199 (1964).

Mailing address: Dr. Harry Sobel, Department of Biochemistry, Colorado State University, Fort Collins, Colorado (USA).

Free Fatty Acids:
Good or Bad for the Heart?

Department of Medicine, University of Florida,
College of Medicine, Gainesville, Florida, USA

METABOLISM OF EXOGENOUS AND ENDOGENOUS FATTY ACIDS IN HEART MUSCLE

M. F. CRASS, III and J. C. SHIPP

Interest in fatty acids as an oxidative substrate and regulator of myocardial metabolism has stemmed from *in vivo* studies using arteriocoronary sinus differences in substrate concentrations across the heart, and from studies in the essentially nonworking Langendorff-perfused heart preparation. A recent and comprehensive review of the literature on fatty acid metabolism in the heart has been published (13, 14). In this paper, the effects of ventricular pressure development on the metabolism of both exogenous and endogenous fatty acids were studied in a more physiologic perfused working rat heart preparation. In addition, metabolic interactions of exogenous and endogenous fatty acids with exogenous glucose and heart glycogen were investigated.

MATERIALS AND METHODS

Animals used in the present experiments were male albino Sprague-Dawley rats (Holtzman Co., Madison, Wis.). All had continuous access to food (Purina rat chow) and water and weighed 250–350 g when used.

The basic perfusion apparatus was described in detail elsewhere (7, 9). The apparatus was modified to a closed system to permit quantitative collection of $^{14}CO_2$ (2).

Rats were anesthetized with sodium pentobarbital (Abbott Laboratories), 60 mg/kg intraperitoneally, or with ether (functionally hepatectomized rats). The chest was opened and the heart excised and mounted on the perfusion apparatus by cannulation of the aorta, and perfusion-washed with a modified Krebs-Henseleit bicarbonate buffer, with or without 5 mM glucose, for 2, 5, or 10 min at 60 mm Hg aortic pressure. During washout, the left atrium was cannulated; following washout, the hearts were perfused by recirculation for varying time intervals. In working hearts, perfusate entered the heart through the left atrium at either 10 or 20 cm H_2O left atrial filling pressure (LAFP) and left the heart via the aorta. In Langendorff-perfused hearts (LAFP = 0 cm H_2O), the medium entered the heart through the aorta in retrograde fashion, at a perfusion pressure of 60 mm Hg. In all cases, the perfusate was held constant at 37°C and was continuously gassed with 95% O_2 – 5% CO_2.

[1]Supported by National Institutes of Health Grants AM 4829 and AM 5444, and by the Florida Heart Association, Grant E69AG6.

Performance of hearts was monitored as described previously (2). Heart rates were similar for working and Langendorff-perfused hearts (range = 260–300 beats/min). Mean aortic peak systolic pressures were 65, 110, and 124 mm Hg at 0 (Langendorff), 10, and 20 cm H_2O LAFP, respectively. Aortic outputs averaged 29 and 43 ml/g wet wt/min in hearts perfused at 10 and 20 cm H_2O LAFP, respectively.

Details of preparation of the fatty acid-albumin complex have been published (2).

Methods used for the determination of perfusate glucose and free fatty acids, in addition to content and radioactivity of heart glycogen and lipid in frozen ventricular tissue (16), were similar to those described previously (2, 3).

Heart lipids were extracted from frozen ventricle in 20 volumes of chloroform : methanol (2:1 v/v). After silicic acid column separation of the neutral lipids + FFA and phospholipids, the neutral lipid fraction was further fractionated by thin-layer chromatography, using a solvent system consisting of petroleum ether : diethyl ether : methanol : acetic acid (90:20:2:3 v/v/v/v).

Palmitic acid (lot no. 721–33) and palmitate-1-^{14}C (lot no. 31–200A-32) were obtained from Applied Science Laboratories, Inc., State College, Pa., and New England Nuclear Corp., Boston, Mass., respectively. Bovine plasma albumin (4 X crystallized, lot nos. 5774, 9641, and 9385) was purchased from Nutritional Biochemicals Corp., Cleveland, Ohio. Glucose-U-^{14}C (lot no. 86–258–12) was obtained from the New England Nuclear Corp.

The technique used for prelabeling heart lipids with palmitate-1-^{14}C *in vivo* was similar to that described by Borgstrom and Olivecrona (1). This procedure consisted in administering 0.5 ml of a palmitate-1-^{14}C serum complex intravenously into functionally hepatectomized rats 30 min prior to excision of the heart and perfusion.

All data were expressed per g, or per 100 mg, dry weight of ventricle.

RESULTS AND DISCUSSION

The uptake and principal fates of 1 mM palmitate-1-^{14}C (complexed to 3% bovine plasma albumin) as the sole exogenous substrate in working as compared with Langendorff-perfused hearts is shown in table I. Working hearts, developing greater ventricular pressures, extracted approximately 30% more labeled fatty acid during a 1 h perfusion. However, $^{14}CO_2$ production was increased by more than two-fold over that observed in Langendorff-perfused hearts. In both perfusion systems, $^{14}CO_2$ was the major metabolic fate. However, the percentage of uptake that was oxidized to $^{14}CO_2$ was substantially greater in working hearts, amounting to approximately 60%, as compared to 35–38% in Langendorff-perfused hearts. Recovery of fatty acid label in heart lipids accounted for 20% or less of the uptake in both systems, although on a percentage basis approximately twice as much label was recovered in tissue lipids

Table I

Metabolism of Exogenous (1 mM) Palmitate-1-^{14}C in Working and Langendorff-Perfused Rat Hearts*

Conditions	Uptake	Fates		
		CO_2	NL + FFA	PL
0 cm H_2O LAFP (6) (59 ± 1)†	100.5 ± 7.4	37.8 ± 2.0	14.7 ± 0.9	3.9 ± 0.2
10 cm H_2O LAFP (7) (109 ± 4)†	144.9 ± 5.1	85.0 ± 2.7	11.9 ± 0.2	3.8 ± 0.3

*Hearts from fed rats were perfused in a recirculated system for 1 h with buffer containing 1 mM palmitate-1-^{14}C. All values are expressed as μmole of palmitate, or C_{16} equivalents, per g dry wt. Each value represents the mean of the number of hearts shown in parentheses ± S.E.M.

† = mean aortic peak systolic pressure.

in the Langendorff group. In both groups, roughly 80% of the ^{14}C-tissue lipid was recovered in the neutral lipid + FFA fraction.

Working hearts were perfused with buffer containing glucose, and the effects of palmitate on various steps in the utilization of 5 mM glucose-U-^{14}C by working hearts are shown in figures 1–4. The results were qualitatively similar in both working and Langendorff hearts, though there were certain quantitative differences that have been alluded to elsewhere (3).

Figure 1 shows the effect of varying concentrations of palmitate on glucose uptake. In the absence of palmitate, the accelerated glucose uptake, which was three to four times greater in working hearts as compared with Langendorff hearts (2, 10), was maximally inhibited in the presence of 0.5 mM palmitate (FFA : albumin molar ratio = 1 . 1). Increasing the exogenous fatty acid concentration to 1.7 mM did not increase the degree of inhibition. The inhibition of glucose uptake at any given concentration was shown to be greater in hearts developing increased ventricular pressures (3, 8). Several observations would suggest that fatty acids directly affect the glucose transport system, and that this effect is dependent on fatty acid oxidation. Randle *et al.* (15) showed that fatty acid inhibition of glucose uptake in Langendorff hearts occurred only under aerobic conditions. Neely *et al.* (8) observed that fatty acids under aerobic, but not anaerobic, conditions reduced the pressure development-accelerated transport of the nonmetabolized sugar, 3-0-methylglucose.

Net mobilization of glycogen occurred in the absence of exogenous fatty acid in the isolated rat heart (fig. 2); this process was accelerated in working hearts developing increased ventricular pressures (2, 10). Addition of fatty acids in physiologic concentrations to the perfusion medium resulted in a marked reduction in net glycogen flux. Virtually complete inhibition of glycogenolysis

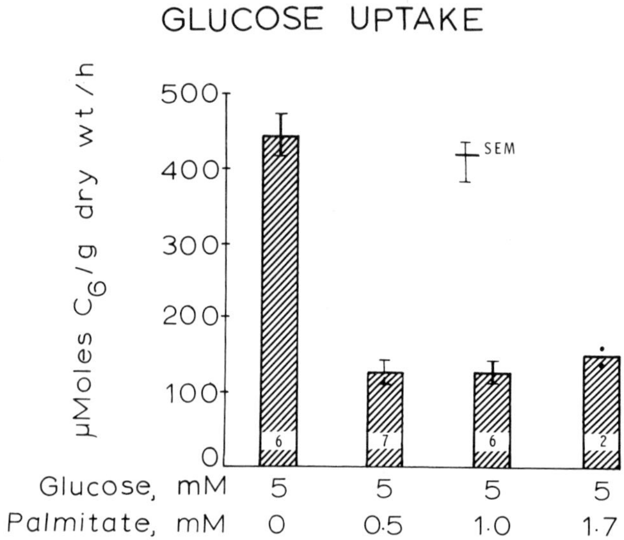

Fig. 1 Effect of different concentrations of palmitate on glucose uptake in the isolated perfused working rat heart (10 cm H_2O LAFP). Hearts from fed rats were perfused in a recirculated system for 1 h with buffer containing 5 mM glucose-U-^{14}C with or without varying concentrations of palmitate complexed to 3% bovine plasma albumin. Each bar represents the mean of the indicated number of hearts± S.E.M.

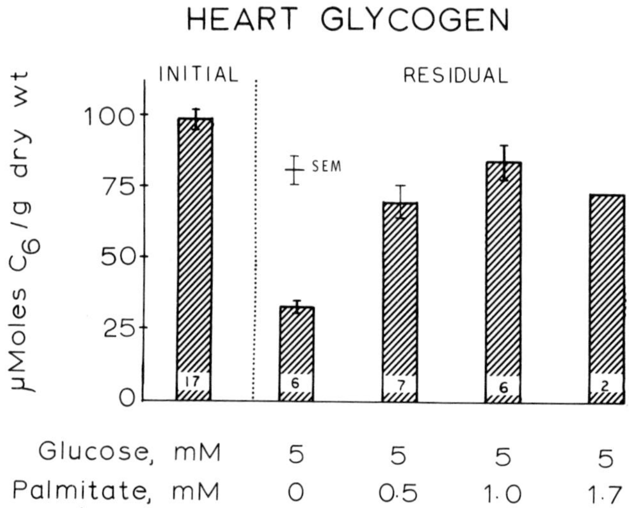

Fig. 2 Effect of different concentrations of palmitate on net glycogenolysis in the isolated perfused working rat heart (10 cm H_2O LAFP). "Initial" and "Residual" = glycogen content before and after a 1 h perfusion, respectively. Other details were the same as described in legend to figure 1. Each bar represents the mean of the indicated number of hearts± S.E.M. Glycogen is expressed as µmole of glucose, or C_6 equivalents, per g dry wt.

in the working heart was obtained at the 1.0 mM level. Under similar conditions, Neely *et al.* (8) observed an increased glucose-6-phosphate concentration and a concomitant decrease in the level of P_i; both factors would favor a reduction in phosphorylase activity.

Figure 3 shows a marked reduction in the rate of glucose oxidation in working hearts perfused with buffer containing fatty acid. Inhibition of $^{14}CO_2$ production was maximal in the presence of 1.0 mM palmitate.

Net lactate production in working hearts was relatively unaltered in the presence of fatty acid (fig. 4). However, lactate was one of the principal fates of glucose with fatty acid present. Lactate accounted for 14–58% of the glucose uptake as exogenous palmitate was increased from 0 to 1.0 mM. Garland and Randle (6) have provided evidence for a direct inhibition of pyruvate conversion to acetyl CoA under similar experimental conditions.

These data would suggest that long-chain fatty acids inhibit glucose utilization at multiple steps in the isolated working heart. Similar results have been obtained in the isolated heart of the squirrel monkey, *Saimiri sciureus* (4).

Recently, effects of pressure development or cardiac work on the utilization of the heart ventricular lipids have been studied in this laboratory. Previous studies in the Langendorff heart perfused with substrate-free buffer (5, 8, 12) have shown that this preparation maintained contractility for periods of 1 h or

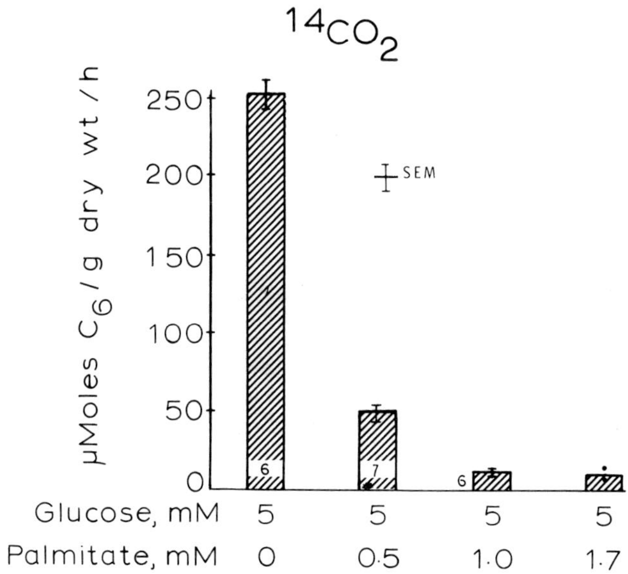

Fig. 3 Effect of different concentrations of palmitate on glucose oxidation in the isolated perfused working rat heart (10 cm H_2O LAFP). Other details were the same as described in legend to figure 1. Each bar represents the mean of the indicated number of hearts ± S.E.M. $^{14}CO_2$ production is expressed as μmole of glucose, or C_6 equivalents, per g dry wt.

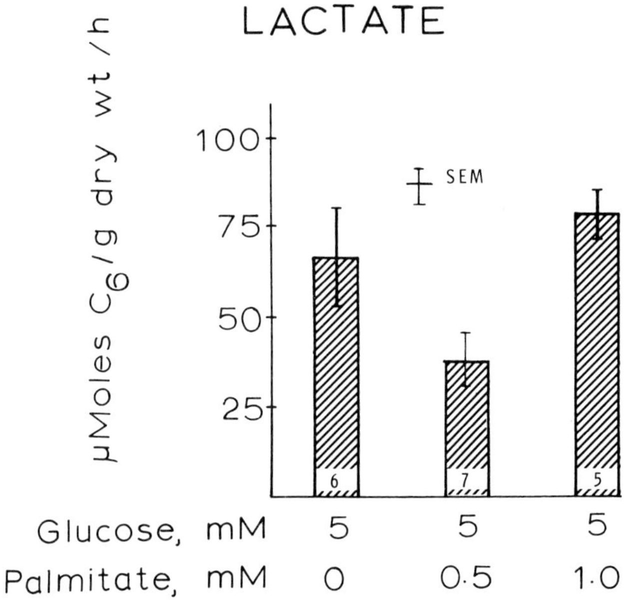

Fig. 4 Effect of different concentrations of palmitate on net lactate production in the isolated perfused working rat heart (10 cm H_2O LAFP). Other details were the same as described in legend to figure 1. Each bar represents the mean of the indicated number of hearts ± S.E.M. Lactate is expressed as µmole of glucose, or C_6 equivalents, per g dry wt.

more, suggesting the possible importance of the endogenous heart lipids as an energy source. The present studies examined the possibility that endogenous lipid metabolism was stimulated in working hearts developing increased ventricular pressures. Figure 5 shows results obtained from hearts that were perfused with substrate-free buffer for 5 or 15 min at 0, 10, and 20 cm H_2O LAFP. Residual endogenous total glycerides were determined. After 5 min of perfusion, residual glyceride content was significantly less in working hearts than in Langendorff hearts. However, after 15 min, residual glyceride contents were similar. Neither perfusion per se nor increased pressure development altered the content of total heart phospholipids (measured as lipid P) (fig. 6).

To obtain a better estimate of endogenous lipid mobilization and oxidation, heart lipids were prelabeled *in vivo* with palmitate-1-^{14}C. Of the label recovered in the heart lipids 30 min after injection, 80% was in triglycerides and 15% in phospholipids. No label was recovered in glycogen. Table II shows the accelerated mobilization of labeled triglyceride fatty acids in working hearts perfused for 10 min with substrate-free buffer. The difference in residual triglyceride fatty acid label in hearts perfused at 0 and 10 cm H_2O LAFP was statistically significant. These observations suggested that the activity of the myocardial lipase system(s) was stimulated in working hearts developing

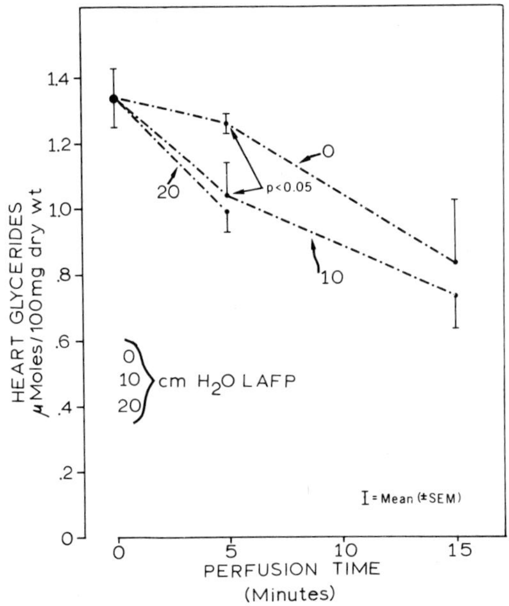

Fig. 5 Effect of pressure development on the mobilization of endogenous glycerides in the isolated working rat heart. Hearts from fed rats were perfused with substrate-free buffer for varying time intervals at 0, 10, or 20 cm H_2O LAFP. Values for residual total heart glycerides are shown. Each point represents the mean of 4–10 hearts ± S.E.M.

increased ventricular pressures. In similar prelabeled perfusion experiments, but for 30 min duration, there was a marked decrease in labeling of the triglyceride fraction in both working and Langendorff-perfused hearts (table III). The rapid transient stimulation of mobilization, shown in figure 5 and table II, was not evident after the longer perfusion. In both working and nonworking hearts, phospholipid labeling increased slightly, suggesting the transfer of acyl groups to the phospholipids from other ^{14}C-lipid fractions.

Endogenous lipid oxidation was stimulated in working hearts, as evidenced by the increased $^{14}CO_2$ production (table III). This effect could have: a) resulted from a direct stimulation of the mitochrondrial fatty acid-oxidase system or enzymes of the citric acid cycle (11); or b) been secondary to stimulation of lipolysis, with a subsequently increased availability of fatty acyl CoA or its carnitine derivatives. A similar stimulation of exogenous fatty acid oxidation was observed in working hearts (table I).

Finally, the effects of various exogenous substrates on endogenous lipid mobilization and oxidation were studied in working hearts prelabeled *in vivo* with palmitate-1-^{14}C (table IV). In hearts perfused with plain buffer (no added substrate), neutral lipid radioactivity decreased markedly, with no concomitant change in labeling of the phospholipid fraction. Virtually the entire

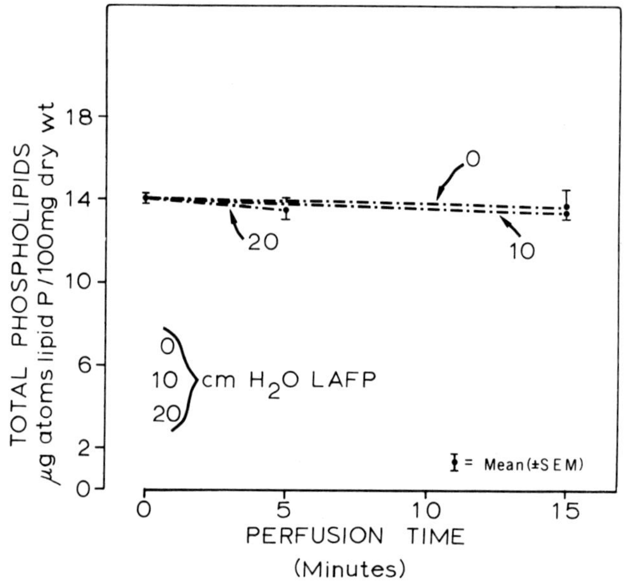

Fig. 6 Effect of pressure development on the mobilization of tissue phospholipids in the isolated working rat heart. Other details were the same as described in legend to figure 5. Values for residual total lipid P are shown. Each point represents the mean of 4–10 hearts ± S.E.M.

Table II

Effect of Ventricular Pressure Development on the Mobilization of Endogenous ^{14}C-triglyceride Fatty Acids*

Conditions	No. of hearts	^{14}C-triglyceride fatty acids
		dpm/100 mg dry wt
Initial†	7	162,534 ± 15,073[a]
0 cm H_2O LAFP‡	4	97,192 ± 15,050[b]
10 cm H_2O LAFP‡	5	58,926 ± 6,651[c]

*Hearts from fed rats, which were prelabeled *in vivo* with palmitate-1-^{14}C, were perfused in a recirculated system with substrate-free buffer. *a vs. b:* $P<0.02$; *a vs. c:* $P<0.001$; *b vs. c:* $P<0.05$.

†Hearts perfusion-washed for 2 min only.

‡Hearts perfused by recirculation for an additional 10 min at the designated LAFP. Each value represents the mean ± S.E.M.

Table III

Mobilization and Oxidation of Endogenous Heart Lipids in Working and Langendorff-Perfused Hearts after 30 min of Substrate-Free Perfusion*

Perfusion conditions	Principal ^{14}C-lipid fractions			
	TG	PL	Total ^{14}C	$^{14}CO_2$
	dpm/100 mg dry wt ± S.E.M.			
Initial† (9)	171,144 ± 7,468	29,385 ± 1,579	221,021 ± 7,343	–
0 cm H_2O LAFP‡ (6)	46,134 ± 4,752	36,853 ± 1,543	97,953 ± 5,703	66,298[a] ± 7,216
10 cm H_2O LAFP‡ (6)	38,155 ± 3,382	40,434 ± 3,362	90,779 ± 6,530	94,545[b] ± 7,378

*Hearts from fed rats, which were prelabeled *in vivo* with palmitate-1-^{14}C, were perfused in a recirculated system with substrate-free buffer. Total ^{14}C refers to sum of labeling in all lipid fractions. *a vs. b*: $P<0.02$.

†Hearts perfusion-washed for 5 min only.

‡Hearts perfused by recirculation for an additional 30 min at the designated LAFP. Each value represents the mean of the number of hearts shown in parentheses ± S.E.M.

Table IV

Effect of Exogenous Substrate(s) on the Utilization of Endogenous Lipids in the Working Rat Heart*

Perfusate Additions	^{14}C-NL + FFA	^{14}C-PL	$^{14}CO_2$
	dpm/100 mg dry wt		
Initial† (14)	181,715 ± 8,945	38,657 ± 2,918	–
Plain buffer (5)	69,570 ± 11,232	42,291 ± 5,119	113,273 ± 12,526
5.5 mM glucose (6)	43,207 ± 7,377	31,381 ± 3,855	98,910 ± 13,638
0.68 mM palmitate (8)	147,372 ± 7,466	30,693 ± 1,707	40,863 ± 3,233
1.3 mM palmitate (6)	123,715 ± 6,657	22,957 ± 1,797	19,056 ± 3,435
0.68 mM palmitate + 5.5 mM glucose (6)	119,048 ± 15,493	26,817 ± 2,530	32,037 ± 5,067

*Hearts from fed rats, which were prelabeled *in vivo* with palmitate-1-^{14}C, were perfused in a recirculated system with buffer containing the specified exogenous substrate.

†Hearts perfusion-washed with substrate-free buffer for 5 min only. All remaining groups of hearts were perfused by recirculation for an additional 30 min at 10 cm H_2O LAFP. Each value represents the mean of the number of hearts shown in parentheses ± S.E.M.

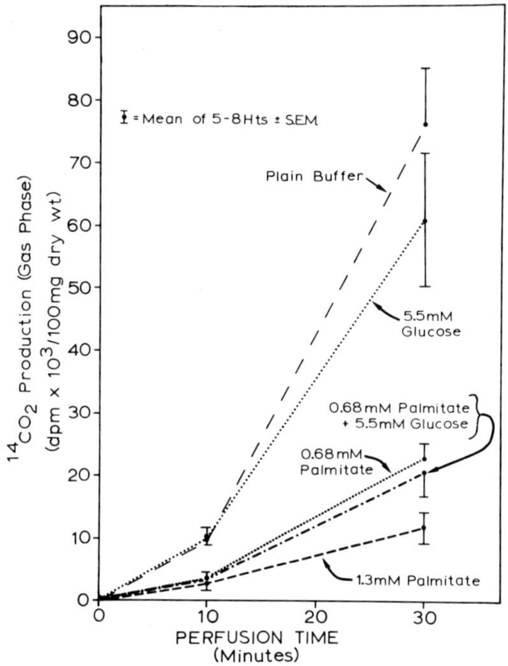

Fig. 7 Effect of different exogenous substrates on the time course of $^{14}CO_2$ production in isolated perfused working rat hearts (10 cm H_2O LAFP) prelabeled *in vivo* with palmitate-1-^{14}C. Other details were the same as described in legend to table IV. Each point represents the mean of 5–8 hearts ± S.E.M.

decrement in tissue lipid label was recovered in $^{14}CO_2$. The presence of 5.5 mM glucose in the perfusion medium did not alter endogenous lipid mobilization or oxidation. However, a physiologic concentration of fatty acids alone, or in combination with glucose, produced a marked reduction in: a) disappearance of endogenous labeled neutral lipid esters; and b) endogenous fatty acid oxidation. Figure 7 shows the time course of endogenous fatty acid oxidation in prelabeled hearts perfused with different exogenous substrates. These observations are clearly suggestive of a preferential utilization of exogenous fatty acids as compared with endogenous neutral lipid esters in the isolated working rat heart. Studies are currently in progress to investigate the nature of the accelerated oxidative metabolism of exogenous and endogenous fatty acids in working hearts, and the apparent substrate regulatory response of endogenous glyceride fatty acids in heart muscle.

SUMMARY

The metabolism of exogenous and endogenous fatty acids, and fatty acid-glucose interactions, were studied in an isolated perfused working rat heart preparation.

Uptake and oxidation of exogenous palmitate (1.0 mM) were accelerated in working hearts developing increased ventricular pressures.

Exogenous palmitate, in physiologic concentrations characteristic of the fed state, inhibited glucose metabolism at multiple sites in working hearts.

Mobilization and oxidation of endogenous fatty acids was increased in working as compared with Langendorff-perfused hearts.

Physiologic concentrations of exogenous fatty acids exerted a sparing effect on the utilization of endogenous heart lipids.

REFERENCES

1. BORDSTRÖM, B. and OLIVECRONA, T.: The metabolism of palmitic acid-1-^{14}C in functionally hepatectomized rats. J. Lipid Res. *2:* 263–267 (1961).
2. CRASS, M. F., III; MCCASKILL, E. S. and SHIPP, J. C.: Effect of pressure development on glucose and palmitate metabolism in perfused heart. Amer. J. Physiol. *216:* 1569–1576 (1969).
3. CRASS, M. F., III; MCCASKILL, E. S. and SHIPP, J. C.: Glucose-free fatty acid interactions in the working heart. J. Appl. Physiol. *29:* 87–91 (1970).
4. CRASS, M. F., III; TULLIS, C. S.; MCCASKILL, E. S. and SHIPP, J. C.: Glucose and free fatty acid metabolism in perfused hearts of normal and hypothyroid monkeys *(Saimiri sciureus).* Comp. Biochem. Physiol. *36:* 201–205 (1970).
5. DENTON, R. M. and RANDLE, P. J.: Concentrations of glycerides and phospholipids in rat heart and gastrocnemius muscles. Effects of alloxan diabetes and perfusion. Biochem. J. *104:* 416–422 (1967).
6. GARLAND, P. B. and RANDLE, P. J.: Control of pyruvate dehydrogenase in the perfused rat heart by the intracellular concentration of acetyl-coenzyme A. Biochem. J. *91:* 6c–7c (1964).
7. MORGAN, H. E.; NEELY, J. R.; WOOD, R. E.; LIEBECQ, C.; LIEBERMEISTER, H. and PARK, C. R.: Factors affecting glucose transport in heart muscle and erythrocytes. Fed. Proc. *24:* 1040–1045 (1965).
8. NEELY, J. R.; BOWMAN, R. H. and MORGAN, H. E.: Effects of ventricular pressure development and palmitate on glucose transport. Amer. J. Physiol. *216:* 804–811 (1969).
9. NEELY, J. R.; LIEBERMEISTER, H.: BATTERSBY, E. J. and MORGAN, H. E.: Effect of pressure development on oxygen consumption by isolated rat heart. Amer. J. Physiol. *212:* 804–814 (1967).
10. NEELY, J. R.; LIEBERMEISTER, H. and MORGAN, H. E.: Effect of pressure development on membrane transport of glucose in isolated rat heart. Amer. J. Physiol. *212:* 815–822 (1967).
11. NEELY, J. R. and WHITFIELD, C. F.: Regulation of fatty acid utilization by isolated rat hearts: Effects of ventricular pressure development. Fed. Proc. *29:* 911 (1970).
12. OLSON, R. E. and HOESCHEN, E. J.: Utilization of endogenous lipid by the isolated perfused rat heart. Biochem. J. *103:* 796–801 (1967).
13. OPIE, L. H.: Metabolism of the heart in health and disease. Part I. Amer. Heart J. *76:* 685–698 (1968).
14. OPIE, L. H.: Metabolism of the heart in health and disease. Part II. Amer. Heart J. *77:* 100–122 (1969).
15. RANDLE, P. J.; NEWSHOLME, E. A. and GARLAND, P. B.: Regulation of glucose uptake by muscle. 8. Effects of fatty acids, ketone bodies, and pyruvate, and of alloxan diabetes and starvation, on the uptake and metabolic fate of glucose in rat heart and diaphragm muscles. Biochem. J. *93:* 652–665 (1964).

16. WOLLENBERGER, A.; RISTAN, O. und SHOFFA, G.: Eine einfache Technik der extrem schnellen Abkuhlung grösserer Gewebestucke. Arch. Ges. Physiol. *270:* 399–412 (1960).

Mailing address: Dr. M. F. Crass, III, Rm. 4038, Basic Science Building, University of Nebraska Medical Center, Omaha, Nebraska 68105 (USA).

Department of Medicine and Pathology, University of Iowa,
College of Medicine, Iowa City, Iowa, USA

EFFECTS OF ACUTE FREE FATTY ACID MOBILIZATION ON THE HEART

J. C. HOAK, E. D. WARNER, and W. E. CONNOR

When the effects of free fatty acids upon the heart are considered as a consequence of acute lipid mobilization, it is appropriate to note the potential effects of unbound long-chain fatty acids. Salts of these fatty acids can exert a detergent effect upon cells (7) and mitochondria (1) and many of their toxic effects may involve this property.

Intravenous injections of sodium stearate into heparinized dogs produce marked electrocardiographic changes (fig. 1) in association with tachycardia, a reduction in systemic arterial pressure, and a decrease in cardiac output (2). The hemodynamic effects of fatty acids are not new discoveries and were known at least as early as 1889, when Munk (6a) observed that fatty acid injections produced a fall in blood pressure and, in large doses, caused shock and death.

Acute injections of fatty acids may also cause extensive morphological changes in mitochondria of the heart, as shown by electron microscopy (fig. 2). Using an isolated perfused heart preparation, Severeid *et al.* (9) found that unbound long-chain fatty acids had a deleterious effect upon myocardial function.

It should be emphasized that the effects just described represent the ultimate in potential toxicity, since long-chain fatty acids are usually transported in the

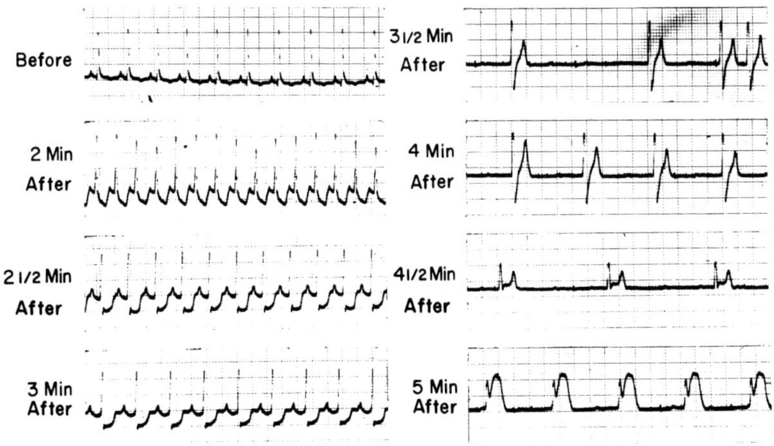

Fig. 1 Electrocardiogram of a dog that received an intravenous infusion of sodium stearate (3.5 mM, pH 8, 10 ml/kg over a 5 min period). Signs of ischemia, myocardial injury, and heart block preceded death.

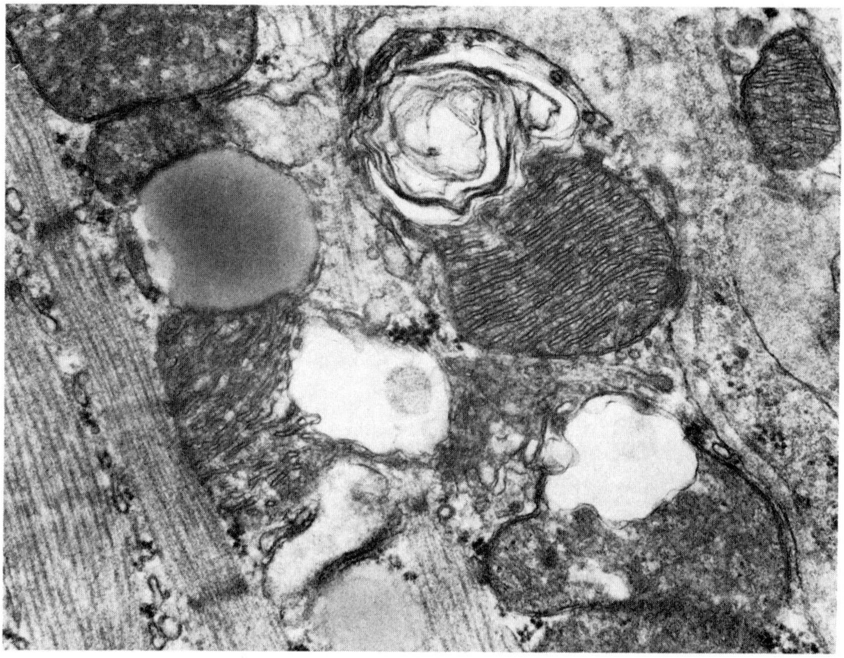

Fig. 2 Electron micrograph of myocardium from a mouse, following an intravenous injection of sodium stearate. Disruption and lysis of mitochondria are seen. Lipid droplets were also frequent findings. ×30,800.

blood bound to albumin. There is little evidence that toxic effects are seen unless the free fatty acid (FFA) concentration exceeds the tight binding sites of albumin, *i.e.*, when the FFA/albumin molar ratio exceeds 2.

In subsequent attempts to study the effects of fatty acids under more physiological conditions, we performed experiments in animals treated with agents that cause high plasma concentrations of FFA as a result of mobilization of endogenous lipid from adipose stores. As shown in figure 3, these agents are thought to stimulate lipolysis in adipose tissue through cyclic AMP by activation of adenyl cyclase.

Using norepinephrine infusions in dogs, we have made observations similar to those reported by other investigators. Acute myocardial necrosis was produced by such treatment, associated with elevated plasma FFA concentrations, elevated serum creatine phosphokinase and glutamic oxaloacetic transaminase activities, and electrocardiographic changes consistent with myocardial infarction (4). The marked increase in the FFA concentration and the development of myocardial lesions and biochemical abnormalities were prevented by treatment with the beta adrenergic blocking agent, propranolol, administered prior to the infusion of norepinephrine.

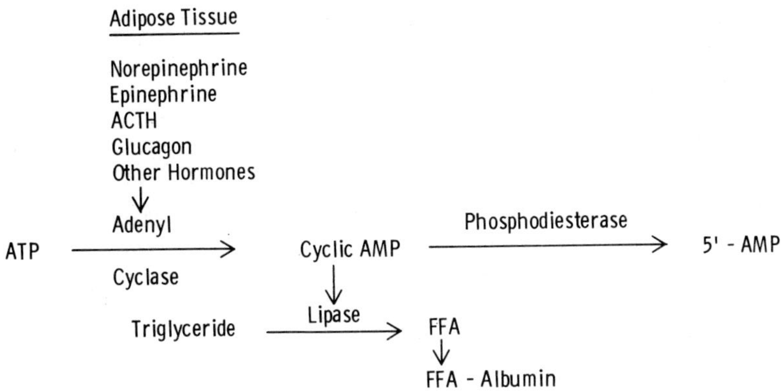

Fig. 3 Mechanism of action by which hormones produce lipolysis in adipose tissue.

Additional experiments were performed in rabbits given ACTH to mobilize endogenous lipid. The effects upon plasma FFA concentrations are shown in figure 4. In 10 rabbits receiving ACTH (50 units/kg, subcutaneously) the mean FFA increased to 1770 μEq/l at 2 h and remained elevated at 1269μEq/l at 5 h.

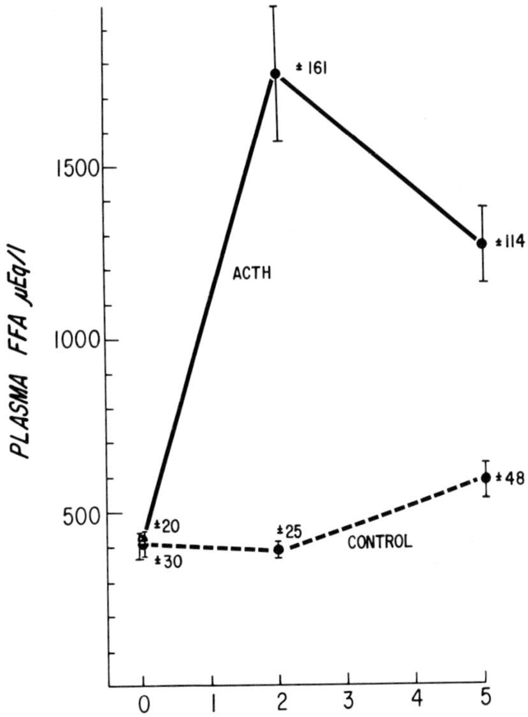

Fig. 4 Effect of ACTH upon FFA in normal rabbits. ACTH gel (50 units/kg, s.c.) caused a sustained increase in plasma FFA.

Control rabbits injected with saline showed only a slight increase in FFA at 5 h. The ACTH-treated rabbits developed toxic signs within 1 h after the injection, and these consisted of an initial increase in respiratory rate followed by generalized weakness, labored respirations, and decreased alertness. None of the control rabbits exhibited toxic signs. Of the 10 animals given ACTH, 4 died during the experiment; pulmonary edema and congestion were common findings at necropsy. Many of the treated rabbits presented electrocardiographic changes. Sinus tachycardia and premature ventricular contractions were common. In others, sinus arrest, supraventricular tachycardia, ventricular tachycardia, and signs of myocardial ischemia were observed (fig. 5). Electron microscopy revealed extensive myocardial changes with destruction of myofibrils, distortion of mitochondrial structure, and osmiophilic intramitochondrial inclusions (fig. 6).

We then turned our attention to mobilization of lipid in birds and produced high plasma FFA in geese by means of intravenous injections of glucagon (3). Birds mobilize lipid to a lesser extent in response to catecholamines or pituitary hormones than do mammals, but show a prompt increase in FFA after glucagon.

Within 10 min of the glucagon injection, toxic signs developed. Characteristic signs included tachycardia, rapid respirations, vomiting, defecation, and

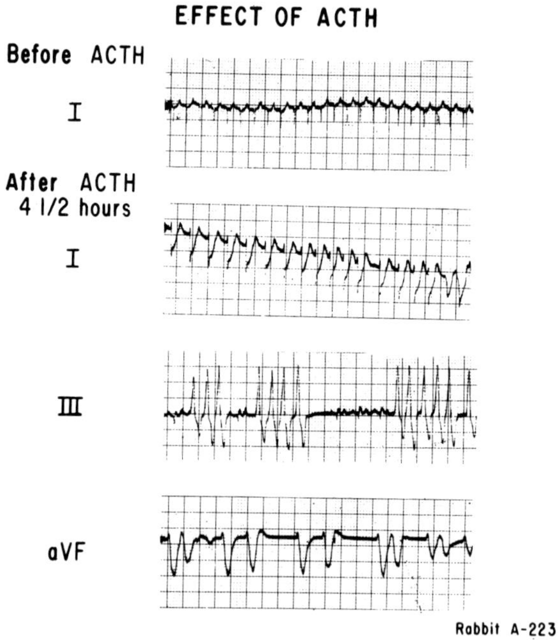

Fig. 5 Electrocardiogram of a rabbit, before and after ACTH. At the time when plasma FFA concentrations were elevated, marked S-T segment depression was observed. Progressive deterioration was associated with the development of ventricular tachycardia.

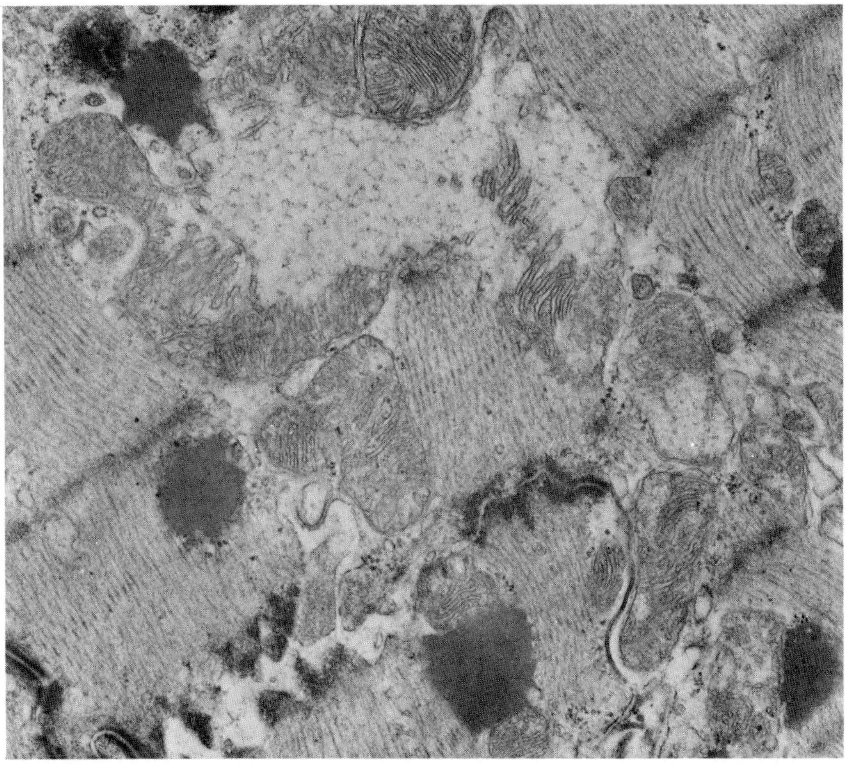

Fig. 6 Electron micrograph of myocardium from a rabbit, after ACTH. Note the area of sarcolysis and disrupted mitochondria. Lipid droplets were increased. × 23,900.

prostration; 2 of 11 geese exhibited progressive deterioration and died prior to completion of the experiment. Glucagon produced a four-fold increase in FFA within 15 min. Mean FFA values increased from 647 to 2686 μEq/l, remained elevated for over 2 h, and returned to normal at 24 h. Control geese did not develop toxic signs or significantly increased FFA values.

The glucagon-treated geese frequently presented electrocardiographic changes; an example of this is shown in figure 7. While sinus tachycardia and premature ventricular contractions were common, sinus arrest, supraventricular tachycardia, ventricular tachycardia, and signs of myocardial ischemia were also observed. At necropsy, myocardial lesions were present in 6 of the 11 geese that had exhibited electrocardiographic abnormalities; in addition, fatty infiltration of the liver was seen in 8, and acute hepatic necrosis in 1 instance.

The typical myocardial lesion on gross examination consisted of a focal hemorrhagic area that was usually located in the ventricular wall. On histological examination, areas of focal hemorrhage were seen. In one of the myocardial lesions, leukocytic infiltration and focal areas of early myocardial necrosis were

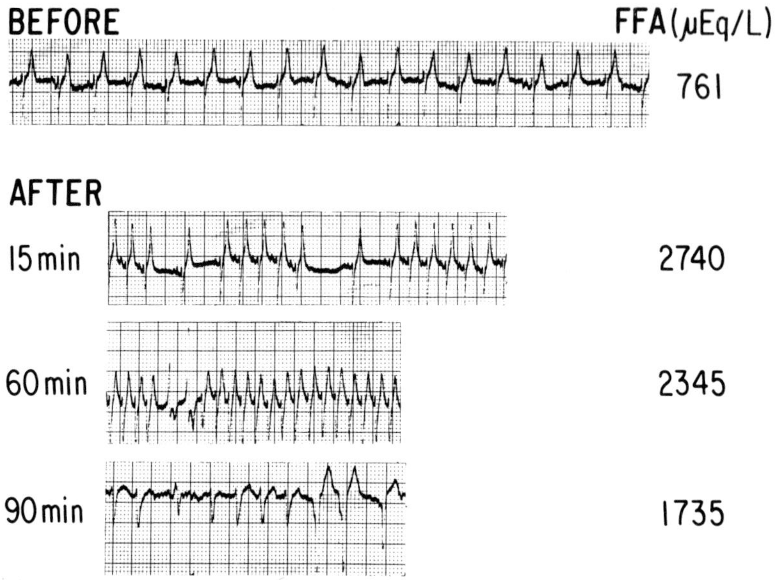

Fig. 7 Electrocardiogram of a goose treated with glucagon. This tracing was taken before and after the administration of glucagon, 0.5 mg/kg. Within 15 min, the FFA increased from 761 to 2740 μEq/l and the electrocardiogram showed an increase in cardiac rate and periods of sinus arrest. Plasma FFA remained high. At 60 min, supraventricular tachycardia was present. By 90 min the rate had slowed, but the bizarre complexes suggested a ventricular rhythm or aberrant conduction with supraventricular arrhythmia. The goose died shortly after the last tracing was taken.

conspicuous features. Small fat droplets were seen in the lesion when the sections were stained with Oil Red O. Electron microscopic examination of the cardiac lesions revealed destructive changes in the myofibrils, distortion of mitochondrial structure, and osmiophilic intramitochondrial inclusions, in addition to the changes observed with light microscopy (fig. 8). For purposes of comparison, an electron micrograph demonstrating myocardium from a normal goose is shown in figure 9.

DISCUSSION

The myocardial lesion produced by glucagon was similar or identical to that produced in dogs by norepinephrine or in rabbits by injection of ACTH. All of these agents were potent stimuli for lipid mobilization in the particular species used.

The relevance of these observations to clinical myocardial disease is speculative, but similar mechanisms may operate in the pathogenesis of myocardial infarction and its complications. The hypothesis that very high plasma concentrations of FFA or a decrease in the rate of oxidation of fatty

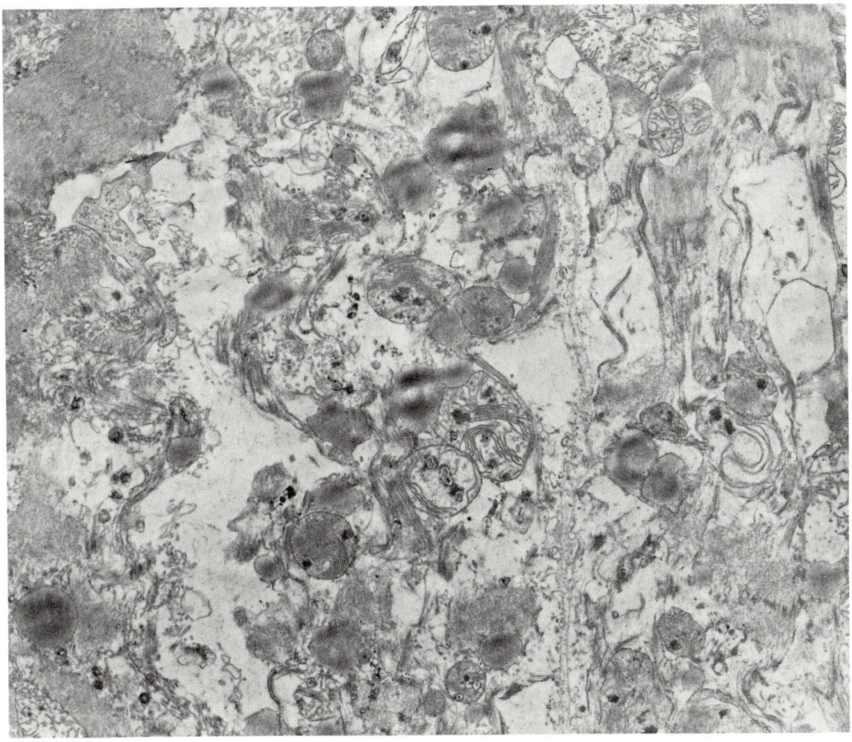

Fig. 8 Electron micrograph of a myocardial lesion that developed in a goose after administration of glucagon. Note the distorted and irregular mitochondria. Intramitochondrial osmiophilic inclusion particles were abundant. Myofibrils showed areas of destruction. Lipid particles were frequently seen. ×6,050.

acids by the cell may result in an accumulation of fatty acids sufficient to prove detrimental to cellular structure and function is an interesting concept that is supported by other experimental studies. Diphtheria toxin has been reported to produce acute myocardial lesions similar to those induced by catecholamines (10). This toxin has also been shown to inhibit the oxidation of long-chain fatty acids by heart homogenate preparations (11). Significant elevation of plasma FFA concentrations were found in patients with acute myocardial infarction (5). Later, patients with serum FFA concentrations above 1200 μEq/l after acute myocardial infarction were reported to have an increased prevalence of serious arrhythmias and a higher mortality rate (8). Elevation of plasma FFA through activation of lipoprotein lipase by heparin was seen to be associated with the development of serious ventricular arrhythmias in dogs with acute myocardial infarction (6). It is unknown whether fatty acids play a causative role or whether they merely reflect other metabolic or hormonal changes in patients with acute

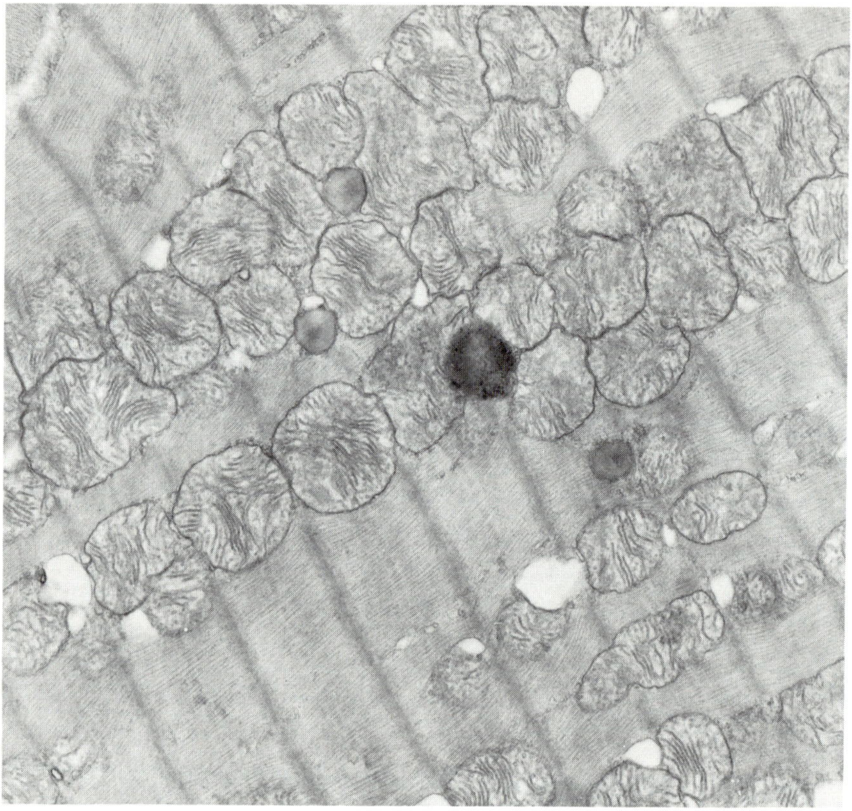

Fig. 9 Electron micrograph of normal goose myocardium. ×12,000.

myocardial infarction. The results reported here, however, do suggest that high plasma FFA concentrations or impaired oxidation of fatty acids by the myocardial cell may be deleterious to cellular function and warrant consideration in the pathogenesis of myocardial disorders.

SUMMARY

Acute lipid mobilization, which resulted in high plasma FFA concentrations, was followed by the development of cardiac rhythm abnormalities, impaired myocardial function and, in some instances, death. Morphological and biochemical abnormalities were associated findings. These results suggest that high plasma FFA concentrations or impaired oxidation of fatty acids by the myocardial cell may play a role in the pathogenesis of myocardial infarction and its complications.

REFERENCES

1. BJÖRNTORP, P.; ELLIS, H. A. and BRADFORD, R. H.: Albumin antagonism of fatty acid effects on oxidation and phosphorylation reactions in rat liver mitochondria. J. Biol. Chem. *239:* 339–344 (1964).

2. HOAK, J. C.; CONNOR, W. E.; ECKSTEIN, J. W. and WARNER, E. D.: Fatty acid-induced thrombosis and death: Mechanisms and prevention. J. Lab. Clin. Med. *63:* 791–800 (1964).

3. HOAK, J. C.; CONNOR, W. E. and WARNER, E. D.: Toxic effects of glucagon-induced acute lipid mobilization in geese. J. Clin. Invest. *47:* 2701–2710 (1968).

4. HOAK, J. C.; WARNER, E. D.; and CONNOR, W. E.: New concept of levarterenol-induced acute myocardial necrosis. Arch. Pathol. *87:* 332–338 (1969).

5. KURIEN, V. A. and OLIVER, M. F.: Serum free fatty acids after acute myocardial infarction and cerebral vascular occlusion. Lancet *ii:* 122–127 (1966).

6. KURIEN, V. A.; YATES, P. A. and OLIVER, M. F.: Free fatty acids, heparin, and arrhythmias during experimental myocardial infarction. Lancet *ii:* 185–187 (1969).

6a. MUNK, V. I.: Berlin Centralblatt fur die medicinischen den Wissenschaften *27:* 514 (1889).

7. NORKIN, S. A.: CZERNOBILSKY, B.; GRIFFITH, E. and DUBIN, I. N.: Effect of albumin and fatty acids on cellular growth in vitro. Arch. Pathol. *80:* 273–277 (1965).

8. OLIVER, M. F.; KURIEN, V. A. and GREENWOOD, T. W.: Relation between serum free fatty acids and arrhythmias and death after acute myocardial infarction. Lancet *i:* 710–714 (1968).

9. SEVEREID, L.: CONNOR, W. E. and LONG, J. P.: The depressant effect of fatty acids on the isolated rabbit heart. Proc. Soc. Exp. Biol. Med. *131:* 1239–1234 (1969).

10. SZAKÁCS, J. E. and MEHLMAN, B.: Pathologic changes induced by 1-norepinephrine. Amer. J. Cardiol. *5:* 619–627 (1960).

11. WITTELS, B. and BRESSLER, R.: Biochemical lesion of diphtheria toxin in the heart. J. Clin. Invest. *43:* 630–637 (1964).

Mailing address: Dr. J. C. Hoak, Department of Medicine University of Iowa College of Medicine, Iowa City, Iowa 52240 (USA).

Myocardial Metabolism Laboratory, Department of Medicine,
Division of Cardiology, Cornell University Medical College,
Institute for Muscle Disease, Inc. New York, USA

EFFECT OF PALMITATE ON HYPOXIC MYOCARDIAL METABOLISM AND CONTRACTILITY [1]

C. S. APSTEIN,[2] R. GMEINER,[3] and N. BRACHFELD[4]

INTRODUCTION

The ultimate fate of the heart that survives myocardial infarction is often determined by the amount and severity of the zone of peri-infarction ischemic tissue that surrounds the infarct. The metabolic and hemodynamic characteristics of this hypoxic myocardium may thus play an essential role in determining the limits that help to define functional adequacy (1, 2).

Although carbohydrate and lipid are the principal myocardial substrates, little is known about their relative effects on myocardial performance. It has been demonstrated that glucose exerts a beneficial effect on hemodynamic performance and tissue survival in the anoxic perfused rat heart (16). In experiments performed on papillary muscles, free fatty acids (FFA) were reported to depress contractility under hypoxic conditions; during aerobic perfusion FFA did not affect contractility (7). Elevated serum levels of FFA have been reported to accompany an increased incidence of dysrhythmias and an enhanced mortality after myocardial infarction in patients (11) and experimental animals (8).

In aerobic experiments, FFA increased myocardial oxygen consumption (4), preserved glycogen stores (15), and had a variable effect on glucose uptake. Conversely, hypoxia enhanced glucose consumption, depressed FFA oxidation, and increased its storage as tissue neutral lipid (14). Although carbohydrate has been generally accepted as the preferential substrate during hypoxia, several theoretical reasons suggest that FFA may prove beneficial during ischemic myocardial metabolism (12).

The present investigation was undertaken to investigate the effect of FFA (palmitate) on carbohydrate metabolism and hemodynamic performance in the

---

[1] Supported by research grants from the National Institutes of Health (PH-43-67-1439 and HE-06216) and the New York Heart Association and the Muscular Dystrophy Assoc. of America, Inc.

[2] Post-doctoral fellow of the National Heart Institute of the National Institutes of Health.

[3] Recipient of the Max Kade Foundation Fellowship during the period of this research.

[4] Recipient of a Career Development Award from the National Advisory Heart Council of the National Institutes of Health.

isolated working rat heart during hypoxia. Hearts were perfused under aerobic and hypoxic conditions in the experimental apparatus originally described by Neely *et al* (9). The hearts perform measurable work and the preparation is felt to be more physiologic than techniques utilizing isolated papillary muscles. Two substrate conditions were provided at each oxygen level: a) 5.5 mM glucose and b) 5.5 mM glucose + 0.4 mM palmitate. Perfusate consisted of Krebs-Ringer bicarbonate gassed with 5% CO_2 and varying concentrations of O_2 and N_2. Bovine serum albumin was present during both substrate conditions. In palmitate perfusates, the palmitate: albumin molar ratio was 5 : 1.

RESULTS

Data are expressed as mean values. The standard error of the mean is recorded in the tables.

Hearts were perfused under aerobic or hypoxic conditions with single or double substrates as follows:

A. Aerobic (pO_2 = 478 ± 4.7 mm Hg)
   Substrate 1: 5.5 mM glucose
   Substrate 2: 5.5 mM glucose + 0.4 mM palmitate

B. Hypoxic (pO_2 = 175 ± 1.5 mm Hg)
   Substrate 1:5.5 mM glucose
   Substrate 2: 5.5 mM glucose + 0.4 mM palmitate

Each group consisted of a series of 4 or 5 hearts.

I. Hemodynamic performance

A. Effect of hypoxia

Hypoxia markedly depressed hemodynamic performance (table I). Heart rate fell 50%, peak aortic systolic pressure decreased 8%, peak left-ventricular systolic pressure fell 29%, maximum dp/dt of the left-ventricular pressure curve decreased 64%, and left-ventricular end-diastolic pressure increased 109%. All above changes were highly significant ($p < 0.001$).

The disparity between the left-ventricular and aortic peak pressures was due to the systolic gradient inherent in our perfusion system. Within this range, an increase in peak aortic systolic pressure of 6 mm Hg represents a change in peak left-ventricular systolic pressure of 30 mm Hg.

B. Effect of palmitate on hemodynamic performance

The presence of 0.4 mM palmitate slightly improved hemodynamic performance at both oxygen levels (table II). Peak aortic systolic pressure was 6.4 mm Hg (9%) higher in the presence of palmitate under aerobic conditions and 5.3 mm Hg (9%) higher with palmitate during hypoxia. Both differences were statistically significant ($p < 0.01$). By extrapolation, this would indicate a 30 mm Hg increase in peak systolic left-ventricular pressure induced by the presence of palmitate at each oxygen level. The presence of FFA had no significant effect on heart rate or cardiac output at either oxygen level.

II. Metabolic characteristics

A. Carbohydrate metabolism: effect of palmitate and hypoxia

Carbohydrate utilization (table III). The presence of 0.4 mM palmitate inhibited total carbohydrate utilization, determined by glucose uptake and net glycogen

Table I

Effect of Hypoxia on Hemodynamic Performance*

State	Heart rate	Peak aortic pressure	Cardiac output	Peak L.V. systolic pressure	L.V. end-diastolic pressure	Max. dp/dt† (units) (L.V. pres.)
	(beat/min)	*(mm Hg)*	*(ml/min)*	*(mm Hg)*	*(mm Hg)*	
Aerobic	304 ± 5	67.8 ± 0.7	39.3 ± 1	95.6 ± 1.4	7.4 ± 0.3	8.5 ± 0.2
Hypoxic	152 ± 9	62.3 ± 0.4	–	68.3 ± 1.2	15.5 ± 0.7	3.1 ± 0.1
% Change	–50.0%	–8.1%	–	–28.7%	+109%	–63.5%
p Value	$p < 0.001$	$p < 0.001$	–	$p < 0.001$	$p < 0.001$	$p < 0.001$

*All experiments were performed with the intra-left ventricular needle. Experiments with both substrates (5.5 mM glucose + 5.5 mM glucose + 0.4 mM palmitate) were grouped together and the mean value expressed. The same over-all effect is observed if each substrate is considered separately.

†Millimeters of peak height read directly from the RC differentiating circuit at a standardized calibration.

Table II

Effect of Palmitate on Hemodynamic Performance*

State	Substrate	Heart rate	Peak aortic pressure	Cardiac output
		(beat/min)	*(mm Hg)*	*(ml/min)*
Aerobic	Glucose	309 ± 10	71.1 ± 1.1	46.2 ± 2.1
	Glucose + palmitate	313 ± 7	77.5 ± 0.9	43.3 ± 0.8
	% Change	+1%	+9%	–6%
	p Value	$p > 0.7$	$p < 0.001$	$p > 0.1$
Hypoxic	Glucose	146 ± 16	61.5 ± 0.9	–
	Glucose + palmitate	132 ± 9	66.8 ± 1.1	–
	% Change	–9.5%	+9%	–
	p Value	$p > 0.4$	$p < 0.01$	–

*All experiments were performed without intra-left ventricular needle. Glucose concentration = 5.5 mM. Palmitate concentration = 0.4 mM.

Table III

Carbohydrate Metabolism: Effect of Hypoxia and Palmitate

State	Substrate	Glucose uptake	Glycogen-olysis†	Lactate production	Total carbohydrate utilization‡	Fractional lactate production §
		(μM/g wet wt./30 min)	μM gluc. eq./g wet wt./30 min			
Aerobic	A. Glucose	36.5 ± 3.8	14.4 ± 1.6	22.4 ± 1.7	50.9 ± 2.8	0.44
	B. Glucose + palmitate	31.0 ± 1.6	9.4 ± 1.2	13.4 ± 1.4	41.4 ± 0.7	0.33
	% Change	−15%	−35%	−40%	−19%	−25%
	p Value	$p > 0.1$	$p < 0.05$	$p < 0.001$	$p < 0.02$	–
Hypoxic	C. Glucose	64.9 ± 0.5	21.6 ± 1.1	53.1 ± 2.6	86.5 ± 1.1	0.61
	D. Glucose + palmitate	33.9 ± 1.1	22.8 ± 1.3	33.9 ± 2.2	56.7 ± 1.2	0.60
	% Change	−48%	+5.6%	−36%	−34%	−2%
Effect of hypoxia	*p* Value	$p < 0.001$	$p > 0.5$	$p < 0.001$	$p < 0.001$	–
A–C	% Change glucose	+78%	+50%	+137%	+70%	+39%
	p Value	$p < 0.001$	$p < 0.01$	$p < 0.001$	–	–
B–D	% Change (glucose + palmitate)	+9.4%	+143%	+153%	+40%	+82%
	p Value	$p > 0.1$	$p < 0.001$	$p < 0.001$	–	–

*All experiments were done without the intra-left ventricular needle. Glucose concentration = 5.5 mM; palmitate concentration = 0.4 mM.

†Decrease in tissue glycogen content.

‡Glucose uptake + glycogenolysis.

§Fraction of total carbohydrate utilized that appeared as lactate.

breakdown, at both oxygen levels. The stimulation of carbohydrate utilization (70%) noted during hypoxic perfusion when glucose was the sole exogenous substrate, was reduced to 40% when palmitate was present, although carbohydrate consumption continued to exceed aerobic controls. In the presence of palmitate, total carbohydrate utilization fell 19% ($p < 0.02$) during aerobic perfusion and 34% ($p < 0.001$) during hypoxia.

It was significant that depression of carbohydrate utilization affected exogenous and endogenous sources of this substrate differently and that the response was related to oxygen availability. Under aerobic conditions, the presence of 0.4 mM palmitate depressed net glycogen breakdown by 35%. Glucose uptake remained fixed and was unaffected by the presence of palmitate.

With hypoxia, this pattern of inhibition by palmitate was reversed. When glucose was the sole substrate, its uptake was markedly stimulated by hypoxia and increased 78%. When palmitate was present, however, glucose uptake was reduced by 48% and hypoxia caused no significant increase in glucose uptake. Nevertheless, glycogenolysis was stimulated by hypoxia and, in contrast with aerobic perfusion, this stimulation persisted despite the presence of palmitate.

The presence of 0.4 mM palmitate markedly inhibited lactate production during both aerobic and hypoxic perfusions (table III). At a high pO_2, palmitate reduced lactate production by 40%. Hypoxia stimulated lactate production. When glucose was the sole exogenous substrate, lactate production exceeded aerobic controls by 137%. In the presence of palmitate, the enhanced lactate production during hypoxia persisted but to a lesser degree (51% greater than aerobic controls).

The over-all effect of hypoxia was to increase carbohydrate utilization and lactate production (Pasteur effect). Palmitate inhibited both of these effects, but the site of the block to utilization was related to the state of tissue oxygenation. Under aerobic conditions, net glycogenolysis was inhibited by palmitate, while exogenous glucose utilization was unaffected; under hypoxic conditions, the reverse was seen. The presence of palmitate served to depress the enhancement of glycolysis induced by hypoxia.

B. Lipid metabolism

Palmitate uptake. FFA uptake was measured by chemical and radioisotope techniques (palmitate-1-^{14}C) (table IV). Mean aerobic palmitate uptake measured chemically was 13.2 μM/g wet weight/30 min; the calculated extraction ratio [(Initial – Final conc.)/Initial conc.] × 100 was 39.0%. Agreement was excellent by isotopic measurement. By this technique, palmitate uptake was 12.9 μM/g wet weight/30 min and the extraction ratio was 35.6%. During

Table IV

Fatty Acid Metabolism: Effect of Hypoxia*

State	Chemical palmitate uptake	Isotopic palmitate uptake	% Chemical extraction	% Isotopic extraction	$^{14}CO_2$ Production (% uptake)	Tissue ^{14}C-lipid deposition (% uptake)
	μM/g wet wt/30 min		%	%	*μM palm. eq./g wet wt /30 min*	
Aerobic	13.2 ± 2.0	12.9 ± 0.4	39.0 ± 4.6	35.6 ± 1.7	8.63 ± 0.69 (66.6%)	2.70 ± 0.23 (21.1%)
Hypoxic	5.8 ± 0.8	7.2 ± 0.4	20.6 ± 2.9	24.6 ± 1.5	1.74 ± 0.18 (24.8%)	5.00 ± 0.19 (71.7%)
% Change	–56%	–44%	–47	–59	–80%	+85%
p Value	$p < 0.01$	$p < 0.001$	–	–	$p < 0.001$	$p < 0.001$

*All experiments were performed without intra-left ventricular needle. The substrate = 5.5 mM glucose + 0.4 mM palmitate in all experiments.

hypoxia, there was a marked decrease in palmitate uptake as measured by both techniques. Uptake fell 50% ($p < 0.01$) and the extraction ratio fell 53% ($p < 0.01$).

Fate of palmitate. The fate of the extracted palmitate-1-^{14}C was determined by radioisotope recovery studies (tables IV and V). The $^{14}CO_2$ produced by the heart was collected and its activity determined. The lipid contents of an aliquot of myocardium were extracted and fractionated by thin-layer chromatography. During aerobic perfusion, $^{14}CO_2$ production was 8.63 μM of palmitate equivalent/ g wet weight/30 min. This represented 66.6% of the labelled palmitate taken up by the heart. During hypoxia, $^{14}CO_2$ production decreased by 80% and was equal to 24.8% of the extracted palmitate-1-^{14}C. Hypoxia thus significantly reduced both palmitate-1-^{14}C uptake and its complete oxidation.

Under aerobic conditions, 2.70 μM palmitate equiv./g wet weight/30 min (21.1% of extracted palmitate-1-^{14}C) was recovered in tissue lipid fractions. With hypoxia, the lipid fractions accumulated 5.00 μM palmitate equiv./g wet weight/30 min (71.7%). Thus, accumulation of palmitate-1-^{14}C into tissue lipid increased by 85% despite a 44% reduction in palmitate-1-^{14}C extraction. In the aerobic state, the primary fate of extracted palmitate was complete oxidation to CO_2; with hypoxia, there was a shift in pathways of utilization; although extraction was somewhat depressed, it persisted, oxidation was inhibited, and the predominant fate of extracted palmitate was tissue deposition.

The fractional distribution of the extracted palmitate-1-^{14}C into the various lipid fractions was similar during both aerobic and hypoxic perfusion. During hypoxia, however, there was a quantitative increase within each fraction, which ranged from 45 to 233%.

During aerobic perfusion, 66.9% of deposited palmitate-1-^{14}C was recovered in the triglyceride fraction (1.81 μM). With hypoxia, accumulation

Table V

Metabolic Fate of Palmitate-1-^{14}C Deposited as Tissue Lipid under Hypoxic and Aerobic Conditions

	Aerobic		Hypoxic	
Lipid fraction	Lipid deposition*	% Total lipid uptake	Lipid deposition	% Total lipid uptake
Total tissue lipid	2.70 ± 0.23	100	5.14 ± 0.27	100
Triglyceride	1.81 ± 0.18	66.9 ± 1.4	3.77 ± 0.24	73.1 ± 1.0
Phospholipid	0.57 ± 0.05	21.1 ± 0.4	0.91 ± 0.07	17.8 ± 1.2
Diglyceride	0.16 ± 0.02	5.8 ± 0.4	0.24 ± 0.01	4.6 ± 0.2
Nonesterified fatty acid	0.13 ± 0.01	5.0 ± 0.6	0.19 ± 0.04	3.9 ± 0.9
Cholesterol ester	0.01 ± 0.003	0.3 ± 0.1	0.03 ± 0.004	0.6 ± 0.1

*μM palmitate equivalent incorporated/g wet wt/30 min ± standard error of the mean.

into triglyceride increased 108% (3.77 μM) ($p < 0.001$) and esterification to triglyceride was the major metabolic fate of the extracted palmitate-1-^{14}C (73%). The increase in tissue triglyceride was roughly proportional to the over-all increase in lipid accumulation. The appearance of ^{14}C label into the other lipid subfractions is shown in table V.

DISCUSSION

Hemodynamic effects

Heart rate, aortic pressure, and cardiac output all decreased markedly with hypoxia, despite the presence of glucose. In contrast, when both glucose and palmitate were present as substrates there was a significant increase in peak aortic pressure, during both aerobic and hypoxic perfusions. This palmitate-induced enhancement of the contractile state of the perfused heart has not been previously described. Since albumin was present in similar concentration in both the single and double substrate perfusion, it seemed unlikely that it was responsible for this effect.

Previous reports of the effects of palmitate on hemodynamic performance are inconclusive. Evans *et al.* (6) reported that, under aerobic conditions, the perfused rat heart can utilize palmitate as the sole substrate and beat for 30 min without "deterioration of visible performance." In contrast, Henderson *et al.* (7), reported that the addition of FFA to the glucose-containing bath fluid depresses myocardial contractility of the isolated papillary muscle under hypoxic conditions. No effect was observed during the aerobic state. Differences in experimental methodology may account for apparently conflicting results. Our experiments utilized a 0.4 mM concentration of palmitate, a 16-carbon saturated fatty acid. Henderson *et al.* (7) used an unsaturated fatty acid in higher concentration (1.0 to 1.75 mM linoleate) and a shorter, 8-carbon, saturated fatty acid (0.5 to 1.75 mM caprylic acid). The rate of metabolic utilization of these differing fatty acids may vary significantly. Henderson *et al.* (7) did not report metabolic measurements. Furthermore, when they evaluated a nonmetabolized fatty acid (pent-4-enoic acid), they also observed a depression of contractility, suggesting that such depression was not due to utilization of the FFA as an energy source, but was ".... mediated directly by the FFA or acyl CoA derivative.... and might be due to a detergent effect of calcium binding by FFA...." (7). Their technique made coronary perfusion impossible. In our experiments, substrate was distributed in a physiologic antegrade fashion via the coronary circulation. Evidence obtained from studies of uptake and distribution of isotope indicated that palmitate was readily metabolized by our preparation and led to the enhancement of contractility.

FFA has not previously been shown to enhance contractility under aerobic conditions. Henderson *et al.* (7) reported no effect on the isolated rat papillary muscle. Other investigators using a working isolated rat heart preparation similar to ours, found no significant difference in contractility when 1 mM palmitate was added to a substrate of 5 mM glucose (5). In these experiments, however, left atrial filling pressure and cardiac output was lower than in our series. Palmitate enhancement of hemodynamic performance may only be apparent at higher filling pressures, or during hypoxic perfusion conditions, as seen in our experiments. A common denominator of both of these states may be a decreased level of intracellular ATP. Under these conditions, the type of substrate provided may prove to be an important factor in maintaining ATP concentration above a critical level.

Metabolic effects

A. Carbohydrate metabolism

Glucose uptake. During aerobic perfusion, glucose uptake was not significantly inhibited by the presence of palmitate. During hypoxic perfusion, however, inhibition was marked and glucose uptake fell 48% ($p<0.001$). Although hypoxic stimulation of glycolysis increased glucose uptake by 78% when it was the sole substrate, the presence of palmitate completely inhibited this response. Total carbohydrate utilization and lactate production fell as well. It seems unlikely that palmitate oxidation is solely responsible for the inhibition of glucose uptake. In the aerobic state, during maximal palmitate oxidation, glucose uptake was not inhibited. Conversely, under hypoxic conditions, when palmitate oxidation was reduced by 80%, inhibition of glucose was marked.

A more likely mechanism is suggested by the probability that the thiokinase and hexokinase reactions compete for a reduced and rate-limiting supply of ATP under hypoxic conditions. In the aerobic state, with high levels of ATP, both reactions may occur with maximal velocity. With hypoxia, intracellular ATP concentration falls rapidly. When glucose is present as the sole exogenous substrate, the hexokinase reaction may utilize much of the available ATP. When both FFA and glucose are present, the substrates compete for available ATP and hexokinase activity is reduced. This competition may also explain the observation that maximal FFA inhibition of glucose uptake occurs at higher work loads since intracellular ATP levels may be decreased under these conditions (10).

Glycogenolysis. Under aerobic conditions, FFA has been shown to exert a glycogen sparing effect (15). Acetate also decreases net glycogenolysis (17). Tissue citrate concentration was increased sufficiently to cause inhibition of phosphofructokinase. The elevated level of glucose-6-phosphate, so induced, is known to enhance glycogen synthetase-D activity and to decrease the activity of phosphorylase-*b* (13). Both mechanisms may explain the decreased rate of net glycogenolysis observed during aerobic perfusion in the presence of acetate, since in the rat heart, 90% of the glycogen synthetase is in the D form (13).

In the present aerobic experiments, net glycogenolysis was decreased by 35% ($p < 0.05$) by the presence of 0.4 mM palmitate. It seems likely that the relative depression of glycogenolysis is induced by a mechanism similar to that induced by acetate.

The failure of palmitate to inhibit glycogen breakdown in the hypoxic state may be a result of lower tissue levels of citrate and glucose-6-phosphate during hypoxia.

Lactate production. Reports of the effect of FFA on lactate production are variable and dependent upon glucose concentration and work load. In the working heart with a left atrial filling pressure of 10 cm H_2O, an aortic pressure "afterload" of 60 mm Hg, and a substrate composed of 15 mM glucose, the addition of palmitate increased lactate production by 54% (10).

The present studies support the concept of tighter substrate control (inhibition of glucose uptake by FFA) at higher work loads. We observed a decrease in lactate production in the presence of palmitate at relatively high work loads. Other investigators report an increased lactate production in the presence of palmitate in experiments conducted at lower work loads (10). The mechanism responsible for inhibition of glucose uptake at higher work loads has not been explained, but may be related to lower cellular ATP levels, as suggested above. The decrease in lactate production is likely a secondary effect of the decrease in glucose uptake.

B. Lipid metabolism

Myocardial extraction of FFA is enhanced by an increase in the FFA to albumin ratio (6) and by an increase in work load (5, 10). In previous studies of the perfused rat heart, hypoxia did not decrease FFA uptake despite a marked reduction in FFA oxidation (80%). Tissue triglyceride deposition was concomitantly increased (3). Similar findings were reported following production of ischemia in the *in vivo* dog heart (14).

In the present experiments, hypoxia caused a reduction in FFA uptake and percent extraction of roughly 50%. Isotopic and chemical determinations were in good agreement. The lack of agreement with the studies above (3, 14) may be related to the relatively high atrial filling pressures used in our experiments.

The fate of the extracted palmitate was markedly affected by hypoxia. $^{14}CO_2$ production fell by 80% and deposition of ^{14}C label into tissue lipid increased 85%. Most of the esterified palmitate was recovered as triglyceride. In general, the pattern of FFA deposition into the various lipid subfractions was proportional to the fractional concentration of endogenous lipid subfractions in control hearts.These findings confirm our earlier observations (14) that hypoxia appears to enhance esterification of the activated fatty acid due to the increased tissue concentration of α-glycerophosphate induced by a stimulation of glycolysis.

Metabolic changes and increased hemodynamic performance

The relationship between the metabolic changes and increased hemodynamic performance seen in the presence of palmitate may be associated with the palmitate-induced inhibition of glycolysis with a concomitant reduction in lactic acid production and cellular acidosis. The depression of contractility associated with acidosis may, therefore, be relieved. The presence of palmitate may also enhance the net yield of high energy phosphate.

SUMMARY

The effect of varying substrate on isolated working rat hearts was studied under aerobic and hypoxic conditions. Either 5.5 mM glucose or 5.5 mM glucose + 0.4 mM palmitate were provided as substrate. Peak aortic systolic pressures were higher ($p < 0.01$) when palmitate was present for both aerobic and hypoxic perfusions. Under aerobic conditions, the presence of palmitate decreased carbohydrate utilization by 19% ($p < 0.02$) and lactate production by 40% ($p < 0.001$). With hypoxia, there was a mean increase of glycolysis of 100%; nevertheless, the presence of palmitate decreased carbohydrate utilization by 35% ($p < 0.001$) and lactate production by 36% ($p < 0.001$). The site of inhibition of carbohydrate utilization varied with the amount of oxygen available; under aerobic conditions, glycogenolysis was reduced by 35% ($p < 0.05$), but glucose uptake was not significantly inhibited; with hypoxia, glucose uptake was inhibited by 48% ($p < 0.001$) in the presence of palmitate, but there was no significant inhibition of glycogen breakdown. Palmitate uptake was reduced 44% ($p < 0.001$) by hypoxia, its oxidation to CO_2 was decreased by 80% ($p < 0.001$), and tissue lipid deposition was increased by 85% ($p < 0.001$). The presence of palmitate decreased the hypoxic stimulation of glycolysis (Pasteur effect) and improved the hemodynamic performance.

REFERENCES

1. BRACHFELD, N.: Maintenance of cell viability. Circulation 39 & 40 (Suppl. IV): p. IV, 202–219 (1969).
2. BRACHFELD, N.: Myocardial metabolism: bioenergetics of the normal and anoxic myocardium; BANYAI, A. and GORDON, B. L. in Advances in Cardiopulmonary Diseases, Vol. IV, p. 66–90, (Yearbook Medical Publishers, Inc., Chicago, 1969).
3. BRACHFELD, N. and KAWADE, M.: Mechanism of triglyceride synthesis by hypoxic myocardium. Amer. J. Cardiol. *21:* 92 (1968).
4. CHALLONER, D. R. and STEINBERG, D.: Effects of fatty acid on the oxygen consumption of perfused rat heart. Amer. J. Physiol. *210:* 280–286 (1966).
5. CRASS, M., III,; MCCASKILL, E. and SHIPP, J. C.: Effect of pressure development on glucose and palmitate metabolism in perfused heart. Amer. J. Physiol. *216:* 1569–1576 (1969).
6. EVANS, J. R.; OPIE, L. H. and SHIPP, J. C.: Metabolism of palmitic acid in perfused rat heart. Amer. J. Physiol. *205:* 766–770 (1963).
7. HENDERSON, A. H. MOST, A. S.; PARMLEY, W. W.; GORLIN, R. and SONNENBLICK, E. H.: Depression of myocardial contractility in rats by free fatty acids during hypoxia. Circulat. Res. *26:* 439–449 (1970).
8. KURIEN, V. A.; YATES, P. A. and OLIVER, M. F.: Free fatty acids, heparin and arrhythmias during experimental myocardial infarction. Lancet *ii:* 185–187 (1969).
9. NEELY, J. R.; LIEBERMEISTER, H.; BATTERSBY, E. J. and MORGAN, H. E.: Effect of pressure development on oxygen consumption by isolated rat heart. Amer. J. Physiol. *212:* 804–814 (1967).

10. NEELY, J. R.; BOWMAN, R. H. and MORGAN, H. E.: Effect of ventricular pressure development and palmitate on glucose transport. Amer. J. Physiol. *216:* 804–811 (1969).
11. OLIVER, M. F.; KURIEN, V. A. and GREENWOOD, T. W.: Relation between serum free fatty acids and arrhythmias and death after myocardial infarction. Lancet *i:* 710–715 (1968).
12. OLSON, R.: Metabolic interventions in the treatment of infarcting myocardium. Circulation 39 & 40 (Suppl. IV): p. IV, 195–201 (1969).
13. OPIE, L. H.: Metabolism of the heart in health and disease, Parts I–III. Amer. Heart J. *76:* 685–698 (1968); *77:* 100–122 (1969); *77:* 383–410 (1969).
14. SCHEUER, J. and BRACHFELD, N.: Myocardial uptake and fractional distribution of palmitate-1-C^{14} by ischemic dog heart. Metabolism *15:* 945–954 (1966).
15. SHIPP, J. C.; OPIE, L. H. and CHALLONER, D.: Fatty acid and glucose metabolism in the perfused heart. Nature *189:* 1018–1019 (1961).
16. WEISSLER, A.; KRUGER, F; BABA, N.; SCARPELLI, D.; LEIGHTON, R. and GALLIMORE, J.: Role of anaerobic metabolism in the preservation of functional capacity and structure of anoxic myocardium. J. Clin. Invest. *47:* 403–416 (1968).
17. WILLIAMSON, J. R.: Glycolytic control mechanisms, I. Inhibition of glycolysis by acetate and pyruvate in the isolated, perfused rat heart. J. Biol. Chem. *240:* 2308–2321 (1965).

Mailing address: N. Brachfeld, M.D., The New York Hospital, Cornell Medical Center, 525 East 68th Street, New York, N.Y. 10021 (USA).

Cardiovascular Division, Peter Bent Brigham Hospital and
Harvard Medical School, Boston, Mass. USA

INFLUENCE OF FREE FATTY ACIDS ON MYOCARDIAL MECHANICS AND OXYGEN CONSUMPTION IN RAT PAPILLARY MUSCLES AND PERFUSED HEARTS, UNDER OXYGENATED AND HYPOXIC CONDITIONS [1]

A. H. HENDERSON[2] and E. H. SONNENBLICK

The relative contributions of different exogenous substrates to total myocardial fuel consumption have been studied extensively, but usually without reference to the mechanical activity of the heart (1). Over 80% of the heart's total energy requirements are needed for mechanical activity (2, 3), however, and it is clear that energy supply must in some way be coupled to mechanical activity. Relatively little attention has been paid to the possible influence of exogenous substrates on mechanical function.Unfortunately, *in vivo* studies of mechanical function are complicated by a multiplicity of interdependent variables. Conventional hemodynamic assessment of the heart's mechanical performance may be influenced by changes in myocardial contractility, series elasticity, external loading throughout the cardiac cycle, or heart rate, and (at suboptimal levels) by alterations in coronary perfusion. Nor can the effects of one exogenous substrate be considered in isolation from the effects of other substrates and hormones, whose circulating levels will be interrelated in the intact animal. On the other hand, any experimental model set up to study one or more of these variables under controlled conditions is, of course, subject to its own artifactual limitations, including the need to consider species differences.

With this in mind, an investigation into the effects of glucose and free fatty acids (FFA) on the mechanical performance of heart muscle was recently carried out, using rat papillary muscle preparations (4). These preparations have the advantage of allowing detailed measurements of mechanical performance at constant preload (resting force) and at constant rate of contraction, while avoiding the complicating factor of variation in coronary perfusion. Adequate oxygenation of core fibers depends upon the inward diffusion of oxygen, so that it is essential to use thin muscles and, when comparing one group of muscles with another under hypoxic conditions, to ensure that the muscles in each group are of comparable mean cross-sectional area. The influence of glucose and FFA

[1] Supported in part by United States Public Health Service Grant 11306 and 5890.

[2] Samuel Levine Fellow of the Massachusetts Heart Association during tenure of this work. Present address: Royal Devon and Exeter Hospital, Exeter, Devon, England.

on mechanical performance was first investigated in well-oxygenated muscles, by alternating the substrate content of the aqueous bicarbonate buffer in which they were suspended. The buffer contained 0.3 mM albumin (to which was bound 0.1 mM FFA) together with 5 mM glucose or various concentrations (0.5 to 1.75 mM) of FFA added as sodium octanoate or linoleate and incubated for 1 h. Pairs of substrates were alternated for 1 h periods in each muscle. Under oxygenated conditions, mechanical performance was found to remain constant whether 5 mM glucose, 1.75 mM FFA, a combination of the two, or no added substrate, was present. The influence of substrates on mechanical performance during hypoxia was then studied. When the pO_2 of the medium is lowered abruptly from 600 to 150 mm Hg, developed force and peak rate of force development (dF/dt) diminish in an approximately exponential and parallel manner, without necessarily reaching a new steady state after 30 min of oxygen deprivation; a change in the pO_2 of the buffer from 600 to < 5 mm Hg results in a similar but greater depression of contractile performance (fig. 1). When these muscles are then reoxygenated, they recover to 55–100% of control values. Groups of muscles were therefore studied under comparably hypoxic conditions with different substrate combinations in the surrounding medium. Isometric force development was measured after 30 min of oxygen deprivation, and expressed as a percentage of the control value obtained under well-oxygenated

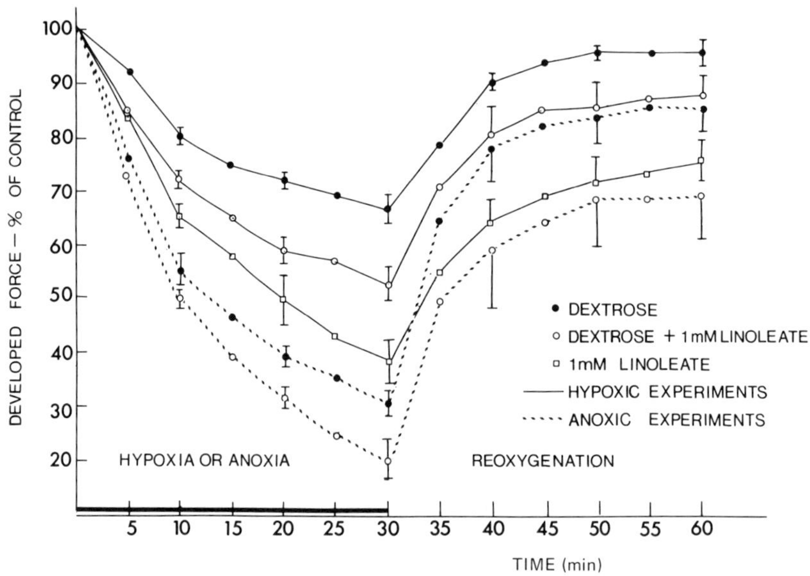

Fig. 1 Time course of isometric force development in five representative groups of rat papillary muscles mounted in buffer containing 5 mM glucose or 1 mM linoleate, or both, and subjected to 30 min hypoxia or anoxia followed by reoxygenation. Mean ± standard error is indicated; N = 6 or 7.

conditions. The presence of 5 mM glucose resulted in significantly less depression of developed force than the absence of any exogenous substrate, while the presence of FFA (0.5 – 1.75 mM) caused significantly more depression (fig. 2). The results with 1.75 mM FFA were similar to those with 0.5 mM FFA. No difference was observed between the effects of octanoate and those of linoleate. When 5 mM glucose was present in addition to 1 mM or 1.75 mM FFA, the extent of the depression was intermediate between that found with either substrate alone. The addition of insulin (50 μU ml) to 5 mM glucose slightly enhanced mechanical performance, although the difference did not reach statistical significance in the small number of muscles studied. Most experiments were conducted under hypoxic conditions, but similar differences were also found with complete anoxia. The extent of the recovery on reoxygenation was generally inversely proportional to the severity of the preceding depression with oxygen deprivation. In summary, the mechanical performance of well-oxygenated muscles was not affected by glucose or by FFA (FFA to albumin molar ratio 6 : 1), but hypoxic or anoxic muscles were depressed by FFA (FFA to albumin 2 : 1). Conversely, glucose enhanced the performance of hypoxic muscles and the additional presence of insulin appeared to augment performance further.

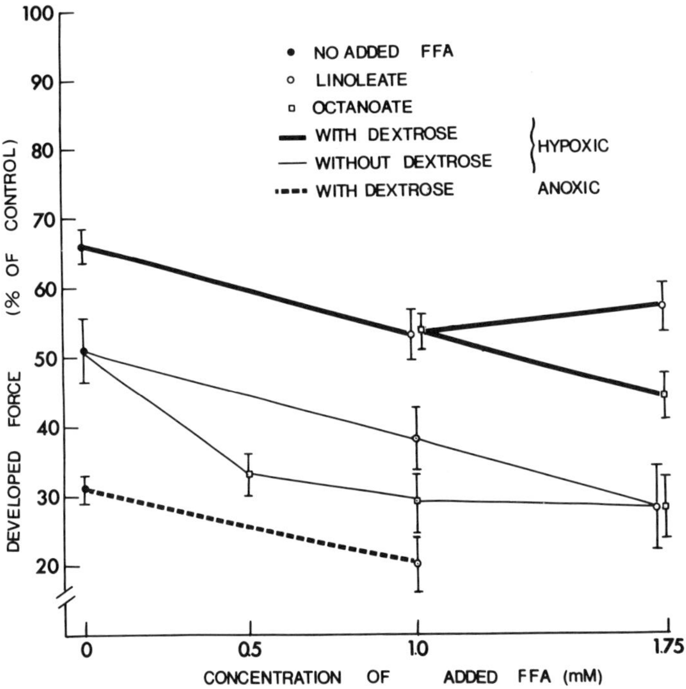

Fig. 2 Influence of external FFA concentration, with and without glucose, on developed force after 30 min hypoxia or anoxia.

Changes in developed force were associated with parallel changes in dF/dt, so that they may be considered to reflect changes in contractility at constant preload, provided that series elasticity remained unchanged. In fact, some muscles developed a reversible increase in resting force at constant resting length (*i.e.*, contracture) (fig. 3). The increase in resting force was generally related to the severity of the depression of developed force. However, FFA appeared to have a disproportionately greater effect in increasing resting force than in depressing developed force (fig. 4). The depressive influence of FFA on

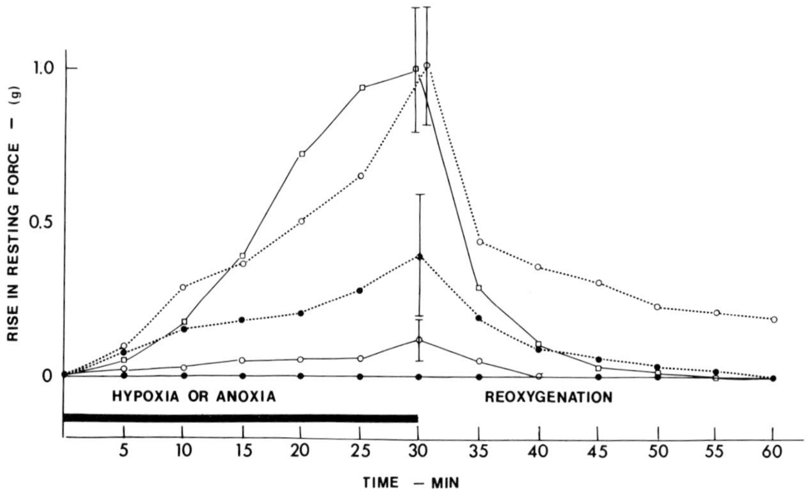

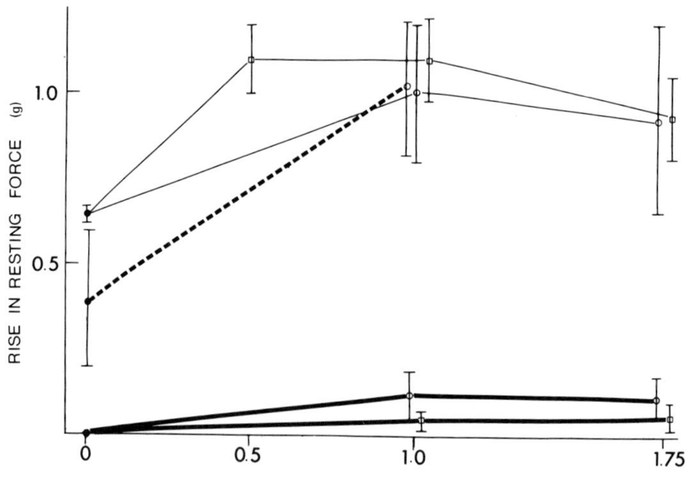

Fig. 3 and 4 As in figures 1 and 2, but showing rise in resting force at constant length (resting force under control conditions, 1 g; muscle mean cross-sectional area 0.84 ± 0.02 mm^2).

contractility may, therefore, have been somewhat underestimated by reference to the changes in developed force and rate of pressure development (dP/dt), since resting force was increased more in the presence of FFA; this effect is likely to have been small, since all experiments were carried out at about the peak of the length-active tension curves of the muscles. In a separate study (5), series elastic compliance was measured during oxygen deprivation and subsequent reoxygenation, and found to be reversibly decreased during the time when contracture occurred (fig. 5). Since contractile element shortening velocity is related to dF/dt through the characteristics of the series elastic element, the depression of contractility induced by FFA would again have been underestimated by the extent of the changes in dF/dt. Thus, the parallel changes in passive and in active length-force relations that occurred in relation to the severity of oxygen deprivation and to the presence of FFA and absence of glucose would tend to have resulted in some underestimation of the extent to which contractility was depressed by FFA as shown by changes in developed force. The observation that FFA tended to favor the development of contracture

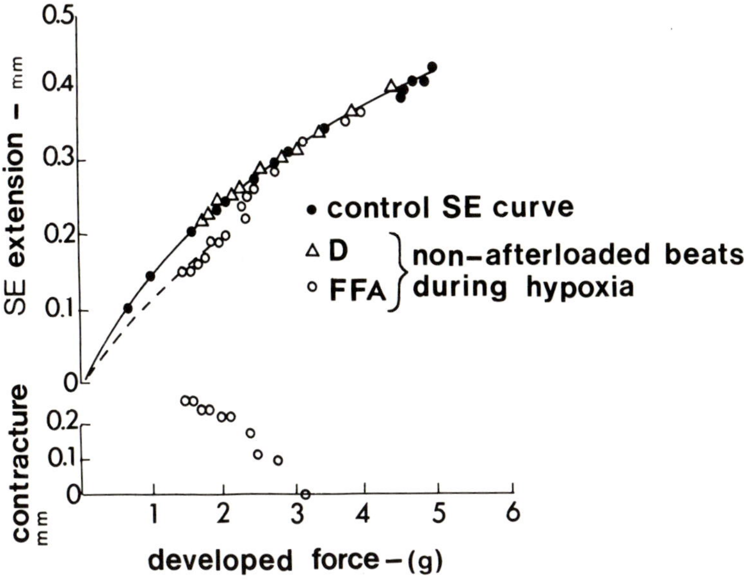

Fig. 5 Continuous line shows series elastic (SE) load-extension curve obtained under oxygenated conditions in representative rat papillary muscle. Superimposed are the values for change in length and force during quick releases during two periods of hypoxia and subsequent reoxygenation, first with 5 mM glucose and then with 1.75 mM linoleate in the medium. With glucose, series elasticity and muscle length did not alter. With linoleate, muscle length shortened as developed force decreased (lower panel), and the values derived from quick-releasing the non-afterloaded muscle deviated from the control curve, indicating an increase in the stiffness of the SE. SE extension curve at 40% maximum developed force is represented by interrupted line. The changes were reversed on reoxygenation.

and so, by implication, to increase the stiffness of the series elastic element, is itself of some intrinsic interest, although it remains completely unexplained at present.

The influence of substrates was studied also in the Langendorff perfused rat heart at 37°C (6). In this preparation, hearts were perfused at constant flow rates to avoid the changes in developed pressure known to be associated with different coronary flow rates, and were paced at constant rate to avoid the changes in contractility associated with differences in rate. Pressure was recorded from a balloon, inserted in the left ventricle and filled with a known volume of water so that the hearts beat isovolumically at constant end-diastolic volumes. The effects of different substrates were then compared in each heart by alternating perfusion with pre-equilibrated oxygenated (pO_2 580 mm Hg) buffer over 20 min periods (fig. 6). The use of buffer containing 0.3 mM albumin, 1.75 mM linoleate, with or without 5 mM glucose, resulted in a significant depression in developed pressure and dP/dt compared with 5 mM glucose alone or the combination of 5 mM glucose and 0.5 mM linoleate. Linoleate, 1 mM, did not depress mechanical performance at steady state, although each change from glucose to FFA was associated with a transient depression of mechanical performance, and each change in the opposite direction was associated with transient augmentation. When no albumin was provided to bind FFA, 0.5 mM

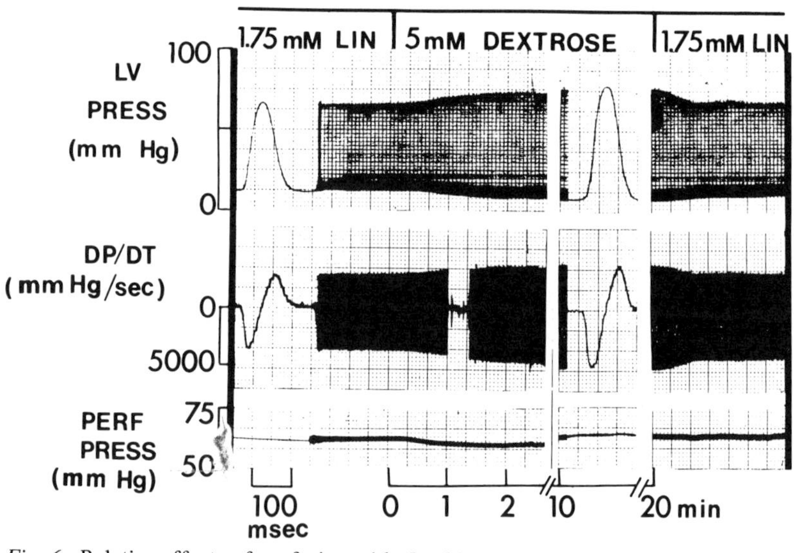

Fig. 6 Relative effects of perfusing with 5 mM glucose and with 1.75 mM linoleate (LIN) in representative rat heart (final wet wt. 680 mg) paced at 300/min. Isovolumic left ventricular pressure, rate of pressure development (DP/DT), and mean perfusion pressure at constant coronary flow (6 ml/min bicarbonate buffer containing 0.3 mM albumin at pO_2 580 mm Hg) are shown.

octanoate depressed the hearts relative to 5 mM glucose or to no exogenous substrate; conversely, 5 mM glucose augmented performance relative to the absence of exogenous substrate, and the addition of insulin (2 mU/ml) resulted in a further significant increase in mechanical performance. A small rise in end-diastolic pressure was also observed with 1.75 mM linoleate. Previous work of others with rat hearts perfused at 32°C has shown that glucose does not improve contractile performance under well-oxygenated conditions, but that it does so when the hearts are hypoxic (7). The findings described here therefore suggest that the hearts may have been inadvertently oxygen-limited, despite a total arterial oxygen supply of 15 ml/min 100 g left ventricle (wet weight). Although the experiments showing enhanced contractility with glucose were conducted with albumin-free buffer and are, therefore, not strictly comparable with those where the perfusate contained albumin, other considerations lend support to the conclusion that all these hearts were probably slightly hypoxic. Thus, the hearts were beating isovolumically at moderate end-diastolic pressure, so that mean wall stress throughout each contraction would be higher than in the so-called working (ejecting) hearts used by other workers; contractility is temperature-related, and the present experiments were carried out at 37°C compared with 32°C in the previously reported study; the increase in developed force with the increase in coronary flow that characterized perfused rat hearts (8) is strongly suggestive of oxygen limitation by analogy with the findings in other species; there is evidence that intrinsic myocardial contractility is unusually high in the rat myocardium (4, 9, 10) and contractility has been shown to be an independent determinant of MVO_2 in dog hearts; and, finally, the oxygen requirements for a given force development appear to be higher in rat than in cat papillary muscles as shown by correlating maximal force development with muscle cross-sectional area (4). It is also possible that the intraventricular balloon might induce artifactual subendocardial underperfusion in these preparations. The apparently discrepant finding that 1.75 mM linoleate depressed contractile performance in the perfused hearts but not in the well-oxygenated papillary muscles would also be explained if, in fact, these perfused rat hearts were oxygen limited. This possibility must, therefore, be borne in mind when interpreting the results of this study. When taken together, the results of the perfused heart and those of the papillary muscle experiments confirm that glucose enhances contractility in hypoxic rat heart muscle (7) and show that the addition of insulin to glucose results in further augmentation of contractile performance. They also indicate that FFA (FFA to albumin $> 2:1$) depress contractility in hypoxic rat myocardial preparations, but not in well-oxygenated preparations.

The influence of substrate on coronary vascular resistance could also be measured in the perfused heart experiments. Perfusion with 1.75 mM linoleate results in an increase of mean coronary pressure at constant flow, relative to perfusion with 5 mM glucose. It is possible that the observed changes in coronary vascular resistance were due to auto-regulation (11), since venous O_2 also rose, whereas in a previous study (12), where venous O_2 fell with FFA, coronary vascular resistance was decreased as shown by a greater flow at constant perfusion pressure. In another study (13), FFA were found to increase coronary vascular resistance at constant perfusion pressure; but in these experiments the buffer containing FFA was also the only buffer to include albumin, so that a separate variable was introduced. In isolated rabbit hearts perfused with aqueous buffer, unbound FFA have been reported to cause an abrupt and critical decrease in coronary flow, while bound FFA have no effect (14). The concentrations of FFA found to influence coronary vascular resistance are all considerably in excess of the tight binding-site capacity of albumin (15) and it would be of interest to determine whether more physiological concentrations of FFA affect coronary vascular resistance in definitely hypoxic hearts. In hearts perfused with blood, it may prove necessary to consider also the influence of FFA on coagulation mechanisms, together with the possibility that raised levels of FFA may predispose to microangiopathy.

Myocardial oxygen consumption (MVO_2) provides a convenient measure of the energy liberated from oxidative metabolism. However, measurements of MVO_2 will present an incomplete picture of the metabolic economy, unless the relevant parameters of mechanical performance (16) are also monitored. FFA have been reported to increase MVO_2 in two previous studies with perfused rat hearts where mechanical activity was not measured. In one (12), they also increased coronary flow at constant perfusion pressure and this, itself, would probably have increased the mechanical demand for oxygen; when coronary flow was held constant, a significant increase in MVO_2 of beating hearts at 37°C was found only with very high concentrations of FFA (FFA to albumin, 10 : 1). In the other study (13), the buffer contained albumin only when FFA were added; we have found that hearts perfused with albumin develop more pressure than those perfused without albumin, and become less edematous as shown by dry-to-wet weight ratios. In our experiments (6), perfusion with 1.75 mM linoleate (FFA to albumin, 6:1) resulted in a lower MVO_2 than with glucose, in association with the depression of contractility (fig. 7). Further analysis of the results showed that, despite the fall in MVO_2, a given amount of work cost slightly more oxygen with this concentration of FFA than with glucose. Under the experimental conditions imposed, any change in contractility would be manifest by an increase in isovolumic pressure development and, thus, in mean wall stress at constant ventricular volume and thickness, so that both these determinants of MVO_2 (16) would be represented in the integrated developed tension per min (or tension time index - TTI), since, in addition, there was no change in the duration of systole. When different values for the TTI (obtained by perfusion at different coronary flow rates)were plotted against MVO_2, the

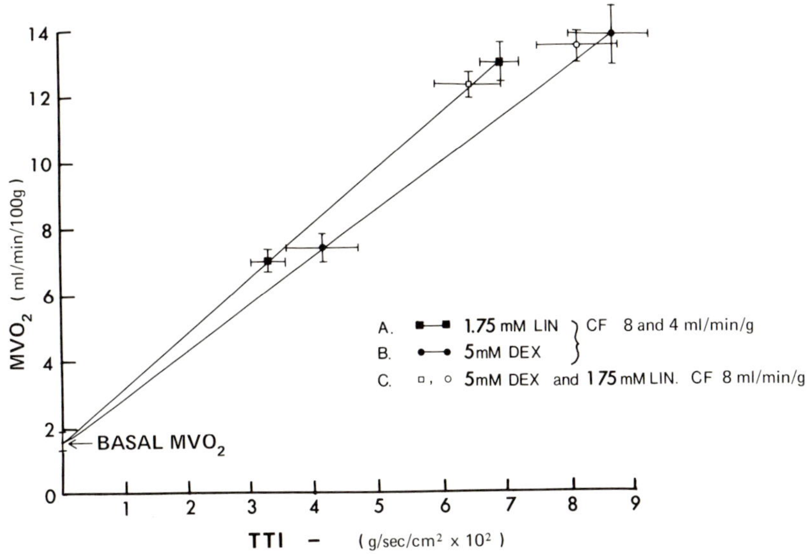

Fig. 7 MVO_2 related to tension time index (TTI) in rat hearts perfused at constant heart rate and left ventricular volume, with either 5 mM glucose (DEX) or 1.75 mM linoleate (LIN) in the perfusate, and at constant coronary flow of either 8 or 4 ml/min/g as indicated. Hearts in group A (N = 8) were perfused with 1.75 mM linoleate and hearts in group B (N = 3) with 5 mM glucose, for alternating 20 min periods at the two different coronary flow rates. Hearts in group C (N = 6) were perfused alternately with 5 mM glucose and 1.75 mM linoleate at constant coronary flow. The ratio of TTI/work MVO_2 (work MVO_2 = total minus basal MVO_2) was significantly less with 1.75 mM linoleate than with 5 mM glucose (paired *t* test).

relationship could be extrapolated to zero TTI to give a value for basal MVO_2; this was found to be similar (1.7 ml/min/100 g) with glucose and with 1.75 mM linoleate. This value was subtracted from the total MVO_2 to give that amount of oxygen consumption associated with mechanical work (work MVO_2). The ratio of TTI/work MVO_2 with 1.75 mM linoleate was significantly lower than that obtained with 5 mM glucose. It is possible that this apparent difference in efficiency would have been due to a difference in the rate of glycolysis and, thus, in the amount of energy provided without oxygen cost, especially since the hearts may have been slightly hypoxic and since the efflux of lactate would be favored by the absence of lactate in the perfusate (17, 18). FFA are known to depress glycolysis in relatively well-oxygenated preparations (19). Without further direct measurement of lactate production, however, it is not possible to exclude an oxygen wasting effect of FFA. However, FFA also decreased the absolute amount of oxygen consumed by the heart; thus, FFA must in addition have depressed oxidative metabolism or contractility directly. FFA (FFA to albumin 4 : 1) have also been reported to increase MVO_2 in hearts arrested by high external concentrations of potassium (12); further experiments suggested that this was due to an increase in mitochondrial non-phosphorylating respiration. Without detracting from the interest of these observations, it should be noted that potassium-arrested hearts may not be in a truly basal state with regard to mechanical and electrical activity. Using a concentration of potassium just sufficient to prevent contractions occurring in response to continued electrical pacing, we observed that diastolic pressure was considerably higher than in the beating hearts,so that energy-requiring active contracture cannot be excluded under these circumstances; moreover, the MVO_2 of such arrested hearts appeared to be much higher than previously reported basal values, and than those obtained by extrapolation from TTI/MVO_2 relations (preliminary unpublished observations). These findings, if confirmed, may of course be related to the fact that the hearts were electrically paced, and this may possibly have increased the energy needs of the sarcolemmal cation pump in the presence of high external potassium concentration.

Even moderate concentrations of FFA appear to depress contractility in the presence of hypoxia, but the underlying mechanism remains to be established. FFA could act either by limiting the supply and translocation of high energy phosphate bonds to the cross-bridges of the myofilaments and to the membrane sites controlling the movement of activator calcium, or by exerting a direct effect upon the contractile mechanism and the control of activator calcium movements. Glycolysis provides the only known anerobic source of energy in the heart to replenish high energy phosphates (20), and it is possible that the differences in contractility observed with different exogenous substrates under hypoxic conditions might reflect associated changes in the rate of glycolysis. In well-oxygenated perfused rat hearts, glucose increases the rate of glycolysis (17), while FFA inhibit glycolysis mainly through inhibition of phosphofructokinase by the accumulation of citrate arising from the oxidative breakdown of FFA (21, 22). In anoxia, the provision of exogenous glucose further increases glycolysis (7); FFA are taken up normally by the cells (23), but they have been reported to cause less accumulation of citrate (24). The influence of FFA on glycolysis under oxygen-limited conditions appears not to have been measured directly, although their inhibitory effect is said to be abolished by anoxia. However, the conclusion that FFA must have exerted a depressant effect on oxidative metabolism or on the contractile mechanism in the perfused heart suggests that their action in depressing contractility in hypoxic papillary muscles was not mediated solely through a reduction in the rate of glycolysis. In an attempt to elucidate the underlying mechanisms a little further, a series of additional experiments were conducted, using the nonmetabolized FFA, pent-4-enoic acid (PA). PA (1 mM), which blocks the carnitine-dependent entry of long-chain FFA into mitochondria (25, 26), was found to have precisely the same effect in depressing the mechanical performance of hypoxic muscles as octanoate or linoleate (1 mM).

Hypoxia results in increased incorporation of label from palmitate-C^{14} into the tissue FFA and triglycerides (TG) (27). While this, in itself, does not necessarily indicate an increase in the absolute amounts of FFA and TG within the cell, recent work suggests that these fractions are indeed increased by hypoxia (28), although Weglicki and Davidson (unpublished observations) were unable to demonstrate any such increase in tissue FFA in perfused rat hearts with severe hypoxia leading to a marked depression of contractility. Raising the extracellular concentration of FFA also increases the intracellular concentration of FFA. Whether FFA ever exceed the binding capacity within the cells, *in vivo*, remains to be confirmed (27), but it seems unlikely that the intracellular concentration of FFA will be increased by the combination of hypoxia and increased circulating levels of FFA. Furthermore, the binding capacity within the cell will be reduced by acidosis associated with hypoxia (29). If FFA do accumulate in excess of binding capacity, the results of *in vitro* experiments showing that unbound FFA exert a detergent effect on cell membranes and mitochondrial respiration (30, 31) are relevant. Linoleate has a stronger detergent effect than octanoate but is more tightly bound to albumin (30, 31), so that these two factors operating in opposite directions may account for our essentially similar findings with these two FFA. Their acyl CoA derivatives will also have detergent properties (30). Thus, unbound FFA may exert a direct toxic effect on the normal processes of cellular metabolism. It is of interest that histological evidence of TG deposition and myocardial damage has been reported following FFA mobilization by catecholamines (32, 33) or, in geese, by glucagon (34). The intracellular accumulation of FFA may also reduce intracellular calcium and hydrogen ion concentrations and so alter enzyme reaction rates. Moderate alkalosis has been shown to enhance the rate of glycolysis (35), suggesting that acidosis may have the reverse effect.

Alternatively, FFA may interfere directly with excitation-contraction coupling. They may affect the normal function of the sarcolemma and sarcoplasmic reticulum, which together probably determine the amount of calcium released with each activation (36); they may bind a small fraction of the activator calcium ions; or they may further lower pH, and acidosis has been shown to decrease contractility independently of its effect on the rate of glycolysis. The hydrogen and calcium ion concentrations in the buffer were not measurably altered by FFA. In muscles depressed by anoxia, developed force could be augmented by raising the external calcium concentration or by paired stimulation (which is thought to act by increasing the amount of activator calcium delivered to the myofilaments), though only to about 50% of the control values obtained under oxygenated conditions. Under control conditions these interventions have only minimal effect (9), probably because rat muscles, with their characteristically negative force-frequency staircase, are already maximally activated at such low frequencies. The fact that contractility could be augmented to a greater extent, both relatively and absolutely, when depressed by anoxia implies that the myofilaments were submaximally activated during anoxia, whereas they were nearly maximally activated when well oxygenated. This suggests that anoxia reduces the amount of calcium delivered to the myofilaments, or that the active sites become less sensitive to calcium, and that the amount of energy available at the active sites, although reduced, is probably not the critical factor limiting contractility in anoxia. Unfortunately, the effect of calcium-mediated interventions in anoxia was not compared in muscles supplied with glucose and with FFA, to determine whether contractility could be augmented to the same level with either substrate present. It is pertinent at this point to note that the influence of FFA on contractility in hypoxic myocardium was studied only in rat-heart preparations, where the control of intracellular calcium kinetics probably differs, at least quantitatively, from that in other mammalian species. For example, caffeine, which is thought to act by altering sarcolemmal permeability to calcium and by decreasing the rate of calcium uptake by sarcoplasmic reticulum, depresses contractility in rat papillary muscles,whereas it augments contractility in cat muscles under comparable conditions (38). Thus, it is possible that the observed depressant effect of FFA, if mediated by a direct action on excitation-contraction coupling, may be peculiar to rat myocardium. The present evidence indicated that FFA exert their effect outside the mitochondria. But it is equally possible that they may act at the surface membrane of the cell. If this were so, it would be necessary to postulate that this action on the cell membrane only affected contractility when the cell was hypoxic. Both in rat and in cat papillary muscles, the time to peak isometric force (TPF) is shortened by hypoxia, and then exhibits a remarkable transient prolongation when the muscles are reoxygenated (fig. 8). In rat muscles, the extent of these changes appears to be related to the severity of oxygen deprivation and to the substrate provided, rather than to the degree of depression of contractility. The differences are most marked in the transient prolongation of TPF that follows reoxygenation: this is much greater following anoxia than hypoxia, and is reduced by the presence of FFA. The association of a negative inotropic effect with shortening of TPF during hypoxia is in marked contrast with the positive inotropic effect associated with shortening of TPF that follows calcium or catechol-mediated interventions. TPF is generally related to the duration of the action potential. It would be of interest to measure the effects of hypoxia and FFA on action potential.

Whatever the underlying mechanisms, it is clear that the results of these experiments could have important clinical implications if confirmed in man. The concentration of FFA found to depress contractility in hypoxic rat-heart preparations (FFA to albumin $> 2:1$) lies within the range of arterial levels found with FFA mobilization in man, although it is true that the binding capacity of blood is probably greater than that of albumin-containing buffer.

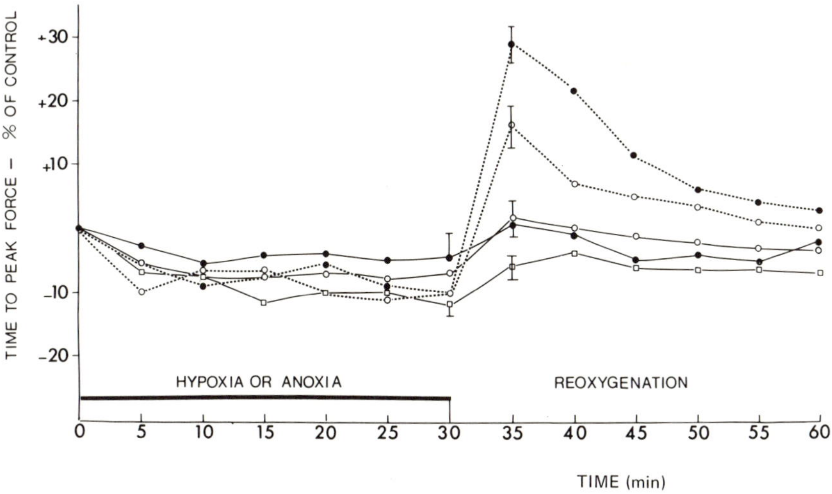

Fig. 8 As in figure 1, but showing time to peak force.

There are conflicting reports about the association between high circulating levels of FFA and the occurrence of arrhythmias and mortality following myocardial infarction (39–41). However, the experimental work of Oliver and his associates does suggest that high levels of FFA are causally associated with arrhythmias in ischemic dog hearts (42). No attempt was made to measure the effect of FFA on rhythm in the experiments described here, but it is certainly possible that arrhythmias and depression of contractility may both be related to a common underlying mechanism. Furthermore, the effect of raised levels of FFA cannot be considered in isolation from the complementary effects of glucose and insulin, about whose beneficial influence in the treatment of myocardial infarction controversy still exists (43–45), perhaps because such a regime has not yet received a proper trial in adequate dosage (46). The recent demonstration of inadequate insulin secretion following myocardial infarction (47, 48) completes the third side of a metabolic triangle. Clearly, the influence of high arterial concentrations of FFA on myocardial function in experimentally induced hypoxia and in patients with ischemic heart disease warrants further study.

SUMMARY

Free fatty acids (FFA) (linoleate, octanoate or, equally, the nonmetabolized FFA pent-4-enoate, in FFA to albumin molar ratios of 2 : 1 to 6 : 1) depressed contractility in hypoxic but not in well-oxygenated rat papillary muscle preparations. FFA also exaggerated the reversible rest contracture and associated increase in series elastic stiffness found with hypoxia. In perfused rat hearts (which were probably slightly hypoxic despite 15 ml O_2 supply/min/100g) beating isovolumically at constant volume, heart rate and coronary flow, contractility was depressed only by high concentrations of FFA (FFA to albumin, 6 :1);

myocardial O_2 consumption (MVO_2) was also decreased, implying that contractility or oxidative metabolism was depressed; but O_2 cost of mechanical work (integrated developed tension/min/total minus basal MVO_2) was greater, possibly because glycolysis was decreased. It is suggested that the contractile mechanisms and/or the supply of energy are influenced directly by FFA or their acyl CoA derivatives, possibly through their detergent effect. FFA concentrations reached *in vivo* are likely to be detrimental only in the presence of hypoxia.

REFERENCES

1. OPIE, L. H.: Metabolism of the heart in health and disease. Amer. Heart J. *76:* 685-698 (1968).
2. CHALLONER, D. R.: Metabolic effect of epinephrine on the QO_2 of the arrested isolated perfused rat heart. Nature *205:* 602-603 (1965).
3. CHALLONER, D. R.: Respiration in myocardium. Nature *217:* 78-79 (1968).
4. HENDERSON, A. H., MOST, A. S.; PARMLEY, W. W.; GORLIN, R. and SONNENBLICK, E. J.: Depression of myocardial contractility by free fatty acids during hypoxia. Circulat. Res. *26:* 439 (1970).
5. HENDERSON, A. H.; PARMLEY, W. W. and SONNENBLICK, E. H.: Series elasticity of heart muscle during hypoxia. Cardiovasc. Res. *5:* 10-14 (1971).
6. HENDERSON, A. H.; CRAIG, R. J.; GORLIN, R. and SONNENBLICK, E. H.: Free fatty acids and myocardial function in perfused rat hearts. Cardiovasc. Res. *4:* 466-472 (1970).
7. WEISSLER, A. M.; KRUGER, F. A.; BABA, N; SCARPELLI, D. G.; LEIGHTON, R. F. and GALLIMORE, J. K.: Role of anaerobic metabolism in the preservation of functional capacity and structure of anoxic myocardium. J. Clin. Invest. *47:* 403-416 (1968).
8. OPIE, L. H.: Coronary flow rate and perfusion pressure as determinants of mechanical function and oxidative metabolism of isolated perfused rat heart. J. Physiol. *180:* 529-541 (1965).
9. HENDERSON, A. H.; BRUTSAERT, D. L.; PARMLEY, W. W. and SONNENBLICK, E. H.: Myocardial mechanics in papillary muscles of the rat and cat. Amer. J. Physiol. *217:* 1273-1279 (1969).
10. HENDERSON, A. H.; CRAIG, R. J.; AMSTERDAM, E. A.; URSCHEL, C. W. and SONNENBLICK, E. H.: Species differences in contractile element performance: adaptation to natural heart rates. Circulation 37/38 (suppl) *6:* 98 (1968).
11. MARTINI, J. and HONIG, C. R.: Direct measurement of intercapillary distance in beating rat heart in situ under various conditions of oxygen supply. Microvasc. Res. *1:* 244 (1969).
12. CHALLONER, D. R. and STEINBERG, D.: Effect of free fatty acid on the oxygen consumption of perfused rat heart. Amer. J. Physiol. *210:* 280-286 (1966).
13. WILLEBRANDS, A. F. and VAN DER VEEN, K. J.: Influence of substrate on oxygen consumption of isolated perfused rat heart. Amer. J. Physiol. *212:* 1529-1535 (1967).
14. SEVEREID, L.; CONNOR, W. E. and LONG, J. P.: The depressant effect of fatty acids on the isolated rabbit heart. Proc. Soc. Exp. Biol. Med. *131:* 1239-1243 (1969).
15. GOODMAN, D. S.: The interaction of human serum albumin with long chain fatty acid anions. J. Amer. Chem. Soc. *80:* 3829-3898 (1958).
16. GRAHAM, T. P.; COVELL, J. W.; SONNENBLICK, E. J.; ROSS, J. and BRAUNWALD, E.: Control of myocardial oxygen consumption: relative influence of contractile state and tension development. J. Clin. Invest. *47:* 376-385 (1968).
17. WILLIAMSON, J. R.: Effects of insulin and diet on the metabolism of L(+)– lactate and glucose by the perfused rat heart. Biochem. J. *83:* 377-383 (1962).

18. HENDERSON, A. H.; CRAIG, R. J.; GORLIN, R. and SONNENBLICK, E. H: Lactate and pyruvate kinetics in isolated perfused rat hearts. Amer. J. Physiol. *217:* 1752-1756 (1969).

19. SHIPP, J. C.; OPIE, L. M. and CHALLONER, D. R.: Fatty acid and glucose metabolism in the perfused heart. Nature *189:* 1018-1019 (1961).

20. SHEUER, J.: Myocardial metabolism in cardiac hypoxia. Amer. J. Cardiol. *19:* 385-392 (1967).

21. NEWSHOLME, E. A.; RANDLE, P. J. and MANCHESTER, K. L.: Inhibition of the phosphofructokinase reaction in perfused rat heart by respiration of ketone bodies, fatty acids and pyruvate. Nature *193:* 270-271 (1962).

22. GARLAND, P. B.; RANDLE, P. J. and NEWSHOLME, E. A.: Citrate as an intermediary in the inhibition of phosphofructokinase in rat heart muscle by fatty acids, ketone bodies, pyruvate diabetes, and starvation. Nature *200:* 169-170 (1963).

23. SHEUER, J. and BRACHFELD, N.: Myocardial uptake and fractional distribution of palmitate-1 C14 by the ischemic dog heart. Metabolism *15:* 945-954 (1966).

24. BOWMAN, R. H.: Effects of diabetes, fatty acids, and ketone bodies on tricarboxylic acid cycle metabolism in the perfused rat heart. J. Biol. Chem. *241:* 3041-3048 (1966).

25. CORREDOR, C.; BRENDEL, K. and BRESSLER, R.: Studies of the mechanism of the hypoglycemic action of 4-pentenoic acid. Proc. Nat. Acad. Sco. *58:* 2299-2306 (1967).

26. RUDERMAN, N. B.; TOEUS, C. J.; LOWY, C.; VREELAND, J. and SHAFRIR, E.: The regulation of hepatic gluconeogenesis: inhibition of gluconeogenesis and fatty acid oxygenation by pent-4-enoic acid. Amer. J. Physiol. *219:* 51-57 (1970).

27. EVANS, J. R.: Importance of fatty acid in myocardial metabolism. Circulat. Res. 14/14 (Suppl. 2) 96-108 (1964).

28. BRACHFELD, N.: Maintenance of cell viability. Circulation 39/40 (Suppl 4) 101-219 (1969).(Amer. Heart. Assoc. Mon. 27. Bondurant, S: Research on acute myocardial infarction).

29. SPECTOR, A. A.; JOHN, K. and FLETCHER, J.: Binding of long-chain fatty acids to bovine serum albumin. J. Lipid. Res. *10:* 56-67 (1969).

30. PANDE, S. V. and MEADE, J. F.: Inhibition of enzyme activities by free fatty acids. J. Biol. Chem. *243:* 6180-6185 (1968).

31. BJORNTORP, P.; ELLS, H. A. and BRADFORD, R. J.: Albumin antagonism of fatty acid effects on oxidation and phosphorylation reactions in rat liver mitochondria. J. Biol. Chem. *239:* 339-344 (1964).

32. ROSENBLUM, I.; WOHL, A. and STEIN, A.: Studies in cardiac necrosis. 3. Metabolic effects of sympathomimetic amines producing cardiac lesions. Toxicol. Appl. Pharmacol. *7:* 344-351 (1965).

33. HOAK, J. C.; WARNER, E. D. and CONNOR, W. E.: New concept of levarterenol-induced acute myocardial necrosis. Arch. Pathol. *87:* 332-338 (1969).

34. HOAK, J. C.; CONNOR, W. E. and WARNER, E. D.: Toxic effects of glucagon-induced acute lipid mobilization in geese. J. Clin. Invest. *47:* 2701-2710 (1968).

35. SCHEUER, J. and BERRY, M. N.: Effect of alkalosis on glycolysis in the isolated rat heart. Amer. J. Phys. *213:* 1143-1148 (1967).

36. EBASHI, S. and ENDO, M.: in BUTLER, J. A. V. and NOBLE, D., Progress in Biophysics and Molecular Biology, pp. 123-182. (Pergamon Press, New York 1968).

37. WEISSLER, A. M.; GELET, T. R.; ALTSCHULD, R. A. and LEIGHTON, R. F.: The disparate effects of audosis on the performance and metabolism of the anoxic heart. J. Clin. Invest. *48i:* 87a-88a (1969).

38. HENDERSON, A. J.; BRUTSAERT, D. L.; FORMAN, R.; MOST, A. S.; PARMLEY, W. W. and SONNENBLICK, E. H.: Effects of caffeine on cardiac muscle. (in preparation).

39. OLIVER, M. F.; KURIEN, V. A. and GREENWOOD, T. W.: Relation between serum-free-fatty-acids and arrhythmias and death after acute myocardial infarction. Lancet *i:* 710-714 (1968).

40. GUPTA, D. K.; YOUNG, R., JEWITT, D. E.; HARTOG, M. and OPIE, L. H.: Increased plasma-free-fatty-acid concentrations and their significance in patients with acute myocardial infarction. Lancet *ii:* 1209-1213 (1969).

41. RUTENBERG, H. L.; PAMINTUAN, J. C. and SOLOFF, L. A.: Serum-free-fatty-acids and their relation to complications after acute myocardial infarction. Lancet *ii:* 559–564 (1969).

42. KURIEN, V. A.; YATES, P. A. and OLIVER, M. F.: Free fatty acids, heparin and arrhythmias during experimental myocardial infarction. Lancet *ii:* 185-187 (1969).

43. SODI-PALLARES, D.; TESTELLI, M. R.; FISHLEDER, B. L.; BISTENI, A.; MEDRANO, E. A.; FRIEDLAND, C. and DEMICHELI, A.: Effects of an intravenous infusion of a potassium-glucose-insulin solution on the electrocardiographic signs of myocardial infarction. A preliminary clinical report. Amer. J. Cardiol. *9:* 166-181 (1962).

44. MITTRA, B.: Potassium, glucose and insulin in treatment of myocardial infarction.Lancet *ii:* 607-609 (1965).

45. MEDICAL RESEARCH COUNCIL: Potassium, glucose and insulin treatment for acute myocardial infarction. Lancet *ii:* 1355–1360 (1968).

46. SODI-PALLARES, D.; DELEON, J. P.; BISTENI, A. and MEDRANO, G. A.: Potassium, glucose and insulin in myocardial infarction. Lancet *ii:* 1315-1316 (1969).

47. ALLISON, S..P.; PROWSE, K. and CHAMBERLAIN, M. J.: Failure of insulin response to glucose load during operation and after myocardial infarction. Lancet *ii:* 478-480 (1967).

48. LEBOVITZ, H. E.; SHULTZ, K. T.; MATTHEWS, M. E. and SCHEEK, R: Acute metabolic responses to myocardial infarction. Changes in glucose utilization and secretion of insulin and growth hormone. Circulation. *39:* 171–181 (1969).

Mailing address: Edmund H. Sonnenblick, M. D., Cardiovascular Unit, Peter Bent Brigham Hospital, 721 Huntington Ave., Boston, Mass. 02115 (USA).

M. R. C. Cardiovascular Research Unit, Department of Medicine,
Royal Postgraduate Medical School, London, England

CATECHOLAMINES AND MYOCARDIAL INFARCTION

R. C. YOUNG, M. THOMAS, M. E. EVANS, and L. H. OPIE

Catecholamines are known to be present in increased amounts in blood and urine after acute myocardial infarction in patients (6, 19). We have been interested in the possibility that these elevated catecholamine levels may have further effects on the myocardium already damaged by infarction.

We approached this experimentally, using a dog heart preparation in which small areas were made ischemic (12). The initial changes in metabolism have been described elsewhere (12) and include loss of lactate (8), phosphate, and potassium into the venous blood draining the ischemic area. By "loss" we mean the situation where the local venous concentration exceeds the arterial level, showing that the ischemic area is releasing the chemical species into the blood. These changes were found as soon as it was possible to take the first venous sample, and were maximal for about the first 30 min. After this initial period, catecholamines were given intravenously as adrenaline (table I) or isoprenaline. The most consistent finding was an exaggeration of the S-T segment elevation on the epicardial electrocardiogram (table I). A few minutes later, there were

Table I

Plasma Adrenaline Concentrations after Adrenaline Infusion in Dogs with Experimental Myocardial Ischemia

Dog no.	Adrenaline infused	Adrenaline measured: Femoral artery	Adrenaline measured: Coronary sinus	Effect on epicardial ECG
	μg/kg/min	*μg/l plasma*		
109	0.12	–	–	–
	0.32	1.6	0.4	S-T elevation
	0	0.2	0	–
	0.70	5.3	3.8	S-T normal deep T wave
114	0.11	–	–	Slight S-T elevation
	0.74–0.90	4.2	0.6	S-T elevation
	0.74–0.90	2.4	0.6	maintained
119	0	0.7	–	
	injected 40 μg into coronary sinus (≃1.4 μg/kg)	0.2	–	No S-T elevation

further losses of phosphate and potassium, suggesting that the "infarct" area had been given a further insult in some ways equivalent to the actual ligation of the artery. The dose of adrenaline required to cause S T segment elevation and metabolic changes was about 0.3–1.0 μg/kg/min, which compares with 0.5 μg/kg/min of adrenaline released spontaneously into the circulation shortly after coronary artery ligation in dogs. (3) Because the lactate loss from the ischemic area was sometimes reversed by exogenous adrenaline, it was not likely that the changes and the loss of potassium and phosphate after adrenaline administration were caused by increased hypoxia in the ischemic area.

Catecholamines are known to mobilize free fatty acids (FFA) (2), which may in certain circumstances be toxic to the heart (11). It seemed possible that FFA were at least contributing to the toxic effects of catecholamines. During catecholamine administration, there is an increased circulating FFA concentration and an increased arteriovenous difference of FFA across the ischemic heart. However, when fatty acids were infused (in the form of triglyceride-heparin combination), there were no comparable changes in the S-T segment of the epicardial ECG, even when conditions were such that the heart rate was similar to that found during the catecholamine infusions. Thus, the suggestion that catecholamines act via FFA alone is not borne out under our experimental conditions, although we cannot exclude the possible mobilization of endogenous triglyceride in the heart by catecholamines.

In the patient studies, we compared urinary catecholamine levels with venous blood FFA levels in two groups of patients. (7). Subjects in both groups had acute myocardial infarction, and were divided up into groups of "uncomplicated" and seriously ill cases. A much clearer distinction between these two groups was evident from a comparison of fasting morning-plasma FFA values than from catecholamine-excretion levels. The mean plasma FFA value day by day could be correlated with the mean adrenaline excretion in the urine day by day, in each group, but there was no good relationship in individual cases. Thus, there is not a simple relationship between venous plasma FFA concentrations and urinary catecholamine excretion levels. One reason for this is the multiplicity of factors that affect the plasma FFA release in patients; such factors include: a) the plasma insulin level, b) the serum growth-hormone level, c) the intravenous intake of dextrose, and possibly d) the plasma cortisol level. Clearly, the circulating FFA concentration in such patients is the end result of a variety of factors that include catecholamine and other hormonal disturbances, the degree of starvation of the patient, the extent of the patient's anxiety, and the severity of the clinical complications of the myocardial infarction.

At present, we are not confident that plasma FFA can be regarded as mediators of any toxic effects that endogenously secreted catecholamines may exert in patients with myocardial infarction. One method of investigating this is to depress the circulating FFA levels in patients with acute myocardial infarction. In our experience (7), the intravenous administration of glucose is an adequate means of approaching this problem and, in 8 patients, the FFA levels were significantly decreased following intravenous glucose administration. These 8 patients include 2 who were severely ill and subsequently died. We do not, however, anticipate that intravenous glucose would be effective in shocked patients in whom there are marked abnormalities of insulin secretion. Nevertheless, intravenous glucose was effective in 1 patient without altering the circulating plasma insulin levels. The problem with this method of controlling plasma FFA concentrations is, of course, that the FFA level in patients with acute myocardial infarction very probably rises immediately there is pain and anxiety at the start of an infarction. Therefore, it would seem likely that the plasma FFA rise could never be totally prevented.

The question remains as to whether catecholamines themselves may be toxic to the patient's heart in some way. The problem of catecholamine toxicity centers around the relevance of doses given to experimental animals and the circulating concentrations found in patients (table II). The relationship between the dose given, the circulating catecholamine concentration, the rate of uptake of catecholamine by the heart, and the ultimate metabolic and mechanical response of the heart requires clarification. We have started work in this direction by investigating the levels of circulating catecholamines associated with various rates of catecholamine infusion in dogs. Measurements of catecholamines in small amounts of blood plasma have been carried out, using a modification of the trihydroxyindole spectrofluorimetric method as established by Valori *et al.* (18). Table I shows that the circulating arterial adrenaline levels required to produce S-T segment elevation on the epicardial electrocardiogram are very approximately 10 times higher than values found spontaneously in dogs with experimental myocardial infarction (table II). From the work of others, it

Table II

Measured Plasma Catecholamine Concentrations in Acute Myocardial Infarction in Man and in Experimental Myocardial Ischemia in Dogs

	Noradrenaline	Adrenaline
A. Patients	*(μg/l plasma)*	
Gazes *et al.* (6)	5	1
McDonald *et al.* (10)	1	0.6
B. Dogs		
Richardson *et al.* (13)	10	0.2
Lammerant *et al.* (9)	noradrenaline + adrenaline = 2	
Young (20)	0.2	0.2
Ceremuzynski *et al.* (3)	adrenaline secretion = 0.5 μg/kg/min	

Table III

Effect of Catecholamine Administration on Myocardium in Experimental Preparations

Author	Dose*	Preparation	Effect
Evans (5)	1 mg *A*	Dog heart-lung	Oxygen-wasting (big effect)
Berne (1)	4–50 μg/min *A* and *NA* (intracoronary)	Dog heart	Oxygen-wasting (big effect)
Sayen *et al.* (15)	0.1–0.2 μg/kg/min *NA;* 0.5 μg/kg *A* (intravenous)	Dog, open chest, regional ischemia	Improved S-T change in ischemic area
Rona *et al.* (14)	2.5 mg/kg Isoprenaline (subcutaneous)	Dog	Necrosis
Sonnenblick *et al.* (17)	0.01–1 μg/kg/min *NA* (into pulmonary artery)	Dog, open chest	Increased V-max
Sobel *et al.* (16)	5 mg/kg *A* (subcutaneous)	Rat	Uncoupled respiration heart mitochondria
Challoner (4)	0.1 μg/ml *A* (above coronary artery)	Isolated rat heart	Increased basal qO_2, 37%

**A* = adrenaline; *NA* = noradrenaline.

appears that very small doses of catecholamines are associated with beneficial effects to the ischemic myocardium (15); it is difficult to compare these experimental doses with ours, but we found sustained S-T segment elevation in the ischemic area with doses in the neighborhood of 0.5μg adrenaline/kg/min, whereas Sayen *et al.* (15) found an improvement toward normal of the raised S-T segment with a single dose of 0.5 μg/kg of adrenaline. Thus, it seems that our "cardiotoxic" dose exceeds the "beneficial" dose of Sayen *et al.* With much larger doses of catecholamines, there may be "oxygen-wasting," uncoupled respiration, and necrosis of the myocardium (table III).

Because the local concentration of catecholamine operative at the mitochondrial level in acute myocardial ischemia is not known, it is at present not profitable to do more than speculate that in some cases of very severe myocardial infarction there may be a greater secretion of catecholamines than is appropriate for the physiological needs of the body.

SUMMARY

Catecholamines, usually in the form of adrenaline, were administered to dogs with acute experimental coronary artery ligation. In a dose of 0.3–0.9 μg/kg/min, the S-T segment of the epicardial electrocardiogram usually showed increased elevation, and there were further losses of potassium and phosphate ions into the local venous blood. This dose is similar to the reported rates of secretion of adrenaline in dogs soon after coronary artery ligation. In considering possible cardiotoxic effects of catecholamines, it is crucial to relate the findings observed to the circulating catecholamine concentration and to the uptake by the heart.

REFERENCES

1. BERNE, R. M.: Effect of epinephrine and norephinephrine on coronary circulation. Circulat. Res. *6:* 644–655 (1958).
2. CARLSON, L. A.: Fatty acid response to trauma, in: Acute myocardial infarction, JULIAN, D. G. and OLIVER, M. F. p. 243 (Livingstone, Edinburgh 1968).
3. CEREMUZYNSKI, L.; STASZEWSKA-BARCZAK, J. and HERBACZYNSKA-CEDRO, K.: Cardiac rhythm disturbances and the release of catecholamines after acute coronary occlusion in dogs. Cardiovasc. Res. *3:* 190–197 (1969).
4. CHALLONER, D. R.: Evidence of uncoupled respiration in thyrotoxic and epinephrine-stimulated myocardium. Amer. J. Physiol. *214:* 365–369 (1968).
5. EVANS, C. L..: The mechanism of cardiac acceleration by warmth and by adrenaline. J. Physiol. (London) *51:* 91–104 (1917).
6; GAZES, P. C.; RICHARDSON, J. A. and WOODS, E. F.: Plasma catecholamine concentrations in myocardial infarction and angina pectoris. Circulation *19:* 657–661 (1959).
7. GUPTA, D. K.; YOUNG, R. C.; JEWITT, D. E.; HARTOG, M. and OPIE, L. H.: Increased plasma free fatty acid concentrations and their significance in patients with acute myocardial infarction. Lancet *ii:* 1209–1213 (1969).
8. HUCKABEE, W. E.: Relationship of pyruvate and lactate during anaerobic metabolism. V. Coronary adequacy. Amer. J. Physiol. *200:* 1169–1176 (1961).
9. LAMMERANT, J.; DE HERDT, P. and DE SCHRYVER, C.: Direct release of myocardial catecholamines in the left heart chambers: the enhancing effect of acute coronary occlusion. Arch. Int. Pharmacodym *163:* 219–226 (1966).
10. MCDONALD, L.; BAKER, C.; BRAY, C.; MCDONALD, A. and RESTIEAUX, N.: Plasma catecholamines after cardiac infarction. Lancet *ii:* 1021–1023 (1969).
11. OLIVER, M. F.; KURIEN, V. A. and GREENWOOD, T. W.: Relation between serum free fatty acids and arrhythmias and death after acute myocardial infarction. Lancet *ii:* 710–715 (1968).
12. OWEN, P.; THOMAS, M.; YOUNG, V. and OPIE, L.: Comparison between metabolic changes in local venous and coronary sinus blood after acute experimental coronary arterial occlusion. Amer. J. Cardiol. *25:* 562–570 (1970).
13. RICHARDSON, J. A. WOODS, E. F. and BAGWELL, E. E.: Circulating epinephrine and norepinephrine in coronary occlusion. Amer. J. Cardiol. *5*: 613–618 (1960).
14. RONA, G.; KAHN, D. S.; and CHAPPEL, C. I.: Studies on infarct-like myocardial necrosis produced by isoproterenol: A review. Rev. Canad. Biol. *22*: 241–255 (1963).
15. SAYEN, J. J.; KATCHER, A. H.; SHELDON, W. F. and GILBERT, C. M.: The effect of levarterenol on polarographic myocardial oxygen, the epicardial electrocardiogram and contraction in non-ischemic dog hearts and experimental acute regional ischemia. Circulat. Res. *8:* 109–128 (1960).
16. SOBEL, B.; JEQUIER, E.; SJOERDSMA, A. and LOVENBERG, W.: Effect of catecholamines and adrenergic blocking agents on oxidative phosphorylation in rat heart mitochondria. Circulat. Res. *19:* 1050–1061 (1966).
17. SONNENBLICK, E. H.; ROSS, J.; COVELL, J. W.; KAISER, G. A. and BRAUNWALD, E.: Velocity of contraction as a determinant of myocardial oxygen consumption. Amer. J. Physiol. *209:* 919–927 (1965).
18. VALORI, C.; BRUNORI, C. A.; RENZINI, V. and COREA, L.: Improved procedure for formation of epinephrine and norepinephrine fluorophors by the trihydroxyindole reaction. Anal. Biochem. *33:* 158–167 (1970).

19. VALORI, C.; THOMAS, M. and SHILLINGFORD, J.: Free noradrenaline and adrenaline excretion in relation to clinical syndromes following myocardial infarction. Amer. J. Cardiol. *20:* 605–617 (1967).

20. YOUNG, R. C.: Unpublished observations.

Mailing address: Lionel H. Opie, M.D., Department of Medicine, University of Capetown and Groote Schuur Hospital (South Africa).

Biochemistry of Cardiac Hypertrophy

Department of Medicine of the Pritzker School of Medicine of the University of Chicago, Department of Biochemistry, and the Argonne Cancer Research Hospital[1]
Chicago, Illinois, USA

SYNTHESIS OF COLLAGEN, MYOSIN, NONCOLLAGENOUS PROTEIN, AND DNA DURING EXPERIMENTAL MYOCARDIAL HYPERTROPHY IN THE RAT[2]

J. L. SKOSEY,[3] V. ASCHENBRENNER,
R. ZAK, and M. RABINOWITZ[4]

A variety of measures are available for the production of experimental cardiac hypertrophy. During and following the development of cardiac hypertrophy, a number of changes have been observed in the composition and metabolism of rodent hearts (2). Workers in a number of laboratories have reported alterations in ventricular DNA. Grove *et al.* (4), for example, noted an increase in the total DNA of the left ventricle following the induction of cardiac hypertrophy in rats by constriction of the thoracic aorta. The DNA increased in proportion to ventricular weight, so that there was no change in the DNA concentration with ventricular hypertrophy. The results of histological studies indicated that the increase in DNA resulted from proliferation of connective tissue cells. No mitotic figures were observed in muscle cells, and increased polyploidy of muscle cells following aortic banding could account for only a minute fraction of the increase. The mitotic indices of connective tissue cells increased in hearts following banding to values ten times greater than control, however. In addition, as shown by Grove *et al.* (5), autoradiography after labeling with tritiated thymidine demonstrated that new DNA synthesis occurred almost exclusively in connective tissue cells.

Collagen is believed to be synthesized only by connective tissue cells. We have undertaken studies of the metabolism of collagen in correlation with cardiac hypertrophy as an index of the changes in the activity of connective tissue cells. Changes in the synthesis of noncollagenous protein (NCP), myosin, and DNA

---

[1] Operated by the University of Chicago for the United States Atomic Energy Commission.

[2] Supported in part by United States Public Health Service Grants HE-09172 and HE-04442 from the National Heart Institute, by the Myocardial Infarction Unit Grant PH-4368133, and by grants from the Chicago and Illinois Heart Association and the E. F. Andrews Foundation.

[3] Arthritis Foundation Postdoctoral Fellow.

[4] Recipient of a U. S. Public Health Service Career Development Award.

have also been determined and compared with estimates of collagen synthesis during hypertrophy.

Some of the properties of collagen facilitate its study. Collagen is, for practical purposes, the only protein in mammalian tissue that contains the imino acid hydroxyproline (hypro). Hypro is derived from proline, which is hydroxylated after it is incorporated into protocollagen, the polypeptide precursor of the collagen peptide chain. Collagen can be extracted quantitatively from tissues into hot trichloroacetic acid (TCA) (3). Thus, heart collagen content and radioactivity were estimated by measuring the content and radioactivity of hypro in hydrolysates of hot TCA extracts of heart (9).

We have confirmed the finding of Grove *et al.* (5) and Bartosova *et al.* (1). The collagen content of the heart does not change during the early phases of hypertrophy induced by aortic banding and, consequently, the concentration falls. Later, the heart collagen content and concentration both increase (fig. 1).

In a series of preliminary experiments, we measured the incorporation of radiolabeled proline into collagen hypro at times ranging from 2 to 30 days after induction of cardiac hypertrophy by aortic banding of 205-210 g female Sprague-Dawley rats (7). Radiolabeled proline was injected intraperitoneally; 3 h later, heart collagen was extracted into hot TCA and its radioactive hypro content measured. Incorporation of radioactivity into NCP (the residue remaining after TCA extraction) was also measured (fig. 2). Collagen hypro radioactivity was increased at each time studied. Enhanced labeling of heart

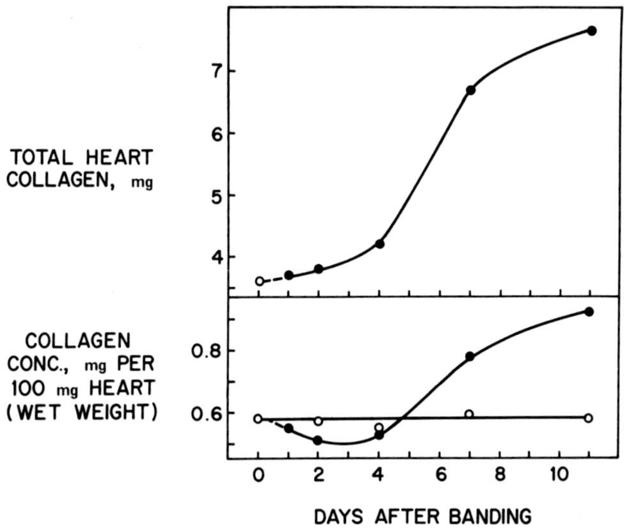

Fig. 1 Effect of banding upon total heart collagen and collagen concentration. Open circles represent means from sham-operated animals; closed circles means from banded animals.

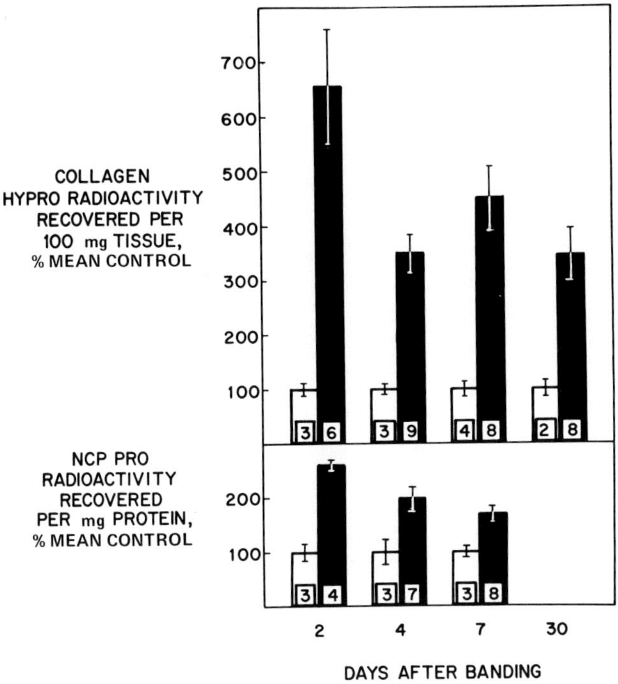

Fig. 2 Effect of banding upon recovery of radioactive proline as collagen hypro and in NCP. Animals were each injected with 10 μCi proline-U-^{14}C on day 7 after banding, and with 50 μCi proline-2,3-^{3}H on all other days. The height of the bars represents mean radioactivity recovered in hearts from sham-operated (white bars) and banded (black bars) animals. The lengths of the vertical lines represent 2 S.E. of the means. The number in each group is given at the base of the bars. Each difference is significant by Student's t test, with a p value of 0.01 or less.

components following aortic constriction could result from an increased delivery of radioactive precursor to the tissue. An increased specific activity of precursor pool and, subsequently, product would result. The percentage increase in labeling of collagen hypro exceeded the increase in labeling of NCP (fig. 2). Thus, it is unlikely that enhanced incorporation of radioactivity into collagen hypro resulted solely from an increase in the specific activity of the precursor pool.

The increase in heart weight resulting from aortic banding varies within a given banding procedure, and between experiments. The extent of collagen labeling varied with heart weight (fig. 3). We wished to study the time course of labeling of collagen and NCP. In order to minimize the variations in the hypertrophic response, a large group of animals was banded on a single day. At varying times after banding, animals were injected with tritiated proline as before. Incorporation into collagen and NCP was measured. A peak of collagen

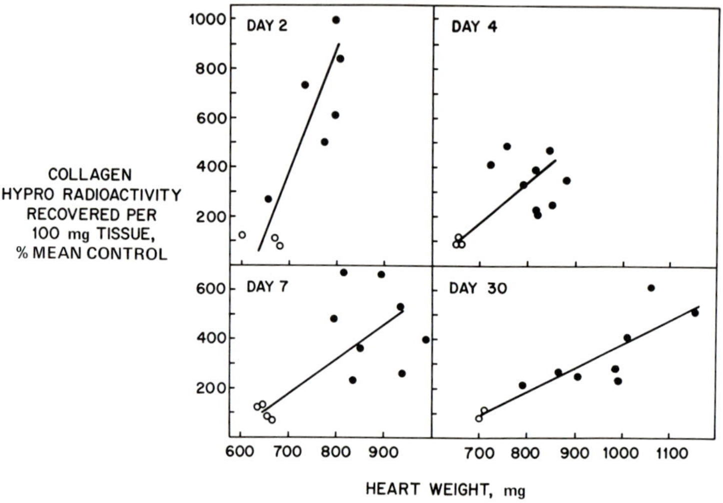

Fig. 3 Correlation of heart weight with collagen synthesis. Individual data from the experiments in figure 2 are plotted. Open circles represent observations in sham-operated animals; closed circles, banded animals.

labeling occurred 4 days after banding (fig. 4). NCP labeling was maximal on about the 2nd day, and remained elevated.

We also explored differences between the metabolism of muscle and connective tissue cells. A group of animals was banded; 2, 4, and 7 days later, members of the group were injected with a mixture containing tritiated thymidine (to measure nuclear DNA labeling), ^{14}C-lysine (to measure the labeling of myosin[5] and noncollagenous protein), and tritiated proline (to measure collagen labeling). The hearts were removed, weighed, and pooled according to whether they exhibited gross hypertrophy, moderate hypertrophy, or were from sham-operated animals. A peak of DNA labeling occurred in both groups of experimental animals 1 week after banding (fig. 5). The curves of myosin and noncollagenous protein synthesis were very similar to each other, reaching a plateau and remaining elevated. Collagen synthesis by the moderately hypertrophied hearts reached a peak on the 4th day after banding, while, in the grossly hypertrophied hearts, it reached a maximum on or before the 2nd day after banding. The time course of labeling of myosin, which is formed in the muscle cells, is distinctly different from that of the labeling of nuclear DNA and collagen, which are formed by the connective tissue cells. Also, changes in collagen labeling by the connective tissue cells in response to banding occur with a different time course than the changes in labeling of DNA.

[5] Myosin was extracted by a procedure to be published elsewhere.

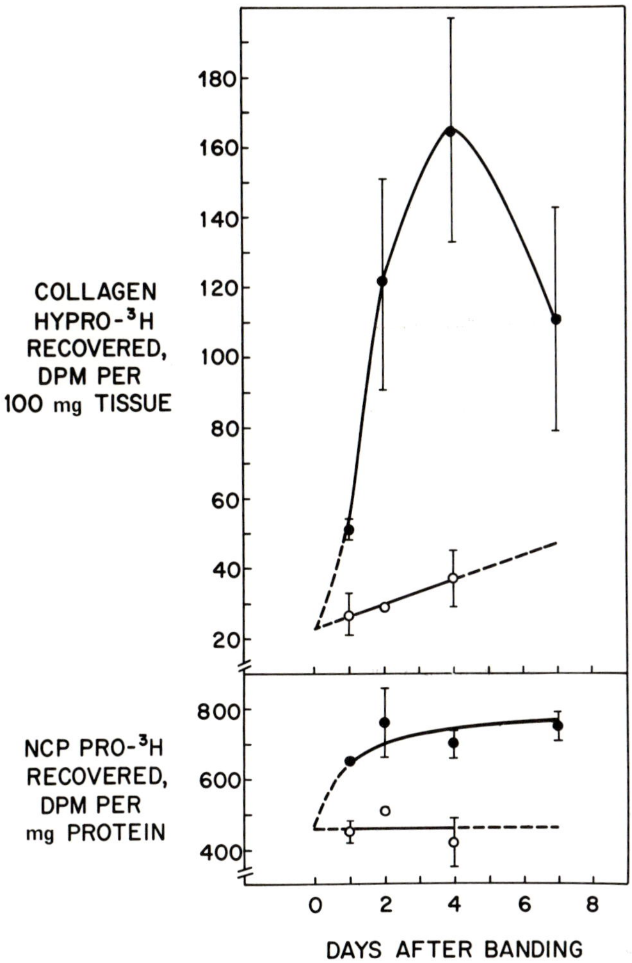

Fig. 4 Time course of collagen and NCP synthesis after banding. The mean radioactivity recovered in sham-operated (open circles) and banded (closed circles) animals is plotted. The lengths of the vertical lines represent 2 S.E. of the means. Where comparisons between banded and sham-operated animals are possible, the differences are significant by Student's t test, with a p value of 0.02 or less.

We can only speculate about the effects of an increased collagen concentration in the hypertrophied heart. It has been suggested that the collagen of blood vessels prevents excessive distension when pressure in them is increased(6). Collagen in the heart might similarly prevent excessive distension, which would result in decreased efficiency of contraction processes (cf. 10).

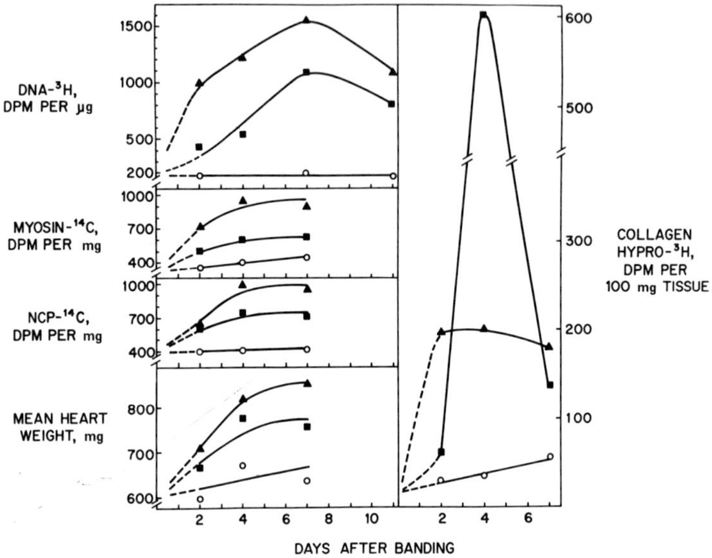

Fig. 5 Time course of synthesis of collagen, myosin, NCP, and DNA, following aortic banding. Animals were injected with a mixture of ^{3}H-proline, ^{14}C-lysine, and ^{3}H-thymidine 2, 4, and 7 days after banding. Hearts were weighed and pooled according to whether they exhibited gross hypertrophy (solid triangles), moderate hypertrophy (solid squares) or no hypertrophy (sham-operated, open circles). The result from each pool of 2–3 hearts is plotted.

SUMMARY

Using appropriate precursors, we have measured the labeling of collagen hydroxyproline (hypro), noncollagenous protein (NCP), myosin, and DNA in rat hearts undergoing hypertrophy as the result of surgically-induced aortic stenosis. A peak of incorporation of radiolabeled proline into collagen hypro occurred at 4 days after aortic banding. The peak of DNA labeling (with ^{3}H-thymidine) occurred at 7 days. Labeling of myosin and NCP was maximal on day 4 and remained elevated. The percentage increase in collagen hypro labeling exceeded the increase in labeling of NCP and myosin. The results indicate that muscle and connective tissue cells are stimulated differently following the induction of cardiac hypertrophy.

REFERENCES

1. BARTOSOVA, D.; CHVAPIL, M.; KORECKY, B.; POUPA, O.; RAKUSAN, K.; TUREK, Z. and VIZEK, M.: The growth of the muscular and collagenous parts of the rat heart in various forms of cardiomegaly. J. Physiol., Lond. *200:* 285–295 (1969).
2. FANBURG, B. L.: Experimental cardiac hypertrophy. New Engl. J. Med. *282:* 723–732 (1970).
3. FITCH, S. M.; HARKNESS, M. L. R. and HARKNESS, R. D.: Extraction of collagen from tissues. Nature, Lond. *176:* 163 (1955).
4. GROVE, D.; NAIR, K. G. and ZAK, R.: Biochemical correlates of cardiac hypertrophy. III. Changes in DNA content; the relative contributions of polyploidy and mitotic activity. Circulat. Res. *25:* 463–471 (1969).

5. GROVE, D.; ZAK, R.; NAIR, K. G. and ASCHENBRENNER, V.: Biochemical correlates of cardiac hypertrophy. IV. Observations on the cellular organization of growth during myocardial hypertrophy in the rat. Circulat. Res. *25:* 473–485 (1969).

6. HARKNESS, R. D.: Mechanical properties of collagenous tissues, in: GOULD, B. S., Treatise on Collagen, Vol. 2, Biology of Collagen, Part A, pp. 247–310 (Academic Press, New York 1968).

7. NAIR, K. G.; CUTILLETTA, A. F.; ZAK, R.; KOIDE, T. and RABINOWITZ, M.: Biochemical correlates of cardiac hypertrophy. I. Experimental model; changes in heart weight, RNA content, and nuclear RNA polymerase activity. Circulat. Res. *23:* 451–462 (1968).

8. PROCKOP, D. J. and KIVIRIKKO, K. I.: Hydroxyproline and the metabolism of collagen, in: GOULD, B. S., Treatise on Collagen, Vol. 2, Biology of Collagen, Part A, pp. 215–246 (Academic Press, New York 1968).

9. PROCKOP, D. J. and UDENFRIEND, S.: A specific method for the analysis of hydroxyproline in tissue and urine. Analyt. Biochem. *1:* 228–239 (1960).

10. SONNENBLICK, E. H.: Myocardial ultrastructure in the normal and failing heart. Hospital Practice *5:* 35–43 (1970).

Mailing address: J. L. Skosey, M.D., Ph.D., Department of Medicine, Pritzker School of Medicine, University of Chicago, 950 East 59th Street, Chicago, Illinois 60637 (USA).

Departments of Medicine and Biochemistry, The University of Chicago, and Argonne Cancer Research Hospital, [1] Chicago, Illinois, USA

INCREASED TURNOVER OF MITOCHONDRIAL CONSTITUENTS IN CARDIAC HYPERTROPHY AND ACUTE HYPOXIA IN THE RAT [2]

V. ASCHENBRENNER, R. ALBIN, R. ZAK, K. G. NAIR, and M. RABINOWITZ

Under normal conditions, cardiac muscle derives its energy almost exclusively from oxidative metabolism localized in mitochondria. Since the mitochondrion is the principle site of myocardial ATP synthesis, changes in mitochondrial functional mass may have important consequences to the metabolism of the rest of the cell. We now know that mitochondria are semiautonomous organelles that have independent systems for the synthesis of nucleic acids and protein (4, 7, 9, 10). Heart mitochondrial components turn over rapidly with a half-life of about 5–6 days (2), indicating a capacity for rapid synthesis. In these experiments, we sought to study the effects of experimentally-produced cardiac hypertrophy and of severe hypoxia on the synthesis and turnover of mitochondria and of other cardiac muscle components. Mitochondria are particularly suitable for these studies, since they are predominantly (over 95%) localized in the cardiac *muscle* cell in whose function we are most interested. Furthermore, as the tissue receptor for oxygen, mitochondria may be particularly sensitive to changes in tissue oxygen tension or to levels of oxygen consumption.

Experimental cardiac hypertrophy was produced by banding the ascending aorta in rats (6). As seen in figure 1, a 20–40% increase in heart weight is achieved within 36–48 h after aortic constriction.

Changes in mitochondrial mass after aortic constriction were estimated from the activity of mitochondrial respiratory enzymes in homogenate and mitochondria. As seen in figure 2, mitochondrial mass appears to increase proportionally greater than total myocardial proteins 1 day after aortic banding, as indicated by the increase in homogenate specific activities of cytochrome oxidase, succinate cytochrome *c* reductase, and succinic dehydrogenase. Total cardiac activities of these enzymes increase to a greater extent than does total cardiac protein

---

[1] Operated by the University of Chicago for the United States Atomic Energy Commission.

[2] Supported in part by Myocardial Infarction Research Unit PH 4368133 and by Grants HE 9172 and HE 4442 from National Heart Institute, National Institutes of Health, Bethesda, and RN–69–15 from Chicago Illinois Heart Association.

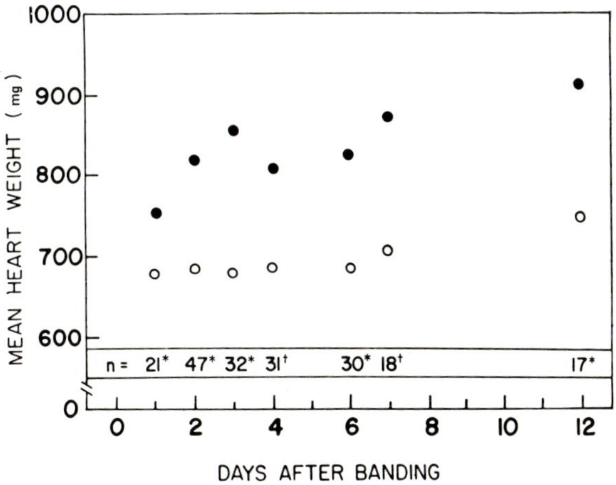

Fig. 1 Changes in heart weight after banding the ascending aorta of rats. From Nair *et al.* (6).

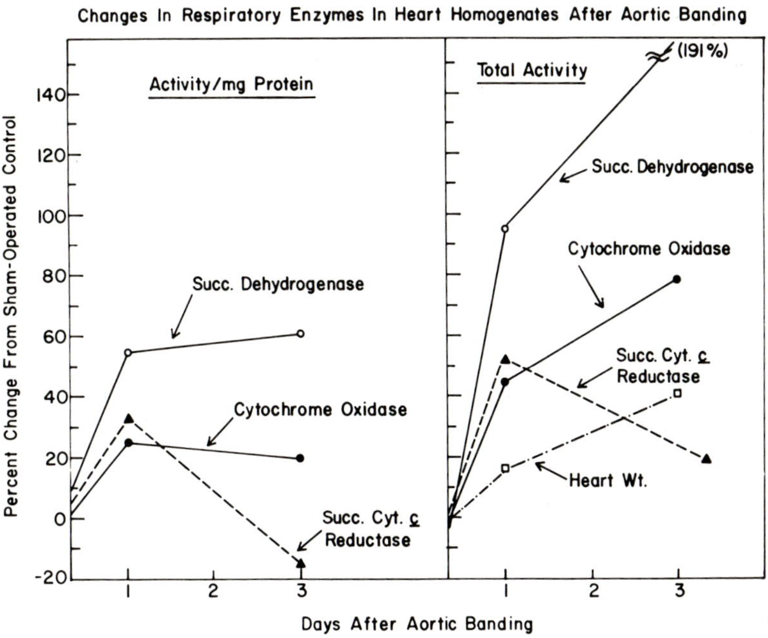

Fig. 2 Effect of banding the ascending aorta of rats on the cardiac respiratory enzyme activities. From Rabinowitz *et al.* (8).

content. Mitochondrial enzymes therefore appear to be synthesized more rapidly than other protein components in early hypertrophy. This preferential synthesis of mitochondrial proteins is less apparent 3 days after banding.

Preferential synthesis of mitochondrial proteins in early cardiac hypertrophy is also indicated by the relative incorporation of ^{3}H-leucine into various myocardial subcellular fractions. Table I shows that 24 h after aortic banding, the increment in incorporation of ^{3}H-leucine into mitochondrial proteins was greater than that into myofibrillar or supernatant proteins. It may be concluded that mitochondrial proteins are rapidly synthesized during the acute phase of cardiac hypertrophy.

To gain insight into the dynamics of synthesis and degradation of mitochondrial proteins, we studied the effects of two experimental procedures: (a) cardiac hypertrophy produced by aortic constriction; and (b) exposure to severe hypoxia, on the turnover of mitochondrial proteins. We will first discuss the effect of hypoxia and will later return to cardiac hypertrophy.

The rationale of the experiments to be described was to label mitochondrial protein or a specific mitochondrial cytochrome, several days before the experimental intervention. The effects of the experimental procedure on the specific radioactivity compared with controls reflects increased synthesis of new unlabeled material. Changes in the rate of destruction are measured by changes in the total cardiac radioactivity of that component.

In the experiment shown in figure 3, mitochondrial proteins were prelabeled with ^{14}C-guanido-arginine. The rapid normal turnover of mitochondrial proteins is apparent. After exposure of rats to a 6 h period of 4–5% hypoxia, the specific activity of mitochondrial proteins was 15–20% less than controls. This decrease is indicative of synthesis of new unlabeled mitochondrial proteins, presumably secondary to increased destruction produced by hypoxia. The slope of the decay curves for hypoxic and control animals is substantially the same, indicating that mitochondria that escape destruction during the hypoxic period are not more susceptible to destruction later.

These conclusions are confirmed by experiments using cytochrome *c* labeled

Table I

Relative Incorporation of ^{14}C-amino Acids into Protein of Subcellular Fractions of Hearts of Banded and Control Animals

Fraction	Sham	Banded	Ratio: $\frac{\text{Band}}{\text{Sham}} \times 100$
Homogenate	1.0	1.0	
Mitochondria	0.95	1.24	130.5
Myofibrils	0.97	0.90	93.0
Supernatant	1.28	1.35	105.5

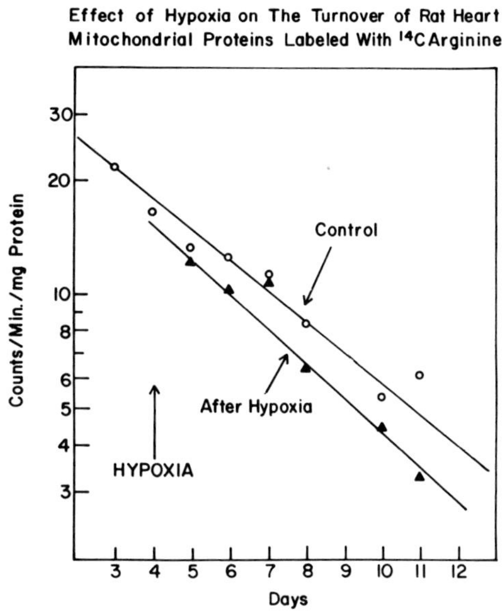

Fig. 3. Effect of a 6 h period of 4–5% O_2 on the specific activities of mitochondrial protein previously labeled with ^{14}C-guanido-arginine. From Rabinowitz *et al.* (8) and Aschenbrenner *et al.* (3).

with ^{3}H-δ-aminolevulinic acid, which is a specific heme precursor and is a non-reutilizable precursor. Accurate measurements of total cytochrome *c* content can also be made, which facilitate interpretation of the results. Cytochrome *c* is known to be bound to the inner mitochondrial membrane, and to turn over at the same rate as the particulate cytochromes *b* and aa_3 (2,5,6).

The specific activity of cytochrome *c* (fig. 4) and also of isolated purified heme aa_3 decreased by 15–30% after exposure to low oxygen atmosphere. No dilution was present at the end of the 6 h hypoxic period, but it was present 6–12 h after return to a normal atmosphere. This indicates that mitochondria may be destroyed during hypoxia, but they are resynthesized only after return to normal oxygen atmosphere. Also seen in figure 4 is that hypoxia causes no decrease in myosin specific radioactivity. This indicates that mitochondria are more sensitive to the effects of hypoxia than is myosin.

Hypoxia leads to a significant decrease in cytochrome *c* content by the end of the 6 h hypoxia period (fig. 5). The level returns to normal 6 h later, indicating a brisk resynthesis. Thus, it seems clear that hypoxia results in a destruction of mitochondrial components that are rapidly synthesized after termination of the hypoxia.

That mitochondrial resynthesis occurs predominantly after return to normal oxygen tension is confirmed by experiments measuring the incorporation of

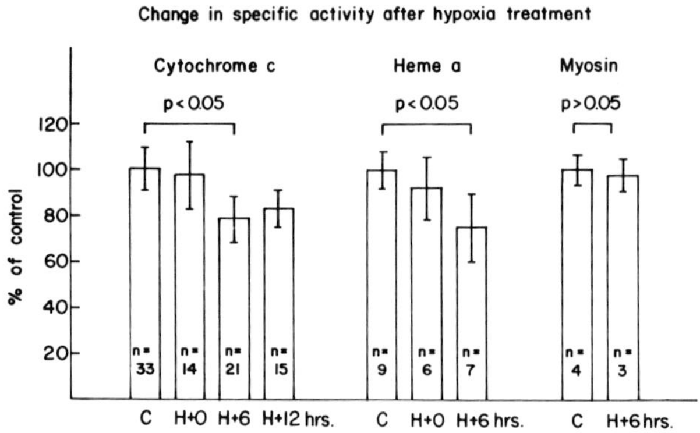

Fig. 4 Effect of a 6 h period of 4–5% O_2 on the specific activities of cytochrome *c*, and heme *a* prelabeled with ^{3}H-δ-aminolevulinic acid, and myosin prelabeled with ^{3}H-leucine. C=control; H+O=end of 6 h of hypoxia; H+6=6 h after return to normal O_2 tensions. From Aschenbrenner *et al.* (3).

^{3}H-leucine into mitochondrial protein during and after hypoxia. Incorporation was at control levels during hypoxia, but increased 200–300% 2 to 3 h after return to normal oxygen tensions (fig. 6). The level of incorporation returns to normal by 24 h.

The effect of aortic constriction in cytochrome *c* turnover was also studied. As seen in figure 7, cytochrome *c* content increases by about 10% after aortic banding. After correction is made for the increase in cytochrome *c* content, a 15–20% dilution of cytochrome *c* specific activity is present. The decrease in specific radioactivity, and in total cardiac cytochrome *c* radioactivity compared

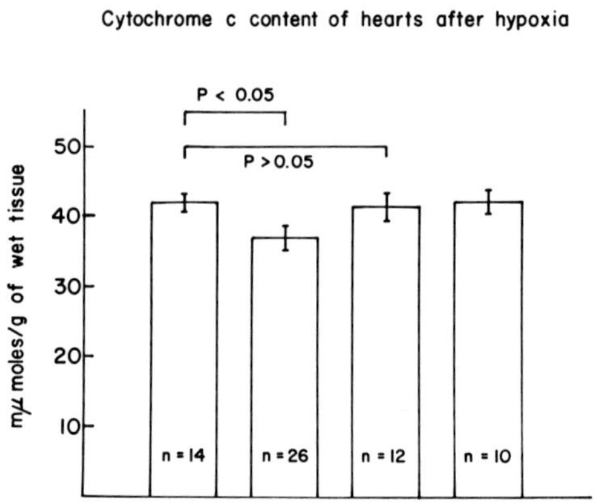

Fig. 5 Cardiac cytochrome *c* content after a 6 h period of hypoxia, and 12 and 18 h after return to a normal O_2 atmosphere. From Aschenbrenner *et al.* (3).

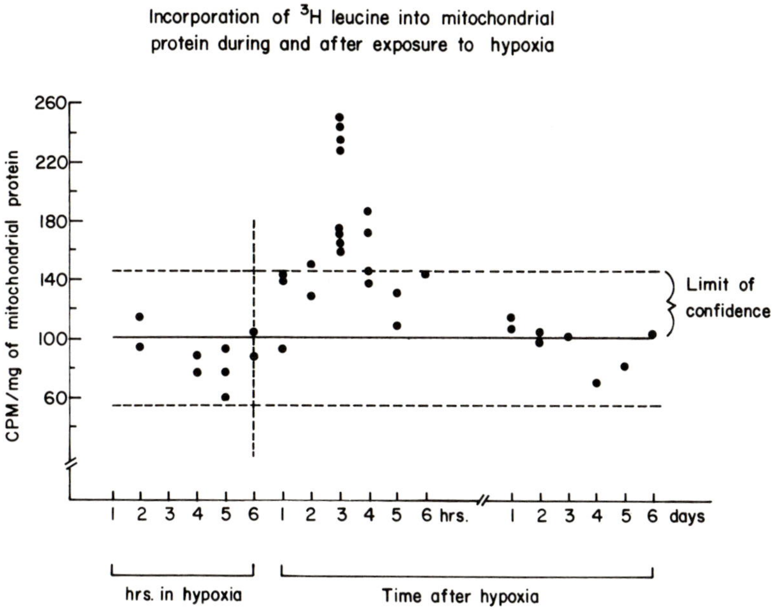

Fig. 6 Incorporation of 1 h pulses of ^{3}H-leucine into mitochondrial protein during the 6 h period of 4–5% O_2 and after return to a normal atmosphere. From Aschenbrenner *et al.* (3).

with controls, indicates an increased turnover immediately after banding. The observed decrease in corrected specific radioactivity was unchanged 3 days after banding, indicating that the rate of destruction had returned toward normal.

From these data and from the turnover rate of cytochrome *c*, estimations of the rate of destruction and synthesis of cytochrome *c* can be made. It appears that rate of destruction has increased considerably during the first 24 h after banding. This increased rate of destruction is compensated by a three-fold increase in synthesis rate. The rate of destruction and synthesis have returned toward normal 3 days after banding.

SUMMARY

Mitochondrial mass increases proportionately greater than other cardiac muscle proteins, 24 h after production of cardiac hypertrophy.

There is an increased rate of mitochondrial destruction 24 h after aortic banding, which returns toward normal thereafter.

Severe hypoxia leads to mitochondrial destruction.

Mitochondrial components are rapidly resynthesized after return to a normal oxygen tension.

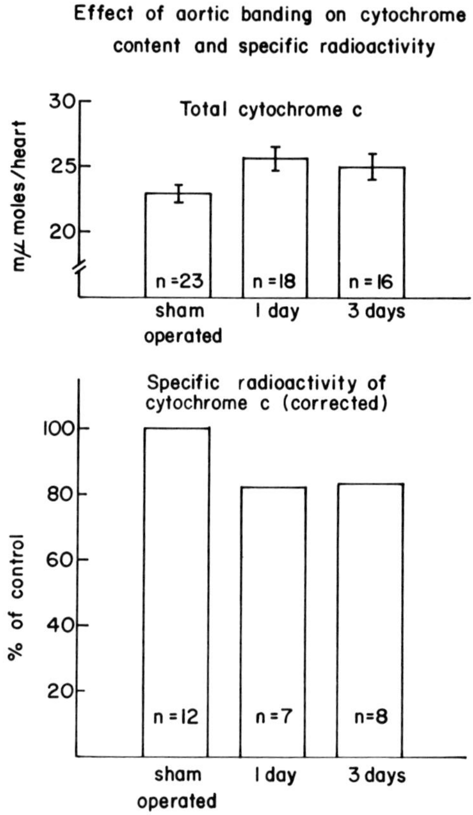

Fig. 7 Effect of banding of the ascending aorta of rats on total cardiac cytochrome *c* content, and on cytochrome *c* specific radioactivity, corrected for changes in cytochrome content. Cytochrome *c* was prelabeled with ^{3}H-δ-aminolevulinic acid 3 days prior to aortic banding. From Albin and Rabinowitz (1).

REFERENCES

1. ALBIN, R. and RABINOWITZ, M.: Unpublished observations.
2. ASCHENBRENNER, V.; DRUYAN, R.; ALBIN, R. and RABINOWITZ, M.: Heme *a*, cytochrome *c* and total protein turnover in mitochondria from rat heart liver. Biochem. J. *119:* 157–160 (1970).
3. ASCHENBRENNER, V.; ZAK, R. and RABINOWITZ, M.: The effect of hypoxia on mitochondrial turnover in rat heart. Am. J. Physiol. (in press).
4. BORST, P. and KROON, A. M.: Mitochondrial DNA: physicochemical properties, replication, and genetic function, in: BOURNE, G. H. and DANIELLI, J. F., International Review of Cytology 26, pp. 108–190. (Academic Press, New York 1969).
5. DRUYAN, R.; DEBERNARD, B. and RABINOWITZ, M.: Turnover of cytochromes labeled with δ-aminolevulinic acid-^{3}H in rat liver. J. Biol. Chem. *244:* 5874–5878 (1969).

6. NAIR, K. G.; CUTILLETTA, A. F.; ZAK, R.; KOIDE, T. and RABINOWITZ, M.: Biochemical correlates of cardiac hypertrophy. 1. Experimental model; changes in heart weight, RNA content, and nuclear RNA polymerase activity. Circulat. Res. *23:* 451–462 (1968).
7. NASS, M. M. K.: Mitochondrial DNA: advances, problems, and goals. Science *165:* 25–35 (1969).
8. RABINOWITZ, M.; ASCHENBRENNER, V.; ALBIN, R.; GROSS, N. J.; ZAK, R. and NAIR, K. G.: Synthesis and turnover of heart mitochondria in normal, hypertrophied and hypoxic rat. in: ALPERT, N., American Heart Association symposium on cardiac hypertrophy, (Academic Press, New York [in press] 1970).
9. RABINOWITZ, M. and SWIFT, H.: Mitochondrial nucleic acids and their relation to the biogenesis of mitochondria. Physiol. Rev. *50:* 376–427 (1970).
10. ROODYN, D. B. and WILKIE, D.: The biogenesis of mitochondria (Methuen and Co. Ltd., London 1968).

Mailing address: Dr. M. Rabinowitz, Department of Medicine and Biochemistry, University of Chicago, 950 East 59th Street, Chicago, Illinois 60637 (USA).

Institute of Circulation Research, German Academy of Sciences,
Berlin-Buch, German Democratic Republic

INCORPORATION OF AMINO ACID INTO TOTAL AND CONTRACTILE PROTEIN OF TISSUE SLICES OF THE HYPERTROPHYING RAT HEART

A. WOLLENBERGER and K. ONNEN

Measurements of the incorporation *in vivo* of radioactive amino acids into the protein of chronically overloaded, hypertrophying hearts have yielded results that were interpreted as indicative of elevated rates of protein synthesis (2, 4). Since amino acid incorporation into the heart depends on the coronary flow and may also be a function of acute cardiac work load (5), we sought a way of studying the relationship between myocardial hypertrophy and amino acid incorporation in the absence of an influence of these variables. Tissue slices prepared from the heart at different stages of cardiac hypertrophy were selected for this purpose.

MATERIALS AND METHODS

Left ventricular hypertrophy was induced in 5- to 6-month-old Wistar rats by constricting the abdominal aorta with a wire spiral (3). Sham-operated and non-operated rats served as controls. The animals were killed on the 2nd, 4th, and 7th day after the operation. The ratio of left ventricular to body weight had increased by 40% at the end of the 7th day, the greatest increase being noted during the first 2 days. Slices of the left ventricle, 0.6 mm in thickness, were incubated under nearly sterile conditions at 37°C in approximately 10 volumes of Krebs-Ringer- bicarbonate solution equilibrated with 95% O_2 and 5% CO_2 and containing glucose and ^{14}C-L-leucine. Total and contractile protein (actomyosin) were isolated for determination of specific radioactivities. The actomyosin was extracted as usual with buffered 0.6 M KCl solution from tissue powder prepared in the presence of liquid nitrogen. It was purified by four-fold precipitation.

RESULTS AND DISCUSSION

The rates of ^{14}C-leucine incorporation into total and contractile protein during incubation were found to be constant for at least 2 h. They were decreased by addition of a mixture of nonradioactive L-amino acids (without leucine) in concentrations corresponding to those present in rat serum. Abolition of respiratory energy production by cyanide and interference with the translation process by puromycin caused an almost complete inhibition of

^{14}C-leucine incorporation. It appears, therefore, that we were dealing with true protein synthesis and not with an energy-independent, nonspecific adsorption or inclusion of radioactive label.

Figure 1 shows that ^{14}C-leucine is incorporated more slowly into the contractile protein of cardiac slices than into the total protein. It is also seen from this figure that the incorporation of the amino acid, especially into the contractile protein, proceeds at a faster pace in slices from the hypertrophying heart than it does in normal cardiac muscle slices. The results obtained at the various times following aortic constriction are summarized in figure 2. The values for the sham-operated amimals do not significantly differ from those of the non-operated rats. By the end of the 2nd day following aortic constriction, incorporation of ^{14}C-leucine into total cardiac tissue protein has risen more than two-fold, incorporation into contractile protein approximately five-fold. The rates of incorporation remain elevated up to the 7th day, when the gain in cardiac mass has come to an end. In this respect the cardiac slices behave differently from isolated heart mitochondria, which in a parallel series of experiments exhibited increased amino acid incorporating ability only during, and not after, completion of compensatory cardiac growth (6).

It is not possible to decide from our data whether the observed increase in the rate of amino acid incorporation into the slices of the hypertrophying and hypertrophied hearts represents an enhancement of the protein synthetic ability of the tissue. The increased incorporation could be due, at least in part, to accelerated entry of labelled amino acid into the cells, such as it has been

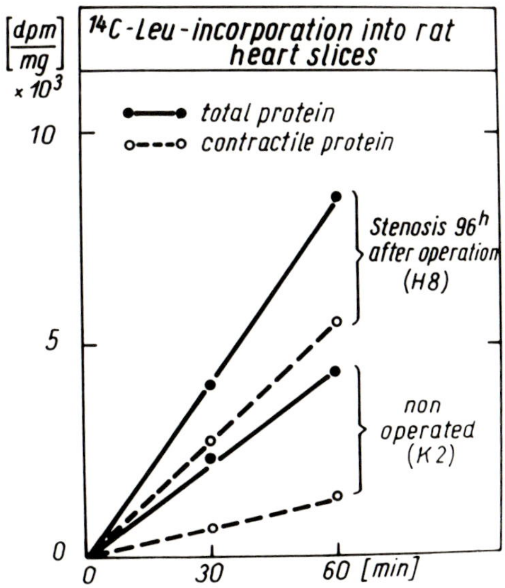

Fig. 1 ^{14}C-leucine incorporation into rat heart slices.

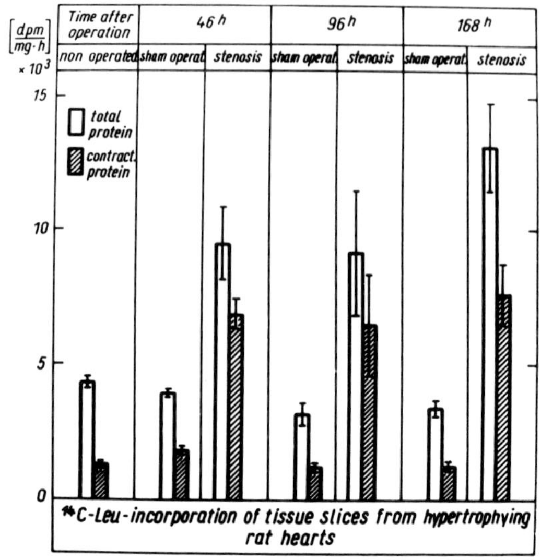

Fig. 2 ^{14}C-leucine incorporation of tissue slices from hypertrophying rat hearts.

observed in hypertrophying and hypertrophied skeletal muscle (1). However, accelerated amino acid uptake could not satisfactorily account for the differential increase in amino acid incorporation into contractile and total protein. These questions are presently under study.

SUMMARY

Compensatory cardiac hypertrophy induced in rats by aortic constriction is associated with an enhancement of the amino acid incorporating ability of the myocardium that is independent of acute influences of hemodynamic and cardiodynamic factors. This effect can be demonstrated in the myocardial tissue *in vitro* under standardized conditions of amino acid and oxygen supply.

REFERENCES

1. GOLDBERG, A. L. and GOODMAN, H. M.: Amino acid transport during work-induced growth of skeletal muscle. Amer. J. Physiol. *216:* 1111–1115 (1969).
2. GUDBJARNASON, S.: TELERMAN, M. and BING, R. J.: Protein metabolism in cardiac hypertrophy and heart failure. Amer. J. Physiol. *206:* 294–298 (1964).
3. KOGAN, A. K.: Novaya prostaya metodika dozirovannogo cusheniya pochechnikh i drugikh arterii y melkikh labratornikh zhivotnikh v khronicheskom eksperimente. Bjull. Eksp. Biol. Med. *52:* (No. 1) 112–115 (1961).
4. MEERSON, F. S. : Hyperfunktion, Hypertrophie und Insuffizienz des Herzens (VEB Verlag Volks und Gesundheit, Berlin, 1969).

5. SCHREIBER, S. S.; ORATZ, M. and ROTHSCHILD, M. A.: Protein synthesis in the overloaded mammalian heart. Amer. J. Physiol. *211:* 314–318 (1966)
6. SHAHAB, L. and WOLLENBERGER, A.: Amino acid incorporation into mitochondria from the hypertrophying heart of rats with aortic constriction. J. Mol. Cell. Cardiol. *1:* 143–155 (1970).

Mailing address: Prof. A. Wollenberger, Institut fur Kreislaufforschung, Deutsche Akademie der Wissenschaften zu Berlin, Lindenberger Weg 70, 1115 Berlin-Buch (Germany/DDR).

Mayo Foundation, Rochester, Minnesota, USA

THE EFFECTS OF HYPERTROPHY ON MYOCARDIAL ENERGY UTILIZATION

J. F. GUNNING and H. N. COLEMAN

INTRODUCTION

It is generally accepted that hypertrophy provides a fundamental mechanism whereby the myocardium compensates for an increased load. In recent studies of papillary muscles from hypertrophied ventricles (12) as well as the intact hypertrophied right ventricles (13), Spann and his co-workers characterized myocardial contractility in terms of force-velocity relationships and demonstrated depressed contractility per unit mass of myocardium. These quantitative measurements of myocardial function confirmed the conclusions reached by Meerson and Pshenikova (10) from earlier investigations. Since the contractile state has been identified as a major determinant of myocardial energy utilization (8, 14), it might be predicted that hypertrophied myocardium would have an altered pattern of energy utilization.

However, no adequate definition of the energy requirements and efficiency of energy utilization in hypertrophied myocardium has been made, owing to the difficulty inherent in characterizing myocardial contractility and correlating changes in mechanical behavior with energy utilization in intact ventricles. The study of isolated papillary muscle allows precise measurement of myocardial mechanics, and the use of a polarographic oxygen electrode allows simultaneous determination of myocardial oxygen consumption ($M\dot{V}O_2$). Accordingly, the purpose of the present investigation is to correlate the mechanics of myocardial contraction and oxygen consumption of normal and hypertrophied right ventricular papillary muscles.

METHODS

Right ventricular hypertrophy without congestive features was produced by chronic constriction of the main pulmonary artery in 10 cats, a procedure described in detail by Spann *et al.* (13). Anesthesia was induced with intraperitoneal pentobarbital (25 mg/kg), 45 to 82 days after the constriction procedure, and the hemodynamic status of the 10 cats in the hypertrophy group and 12 cats in a control group was determined. A cannula was placed in the descending aorta via the femoral artery, the electrocardiogram was monitored, and a small catheter was placed in the right ventricle via the right external jugular vein. Aortic and right ventricular pressures as well as arteriovenous oxygen differences were determined. Cardiac output was measured by the injection of 1.25 mg of indocyanine dye into the right ventricle with aortic sampling. The weights of the right ventricle and liver were recorded.

Papillary muscles were excised from the right ventricle and mounted in a polarographic oxygen electrode muscle bath for the simultaneous determination of oxygen consumption and mechanical activity. The details of the bath, electrodes, lever system, and procedures have been published previously (7). Papillary muscles from normal cats ranged from 4 to 12 mm in length and 0.44 to 1.10 mm^2 (mean 0.90 mm^2) in cross-sectional area. The muscles from hypertrophied ventricles ranged from 4 to 8 mm in length and 0.55 to 1.10 mm^2 (mean 0.92 mm^2) in cross-sectional area. The results from this study were normalized with shortening as muscle lengths, velocity as muscle lengths/sec, tension and g/mm^2, and oxygen consumption as μL/mg dry tissue/contraction $\times 10^{-3}$.

RESULTS

Assessment of hypertrophy state

The findings documenting right ventricular hypertrophy in the operated cats are shown in figure 1. Right ventricular weight to body weight ratio was increased from 0.54 ± 0.03 SEM g/kg in normal cats to 0.93 ± 0.07 g/kg in the

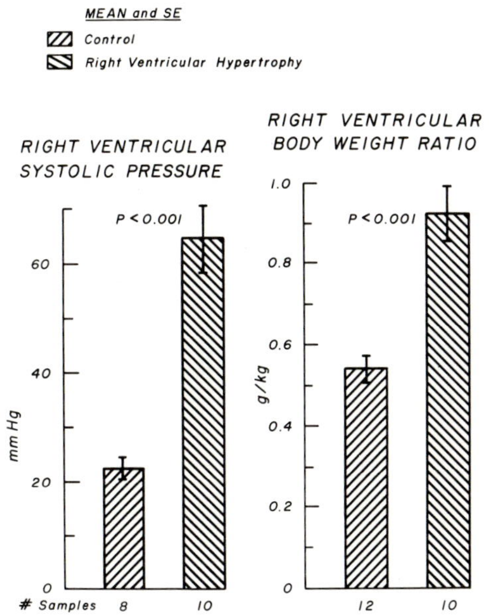

Fig. 1 Average values ± 1 SEM for right ventricular systolic pressure and right ventricular body weight ratio, obtained from 12 control cats and 10 cats with right ventricular hypertrophy following pulmonary artery constriction.

operated cats ($p < 0.001$). Right ventricular systolic pressure increased from 22 ± 2 mm Hg to 64 ± 6 mm Hg ($p < 0.001$). The lack of evidence of a congestive heart failure syndrome is seen in figure 2. There is no significant difference between normal and operated cats of arterio-mixed venous oxygen difference, right ventricular end diastolic pressure, or the ratio of liver weight to body weight. No ascites, pleural effusions, or enlarged systemic veins are found in the pulmonary-artery-constricted cats.

Characterization of contractility of right ventricular papillary muscles from hypertrophied hearts

The mechanics of contraction of the 10 hypertrophied and 12 normal muscles of similar average cross-sectional area (0.90 and 0.92 mm^2) was assessed from isotonic force velocity curves, and isometric contractions at various muscle lengths.

Isotonic force-velocity relations from the two groups of muscles are shown in figure 3. The average preloads, determining initial muscle lengths, are similar (0.44 ± 0.02 g/mm^2 control and 0.46 ± 0.02 g/mm^2 hypertrophied). In the upper panel the force velocity curve from the hypertrophied muscles is depressed downwards and to the left with decreased velocity at all loads. The maximum measured velocity at preload is significantly depressed from 1.35 ± 0.07 to 0.94 ± 0.08 muscle lengths/sec ($p < 0.01$). In the lower panel,

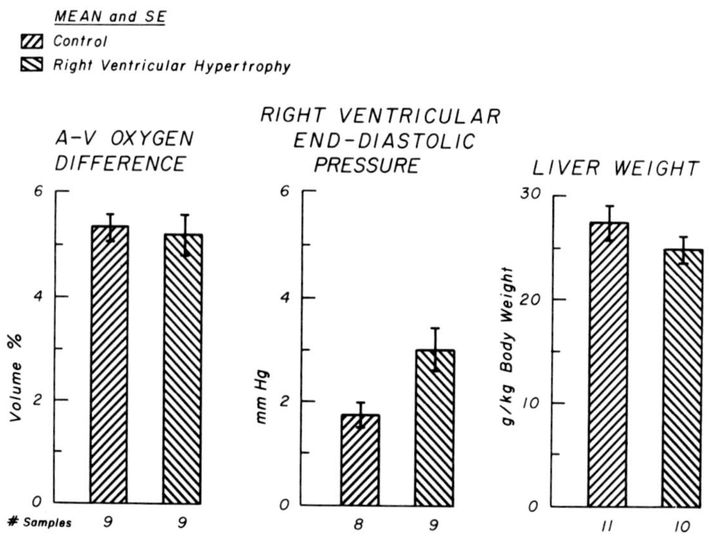

Fig. 2 Average values ± SEM for arterio-venous oxygen difference, right ventricular and diastolic pressure, and liver weight, obtained from 12 control cats and 10 cats with right ventricular hypertrophy following pulmonary artery constriction.

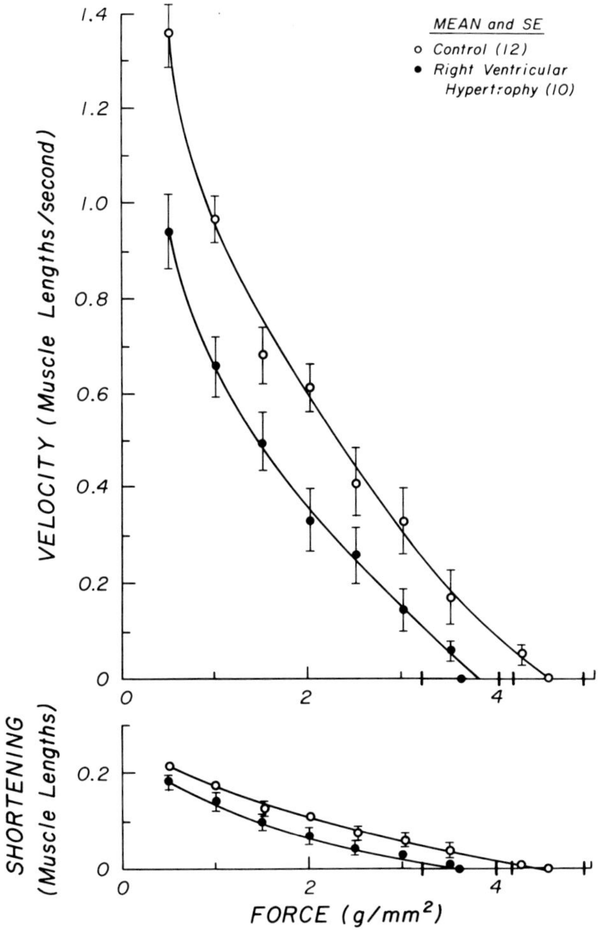

Fig. 3 Average force-velocity relationships ± 1 SEM, obtained from 12 control cats and 10 cats with right ventricular hypertrophy following pulmonary artery constriction. Control preload averaged 0.44 ± 0.01, while the preload in right ventricular hypertrophy muscles averaged 0.46 ± 0.02 g/mm².

shortening and thus the product of shortening X load, external work, was not significantly depressed in the hypertrophied muscles.

In figure 4, the $M\dot{V}O_2$ values corresponding to the afterloaded contractions in figure 3 are depicted. In hypertrophied muscles, the average oxygen consumption for any given level of load is within normal limits.

Average length tension curves for control and hypertrophied muscles are shown in figure 5. The average resting length tension curves for the two groups of muscles are similar, with a resting tension at Lmax of 0.87 ± 0.14 and 1.00 ± 0.12 g/mm^2 respectively. Average developed tension at Lmax was decreased in

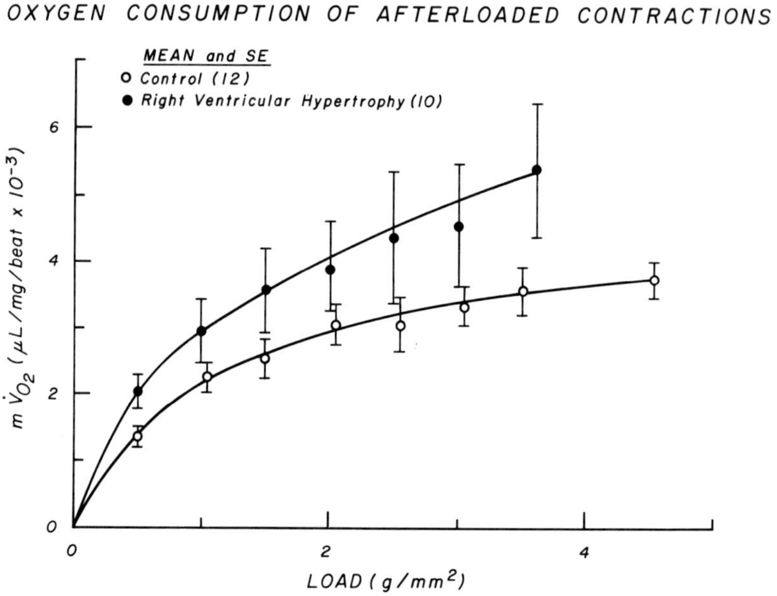

Fig. 4 Average oxygen consumption values ± 1 SEM for the control muscles (12) and from right ventricular hypertrophy muscles. Load is normalized as g/mm², while $M\dot{V}O_2$ is expressed as μL/mg dry tissue/beat × 10^{-3}.

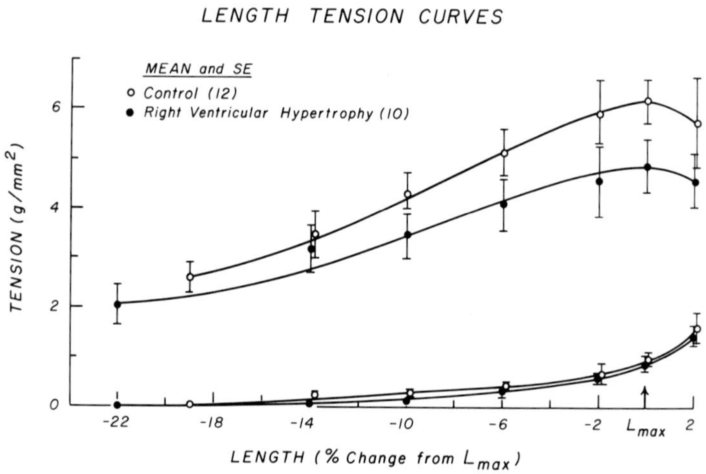

Fig. 5 Average length tension curves ± 1 SEM of papillary muscles obtained from 12 control cats and 10 cats with right ventricular hypertrophy following pulmonary artery constriction. Abscissa = length % change from L_{max}; ordinate = tension g/mm².

the hypertrophied muscles, but the difference was not statistically significant (6.19 ± 0.55 and 4.90 ± 0.54 g/mm^2, $p > 0.1$). The time course of isometric tension development for the two groups of muscles was analyzed in terms of the maximum rate of tension development and time from onset of tension development to peak tension at L_{max}. The maximum rate of rise of tension at L_{max} was significantly depressed in the hypertrophied muscles from 34.1 ± 3.3 g/mm^2/sec to 22.3 ± 2.3 g/mm^2 sec ($p < 0.01$). The time to peak tension was significantly greater in the hypertrophied muscles (354.5 ± 9.4 msec compared with 302.4 ± 5.4 msec, $p < 0.01$), despite the fact that less tension was being developed.

Oxygen consumption was determined for five levels of tension development on each length tension curve for 12 control and 10 hypertrophied muscles. The relationship between tension development and $M\dot{V}O_2$ in these isometric contractions is presented in figure 6. As shown in this figure, the relationship between $M\dot{V}O_2$ and isometric developed tension for normal muscles is described by the least squares linear regression equation $Y = 0.13 + 0.56\ x$ (r = 0.68, number of points = 55) and for the hypertrophied muscles $Y = -0.68 + 1.4\,x$ ($r = 0.79, N = 52$) (fig. 6). The following statistical analysis was applied. The slopes of the regression lines relating $M\dot{V}O_2$ and isometric developed tension for the individual muscles of each group were calculated. The two groups of slopes obtained were then compared, using both a Student t test and the nonparametric Rank Sum Test. The mean slope for the normal muscles was 0.63 ± 0.08 and for the hypertrophied muscles 1.28 ± 0.27 μL O_2/mg/beat $\times 10^{-3}$/g/mm^2 of tension. This difference is significant ($p < 0.01$). Thus, the oxygen cost of unit tension development is significantly increased in the hypertrophied muscles. Further evidence that the $M\dot{V}O_2$ of isometric contraction was greater in hypertrophied muscles is seen when the $M\dot{V}O_2$ per g of tension developed at the peak of the length tension curves was compared. The $M\dot{V}O_2$/g/mm^2 of tension developed was 1.13 μL/mg/beat $\times 10^{-3}$ for the hypertrophied muscle and 0.61 for normal muscle ($p < 0.01$).

Resting oxygen consumption

Resting oxygen consumption was determined in 11 normal and 8 hypertrophied muscles. The $M\dot{V}O_2$ was increased from 2.07 ± 0.19 SEM μL/mg/h in the normal muscles to 3.42 ± 0.45 μL/mg/h in the hypertrophied muscles ($p < 0.01$).

DISCUSSION

The effects of hypertrophy on the function of the ventricle have interested numerous investigators (1, 2, 4). In initial studies, a definitive answer to this question was precluded by the lack of effective means of characterizing contractile state and comparing different ventricles. However, the studies of

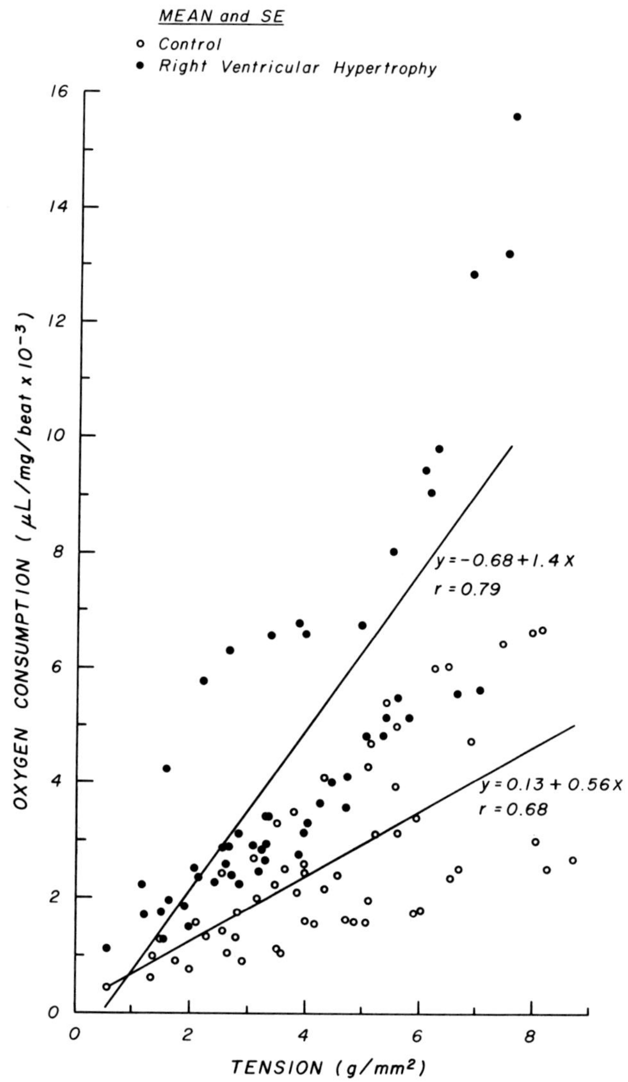

Fig. 6 Relationship between peak isometric tension development and $M\dot{V}O_2$. Abscissa = tension g/mm^2; ordinate = $M\dot{V}O_2$ μL/mg/beat × 10^{-3}. The relation between tension and $M\dot{V}O_2$ for control muscles is described by the least squares regression equation $Y = 0.13 + 0.56\,x$; for hypertrophied muscles $Y = -0.68 + 1.4\,x$.

Meerson and Pshenikova (10) suggested depressed contractile function and more recent studies by Spann *et al.* and Bing *et al.* (3) have demonstrated clear depression of myocardial contractility of hypertrophied nonfailing myocardium. The results from the present study confirm these findings of depressed myocardial contractility associated with the hypertrophy state. This decrease in contractility was characterized by the depression of the force-velocity relationship in hypertrophied muscles, with a depression of the maximal measured velocity of contraction at similar preloads. Although contractility was decreased, equivalent amounts of external work were performed at any level of tension development. The contractile state has recently been identified as one of the major factors determining $M\dot{V}O_2$ (8, 14), with acute depression of contractility being associated with decreased $M\dot{V}O_2$ (9). Thus, in afterloaded contractions with equal levels of tension and external work, the decrease in velocity of contraction would be expected to result in a decrease in $M\dot{V}O_2$. However, the average $M\dot{V}O_2$ at any given load in hypertrophied muscles was increased over the $M\dot{V}O_2$ average of normal muscles (fig. 4).

In isometric contractions, there was a depression of the maximum rate of tension development, dP/dt, consistent with the decrease in velocity of contraction noted in figure 3. The tension developed at the apex of the length tension curve was only slightly decreased, with the result that the resting length tension curve for hypertrophied muscles was normal (fig. 5). Since the level of tension is a function of the rate of tension development and the time to peak tension development (TTP) (5), it may be assumed that the increase in TTP partially compensated for the moderate decrease in dP/dt, with the result that peak developed tension at L_{max} was maintained. The average developed tension of 4.9 ± 0.5 seen in hypertrophied muscles in this study is similar to the value of 5.3 ± 0.6 g/mm^2 found by Spann *et al.* (13) utilizing a similar procedure to produce hypertrophy.

In a previous investigation, the $M\dot{V}O_2$ per g of tension from isometric contractions at the apex of the length tension curve was found to be 0.57 ± 0.09 (6). The results from the control group of the present study are quite similar at 0.61 ± 0.08 μL/mg/contraction/g/mm^2 tension. The major finding of this study is the increased $M\dot{V}O_2$ per g of tension development in the hypertrophied muscles. This is demonstrated both at the apex of the length tension curve (1.13 ± 0.08 μL/mg/contraction × 10^{-3}) and when the relationships between developed tension and $M\dot{V}O_2$ are plotted for all isometric contractions examined during definition of the length tension relationship. This relationship is expressed by the equation $Y = -0.68 \pm 1.40\,x$ and is contrasted to the relation in normal muscles of $Y = 0.13 \pm 0.56\,x$. Since force development in active muscle may be considered to be the result of an interaction between an active contractile element and a passive series elastic component (SEC), extension of the SEC during isometric contraction results in the performance of "internal work" (7). With the evidence that SEC is unaffected by the process of hypertrophy (11), it is then possible to conclude that hypertrophied myocardium is inefficient in the conversion of oxygen to "internal work" (tension).

This finding offers an explanation for the failure to find a decrease in $M\dot{V}O_2$ associated with depression of contractile state in afterload contractions. In afterloaded contractions, $M\dot{V}O_2$ is the sum of $M\dot{V}O_2$ associated with external work, tension development, the contractile state, and activation factors (7). It can be assumed that the increased $M\dot{V}O_2$ per g in hypertrophied muscles has cancelled a diminution of $M\dot{V}O_2$ expected from the decrease in contractile state. Thus, the net result of all factors affecting $M\dot{V}O_2$ for afterloaded contractions was to normalize $M\dot{V}O_2$ to levels found in control muscles.

In addition to the increased $M\dot{V}O_2$ during contraction, the resting $M\dot{V}O_2$ of hypertrophied muscles was found to be significantly increased. The significance of this finding awaits further investigation.

These preliminary data demonstrate that experimental cardiac hypertrophy is associated with a decrease in myocardial contractility and inefficiency in conversion of oxygen to tension development.

SUMMARY

The contractile state and oxygen consumption ($M\dot{V}O_2$) of papillary muscles from hypertrophied right ventricles of cats with pulmonary artery constriction were examined. Myocardial contractility as characterized by force velocity relations was depressed in hypertrophied muscles. Inefficiency of conversion from energy to tension development was demonstrated by an increase in $M\dot{V}O_2$ per g of tension development at the apex of the length tension curve from 0.61 to 1.13 μL/mg/contraction $\times 10^{-3}$.

REFERENCES

1. ALEXANDER, N.; GOLDFARB, T. and DRURY, D. R.: Cardiac performance of hypertensive aorta-constricted rabbits. Circulat. Res. *10:* 11-16 (1962).
2. BEZNAK, M.: Cardiac output in rats during the development of cardiac hypertrophy. Circulat. Res. *6:* 207-212 (1958).
3. BING, O. H. L.; MATSUSHITA, S.; FANBURG, D. L. and LEVINE, H. J.: Mechanical properties of rat cardiac muscle during experimental hypertrophy. Circulat. Res. (in press 1970).
4. BRETSCHNEIDER, H. J.; BUCHERL, E.; FRANK, A. und HUSTEN, M.: Uber die energitek des normalen and hypertrophischen isolierten Saugchirtierzens. Arch. gest. Physiol *254:* 458-486 (1952).
5. BUCCINO, R. A.; SONNENBLICK, E. H.; SPANN, J. F., Jr. and BRAUNWALD, E.: Interactions between changes in intensity and duration of active state in the characterization of inotropic stimuli on heart muscle. Circulat. Res. *21:* 857–867 (1968).
6. COLEMAN, H. N.; DEMPSEY, P. J. and COOPER, T.: Myocardial oxygen consumption following chronic cardiac denervation. Amer. J. Physiol. *218:* 475–478 (1970).
7. COLEMAN, H. N.; SONNENBLICK, E. H. and BRAUNWALD, E.: Myocardial oxygen consumption associated with external work: The Fenn effect. Amer. J. Physiol. *217:* 291–296 (1969).
8. GRAHAM, T. P., Jr.; COVELL, J. W.; SONNENBLICK, E. H.; ROSS, J. and BRAUNWALD, E.: Control of myocardial oxygen consumption: Relative influence of contractile state and tension development. J. Clin. Invest. *47:* 375-385 (1968).

9. GRAHAM, T. P., Jr.; ROSS, J., Jr. and COVELL, J. W.: Myocardial oxygen consumption in acute experimental cardiac depression. Circulat. Res. *21:* 123-138 (1967).

10. MEERSON, F. Z. and PSHENIKOVA, M. G.: Effect of hypertrophy on the contractile function of the heart. Bull. Eksperim. Biol. Med.

11. PARMLEY, W. W.; SPANN, J. F., Jr.; TAYLOR, R. R., and SONNENBLICK, E.: The series elasticity of cardiac muscle in hyperthyroidism, ventricular hypertrophy, and heart failure. Proc. Soc. Exp. Biol. Med. *127:* 606–609 (1968).

12. SPANN, J. F., Jr.; BUCCINO, R. A.; SONNENBLICK, E. H. and BRAUNWALD, E.: Contractile state of cardiac muscle obtained from cats with experimentally produced ventricular hypertrophy and heart failure. Circulat. Res. *21:* 341-354 (1967).

13. SPANN, J. F., Jr.; COVELL, J. W.; ECKBERG, D. L.; SONNENBLICK, E. H.; ROSS, J., Jr. and BRAUNWALD, E.: Myocardial contractile state in hypertrophy and congestive failure. Studies in intact heart and isolated muscle. Circulat. *36:* (Suppl. 2) 241 (1967).

14. SONNENBLICK, E. H.; ROSS, J., Jr.; COVELL, J. W.; KAISER, G. A. and BRAUNWALD, E.: Velocity of contraction as a determinant of myocardial oxygen consumption. Amer. J. Physiol. *209:* 919-927 (1965).

Mailing address: Dr. H. N. Coleman, Section of Physiology, Mayo Clinic, Mayo Foundation, Rochester, Minnesota (USA).

Department of Experimental Pathology, School of Medicine,
University of Komensky, Bratislava, Czechoslovakia

MYOCARDIAL METABOLIC CHANGES IN CARDIAC HYPERTROPHY AND HEART FAILURE

A. FIZELOVA and A. FIZEL

At the present time, there exist two points of view on the possible metabolic cause of heart failure. One school of thought (6, 22, 26) holds that energy production is at fault, while the other group (12, 18, 23) thinks that the defect lies in the organs of energy utilization, in the contractile proteins. These opinions are based on the metabolic disturbances observed in the failing heart. As heart failure already represents the final stage of the progressive process that starts with overload of the heart (valve defects, hypertonia), we studied changes in myocardial protein fractions and energy stores in several stages, from the beginning of overload of the heart until its failure, in a model of chronic heart failure resulting from aortic insufficiency.

MATERIAL AND METHODS

Aortic insufficiency was induced in rabbits by perforation of the aortic valves from the right carotid artery (9). Subsequently, the changes in the size of the heart on X-ray were studied daily for the first week and then at three-weekly intervals. On the basis of the changes occurring in the size of the heart on X-ray, in the heart weight, and in the nucleic acid concentrations (fig. 1), six different time periods were chosen at which to make the biochemical determinations.

Groups I, II, and III were examined on the 2nd, 7th, and 30th day after operation, at the stage of *developing cardiac hypertrophy* (DCH), when the size of the heart on X-ray, the heart weight, and RNA concentration were increased (fig. 1). Group IV was examined during the 5th to 6th month, at the stage of *relatively stabilized cardiac hypertrophy* (RSCH), when the size of the heart did not change markedly, the heart weight approximately doubled, the RNA concentration returned to the control values, and the DNA concentration decreased. Group V was examined during the 7th – 10th month, at the stage of *relative cardiac atrophy* (RCA), at the time when the heart weight, the size of the heart on X-ray, and the nucleic acid concentrations were decreased as compared with the previous period. Group VI was examined during the 8th – 12th month, at the stage of *heart failure*, when the heart weight, its size on X-ray, and the RNA concentration again increased. The changes in the size of the heart on X-ray in the course of aortic insufficiency are presented in detail elsewhere (9).

In the first part of the experiment, after taking the total heart weight and the left ventricular weight, the nucleic acid and protein fractions were examined in the myocardium of the left ventricle. Extraction of the individual protein fractions was carried out with neutral salt solutions of a different ionic strength that did not cause the proteins to denature, thus maintaining their biological activity.

The *sarcoplasmic* and *myofibrillar* proteins were determined by the method of Ivanov *et al.* (14): after removing the fat tissue, the left ventricle was minced with scissors in a cold temperature, diluted with a tenfold volume of phosphate buffer (0.03 M pH 7.4), and homogenized in a Potter-Elvehjem homogenizer, using a high speed electric motor. The homogenate was centrifuged at 2,000 × G for 10 min, and the residue obtained was again homogenized (twice in succession) with the same phosphate buffer, and recentrifuged. The combined supernatant fluid contains the proteins soluble in neutral salt solution of low ionic strength ($\mu < 0.2$), *i.e.*, the *sarcoplasmic* proteins, including the proteins of the granules. Following this procedure, the fraction belonging to the proteins of the sarcoplasm was not observed at the electrophoretic division of the residue.

After separation of the sarcoplasmic proteins, the proteins soluble in the high ionic strength ($\mu > 0.5$), *i.e.*, the *myofibrillar* proteins, were extracted from the residue: at first, by a sixfold volume of Weber's solution (0.6 M KCl, 0.04 M $NaHCO_3$, 0.01 M Na_2CO_3, pH 9.33) in the course of 18 h; then, after addition of ATP (2 mg/g tissue), by a threefold volume of 0.6 M KCl (2 h); and finally, by a threefold volume of Weber's solution (1 h). The individual extracts were pooled. The electrophoretic division did not show the presence of myofibrillar proteins in the residue. After separation of the sarcoplasmic and myofibrillar proteins, the stroma proteins remain in the residue.

From the myofibrillar protein extract so-obtained, *actomyosin* was separated (1). A part of the extract was diluted with a fifteenfold volume of ice-cold, glass-distilled water and adjusted to pH 6.8 with acetic acid. The precipitated actomyosin was immediately centrifuged at 2,000 × G for 10 min at 0°C. It was stored in the refrigerator overnight and centrifuged at 14,000 × G for 15 min at 0°C.

The *mitochondrial* proteins were estimated by Hogeboom's method (13). The muscle of the left ventricle was homogenized in a twentyfold volume of sucrose solution with EDTA (0.25 M sucrose, 0.002 M EDTA, pH 7.4). The homogenate was centrifuged at 600 × g for 15 min in a cold centrifuge and the residue was homogenized with the same sucrose solution and again centrifuged at 600 × G. The mitochrondrial proteins were obtained from the pooled supernatants by centrifugation at 10,000 × G in the course of 30 min and washed with the same sucrose solution. This fraction was verified electronmicroscopically.

A 1g sample of muscle was homogenized in a sevenfold volume of Weber's solution, diluted by the same volume of 0.6 M KCl and, after addition of ATP (0.02 ml of 10 % ATP), extracted for 24 h. The volume of the homogenate was completed with 0.6 M KCl at a ratio of 1 : 30, pH 8.7 – 8.9 (1). A part of this homogenate was used for determination of the *total protein.* Another part was diluted by a fivefold volume of 0.05 N NaOH, and centrifuged after a 24 h storage. The *noncollagen* protein was determined in the supernatant by the procedure of Lilienthal *et al.* (16). The *collagen* protein was determined from the difference of the total and the noncollagen protein.

Homogenization of the muscle tissue as well as extraction and centrifugation of the individual protein fractions were carried out at a temperature of 0–+2° C. The total protein and protein fractions were determined as nitrogen by the Kjeldahl micromethod using the digestive mixture according to Perrin (19) and, after calculation by factor 6.25, expressed as mg protein/g wet weight of the left ventricle.

The *nucleic acids* were determined by the Schmidt and Thannhauser procedure (21) modified by Canev and Markov (4). The optical density of the samples was measured by a spectrophotometer, using the double-wavelength method.

In the second part of the experiment, phosphates and glycogen were studied in the myocardium of the left ventricle. Before taking the muscle samples for biochemical assay, the rabbits were anesthetized with thiopental and artificially respirated with a respirator for 10 min. Their chests were opened widely by cutting through 3 to 4 ribs on both sides of the sternum. The central portion of the left ventricular myocardium was cut free with a pair of scissors and instantly frozen by means of the Wollenberger tongs (28) precooled in liquid nitrogen. The maximal thickness of the frozen muscle samples was 1.0 – 1.5 mm. The time interval between removal and freezing of the ventricle was less than 3 sec. Utilizing a mortar and pestle precooled with liquid nitrogen, the frozen portion of the ventricle was pulverized to a fine powder, under a layer of liquid nitrogen. This finely ground powder was utilized for determination of the phosphate and glycogen concentrations.

Adenosine triphosphate (ATP) and *adenosine diphosphate* (ADP) were determined in the trichloroacetic extract (from 400 mg muscle powder), following their separation on paper chromatography according to Fleckenstein *et al.* (11). The detected areas of paper were ashed and estimated as phosphate by the Berenblum and Chain method (2). Perchloric acid extract (from 200 mg muscle powder) was utilized for the determination of *inorganic phosphate* (Pi) and *creatine phosphate* (CP), according to the method of Wollenberger *et al.* (27). Inorganic phosphate, after its reaction with the acid-molybdate reagent, was transported into isopropylacetate and the development of color reaction was carried out by means of $SnCl_2$. The total phosphate (Pi plus CP) was determined on a second aliquot of the muscle extract, after incubation with the acid-molybdate reagent for 30 min at room temperature, using the same technic as for the Pi. CP was estimated as the difference between the total and inorganic phosphates. According to Wahler and Wollenberger (25), there are no acid-molybdate-labile phosphates in the heart muscle other than creatine phosphate. The *glycogen* concentration was determined colorimetrically, according to Carroll *et al.* (5). Glycogen was precipitated from trichloroacetic acid filtrate of muscle powder by ethanol and determined by an anthrone procedure.

In both parts of the study, 10 rabbits were used in each experimental group and 15 served as controls. The results obtained were examined statistically with the *t*-test.

RESULTS

Changes in heart weight

The total weight of the heart increases rapidly after the formation of aortic insufficiency (table I and fig. 1). At the stage of DCH, it increases on the 7th day (group II) by 38.2 % and on the 30th day (group III) by 60.1 %. During the further course of aortic insufficiency, the weight of the heart increases more slowly. At the stage of RSCH (group IV), it approximately doubles as compared with the control value. At the time of RCA (group V), the weight of the heart decreases by 16.2 % in comparison with group IV, but it remains increased compared with the controls. At the time of heart failure (group VI), the heart weight increases by 11.1 % as compared with the stage of RCA.

Table I

Changes in Heart Weight and Nucleic Acids in the Course of Aortic Insufficiency

Group*	Heart weight	Left ventricle	RNA	DNA
	g/kg body wt.		*mg % P*	
Control	2.002 ± 0.107	0.787 ± 0.068	18.52 ± 2.24	9.26 ± 1.18
II. Developing cardiac hypertrophy, 7th day	2.791 ± 0.487[b]	1.057 ± 0.193[b]	26.12 ± 2.38[a]	9.53 ± 1.24
III. Developing cardiac hypertrophy, 30th day	3.239 ± 0.302[a]	1.319 ± 0.131[a]	22.52 ± 2.79[c]	8.52 ± 1.60
IV. Relatively stabilized cardiac hypertrophy	3.964 ± 0.479[a]	1.646 ± 0.243[a]	17.38 ± 1.78	6.18 ± 1.09[a]
V. Relative cardiac atrophy	3.322 ± 0.312[a]	1.321 ± 0.149[a]	15.84 ± 2.08[c]	6.09 ± 1.59[a]
VI. Heart failure	3.701 ± 0.332[a]	1.384 ± 0.162[a]	22.95 ± 2.18[b]	6.72 ± 1.26[a]

All values expressed as mean ± SD; [a] $p < 0.001$; [b] $p < 0.01$; [c] $p < 0.05$.
*The heart weight and nucleic acids were not examined in group I.

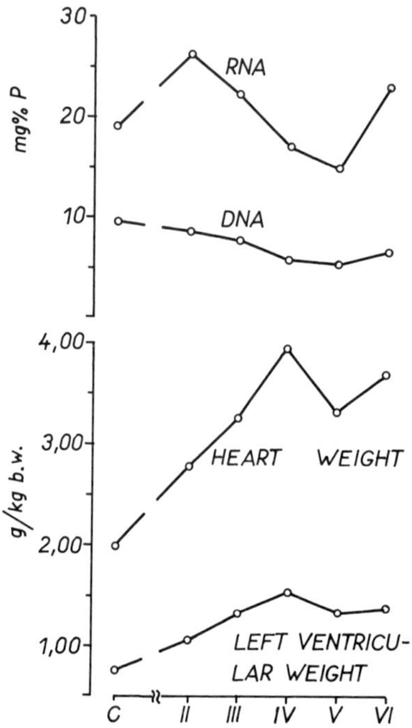

Fig. 1 Changes in heart weight, left ventricular weight, and concentration of nucleic acids in the course of aortic insufficiency. C: controls; II: developing cardiac hypertrophy (7th day); III: developing cardiac hypertrophy (30th day); IV: relatively stabilized cardiac hypertrophy; V: relative cardiac atrophy; VI: heart failure.

The weight changes of the individual parts of the heart do not show the same trend in the course of aortic insufficiency. At the stage of DCH all parts of the heart participate relatively evenly in the increase of its total weight. At the stage of RSCH as well as of RCA, the changes of the total heart weight are caused mainly by changes of the left ventricle and septum, whereas the weight of the right ventricle and the auricles does not change substantially as compared with group III. On the other hand, however, at the time of heart failure, the repeated increase in the total heart weight is caused mainly by the weight increase in the right ventricle and auricles. The changes of the individual parts of the heart are presented in detail elsewhere (9).

Changes in nucleic acids

Parallel with changes in the heart weight, changes in the concentration of nucleic acids also occurred (table I and fig. 1). At the time of DCH, the concentration of RNA increased by 39.9 % (group II) and by 21.4 % (group III); the concentration of DNA did not change. At the time of RSCH (group IV), the concentration of RNA returned to the control values and the DNA concentration decreased by 33.0 %. At the time of RCA (group V), there was a decrease in the concentration of RNA and DNA (by 14.5 % and 34.0 %, respectively). At the time of heart failure (group VI), the concentration of RNA increased again by 24.0 %, while the concentration of DNA remained decreased.

Changes in myocardial proteins

At the stage of DCH (groups II and III), the concentration (expressed in mg protein/g wet muscle weight) of actomyosin decreased by 23.8 % and 19.0 %, and the concentration of mitochondrial proteins increased by 44.0% and 69.8%. The concentration of total and sarcoplasmic proteins as well as of collagen was not changed significantly, except for the concentration of the total protein on the 7th day (group II), which was decreased by 10.8 %. No significant changes in protein concentrations were observed at the stage of RSCH, except for the mitochondrial proteins, which were still increased. At the stage of RCA (group V) the opposite changes were observed as compared with the stage of DCH: the actomyosin concentration was increased by 20.0 % and the mitochondrial protein concentrations were decreased by 25.6 %. The concentration of other protein fractions did not change. At the time of heart failure, the actomyosin concentration markedly decreased (by 28.1 %); on the other hand, the concentration of collagen increased by 27.1 % and that of mitochondrial proteins returned to the control values (table II and fig. 2).

Table II

Changes in the Concentration of Protein Fractions

Group	Total protein	Sarco-plasmic proteins	Mitochon-drial proteins	Actomyosin	Collagen
			mg/g wet wt.		
Control	141.3 ± 10.6	54.3 ± 5.8	4.3 ± 0.6	41.3 ± 3.4	21.7 ± 2.8
I. Developing cardiac hypertrophy, 2nd day	–	–	4.9 ± 0.6	–	–
II. Developing cardiac hypertrophy, 7th day	126.4 ± 9.3^b	50.3 ± 4.5	6.2 ± 0.7^b	31.4 ± 3.7^a	18.5 ± 6.5
III. Developing cardiac hypertrophy, 30th day	134.8 ± 12.5	55.5 ± 6.1	7.3 ± 0.8^a	33.4 ± 5.5^c	18.0 ± 5.1
IV. Relatively stabilized cardiac hypertrophy	134.1 ± 8.2	50.0 ± 5.4	5.0 ± 0.5^c	41.5 ± 4.6	20.7 ± 4.5
V. Relative cardiac atrophy	142.8 ± 9.9	50.8 ± 3.4	3.2 ± 0.7^b	49.6 ± 4.0^a	22.6 ± 4.8
VI. Heart failure	134.3 ± 11.8	52.4 ± 6.7	3.8 ± 0.7	29.7 ± 4.0^a	27.6 ± 4.9^b

All values expressed as mean ± SD; $^a\, p < 0.001$; $^b\, p < 0.01$; $^c\, p < 0.05$.

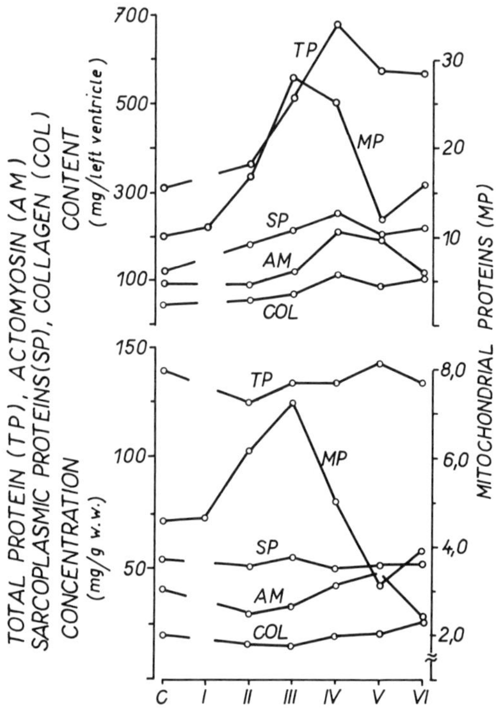

Fig. 2 Changes in concentration and content of the total protein and its fractions in the course of aortic insufficiency. C: controls; I: developing cardiac hypertrophy (2nd day); II: developing cardiac hypertrophy (7th day); III: developing cardiac hypertrophy (30th day); IV: relatively stabilized cardiac hypertrophy; V: relative cardiac atrophy; VI: heart failure.

The water concentration was significantly increased only at the stage of DCH on the 7th day (by 1.8 %); in the further course of aortic insufficiency, it did not change significantly.

The total content of all protein fractions (expressed in mg protein/ myocardium of the left ventricle), except that of actomyosin, was increased at the stage of DCH on the 7th day (group II); the content of actomyosin was not changed. On the 30th day (group III), with an increase of myocardial mass, the content of all protein fractions, including actomyosin, increased further. At the stage of RSCH, when the left ventricular weight increased by 109.9 %, the content of total protein and its fractions approximately doubled. At the time of RCA (group V), when the left ventricular mass decreased as compared with the period of RSCH, the content of actomyosin did not change and the content of other protein fractions decreased. At the stage of heart failure, the content of actomyosin markedly decreased; on the other hand, other protein fractions, predominantly collagen, increased (table III and fig. 2).

Table III

Changes in the Total Content of Protein Fractions (mg/myocardium of the left ventricle)

Group	Total protein	Sarcoplasmic proteins	Mitochondrial proteins	Actomyosin	Collagen	Weight of left ventricle
						g
Control	324.0 ± 42.8	125.4 ± 6.9	9.8 ± 1.6	94.9 ± 14.2	49.8 ± 14.4	2.30 ± 0.21
I. Developing cardiac hypertrophy, 2nd day	–	–	13.0 ± 3.2	–	–	2.54 ± 0.42
II. Developing cardiac hypertrophy, 7th day	365.7 ± 37.7^c	146.2 ± 26.3	17.0 ± 2.1^a	90.0 ± 19.3	55.0 ± 27.5	2.89 ± 0.22^b
III. Developing cardiac hypertrophy, 30th day	513.2 ± 113.1^a	211.8 ± 55.8^a	28.1 ± 3.4^a	125.6 ± 31.2^b	69.3 ± 25.5^c	3.76 ± 0.37^a
IV. Relatively stabilized cardiac hypertrophy	683.1 ± 109.3^a	254.3 ± 59.8^a	25.9 ± 2.8^a	213.2 ± 50.5^a	109.4 ± 54.1^b	5.22 ± 0.78^a
V. Relative cardiac atrophy	576.9 ± 78.4^a	204.8 ± 39.2^a	12.9 ± 3.2	201.4 ± 42.1^a	91.8 ± 20.8^a	3.96 ± 0.40^a
VI. Heart failure	571.2 ± 116.2^a	219.3 ± 37.5^a	16.2 ± 2.6^a	126.2 ± 33.6	116.1 ± 27.1^a	4.21 ± 0.49^a

All values expressed as mean ± SD; ^{a} $p < 0.001$; ^{b} $p < 0.01$; ^{c} $p < 0.05$.

Changes in high-energy phosphate and glycogen stores

At the stage of DCH on the 2nd day after cardiac overload (group I), the high-energy phosphate concentrations were significantly reduced (CP by 47.5 %; ATP by 23.8 %; and ADP by 18.5 %). In contrast, the glycogen concentration was increased by 34.0 %. On the 7th and 30th days (groups II and III) at a time when mitochondrial protein contractions were increased, the ATP, ADP, and glycogen concentrations returned to the control values, but the concentration of CP still remained decreased by 18.9 % and 19.9 %. When the left ventricular mass had approximately doubled (group IV), the concentration of ATP increased by 21.4 % and that of glycogen by 49.9 %, while the concentrations of ADP and CP did not differ as compared with the controls. However, when compared with the stage of DCH (group III), the CP concentration was seen to be increased by 20.1 %. The concentration of mitochondrial proteins then began to decrease. At the stage of RCA (group V), the concentration of ATP was increased by 11.4 %, that of CP by 32.0 %, and that of glycogen by 33.8 %. At the time of heart failure, opposite changes occurred in the energy stores. The concentration of ATP, CP, and glycogen decreased, but these changes were not significant as compared with the controls (table IV and fig. 3).

DISCUSSION

The observations that the changes occurring in myocardial energy stores and proteins do not parallel the individual stages of aortic insufficiency, although the defect of the aortic valves remains constant, raises the question of the nature of the underlying mechanism of these changes. From the results of the present experiments, we cannot give a conclusive answer to this problem. We suggest

Table IV

Changes in the High-Energy Phosphate and Glycogen Concentrations

Group	ATP	ADP *µmoles/g wet wt.*	CP	Glycogen *mg/100g wet wt.*
Control	4.75 ± 0.44	0.92 ± 0.13	7.71 ± 0.58	235.8 ± 67.2
I. Developing cardiac hypertrophy, 2nd day	3.62 ± 0.49^a	0.75 ± 0.10^a	4.05 ± 0.83^a	315.0 ± 52.0^c
II. Developing cardiac hypertrophy, 7th day	4.87 ± 0.45	0.87 ± 0.16	6.26 ± 0.86^b	199.8 ± 71.9
III. Developing cardiac hypertrophy, 30th day	4.99 ± 0.48	0.89 ± 0.14	6.18 ± 0.85^b	242.1 ± 68.9
IV. Relatively stabilized cardiac hypertrophy	5.77 ± 0.65^b	0.81 ± 0.19	7.46 ± 0.61	352.7 ± 87.8^b
V. Relative cardiac atrophy	5.34 ± 0.60^c	0.85 ± 0.18	10.18 ± 1.87^b	314.7 ± 84.0^c
VI. Heart failure	4.47 ± 0.52	0.83 ± 0.16	6.97 ± 0.90	194.2 ± 60.2

All values expressed as mean ± SD; $^a\ p < 0.001$; $^b\ p < 0.01$; $^c\ p < 0.05$.

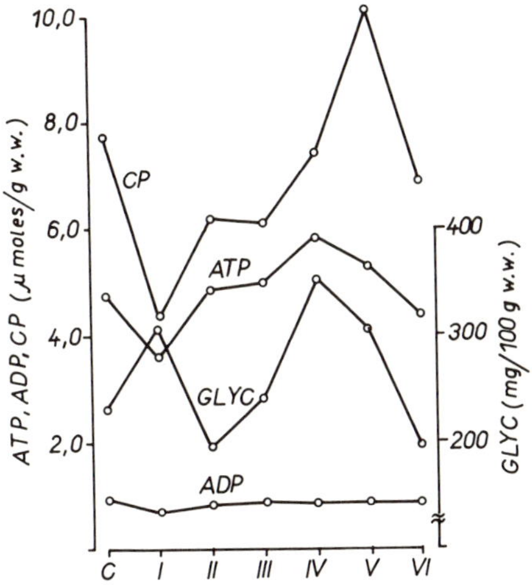

Fig. 3 Changes in the myocardial energy stores in the course of aortic insufficiency. ATP (adenosine triphosphate); ADP (adenosine diphosphate); CP (creatine phosphate); GLYC (glycogen). For other symbols, see legend to figure 2.

that these changes may be connected with a different quantity of work that is performed by each unit of the myocardium at the individual stages of aortic insufficiency.

Immediately after the onset of cardiac overload, each unit of the myocardium of the left ventricle must perform more work than under normal conditions. This leads to a rapid decrease of the myocardial high-energy stores and to activation of nucleic acid and protein synthesis, initially to increased synthesis of the sarcoplasmic and mitochondrial proteins and, later, of the myofibrillar proteins also. This process results in a rapid growth of the myocardial mass. The development of cardiac hypertrophy leads to the distribution of the increased work within its augmented mass; hence, each unit of the myocardium performs relatively less work, an observation that is supported by the experiments of numerous authors (15, 24). The energy requirements for each unit of the myocardium decrease and, therefore, the myocardial energy stores increase. As a result of this, there occurs a decrease of nucleic acids and protein synthesis, mainly sarcoplasmic and mitochondrial proteins. The myocardial mass gradually starts to diminish. In consequence of the progressive reduction of the left ventricular mass, there arises a situation similar to the one observed at a very early stage of the ventricular overload. Each muscle weight unit must once again perform increased work. The energy requirements per unit weight increase and the myocardial energy stores show a relative fall. The protein synthesis of the energy-producing structures is activated, but the synthesis of the contractile proteins does not increase; on the contrary, a marked reduction of the actomyosin content and of the ATPase activity is observed (10). Consequently, left ventricular failure develops. Because the high-energy phosphate stores are sufficient in the failing heart, the reduction of contractile proteins and their ATPase activity may be in a causal relationship to the impaired myocardial contractility rather than to a relative decrease of energy stores.

The changes in the myocardial energy stores as well as in nucleic acid and protein synthesis at the various stages of aortic insufficiency take place simultaneously with the changes of sympathoadrenal and adrenocortical activity (9). Since the catecholamines and cortical hormones modify the mechanical activity of the heart (3) and myocardial metabolic processes (7, 8, 17, 20), it is possible that the changes observed in the sympathoadrenal and adrenocortical activity may take part in the changes of myocardial energy stores.

Further investigations are required before we can precisely determine the nature of the underlying mechanism of the changes that occur in the myocardial protein synthesis and energy stores in the course of aortic insufficiency.

SUMMARY

Cardiac overload induced by aortic insufficiency leads to a decrease of the high energy phosphate stores and to activation of nucleic acid and protein synthesis at the initial stage of the developing cardiac hypertrophy. At the stage of relatively stabilized cardiac hypertrophy, the nucleic acid and protein synthesis are normalized: the concentrations of ADP and CP return to the control values and the concentration of ATP increases. At the stage of relative cardiac atrophy, nucleic acid and protein synthesis are reduced; however, the ATP and CP concentrations are increased. At the stage of heart failure, a reduction of the energy stores to the control values takes place and nucleic acid and protein synthesis, except for actomyosin, increases.

REFERENCES

1. BENSON, E. S.; HALLAWAY, B. E. and FREIER, E. F.: Extraction and quantitative characteristics of actomyosin in canine heart muscle.Circulat. Res. *3:* 215-220 (1955).
2. BERENBLUM, I. and CHAIN, E.: An improved method for the colorimetric determination of phosphate. Biochem. J. *32:* 295-298 (1938).
3. BRAUNWALD, E.: The control of ventricular function in man. Brit. Heart J. *27:* 1-8 (1965).
4. CANEV, R. G. and MARKOV, G. G.: A contribution to quantitative spectrophotometric determination of nucleic acids (In Russian). Biochimija *25:* 151-159 (1960).
5. CARROLL, N. V.; LONGLEY, R. W. and ROE, J. H.: The determination of glycogen in liver and muscle by use of anthrone reagent. J. Biol. Chem. *220:* 583-593 (1956).
6. FEINSTEIN, M. B.: Effects of experimental congestive failure, ouabain, and asphyxia on the high energy phosphate and creatine content of guinea pig heart. Circulat. Res. *10:* 333-346 (1962).
7. FIZEL, A. und FIZELOVA, A.: Studium einiger metabolischer Prozesse des versagenden Herzens im Experiment. X. Z. Kreislaufforschg. *55:* 614-627 (1966).
8. FIZEL, A. und FIZELOVA, A.: Studium einiger metabolischer Prozesse des versagenden Herzens im Experiment. XI. Z. Kreislaufforschg. *55:* 804-818 (1966).

9. FIZEL, A. and FIZELOVA, A.: Changes of the sympathoadrenal and adrenocortical activity in the course of cardiac hypertrophy and heart failure. Z. Kreislaufforschg. *58:* 1115-1128 (1969).

10. FIZEL, A; FIZELOVA, A. and BOHUNICKA, E.: Changes in the myocardial protein fractions in the cardiac hypertrophy and heart failure. (In Slovak). Cs.fysiol. *17:* 197-197 (1968).

11. FLECKENSTEIN, A.: JANKE, J. und ELKE, M.: Papierchromatographische Trennung von Adenosinetriphosphat (ATP), Adenosinediphosphat (ADP) und von anderen Phosphorverbindungen bei der Koffeinkontraktur des Frosch-und Kroten-Rectus. Arch. exp. Path. Pharmak. *221:* 404-417 (1954).

12. GORDON, M. D.; BROWN, A. L. and BROWN, A. L., Jr.: Myofibrillar adenosine triphosphatase activity of human heart tissue in congestive heart failure: Effects of ouabain and calcium. Circulat. Res. *18:* 534-542 (1966).

13. HOGEBOOM, G. H.: Fractionation of cell components of animal tissues. In: Methods in Enzymology, Vol. 1: pp. 16-19, (Academic Press, New York 1955).

14. IVANOV, I. I.; ZACHOVA, Z. N.; ZINOVJEVA, I. P.; MIROVIC, N. I.; MOISEEVA, V. P.; PARSINA, E. A.; TUKACINSKIJ, S. E. and JURJEV, V. A.: The composition of protein fractions and the contractile function of various types of muscles (In Russian). Biochima ja *24:* 451-458 (1959).

15. KRAMES, B. and NORTUP, D. W.: Isometric tension development in the hypertrophied heart. Fed. Proc. *23:* 358-358 (1964).

16. LILIENTHAL, J. L.; JR.; ZIERLER, K. L.; FOLK, B. P.; BUKA, R. and RILEY, M. J.: A reference base and system for analysis of muscle constitutents. J. Biol. Chem. *182:* 501-505 (1950).

17. MARCHETTI, G.: MERLO, L. and NOSEDA, V.: Myocardial uptake of free fatty acids and carbohydrates after beta adrenergic blokade. Amer. J. Cardiol. *22:* 370-374 (1968).

18. OLSON, R. E.: Myocardial metabolism in congestive heart failure. J. Chron. Dis. *9:* 442-464 (1959).

19. PERRIN, C. H.: Rapid modified procedure for determination of Kjeldahl nitrogen. Anal. Chem. *25:* 968-972 (1953).

20. RAAB, W.: The adrenergic-cholinergic control of cardiac metabolism and function. Adv. Cardiol. *1:* 65-152 (1956).

21. SCHMIDT, G. and THANNHAUSER, S. J.: A method for the determination of desoxyribonucleic acid, ribonucleic acid and phosphoproteins in animal tissues. J. Biol. Chem. *161:* 83-89 (1945).

22. SCHWARTZ, A. and LEE, T. S.: Study of heart mitochondria and glycoglytic metabolism in experimentally induced cardiac failure. Circulat. Res. *10:* 321-332 (1962).

23. SOBEL, B. E.; SPANN, J. F.; POOL, P. E.; SONNENBLICK, E. H. and BRAUNWALD, E.: Normal oxidative phosphorylation in mitochondria from the failing heart. Circulat. Res. *21:* 355-361 (1967).

24. SPANN, J. F.; Jr.; BUCCINO, R. A.; SONNENBLICK, E. H. and BRAUNWALD, E.: Contractile state of cardiac muscle obtained from cats with experimentally produced ventricular hypertrophy and heart failure. Circulat. Res. *21:* 341-354 (1967).

25. WAHLER, B. E. und WOLLENBERGER, A.: Zur Bestimmung des Ortophosphats neben säauremolybdat-labilen Phosphosäureverbindungen. Biochem. Z. *329:* 508-520 (1958).

26. WOLLENBERGER, A.; KLEITKE, B. and RAABE, G.: Some metabolic characteristics of mitochondria from chronically overloaded, hypertrophied hearts. Exp. Molec. Path. *2:* 251-260 (1963).

27. WOLLENBERGER, A.: KRAUSE, E. G. und WAHLER, R. E.: Uber den tatsächlichen Ortophosphat- und N-Phosphorylkreatingehalt des Säugetierherzens. Pflügers Archiv *270:* 413-427 (1960).

28. WOLLENBERGER, A.; RISTAU, O. und SCHOFFA, G.: Eine einfache Technik der extrem schnellen Abkühlung grösserer Gewebstücke. Pflügers Archiv *270:* 399-412 (1960).

Mailing address: A. Fizelova, Ph.D., Department of Experimental Pathology, School of Medicine, Sasinkova 4, Bratislava (Czechoslovakia).

Institute of Circulation Research, German Academy of Sciences, Berlin-Buch, German Democratic Republic

RESPONSES OF THE HEART MITOCHONDRIA TO CHRONIC CARDIAC OVERLOAD AND PHYSICAL EXERCISE

A. WOLLENBERGER

The heart as an uninterruptedly working muscle must rely for the performance of its function on a steady and rapid supply of energy in the form of ATP, which is furnished, in the presence of oxygen, mainly by oxidative phosphorylation in the mitochondria. The magnitude of the task that the mitochondria have to perform in the heart is attested to by the fact that, in the muscle cells of mammalian myocardium, these organelles constitute approximately one third of the total cellular mass.

CHRONIC CARDIAC OVERLOADING

Some years ago, when one of the main interests in our laboratory was the mechanism of congestive heart failure, we decided to take a closer look – biochemically and with the electron microscope – at the chronically overloaded heart during the stage preceding the onset of manifest heart failure, which is the stage of compensatory cardiac hypertrophy not yet associated with overt functional decompensation. We thought it might be possible that there are cellular and subcellular changes that may predispose the hypertrophied heart to failure. A first series of experiments was performed on adult dogs whose ascending aorta was permanently constricted with a nylon band to an extent just tolerated by the animals. Several months afterwards when there was X-ray evidence of cardiac enlargement, and at later dates, the dogs were sacrificed. At these times the animals exhibited no signs and symptoms of heart failure. End diastolic intraventricular pressures were normal and an exercise tolerance test, consisting of a 2-mile run at moderate speed, was performed without apparent effort.

Examination of the heart muscle revealed rather pronounced changes affecting the mitochondria (21). There was a deterioration in the structure of the organelles, mainly a partial loss of their inner membranes, the cristae. Histometric measurements showed that the mitochondrial compartment in the heart muscle cells of the hypertrophied left ventricle had diminished by about 30%. Figure 1, series I, shows that the hypertrophy of the left ventricle or, to be more specific, the hypertrophic state at the time of sacrifice, was due largely to an increase in the mass (volume) of the myofibrils of the myocardial cells,

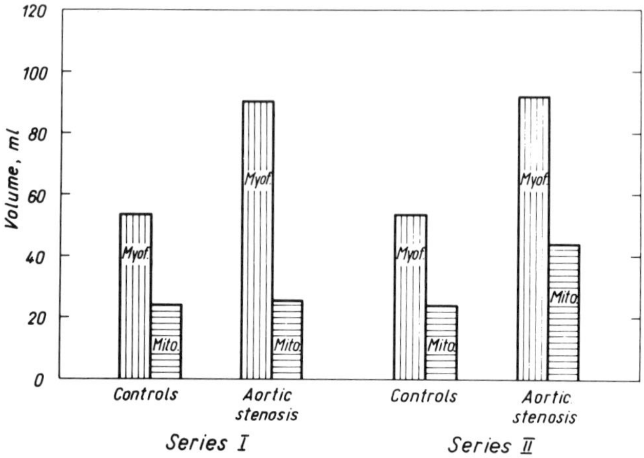

Fig. 1 Contribution of the myofibrils and the mitochondria to hypertrophy of the left ventricle of dogs with experimental aortic stenosis. The columns represent the total myofibrillar and mitochondrial volumes in the heart muscle cells of the left ventricle, whereby the total heart muscle cell volume of the control ventricles is arbitrarily set at 100 ml. It is assumed that the increase in the volume of the heart muscle cells due to aortic stenosis is proportional to the increase in the mass of the ventricle. From Wollenberger (21).

whereas the total mass of the mitochondria in these cells was only slightly increased. Similar changes had been observed earlier in cardiac muscle of patients with valvular heart disease and heart failure (8, 15).

Our measurements were made prior to heart failure, but several months to 2 ½ years after constriction of the aorta. From the results of more recent experiments on rabbits and rats conducted by Meerson *et al.* (13) and in our laboratory (17), it seems that in the first week or first 2 weeks following aortic constriction, the relative mass of the mitochondria in the hypertrophying heart increases. In the isolated mitochondria of such hearts, we have noted marked elevations in the rate of incorporation of amino acids into protein (19). As is well known, mitochondria possess a limited degree of genetic autonomy, *i.e.*, they contain a small amount of DNA and are capable of coding and synthesizing part of their protein. In our dog experiments, mitochondria were also isolated, not for the purpose of measuring protein synthesis, but for the measurement of respiration and phosphorylation. In agreement with results obtained by Schwartz and Lee (18) in guinea pigs with aortic stenosis, the dog heart mitochondria exhibited decreases in the rate of O_2 consumption and in the efficiency of phosphorylation (21). Taken in conjunction with the reduction in mitochondrial mass per g heart, this meant a rather serious curtailment of the capacity of the hypertrophic myocardium to synthesize ATP (see table I, series I).

In a second series of dog experiments, a more sophisticated technique was

Table I

Phosphorylative Capacity of Mitochondria from Normal and Hypertrophied Dog Hearts (Calculated per g Heart)

Experimental Series	Group	Left ventricle (g)/body wt. (kg)	Mitochondrial protein, mg/g left ventricle	Rate of phosphorylation (μmoles phosphate/ min/g)*
I	Control	3.68	39.8	25
	Aortic stenosis	5.69	27.4	9
II	Control	4.01	43.6	29
	Aortic stenosis	6.87	47.7	32

*In medium containing glutamate, 30°C. From Wollenberger and co-workers (21).

used for producing aortic stenosis. A nylon band was placed around the ascending aorta of puppies without narrowing the vessel. As the animals grew, a stenosis developed at the banding site. By the time the dogs were fully grown, their hearts were markedly hypertrophied, to almost double the normal size. But there was no decrease in the mass of mitochondria per g myocardium (fig. 1, series II) nor was there any major abnormality in cardiac mitochondrial structure and metabolism (21) (see table I, series II). Similar conclusions were reached later in studies on mitochondria from the failing heart of animals with chronic outflow obstruction (20) and on patients with valvular heart disease (4). In view of the disparity of the results that I have discussed, all we can say at present is that, depending on the experimental conditions or clinical state, chronic cardiac overloading leading to heart hypertrophy and failure may or may not be associated with deficiencies in myocardial mitochondrial mass, structure, and metabolic activity.

ENDURANCE TRAINING

Compensatory cardiac growth is not only an adaptive response to a permanent cardiac overload resulting from pathologic conditions such as aortic and pulmonary stenosis, hypertension, and valvular insufficiency; it is also an adaptive reaction to intermittent mechanical stress of the heart caused by repeated dynamic physical exercise of a certain intensity, duration, and frequency. It is precisely this type of exercise that is being recommended by some experts as affording a good chance of protection against ischemic heart disease. I, personally, have the feeling that this may be true and I am an advocate of regular vigorous exercise. But the question may be raised whether in following such an exercise program of presumptive cardioprotective value one is not incurring a risk in the sense that acquisition of an enlarged and hypertrophic heart may perhaps be a step in the direction of an eventual heart failure. I have spoken to a number of cardiologists who believe in this danger, evidently because they are impressed by the fact that the vast majority of patients with congestive heart failure whom they see have enlarged hearts and that, *vice versa,* the vast majority of people free of manifest cardiac disease do not have enlarged hearts. But arguments based on feelings and impressions will not help to resolve this problem: An objective approach is needed. The characterization of cardiac hypertrophy due to exercise – morphologically, biochemically, and functionally – the prognosis of this hypertrophy, and the analysis of the responses of the heart and its cells to acute exercise that, when repeated, may induce compensatory hypertrophy of the heart, thus become important tasks for those wishing to know more about the presumed beneficial effect of physical exercise on cardiovascular health.

I mentioned earlier, the changes in mitochondrial mass per g myocardium in animals subjected to chronic overload of the heart are apparently of biphasic nature: there is first an increase, and then a decrease in this parameter. The same appears to be the case in animals subjected to more or less vigorous physical training. This is seen in figure 2, which illustrates and compares the changes in the ratio of mitochondrial to myofibrillar mass in the muscle cells of the heart during extended periods of physical training and of chronic cardiac overloading. The myofibrils and the mitochondria are, quantitatively speaking, the major subcellular structure of the myocardial cells, normally accounting, for approximately 1/2 and 1/3, respectively, of the total cellular mass. The curve constructed from the data of Bozner and Meesen (2) represents rats trained to swim 5–6 h uninterruptedly each day. A great increase in the ratio during the first 4 days, reflecting an increase in the mitochondrial compartment and not a disappearance of myofibrils, was followed by a decline that apparently had not

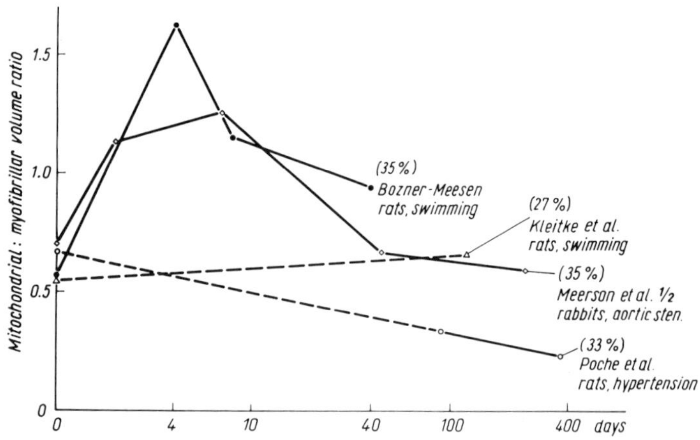

Fig. 2 Changes in the mitochondrial to myofibrillar mass ratio in the heart muscle cells of left ventricle in cardiac hypertrophy due to various causes. The figures in parentheses denote the degree of hypertrophy at the end of the experiments. The ratios determined by Meerson *et al.* (13) are divided by 2.

come to an end after 40 days, when the experiments were terminated. A similar time course was observed in the experiments of Meerson *et al.* (13) on rabbits with permanent aortic constriction. The figures in parentheses denote the degree of hypertrophy at the end of the experiments, expressed as the percentage of increase in the ratio of heart weight to body weight. The curve constructed from the data of Poche *et al.* (16) shows the reductions in the mitochondrial to myofibrillar mass ratio in the heart of a group of rats after 3 months and 1 year of experimental renal hypertension. A disappearance of mitochondria was observed. The first part of the curve is drawn as a broken line, because it is not certain that there was not an initial increase in the ratio. The other broken line in figure 2 represents a group of swimming rats studied in our laboratory (9). The training schedule, here, was less exacting than that of Bozner and Meesen's (2) rats, namely, 60–90 min of swimming each day. The mitochondrial: myofibrillar mass ratio had increased after 4 months of training; but again we do not know whether it may not perhaps have increased more strongly in the early part of the experiments. In all cases, there was a moderate hypertrophy of the heart. On the whole, these results indicate that, at the cellular level, compensatory cardiac growth is not a uniform process, the mitochondria apparently responding earlier to the stimulus of an increased mechanical demand than do the myofibrils. The difference might be related to the fact that the turnover of mitochondrial proteins is more rapid than that of the myofibrillar proteins.

Arcos *et al.* (1) compared the electron microscopic appearance of cardiac muscle of normal laboratory rats with that of rats that swam up to 6 h daily for about 40 days. The mitochondria were enlarged in the swimming rats, with otherwise normal, orthodox-like appearance. The mitochondrial mass per g heart was increased by 50%; the hearts were moderately hypertrophied. The mitochondria seemed to be in good functional condition, as indicated by a relatively high respiratory control index, showing that they were tightly coupled. Continuation of this strenuous training regimen for another 40 days led to a reduction in the density of packing of the mitochondrial cristae in the heart. This change was associated with a slackening of respiratory control, indicative of at least a mild degree of uncoupling of respiration and phosphorylation, with a reduction of the mitochondrial mass per g heart to the starting value, and with focal decreases in the activity of succinic dehydrogenase, a typical mitochondrial enzyme. The mitochondrial shrinkage and condensation of cristae that accompany uncoupling in heart muscle cells (3) may possibly have played a part in the observed morphologic changes. At this stage of training, the rats were no longer able to swim continuously for 6 h. They were evidently overworked and overtrained, and this fact was reflected, and had perhaps one of its causes, in the state of the cardiac mitochondria. In our experiments (9; see above), in which the rats swam daily for only 60 to 90 min, the animals showed no signs of fatigue, the mitochondria of the heart muscle cells had a normal appearance in the electron microscope, the cristae were very densely packed, and succinic dehydrogenase activity was not reduced, but was moderately augmented, in proportion to the increase in mitochondrial content of the tissue.

IMMEDIATE EFFECTS OF STRENUOUS EXERCISE

In the experiments of Arcos *et al.* (1) and in ours also (9), the rats were killed 18–24 h following the last swim, after they had had a chance to recover completely or partly from the stress of exercise. Now, what is the immediate effect of exercise on the cardiac mitochondria? This question was first studied by Laguens and his co-workers (11, 12), who let untrained dogs and rats swim to exhaustion or near-exhaustion, usually for 1/2 to 2 h. The animals were immediately killed, and samples of the myocardium were fixed and processed for electron microscopy. The electron micrographs revealed enlargement of the mitochondria in the heart muscle cells, swelling to the point of disruption, and apparent fusion of mitochondria, leading to the formation of greatly elongated structures. Strikingly similar changes had been observed several years earlier in our laboratory (22) in the hypertrophied failing heart of dogs with aortic constriction (fig. 3).

Planimetric evaluation of the electron micrographs of the myocardium of the animals killed after a single exhaustive swim indicated, in agreement with findings reported by other authors (2, 14), an increase in total mitochondrial mass (11). Laguens and Dumm (10) suggest that new mitochondria were formed in the heart during swimming, since they obtained autoradiographic evidence of DNA synthesis inside the heart mitochondria of the swimming animals. It was subsequently found (7) that swimming to near exhaustion also increases the capacity of the heart mitochondria to synthesize protein. As noted above, a similar phenomenon has been observed in mitochondria isolated from the heart of rats with aortic constriction, at the time that cardiac hypertrophy begins to develop, *i.e.*, less than 24 h after constriction (19).

COMMENTS

It is not yet possible to say how cardiac overloading and exercise bring about the effects just discussed. The reported stimulation of DNA synthesis and the increased protein synthetic capacity of the heart mitochondria under these conditions are probably related to and may partly account for, the increase in total and relative mitochondrial mass in the heart early in cardiac hypertrophy, whether the latter is induced by repeated vigorous exercise or by aortic constriction or other pathologic changes. Enhanced macromolecular synthesis in heart mitochondria may also be of relevance to the recovery of the tissue from the acute effects of strenuous exercise. This recovery has not yet been studied

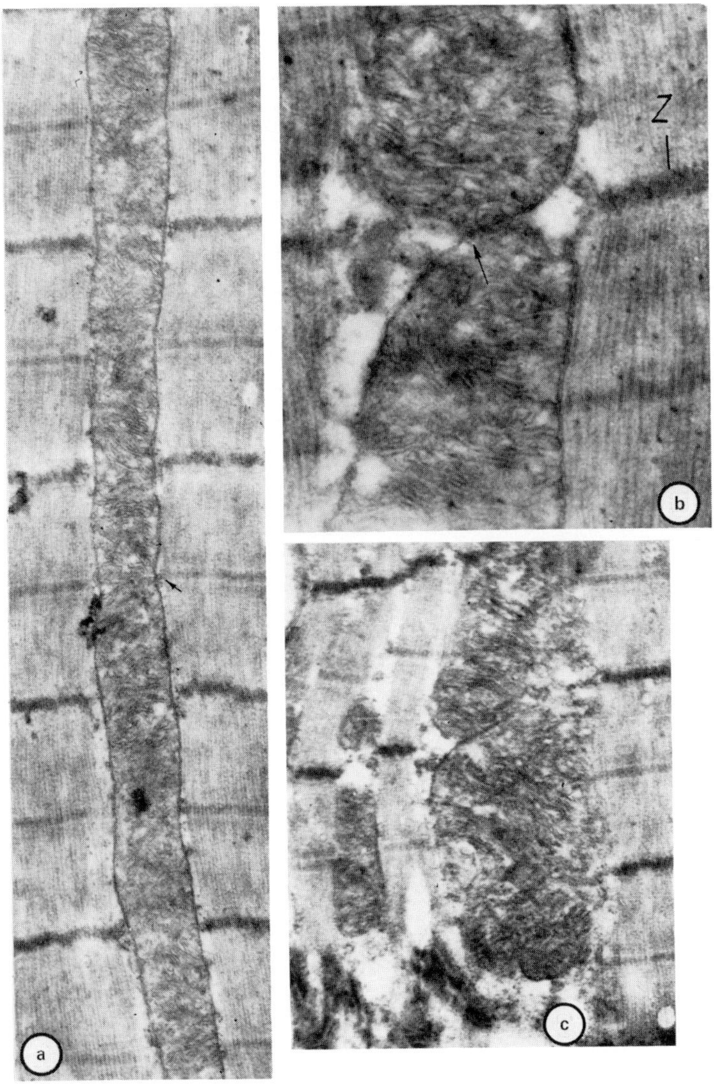

Fig. 3 Electron micrographs of myocardium of dogs with aortic constriction and cardiac hypertrophy, showing mitochondrial changes, *i.e.,* elongation (a), enlargement (c), and apparent fusion (b), resembling those seen in the tissue immediately following an exhaustive swimming bout (12). × 21,700 (a); × 42,000 (b); × 17,850 (c). From Wollenberger (21) and Wollenberger and Schulze (22).

biochemically and cytologically in heart muscle. Figure 4, which is taken from a recent paper by Gollnick and King (6), shows electron micrographs of fibers of the gastrocnemius muscle of trained rats immediately after, and 24 h after, a run to exhaustion. Right after the run, the mitochondria look swollen, the cristae are

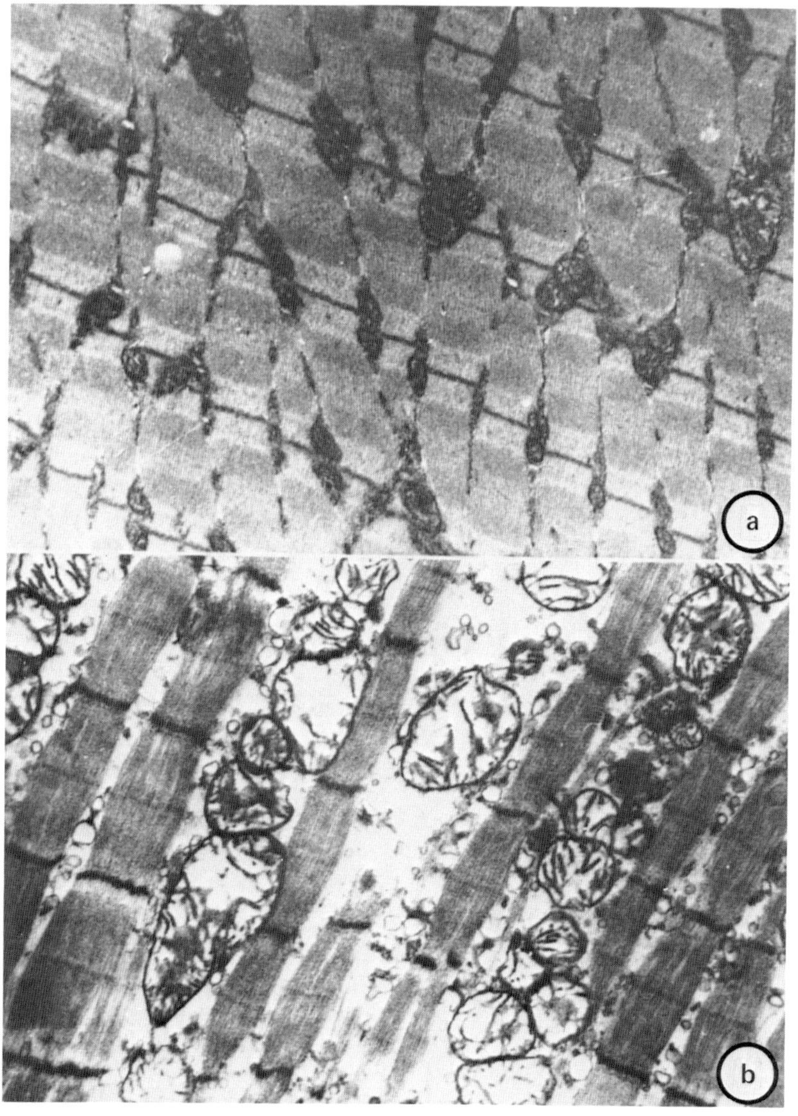

Fig. 4 Electron micrographs of skeletal muscle of trained rats immediately (b) and 24 h (a) after an exhaustive run. × 10,500. From Gollnick and King (6).

widely separated and have partly disappeared. These changes are fully reversible; after 24 h of rest, the muscle fibers have regained their normal appearance. Since the mitochondrial alterations immediately after running resemble those produced by hypoxia, the authors suggest that hypoxia may have been a factor responsible for these alterations.[1]

Exercise physiologists maintain that one of the most characteristic attributes of trained animals and athletes is their ability to recover comparatively quickly from the effects of fatiguing exercise. One therefore feels justified in predicting that the return from an exhausted, hypoxia-like cellular state (cf. fig. 4b) to the normal resting condition, a process in which mitochondrial repair presumably involving macromolecular syntheses appears to be a most notable feature, should occur more rapidly and easily in trained than it would in untrained muscle. Similar differences may conceivably exist between trained and untrained heart muscle and with respect to recovery from genuine hypoxia. This line of speculation leads us right to the question whether prior physical training may not have a favourable influence on the repair and recovery processes in hypoxia-injured cardiac tissue. If this should prove to be the case, we would perhaps know one of the reasons why physically active men have a higher rate of survival after myocardial infarction than do inactive men (5).

SUMMARY

The effects of chronic cardiac overload and of vigorous physical exercise and training upon the mass, structure, protein synthetic ability, and phosphorylative capacity of the heart mitochondria are compared and discussed, especially in relation to the compensatory hypertrophy of the heart associated with these conditions.

REFERENCES

1. ARCOS, J. C.; SOHAL, R. S.; SUN, S.-C.; ARGUS, M. F. and BURCH, G. E.: Changes in ultrastructure and respiratory control in mitochondria or rat heart hypertrophied by exercise. Exp. Mol. Path. *8:* 49–65 (1968).
2. BOZNER, A. und MEESEN, H.: Die Feinstrucktur des Herzmuskels der Ratte nach einmaligem und nach wiederholtem Schwimmtraining. Virchows Arch. Zellpath. Abt. B *3:* 248–269 (1969).
3. BUFFA, P.; GUARRIERRA-BOBYLEVA, V.; MUSCATELLO, U. and PASQUALI-RONCHETTI, I.: Conformational changes of mitochondria associated with uncoupling of oxidative phosphorylation *in vivo* and *in vitro*. Nature *226:* 272–274 (1970).
4. CHIDSEY, C. A.; WEINBACH, E. G.; POOL, P. E. and MORROW, A. G.: Biochemical studies of energy production in the failing human heart. J. Clin. Invest. *45:* 40–50 (1966).
5. FRANK, C. W.; WEINBLATT, E.; SHAPIRO, S. and SAGER, R. W.: Physical inactivity as a lethal factor in myocardial infarction among men. Circulation *34:* 1022–1033 (1966).

---

[1] Addendum: After completion of this manuscript, King and Gollnick (23) reported that they observed similar mitochondrial changes in the myocardium of rats exercised to the point of exhaustion.

6. GOLLNICK, P. D. and KING, D. W.: The immediate and chronic effect of exercise on the number and structure of skeletal muscle mitochondria; in POORTMANS, J. R., Biochemistry of Exercise, Medicine and Sport, Vol. 3, pp. 239–244 (S. Karger, Basel 1969).
7. HAMBERGER, A.; GREGSON, N. and LEHNINGER, A.: The effect of exercise on amino acid incorporation into mitochondria of rabbit tissues. Biochim. Biophys. Acta *186:* 373–383 (1969).
8. KISCH, B.; CAVUSOGLU, M. and MARANGONI, B. A.: Electron microscopic changes in the human heart in cardiac failure. Exp. Med. Surg. *17:* 85–100 (1959).
9. KLEITKE, B.; SCHULZE, W. und WOLLENBERGER, A.: Uber das Verhalten der Mitochondrien des Rattenherzens bei Belastung durch experimentelle Aortenkonstriktion und korperliches Training. Acta Biol. Med. German. *17:* 343–348 (1966).
10. LAGUENS, R. P. and DUMM, C. L. G.: Deoxyribonucleic acid synthesis in the heart mitochondria after acute and exhaustive exercise. Experientia *24:* 163–164 (1968).
11. LAGUENS, R. P. and GOMEZ-DUMM, C. L. A.: Fine structure of myocardial mitochondria in rats after exercise for one-half to two hours. Circulation Res. *21:* 271–279 (1967).
12. LAGUENS, R. P.; LOZADA, B. B.; DUMM, C. G. and BERAMENDI, A. R. R.: Effect of acute and exhaustive exercise upon the fine structure of heart mitochondria. Experientia *22:* 244–246 (1966).
13. MEERSON, F. Z.; ZALETAYEVA, T. A.; LAGUTCHEV, S. S. and PSHENNIKOVA, M. G.: Structure and mass of mitochondria in the process of compensatory hyperfunction and hypertrophy of the heart. Exp. Cell Res. *36:* 568–578 (1964).
14. PELOSI, G. and AGLIATI, G.: The heart muscle in functional overload and hypoxia. Lab. Invest. *18:* 86–93 (1968).
15. POCHE, R.: Submikroskopische Beitrage zur Pathologie der Herzmuskelzelle bei Phosphorvergiftung, Hypertrophie, Atrophie und Kaliummangel. Virchows Arch. Path. Anat. *331:* 165–248 (1958).
16. POCHE, R.; MATTOS, C. M. D. M.; REMBARZ, H.-W. und STOELPEL, K.: Uber das Verhaltnis Mitochondrien: Myofibrillen in den Herzmuskelzellen der Ratte bei Druckhypertrophie des Herzens. Virchows Arch. Path. Anat. Abt. A *344:* 100–110 (1968).
17. SCHULZE, W.: Unpublished observations.
18. SCHWARTZ, A. and LEE, K. S.: Study of heart mitochondria and glycolytic metabolism in experimentally induced cardiac failure. Circulation Res. *10:* 321–332 (1962).
19. SHAHAB, L. and WOLLENBERGER, A.: Amino acid incorporation into mitochondria from the hypertrophying heart of rats with aortic constriction. J. Mol. Cell. Cardiol. *1:* 143-155 (1970).
20. SOBEL, B. E.; SPANN, J. F.; POOL, P. E.; SONNENBLICK, E. H. and BRAUNWALD, E.: Normal oxidative phosphorylation in mitochondria from the failing heart. Circulation Res. *21:* 355-363 (1967).
21. WOLLENBERGER, A.: Die Mitochondrien im hypertrophen und insuffizienten Herzen; in WOLLHEIM, E. and SCHNEIDER, K. W., Eds., Herzinsuffizienz: Hämodynamik und Stoffwechsel, pp. 202–217 (Thieme-Verlag, Stuttgart 1964).
22. WOLLENBERGER, A. and SCHULZE, W.: Mitochondrial alterations in the myocardium of dogs with aortic stenosis. J. Biophys. Biochem. Cytol. *10:* 285–288 (1961).
23. KING, D. W. and GOLLNICK, P. D.: Ultrastructure of rat heart and liver after exhaustive exercise. Amer. J. Physiol. *218:* 1150–1155 (1970).

Mailing address: Prof. A. Wollenberger, Institut fur Kreislaufforschung, Deutsche Akademie der Wissenschaften zu Berlin, Lindenberger Weg 70, 115 *Berlin–Buch* (German Democratic Republic).

Hereditary Cardiomyopathy: A New Disease Model

Thorndike Memorial Laboratory and the Harvard Medical Unit,
Boston City Hospital, and Department of Medicine,
Harvard Medical School, Boston, Massachusetts, USA

HEMODYNAMICS OF THE HEREDITARY CARDIOMYOPATHY OF SYRIAN HAMSTERS [1]

W. H. ABELMANN, F. E. JEFFREY, and R. WAGNER

Congestive heart failure is the clinical syndrome common to the majority of diseases of the heart and constitutes one of the most frequently encountered problems in clinical practice. Yet, the underlying causes remain incompletely understood. Recent experimental work has focussed largely upon the myocardium itself, by means of studies of ultrastructure, histochemistry, and muscle mechanics. The paucity of animal models of naturally occurring chronic congestive heart failure (11) has limited investigation of the pathophysiology of this condition in the intact organism largely to man.

In the hereditary cardiomyopathy of Syrian hamsters, recently described by Bajusz *et al.* (3), a predictable model of chronic congestive heart failure has become available. It is the purpose of this communication to describe methods that have been developed to permit hemodynamic study of intact hamsters, to report normal values, and to present the hemodynamic profile of the decompensated stage of the hereditary cardiomyopathy of the Syrian hamster.

MATERIAL AND METHODS

A total of 13 male Syrian golden hamsters with hereditary cardiomyopathy of the BIO 14.6 strain (Bio-Research Institute, Cambridge, Massachusetts), 240- to 380-days old, and 15 age-matched, male, random-bred, healthy hamsters were studied. The animals were fed *ad libitum* until the time of study. At subsequent necropsy, all cardiomyopathic hamsters showed evidence of advanced congestive heart failure, with cardiac dilatation, hepatomegaly, ascites, pleural effusions, and generalized subcutaneous edema.

All animals were anesthetized with sodium pentobarbital 40-60 mg/kg given intraperitoneally. The electrocardiogram was recorded with needle electrodes applied to the extremities. The trachea was incised and cannulated, and the animals were ventilated with ambient air by means of a rodent respirator (model 680, Harvard Apparatus Company, Millis, Mass.). A 1 cm long left parasternal incision was made, and the pectoral muscles were retracted, exposing the rib cage. Both right and left ventricles were then entered by transthoracic puncture

[1] Supported by Grants HE 5244 and HE 10539 from the National Heart and Lung Institute, United States Public Health Service.

with no. 26 needles. Left ventricular blood was analyzed for pO_2 and pH in 8 normal and 7 myopathic animals, using an ultra-micro gas analyzer (model 113-S1, Instrumentation Laboratory, Lexington, Mass.). Arterial oxygen tension was 72.1 ± 10.9 mm Hg (mean ± S.D.) in normal and 75.7 ± 17.3 mm Hg in myopathic hamsters; pH was 7.38 ± 0.12 in normal and 7.49 ± 0.13 in myopathic hamsters. The differences between the two groups of hamsters were not statistically significant.

Oxygen consumption was calculated from the oxygen content of expired air, measured paramagnetically (Oxygen Analyzer, type OA 101 Mk II, Servomex Controls Ltd., Crowborough, Sussex, England) and the minute ventilation obtained by calibrating the respirator pump. A factor correcting for CO_2 produced was determined by analysis of expired air from several animals in each group for CO_2 by means of the micro-Scholander method. The arterio-venous oxygen difference was derived from the oxygen content of right and left ventricular blood, measured by means of the micromanometric method of Van Slyke and Plazin (17). The cardiac output was calculated, according to the Fick equation, from oxygen consumption and arterio-venous oxygen difference obtained simultaneously. The hematocrit of left ventricular blood was measured by the micromethod.

Right and left ventricular pressures were measured with a high fidelity fiberoptic manometer (American Optical Corporation, Framingham, Mass.) as described previously (14), and recorded on a direct writing oscillograph (Sanborn 964, Hewlett-Packard Company, Waltham, Mass.). The experimental arrangement is illustrated schematically in figure 1.

Measurements are given as means ± standard deviations. The significance of differences was tested by means of Student's *t* test.

RESULTS

Body weight, heart weight, and hematocrit

The body weight of cardiomyopathic hamsters (151.6 ± 13.7 g) significantly ($p < 0.001$) exceeded that of normal animals (114.3 ± 10.3 g). This represented an increase of 33%. Hearts of myopathic animals weighed 747 ± 110 mg compared with control weights of 418 ± 46 mg, a difference of 79% ($p < 0.001$).

The hematocrit of myopathic hamsters was 37.1 ± 5.0% compared with 42.4 ± 3.0% for normal animals ($p < 0.005$).

Heart rate and ventricular pressures

Cardiomyopathy was associated with a significant bradycardia of 288 ± 35 beats/min compared with a control heart rate of 389 ± 60 beats/min (fig. 2).

In the presence of decompensated cardiomyopathy, left ventricular systolic pressure was significantly decreased from the normal 121.5 ± 26 mm Hg to

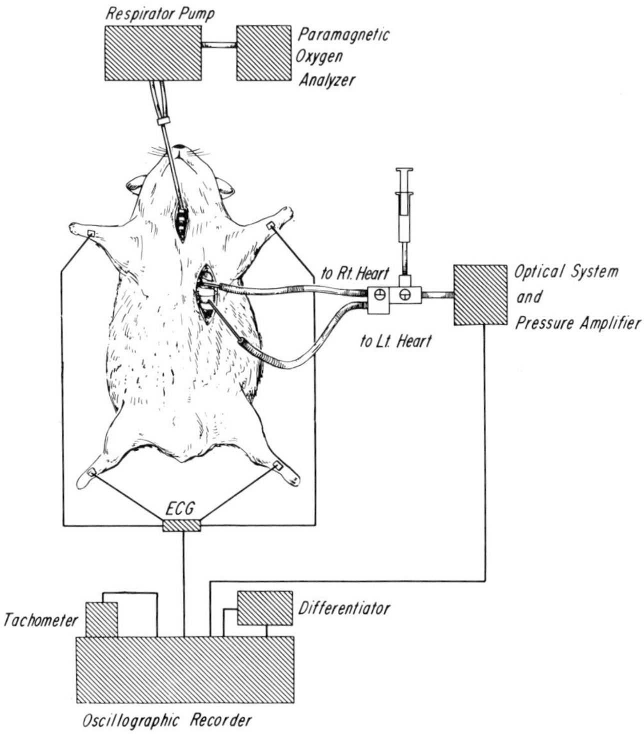

Fig. 1 Schema of the experimental arrangement.

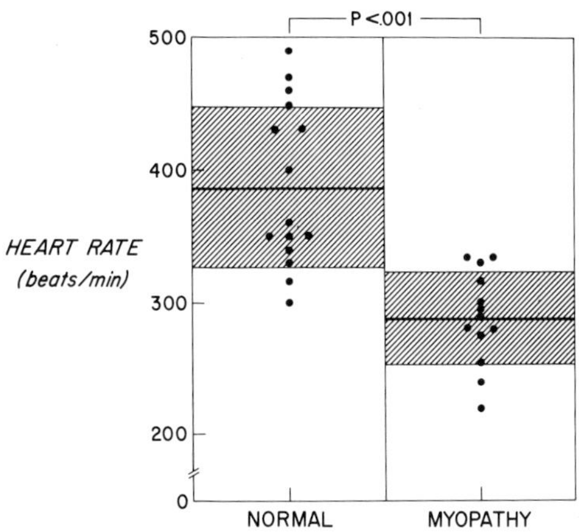

Fig. 2 The heart rate of 13 myopathic hamsters is compared with that of 15 normal control animals. Individual values are plotted, means indicated by horizontal lines, and standard deviations of the mean by shaded areas. The significance of the difference between the means is indicated.

103.0 ± 20 mm Hg ($p < 0.005$), and left ventricular end-diastolic pressure significantly increased from a control of 10.1 ± 4.4 mm Hg to 24.8 ± 9.7 mm Hg ($p < 0.001$) (fig. 3). The myopathic right ventricle was found to have a systolic pressure increased from 32.3 ± 8.3 mm Hg to 42.3 ± 9.0 mm Hg ($p < 0.001$) and an end-diastolic pressure elevated from a control value of 4.3 ± 3.6 mm Hg to 13.0 ± 4.5 mm Hg ($p < 0.001$) (fig. 4).

Oxygen consumption, arterio-venous oxygen difference, cardiac output, and stroke volume (figs. 5 and 6)

Oxygen consumption in the myopathic animals was identical to that in the controls, *i.e.*, 1.31 ± 0.48 and 1.26 ± 0.31 ml/min STP, respectively. The arterio-venous oxygen difference was 7.34 ± 2.49 vol% in the myopathic hamsters, a figure significantly less than the control value of 9.32 ± 2.34 vol% (fig. 5). Cardiac output, therefore, was numerically greater in the heart failure animals than in the controls, 18.3 ± 5.7 ml/min compared with 14.7 ± 6.4 ml/min, but this difference was not statistically significant. Stroke volume, however, was significantly greater in the myopathic animals than in controls, *i.e.*, 0.063 ± 0.014 ml/beat compared with 0.038 ± 0.016 ml/beat ($p < 0.001$) (fig. 6).

Peripheral vascular resistance

In the absence of measurements of systemic arterial pressure, exact calculation of peripheral vascular resistance is not possible. However, approximations of this parameter have been made by means of the following formula and

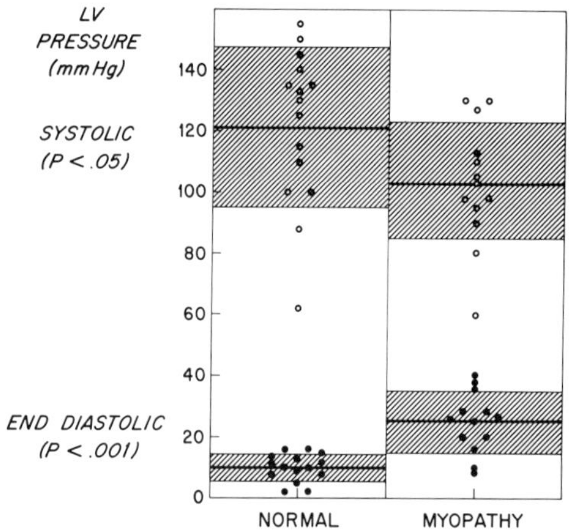

Fig. 3 Left ventricular systolic (circles) and end-diastolic (dots) pressures of normal and myopathic hamsters. Presentation as in figure 2.

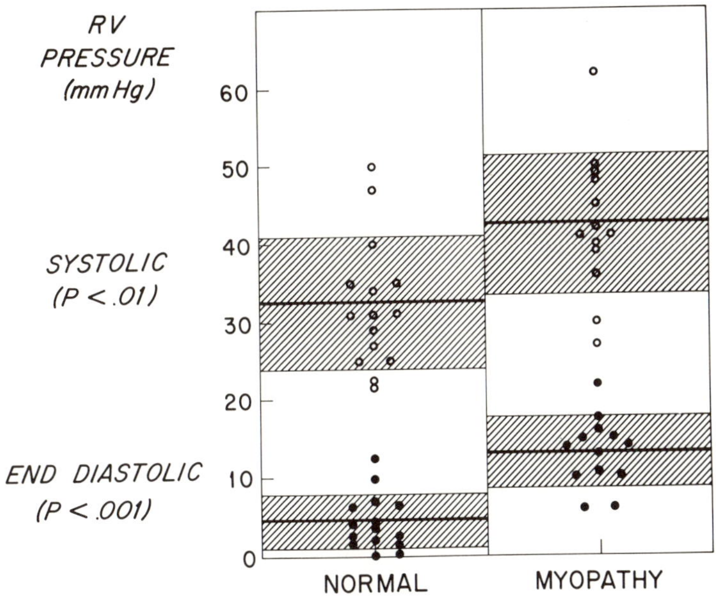

Fig. 4 Right ventricular systolic (circles) and end-diastolic (dots) pressures of normal and myopathic hamsters. Presentation as in figure 2.

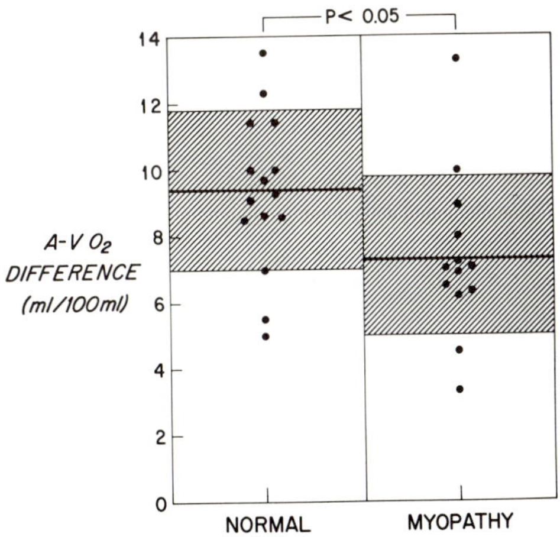

Fig. 5 Arterio-venous oxygen difference of normal and myopathic hamsters. Presentation as in figure 2.

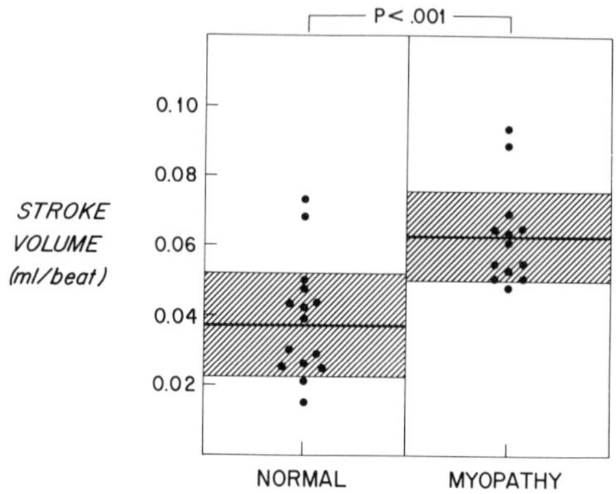

Fig. 6 Stroke volume of normal and myopathic hamsters. Presentation as in figure 2.

expressed as an index, the mean value obtained for normal control hamsters being set at 100.

Index of peripheral vascular resistance

$$= \frac{\begin{pmatrix}\text{Left ventricular}\\ \text{systolic pressure}\end{pmatrix} - \begin{pmatrix}\text{Right ventricular}\\ \text{diastolic pressure}\end{pmatrix}}{\text{Cardiac output}}$$

The index of peripheral vascular resistance derived in this manner for myopathic hamsters was 58 ± 14, *i.e.*, significantly depressed by 42% ($p < 0.005$).

DISCUSSION

Hemodynamics of the normal hamster

The measurements of left ventricular pH and pO_2, which according to the oxygen dissociation curve determined for the Syrian golden hamster by Ulrich *et al.* (16) correspond to a saturation for oxygen of about 95%, indicate that the animals were adequately ventilated. The heart rates measured in the present study closely approach the value of 400±25 (S.D.) beats/min reported by Hoenig and Mohr (8), also obtained under pentobarbital anesthesia. The present values for left and right ventricular pressure are essentially identical with those obtained in this laboratory by direct ventricular puncture at open thoractomy (9).

The oxygen consumption of 11.0 ml/kg/min in anesthetized hamsters of the present series is somewhat lower – as expected – than the resting oxygen consumption of 15.2 ml/kg/min reported for unanesthetized golden hamsters by Adolph and Richmond (2). To our knowledge, neither arterio-venous oxygen difference nor cardiac output have ever been measured in the hamster. In comparison with those recorded in other mammals, the arterio-venous oxygen differences measured here appear rather large. It must be remembered, however, that the animals were under barbiturate anesthesia, which may depress cardiac function.

The cardiac output of 0.13 l/kg/min is significantly below the 0.26 l/kg/min value measured in the anesthetized rat (4), but approaches the 0.15 l/kg/min and 0.12 l/kg/min values reported for the ferret (10) and rabbit (13), respectively.

Hemodynamics of the myopathic hamster

The late stage of the hereditary cardiomyopathy of the golden Syrian hamster was characterized by a dilated heart of increased weight, visceral congestion, pleural effusion, ascites and subcutaneous edema, and was associated with a weight gain of 32.6% and a fall in hematocrit of 12.5% as compared with controls. Both changes are probably entirely attributable to fluid retention associated with congestion. Although the significant bradycardia is at first surprising inasmuch as congestive heart failure is usually associated with tachycardia, we have previously reported significant bradycardia in another experimental model of cardiomyopathy in rodents, namely subacute myocarditis due to *Trypanosoma cruzi* (1).

The left and right ventricular pressures, both systolic and end-diastolic, all differed significantly from the control studies. Elevated right and left ventricular filling pressures associated with decreased left ventricular systolic and increased right ventricular systolic pressure are characteristic of biventricular heart failure. These results are fully compatible with the gross and microscopic appearance of the lungs, liver, and other viscera at this stage (3).

The vast majority of clinical states of biventricular congestive heart failure are associated with an abnormally low cardiac output and widened arterio-venous oxygen difference. In the present model of congestive heart failure, however, cardiac output not only failed to be below the normal control, but actually exceeded the control values by 25%, although this difference was not statistically significant. This normal to elevated cardiac output was associated with an unchanged oxygen consumption, an arterio-venous oxygen difference significantly narrowed by 21%, and a stroke volume significantly increased by 55%. This is the classic hemodynamic profile of so-called high output failure, a syndrome often associated with pathophysiologic alterations in the systemic circulation (7). Indeed, the significantly decreased index of peripheral vascular resistance suggests either dilatation of systemic resistance vessels or perfusion of an increased number of such vessels.

On the basis of the present data, high output heart failure due to anoxia, anemia, and hypermetabolism may be considered ruled out as the pathogenetic mechanism. It is difficult to exclude a possible role that the diffuse skeletal myopathy itself might play in peripheral vasodilatation and increased systemic blood flow; but evidence for such a mechanism is lacking.

The possibility must now be considered that the progressive expansion of extracellular fluid volume (5) – and presumably also of the intravascular and circulating blood volumes – may play a pathogenetic role in the congestive heart failure associated with the hereditary cardiomyopathy of the hamster, rather than represent a pathophysiologic consequence. In support of this possibility, Bajusz and his associates (3) failed to find a clear relationship between severity of myocardial lesions and heart failure and, furthermore, demonstrated that parabiosis between myopathic and healthy hamsters prevented the development of congestive heart failure without influencing the pathologic lesions of cardiac and skeletal muscle. A component of circulatory congestion (6), independent of myocardial disease, may thus be operant in this disease of hamsters and may precipitate or enhance cardiac failure. It is also of interest that recent serial studies of muscle mechanics of the myopathic hamster by Lochner *et al.* (12) demonstrated depressed contractility and development of myocardial tension, which appeared late in the disease, long after the appearance of histopathologic abnormalities.

Finally, it is noteworthy that in progressive muscular dystrophy in man, cardiac output has been found elevated (15). Thus, further investigation of the pathophysiology of congestive heart failure in the myopathic hamster, a unique model of high output heart failure, may also have implications for human myopathies.

SUMMARY

The hemodynamics of the hereditary cardiomyopathy of Syrian hamsters of the BIO 14.6 strain were assessed in 13 intact myopathic males in chronic congestive heart failure and compared with 15 healthy control animals. Under pentobarbital anesthesia, mechanical ventilation assured normal arterial oxygen tension and pH. Transthoracic needles were placed into the left and right ventricles. Ventricular pressures were measured with a high fidelity fiberoptic manometer, oxygen contents of right and left ventricular blood by the micromanometric method of Van Slyke, and oxygen consumption by paramagnetic analysis of expired gas.

Congestive heart failure in the myopathic hamster was characterized by significant bradycardia, elevation of the end-diastolic pressures of both ventricles, decrease in left ventricular systolic pressure and increase in right ventricular systolic pressure. Whereas the oxygen consumption did not differ between cardiomyopathic and normal hamsters, the arterio-venous oxygen difference was significantly narrowed in the former, resulting in a cardiac output that was normal to slightly increased. Stroke volume was significantly increased in the cardiomyopathic animals. An index of peripheral vascular resistance was found to be decreased 42% below the norm.

It is concluded that the chronic congestive heart failure associated with hereditary cardiomyopathy of the Syrian hamster is of the high output type. Vasodilatation or perfusion of an increased number of resistance vessels in the systemic circulation is postulated.

ACKNOWLEDGEMENTS

The authors are most grateful to Drs. F. Homburger and E. Bajusz for supplying the hamsters used in this study, and to Miss Nancy W. Hewett, Mr. Curtis T. Jones, and Mr. David Rhodes for technical assistance.

REFERENCES

1. ABELMANN, W. H.; KUMAR, R. and WAGNER, R.: Biventricular heart failure in experimental Chagasic myocarditis. Circulation *40* (Suppl. III): 33 (1969).
2. ADOLPH, E. F. and RICHMOND, J.: Adaptation to cold in the golden hamster and ground squirrel measured chiefly by rates of body cooling. J. Appl. Physiol. *9:* 53–58 (1956).
3. BAJUSZ, E.; BAKER, J. R.; NIXON, C. W. and HOMBURGER, F.: Spontaneous, hereditary myocardial degeneration and congestive heart failure in a strain of Syrian hamsters. Ann. N.Y. Acad. Sci. *156:* 105–129 (1969).
4. BLOOD, F. R.; SMITH, D. L. and D'AMOUR, F. E.: Cardiac output in the rat at normal and at high altitudes and its relationship to gas embolism. Amer. J. Physiol. *163:* 268–271 (1950).
5. CANTIN, M., and BAJUSZ, E.: Juxtaglomerular apparatus and adrenal cortex in hamsters with heart failure. Relationship between morphologic changes, serum electrolytes, and extracellular fluid volume. Arch. Path. *87:* 626–636 (1969).
6. EICHNA, L. W.: Circulatory congestion and heart failure. Circulation *22:* 864–886 (1960).
7. GUYTON, A. C.: Circulatory Physiology: Cardiac Output and Its Regulation. (Saunders, Philadelphia 1963).
8. HOENIG, W. und MOHR, W.: Elektrokardiographische Studien beim Goldhamster, einem neuerdings viel verwendeten Laboratoriumstier. Z. Tropenmed. Parasit. *4:* 117–121 (1953).
9. JEFFREY, F. E.; WAGNER, R. and ABELMANN, W. H.: Hemodynamics of biventricular heart failure in a natural animal model of cardiomyopathy. Clin. Res. *18:* 115 (1970).
10. KEMPF, J. E. and CHANG, H. T.: The cardiac output and circulation time of ferrets. Proc. Soc. Exp. Biol., N.Y. *72:* 711–714 (1949).
11. KUMAR, R.; KLINE, I. K. and ABELMANN, W. H.: Experimental *Trypanosoma cruzi* myocarditis. Relative effects upon the right and left ventricles. Amer. J. Path. *57:* 31–48 (1969).
12. LOCHNER, A.; BRINK, A. J. and VAN DER WALT, J. J.: The significance of biochemical and structural changes in the development of the myocardiopathy of the Syrian hamster. J. Mol. Cell. Cardiol. *1:* 47–64 (1970).
13. ODAIRA, T.: Studien uber Gasstoffwechsel und Minutenvolum. I. Mitteilung. Die Beziehung des Gasstoffwechsels und Minutenvolums zur inneren Sekretion. Tohoku J. Exp. Med. *6:* 325–366 (1925).
14. RAMIREZ, A.; HOOD, W. B.; POLANYI, M.; WAGNER, R., YANKOPOULOS, N. A. and ABELMANN, W. H.: Registration of intravascular pressure and sound by a fiberoptic catheter. J. Appl. Physiol. *26:* 679–683 (1969).
15. SUNDERMEYER, J. F.; GUDBJARNASON, S.; WENDT, V. E.; DEN BAKKER, P. B. and BING, R. J.: Myocardial metabolism in progressive muscular dystrophy. Circulation *24:* 1348–1355 (1961).

16. ULRICH, S.; HILPERT, P. und BARTELS, H.: Uber die Atmungsfunktion des Blutes von Spitzmäusen, weissen Mäusen und syrischen Goldhamstern. Pflügers Arch. ges. Physiol. *277:* 150–165 (1963).

17. VAN SLYKE, D. D. and PLAZIN, J.: Micromanometric Analysis (Williams and Wilkins, Baltimore, 1961).

Mailing address: Walter H. Abelmann, M. D., Thorndike Memorial Laboratory, Boston City Hospital, 818 Harrison Avenue, Boston, Massachusetts 02118 (USA).

Division of Myocardial Biology, Baylor College of Medicine, and the
Fondren-Brown Cardiovascular Research and Training Center,
Methodist Hospital, Houston, Texas, and
Bio-Research Institute, Boston, Massachusetts, USA

SEVERAL BIOCHEMICAL CHARACTERISTICS OF THE CARDIOMYOPATHIC SYRIAN HAMSTER [1]

A. SCHWARTZ,[2] L. A. SORDAHL, C. A. CROW, W. B. McCOLLUM,
S. HARIGAYA, and E. BAJUSZ

Two important biochemical aspects of cardiac muscle obtained from the cardiomyopathic Syrian hamster (strain BIO 14.6) were studied, one representing energy liberating and storage mechanisms and the other, energy utilization. These include mitochondria and the cardiac relaxing system (CRS) (probably representing elements of the sarcoplasmic reticulum). With respect to mitochondria, both ultrastructural and oxidative phosphorylation measurements were carried out, particularly related to the development of the disease. Mitochondria were isolated in a KCl-EDTA medium, which appears to be optimal for the maintenance of both structure and function in heart mitochondria. Both control (LSH strain) and experimental animals, ranging in age from 20 to 300 days, were employed.

Ultrastructurally, the tissue from both right and left ventricle, as well as apex of hearts taken from 20-day-old Syrian hamsters, revealed an increase in the number of mitochondria as compared with controls. In addition, the size of mitochondria was smaller. At 80 days, the experimental hearts were in an early hypertrophic phase and exhibited peak lesion development. Mitochondria were still numerous, but many were already degenerating with respect to cristae. At 126-127 days, the hearts exhibited peak hypertrophy with very numerous mitochondria, the latter tending to collect in areas or clumps, rather than exhibiting the usual even distribution observed in normal hearts. In addition, elements of the sarcoplasmic reticulum were significantly dilated. Numerous mitochondria were still evident, although many were in degenerative phases; healed lesions were evident and definite myofibrillar degeneration was noted (figs. 1 to 4).

Isolated mitochondria from all phases of development exhibited biochemical characteristics that were consistent with the ultrastructural observations described. For example, at 20-30 days, mitochondria showed higher oxygen consumption and respiratory control than similarly isolated normal mitochon-

[1] Supported by United States Public Health Service grants HE 05435, HE 07906, and HE 09791.

[2] Recipient of a Career Development Award (K_3 HE 11,875).

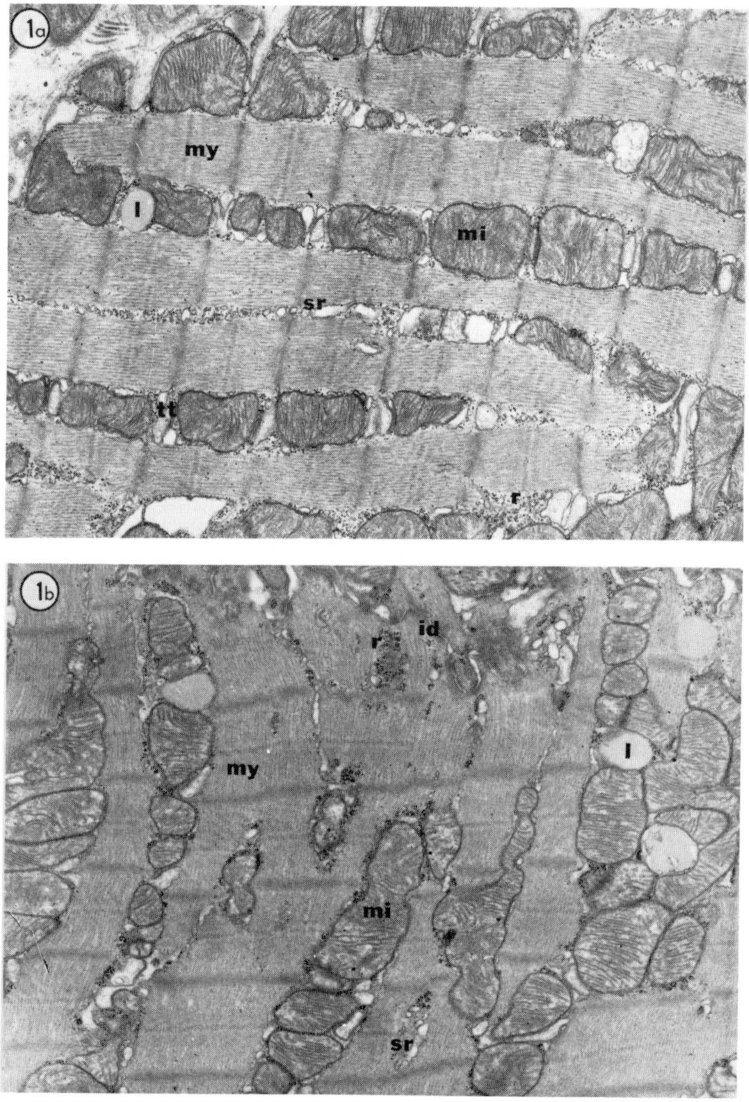

Fig. 1 a. Control hamster, 20 days. Mitochondria (mi) occur in rows between myofibrils (my) and are aggregated at the pole of the nucleus. × 9,700. b. Cardiomyopathic hamster, 20 days. Mitochondria are more numerous than in control myocardium and occur in rows and small clumps between the myofibrils. × 10,050. sr = sarcoplasmic reticulum; tt = transverse tubule; l = lipid droplet; id = intercalated disc; r = ribosomes.

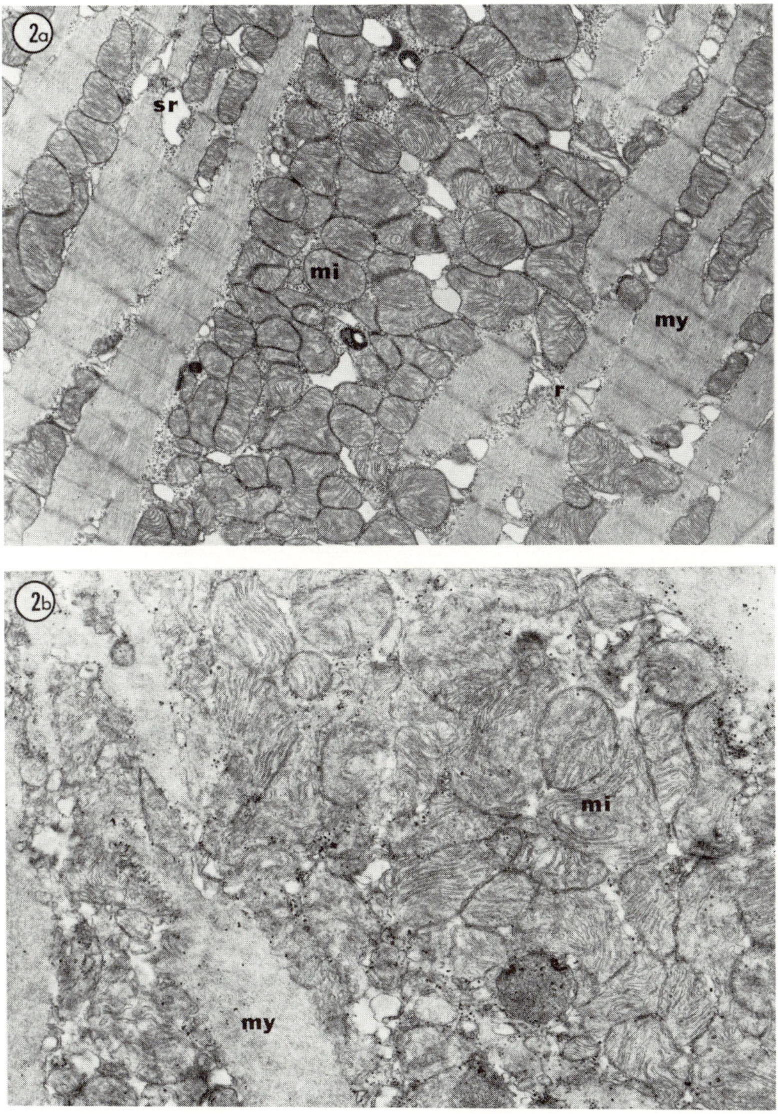

Fig. 2 a. Cardiomyopathic hamster, 80 days. During the early phase of hypertrophy, clumps of mitochondria (mi) grow larger. In most cells, the cristae and matrix appear normal. Elements of the sarcoplasmic reticulum (sr) are dilated. Ribosomes (r) are abundant. × 6,075. b. Cardiomyopathic hamster, 80 days. In degenerating areas of the myocardium, mitochondria and myofibrils are disrupted. × 15,400.

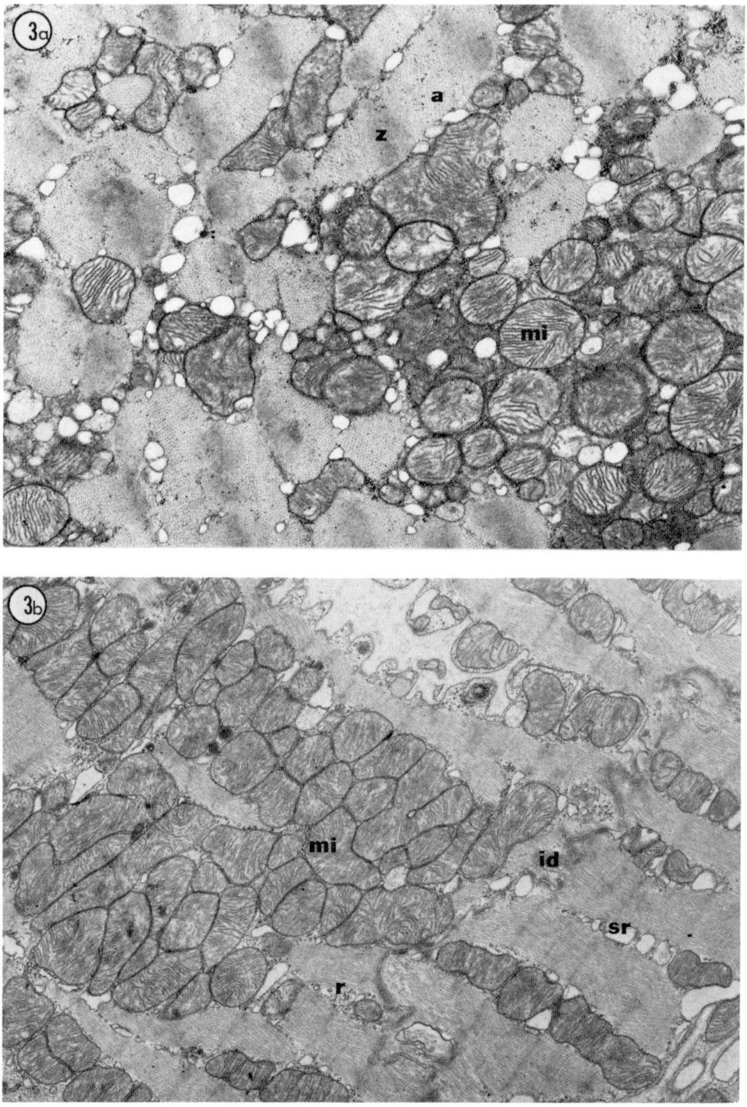

Fig. 3 a. Cardiomyopathic hamster, 80 days. Alterations in the pattern of the cristae and in the density of the matrix occur in the mitochondria (mi) of some cells during early hypertrophy. × 9,300. b. Cardiomyopathic hamster, 127 days. During peak hypertrophy, large clumps of mitochondria, abundant ribosomes (r) and dilated elements of the sarcoplasmic reticulum (sr) are characteristic features. × 7,400.

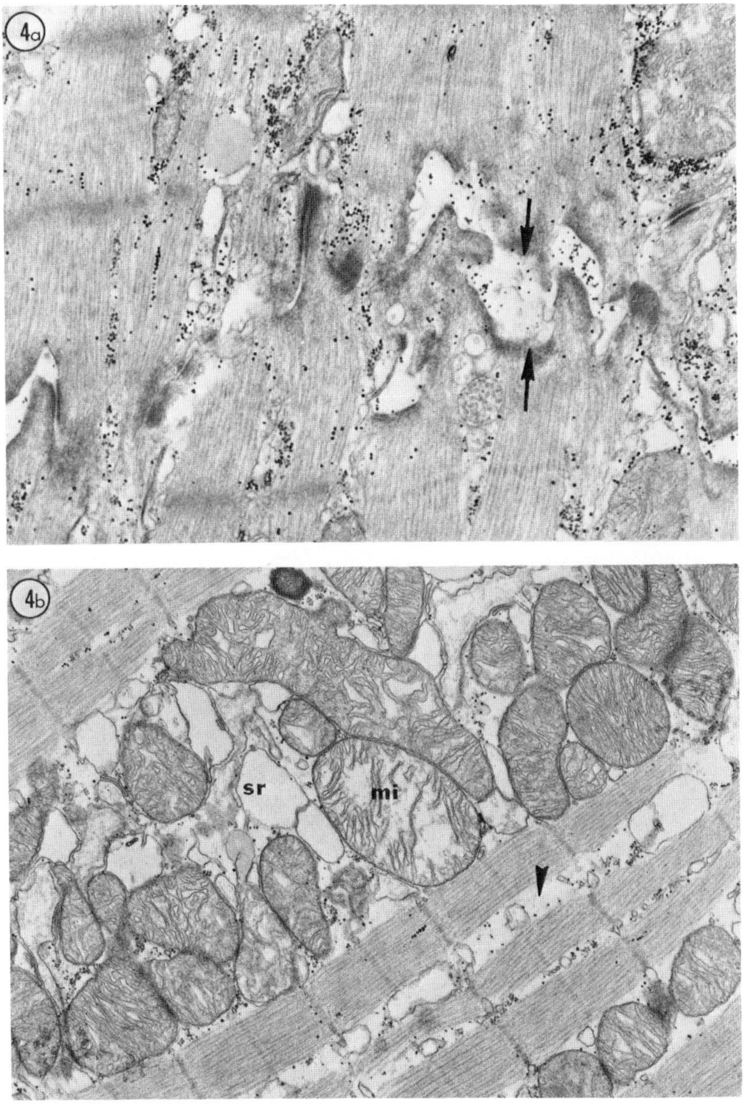

Fig. 4 a. Cardiomyopathic hamster, 303 days. During the stage of dilatation and terminal failure, alterations in many cells resemble those caused by hypoxia. The membranes of many intercalated discs are separated (arrows). × 16,200. b. Cardiomyopathic hamster, 303 days. Elements of the sarcoplasmic reticulum (sr) are significantly dilated. Mitochondria (mi) often show swelling and loss of matrix density. Intracellular edema is evident (arrowhead). × 13,200.

dria from the same age group. At 80 days, both of these parameters were depressed; at 126-127 days, there appeared to be a return to the normal state and, with progressive congestive heart failure, mitochondria showed depressed oxygen consumption and respiratory control. There was little or no change in ADP : O ratio in any of the stages. The most dramatic changes occurred with NADH-linked substrates. Ultrastructural and biochemical characteristics of human cardiomyopathy hearts resembled those described for the Syrian hamster. Isolated sarcoplasmic reticulum vesicles (CRS) from hearts in congestive heart failure exhibited diminished *rate* of calcium binding, with respect to controls. This finding is consistent with the ultrastructural observations reported above.

In addition to the cardiomyopathic strain BIO 14.6, two new strains were employed: BIO 82.62 and BIO 40.54, both characterized by a more rapidly developing congestive heart failure. Table I shows quite clearly that both ATP-dependent *binding*, as well as uptake (in the presence of oxalate), of all three cardiomyopathic strains (BIO. 82.62, BIO 40.54, and BIO 14.6) were significantly depressed with respect to controls of the same age group. The birth date of all animals employed in this study is August 26, 1969. All experimental animals exhibited significant congestive heart failure characterized by a marked increase in heart weight/body weight ratio (table II), and varying degrees of ascites, hydrothorax and/or subcutaneous edema. Employing a murexide-spectrophotometric assay (1), the rate of calcium binding was found to be significantly lower in all of the cardiomyopathic strains as compared with the controls (table III). Some typical direct tracings are shown in figure 5 and a comparison of binding and release is exhibited in figure 6.

Before any definitive conclusions can be drawn, it should be emphasized that

Table I

Maximal Calcium Binding and Uptake in Isolated Cardiac Relaxing System (Sarcoplasmic Reticulum) from Three Strains of Cardiomyopathic Syrian Hamsters

	BIO-RB (control)	BIO 14.6	BIO 82.62	BIO 40.54
		nmoles Ca^{++}/mg protein		
ATP-dependent binding	49 (17)	30 (18)	36 (16)	36 (18)
ATP-dependent uptake	3700 (17)	2800 (18)	2850 (16)	3350 (18)

Assay and incubation methods and techniques were as described previously by Harigaya and Schwartz (1). Temperature for binding = 25°C; time = 5 min. Temperature for uptake = 32°C; time = 30 min. pH = 6.8. The numbers in parentheses refer to the number of animals employed in each group. Calcium content determined by atomic absorption spectrophotometry, in CRS fractions of each group, BIO-RB, BIO 14.6, BIO 82.62, and BIO 40.54, respectively (in nmoles/mg protein) = 10, 14, 13, and 12.

Table II

Some Characteristics of Cardiomyopathic Hamsters

Strain	Body weight	Heart weight	HW/BW × 10^3
BIO-RB (Control	128.3 ± 2.8	0.49 ± 0.02	3.83
BIO 14.6	141.2 ± 2.6	0.77 ± 0.04	5.52
BIO 82.62	124.1 ± 2.2	0.78 ± 0.05	6.22
BIO 40.54	133.3 ± 2.4	0.63 ± 0.02	4.72

The values ± SEM are listed.

Table III

The Rate of Calcium Binding of CRS from Cardiomyopathic Syrian Hamsters

Strain	Binding (nmoles calcium/mg protein/3 sec)
BIO-RB (Control)	17.1 ± 0.7
BIO 14.6	10.0 ± 0.5
BIO 82.62	8.3 ± 1.4
BIO 40.54	12.3 ± 1.9

A murexide dual-beam spectrophotometric procedure (1) was employed. Values ± SEM are listed.

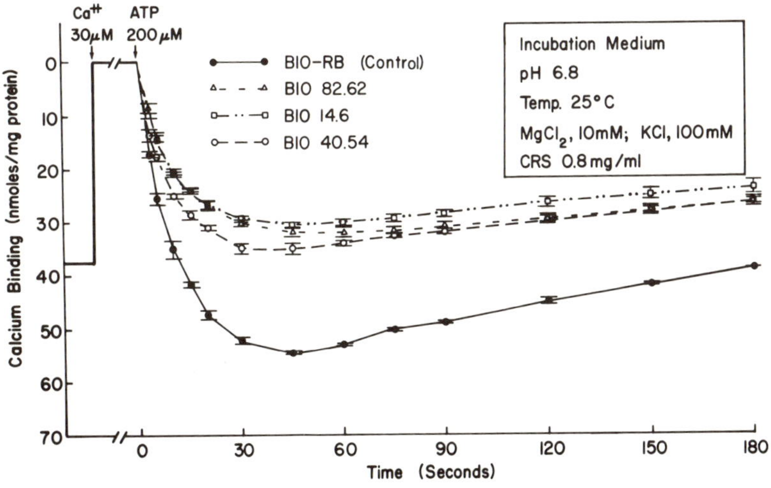

Fig. 5 Calcium binding by CRS from control and cardiomyopathic hamsters–dual beam spectrophotometer.

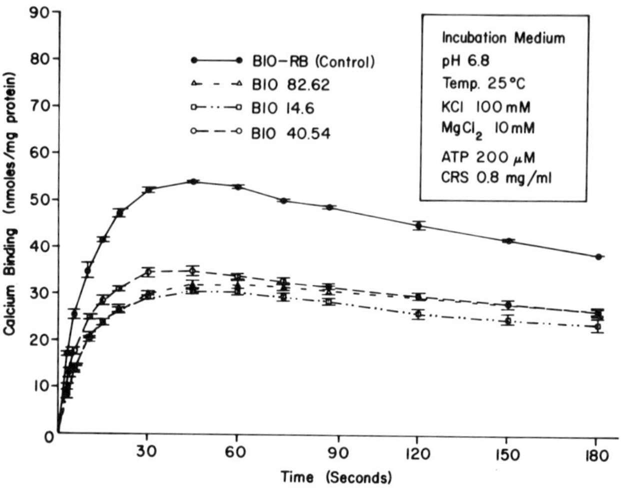

Fig. 6 Calcium binding by CRS from control and cardiomyopathic hamsters.

complete characterization of the CRS fraction, both biochemical and morphological, is most important.

SUMMARY

In three strains of cardiomyopathic Syrian hamsters (BIO 14.6, BIO 82.62 and BIO 40.54), isolated sarcoplasmic reticulum cardiac relaxing system showed significant depression of calcium binding and release as compared with controls. The dependence of calcium binding appeared to correlate with the failing process. These results were found both in the hypertrophic phase and in the severely failing animal. Mitochondria were isolated from cardiomyopathic Syrian hamsters in various stages of hypertrophy and congestive failure. During the early hypertrophic phase, there was a demonstrable and significant increase in numbers of mitochondria as well as a significant increase in respiratory control and oxygen consumption. Both electron microscopic studies and biochemical experiments were employed to determine these results. During the later stage of hypertrophy and early failure, degeneration of mitochondria was noted, accompanied by a decrease in phosphorylating oxygen consumption. During severe congestive failure, respiratory control of mitochondria as well as phosphorylating oxygen consumption were significantly depressed as compared with controls.

REFERENCES

1. HARIGAYA, S. and SCHWARTZ, A.: The rate of calcium binding and uptake in animal and human cardiac muscle: Membrane vesicles ("relaxing system") and mitochondria. Circulation Res. *25:* 781–794 (1969).

Mailing address: Dr. Arnold Schwartz, Division of Myocardial Biology, Baylor College of Medicine, Texas Medical Center, Houston, Texas, 77025 (USA).

Cardiovascular Division, Harvard Medical School, Peter Bent Brigham Hospital, and Bio-Research Institute, Boston, Massachusetts, USA

A BIOCHEMICAL DEFECT IN THE FUNCTION OF THE SARCOPLASMIC RETICULUM IN THE HEREDITARY CARDIOPATHY OF THE SYRIAN HAMSTER [1]

E. W. GERTZ, A. STAM, Jr., E. BAJUSZ, and E. H. SONNENBLICK

The current concept of excitation-contraction coupling casts the sarcoplasmic reticulum in the role of the principal controller of contractility. Under physiologic conditions the sarcoplasmic reticulum assumes a major role in supplying calcium for contraction while the ATP dependent calcium pump of the sarcoplasmic reticulum is thought to bring about relaxation by sequestering the calcium from the myofibrils (11, 14, 16). In previous studies, it was suggested that a depression of the sarcoplasmic reticular ATP dependent calcium pump may play a role in the etiology of myocardial failure (3, 5). It has been theorized that were this pump to be depressed, it would pump less calcium per unit of time, leading to a loss of calcium from the sarcoplasmic reticulum and, eventually, to less calcium for contraction.

Although no direct cause and effect relationship has been demonstrated, certain drugs, such as barbiturates and local and general anesthetics, which cause depression of contractility of cardiac and skeletal muscles (1, 12, 13), have been shown to depress the ATP dependent calcium pump of skeletal and/or cardiac sarcoplasmic reticulum (3, 4, 8, 10, 14). Further, in acute heart failure in dogs (5) and chronic heart failure in cattle (18) and humans (6), a depression of the ATP dependent calcium pump of the sarcoplasmic reticulum has been demonstrated.

The BIO 14.6 strain of Syrian hamster offers a unique opportunity to study the function of the sarcoplasmic reticulum in a naturally occurring form of myocardial failure (7). These animals have been bred with an autosomal recessive gene that produces a spontaneous cardiomyopathy. Myolytic lesions with minimal inflammatory reaction occur between the 30–40th day of life and reach their peak at approximately 60 days (2). Thereafter, healing occurs with minimal proliferation of connective tissue. At approximately this time, hypertrophy develops and continues at varying rates. By 200 days, hypertrophy is well advanced and signs of congestive heart failure begin to appear, characterized by marked fluid accumulation with ascites and pleural effusions and hepatomegaly.

---

[1] Supported in part by United States Public Health Service Grant HE 11306-04, ITI HE 5890, IFO2 HE 44215-01, and HE 09791, and a grant from The Hartford Foundation.

With these findings in mind, BIO 14.6 hamsters and appropriate normal controls have been studied with specific attention to the *quantity* of sarcoplasmic reticulum in these hearts, and the *quality* or performance of the sarcoplasmic reticulum as characterized by the calcium oxalate capacity and calcium oxalate rates at different stages of the disease.

METHODS

Animals of both sexes were sacrificed either by decapitation with a guillotine or by cervical fracture. Two groups of hearts were obtained: one for determination of wet and dry weights and the other for the calcium studies.

The hearts from the first group were removed from the animals, trimmed of great vessels and atria, blotted and weighed. They were then dried to a constant weight and reweighed.

The hearts for the second group were quickly removed from the animals, placed in an ice-cold sucrose solution, trimmed as above, and minced and weighed in groups of from 10-35 hearts. All steps were carried out at as close to 0°C as possible. Homogenization was carried out in a Sorvall Omnimixer with a setting of 120 on a powerstat for 40 sec. Blending was done in a 4 : 1 volume to weight ratio of a solution containing 0.3 M sucrose and 10 mM imidazole (pH 7.0 at 0°C). Isolation of the fragmented sarcoplasmic reticulum was carried out using the procedures previously described (4), except that 0.6 M KCl was added to the final wash and centrifuges and rotors used were from International Equipment Company.

The Millipore filtration technique previously described was used for measuring fragmented sarcoplasmic reticular calcium uptake rates and capacity (4). When capacity was being determined, 0.64 mM $CaCl_2$, 1.6 mM EGTA, 6 mM creatine phosphate, and 5 mM azide were included in the incubation bath. All experiments were carried out at 37°C. The technique of Solero *et al.* (15) was used for estimating the quantity of sarcoplasmic reticulum in the myocardium. Proteins were determined by the method of Lowry *et al.* (9). All water used was first glass-distilled and then run through deionizing and charcoal beds.

RESULTS

The failing hearts utilized in this study were obtained from animals with a naturally occurring cardiomyopathy. Neither surgical interventions (18) nor pharmacological agents (6) that might produce changes in and of themselves were introduced.

Figure 1 represents wet and dry weights of the left and right ventricles as well as percent water of the ventricles. Each bar represents the mean of at least 20 determinations. In this series of animals, the hearts begin to hypertrophy

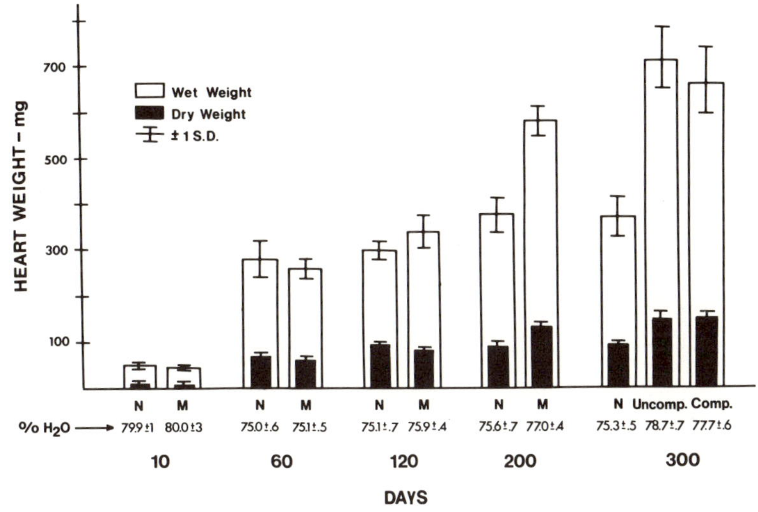

Fig. 1 The wet and dry weights and percent water of hamster ventricles. Each bar represents the mean of 20 determinations; N = normal; M = myopathic; Days = age of animal; Uncomp. = anasarca; Comp. = without anasarca.

sometime after 120 days. The difference in weights at 200 days is significant and by 300 days the weight of the dystrophic hearts is approximately twice that of the controls. The total heart water is statistically different from controls at 200 and 300 days; the absolute values are not appreciably different, *i.e.*, 75% for controls, 77% for 200 days, 77.7% for 300 days compensated, and 78.7% for animals in gross congestive heart failure. Therefore, these hearts double their weights, beginning sometime after 120 days, and this increase in weight is associated with an increase in the solid constituents, not primarily with edema formation.

Figure 2 illustrates the calcium oxalate pumping rates of the isolated sarcoplasmic reticulum. Each point on this figure and subsequent figures represents the mean of three experiments. Each experiment contained material pooled from between 10 and 30 hearts. The rate of pumping is expressed in μmoles of calcium per mg fragmented sarcoplasmic reticular protein per min. The rate in normals of all ages averaged 1.08 ± 0.06 μmole/mg/min. In the myopathic animals, the rate was unchanged from controls at 10 days, reduced by 25% at 60 days, not different from controls at 120 days, reduced 22% at 200 days, 30% at 300 days when the animals were compensated, and 77% when they were in marked congestive heart failure.

The calcium oxalate capacity of fragmented sarcoplasmic reticulum of normal and myopathic hearts is illustrated in figure 3. The capacity averaged 5.6 μmoles of calcium per mg of fragmented sarcoplasmic reticular protein in the normals.

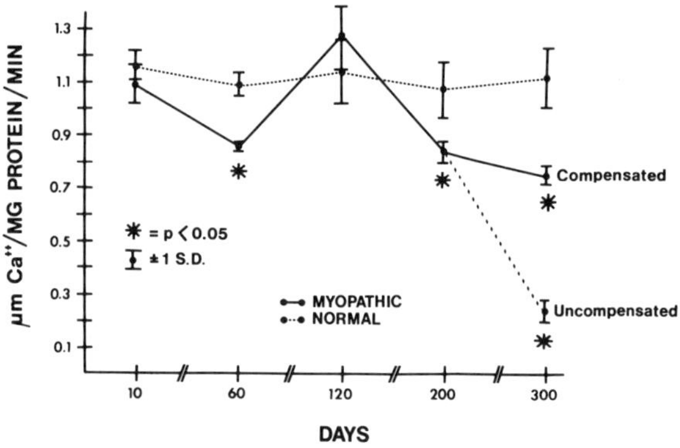

Fig. 2 Calcium uptake by fragmented sarcoplasmic reticulum of Syrian hamsters. See "Methods" section and reference 9 for experimental conditions; each point represents the mean of three different experiments; each experiment contained material pooled from between 10–35 hearts; •-• myopathic; •---• normal; *$p. < 0.05$; ± 1 standard deviation.

This capacity was not significantly different in the myopathic material except at 60 days, when the capacity was 25% lower than the appropriate control.

After the calcium oxalate capacity of the reticulum is determined, the relative quantity of sarcoplasmic reticulum in the heart can be determined by the calcium oxalate capacity of the heart homogenate. It is our contention that if the sarcoplasmic reticular calcium oxalate capacity is unchanged in the myopathic hearts from controls, any change in the whole heart calcium oxalate capacity reflects a change in the amount of reticulum in the heart.

Calcium oxalate capacity of whole heart homogenate with mitochondrial

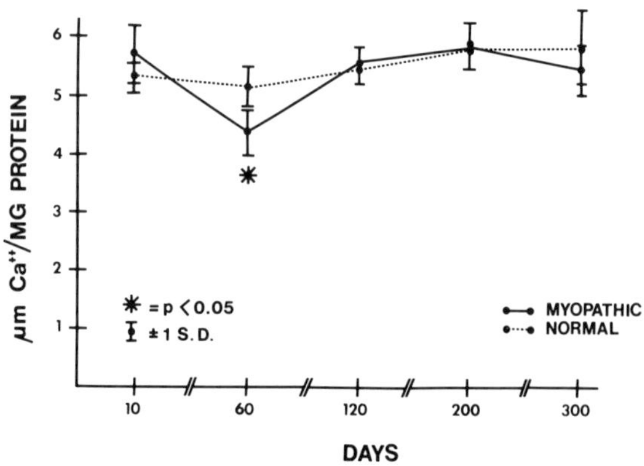

Fig. 3 Calcium oxalate capacity of Syrian hamster fragmented sarcoplasmic reticulum. See "Methods" section and reference 16 for experimental conditions; each point represents the mean of three different experiments; each experiment contained material pooled from between 10-35 hearts; •-• myopathic; •---• normal; * $p < 0.05$; ± 1 standard deviation.

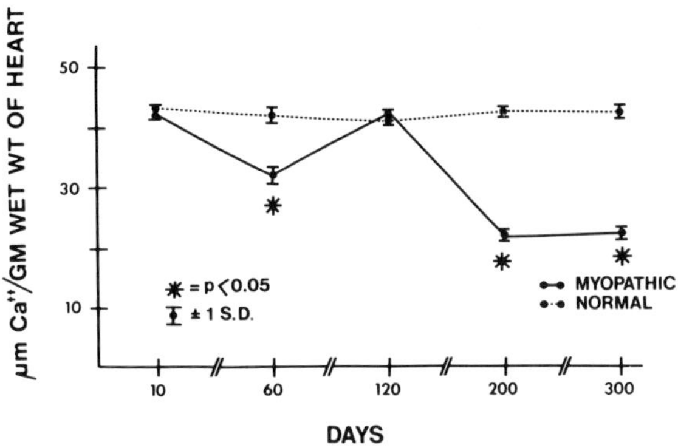

Fig. 4 Calcium oxalate capacity of whole heart expressed as μmoles calcium per unit of heart. See "Methods" section and reference 16 for experimental conditions; each point represents the mean of three different experiments; each experiment contained material pooled from between 10-35 hearts; •-• myopathic; •—• normal; * $p < 0.05$; ± 1 standard deviation.

uptake blocked by azide is illustrated in figure 4. Here, the capacity is expressed in μmoles of calcium per wet weight of heart. At 200 and 300 days, there is a 50% reduction in calcium capacity in the myopathic hearts. At 60 days, there is a 25% reduction. This reduction may represent only the reduction of calcium capacity of the fragmented sarcoplasmic reticulum and not a decrease in the actual amount of reticulum.

Figure 5 represents the calcium capacity of the heart expressed as μmoles of calcium per whole heart. Data expressed in this manner show that, except at 60 days, the total calcium oxalate capacity of the myopathic heart is not different

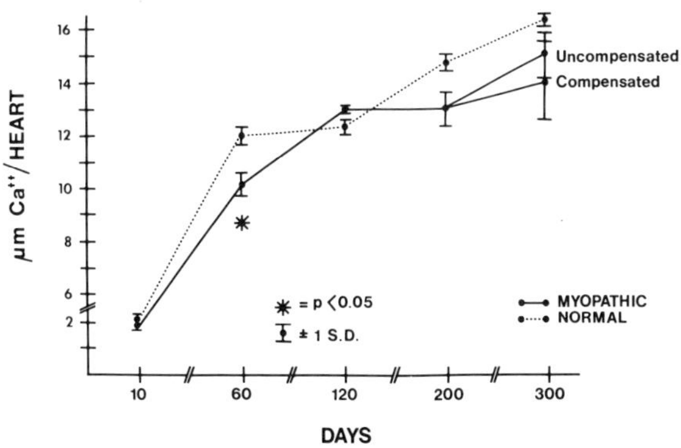

Fig. 5 Calcium oxalate capacity of whole heart expressed as μmoles calcium per whole heart. See "Methods" section and reference 16 for experimental conditions; each point represents the mean of three different experiments; each experiment contained material pooled from between 10-35 hearts; •-• myopathic; •—• normal; * $p < 0.05$; ± 1 standard deviation.

from the controls. Therefore, if the calcium capacity per unit heart drops by approximately 50% and the whole heart capacity stays the same, one must postulate a dilution of the sarcoplasmic reticulum by the resulting hypertrophying muscle.

DISCUSSION

Briggs *et al.* (3) have suggested that a depression of the sarcoplasmic reticular calcium pump might be causally related to myocardial failure. This hypothesis was based on the direct action of various pharmacologic agents to depress calcium pumping by preparations of sarcoplasmic reticulum. Other studies of the spontaneously failing heart-lung preparations later lent further support to this view (5). Recently, others have shown a similar depression of the ATP calcium pumping in association with myocardial failure. One of these studies (18) was performed on hearts obtained from animals after surgical interventions and the other (6) on human hearts obtained only after multiple pharmacologic interventions. The animals used in this study, BIO 14.6 Syrian hamsters, are unique in that their disease process is one that is initiated without external manipulation. Further, the process continues in a predictable and reproducible fashion, so that biochemical studies can be performed at selected intervals in the course of the disease.

As myocardial dilation and hypertrophy develops in these animals, failure occurs. The calcium oxalate capacity of the fragmented sarcoplasmic reticulum per milligram of reticular protein remains normal throughout. The calcium pumping rate of the reticulum, in contrast, falls as failure progresses. The calcium oxalate capacity of whole heart homogenates is reduced as the hearts hypertrophy, attesting to the fact that the quantity of sarcoplasmic reticulum per unit of muscle mass is reduced. Thus, during the hypertrophy in the present pathologic condition, there appears to be a "dilution" of sarcoplasmic reticulum by the hypertrophying myofibrillar elements. Were this the case, one would reach a point where there would be an inadequate amount of sarcoplasmic reticulum to store the calcium needed to activate the myofilaments, and failure might ensue. Thus, in this cardiomyopathy, in addition to a *qualitative* defect in calcium pumping reflected by reduced rates, there appears to be a *quantitative* reduction in the calcium capacity per unit of muscle due to dilution of the sarcoplasmic reticulum by hypertrophy.

The present study provides substantial support for the view that there is a depression of the rate of calcium pumping by the sarcoplasmic reticulum in the hereditary cardiomyopathy of the Syrian hamster and that this depression increases as evidence of congestive heart failure progresses. However, the pertinence of these observations to the etiology of heart failure in clinical situations remains speculative. While this form of cardiomyopathy may not be representative of all forms of myocardial failure, the consistent nature of the defect in association with reduced myocardial function provides an intriguing model.

Studies are now underway to further elucidate the sequence of events that occur in this cardiomyopathy and to characterize the non-oxalate ATP dependent calcium binding and calcium stimulated "extra" ATPase activity of the sarcoplasmic reticulum. In addition, studies to determine the collagen content of these tissues are also in progress. Moreover, the question as to whether the apparent "dilution" of the SR by myofibrillar hypertrophy occurs in other forms of compensatory hypertrophy is presently being explored. Should this be the case, it might help to explain the reduction in myocardial contractility that has been found in the presence of right ventricular hypertrophy in cats with experimental pulmonary artery stenosis (17).

SUMMARY

The etiology of the hereditary cardiomyopathy with compensatory hypertrophy and congestive failure that develops in the Syrian hamster is undefined. Accordingly, the fragmented sarcoplasmic reticulum (SR) has been isolated and studied in normal (N) and myopathic (CM) hamsters in terms of rate of ATP dependent calcium trapping, calcium capacity per unit SR, and SR quantity per unit of myocardium. The ATP dependent Ca^{++} oxalate pumping of the SR was not different from N in CM at 10 days, reduced 22.4% at 200 days, 30.5% at 300 days with moderate failure, and 77.5% at 300 days with severe failure. The quantity of SR in the tissue as judged by the Ca^{++} oxalate capacity of homogenate was unchanged at 10 days but markedly reduced at 200-300 days. Nevertheless, the Ca^{++} oxalate capacity per unit SR was unchanged at all ages studies, suggesting a dilution of the SR by the resultant hypertrophy.

REFERENCES

1. ASHER, M. and FREDERICKSON, E. L.: Halothane versus chloroform: the dose response using the isolated rabbit heart. Anesth. Analg. Curr. Res. *41:* 429–434 (1962).
2. BAJUSZ, E.; BAKER, J. R.; NIXON, C. W. and HOMBURGER, F.: Spontaneous, hereditary myocardial degeneration and congestive heart failure in a strain of Syrian hamster. Ann. New York Acad. Sci. *156:* 105–129 (1969).
3. BRIGGS, F. N.; GERTZ, E. W. and HESS, M. L.: Calcium uptake by cardiac vesicles: inhibition by amytal and reversal by ouabain. Biochim. Zeit. *345:* 122–131 (1966).
4. FUCHS, F.; GERTZ, E. W. and BRIGGS, F. N.: The effect of quinidine on calcium accumulation by isolated sarcoplasmic reticulum of skeletal and cardiac muscle. J. Gen. Physiol. *52:* 955–968 (1968).
5. GERTZ, E. W.; HESS, M. L.; LAIN, R. F. and BRIGGS, F. N.: Activity of the vesicular calcium pump in the spontaneously failing heart-lung preparation. Circ. Res. *20:* 477–484 (1967).
6. HARIGAYA, S. and SCHWARTZ, A.: Rate of calcium binding and uptake in normal animal and failing human cardiac muscle: Membrane vesicles (Relaxing System) and mitochondria. Circ. Res. *25:* 781–794 (1969).
7. HOMBURGER, P.; NIXON, C. W.; HARROP, J.; WILGRAM, G. and BAKER, J. R.: Further morphologic and genetic studies on dystrophy-like primary myopathy of the Syrian hamster. Fed. Proc. *22:* 195 (1963).
8. LAIN, R. F.; HESS, M. C.; GERTZ, E. W. and BRIGGS, F. N.: Calcium uptake activity of canine myocardial sarcoplasmic reticulum in the presence of anesthetic agents. Circ. Res. *23:* 597–604 (1968).

9. LOWRY, O. H.; ROSENBROUGH, N. J.; FARR, A. L. and RANDALL, R. J.: Protein measurement with the Folin phenol reagent. J. Biol. Chem. *193:* 265–275 (1951).

10. MARTONOSI, A. and FERETOS, R.: Sarcoplasmic reticulum: II Correlation between adenosine triphosphate activity and calcium uptake. J. Biol. Chem. *239:* 659–668 (1964).

11. NAYLOR, W. G.: The significance of calcium ions in cardiac excitation and contraction. Amer. Heart J. *65:* 404–411 (1963).

12. PARMLEY, W. W. and BRAUNWALD, E.: Comparative myocardial depressant and anti-arrhythmic properties of d-propranolol, dl-propranolol and quinidine. J. Pharmacol. Exptl. Therap. *158:* 11–21 (1967).

13. PRICE, H. L. and HELRICH, M.: The effect of cyclopropane, diethyl ether, nitrous oxide, thiopental, hydrogen ion concentration on myocardial function of the dog heart-lung preparation. J. Pharmacol. Exp. Therap. *115:* 206–216 (1955).

14. SANDOW, A.: Excitation-contraction coupling in skeletal muscle. Pharmacol. Rev. *17:* 265–320 (1965).

15. SOLERO, R. J.; GERTZ, E. W. and BRIGGS, F. N.: The volume of sarcoplasmic reticulum in skeletal and cardiac muscle, in: Cardiac Hypertrophy (Academic Press, 1970).

16. SONNENBLICK, E. H. and STAM, A. C.: Cardiac muscle; activation and contraction. Ann. Rev. Physiol. *31:* 647–674 (1969).

17. SPANN, J. F.; BUCCINO, R. A.; SONNENBLICK, E. H. and BRAUNWALD, E.: Contractile state of cardiac muscle obtained from cats with experimentally produced ventricular hypertrophy and heart failure. Circ. Res. *21:* 341–345 (1967).

18. SUKO, J.; YOSHIHISA, I.; VOGEL, J. and CHIDSEY, C. Abnormal calcium uptake and ATPase in sarcoplasmic reticulum of failing heart. Circulation *40:* (Suppl. III) 197 (1969).

Mailing address: E. W. Gertz, M.D., Cardiovascular Division, Peter Bent Brigham Hospital, 721 Huntington Ave., Boston, Massachusetts 02115 (USA).

Departments of Pathology and Physiology,
University of Ottawa, Faculty of Medicine,
Ottawa, Ontario, Canada

EARLY ULTRASTRUCTURAL AND BIOCHEMICAL CHANGES IN THE MYOPATHIC HAMSTER HEART [1]

B. B. NADKARNI, B. HUNT, and H. A. HEGGTVEIT

The histopathology of the myocardium of the BIO 14.6 strain of Syrian golden hamster has been thoroughly studied (1, 2, 4). To date, however, there have been only a few preliminary reports on myocardial ultrastructure in the dystrophic hamster (3, 10, 11). This paper documents our electron microscopic observations on the failing and pre-failing hearts of these myopathic animals and provides some biochemical data on the electrolyte composition of the myocardium and serum.

METHODS

Myopathic hamsters (Bio-Research Institute, Cambridge, Mass.) and normal animals were examined at 3-day and weekly intervals from birth. Arterial blood for biochemical studies was collected from the abdominal aorta, after which the heart was removed. The apex of the left ventricle was excised and minced in phosphate-buffered glutaraldehyde, post-fixed in osmium tetroxide, and embedded in Epon. Thin sections were cut on a Reichert OM-U2 ultramicrotome, stained with uranyl acetate and lead citrate, and examined on a Philips EM 300 electron microscope.

The remaining portions of the hearts were divided in half for histological and biochemical studies. Serum and tissue magnesium and calcium levels were measured by atomic absorption spectrophotometry, following a method described in detail elsewhere (8).

RESULTS

Ultrastructural observations

The fine structure of the normal young hamster heart is shown in figure 1. Myocardial cells of newborn myopathic hamsters present a fairly uniform immature or primitive appearance. Although Z-bands are well formed, they are

[1] Supported in part by a grant from the Ontario Heart Foundation.

Fig. 1 Myocardium of a 1-month-old normal hamster, showing regular sarcomere band patterns and intact mitochondria. ×38,900.

disoriented and the intervening sarcomeres remain, for the most part, structureless and empty. Myofilaments arise from either side of the Z-bands and radiate bicollaterally. By 10 days, large numbers of myofilaments have become organized into myofibrils, but these structures, for the most part, are still incompletely formed and in a state of disarray (fig. 2). At this stage, there is a prominence of the sarcoplasmic reticulum and many mitochondria are distended and vacuolated.

Even as new myofibrils are developing, previously-formed contractile units show evidence of focal myolysis, with loss of filamentous and Z-band material although the T-tubules remain initially visible (fig. 3). Myolytic cells are interspersed, either singly or in groups, with reasonably intact myocardial cells. The mitochondria of the latter, however, frequently show hypertrophy and distortion of their internal fine structure (fig. 3). The glycogen content of the myocardium is not appreciably diminished at this stage. As the myolytic cells collapse and disappear, adjacent myocardial cells undergo some degree of enlargement. The appearance of new myolytic foci decreases with age and, prior to the onset of cardiac failure (between 45 and 65 days of age in our animals), most myofibrils have a relatively normal sarcomere band pattern (fig. 4). Mitochondria, on the other hand, continue to show hypertrophy and vesiculation and begin to increase in number.

With the onset of overt cardiac failure, the heart shows gross evidence of hypertrophy and dilatation. This is reflected at the ultrastructural level by hypertrophy of myofibrils through the addition of more filaments. Progression of heart failure is accompanied by fraying and separation of the myofilaments with blurring of Z-bands and disappearance of the I-band regions (fig. 5). Mitochondrial alterations become more pronounced with increased numbers of enlarged, bizarre forms, distorted cristal patterns, and apparent budding. Excessive fluid uptake is evidenced by prominent pinocytotic activity on the part of capillary endothelial cells and sarcolemmal membranes. Endothelial cells and subendothelial smooth muscle cells of the endocardium also show exaggerated numbers of water transport vesicles (fig. 6). The subsarcolemmal spaces are distended in a bleb-like fashion. Glycogen granules rapidly disappear, while lipid droplets and lipochrome pigment accumulate within the failing myocardial cells.

Biochemical data

The results of the plasma and myocardial calcium and magnesium determinations in 2 groups of dystrophic hamsters and age-matched controls are given in table I and figure 7.

Fig. 2 Myocardium of a 10-day-old myopathic hamster, showing incomplete formation and disorientation of myofibrils, enlargement and vacuolization of mitochondria, and dilatation of sarcoplasmic reticulum. × 17,000.

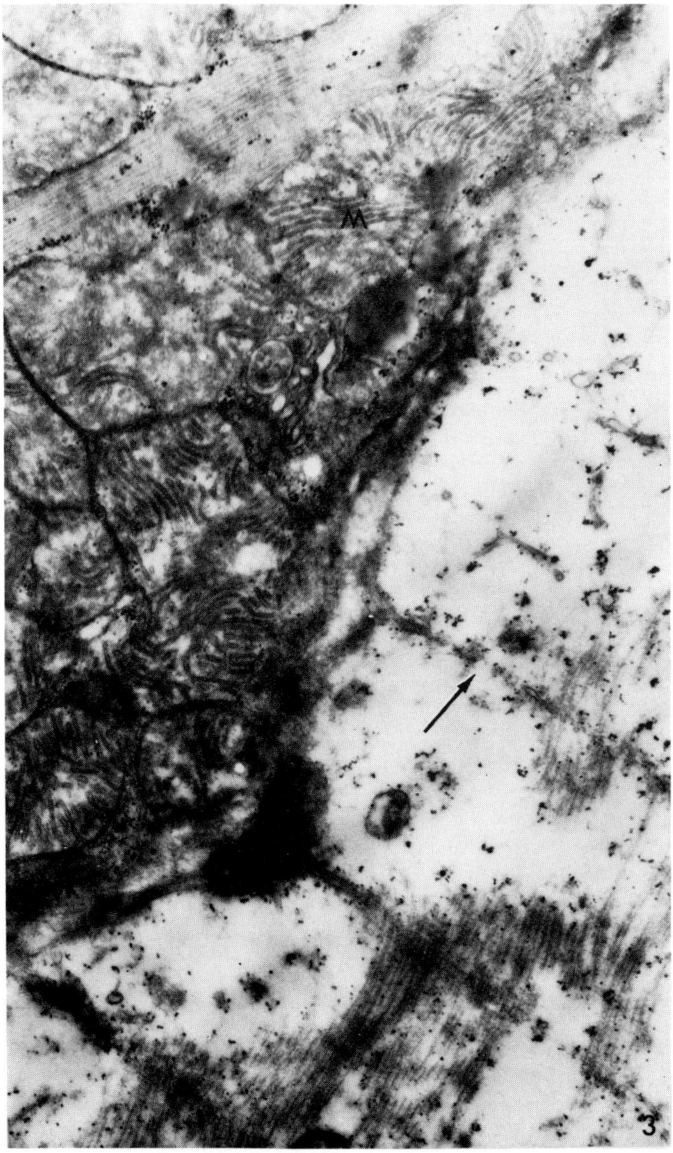

Fig. 3 Myocardium of a 25-day-old myopathic hamster. One of the two adjacent myocardial cells shows myolysis with disappearance of myofilaments, although the T-system can be seen (arrow). The mitochondria of the other cell are hypertrophic with some distortion of their cristae. Glycogen granules are abundant. ×50,200.

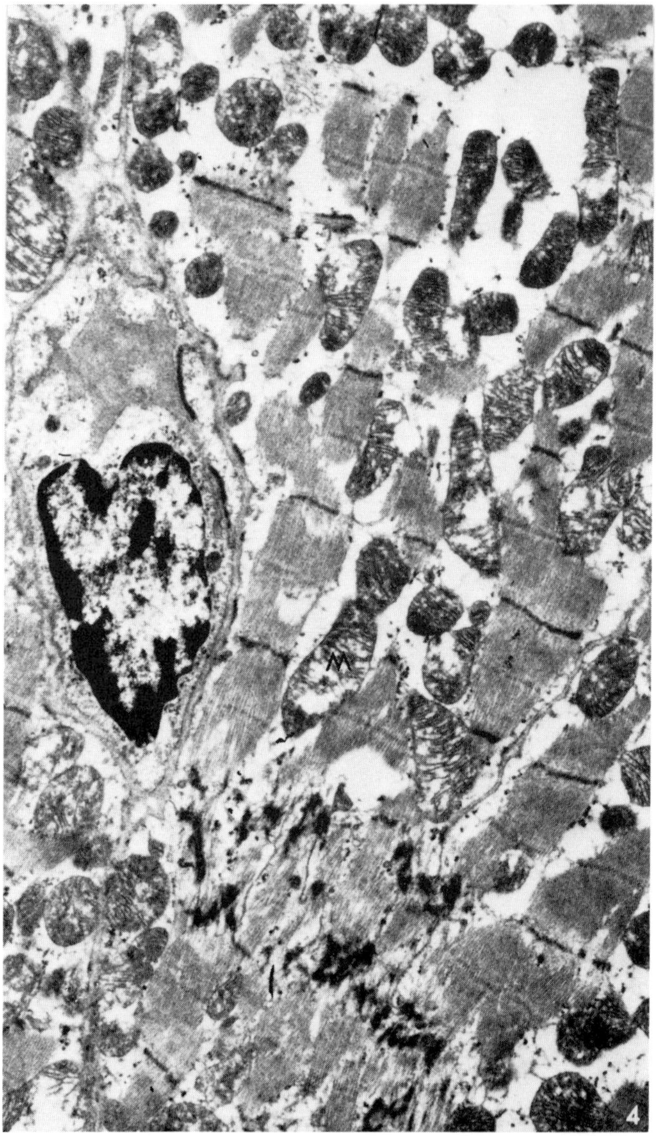

Fig. 4 Myocardium of a 2-month-old myopathic hamster not yet in heart failure. Myofibrils are relatively normal, but mitochondria exhibit prominent swelling and vacuolization. ×13,000.

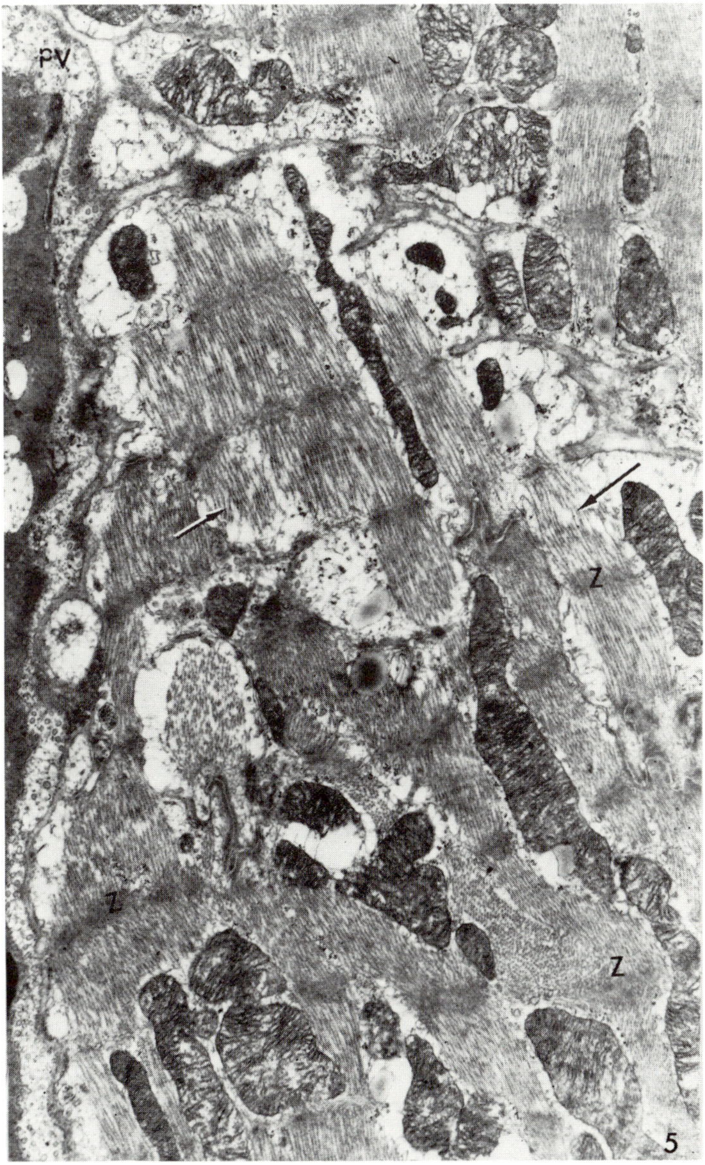

Fig. 5 Myocardium of a 65-day-old myopathic hamster with heart failure. Myofilaments disclose focal fraying (arrows) and there is blurring of the Z-bands. Mitochondria are hypertrophied and irregular. Subsarcolemmal spaces are prominent and devoid of glycogen. Capillary endothelium contains many pinocytotic vesicles. ×24,300.

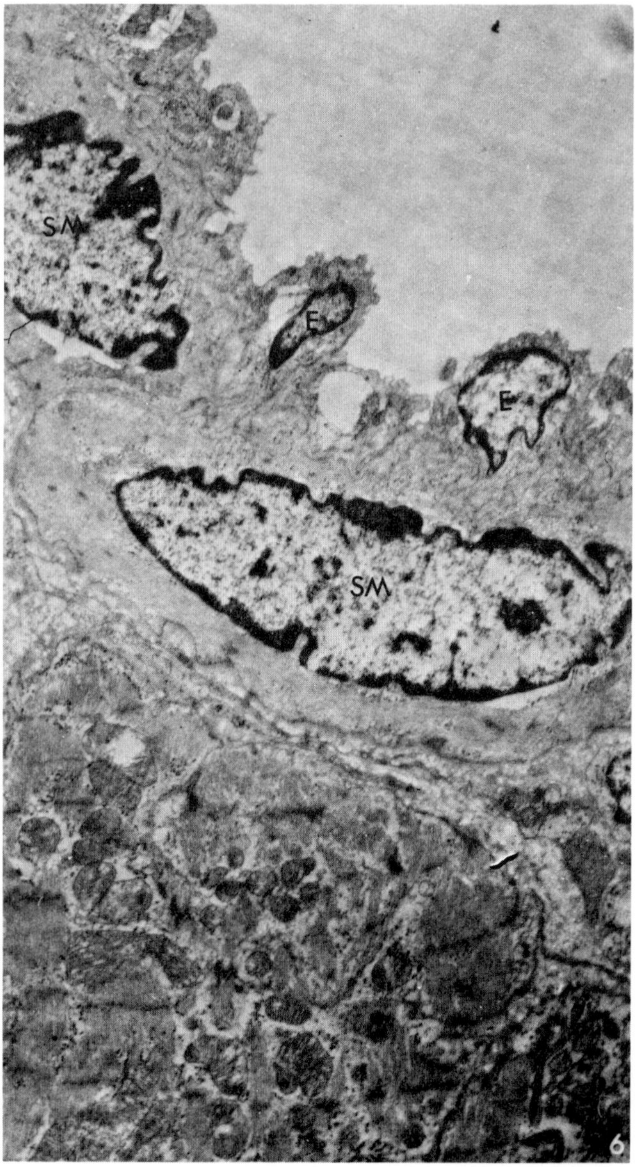

Fig. 6 Endocardium of a 65-day-old myopathic hamster in heart failure. The endothelium (E) and subendothelial smooth muscle (SM) cells show increased pinocytotic activity. × 11,300.

Table I

Plasma and Myocardial Calcium and Magnesium Levels*

	Plasma		Heart muscle	
	Control	Myopathic	Control	Myopathic
	mg/100 ml		*mg/100 g dry tissue*	
Calcium	10.9 (± 0.30)	10.7 (± 0.54)	20.8 (± 1.2)	267.8 (± 56.0)
Magnesium	2.54 (± 0.13)	2.6 (± 0.10)	103 (± 1.8)	107 (± 2.1)

*Figures in parenthesis represent the S.E. of the mean value.

DISCUSSION

Delayed or defective genesis of the contractile elements in the newborn myopathic hamster is followed by myolytic changes in the first weeks of life. These alterations subside until heart failure ensues, when the myofibrils again manifest fine structural damage. Mitochondrial abnormalities are present at all ages studied, suggesting that a defective energetic mechanism may be a significant pathogenetic mechanism in this disorder.

Uncoupling of oxidative phosphorylation has been demonstrated in both cardiac and skeletal muscles of dystrophic hamsters (10, 13), although the data of Blanchaer and Wrogemann (5) are at variance with these findings. Studies of cardiac muscle in experimental magnesium deficiency (a condition known to induce uncoupling of oxidative phosphorylation) suggest that cellular depletion of magnesium may play a fundamental part in various disorders of the myocardium (6–8). Alterations in mitochondrial and cellular membrane permeability, with ionic shifts and defective energy production or utilization, may be a common denominator in the subsequent development of overt necrotic lesions. This observation has some experimental support but remains unproven.

Our initial data, derived from several 30- and 60-day-old hamsters, showed that blood and tissue levels of magnesium were decreased and calcium levels were markedly increased (fig. 7). These findings are in agreement with the observations of Bajusz and Lossnitzer (4). Subsequent determinations carried out on older animals failed to confirm a significant difference in serum or myocardial magnesium between the control and dystrophic animals (table I). Calcium levels and tissue water, however, were significantly elevated in the myopathic hamsters, even before the onset of overt heart failure. It is possible that, although magnesium levels are essentially normal, the intracellular distribution and availability of magnesium ions may be altered.

The relative roles of the biochemical and ultrastructural changes in the myopathic hamster heart require further elucidation. One of the major

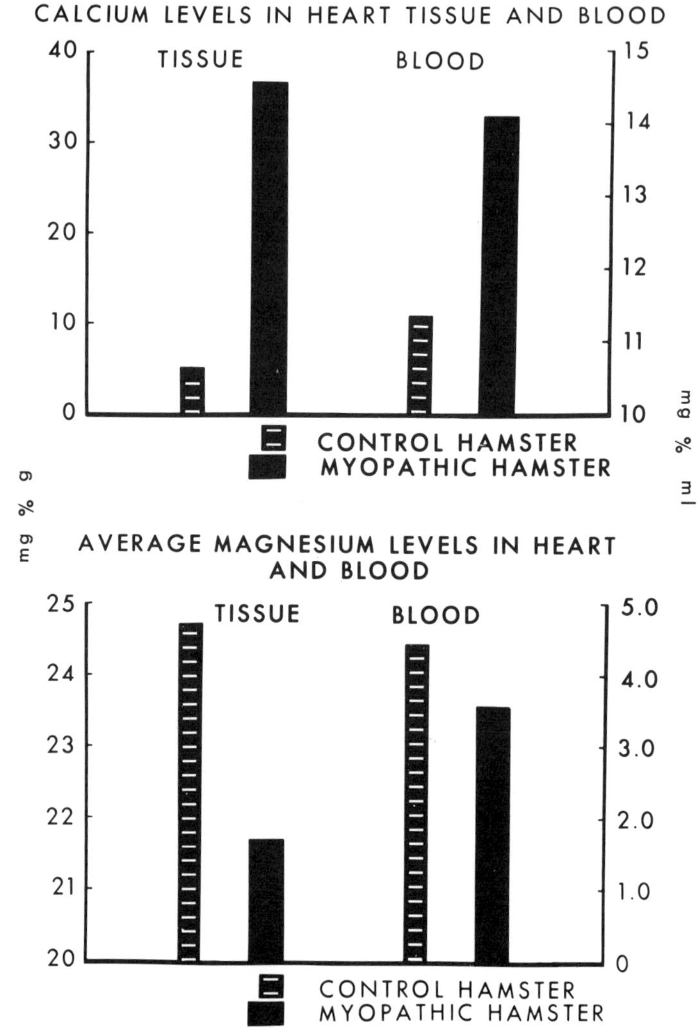

Fig. 7 Calcium and magnesium levels in hamsters.

difficulties, in this regard, is the separation of secondary or compensatory mechanisms from primary events.

SUMMARY

Hearts of myopathic hamsters of BIO 14.6 strain and of normal animals were examined at weekly intervals from birth. Arterial blood collected from the lower abdominal aorta was analyzed by atomic absorption spectrophotometry for magnesium and calcium levels.

Histopathological, ultrastructural, and biochemical studies were performed on myocardial tissue.

Delayed or defective genesis of the contractile elements in the newborn myopathic hamster is followed by myolytic changes in the first few weeks of life. These pathological alterations subside until cardiac failure ensues, when the myocardium again reveals ultrastructural damage. Consistent mitochondrial abnormalities seen at all ages suggest a defective energetic mechanism in the pathogenesis of this cardiomyopathy. Serum and tissue calcium levels are significantly elevated. However, the magnesium contents, which are lower in younger animals, show no significant difference in the older ones. Further studies on the possibility of intracellular redistribution and altered availability of magnesium ions are being conducted.

REFERENCES

1. BAJUSZ, E.: Hereditary cardiomyopathy: A new disease model. Amer. Heart J. *77:* 686–696 (1969).
2. BAJUSZ, E.; BAKER, J. R.; NIXON, C. W. and HOMBURGER, F.: Spontaneous hereditary myocardial degeneration and congestive heart failure in a strain of Syrian hamsters. Ann. N. Y. Acad. Sci. *156:* 105–129 (1969).
3. BAJUSZ, E.; BÜCHNER, F.; ONISHI, S. und RICKERS, K.: Hypertrophie des Herzmuskels bei erbbedingter Myopathie. Naturwissenschaften *56:* 568–569 (1969).
4. BAJUSZ, E. and LOSSNITZER, K.: A new disease model of chronic congestive heart failure: Studies on its pathogenesis. Trans. N. Y. Acad. Sci. *30:* (Ser. II): 939–948 (1968).
5. BLANCHAER, M. C. and WROGEMANN, K.: Oxidative phosphorylation by mitochondria isolated from hearts of BIO 14.6 myopathic hamsters. Trans. N. Y. Acad. Sci. *30:* (Ser. II):949–950(1968).
6. HEGGTVEIT, H. A.: The cardiomyopathy of magnesium deficiency; in BAJUSZ, E., electrolytes and cardiovascular diseases, vol. 1, pp. 204-220 (Karger, Basel 1965).
7. HEGGTVEIT, H. A.; HERMAN, L. and MISHRA, R. K.: Cardiac necrosis and calcification in experimental magnesium deficiency: A light and electron microscopic study. Amer. J. Path. *45:*757–782(1964).
8. HEGGTVEIT, H. A. and NADKARNI, B. B.: Ultrastructural pathology of the myocardium in: BAJUSZ, E. and JASMIN, G. Methods and achievements in experimental pathology, vol. 5, pp. 474–517 (Karger, Basel 1971).
9. HUNT, B.: The estimation of magnesium in plasma, muscle and bone by atomic absorption spectrophotometry. Clin. Chem. *15:*979–996(1969).
10. LOCHNER, A. and BRINK, A. J.: Oxidative phosphorylation and glycolysis in the hereditary muscular dystrophy of the Syrian hamster. Clin. Sci. *33:* 409 (1967).
11. NADKARNI, B. B.; HUNT, B. and HEGGTVEIT, H. A.: Ultrastructural studies of myopathic hamster heart. Proc. Canad. Fed. Biol. Soc. *13:* 80 (1970). Abstract.
12. NAKAO, K.; OKA, M.; CHEN, C.; BAJUSZ, E. and ANGRIST, A.: Myocardial and valvular studies in myocardiopathy in the BIO 14.6 strain hamster. 10th Conf. of the Internat. Soc. Geograph. Path., Jerusalem, September 1969. Path. Microbiol. *35:* 118–124 (1970).
13. OPIE, L. H.; LOCHNER, A.; BRINK, A. J.; HOMBURGER, F. and NIXON, C. W.: Oxidative phosphorylation in hereditary myocardiopathy in the Syrian hamster. Lancet *ii:* 1213–1214 (1964).

Mailing address: H. Alexander Heggtveit, M.D., Department of Pathology, University of Ottawa, Faculty of Medicine, Ottawa 2 (Canada).

Department of Physiology and Biophysics,
Hahnemann Medical College, Philadelphia, Pa., and
Bio-Research Institute, Boston, Massachusetts, USA

ADRENERGIC NEUROHUMORS IN THE HEART OF HAMSTERS WITH HEREDITARY MYOPATHY DURING CARDIAC HYPERTROPHY AND FAILURE [1]

E. T. ANGELAKOS, L. C. CARBALLO, J. B. DANIELS,
M. P. KING, and E. BAJUSZ

Evidence accumulated in recent years indicates that in congestive heart failure there are significant changes in the function of the adrenergic nerve terminals to the heart. In the human disease as well as in experimentally induced heart failure, there is a marked decrease in the cardiac norepinephrine (NE) concentration during heart failure (2, 9, 10, 12, 15, 21, 25, 27, 28). In experimentally induced failure, it could be shown that the decrease in cardiac NE concentration is greater than what might be expected from concurrent cardiac hypertrophy and, therefore, is believed to reflect a true decrease in the cardiac NE stores. In addition, there is now evidence that at least in some forms of congestive heart failure there is: a) a reduction in the chronotropic and inotropic responses of the heart to cardiac sympathetic nerve stimulation (13); b) a reduction in the cardiac NE stores available for release by tyramine (12); and c) a reduction in tyrosine hydroxylase activity, the rate limiting enzyme in the synthesis of NE from tyrosine (23).

However, in both clinically observed and experimentally induced heart failure it is often not clear whether the changes observed in the function of the cardiac adrenergic nerve terminals precede the development of failure or are simply a direct consequence of the pathophysiologic alterations produced by the condition of congestive failure. Interpretation of the results from experimentally induced failure is similarly limited, since the interventions necessary to induce failure produce marked hemodynamic changes from the beginning and, therefore, do not provide a reliable approach to differentiating primary from secondary effects.

In recent years, a strain of hamsters (BIO 14.6) with hereditary myopathy was developed. These animals are normal at birth but develop progressive myocardiopathy leading to spontaneous heart failure at an early age (4 – 9 months). The development of myocardial lesions and the course of the hereditary disease in these animals have been studied extensively (6, 17, 10, 22).

[1] This work was supported by grants from United States Public Health Service, National Institutes of Health: HE 09616, HE 05680, and HE 09791.

This animal model provides a means for investigating cardiac adrenergic mechanisms before and during the development of heart failure and could, therefore, help in the differentiation between primary and secondary effects. Furthermore, the spontaneous nature of the disease avoids the complication inherent in the production of experimental heart failure.

Animals of this strain have been used in extensive studies performed over the past three years to evaluate the function of the adrenergic nerve terminals at various stages of the disease. An over-all review of the results obtained are presented here. Some preliminary findings have been reported previously (3,4) and more detailed reports will appear elsewhere.

METHODS

Animals of the BIO 14.6 strain were obtained from the Bio-Research Laboratories Inc. (Boston). Studies were made in animals ranging from 1 to 9 months of age. In each case, animals of the same age from a similar strain without the hereditary disease (LSH strain) were obtained from the same source and were used as controls.

For the studies on the functional status of the heart, the animals were anesthetized with pentobarbital, a thoracotomy was performed during artificial respiration, and the heart was attached to a transducer by a thread sutured at the apex. With the heart *in situ,* progressive tension was applied and the force of cardiac contraction was recorded at each stage of tension using conventional carrier amplifiers and direct writing recorders. In this manner, an active length-tension relationship was obtained with the heart *in situ.* The results shown in figure 2 represent the maximum active tension developed by the heart under these conditions. However, in all cases, the entire length-tension curve was shifted. Further details of these studies will be presented elsewhere.

Chemical analyses for cardiac catecholamines were performed fluorometrically, using the trihydroxyindole method in pools of 5 – 6 hearts. Details of the methodology used have been published in connection with other studies on cardiac catecholamines made in this laboratory (1).

Histochemical studies were made using the formaldehyde fluorescence histochemical technique of Falck and Hillarp as previously described (3, 14).

NE synthesis and turnover were evaluated from the concentration of C^{14} NE found in the heart 1 h after the intravenous injection of C^{14} tyrosine. The animals were anesthetized with pentobarbital, and 10 μc of C^{14} 1-tyrosine (367 to 395 mC/mM specific activity) was injected into a catheterized vein. The animals were sacrificed 60 min after the injection and the hearts were removed rapidly, weighed, and processed for extraction of NE. Total tyrosine, C^{14} tyrosine, total NE, C^{14} NE, and C^{14} dopamine were determined; 6 hearts were pooled in each case. Initial extraction of catecholamines and isolation by adsorption on alumina were performed as previously described for the determination of NE. In addition, in these studies, the alumina eluates were passed through a Dowex 50 column to separate NE from DOPA and dopamine. DOPA was isolated in the initial filtrate of the Dowex column, NE was eluted with 1 N HCl, and dopamine with 2 N HCl. Tyrosine was extracted from the alumina filtrates and determined fluorometrically as described by Udenfriend (29). C^{14} activity was measured in a liquid scintillation counter using Bray's solution. The counts per min were corrected in each case for quenching effects of various solutions and for tyrosine impurities, *etc.*, so that all counts are comparable.

RESULTS

Studies were performed in animals with the hereditary disease, whose age ranged from 35 to 380 days. Animals of the same age from a similar strain, without the hereditary disease, served as controls. In each case, equal numbers of animals from each group (control and diseased) were sacrificed at the same time and parallel studies were made.

For the purposes of this presentation, the results have been pooled into 4 groups corresponding to four stages in the development of myocardial disease in the dystrophic strain. The first group consists of animals at an early age (1 month old), when the dystrophic strain shows no evidence of heart disease histologically or functionally. The second group consists of animals 2 – 3 months of age, when in the dystrophic strain there is some histologic evidence of focal myocardiopathy as described previously (6). These animals still do not exhibit cardiac hypertrophy or any functional signs of heart disease. The third group consists of animals 3 – 4 months of age, when in the dystrophic strain there are distinct myocardial lesions and significant cardiac hypertrophy. These animals show some impairment of cardiac function when tested (see below) but no overt signs of heart failure. The last group includes animals 4 – 9 months of age, when the dystrophic strain is in the final stages of the disease. These animals show extensive myocardial lesions, considerable cardiac hypertrophy, and overt signs of congestive heart failure including cardiac enlargement and extensive pulmonary edema, ascites, and peripheral edema.

Division into these pathophysiologic stages is essential for comparison of the results between different cohorts of animals, since in the dystrophic animals of the same strain the progress of the disease does not follow the same time course in each cohort. Therefore, the age of different cohorts may not correlate exactly with a given stage of the disease. In all cases, no evidence of myocardiopathy is apparent prior to 1 month of age. However, the development and subsequent course may vary with each cohort; consequently, some cohorts may develop overt signs of heart failure as early as 4 months of age, while in others the entire progress of the disease may be slower so that signs of overt failure do not occur until 8 – 10 months of age. The genetic or other factors responsible for this are not entirely clear.

The results presented here include studies on groups of animals obtained from several cohorts of the dystrophic strain over a period of 3 years. It was therefore essential to group the results according to the pathophysiologic status of the animals studied in each case rather than strictly on the basis of age.

Cardiac hypertrophy

The results of total wet heart weights at each of the four stages of the disease are shown in figure 1. A moderate increase in heart size is observed in the controls, consistent with body growth in the first few months of age. The dystrophic strain begins to show signs of cardiac hypertrophy as early as 3 months of age, becoming progressively greater as the disease advances, although in some cohorts it may not be evident until 4 months of age or later. At the terminal stages of heart failure, the large heart weights found may, in part, reflect cardiac edema. However, in both cases, histologically there is distinct myocardial fiber enlargement. Furthermore, previous studies of dry heart weights in dystrophic animals indicate little or no increase in water content of cardiac tissue at various stages of the disease (7).

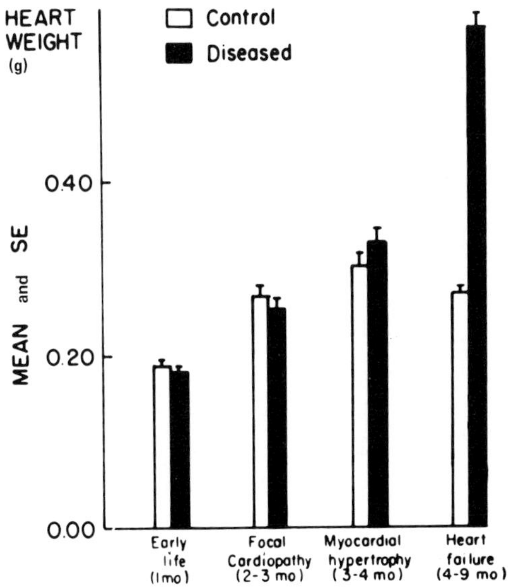

Fig. 1 Total wet weight of the heart from diseased and control strains at various ages. Means ± SE from 55 to 90 hearts per group.

Functional status

The force of contraction of the heart beating *in situ* was determined as outlined under Methods. Details of these studies will be presented elsewhere. A summary of the results obtained for maximum tension developed at each stage is shown in figure 2. This information is pertinent in the over-all evaluation of the stage of the disease in each of the groups studied. Results are expressed both in terms of maximum tension (g/heart) measured, as well as tension developed per unit *dry* heart weight (g/g dry heart weight). In the dystrophic group, the force/heart indicates the maximum capacity of the hypertrophied heart, while the force/g dry weight of heart shows the impairment in heart function per unit mass. Note that there is no significant difference between control and dystrophic animals in the early stages of myocardiopathy. Therefore, studies at an earlier stage did not appear to be necessary in view of the many technical difficulties involved in working with very small animals. During the stage of moderate hypertrophy, in the absence of overt heart failure, there is considerable impairment of cardiac function in the dystrophic group in terms of force/unit heart mass. However, the force/total heart is only moderately reduced compared to the control. Thus, during this stage, total cardiac function is largely compensated by myocardial hypertrophy. This is consistent with the absence of signs of heart failure.

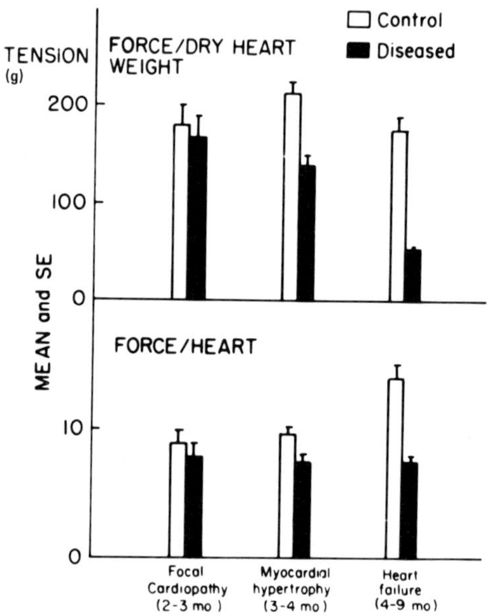

Fig. 2 Peak active tension (g) developed by the heart *in situ*. Means ± SE from 10 hearts per group.

In animals with heart failure, both force/heart and force/unit heart mass, are markedly decreased. This group consisted of animals with overt but moderate heart failure. Animals with advanced failure (excessive edema) could not be studied, as they did not tolerate well the anesthesia and thoracotomy necessary for these studies.

Cardiac catecholamines

The concentration of catecholamines in the heart was determined in several series of experiments in control and diseased animals between 1 and 10 months of age. A total of 100 determinations were made in pools of 5 to 6 hearts (552 hearts total). Some preliminary results have been reported previously (2). A summary of all the results obtained to date on the NE concentrations is shown in figure 3. Chemical determinations for epinephrine were also made. However, in all cases, epinephrine was found only in very small concentrations (less than 5% that of NE) and there were no significant differences between control and dystrophic animals.

In general, the average results of all the studies made (fig. 3) indicate that early in life (1 month of age) there are no significant differences between control and dystrophic animals in the NE stores in the heart. However, during the next stage of focal cardiopathy (2 – 3 months of age) the NE stores were found to be somewhat higher in the dystrophic than in the control groups of the same age. This difference, although small (10 – 20%) was found consistently in all the studies made and was statistically significant in each case. During the stage of moderate hypertrophy, prior to the development of heart failure, cardiac NE stores in the dystrophic group decrease; at this stage, therefore, they are not significantly different from the controls. The decrease in NE stores becomes progressively more marked as the disease advances so that, at the final stage, in the presence of congestive heart failure, NE stores are significantly and markedly lower in the dystrophic than in the control group. At this stage, cardiac NE concentrations are reduced by 50% or more as compared with those found in the control group of the same age or in the dystrophic group at earlier stages of the disease.

Since there is significant hypertrophy in the last two stages, it is pertinent to examine NE stores also in terms of total NE per heart. These results are also shown in figure 3. It is clear from this latter analysis that, in the dystrophic group, total cardiac NE increased progressively throughout the period of cardiac hypertrophy and that much of the decrease in NE concentration observed at the last stage is due to cardiac hypertrophy. Nevertheless, there is also a net decrease in total NE during the final stage. This becomes more significant, because the controls show a small but progressive increase in total cardiac NE as the heart size increases during the growth period.

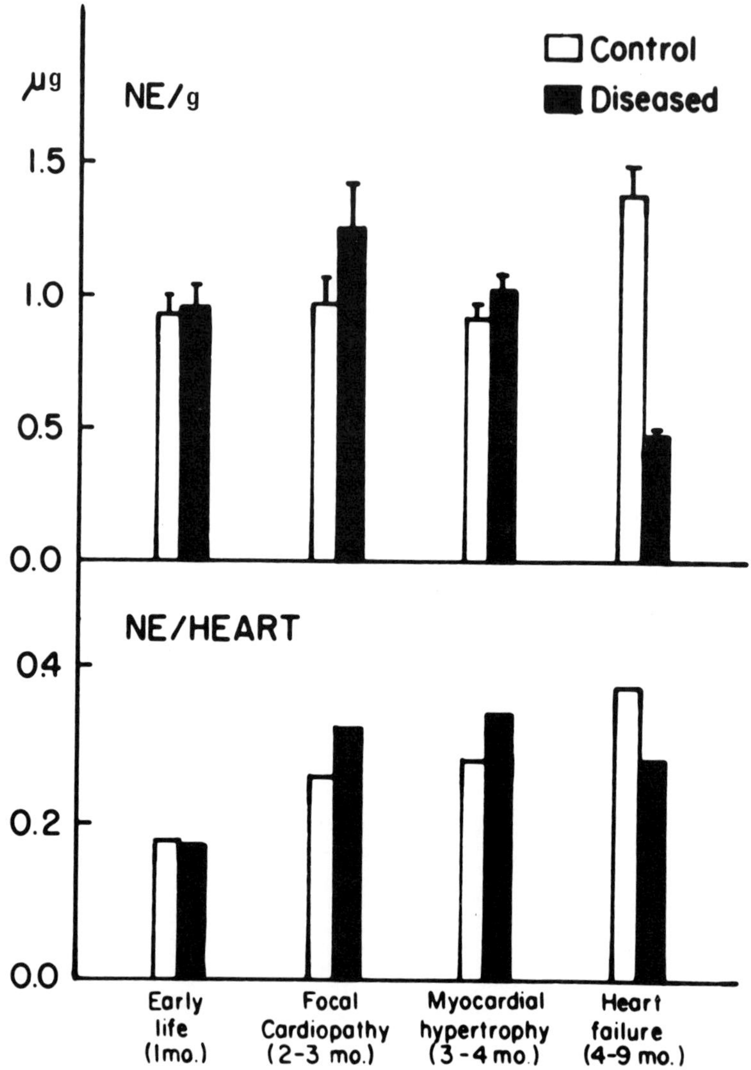

Fig. 3 Norepinephrine (NE) in the heart of control and diseased groups. Means ± SE from 55 to 90 hearts per group.

Histochemical studies

Ventricular sections from control and dystrophic animals at all stages of the disease were examined with the fluorescence histochemical technique for catecholamines (see "Methods"). In the early stages of the disease (2 – 4 months of age), the fluorescence intensity of adrenergic nerve fiber terminals was distinctly greater in the dystrophic than in the control groups. Even in the last stage of the disease during congestive failure, in hearts in which the concentration of amines determined chemically was distinctly reduced, histochemical observation again showed increased fluorescence intensity of adrenergic nerve terminals. Only in the very final stages of the disease, in the presence of advanced heart failure, did the fluorescence of the adrenergic nerve terminals appear to be less in the diseased than in the control group.

In the preparations from the dystrophic group after the first stage, distinct focal areas could be observed where no adrenergic fibers were present. These appeared to correspond to the areas of focal necrosis and became more numerous and more extensive as the disease progressed. They were very striking in the last stage of the disease.

An indication of the difference in fluorescence intensity could be arrived at from the difference in photographic exposure time necessary to take optimal microphotographs of the nerve terminals in each case. Figure 4 shows such microphotographs, which, as indicated in the figure, require a greater exposure time in the control than in the dystrophic animal. The latter was at an advanced but not terminal stage of the disease.

NE synthesis and turnover

The concentration of C^{14} NE found in the heart 1 h after the intravenous administration of C^{14} tyrosine is shown in figure 5. The extraordinary cost of these studies precluded determinations in more than one point in time. Although the actual rate of NE synthesis cannot be calculated from these data, it is clear that a higher concentration of C^{14} NE reflects a higher rate of NE synthesis from tyrosine. This value appears to be greater in the dystrophic than in the control animals in all stages (fig. 5). This increase is also found early in life, before there is any other evidence of disease in the dystrophic group.

Since early in life the total cardiac NE concentrations are not different between control and dystrophic groups (fig. 3), a high synthesis rate suggests that the turnover of NE is higher in the dystrophic than in the control group at this early stage. This is shown when the specific activity of C^{14} NE found in the heart is calculated (fig. 5). The NE specific activity appears to be similar in the control and diseased groups in the later stages of the disease, *i.e.*, when cardiac function is fully compensated by cardiac hypertrophy. However, NE specific activity again increases substantially in the last stage of the disease so that, in the presence of heart failure, NE turnover appears to be markedly increased.

In addition to C^{14} NE, the amount of C^{14} dopamine (DA) was also determined following administration of C^{14} tyrosine (TY). Calculations of the C^{14} ratios in NE/TY, NE/DA, and DA/TY again indicated that, in the dystrophic

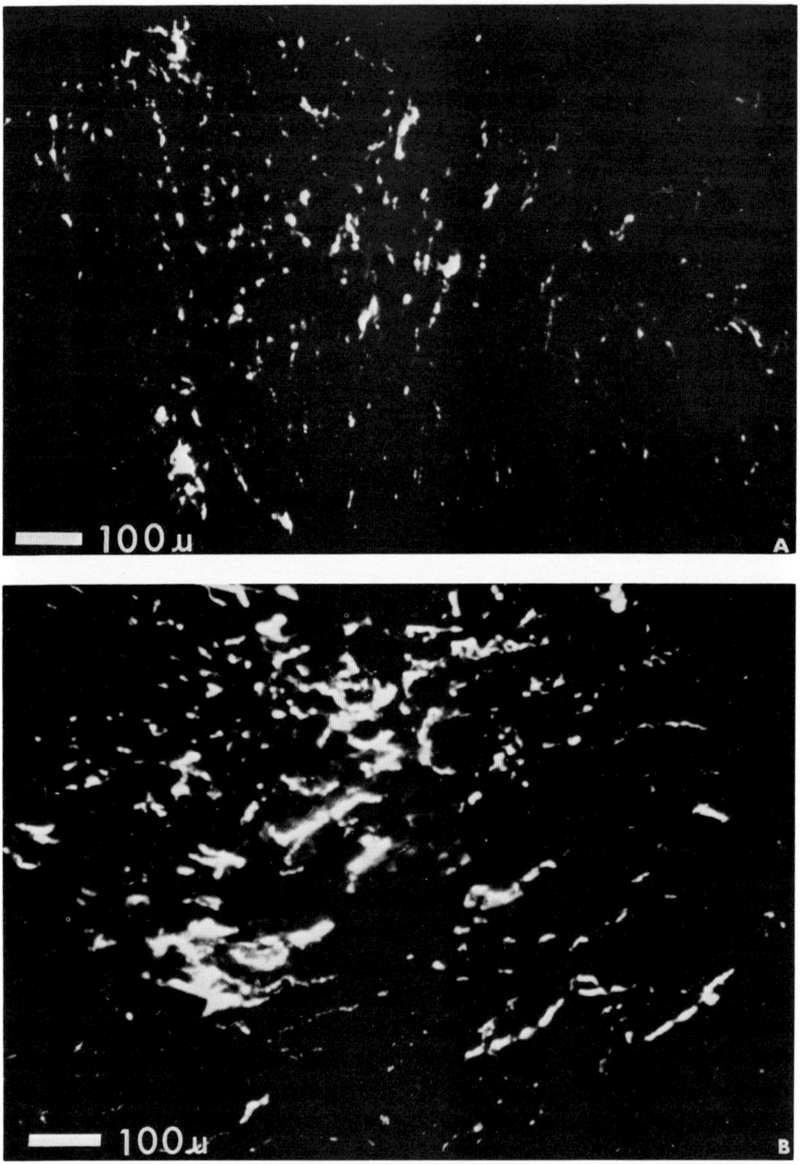

Fig. 4 Fluorescence microphotographs of cardiac adrenergic nerve terminals: (A) from control strain, 4 months of age. Photographic exposure 16 min on Kodak Ektachrome; (B) from dystrophic strain, 4 months of age, with marked hypertrophy and moderate degree of heart failure. Photographic exposure 8 min on Kodak Ektachrome.

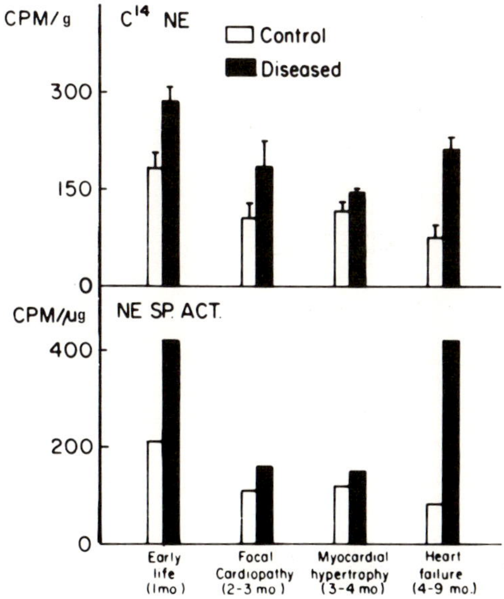

Fig. 5 C^{14} norepinephrine (NE) found in the heart 1 h after administration of C^{14} tyrosine. Means ± SE from 5 pools (6 hearts each) for each group.

animals, there was an increase of synthesis from TY to NE and that this was comparable from early life to the early stages of the disease (fig. 6).

Similarly, at all ages and stages, the synthesis of NE from DA was faster in the dystrophic than in the control groups (fig. 6). The increase was more prominent early in life and again during the stage of congestive failure.

The synthesis of DA from TY was not significantly different in the dystrophic animals as compared to the controls at all stages except at the last stage. At the stage of congestive failure, there was a marked drop in the concentration of C^{14} DA in the diseased hearts and, hence, the DA/TY ratio was much lower (fig. 6).

The above calculations are based to some extent on the assumption that the C^{14} TY injected reached the same concentration in the adrenergic nerve terminals of the heart in both the control and dystrophic animals. While this cannot be tested directly, the total cardiac concentration of TY as well as the amount of C^{14} TY, and hence the specific activity of C^{14} TY, were determined in each group (fig. 7).

The amount of total TY/g tissue in the heart of control and dystrophic animals was similar at all stages and not significantly different during the early and mid stages. However, there was a small but significant increase in the total TY/g in the hypertrophied hearts (third and fourth stages) of the dystrophic group. Thus, although the C^{14} TY found in the heart was similar in all groups,

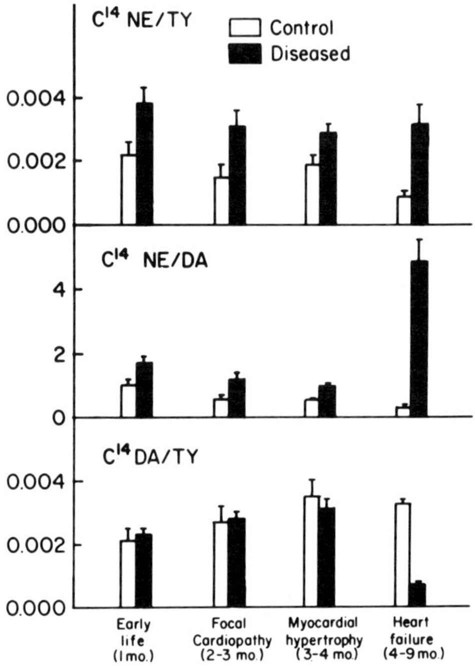

Fig. 6 Ratios of C^{14} counts as norepinephrine (NE), tyrosine (TY) and dopamine (DA) found 1 h after administration of C^{14} TY. Means ± SE from 5 pools for each group.

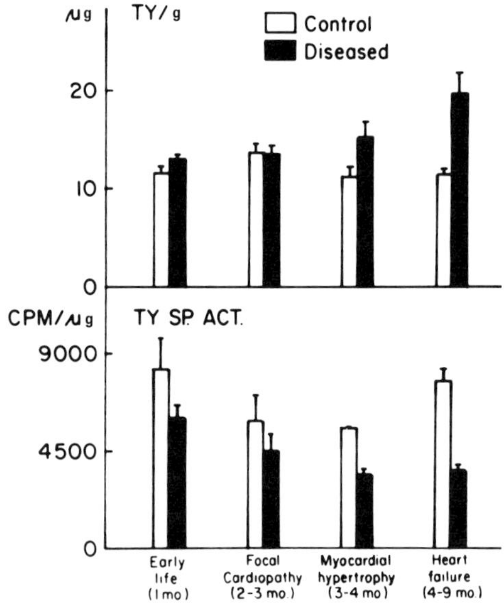

Fig. 7 Total (cold) tyrosine (TY) and C^{14} TY specific activity found 1 h after administration of C^{14} TY. Means ± SE from 5 pools for each group.

the specific activity of the C^{14} TY was distinctly and significantly lower in the hypertrophied hearts, *i.e.*, third and fourth stages (fig. 7). This seems to be related to cardiac hypertrophy rather than to congestive heart failure, since the levels in the hypertrophy groups without heart failure were similar to those in the last group with hypertrophy and heart failure.

DISCUSSION

All available evidence indicates that the strain of animals studied develops progressive cardiac myopathy, leading to an early death from congestive heart failure. While at an early stage there is no myocardial pathology, the disease advances through a progressive focal cardiopathy. At first, this is apparently compensated by myocardial hypertrophy, but it eventually leads to the stage of congestive failure (6).

The method used in the present studies for evaluating the functional status of the heart *in situ* can only provide a gross estimate of myocardial contractility and is undoubtedly affected by other factors. Nevertheless, when control and diseased animals are compared under the same conditions, the results show that the force per unit heart weight is nearly the same during the early life and becomes progressively less as the disease progresses.

Similar conclusions were reached by Brink and Lochner (8), who performed a more detailed evaluation of myocardial contractility in this strain of animals before and after the development of heart failure. More precise studies of the cardiac dynamics in this strain of animals have been carried out recently by Abelmann *et al.* and by Forman *et al.* and are presented in other parts of this book.

When heart failure is fully developed, cardiac NE concentrations are reduced, as was reported previously (2). This is consistent with observations made in man and in a variety of forms of experimental heart failure (9, 12, 15, 28). In the present case, as in other forms of heart failure studied previously, the decrease in cardiac NE concentrations is only partially due to a 'dilution' effect of myocardial hypertrophy, since the total amount of NE per heart is also significantly reduced.

During the progress of the disease and prior to heart failure there is some indication that cardiac NE is higher in the diseased than in the control group. The increase is more apparent when NE/heart is considered, which discounts the effect of myocardial hypertrophy. This difference although small was consistently found in several experiments with different cohorts of animals and appears to be of some significance in the light of subsequent observations on NE turnover.

The histochemical observations provide an additional insight into the changes occurring at the adrenergic nerve terminals. The increase in nerve terminal fluorescence in the latter stages of the disease, including the early phases of heart failure, suggests that the intraneuronal concentration of NE is augmented. This is apparently reflected in the moderate increase in total NE observed prior to heart failure.

It appears, therefore, that in the early stages of heart failure the intraneuronal concentration of NE remains high. The decrease in total heart NE could be due to progressive focal necrosis, with destruction of nerve terminals in the necrotic areas. The subsequent decrease in fluorescence intensity observed in animals with advanced heart failure nearing the terminal stage may well reflect changes in intraneuronal NE concentration that were secondary to the marked hemodynamic alterations.

The 'dilution' effect of necrotic or fibrotic tissue has also been suggested as a contributing factor in the low concentrations of atrial NE found in tissues taken from patients with congestive failure (10). In general, it must be emphasized that, in the presence of extensive tissue damage, total cardiac NE concentrations are not a reliable index of intraneuronal NE concentrations at the adrenergic terminals.

If the formation of C^{14} NE following administration of C^{14} TY is taken as an indication of the rate of NE synthesis, the results show a consistently higher NE synthesis in the diseased than in the control groups in all stages. Measurement of the actual NE synthesis rate would require measurements of C^{14} NE formation at several points in time following administration of C^{14} TY. Such studies at several age groups were precluded in this project because of their extremely high cost. In fact, precise measurements of NE synthesis rates would also require a knowledge of the concentration of C^{14} TY at the site of NE synthesis, *i.e.*, within the adrenergic nerve terminals. This cannot be obtained experimentally and, therefore, only relative rates of NE synthesis can be determined by any of the approaches available.

In any event, the present results are taken as indicating that there is a higher rate of NE synthesis in the diseased strain at all ages as compared to controls of the same age. The difference was found to be more marked early in life and again during heart failure. This interpretation assumes that in both groups of animals: a) the pathways for NE synthesis are the same; b) the concentrations of C^{14} TY reached at the cardiac adrenergic terminals are the same; and c) the time at which peak intraneuronal C^{14} NE contraction is reached is not markedly different.

In connection with the second assumption, it is of interest that total TY was seen to be elevated in the diseased group during the last two stages. It is not clear whether this finding is related to the presence of extensive myocardial lesions, cardiac hypertrophy, or other factors. Since the same amount of C^{14} TY was given in both groups, the presence of high total TY results in lower values of calculated TY specific activity in the diseased groups. However, there is no reason to believe that the intraneuronal concentration of TY at the adrenergic terminals was actually lower in the diseased than in the control groups. If this were in fact the case, then the NE synthesis would be even higher in the diseased group than is indicated from the C^{14} NE levels.

Similar conclusions are reached when the ratios of C^{14} NE/TY, NE/DA, and DA/TY are examined. In this analysis, any discrepancies in the total NE or TY levels are not involved. All these ratios again suggest an increased synthesis of NE from TY in the diseased strain at all stages of the disease.

At the intermediate stages of NE synthesis, the results suggest that the rate of formation of NE from DA is also somewhat higher at the first stages of the disease but becomes much greater during the stage of congestive failure. During this stage, the rate of formation of NE from DA apparently becomes relatively greater than that of the formation of DA from TY. The same conditions would occur if there were a net loss of cardiac DA during this stage. However, this could not be tested, since the total concentration of DA in the heart, which is very low, could not be measured with any degree of reliability.

The results show that NE synthesis is markedly elevated during the stage of congestive failure. Therefore, at this stage, the low total NE levels observed are not a result of reduced synthesis but, rather, are due to a high rate of NE release leading to a net loss.

These findings are not in agreement with conclusions reached by others in studies on NE synthesis in certain forms of experimental heart failure. Spann *et al.* (26) reported that, in the congestive failure occurring in guinea pigs following chronic aortic constriction, there is no change in NE turnover as measured by the rate of disappearance of cardiac C^{14} NE following administration of the same compound. The disparity between the results of Spann *et al.* and those reported here may be due to a basic difference in the type of heart failure. On the other hand, the measurements of NE turnover in the two studies involved different intraneuronal pools of NE. The pool of NE obtained from administration of the amine has been found in several studies to behave differently in many respects from the NE formed from TY.

The present findings on NE synthesis in heart failure are also in contrast to the observations of Pool *et al.* (23). In experimentally induced heart failure in the dog, these authors found a decrease in tyrosine hydroxylase activity, the rate limiting enzyme in the formation of NE from TY. These findings suggest a defect in the NE synthesis from TY in that form of heart failure, in contrast to the results obtained in the present study. Whether this reflects differences in the type of heart failure or other aspects involved cannot be determined at present.

The studies of Alousi and Beards (reported elsewhere in this book) are of interest, since these workers did not find a significant decrease in total cardiac NE during the stage of heart failure in another group of hamsters from the same strain. As the present results indicate (fig. 3), there exists a stage when the total NE concentration of the heart of the diseased group changes from being significantly higher to significantly lower that that of the controls. Hence, at the early stages of heart failure, there may be no significant differences in the total cardiac NE between the control and diseased groups. Alousi and Beards also found that the NE turnover of the diseased hearts measured in heart homogenates *in vitro* was not significantly different from the controls. When this is compared with our findings *in vivo,* where a marked increase in NE turnover was observed, the two sets of results indicate that the hearts of the diseased animals had a normal ability to synthesize NE and, hence, the increased turnover observed *in vivo* must reflect an increased 'drive' of the synthetic apparatus, presumably through an increased activity of the adrenergic nerves to the heart.

It has been demonstrated that enhanced adrenergic nerve activity results in increased NE synthesis (16). Thus, an increase in the latter parameter *in vivo* may be taken as suggesting enhanced activity of the former. Increased NE synthesis in the presence of stable or low NE stores indicates increased NE turnover and is reflected by enhanced NE specific activity. On this basis, it may be concluded that, in the diseased strain under study, there is an increased cardiac adrenergic nerve activity throughout life, which is maintained and intensified during the stage of heart failure. In addition, on the basis of the histochemical observations, it appears that the intraneuronal concentration of NE is also high in this strain throughout life.

Of particular interest is the finding that there is a marked difference in cardiac NE synthesis and turnover in the diseased strain early in life before any signs of cardiac disease are apparent. This can be taken as suggesting that even at this early stage, there is some functional compensation in the heart through an increase in adrenergic nerve activity. Alternatively, it may be that high adrenergic nerve activity with an increased release of NE is responsible for the subsequent focal myocardial pathology and, hence, the development of myocardial disease. Certainly, focal myocardial necrosis has been observed to occur following the administration of large doses of catecholamines (24). In this connection, it is noteworthy that recent studies by Bajusz *et al.* (5) indicate that chronic administration of catecholamines in the diseased strain of hamsters hastens the progress of the disease and leads to the development of myocardial lesions earlier in life.

In general, it appears that marked alterations in the cardiac adrenergic mechanisms exist in this strain of animals prior to the onset of any signs of heart disease. Whether this has a counterpart in any form of heart disease in man remains to be elucidated.

SUMMARY

Cardiac norepinephrine (NE) stores, synthesis, and turnover were evaluated prior to and during the development of congestive heart failure in a strain of hamsters (BIO 14.6) with hereditary myopathy. The NE stores at the cardiac adrenergic nerve terminals were studied chemically and histochemically, while NE synthesis was evaluated from measurements of C^{14} NE following administration of C^{14} tyrosine (TY).

The results show that during early life, prior to the development of myocardial lesions, NE formation occurs at a greater rate in the diseased strain as compared to a control group of the same age. Since NE stores are essentially the same in both groups during this stage, it is concluded that NE turnover is higher in the diseased strain and this probably results from increased cardiac sympathetic nerve activity.

Similar findings, but to a lesser degree, were obtained at later stages of the disease, at a time when myocardial pathology is present but is largely compensated by myocardial hypertrophy. During these intermediate stages, there is a small but significant increase in cardiac NE stores as well as an increase in C^{14} NE formation.

During the last stage, in the presence of overt heart failure, cardiac NE concentration is markedly reduced. This is due, in part, to a 'dilution' effect of the prominent myocardial hypertrophy that is present at this stage and, in part, to a similar 'dilution' effect produced by the presence of extensive focal myocardial necrosis and fibrosis with focal destruction of adrenergic nerve terminals. Histochemical observations indicate that, at least during the early phases of heart failure, the intraneuronal concentration of NE at the cardiac adrenergic terminals is elevated. At the same time and throughout the period of heart failure, NE formation increases markedly, which suggests an increase in NE turnover.

Similarities and differences between these results and observations made in various forms of clinically observed or experimentally induced heart failure are discussed.

REFERENCES

1. ANGELAKOS, E. T.: Regional distribution of catecholamines in the dog heart. Circulat. Res. *16:* 39–44 (1965).
2. ANGELAKOS, E. T.: Electrocardiographic (ECG) changes and cardiac catecholamine concentrations in the dystrophic hamster strain (Bio 14.6 line) at various ages. Trans. N. Y. Acad. Sci. *30:* 955–956 (1968).
3. ANGELAKOS, E. T. and KING, M. P.: Fluorescence histochemistry of the biogenic amines. in EFRON, D. H. Psycopharmacology, pp. 27–37 (Public Health Service Publication No. 1836, Washington, D. C. 1968).
4. ANGELAKOS, E. T.; KING, M. P. and MILLARD, R. W.: Regional distribution of catecholamines in the hearts of various species. Ann. N. Y. Acad. Sci. *156:* 219–240 (1969).
5. BAJUSZ, E.; HOMBURGER, F.; BAKER, J. R. and BOGDONOFF, P.: Dissociation of factors influencing myocardial degeneration and generalized cardiocirculatory failure. Ann. N. Y. Acad. Sci. *156:* 396–420 (1969).
6. BAJUSZ, E.; HOMBURGER, F.; BAKER, J. R. and OPIE, L. H.: The heart muscle in muscular dystrophy with special reference to involvement of the cardiovascular system in the hereditary myopathy of the hamster. Ann. N. Y. Acad. Sci. *138:* 213–231 (1966).
7. BAJUSZ, E. and LOSSNITZER, K.: A new disease model of chronic congestive heart failure: Studies on its pathogenesis. Trans. N. Y. Acad. Sci. *30:* 939–948 (1968).
8. BRINK, A. J. and LOCHNER, A.: Work performance of the isolated perfused beating heart in the hereditary myocardiopathy of the Syrian hamster. Circulat. Res. *21:* 391–401 (1967).
9. CHIDSEY, C. A.; BRAUNWALD, E. and MORROW, A. G.: Catecholamine excretion and cardiac stores of norepinephrine in congestive heart failure. Amer. J. Med. *39:* 442–451 (1964).
10. CHIDSEY, C. A.; BRAUNWALD, E.; MORROW, A. G. and MASON, D. T.: Myocardial norepinephrine concentration in man. Effects of reserpine and of congestive heart failure. New Eng. J. Med. *269:* 653–658 (1963).
11. CHIDSEY, C. A.; KAISER, G. A. and BRAUNWALD, E.: Biosynthesis of norepinephrine in isolated canine heart. Science *139:* 828–829 (1963).

12. CHIDSEY, C. A.; KAISER, G. A.; SONNENBLICK, E. H.; SPANN, J. F. and BRAUNWALD, E.: Cardiac norepinephrine stores in experimental heart failure in the dog. J. Clin. Invest. *43:* 2386–2393 (1964).
13. COVELL, J. W.; CHIDSEY, C. A. and BRAUNWALD, E.: Reduction of the cardiac response to postganglionic sympathetic nerve stimulation in experimental heart failure. Circulat. Res. *19:* 51–56 (1966).
14. FALCK, B.: Observations on the possibilities of the cellular localization of monoamines by a fluorescence method. Acta. Physiol. Scand. *56:* Suppl. *197:* 1–25 (1962).
15. FISCHER, J. E.; HORST, W. D. and KOPIN, I. J.: Norepinephrine metabolism in hypertrophied rat hearts. Nature *207:* 951–953 (1965).
16. GORDON, R.; REID, J. V. O.; SJOERDSMA, A. and UDENFRIEND, S.: Increased synthesis of norepinephrine in the rat heart on electrical stimulation of the stellate ganglia. Mol. Pharmacol. *2:* 606–613 (1966).
17. HOMBURGER, F.; BAKER, J. R.; NIXON, C. W. and WHITNEY, R.: Primary, generalized polymyopathy and cardiac necrosis in an inbred line of Syrian hamsters. Med. Exp. *6:* 339–345 (1962).
18. HOMBURGER, F.; BAKER, J. R.; NIXON, C. W. and WILGRAM, G.: New hereditary disease of Syrian hamsters. Arch. Int. Med. *110:* 660–662 (1962).
19. HOMBURGER, F.; BAKER, J. R.; NIXON, C. W.; WILGRAM, G. and HARROP, J.: The early histopathological lesions of muscular dystrophy in the Syrian golden hamster. J. Path. Bact. *89:* 133–138 (1965).
20. HOMBURGER, F.; NIXON, C. W.; EPPENBERGER, M. and BAKER, J. R.: Hereditary myopathy in the Syrian hamster: studies on pathogenesis. Ann. N. Y. Acad. Sci. *138:* 14–27 (1966).
21. MEERSON, F. Z.; MANUKHIN, B. N.; PSHENICHNIKOVA, M. G. and ROZANOVA, L. S.: Mediator metabolism in myocardium in compensatory hyperfunction and hypertrophy of heart. Pat. Fiziol. Eksp. Ter. *7:* 32 (1963). Cited in Fed. Proc. (Tran l. Suppl.) *23:* T593–T595 (1964).
22. OPIE, L. H.; LOCHNER, A.; BRINK, A. J.; HOMBURGER, F. and NIXON, C. W.: Oxidative phosphorylation in hereditary myocardiopathy in the Syrian hamster. Lancet *ii:* 1213 (1964).
23. POOL, P. E.; COVELL, J. W.; LEVITT, M.; GIBB, J. and BRAUNWALD, E.: Reduction of cardiac tyrosine hydroxylase activity in experimental congestive heart failure. Circulat. Res. *20:* 349–353 (1967).
24. RAAB, W.: The nonvascular metabolic myocardial vulnerability factor in "coronary heart disease". Amer. Heart J. *66:* 685–706 (1963).
25. SPANN, J. F.; CHIDSEY, C. A. and BRAUNWALD, E.: Reduction of cardiac stores of norepinephrine in experimental heart failure. Science *145:* 1439–1441 (1964).
26. SPANN, J. F.; CHIDSEY, C. A.; POOL, P. E. and BRAUNWALD, E.: Mechanism of norepinephrine depletion in experimental heart failure produced by aortic constriction in the guinea pig. Circulat. Res. *17:* 312–321 (1965).
27. TOMOMATSU, T.; UEDA, Y.; MATSUMOTO, T.; ISOMA, T. and KONDO, Y.: Catecholamine in congestive heart failure. Jap. Heart J. *4:* 13–24 (1963).
28. TORSTI, P.; PENTTILA, O. and TORSTI, E.: The catecholamine content of the failing rabbit heart. Ann. Med. Exp. Fenn. *42:* 33–36 (1964).
29. UDENFRIEND, S.: Fluorescence Assay in Biology and Medicine (Academic Press, New York 1962).

Mailing address: E. T. Angelakos, M.D., Ph.D., Department of Physiology and Biophysics, Hahnemann Medical College and Hospital, 235 North Fifteenth Street, Philadelphia, Pa. 19102 (USA).

Geigy Pharmaceuticals, Division of Geigy Chemical
Corporation, Ardsley, New York, U S A

CATECHOLAMINE, PROTEIN, AND RNA CONTENT IN ADVANCED CONGESTIVE HEART FAILURE IN THE SYRIAN HAMSTER

A. A. ALOUSI and J. A. BEARDS

INTRODUCTION

The exact molecular alterations that lead to the mechanical impairment of cardiac muscle in patients with congestive heart failure remain unknown. However, there is ample evidence to indicate that the activity of the sympathetic nervous system is modified in such patients (7). The influence of the sympathetic nervous system on the mechanical and electrical properties of the normal heart has been recognized for many years, and it is now very well documented that this system plays a very important compensatory role in the cardiovascular readjustment in subjects with heart failure (4, 10, 12). Hyperactivity of the sympathetic nervous system in cardiac patients and experimental animals has been reflected in an elevation of plasma norepinephrine (NE) level at rest and during exercise, an increase in urinary excretion of NE, and depletion of cardiac NE (8, 10, 17).

The present report consists of a study of some cellular constituents known to affect myocardial contraction and, hence, performance; these are protein and catecholamines. Cardiomyopathic (Bio 14.6) Syrian hamsters in a late stage of cardiac decompensation have been used as a model for the study of congestive heart failure (1, 5).

METHODS

The Langendorff preparation

Male myopathic Syrian hamsters (Bio 14.6) weighing 150-200 g, obtained from the Bio-Research Institute in Boston, were used in this study. The hamsters were 8-9 months old and in a late stage of cardiac decompensation. Normal, healthy hamsters from an unrelated strain–London School of Hygiene–were used as controls.

The hamster was sacrificed by decapitation, the thoracic cavity opened, and a PE 60 catheter inserted and tied into the ascending aorta 2.5 mm above the heart; the catheter was being perfused through a side arm of the Langendorff apparatus. The heart was then removed from the hamster and the catheter cut to 1 inch in length and placed on a 20-gauge needle at the base of the Langendorff apparatus. The heart was then perfused at a constant perfusion pressure with Chenoweth's solution of the following composition: 14 g NaC1, 0.84 g KC1, 0.48 g $CaCl_2$, 0.85 g $MgCl\cdot 6H_2O$, 3.6 g dextrose, 16 mg disodium EDTA, and 4.2 g $NaHCO_3$. All salts were dissolved in 2 liters of glass distilled water and equilibrated with 5% CO_2, 95% O_2 gas mixture. The perfusate was recirculated using an air pump.

One end of a cotton suture was attached to the apex of the heart and the other end passed under a pulley to a strain gauge calibrated so that 1 g of tension deflected the pen of a Beckman Dynograph 0.5 cm. The lower chamber of the Langendorff apparatus was then raised and the diastolic tension on the heart adjusted to 1.5 g. The heart was allowed to equilibrate for 15 min before any drug treatment was applied. All drugs and radioactive isotopes were added to the perfusate in the concentrations indicated under individual experimental results. At the end of the perfusion time, the heart was blotted dry, weighed, and homogenized in ice-cold 10% trichloroacetic acid (TCA). The acid homogenate was centrifuged, the supernatant decanted and saved, and the precipitate washed with 5% TCA and saved until further testing.

Studies in heart homogenates

The hamster was sacrificed by decapitation, the chest opened, and the heart quickly removed. After being dipped for a few seconds in ice-cold saline solution, the heart was homogenized in 0.1 M sodium phosphate buffer (pH 7.4) containing 1 mM disodium EDTA and 1mM Pargyline.

An aliquot of the homogenate was transferred to a 20 ml beaker and 10 μc H^3-L-tyrosine (specific activity 35.1 c/mM) or 1 μc L-C^{14}-alanine (specific activity 0.125 mc/mM) and 10 μc H^3-uridine (specific activity 26.7 c/mM) was added. The homogenate was incubated at 37°C in a Dubnoff metabolic shaker under O_2. At the end of the incubation period, the reaction was stopped by the addition of 10% TCA and the mixture was transferred to a centrifuge tube and centrifuged. The acid extract contained catecholamines and other acid soluble compounds and the precipitate contained nucleic acids, proteins, and lipids.

Ribonucleic acid (RNA) determination

A modification of the original Schneider procedure for RNA was adopted (see Munro [15]). The TCA precipitate obtained, as described above, was washed three times with cold 5% TCA. The pellet was then hydrolyzed and centrifuged. The supernatant was transferred quantitatively to a centrifuge tube and total RNA was determined by measuring ribose content by the orcinol method. H^3-uridine incorporation into heart RNA was determined by liquid scintillation spectrometry.

Protein determination

The residual pellet was washed with water, acetone, ethanol-ether (3 : 1), and ether to remove all the lipids. The pellet was then hydrolyzed in 1 N NaOH at 90°C for 30 min. Total protein was determined in alkaline hydrolysate by the biuret method. C^{14}-alanine incorporation into heart protein was determined by liquid scintillation spectrometry.

Catecholamine determination

Norepinephrine content: Total NE in the heart was determined by the trihydroxy-indole method as described by Crout (11).

Isolation of catecholamines: NE and dopamine were isolated on a Dowex-50-Na^+ column (3).

RESULTS

Physical appearance and gross pathology

The Bio 14.6 Syrian hamster in a late stage of cardiac decompensation exhibits all the clinical symtoms characteristic of generalized myopathy and congestive heart failure. These include muscular weakness and lack of coordination, hyperpnea, tachycardia, and generalized subcutaneous edema.

At autopsy, generalized venous congestion, ascites, hydrothorax, and hydropericardium are found.

Enlargement of the heart

The gross pathological signs of cardiac insufficiency are demonstrated by the large volume of the heart, with dilatation of the ventricles and auricles and the presence of fluid and thrombi in both chambers. Table I summarizes the results of comparative studies of body weight, heart weight, and heart protein content

Table I

Cardiac Enlargement in Myopathic Hamster*

Experimental conditions	Body wt. (g)	Heart wt. (g)	Heart wt. (g) / Body wt. (100g)	Protein conc. (mg/heart)	Protein conc. (mg/g heart)
Myopathic hamsters	153.80 ± 5.70 (13)	1.03 ± 0.04 (13)	0.68 ± 0.03 (13)	113.0 ± 5.38 (13)	109.03 ± 3.48 (13)
Normal hamsters	124.80 ± 9.20 (15)	0.56 ± 0.02 (15)	0.46 ± 0.01 (15)	80.5 ± 3.45 (15)	142.02 ± 3.57 (15)
	$p < 0.02$	$p < 0.01$	$p < 0.01$	$p < 0.01$	$p < 0.01$

*Values are means ± standard error of mean; numbers in parentheses refer to the number of experiments; p values are based on Student t test.

in the myopathic and normal hamsters. Hamsters in advanced stages of cardiomyopathy are heavier in body weight than normal hamsters of the same age and sex. The hearts of myopathic hamsters are significatly heavier than those of the controls. However, the ratio of heart weight to body weight remains higher in the myopathic hamsters, signifying the severity of cardiac enlargement.

In order to determine whether the increase in heart weight of myopathic hamsters is due to or accompanied by an increase in protein content of the myocardium, total protein was determined in hearts of myopathic and normal hamsters. As shown in table I, the myopathic hearts contain more protein per heart than the normal hearts. However, the concentration of protein per gram heart is significantly lower in myopathic hearts than in the control hearts.

Protein and RNA synthesis

Heart homogenates from myopathic and normal hamsters were incubated in the presence of 1μc L-C^{14}-alanine and 10 μc H^3-uridine for 1 h. At the end of the incubation period, the reaction was stopped and protein and RNA contents as well as their specific activities were determined. Table II summarizes our results. The incorporation of C^{14}-alanine into heart protein and H^3-uridine into heart RNA was significantly reduced in myopathic hearts. Although the total amount of RNA per heart is more in the myopathic heart, the amount of RNA per gram protein is the same as in the controls.

Catecholamines and the negative inotropy and chronotropy of the myopathic heart

In the Langendorff preparation, the isolated perfused heart of a myopathic hamster in a late stage of cardiac decompensation exhibits negative inotropy and chronotropy as compared with the normal heart (fig. 1, a and b).

In view of the importance of the sympathetic nervous system in stimulating

Table II

Incorporation of Radioactive Precursors into Heart Protein and RNA*

Experimental conditions	Protein specific activity (cpm/mg protein)	RNA concentration (μg/g protein)	RNA specific activity (cpm/μg RNA)
Myopathic hearts	130.1 ± 6.09 (4)	17.26 ± 2.01 (4)	93.66 ± 8.20 (4)
Normal hearts	224.0 ± 2.60 (4)	14.75 ± 0.88 (4)	118.26 ± 7.31 (4)
	$p < 0.01$	$p > 0.1$	$p < 0.02$

*Values are means ± standard error of mean; numbers in parentheses refer to the number of experiments; p values are based on Student t test.

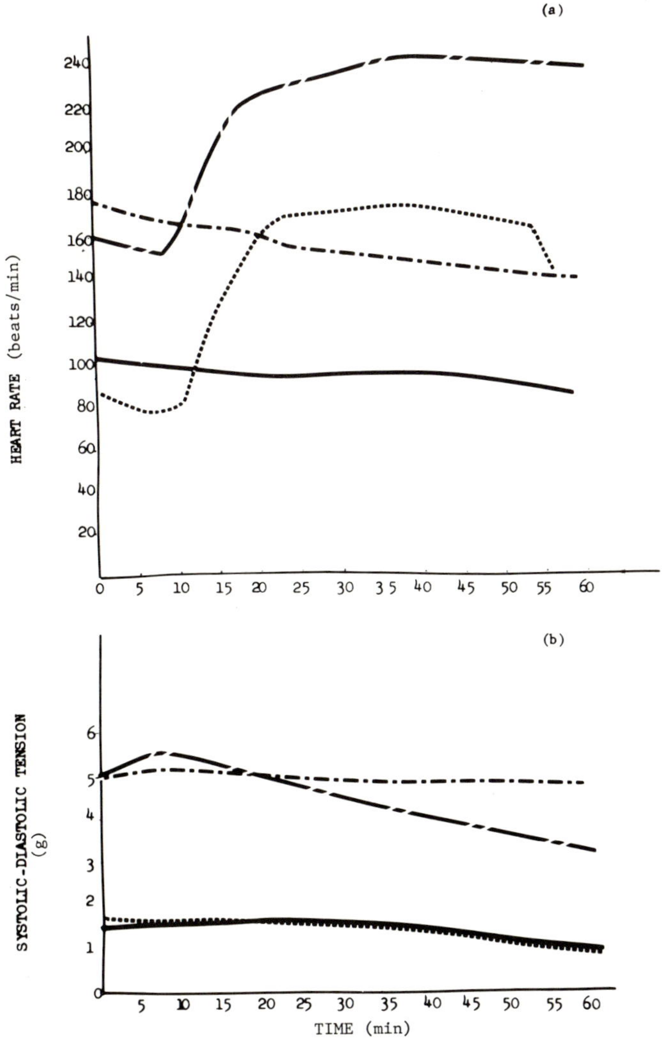

Fig. 1 a and *b*, Effect of dihydroxyphenylalanine on the heart rate and force of contraction of the isolated perfused hamster heart. Measurements were made every 5 min for 60 min. The values are averages of 9 hearts per experiment. —— Myopathic controls; .... Myopathic + DOPA; —·—· Normal controls; - - - Normal + DOPA.

the rate and force of contraction of the normal heart, a study of total NE content and NE concentration in the myopathic heart was conducted. Table III shows that in the myopathic heart the amount of NE per heart is the same as in the normal heart. However, the amount of NE per gram protein is significantly lower in the myopathic heart than in the normal heart.

Table III

Norepinephrine Content of the Isolated Perfused Heart*

Experimental conditions	NE concentration (μg/ heart)	NE concentration (μg/g protein)
Myopathic hearts	0.51 ± 0.057 (8)	4.42 ± 0.395 (8)
Normal hearts	0.57 ± 0.076 (9)	7.42 ± 1.241
	$p > 0.1$	$p < 0.05$

*Values are means ± standard error of mean; numbers in parentheses refer to number of experiments; p values are based on Student t test.

Table IV summarizes our results on the study of the rate of H^3-tyrosine incorporation into heart NE. Heart homogenates from myopathic and normal hamsters were incubated in the presence of 10 μc H^3-tyrosine for 1 h; total NE and NE specific activity were measured at the end of the incubation time. As shown in table IV, there is less NE per gram protein in the myopathic heart than in the normal heart, while NE specific activity is about the same in both hearts.

If the low NE concentration of the myopathic heart is responsible for the negative inotropy and chronotropy, it should be possible to correct this defect by increasing the NE concentration of the myopathic heart.

Perfusing the heart for 1 h with a perfusate containing NE precursor DOPA in a concentration of 1×10^{-3} M caused a significant increase in heart rate of myopathic and normal hearts. DOPA infusion caused no change in the force of contraction of either heart (fig. 1, a and b).

NE synthesis in the myopathic and normal heart increased markedly after DOPA infusion. This increase was significantly greater than in the control hearts, as is shown in table V.

Table IV

Norepinephrine Synthesis in Heart Homogenates*

Experimental conditions	NE concentration (μg NE/g protein)	NE specific activity (cpm $\times$ 10^3/μg NE)
Myopathic hearts	5.54 ± 0.788 (7)	1627 ± 23.5 (7)
Normal hearts	8.50 ± 0.897 (7)	1940 ± 30.3 (7)
	$p < 0.02$	$p > 0.1$

*Values are means ± standard error of mean; numbers in parentheses refer to the number of experiments; p values are based on Student t test.

Table V

Effect of Dihydroxyphenylalanine on Norepinephrine Synthesis in the Isolated Perfused Heart*

	Control		1×10^{-3} M DOPA	
Experimental conditions	NE concentration (μg/heart)	NE concentration (μg/g protein)	NE concentration (μg/heart)	NE concentration (μg/g protein)
Myopathic hearts	0.51 ± 0.057 (8)	4.42 ± 0.395 (8)	6.77 ± 0.823 (5)	60.62 ± 9.550 (5)
Normal hearts	0.57 ± 0.076 (9)	7.42 ± 1.241 (9)	3.32 ± 0.95 (6)	41.17 ± 2.130 (6)
	$p > 0.1$	$p < 0.05$	$p < 0.01$	$p < 0.1$

*Values are means ± standard error of mean; numbers in parentheses refer to the number of experiments; p values are based on Student t Test.

Response of the myopathic heart to the cardiotonic effect of exogenous NE and ouabain

In order to test the response of the isolated perfused heart to the positive inotropic effect of exogenous NE, 0.01 μg/ml NE was added to the perfusate in the Langendorff preparation, using myopathic and normal hearts. As shown in figure 2, NE caused a marked increase in the force of contraction of both hearts.

Ouabain, added to the perfusate in a concentration of 1×10^{-4} M, caused a marked increase in the force of contraction of both hearts (fig. 2).

DISCUSSION

It is evident that the myopathic Syrian hamster, in a late stage of cardiac decompensation, provides an excellent model for the study of the cellular alterations responsible for the development of congestive heart failure.

The increase in total protein and RNA of the myopathic heart is a clear indication of a compensatory hypertrophy of the heart during early stages of the disease. Sumner and McIntosh (18) reported a greater increase in protein and RNA than in DNA during the hypertrophy of the heart. They suggested an increase in cell size rather than cell number. The significant reduction in protein concentration per gram heart in the myopathic hamster is due to the accumulation of other constituents, mainly extracellular fluid (6). The low specific activity of protein and RNA in our isotope incorporation studies indicates a low rate of synthesis of these macromolecules during heart failure (14).

A considerable amount of controversy exists regarding the major cause of heart failure. Among the many factors being considered are: alteration in contractile protein; the level of ATP and creatine phosphate; mitochondrial

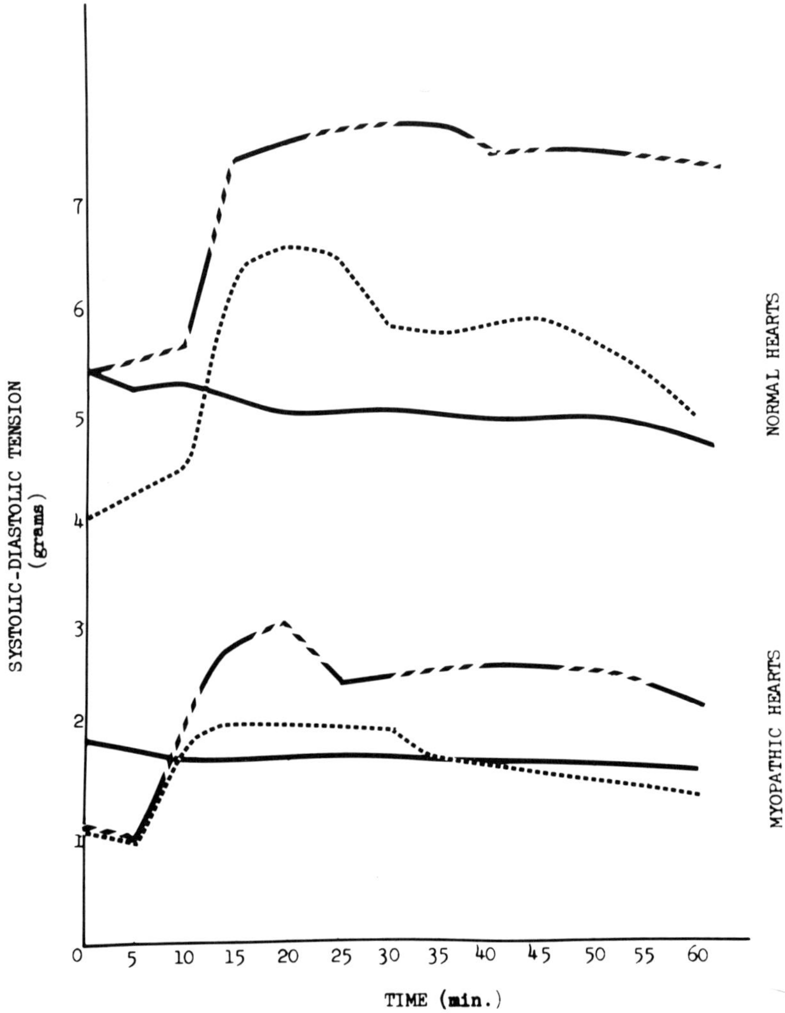

Fig. 2 The effect of norepinephrine and ouabain on the force of contraction of the isolated perfused hamster heart. Measurements were made every 5 min for 60 min. Values are averages of 6 hearts per experiment. ——Controls; ------ + NE; – · · · · – + Quabain.

function and ATP-ase activity. The results of these studies are very variable and most confusing (7, 13, 16).

Covell *et al.* (10) have shown that experimental heart failure is associated with a marked reduction in inotropic and chronotropic response to electrical stimulation of the sympathetic nerve to the heart. In view of our interest in the effect of NE and stress on the metabolism of the heart (2), NE metabolism in the myopathic heart was studied. Low levels of NE per gram protein were observed in the myopathic heart. This could not be attributed to NE depletion, owing to the fact that the total amount of NE per heart is the same as in the normal heart. It was, therefore, concluded that the increase in total protein in the myopathic heart without a corresponding increase in the adrenergic innervation or NE content of the existing nerves is responsible for the low level of NE per gram protein. This is contrary to the observations of Spann *et al.* (17) and Chidsey *et al.* (9). who reported a depletion of NE in hypertrophied hearts and ruled out the possibility of NE dilution by the increased mass of the heart. The disparity in our results could be due to the use of different animal species at a different stage of cardiac stress.

The rate of NE synthesis from tyrosine in the myopathic heart is comparable to that of the normal heart. Dihydroxyphenylalanine perfusion caused a more than ten-fold increase in NE synthesis in the myopathic heart and a six-fold increase in NE synthesis in the normal heart. To our surprise, the marked increase in NE level after DOPA perfusion had no effect on the force of contraction of either heart. However, the addition of a small amount of exogenous NE or ouabain caused a marked increase in the force of contraction of both myopathic and normal hearts. The possibility of changes in the rate of release of NE or in the response of adrenergic receptors in the heart has still to be investigated.

SUMMARY

The myopathic Syrian hamster in a late stage of cardiac decompensation exhibits all the clinical symptoms characteristic of congestive heart failure.

The hypertrophy of the myopathic heart is accompanied by an increase in protein and RNA content and a decrease in protein and RNA concentration.

The rate of protein and RNA synthesis is lower in the myopathic heart than in the normal heart.

The isolated perfused myopathic heart exhibits negative inotropy and chronotropy.

Total amount of NE per myopathic heart is equal to that of the normal heart. However, NE concentration per gram protein is lower in the myopathic heart.

NE synthesis in the myopathic heart is comparable to that of the normal heart.

Dihydroxyphenylalanine perfusion causes a marked increase in NE concentration and heart rate of the myopathic and normal heart without any significant effect on the force of contraction.

The myopathic heart responds positively to the inotropic effect of exogenously added NE or ouabain.

REFERENCES

1. ALOUSI, A. A.; BEARDS, J. and TANZ, R. D.: Catecholamine, protein and RNA content in advanced congestive heart failure. Pharmacologist *11:* 288 (1969).
2. ALOUSI, A. A. and MALLOV, S.: Effect of hyperthyroidism, epinephrine, and diet on heart lipoprotein lipase activity. Amer. J. Physiol. *206:* 603–609 (1964).
3. ALOUSI, A. A. and WEINER, N.: The regulation of norepinephrine synthesis in sympathetic nerves: Effect of nerve stimulation, cocaine, and catecholamine-releasing agents. Proc. Nat. Acad. Sci. *56:* 1491–1496 (1966).

4. ANZOLA, J. and RUSHMER, R.: Cardiac responses to sympathetic stimulation. Circulat. Res. *4:* 302–307 (1956).
5. BAJUSZ, E.; HOMBURGER, F.; BAKER, J. R. and OPIE, L. H.: The heart muscle in muscular dystrophy with special reference to involvement of cardiovascular system in hereditary myopathy of the hamster. Ann. N. Y. Acad. Sci. *138:* 213–231 (1966).
6. BAJUSZ, E. and LOSSNITZER, K.: A new disease model of chronic congestive heart failure: Studies on its pathogenesis. Trans. New York Acad. Sci. *30:* 939–948 (1968).
7. BRAUNWALD, E.; ROSS, J. and SONNENBLICK, E. H.: Mechanisms of contraction of the normal and failing heart. New Eng. J. Med. *277:* 794–1022 (1967)
8. CHIDSEY, C. A.; BRAUNWALD, E. and MORROW, A. G.: Catecholamine excretion and cardiac stores of norepinephrine in congestive heart failure. Amer. J. Med. *39:* 442–451 (1965).
9. CHIDSEY, C. A.; KAISER, G. A.; SONNENBLICK, E. H.; SPANN, J. F. and BRAUNWALD, E.: Cardiac norepinephrine stores in experimental heart failure in the dog. J. Clin. Invest. *43:* 2386–2393 (1964).
10. COVELL, J. W.; CHIDSEY, C. A. and BRAUNWALD, E.: Reduction of cardiac response to postganglionic sympathetic nerve stimulation in experimental heart failure. Circulat. Res. *19:* 51–56 (1966).
11. CROUT, J. R.: Catecholamines in urine, in: Standard Methods in Clinical Chemistry *3:* 62–80 (1961).
12. GAFFNEY, T. E. and BRAUNWALD, E.: Importance of adrenergic nervous system in support of circulatory function in patients with congestive heart failure. Amer. J. Med. *34:* 320–324 (1963).
13. GORDON, M. S. and BROWN, A. L.: Myofibrillar adenosine triphosphatase activity of human heart tissue in congestive heart failure: Effect of ouabain and calcium. Circulat. Res. *18:* 534–542 (1966).
14. MEERSON, F. Z.: On mechanism of compensatory hyperfunction and insufficiency of heart. Cor Vasa *3:* 161 (1961).
15. MUNRO, H.: The determination of nucleic acids. Meth. Biochem. Anal. *14:* 113–176 (1966).
16. NEBEL, M. L. and BING, R. J.: Contractile proteins of normal and failing human heart. Arch. Int. Med. *111:* 190–195 (1963).
17. SPANN, J. F.; CHIDSEY, C. A.; POOL, P. E. and BRAUNWALD, E.: Mechanism of norepinephrine depletion in experimental heart failure produced by aortic constriction in the guinea pig. Circulat. Res. *17:* 312–321 (1965).
18. SUMNER, R. G. and McINTOSH, H. D.: Nucleic acid studies in experimental cardiomegaly. Circulat. Res. *12:* 170–175 (1963).

Mailing address: Dr. A. A. Alousi, Geigy Research, Ardsley, New York 10502 (USA).

Department of Biochemistry, University of Manitoba,
Winnipeg, Manitoba, Canada

ABNORMAL OXIDATIVE PHOSPHORYLATION IN SKELETAL MUSCLE MITOCHONDRIA OF THE BIO 14.6 DYSTROPHIC SYRIAN HAMSTER[1]

K. WROGEMANN,[2] M. C. BLANCHAER, and B. E. JACOBSON

Abnormal oxidative phosphorylation has been reported in muscle mitochondria of dystrophic hamsters of the BIO 1.50 strain by Lochner and Brink (6) and decreased utilization of ^{14}C-labeled palmitate, pyruvate and acetyl-carnitine has been found by Strickland and his co-workers (4, 5) in muscle mitochondria from dystrophic mice. In the present work, we describe two distinct defects in these processes in skeletal muscle mitochondria of dystrophic BIO 14.6 hamsters. One defect occurs in older hamsters and resembles that found in mice. The other, which is limited to young hamsters, is unusual in that it is ameliorated by magnesium.

Mitochondria were prepared in two ways. In one procedure, resembling that of Lochner and Brink (6), minced muscle was homogenized in 10 volumes of medium (0.25 M sucrose, 1 mM EDTA, pH 7.4), using a Kontes D glass-on-glass homogenizer, and the mitochondria isolated by differential centrifugation. Mitochondria were also prepared, using the proteinase Nagarse in a medium containing 0.21 M mannitol, 0.07 M sucrose, 10 mM Tris-phosphate, and 0.1 mM EDTA. These procedures and other methods have been described elsewhere (2, 3, 8).

With the Nagarse procedure for isolating mitochondria, we were previously unable to detect any abnormalities in organelles from BIO 14.6 hamsters aged 97–124 days (3). In the present work with older animals, respiration and oxidative phosphorylation by mitochondria isolated without Nagarse proved defective with pyruvate or palmitate as substrate (table I). In another larger series, these abnormalities were also demonstrated with palmityl-L-carnitine and acetyl-L-carnitine as substrates. However, succinate and NADH were oxidized at normal rates, suggesting that the defect did not lie in the respiratory chain. From these findings, it was concluded (3) that the primary defect probably stemmed from a failure of the tricarboxylic acid cycle to deliver reducing equivalents to the respiratory chain. The loose coupling of phosphorylation apparent in table I seems to be due to a separate defect. Mitochondria isolated with the help of Nagarse from old dystrophic hamsters also exhibited lowered mean respiratory

---

[1]Supported by grants from the Muscular Dystrophy Association of Canada and the Medical Research Council of Canada.

[2]Post doctoral Fellow of the Muscular Dystrophy Association of Canada.

Table I

Skeletal Muscle Mitochondria Isolated without Nagarse from Old Hamsters*

Hamsters	Pyruvate			Palmitate
	O_2 rate†	P/O	P rate†	O_2 rate†
Normal (N = 7)	204 ± 9	2.08 ± 0.04	849 ± 51	82 ± 5
Dystrophic (N = 6)	124 ± 5	1.89 ± 0.05	473 ± 26	40 ± 4

*Means ± S.E. for hamsters 220 ± 16 days old. All means were significantly greater in the normal animals ($p < 0.02$). A manometric test system was used (3).

†μmole/g mitochondrial protein/min.

rates, P/O's and phosphorylation rates, but these did not differ significantly from the values found with normal muscle mitochondria.

In contrast to the above findings, we recently observed that mitochondria from younger dystrophic hamsters exhibited more marked abnormalities when isolated with Nagarse than without this proteinase (table II). In the most severe cases, phosphorylation was completely uncoupled. Typically, the respiratory rate was 25–50% of normal and decreased progressively during the experiment. Interestingly, addition of glucose, hexokinase and $MgCl_2$ to trap ATP as it was generated restored to normal levels respiration and phosphorylation by dystrophic mitochondria isolated with Nagarse. Of the constituents of the ATP trap, magnesium alone had a beneficial effect. Addition of 3 mM $MgCl_2$ to the abnormal mitochondria at the beginning of a polarographic experiment resulted in a doubling of the initial O_2 rate and a restoration of phosphorylative coupling to yield ADP/O and respiratory control ratios (R.C.R.) approaching those obtained with normal organelles. However, addition of magnesium later in the experiment had no beneficial effect.

A review of the necropsy findings indicated that the slowest respiration and complete uncoupling occurred most frequently in 50- to 80-day-old hamsters with numerous longitudinal white streaks in the muscles. Such streaks in stained sections proved to be regions of massive muscle necrosis and myolysis in which there was extensive phagocytic infiltration. Since such muscle was found in unpublished experiments to contain increased amounts of the acid hydrolases known to uncouple phosphorylation (7), it seemed possible that their release might have damaged the mitochondria during the isolation procedure. However, when equal portions of minced normal and dystrophic muscle were mixed, the behaviour of the mitochondria isolated from the mixture gave no indication that the hydrolases of the dystrophic muscle have a deleterious effect on the organelles of the normal tissue.

Table II

Effect of an ATP Trap on Skeletal Muscle Mitochondria Isolated in the Presence and Absence of Nagarse*

			No ATP trap				ATP trap present		
Nagarse	Hamsters	N	O_2 rate	R.C.R.	ADP/O	P rate	O_2 rate	P/O	P rate
Absent	Normal	6	117 ± 18	4.3 ± 0.7	1.88 ± 0.31	487 ± 96	186 ± 19	2.19 ± 0.09	827 ± 108
	Dystrophic	6	108 ± 19	4.0 ± 0.8	1.85 ± 0.37	471 ± 82	163 ± 41	2.22 ± 0.11	767 ± 224
	p		N.S.	N.S.	N.S.	N.S.	N.S.	N.S.	N.S.
Present	Normal	6	176 ± 13	2.5 ± 0.1	1.90 ± 0.07	679 ± 78	177 ± 5	1.92 ± 0.09	673 ± 38
	Dystrophic	5	96 ± 21	1.6 ± 0.4	0.70 ± 0.43	193 ± 121	195 ± 8	2.08 ± 0.04	813 ± 45
	p		< 0.01	< 0.05	< 0.02	< 0.01	N.S.	N.S.	< 0.05

*Polarographic test system (3).

Table III

Mitochondrial Respiration and Magnesium Content

Hamsters	N	O_2 rate	Magnesium*
Normal	10	187 ± 10	0.89 ± 0.05
Dystrophic			
Coupled	8	161 ± 18†	0.93 ± 0.09†
Uncoupled	8	43 ± 5	1.00 ± 0.09†

*μg/mg mitochondrial protein.
†Means not significantly different from normal ($p > 0.05$).

The white streaks associated with muscle destruction decreased in frequency in dystrophic hamsters older than 80 days. This presumably accounts for the absence of the magnesium-responsive defect in older animals (3). A similar relationship between age and the incidence and severity of myocardial lesions has been described by Bajusz and Lossnitzer (1). It may be significant that these workers also found that magnesium levels in both serum and myocardium were decreased in young BIO 14.6 hamsters but returned to normal before the animals attained an age of 100 days. The latter finding, together with the present observations, are suggestive of a transient skeletal muscle deficiency of magnesium. If this does occur, it clearly is not associated with a fall in the magnesium content of the skeletal muscle mitochondria (table III).

SUMMARY

Abnormal respiration and oxidative phosphorylation have been detected in skeletal muscle mitochondria isolated without the proteinase Nagarse from old (220 ± 16 days) dystrophic hamsters of BIO 14.6 strain. Another type of disturbance of these functions was detected in muscle mitochondria isolated with the help of Nagarse from younger (67 ± 4 days) animals of the same strain. Although this defect was ameliorated by *in vitro* addition of magnesium, the mitochondrial content of magnesium was normal.

REFERENCES

1. BAJUSZ, E. and LOSSNITZER, K.: A new disease model of chronic congestive heart failure: studies on its pathogenesis. Trans. N. Y. Acad. Sci. Series II. *30:* 939–948 (1968).
2. BLANCHAER, M. C.; LUNDQUIST, C. G. and GRIFFITH, T. J.: Factors influencing the utilization of reduced nicotinamide adenine dinucleotide by pigeon heart mitochondria. Canad. J. Biochem. *44:* 105–117 (1966).
3. JACOBSON, B. E.; BLANCHAER, M. C. and WROGEMANN, K.: Defective respiration and oxidative phosphorylation in muscle mitochondria of hamsters in the late stages of hereditary muscular dystrophy. Canad. J. Biochem. *48:* 1037–1042 (1970).
4. LIN, C. H.; HUDSON, A. J. and STRICKLAND, K. P.: Fatty acid metabolism in dystrophic muscle in vitro. Life Sci. *8:* 21–26 (1969).
5. LIN, C. H.; HUDSON, A. J. and STRICKLAND, K. P.: Palmitic acid-1-^{14}C oxidation by skeletal muscle mitochondria of dystrophic mice. Canad. J. Biochem. *48:* 566–572 (1970).

6. LOCHNER, A. and BRINK, A. J.: Oxidative phosphorylation and glycolysis in the hereditary muscular dystrophy of the Syrian hamster. Clin. Sci. *33:* 409–423 (1967).

7. MELLORS, A.; TAPPEL, A. L.; SAWANT, P. L. and DESAI, I. D.: Mitochondrial swelling and uncoupling of oxidative phosphorylation by lysosomes. Biochim. Biophys. Acta. *143:* 299–309 (1967).

8. WROGEMANN, K. and BLANCHAER, M. C.: Respiration and oxidative phosphorylation by muscle and heart mitochondria of hamsters with hereditary myocardiopathy and polymyopathy. Canad. J. Biochem. *46:* 323–329 (1968).

Mailing address: M. C. Blanchaer, M.D., Department of Biochemistry, University of Manitoba, Winnipeg 3, Manitoba (Canada).

Institut de Médecine et de Chirurgie expérimentales,
Université de Montréal, Montréal, Canada, and
Bio-Research Institute, Boston, Massachusetts, USA

STUDY OF THE HYPOTHALAMIC-HYPOPHYSEAL SYSTEM IN HEREDITARY CARDIOMYOPATHIC HAMSTERS [1]

E. HORVATH, E. BAJUSZ, and K. KOVACS

INTRODUCTION

Investigations on the different aspects of hereditary cardiomyopathy in pedigree hamsters (BIO 14.6 strain) (2, 3, 8), have shown the concurrent presence of endocrine abnormalities, such as an increase in the thickness but not in the volume of the zona glomerulosa as well as a striking atrophy of the inner layers of the adrenal cortex (4). On the basis of these observations, we undertook a morphologic examination of the hypothalamic-hypophyseal system of cardiomyopathic hamsters in order to determine whether it is also involved in the development of endocrine alterations.

MATERIALS AND METHODS

A total of 42 hamsters were divided into four groups: 1) 100-day-old cardiomyopathic (BIO 14.6 strain) hamsters without signs of congestive heart failure (10 animals); 2) 100-day-old control (RB strain) hamsters (10 animals); 3) 210- to 420-day-old (BIO 14.6 strain) cardiomyopathic edematous hamsters (11 animals); and 4) 210- to 420-day-old (RB strain) control hamsters (11 animals). The animals of groups 1 and 2 were males; those of groups 3 and 4 were of both sexes. All the animals were killed by decapitation, the skulls were opened, and the entire heads fixed in 6% neutral formalin. After fixation, the brains and pituitaries were excised; the pituitary glands were then weighed on an analytical balance. The hypothalamic areas (including the nucleus supraopticus and nucleus paraventricularis) were removed. The hypothalami and the pituitaries were embedded in paraffin and consecutive sections were stained as follows: hematoxylin-phloxine; Gomori-hematoxylin (6); aldehyde-fuchsin (7); and aldehyde-thionin (10); in addition, the pituitaries were also examined following staining by the PAS technique and the Goldberg-Chaikoff trichrome method (5).

From the pituitaries stained with trichrome, 80 unselected fields were

[1] Supported by grants from le Ministère de la Santé, Québec, the Medical Research Council of Canada (Block Term Grant MT-1829), the Quebec Heart Foundation, Montreal, and US Public Health Service Grant No. HE-09791 (National Heart Institute).

counted, using a magnification of 1,875 (2,300–3,200 glandular cells); the proportion of basophil, acidophil, and chromophobe cells was then calculated as a percentage of all the counted cells.

RESULTS

Macroscopic findings

No abnormalities were observed in the hypothalami. The pituitaries of the 100-day-old cardiomyopathic hamsters (group 1) appeared to be normal, although the glands were considerably smaller than those of the controls (group 2). In the hamsters showing signs of congestive heart failure (group 3), the adenohypophyses were strikingly smaller than those of the controls (group 4), forming a tiny, almost transparent rim around the posterior lobe. The neurohypophyses were markedly enlarged and swollen. As shown in table I, the pituitary weight was significantly less in both cardiomyopathic groups. (The absolute pituitary weights are given, because, owing to the edema of the cardiomyopathic animals (4), the body weight does not provide a reliable basis for calculation.)

Microscopic findings

Hypothalamic-neurohypophyseal system

Following staining with Gomori-hematoxylin, aldehyde-fuchsin, and aldehyde-thionin, the nucleus paraventricularis of each group showed similar characteristics: the round or oval ganglionic cells were relatively pale, and

Table I

Pituitary Weight in Control and Cardiomyopathic Hamsters

Group	No. of animals	Pituitary weight *mg*
1. Cardiomyopathic (BIO 14.6, 100-day-old) hamsters	10	3.3 ± 0.13†
2. Controls (RB)	10	4.9 ± 0.29
3. Cardiomyopathic (BIO 14.6, edematous) hamsters	11	4.7 ± 0.27*
4. Controls (RB)	11	6.1 ± 0.30

*$p < 0.005$.
†$p < 0.001$.

contained a moderate, but variable amount of secretory granules in their cytoplasm.

The supraoptic nucleus in groups 1, 2, and 4 was normal, consisting of closely-situated oval or polygonal ganglionic cells; a considerable amount of evenly-distributed neurosecretory material was found in their cytoplasm (fig. 1). Only in the hamsters with congestive heart failure were several cells of the nucleus supraopticus enlarged, exhibiting a pale cytoplasm and containing only a few randomly-distributed neurosecretory granules (fig. 2). The neurohypophyses and the nucleus supraopticus showed similar changes. The posterior lobes of hamsters of groups 1, 2, and 4 contained a considerable amount of neurosecretory material that stained well with Gomori-hematoxylin, aldehyde-fuchsin, and aldehyde-thionin (fig. 3). Only a small amount of moderately-stained neurosecretory granules could be seen in the enlarged, sponge-like neurohypophyses of the old (edematous) cardiomyopathic hamsters (fig. 4).

Anterior lobe

Control animals—As judged by the results of the PAS and Goldberg-Chaikoff techniques, a striking cytologic difference was found between the normal males

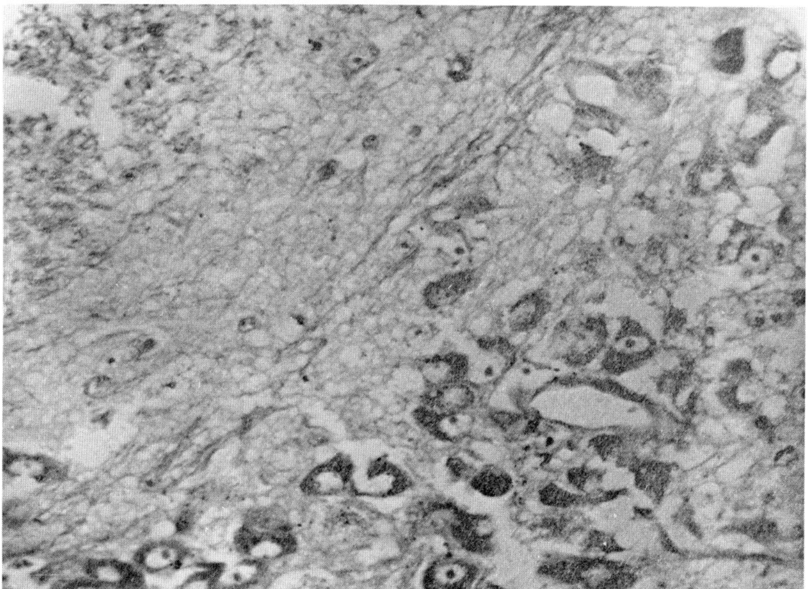

Fig. 1 Nucleus supraopticus of a control hamster. The cytoplasm of the ganglionic cells is tightly packed with intensely stained neurosecretory granules. Aldehyde-thionin stain, × 275.

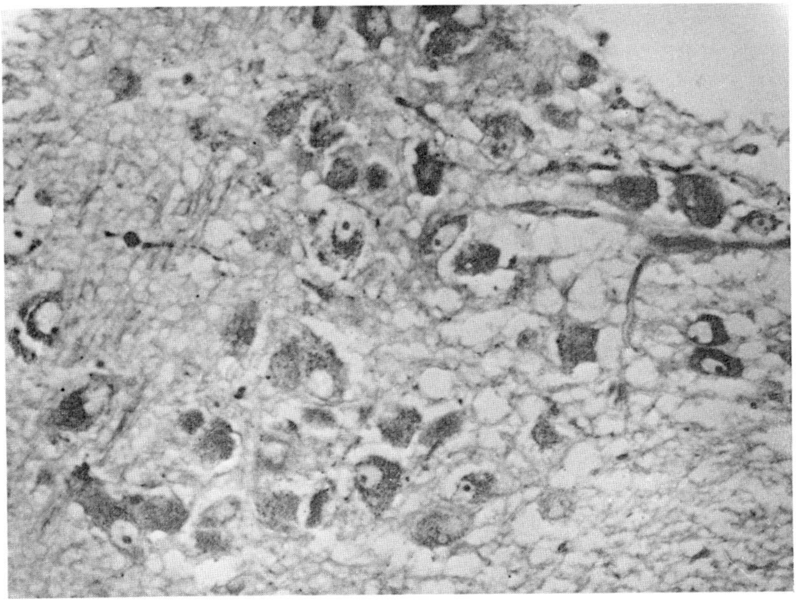

Fig. 2 Nucleus supraopticus of an old (edematous) cardiomyopathic hamster. The amount of neurosecretory material is decreased in several ganglionic cells. Aldehyde-thionin stain, × 275.

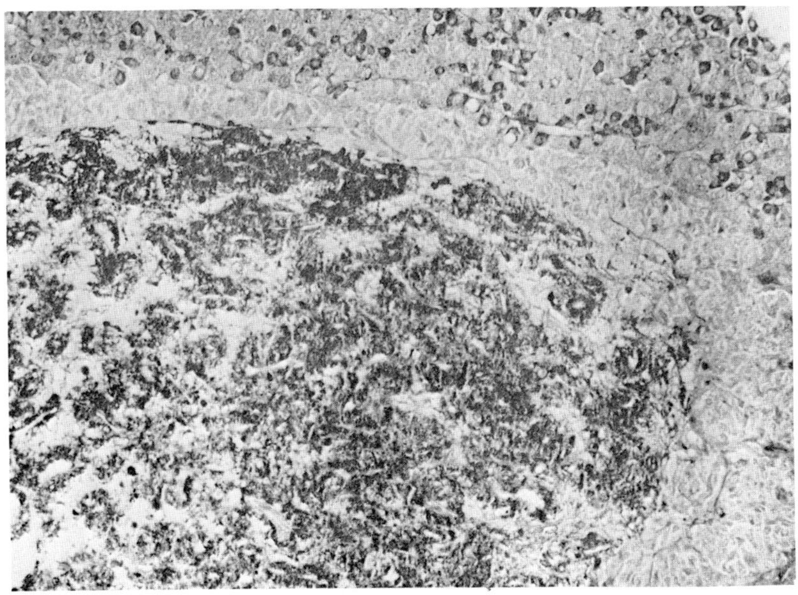

Fig. 3 Neurohypophysis of a control hamster. The posterior lobe contains a large amount of neurosecretory material. Aldehyde-thionin stain, × 100.

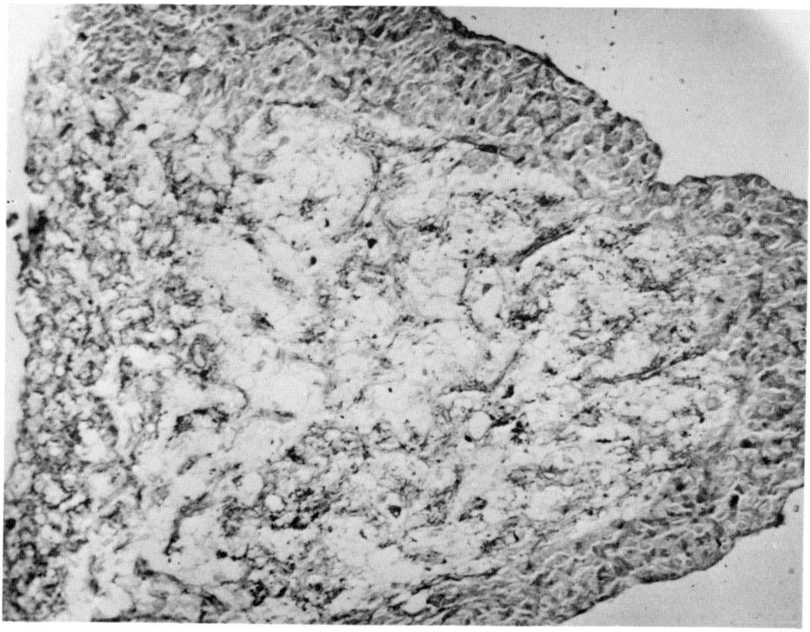

Fig. 4 Neurohypophysis of an old (edematous) cardiomyopathic hamster. Note the almost complete depletion of neurosecretory material. Aldehyde-thionin stain, × 100.

and females: the pituitaries of the males contained a high percentage of basophils (in some instances, over 40%) and relatively few acidophils (fig. 5). In females, the acidophil cells were dominant (table II). The relative proportion of the different basophils could not be determined since, in hamsters, not only the thyrotroph cells, but also the majority of gonadotrophs were stained with aldehyde-fuchsin and aldehyde-thionin. It is noteworthy that colloid droplets, as well as a great number of vacuoles partly filled with basophil colloid, were found mainly in the anterior lobe of old males, while the pituitaries of females were usually devoid of these formations. Occasionally, real cysts, lined with ciliated epithelium, were also formed.

Cardiomyopathic hamsters–There was no significant difference between the quantitative distribution of various cell types in 100-day-old cardiomyopathic hamsters and in controls of the same age. However, signs of degranulation could be observed in the early stage of cardiac disease also. On the other hand, striking changes were observed in the pituitaries of old hamsters showing signs of congestive heart failure. The number of basophils decreased to one-third of the control values in males and about one-half in females, with a corresponding increase in the number of chromophobes. Furthermore, only a few of the remaining basophils could be considered normal; the majority of them were pale and sparsely granulated (fig. 6). Surprisingly, the colloid vesicles could not be

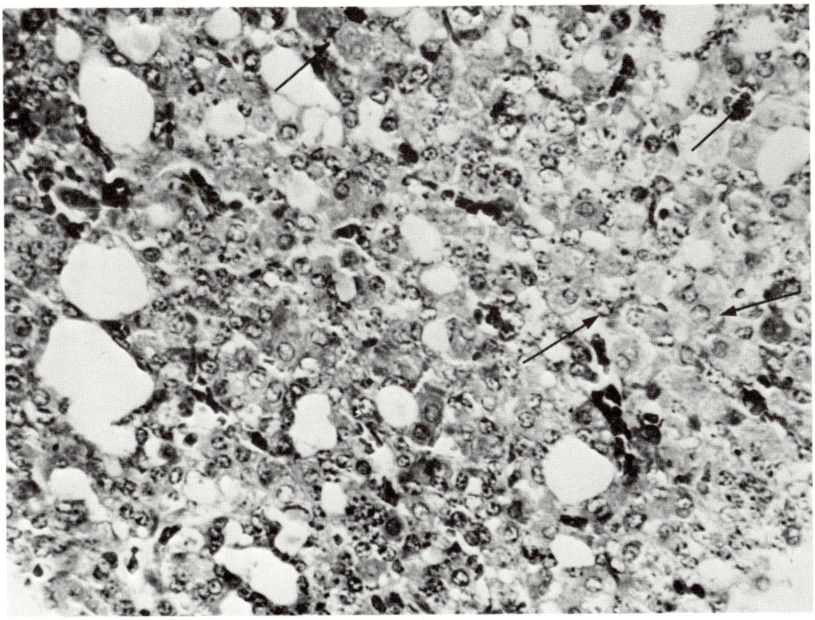

Fig. 5 Adenohypophysis of a control male hamster. Here, numerous basophils (arrows) are seen. Goldberg-Chaikoff trichrome method, × 275.

observed in these pituitaries. No abnormalities were found in the acidophils and chromophobes.

DISCUSSION

The results of the present study show that the hereditary cardiomyopathy of BIO 14.6 hamsters is accompanied by multiple abnormalities in the hypo-

Table II

Distribution of Various Cell Types in the Adenohypophysis of Control and Cardiomyopathic Hamsters

Group	Sex	Basophils	Acidophils	Chromophobes
		%	%	%
Control (RB)	♂	37.6 (36.4–43.0)	25.6 (20.1–32.9)	36.8 (29.8–43.5)
Edematous (BIO 14.6)	♂	12.6 (4.2–32.5)	20.2 (11.7–23.2)	67.2 (39.8–78.4)
Control (RB)	♀	11.8 (11.1–12.5)	49.3 (47.4–51.2)	38.9 (37.7–40.1
Edematous (BIO 14.6)	♀	6.5 (5.9–7.2)	50.4 (49.0–51.8)	43.1 (41.0–45.1)

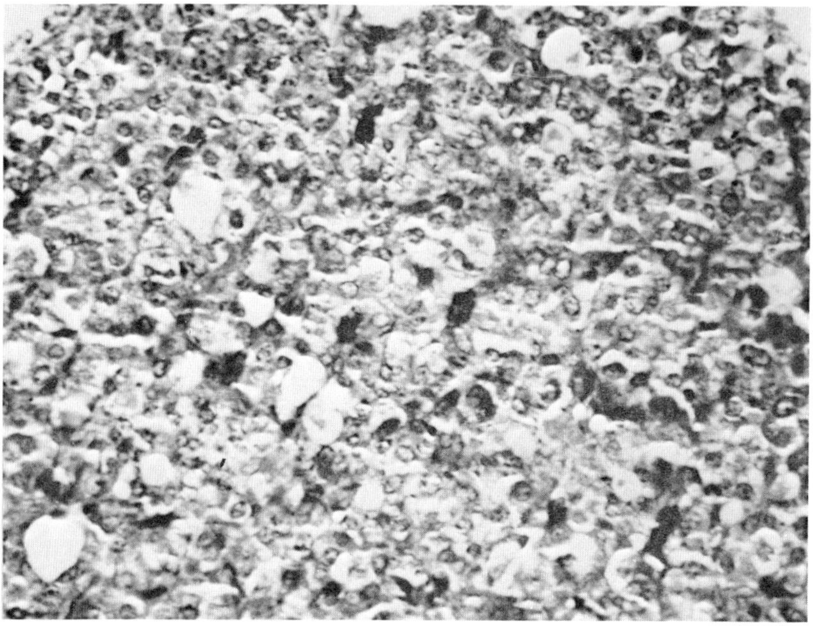

Fig. 6 Anterior pituitary of an old (edematous) male cardiomyopathic hamster. The gland is composed mainly of chromophobes and a few small acidophil cells. Goldberg-Chaikoff trichrome stain, × 275.

thalamic-hypophyseal system. In the advanced (edematous) phase of the disease, an enlargement of the neurohypophyses was found, concurrently with a depletion of neurosecretory material both in the posterior lobe and in the nucleus supraopticus. The anterior pituitary of cardiomyopathic animals was significantly smaller even in the early (nonedematous) stage of the disease; later, an extensive degranulation of basophil cells was also observed.

The morphologic and histochemical changes observed in the hypothalamic-neurohypophyseal system corresponded to those found following the over-production and release of the so-called posterior lobe hormones caused by such stimuli as dehydration or hyperosmosis of the plasma. However, the cardiomyopathic hamsters were not in a state of dehydration or hyperosmosis. A consistent decrease of hematocrit values, expansion of the extracellular fluid space (3, 4), and a hyponatremic tendency (1) have been observed during the development of cardiac disease. Despite these alterations, the morphologic changes of the nucleus supraopticus and of the posterior lobe suggest an increased release of ADH. This pattern resembles manifestations of the Schwartz-Bartter syndrome in humans, in which the hypothalamic-hypophyseal system is responsible for the "inappropriate secretion of antidiuretic hormone" (11). If this were the case here, the over production of ADH could be considered an aggravating factor leading to the expansion of the extracellular fluid space and the mechanical overload of the already-damaged heart muscle. Nevertheless, the morphologic alterations, as well as the laboratory data found earlier, suggest the possibility of an inappropriate secretion of ADH. Direct measurements of the antidiuretic activity are needed to support this assumption and, further, to clarify whether the changes of the hypothalamic-neurohypophyseal system are connected with congestive heart failure or represent another genetic defect unrelated to the cardiopathy.

It is very difficult to interpret the changes found in the adenohypophysis, in terms of function, without direct hormone-determinations. It is not clear whether the diminished size and weight of the pituitaries is attributable to a primary hypoplasia of the gland caused by unidentified factors or whether it is the consequence of cardiomyopathy leading to an inadequate blood supply. However, the latter possibility is unlikely, since adenohypophyseal atrophy was evident in the early phase of the cardiac disease also, even prior to the manifestations of severe circulatory disturbances. Moreover, it is well known that chronic congestive heart failure in humans does not lead to atrophy of the anterior lobe.

Evaluation of the functional significance of the cytologic changes is also difficult. Undoubtedly, these could not indicate a primary defect, since the abnormalities were seen only in the later (edematous) phase of the cardiomyopathy. The degranulation of the basophil cells might be the result of prolonged stress due to the cardiac condition. It is known that similar changes occur also in the course of various severe nonspecific injuries. Interestingly, despite the reduced size of the inner layers of the adrenal cortex, the blood cortisol level is significantly increased in the cardiomyopathic hamster (9). However, further investigations are needed to clarify the exact relationship between the cytologic changes in the anterior lobe and the morphologic and functional alterations of the adrenal cortex.

SUMMARY

Morphologic and histochemical studies of the hypothalamic-hypophyseal system of hamsters suffering from hereditary cardiomyopathy gave rise to the following observations:

1. A considerable decrease in the pituitary weight, which was statistically significant even in the early phase of the disease.

2. Macroscopically, the appearance of the pituitaries of 100-day-old animals was normal, in contradistinction to the older (edematous) animals in which the posterior lobe was enlarged.

3. No histologic or histochemical difference was detectable in the hypothalamic-neurohypophyseal system of 100-day-old cardiomyopathic as compared with the control hamsters. In the later phase of the disease, the neurohypophysis was swollen and its neurosecretory material depleted. The latter change was also observed in the cells of the nucleus supraopticus.

4. The anterior lobe basophils were remarkably degranulated in the old (edematous) cardiomyopathic hamsters.

These findings suggest an enhanced production and release of posterior lobe hormones in the later phase of the cardiomyopathy. However, the functional significance of the adenohypophyseal changes is not yet clear.

REFERENCES

1. BAJUSZ, E.: Unpublished data.
2. BAJUSZ, E.; BAKER, J. R.; NIXON, C. W. and HOMBURGER, F.: Spontaneous hereditary myocardial degeneration and congestive heart failure in a strain of Syrian hamsters. Ann. N. Y. Acad. Sci. *156:* 105–129 (1969).
3. BAJUSZ, E. und LOSSNITZER, K.: Ein neues Krankheitsmodell: Erbliche nichtvaskuläre Myokarddegeneration mit Herzinsuffizienz. Munch. Med. Wchschr. *110:* 1756–1768 (1968).
4. CANTIN, M. and BAJUSZ, E.: Juxtaglomerular apparatus and adrenal cortex in hamsters with heart failure. Relationship between the morphologic changes, serum electrolytes, and extracellular fluid volume. A. M. A. Arch. Path. *87:* 626–636 (1969).
5. GOLDBERG, R. C. and CHAIKOFF, I. L.: On occurrence of 6 cell types in dog anterior pituitary. Anat. Rec. *112:* 265–274 (1952).
6. GOMORI, G.: A differential stain for cell types in the pancreatic islets. Amer. J. Path. *15:* 497–499 (1939).
7. HALMI, N. S.: Differentiation of 2 types of basophils in adenohypophysis of rat and mouse. Stain Techn. *27:* 61–64 (1952).
8. HOMBURGER, F.; BAKER, J. R.; WILGRAM, G.; CAULFIELD, J. and NIXON, C. W.: Hereditary dystrophy-like myopathy: The histopathology of hereditary dystrophy-like myopathy in Syrian hamsters. A. M. A. Arch. Path. *81:* 302–307 (1966).
9. KNIGGE, K. M.: Personal communication.
10. PAGET, G. E. and ECCLESTON, E.: Simultaneous specific demonstration of thyrotroph, gonadotroph and acidophil cells in the anterior hypophysis. Stain Techn. *35:* 119–122 (1960).
11. SCHWARTZ, W. B.; TASSEL, D. and BARTTER, F. C.: Further observations on hyponatremia and renal sodium loss probably resulting from inappropriate secretion of antidiuretic hormone. New Engl. J. Med. *262:* 743–748 (1960).

Mailing address: K. Kovacs, M.D., Ph.D., Department of Pathology, St. Michael's Hospital, 30 Bond Street, Toronto 2, Ontario (Canada).

Département de Pathologie, Université de Montréal, Montréal, Québec, Canada;
Laboratorio di Fisiologia Clinica del CNR, Pisa, Italy; and
Bio-Research Institute, Boston, Massachusetts, USA

PARTICIPATION OF RENAL ELECTROLYTES IN THE MECHANISM OF EDEMA IN CONGESTIVE HEART FAILURE: STUDIES ON CARDIOMYOPATHIC HAMSTERS [1]

M. CANTIN, A. LEONE, and E. BAJUSZ

Earlier studies on the mechanism of edema formation in BIO-14.6 cardiomyopathic hamsters failed to establish a clear-cut relationship between variations in electrolytes and the activation of the juxtaglomerular apparatus-zona glomerulosa system during the evolution of congestive heart failure (6). Only in the severely edematous animals was there an evident relationship betwen the serum electrolytes (increase in sodium and decrease in potassium levels) and activation of the renin-angiotensin-aldosterone system as evidenced by a significant increment of the juxtaglomerular granulation index (JGI) and of the thickness of the zona glomerulosa. Since, during congestive heart failure, the serum levels of sodium and potassium are poor indicators of hormonal influence because of rather wide fluctuations (11, 24), it was decided to undertake a systematic study of various electrolytes, from early infancy (10-20 days) to the time of maximal edema development (280-300 days), not only in serum, but also in renal tissue and in urine. In order to obtain further evidence concerning the participation of the kidney and adrenal cortex in the pathophysiology of congestive heart failure, the histochemistry of the macula densa and of the adrenal cortex as well as the ultrastructure of the kidney were also investigated.

MATERIALS AND METHODS

Five experimental series were performed on a total of 195 cardiomyopathic hamsters (BIO-14.6) and on an equal number of healthy control hamsters (BIO-RB) of the same age. Unless otherwise indicated, the groups consisted of 10 animals each (5 males and 5 females). All the hamsters were kept under identical, controlled-housing conditions in air-conditioned rooms; tap water and hamster-chow (Guilford) supplemented by oats and wheat germ was offered ad libitum. Because the severity of the cardiopathy is time-dependent, the diseased animals and their respective controls were divided into various groups according to age. In most experiments, heart, liver, and lungs were taken

[1] Supported by Medical Research Council of Canada Grant MT-1973; National Heart Institute, Bethesda, Grant HE-09791; and The Hartford Foundation, New York.

for histological examination. To estimate the severity of congestive heart failure, the degree of generalized edema and of pleural and abdominal effusions was assessed and complemented by histopathological studies of cardiac, hepatic, and pulmonary alterations. These observations were combined in a previously established (1, 6) grading scale (0–3) for each animal and then averaged for the group (tables I, V, VI).

The first experimental series (tables I to IV) was performed on 60 cardiomyopathic hamsters (BIO-14.6) and on an equal number of healthy control animals (BIO-RB) of the same age. All the electrolytes (sodium, potassium, magnesium, and calcium) were measured, at the intervals indicated in tables I to IV, with an atomic absorption spectrophotometer (Perkin Elmer, model 290) as described in detail elsewhere (2). Blood samples for serum electrolytes (table II) were obtained from the jugular vein. Urine samples for sodium determinations were taken twice daily (between 9 a.m. and 10 a.m. and between 4 p.m. and 5 p.m.) during 2 consecutive days. The average value of these four measurements was taken to indicate the urinary sodium concentration of a given animal (table IV). For renal electrolyte determinations (table II), the animals were killed with chloroform and the kidneys were immediately removed and weighed on an analytical balance. They were dried for 24 h under vacuum and digested with a mixture of nitric and perchloric acid in an oven for 1 h. The residue was then diluted to final volume with deionized water (2).

The second series was an exact replica of the first as regards the experimental design. The ureters were ligated under ether anesthesia; immediately afterwards, labeled inulin: 0.2 ml (0.6264 μc) of a 100 mg/100 ml solution of inulin carboxyl-C14 in pH 7.4 phosphate buffer, was injected via the jugular vein. Exactly 60 min later, the animals were killed with chloroform; the kidneys were removed, and their radioactivity as well as the water content were measured. From these data, the distribution of intra-and extracellular sodium in renal tissue (table III) was determined.

The third experimental series consisted of 50 cardiomyopathic hamsters (BIO-14.6) and of an equal number of healthy control animals (BIO-RB)selected so as to ensure as much as possible a degree of congestive heart failure comparable to that of the experimental animals of the first and second series. All these animals were killed with a guillotine at the intervals indicated in table V. The left kidney was processed for measurement of the juxtaglomerular granulation index (JGI) according to a previously described technique (6). The right kidney was cut longitudinally through the hilus, each half being immediately frozen on chucks in dry ice. Sections 10μ thick were immediately cut in a cryostat at -20°C and processed according to a technique described elsewhere (7) for the demonstration of glucose-6-phosphate dehydrogenase activity at the level of the macula densa. The intensity of the reaction was estimated semi-quantitatively by the same technique used for the calculation of

Table I

Summary of Gross Pathology and Clinical State of Cardiomyopathic Hamsters Listed in Tables II, III, and IV

Group*	Age (days)	Cardiac hypertrophy	Liver enlargement (congestion)	Ascites and/or hydrothorax	Generalized subcutaneous edema	Assessment of congestive heart failure
2	10–25	0 to +	0	0	0	0
4	40–60	+	0 to +	0	0	0
6	70–90	++	+	0	0	0.6
8	100–120	++	+ to ++	0	0	1.0
10	180–200	+++	++ to +++	+ to +++	+ to +++	2.5
12	280–330	+++†	+++	+++	+++	3.0
	0 = none		+ = slight	++ = moderate		+++ = severe

*Each group consisted of 10 animals (5 males and 5 females).
†Extreme cardiac dilatation.

Table II

Concentration of Serum and Renal Electrolytes in Cardiomyopathic (BIO-14.6) and Healthy Control (BIO-RB) Hamsters at Various Ages*

Group*	Strain	Age	Serum Electrolytes†			Renal Electrolytes†			
			Ca	K	Mg	Ca	K	Mg	Na
		days		*mg %*			*mg/g dry wt.*		
1	Control (BIO-RB)	10–25	9.5 ± 0.21	17.1 ± 0.83	3.4 ± 0.19	0.486 ± 0.05	14.01 ± 0.58	0.996 ± 0.05	7.86 ± 0.19
2	Cardiomyopathic (BIO-14.6)		*11.3* ± 0.18	16.7 ± 1.56	3.7 ± 0.22	0.284 ± 0.02	13.97 ± 0.73	0.987 ± 0.02	*8.42* ± 0.17
3	Control	40–60	9.8 ± 0.31	15.8 ± 1.33	3.1 ± 0.18	0.512 ± 0.02	13.02 ± 0.29	0.921 ± 0.11	7.37 ± 0.26
4	Cardiomyopathic		*12.3* ± 0.23	16.3 ± 0.77	3.5 ± 0.21	*0.302* ± 0.37	*15.12* ± 0.37	0.942 ± 0.09	*8.98* ± 0.37
5	Control	70–90	10.2 ± 0.53	16.0 ± 0.47	3.6 ± 0.17	0.426 ± 0.03	14.12 ± 0.81	1.005 ± 0.07	7.25 ± 0.27
6	Cardiomyopathic		10.1 ± 0.27	15.2 ± 2.13	3.3 ± 0.08	0.413 ± 0.05	14.76 ± 0.46	0.996 ± 0.13	*9.42* ± 0.19
7	Control	100–120	9.9 ± 0.35	14.8 ± 2.13	3.5 ± 0.24	0.408 ± 0.07	13.53 ± 0.19	0.912 ± 0.05	8.03 ± 0.27
8	Cardiomyopathic		9.4 ± 0.72	14.0 ± 0.56	3.4 ± 0.17	0.431 ± 0.11	13.81 ± 0.38	0.934 ± 0.05	*10.1* ± 0.21
9	Control	180–200	10.2 ± 0.31	16.4 ± 1.48	3.4 ± 0.23	0.382 ± 0.06	15.17 ± 0.57	0.988 ± 0.09	8.12 ± 0.17
10	Cardiomyopathic		*9.3* ± 0.26	19.8 ± 1.06	*3.9* ± 0.11	*0.426* ± 0.02	15.01 ± 0.41	1.003 ± 0.08	*10.92* ± 0.31
11	Control	280–330	9.8 ± 0.42	18.1 ± 0.78	3.1 ± 0.25	0.386 ± 0.03	14.80 ± 0.63	0.903 ± 0.06	8.9 ± 0.24
12	Cardiomyopathic		*8.1* ± 0.19	*21.3* ± 1.97	*4.0* ± 0.22	*0.458* ± 0.06	*11.26* ± 0.37	0.939 ± 0.07	*11.87* ± 0.72

*Each group consisted of 10 animals (5 males and 5 females).

†Values represent mean ± S.E. All the italicized figures are significant in comparison with their respective control ($p < 0.001$). A $p \geqslant 0.05$ was taken as insignificant.

Table III

Distribution of Intra- and Extracellular Sodium in Renal Tissues of Cardiomyopathic (BIO-14.6) and Healthy Control (BIO-RB) Hamsters (Derived Data)

Group	Strain	Age	Renal Na Concentration* Extracellular	Intracellular
		days	*mg/g dry wt.*	
1	Control (BIO-RB)	10–25	4.98	2.88
2	Cardiomyopathic (BIO-14.6)		4.72	3.70 ↑↑
3	Control	40–60	5.31	2.60
4	Cardiomyopathic		4.77	4.21 ↑↑
5	Control	70–90	4.24	3.01
6	Cardiomyopathic		4.96	4.46 ↑↑
7	Control	100–120	5.80	2.23
8	Cardiomyopathic		5.38	4.72 ↑↑
9	Control	180–200	6.01	2.11
10	Cardiomyopathic		6.86	3.06 ↑
11	Control	280–330	4.73	3.17
12	Cardiomyopathic		7.91	3.96 ↑

* ↑ = moderate and ↑↑ = marked increase, when compared with control value.

Table IV

Sodium Concentration in Urine of Cardiomyopathic (BIO-14.6) and Healthy Control (BIO-RB) Hamsters at Various Ages

Group*	Strain	Age	Na in Urine†
		days	
1	Control (BIO-RB)	10–25	1008.0 ± 31.2
2	Cardiomyopathic (BIO-14.6)		986.5 ± 28.6
3	Control	40–60	997.3 ± 36.8
4	Cardiomyopathic		969.0 ± 41.5
5	Control	70–90	932.7 ± 20.4
6	Cardiomyopathic		958.1 ± 34.3
7	Control	100–120	801.3 ± 25.6
8	Cardiomyopathic		*912.1* ± 14.3↑
9	Control	180–200	745.5 ± 18.5
10	Cardiomyopathic		*326.4* ± 16.1↓
11	Control	280–330	752.2 ± 23.8
12	Cardiomyopathic		*237.9* ± 11.6↓

*Each group consisted of 10 animals (5 males and 5 females).

†Urine samples were taken twice daily (between 9 a.m. and 10 a.m. and between 4 p.m. and 5 p.m.) during 2 consecutive days. The average value of these four measurements was taken to indicate the urinary Na concentration of a given animal. All the italicized figures are significant in comparison with their respective control ($p < 0.001$).

Table V

Variation of the Juxtaglomerular Granulation Index (JGI) and of the Macula Densa Histochemical Index (MDHI) of Cardiomyopathic (BIO-14.6) and Healthy Hamsters (BIO-RB) at Different Ages

Group*	Severity of congestive heart failure (grades 0–3)	Age	JGI†	MDHI†
	Control (BIO-RB)	*days*		
1	0	31	7.7 ± 2.51	13.1 ± 2.51
2	0	71	8.9 ± 2.02	11.2 ± 2.20
3	0	100	9.1 ± 2.22	9.1 ± 1.41
4	0	153	8.4 ± 1.51	10.5 ± 1.76
5	0	280	10.2 ± 1.43	14.3 ± 2.20
	Cardiomyopathic (BIO-14.6)			
6	0	31	12.2 ± 3.01	12.4 ± 1.81
7	0.6	71	13.3 ± 3.10	12.1 ± 2.02
8	0.8	100	14.3 ± 3.43	10.0 ± 1.43
9	1.9	153	*15.1* ± 2.09	10.3 ± 1.98
10	2.8	280	*17.2* ± 2.05	15.1 ± 3.10

*Each group consisted of 10 animals (5 males and 5 females).

†Values represent mean ± S.E. All the italicized figures are significant in comparison with their respective control ($p < 0.01$). A $p \geqslant 0.05$ was taken as insignificant.

the JGI, and expressed as the macula densa histochemical index (MDHI) (table V).

The right adrenal was fixed in formol calcium, cut at 5μ through the center of the gland in a cryostat, and stained with Oil red O and hematoxylin; the widths of both the zona glomerulosa and the inner zones were then determined as in our earlier studies (5, 6). The left adrenal was frozen on chucks in dry ice, sectioned at 10μ in a cryostat at -20°C and processessed as described elsewhere (7) for the demonstration of glucose-6-phosphate dehydrogenase, 6-phosphogluconic dehydrogenase, and $\triangle$ 5-3 β hydroxysteroid dehydrogenase.

The fourth experimental series (Table VI) consisted of two groups of male and female BIO-14.6 hamsters 5 and 20 days old, respectively, and of two control groups (BIO-RB) of the same sex and age. These animals were killed with chloroform. The left kidney was processed for the estimation of the JGI, while the adrenals were fixed in formalin and sectioned at 5μ after paraffin embedding. After staining with hematoxylin-phloxine or according to the PAS technique, the widths of the zona glomerulosa and of the inner zones (fasciculata and reticularis) were measured.

An electron microscopic study of the kidney was performed in 5 hamsters (2 males and 3 females) of the BIO-14.6 line in severe congestive heart failure (280 days) and in 5 control animals (BIO-RB) of the same sex and age. In this fifth

Table VI

Juxtaglomerular Granulation Index (JGI) and Width of Adrenocortical Zones of Cardiomyopathic (BIO-14.6) and Healthy (BIO-RB) Hamsters at the Postnatal Period

Group*	Severity of congestive heart failure (grades 0–3)	Age	JGI	Zonal width (μ)	
				Zona glomerulosa	Inner zones
		days			
	Control (BIO-RB)				
1	0	5	1.6 ± 0.45†	29.9 ± 0.97	83.1 ± 9.18
2	0	20	0.78 ± 0.23	23.9 ± 0.94	216.4 ± 11.39
	Cardiomyopathic (BIO-14.6)‡				
3	0	5	0.52 ± 0.28	29.2 ± 1.13	83.0 ± 3.73
4	0	20	1.3 ± 0.34	24.5 ± 0.09	195.7 ± 9.10

*Each group consisted of 10 animals (5 males and 5 females).

†Values represent mean ± S.E.

‡None of the values recorded in cardiomyopathic animals are significant ($p > 0.05$) in comparison with the respective controls.

experimental series, the animals were lightly anesthetized with ether and the left kidney was perfused through the abdominal aorta with 2% glutaraldehyde buffered with cacodylate-HCl (0.1 M at pH 7.1) according to a previously described technique (20). After perfusion, thin fragments of renal cortex were placed in the same fixative for 1 h and then washed for three periods of 15 min each in cacodylate buffer to which 2% sucrose was added. Subsequently, the specimens were left in buffer for 12 h, post-fixed in osmium tetroxide buffered with veronal acetate for 1 h, and embedded in Epon. Thick sections were stained with toluidine blue for orientation and thin sections were then cut with a diamond knife in an LKB ultramicrotome; these latter were mounted on grids coated with formvar and carbon, stained with uranyl acetate and lead citrate, and examined with the Philips 200 electron microscope.

RESULTS

Serum electrolytes

The results summarized in table II show that a significant increase in serum calcium was already present in the youngest animals (10-25 days) and persisted up to the 60th day. In contrast, the oldest animals were found to have significantly lower values for this electrolyte. A significant change, *i.e.*, an increase, in serum magnesium was present only in the two oldest groups (180-200 days and 280-330 days). On the average, the serum potassium appeared to be elevated in the two last-mentioned groups. However, this parameter showed great variations from animal to animal; in fact, the serum potassium level in some hamsters with severe peripheral manifestations of congestive heart failure was markedly lower than in the appropriate controls. This observation, together with our earlier findings (6), seems to indicate that during advanced stages of congestive heart failure of the hamster, an increase or decrease in serum potassium concentration may occur equally, depending upon conditions (endocrine and/or metabolic) that are not yet understood.

Renal electrolytes

In hamsters with the hereditary cardiomyopathy, the renal concentration of sodium (table II) became elevated during the very early (*i.e.*, subclinical) stages of congestive heart failure. A further progressive increase paralleled the onset and progression of the peripheral manifestations of congestive heart failure, reaching its highest values during the advanced stages of generalized venous congestion in 300-day-old hamsters. The variations in the renal concentration of potassium proved to be significant in two groups of animals: it was increased in 50-day-old and decreased in 300-day-old animals; the magnesium concentration in the kidney remained unchanged during the entire course of the chronic cardiac disease. Interestingly enough, young cardiomyopathic hamsters were characterized by low renal calcium concentrations in spite of their elevated serum calcium levels. However, during the terminal phases of congestive heart failure, the calcium values in the kidneys of cardiomyopathic hamsters were higher than in appropriate healthy controls of the same sex and age, despite lower serum calcium values.

Distribution of intra- and extracellular sodium

Except in the most severely affected cardiopathic animals (300 days), where the localization of renal sodium appeared to be mainly extracellular (table III), the derived data indicate an intracellular location for the increased load of sodium detected in the kidneys of animals in congestive heart failure.

Urinary sodium

There was a marked reduction in the urinary excretion of sodium (table IV) only in the late stages of congestive heart failure, *i.e.*, in 190-day-old and 300-day-old animals. Paradoxically, the 110-day-old hamsters showed an increased urinary excretion of sodium.

Morphology of the kidney

Juxtaglomerular granulation index (JGI)

The degree of granulation of the juxtaglomerular apparatus was extremely low in very young (5- and 20-day-old) control animals (table V). It was increased somewhat in older animals (31 days) and remained at practically the same level thereafter (table VI). The JGI of cardiomyopathic animals (fig. 1) followed the same pattern, but at a slightly higher level for most of the periods at which it was measured. Only in edematous animals (153- and 280-day-old) did it become significantly elevated (tables V and VI).

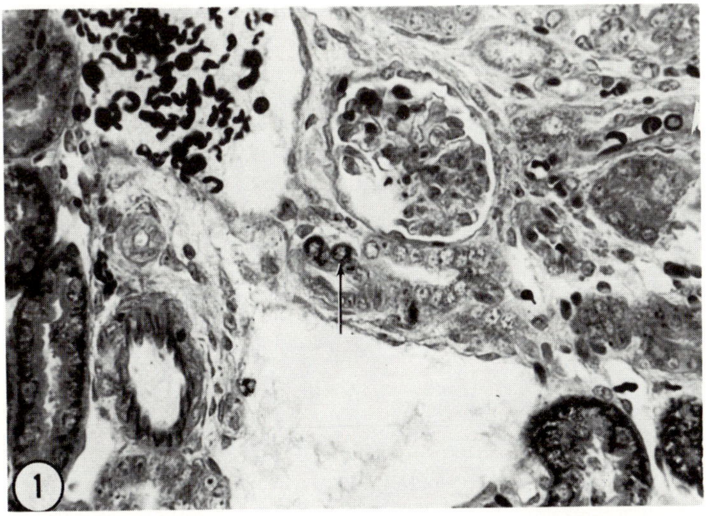

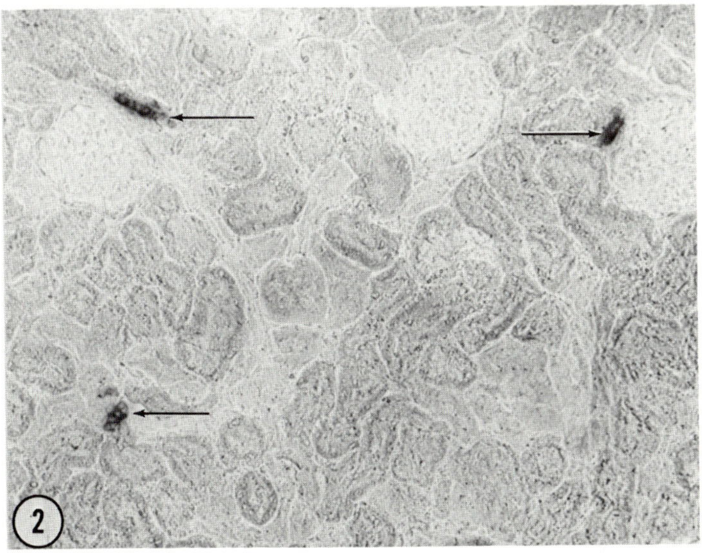

Fig. 1 Juxtaglomerular cells (arrow) located in the wall of an arteriole afferent to a glomerulus (above) in a 5-day-old cardiomyopathic hamster. Bowie, × 400.

Fig. 2 Frozen section of renal cortex stained for demonstration of glucose-6-phosphate dehydrogenase activity in the macula densa (arrows) of cardiomyopathic hamster in severe congestive heart failure. × 100.

Macula densa histochemical index (MDHI)

The evaluation of the activity of glucose-6-phosphate dehydrogenase at the level of the macula densa (figs. 2 and 3) revealed rather wide individual variations; but as can be seen in table V, the differences were not significant at any of the age groups studied.

Histology of the adrenal cortex

The data summarized in table VI indicate that in very young (5- and 20-day-old) hamsters there were no significant differences in the thickness of either the zona glomerulosa or of the inner zones between cardiomyopathic and control animals (figs. 7 and 8). In older cardiomyopathic animals (31-280 days), the adrenal cortex showed changes similar to those already reported (6), *i.e.*, a progressive decrease in thickness of the inner zones, with eventual signs of atrophy, and a progressive but small increment in the width of the zona glomerulosa that became significant after 153 and 280 days of life.

Histochemistry of the adrenal cortex

The activity of glucose-6-phosphate dehydrogenase and of 6-phosphogluconate dehydrogenase in the adrenocortical cells could be foreseen from previous histological studies. There was a decrease of activity for both enzymes

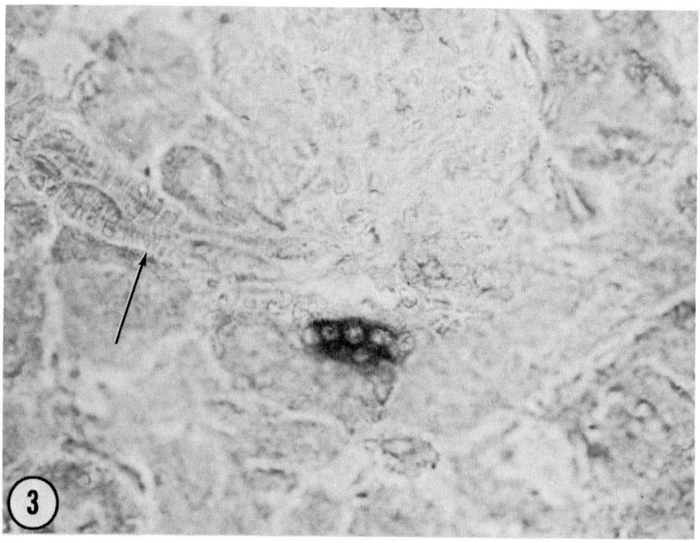

Fig. 3 Glucose-6-phosphate dehydrogenase activity in the macula densa of a control hamster. The afferent arteriole (arrow) is situated between the macula densa (below) and the glomerulus (above). × 600.

in the inner region of the inner zones, starting in 100-day-old cardiomyopathic animals and becoming more evident in markedly edematous animals. The outer region of the same zone showed no change in activity at any period studied. In contrast, the activity of glucose-6-phosphate dehydrogenase and 6-phosphogluconate dehydrogenase in the zona glomerulosa was seen to be slightly augmented as early as at 100 days in the cardiomyopathic animals and became more and more pronounced with increasing age (figs. 4 and 5). There was no change in the activity of Δ5-3 β hydroxysteroid dehydrogenase at the level of the zona glomerulosa in cardiomyopathic animals as compared with controls (fig. 6). The activity of this enzyme was decreased in the inner region of the inner zones and normal in the outer region.

Ultrastructure of the kidney

The study of the fine structure of glomeruli and cortical tubules revealed marked differences betwen the severely edematous hamsters (280-day-old) and control animals of the same age.

Glomeruli

The most striking finding of the present investigation was the focal and irregular thickening of the glomerular basement membranes. Their normal structure was replaced by empty, poorly delimited, irregular spaces or by a fine, irregular, osmiophilic network enclosing a pale matrix. Electron-dense, granular, intramembranous deposits as well as small vesicles were also present. There were no extramembranous or subendothelial deposits. The visceral epithelial cells showed minimal villous transformation and their cytoplasm contained an abundant rough-surfaced endoplasmic reticulum as well as free ribosomes. In several of these cells, cytofilaments were prominent, as were a number of dense bodies; the foot processes were often blunted and fused. The parietal cells appeared slightly hyperactive. The capillary loops were patent in the 200 glomeruli examined. The cytoplasm of endothelial cells was sometimes swollen and the number of the cytoplasmic organelles increased. Several mesangial cells appeared strikingly vacuolated. None of these changes were present in the glomeruli of control hamsters.

Proximal convoluted tubules

In the cardiomyopathic hamsters, there was a net increase in the number of apical tubular invaginations, vesicles, and vacuoles. The most evident change was, however, a widening of the extracellular spaces between the basilar interdigitating processes. Whereas in the tubules of control animals (fig. 9) these spaces were virtual, in edematous hamsters they were enlarged and, in some

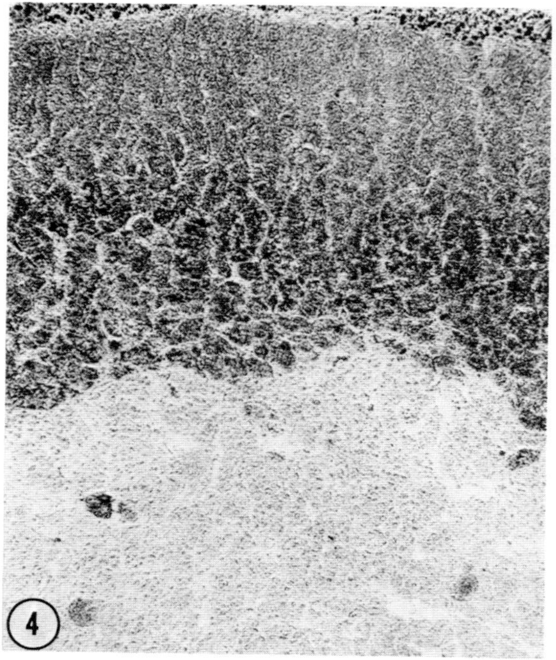

Fig. 4 Frozen section of adrenal cortex stained for demonstration of 6-phosphogluconate dehydrogenase, in control hamster. Note weak reaction in zona glomerulosa (top) and stronger reaction in the inner zones (middle area). The medulla (below) contains a few reactive (dark) adrenocortical cells. × 250.

Fig. 5 In a cardiomyopathic hamster with moderate heart failure, the activity of 6-phosphogluconate dehydrogenase is increased in zona glomerulosa (top) and outer region of inner zones (middle area), while it is decreased in the inner region of inner zones. In this last-mentioned region, the cells are separated by dilated vascular channels. × 250.

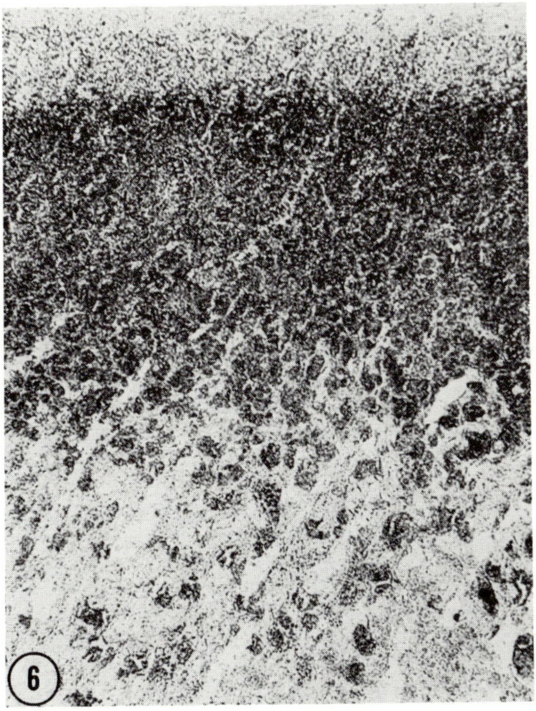

Fig. 6 Portion of the adrenal cortex of cardiomyopathic hamster with incipient heart failure, stained for demonstration of Δ5-3 β hydroxysteroid dehydrogenase. Note slight reaction in the zona glomerulosa (top) and strong reaction in inner zones (middle area). Several adrenocortical cells (dark) are located in the medulla. × 250.

areas, extended deeply into the cytoplasm (fig. 11). The intercellular spaces, however, were virtual in both groups of animals. There was no difference in either number or shape of various cellular organelles, such as mitochondria, cytosomes, dense bodies, and microbodies.

Distal convoluted tubules

In the cardiomyopathic animals, this part of the nephron showed marked edema, particularly in the subnuclear region towards the basal portion of the cells (fig. 12). Apical coated and uncoated vesicles and vacuoles were more numerous in cardiomyopathic than in control hamsters. The extracellular spaces between the basilar interdigitating processes became extremely enlarged in edematous hamsters (figs. 10 and 12). Some of these spaces were filled with slightly electron-opaque material. The intercellular spaces were, however, only rarely enlarged. In some cases, the mitochondria located in these processes took on

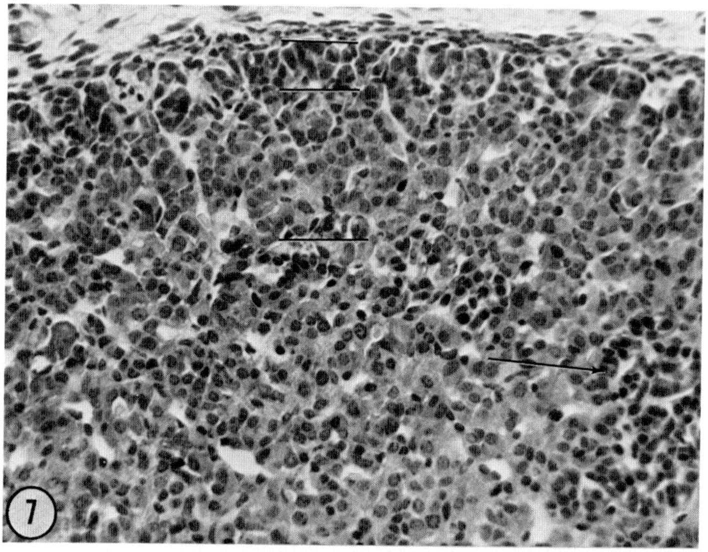

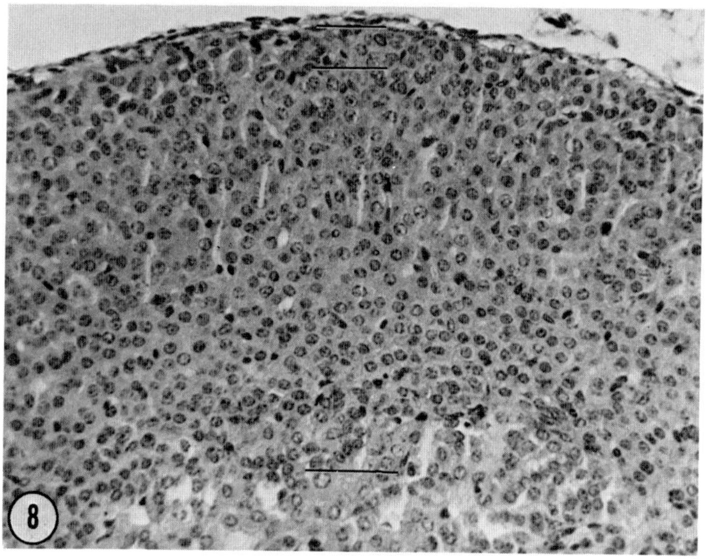

Fig. 7 Portion of the adrenal cortex of a 5-day-old cardiomyopathic hamster, showing (from top to bottom) zona glomerulosa and thin inner zones. The medulla (arrow) is still poorly structured and contains numerous adrenocortical cells. Both glomerulosa and inner zones are delimited by parallel lines. HP, × 125.

Fig. 8 Portion of adrenal cortex of 20-day-old cardiomyopathic hamster. The zona glomerulosa is compact. The inner zones are much larger than in the 5-day-old animal. Portion of medulla is visible at bottom. Both zona glomerulosa and inner zones are delimited by parallel lines. PAS, × 100.

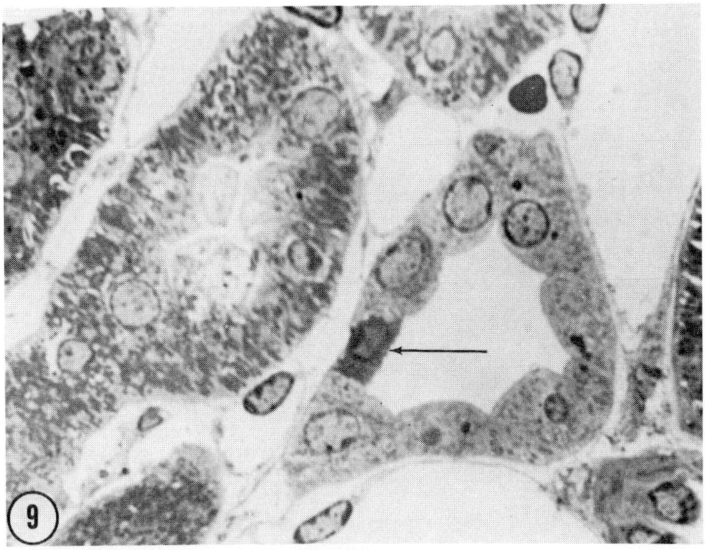

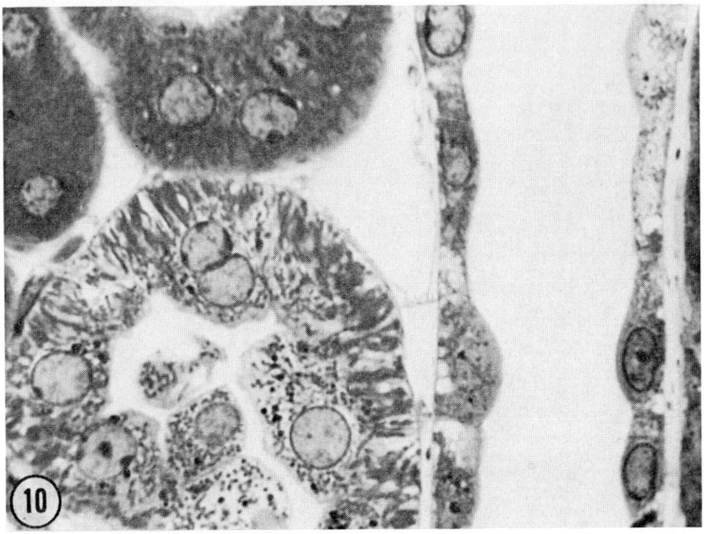

Fig. 9 In normal hamster, both the cortical collecting duct (right) and the distal convoluted tubule (left) show a relatively compact structure. Note dark cell (arrow) in collecting duct. Epon embedding; toluidine blue stain, × 1,000.

Fig. 10 In cardiomyopathic hamster in severe heart failure, both cortical collecting duct (longitudinal section at right) and distal convoluted tubule (bottom left) have a spongy appearance, owing to slits and vacuoles located at the basal pole of the cells. Part of proximal convoluted tubules (upper left) shows little change. Epon embedding; toluidine blue stain, × 1,000.

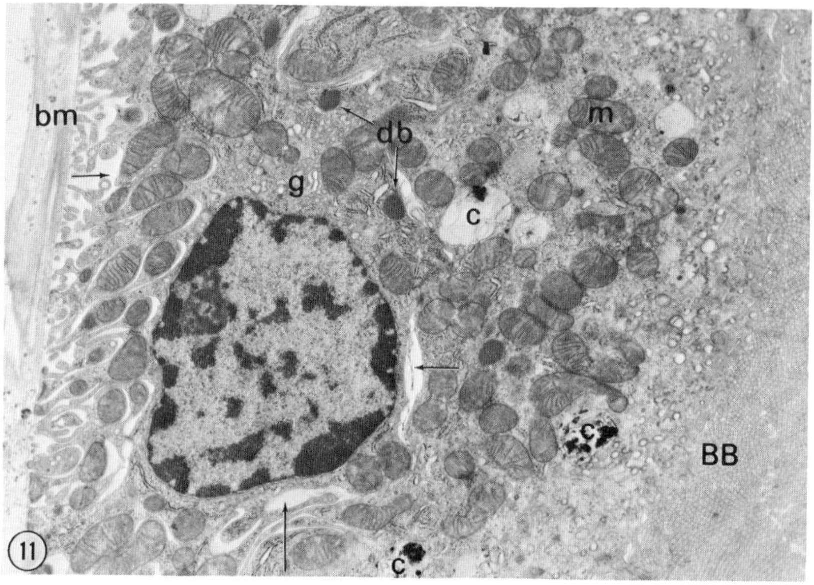

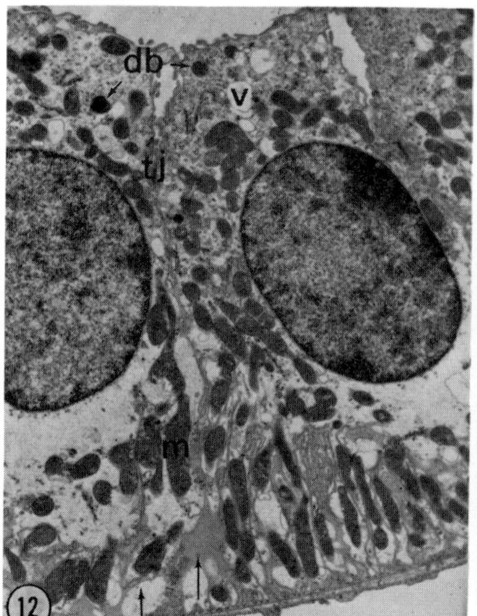

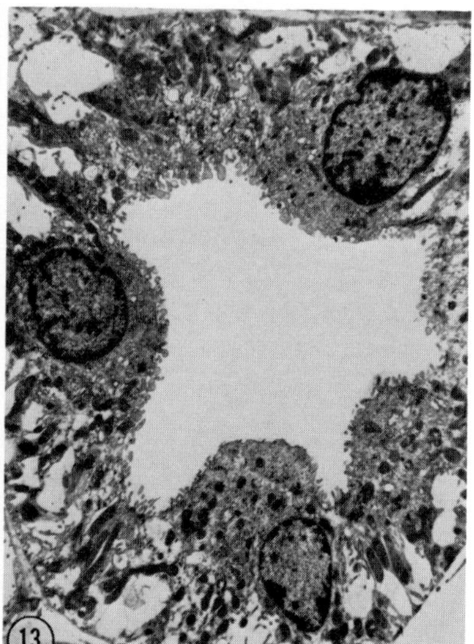

Fig. 11 Proximal convoluted tubular cell of cardiomyopathic hamster in severe congestive heart failure. The basilar interdigitating processes delimit large spaces (arrows), not only near the basement membrane (bm), but also in the perinuclear cytoplasm. Note large number of vacuoles and vesicles at the apex of the cell immediately under the brush border (BB). A Golgi zone (g), dense bodies (db), cytosomes (c), and mitochondria (m) can be seen. × 5,000.

unusual, irregular, elongated shapes. Here again, the nuclei as well as the other cellular organelles appeared unchanged.

Cortical collecting ducts

While in control animals the intercellular spaces of cortical collecting ducts were mostly virtual and only little space existed between the basilar interdigitating processes (fig. 9), in cardiomyopathic animals, both the intercellular spaces and the basilar spaces were extensively enlarged, giving the cells a characteristic spongy appearance (figs. 10 and 13). In several cells, the mitochondria were longer and larger than in controls and, in certain cases, they assumed abnormal shapes. These changes were present in clear as well as in dark cells.

DISCUSSION

Morphological and biochemical studies have implicated the renin-angiotensin-aldosterone system in the development and maintenance of edema during the course of congestive heart failure (6, 11). However, various reports have also stressed the absence of hyperaldosteronism in several cases of congestive heart failure with edema and most morphological and biochemical results indicate that, at best, stimulation of the system is far from maximal (1, 3, 6, 18). The present study indicates a close temporal correlation between the increase of the JGI, the enhanced enzymatic activity of the zona glomerulosa, the urinary retention of sodium, and the occurrence of clinically evident peripheral edema, of ascites and of hydrothorax. The absence of increased activity of $\Delta 5$-3 β hydroxysteroid dehydrogenase at the level of the zona glomerulosa seems to be a particularity of the hamster, since the same phenomenon has been observed in this species during sodium deficiency when the activity of both glucose-6-phosphate dehydrogenase and 6-phosphogluconate dehydrogenase was maximal (7). Morphologically at least, the renin-angiotensin-aldosterone system does not seem to participate in the very early (35 days of age) and progressive increase of the extracellular fluid volume characteristic of cardiomyopathic hamsters (6). In the pre-edematous stages, the JGI and the thickness and enzymatic acticity of the zona glomerulosa are not significantly different from

---

Fig. 12 Part of two cells of distal convoluted tubules of cardiomyopathic hamsters in severe congestive heart failure. Note large spaces (arrows) between basilar interdigitating processes, some of which are filled with slightly electron-opaque material. The mitochondria (m) in this area are large and irregularly shaped. The cytoplasm is edematous in the perinuclear regions towards the basal portion of the cells. The apex shows several vacuoles (V), a tight junction (tj), and dense bodies (db). × 3,960.

Fig. 13 Spongy aspect of cortical collecting duct due to widening of spaces between interdigitating processes and between cells, in a cardiomyopathetic hamster in severe heart failure. × 1, 980.

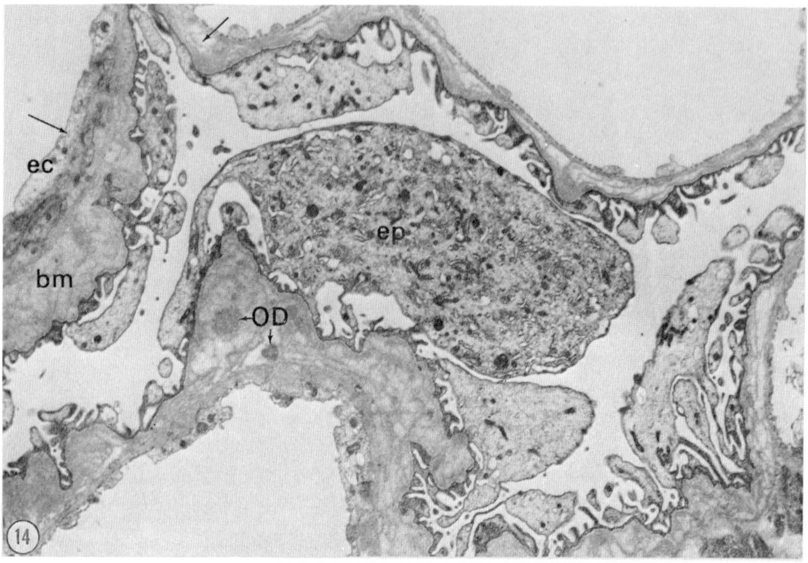

Fig. 14 Glomerulus of cardiomyopathic hamster in severe heart failure. The basement membranes (bm) are irregularly thickened. Some of them contain empty, irregular spaces (arrow). Most show complete disappearance of normal structure, which is replaced by alternating light and dark undulating bands. Osmiophilic deposits (OD) are present. The endothelial cell (ec) cytoplasm is sometimes swollen (arrow). The epithelial cell (ep) podocytes are frequently fused. × 5,230.

those of control animals and the absence of urinary sodium retention precludes the presence of hyperaldosteronism.

Several investigators have shown that the activity of glucose-6-phosphate dehydrogenase at the macula densa level increases under conditions of hyperactivity of the juxtaglomerular complex (ischemic kidney, sodium deficiency, adrenalectomy) and decreases in the reverse circumstances (contralateral, nonischemic kidney, overdose of DOCA and sodium). It has been postulated that the activity of the macula densa may reflect osmotic changes in the tubular fluid at that level. The macula densa would then transmit appropriate signals to the juxtaglomerular cells for renin secretion (4, 14). The absence of hyperactivity of glucose-6-phosphate dehydrogenase even in severely edematous hamsters, where the JGI is significantly elevated, is possibly attributable to the high sodium content of the kidney, since biochemically and histochemically the effect of sodium on the activity of this enzyme is inhibitory (22); it is not due to any inherent lack of reactivity of the macula densa in the hamster, because sodium deficiency in the same species induces a significant increase in the activity of the enzyme in this structure (7).

The early and progressive increase of the renal sodium concentration that occurs long before any morphological sign of activation of the juxtaglomerular

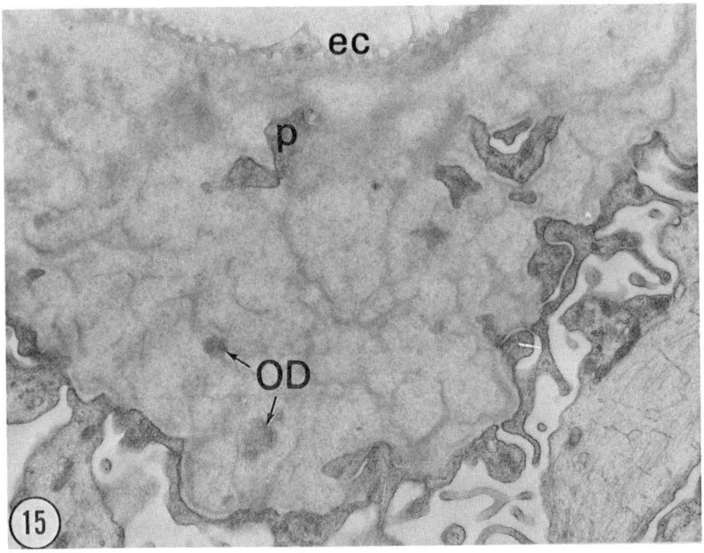

Fig. 15 Detail of glomerulus in severely edematous cardiomyopathic hamster. The fenestrated endothelium (ec) appears normal. The thickened basement membrane is composed of light material interspersed with thin, wavy, dark bands. Osmiophilic deposits (OD) and portions of podocytes (p) surrounded by the basement membrane are visible. The foot processes are completely fused. × 11,250.

apparatus-zona glomerulosa system does not seem to depend on hyperaldosteronism. Extrapolated data suggest that the increase in renal sodium concentration is mainly intracellular. However, the technical problems encountered in measuring, *in vivo,* the extracellular (inulin) space of the kidney do not allow us to make definite conclusions at this time. It is worth mentioning that, during congestive heart failure, the increase in sodium concentration is not limited to renal tissue but is also present in the arterial walls (16, 25), seemingly without a concomitant increase in their water content (16). The sodium content of the adrenal cortex is, however, not changed by cardiac decompensation (24).

The morphological changes in glomeruli observed during the late stage of congestive heart failure are not unexpected, since it is known that a nephrotic syndrome with massive proteinuria can occur during cardiac decompensation, constrictive pericarditis, or tricuspid valve disease in humans (23). Although basement membrane thickening has been observed with the light microscope in a few cases of heart failure (23), no ultrastructural study of this condition has, to our knowledge, been reported. It seems likely that the glomerular basement membrane alterations found in cardiomyopathic hamsters are due to either increased intravascular pressure or hypoxia and stasis, or both (10). In these animals, the liver is extremely congested and the main venous channels, such as the vena cava and renal veins, are enormously distended. The renal venous pressure, however, is probably not as increased as in renal vein thrombosis, where tubular atrophy, interstitial edema, fibrosis, and dilatation of venous channels are present (23). None of these changes were prominent in the present study. The glomerular pathology of renal vein occlusion is likewise different from the findings reported here. It resembles alterations characteristic of membranous glomerulonephritis and consists mainly of irregular thickening of the basement membrane, subepithelial, intramembranous and subendothelial osmiophilic deposits of varying size and, in a few cases, of spurs of lamina densa projecting towards the visceral epithelial cells. Fusion of the foot processes is generally much more extensive, and villous hyperplasia of epithelial cells is sometimes prominent (23). It is possible that, had the disease process been allowed to proceed further, analogous changes that seem to be partly a consequence of proteinuria (23) would have been found in cardiomyopathic hamsters. Further studies involving longer periods of observation as well as measurement of venous pressure and of proteinuria are needed to elucidate these questions. The increased pressure in the vena cava affluents may also have contributed to the renal retention of sodium observed in the late stages of congestive heart failure (12).

The widening of basilar interdigitating processes observed in proximal and distal convoluted tubules during the present experiments is possibly due to increased flow of water and ions, particularly sodium, in these structures. Although several reports (9) have shown that the fine structure of the renal tubules can vary depending on a host of factors related to fixation procedures, the use of the same rigid techniques by the same operator throughout the whole course of this study makes it unlikely that the alterations found in severely edematous hamsters are artefactual, particularly since such changes were not found in control animals. The widening of basilar interdigitating processes has been observed, not only in kidney tubular cells, but also in several structures, specialized in water transport, such as salivary gland epithelium, choroid plexus, ciliary body, salt gland, and gallbaldder during periods of increased flow (8, 9, 15, 19). In both proximal and distal renal tubules, analogous changes have been described after administration of electrolyte solutions (8) or of desoxy-corticosterone and sodium to experimental animals (15). In distal tubules, the presence of widened basilar interdigitating processes was accompanied by cellular edema and mitochondrial alterations (15). The prominence of the vacuolar system in the apical portion of proximal and distal tubular cells might be related to enhanced endocytosis (9).

Finally, the changes observed in cortical collecting ducts are similar to those described *in vitro* in the same structures during periods of vasopressin-induced osmotic water transport (13). This finding is consistent with the increase in antidiuretic hormone secretion observed in congestive heart failure (3) and with the morphological evidence of increased release of neurosecretory material noted in cardiomyopathic hamsters (17). Further studies are needed to determine if these changes in cortical collecting duct morphology are present during the early stages of cardiac decompensation in the hamster. If so, the increased vasopressin secretion would then explain the increment in extracellular fluid volume observed at these periods, when the sodium balance is positive and the renin-angiotensin-aldosterone system does not appear to be stimulated.

SUMMARY

The participation of the renin-angiotensin-aldosterone system in the electrolyte shifts and the development of extracellular fluid volume expansion characteristic of congestive heart failure seems to be a late event mainly limited to the period when the cardiomyopathic hamsters present clinically evident peripheral edema, ascites, and hydrothorax. In these animals, there was no parallelism between the increase of the juxtaglomerular granulation index and the enzymatic activity of the macula densa. The renal concentration of sodium became significantly elevated during the very early (i.e.,"subclinical") stages of congestive heart failure and further progressive increases paralleled the onset and progression of the disease. Derived data seem to indicate an intracellular location for the increased load of sodium detected in the kidney. The variations in the renal concentration of potassium proved to be significant only during the advanced stages of cardiocirculatory insufficiency, while the magnesium concentration of the kidney remained unchanged. Young cardiomyopathic hamsters were characterized by low renal calcium concentrations in spite of elevated serum calcium levels, During the terminal phases of congestive heart failure, however, the calcium values in the kidneys of cardiomyopathic hamsters were higher despite lower serum calcium values. In severe congestive heart failure, electron microscopy revealed glomerular basement membrane thickening with fusion of epithelial foot processes and vacuolation of mesangial cells. Proximal and distal convoluted tubules as well as cortical collecting ducts were characterized by marked widening of extracellular spaces between the basilar interdigitating processes. Distal convoluted tubules, moreover, showed edema and mitochondrial alterations, while the intercellular spaces of cortical collecting ducts were also widened. These tubular changes have been interpreted as indicating an increased flow of water and ions.

REFERENCES

1. BAJUSZ, E.: The role of the adrenal cortex in the pathogenesis of a spontaneous hereditary congestive heart failure, in: JASMIN, G., (ed.) Endocrine aspects of disease processes, p. 388–406 (Warren H. Green Inc. St. Louis 1968).
2. BAJUSZ, E. and LOSSNITZER, K.: A new disease model of chronic congestive heart failure: studies on its pathogenesis. Trans. N. Y. Acad. Sci. *30:* 939–948 (1968).
3. BELLEAU, L.; MION, H.; SIMARD, S.; GRANGER, P.; BERTRANOU, E.; NOWACZYNSKI, W., BOUCHER, R. and GENEST, J.: Studies on the mechanisms of experimental congestive heart failure in dogs. Canad. J. Physiol. Pharmacol. *48:* 450–456 (1970).
4. BUNAG, R. D.; VANDER, A. J. and KANEKO, Y.: Control of renin release, in: PAGE, I. H. and MC CUBBIN, J. W., (eds.) Renal hypertension, p. 100–117 (Year Book Medical Publishers, Inc. Chicago 1968).
5. CANTIN, M.: Relationship of juxtaglomerular apparatus and adrenal cortex to biochemical and extracellular fluid volume changes in magnesium deficiency. Lab. Invest. *22:* 558–568 (1970).
6. CANTIN, M. and BAJUSZ, E.: Juxtaglomerular apparatus and adrenal cortex in hamsters with heart failure. Arch. Path. *87:* 626–636 (1969).
7. CANTIN, M.; FORTHOMME, D. and DUMOUCHEL, J. C.: Influence of dietary sodium and desoxycorticosterone on the histochemistry of the juxtaglomerular apparatus,

the adrenal cortex and the salivary glands in the hamster. Rev. Canad. Biol. *30:* 11–26 (1971).

8. CAULFIELD, J. B. and TRUMP, B. F.: Correlation of ultrastructure with function in the rat kidney. Amer. J. Path. *40:* 199–211 (1962).
9. CHATELANAT, F. and SIMON, G. T.: Ultrastructural pathology of the tubules and interstitial tissue, in: ROUILLER, C. H. and MÜLLER, A. F. (eds.) The Kidney, Vol. 1, pp. 449–530 (Academic Press, New York 1969).
10. CORNOG, J. L.; RAWSON, A. J.; KARP, L. A. and ARVAN, D. A.: Immunofluorescent and ultrastructural study of the renal glomerulus in renal vein thrombosis. Lab. Invest. *22:* 101–111 (1970).
11. DAVIS, J. O.: Mechanisms of salt and water retention in congestive heart failure: The importance of aldosterone. Amer. J. Med. *29:* 486–507 (1960).
12. DAVIS, J. O.; JOHNSTON, C. I.; HOWARDS, S. S. and WRIGHT, F. S.: Humoral factors in the regulation of renal sodium excretion. Fed. Proc. *26:* 60–69 (1967).
13. GANOTE, C. E.; GRANTHAM, J. J.; MOSES, H. L.; BURG, M. B. and ORLOFF, J.: Ultrastructural studies of vasopressin effect on isolated perfused renal collecting tubules of the rabbit. J. Cell. Biol. *36:* 355–367 (1968).
14. HATT, P. Y.: The juxtaglomerular apparatus, in: A. J. DALTON and F. HAFUENAU (eds.), Ultrastructure of the kidney, p. 101–141. (Academic Press, New York 1967).
15. HATT, P. Y.; PERCHERON, F.; ALLAIRE, Y.; KERNEC, H.; LAFON, Y. et ROQUES, B.: Contribution de la microscopie électronique à l'étude du mécanisme de l'hypertension artérielle expérimentale d'origine rénale. Path. Biol. *11:* 417–137 (1963).
16. HAYDUK, K.; BRECHT, H. M.; VLADUTIU, A.; SIMARD, S.; ROJO-ORTEGA, J. M.; BELLEAU, L.; BOUCHER, R. and GENEST, J.: Renin activity and norepinephrine, cation, and water content of cardiovascular tissue of dogs with congestive heart failure and ascites. Canad. J. Physiol. Pharmacol. *48:* 463–468 (1970).
17. HORVATH, E.; BAJUSZ, E. and KOVACS, K.: Study of the hypothalamic-hypophyseal system in hereditary cardiomyopathic hamsters. This Book, pp. 294–302.
18. KAUFMAN, W.; STEINER, B.; DURR, F.: MEURER, K. A. und BEHN, C.: Induzierter Aldosteronismus bei hydropischer Herzinsuffizienz. Klin. Wochenschr. *47:* 16–25 (1969).
19. KAYE, G. I.; WHEELER, H. O.; WHITLOCK, R. T. and LANE, N.: Fluid transport in the rabbit gallbladder. J. Cell Biol. *33:* 237–268 (1966).
20. MANDALENAKIS, N.; CANTIN, M. et DUMONT, A.: L'appareil juxtaglomerulaire du rein endocrinien. Etude histologique et ultrastructurale. Path. Biol. *18:* 233–240 (1970).
21. PEASE, D. C.: Infolded basal plasma membranes found in epithelia noted for their water transport. J. Biophys. Cytol. *2:* (4) Suppl., 203–208 (1956).
22. REEVES, G. and SOMMERS, S. C.: Sensitivity of the renal macula densa to urinary sodium. Proc. Soc. Exp. Biol. Med. *120:* 324–326 (1965).
23. ROSENMANN, E.; POLLACK, V. E. and PIRANI, C. L.: Renal vein thrombosis in the adult. A clinical and pathologic study based on renal biopsies. Medicine *47:* 269–335 (1968).
24. YANKOPOULOS, N. A.; DAVIS, J. O.; COTLOVE, E. and TRAPASSO, M.: Mechanism of myocardial edema in dogs with chronic congestive heart failure. Amer. J. Physiol. *199:* 603–608 (1960).
25. ZELIS, R.; DELEA, C. S.; COLEMAN, H. N. and MASON, D. T.: Arterial sodium content in experimental congestive heart failure. Amer. J. Cardiol. (abstr), *23:* 144–145 (1969).

Mailing address: Dr. Marc Cantin, Départment de Pathologie, Université de Montréal, Case Postale 6128, Montreal 101, Qué. (Canada).

Cardiac Morphology: Structural-Functional Relationships

Department of Pathology, Richardson Laboratories,
Queen's University and Kingston General Hospital,
Kingston, Ontario; and Pathology Institute, Department of
Pathology, McGill University, Montreal, Quebec, Canada

SPECIFIC GRANULES IN HUMAN AND NONHUMAN VERTEBRATE CARDIOCYTES

S. A. BENCOSME and J. M. BERGER

INTRODUCTION

Mammalian cardiac muscle cells (cardiocytes) contain diverse populations of granules. The best known of these granular structures are lipofuscin granules, which at the fine structural level have been variously termed Cytosoma C (17), lysosomal bodies (13), lipofuscin granules (31), and residual bodies (14). They have been correlated at the light and electron microscopic levels both in situ (7, 19) and in vitro (7), and may appear as large, oblate, or irregularly shaped electron-dense bodies with a variety of inclusions such as fat, ferritin, membranes, organelles, or coarse particulate granules (6). As a group, these latter oblate or irregularly shaped granules have been recently classified tentatively as *C-granules* (5, 6), following the designation of Lindner, who called these *Cytosoma C* (17). The C-granules may appear in substantial numbers in both atrial and ventricular cardiocytes and measure about 450–1900 mμ along the long axis. They increase in number with senescence and in pathologic hearts (13, 43). Wheat (43) has shown lipofuscin granules in ventricle to measure 0.5–1.5 μ in diameter. Their cores have been shown to be bound by a unit membrane (14) and to be acid-phosphatase positive (14). In certain cases, the diameters and general appearance of C-granules may resemble another class of granules that, though less well known, is far more intriguing. *This other class of granules is known as "specific granules."*

By "specific granules," we are referring to membrane-bound granules with homogeneously electron-dense cores, morphologically very similar to secretory granules in endocrine cells, but present abundantly in mammalian atrial, and nonmammalian vertebrate atrial and ventricular cardiocytes. The specific granules have, in the past, been called *corps denses* (8), microbodies (16), secretory granules (30), dense bodies (35), *dichte Körper* (31), atrial granules (13), and specific granules (14, 40, 41). Recently, a nomenclature based on certain differences in the size and texture of the cores among specific granules has been proposed in connection with possible transitional forms (5, 6). On the basis of their numbers and distribution in cardiocytes, it appeared that striated muscle cells in heart may have subsumed secretory as well as contractile functions.

Specific granules were first described by Kisch (15) and later by Bompiani *et al.* (8), both of whom recognized that these granules were characteristic of atrium in mammals. Jamieson and Palade (14) showed that mammalian atrial specific granules were acid-phosphatase negative and did not appear to take up ^{3}H-dopamine, or to appreciably bind ^{3}H-norepinephrine in situ.

Ultrastructural studies were undertaken in our laboratories on human and nonhuman vertebrate atrial and ventricular cardiocytes in order to: (1) assess the distribution of specific granules in heart from representative species; (2) determine the possible relationship of specific granules to lysosomal bodies (C-granules); and (3) ascertain in what respects the specific granules could be related to morphologically similar granules in endocrine cells.

MATERIAL AND METHOD

Samples from atrium and ventricle of species from a wide variety of animal Classes, including man, were processed for light and electron microscopy. Tissues were fixed in formol, Zenker-formol, or osmium iodide fixative of Champy. A modified Gomori-acid-phosphatase technique for electron microscopy (22) was used in specific cases. The method described by Hibbs and Ferrans (12) using differentiation with 95% ethanol was employed on toluidine-blue-stained Epon sections. For electron microscopy, samples were fixed in buffered osmium tetroxide or 3% glutaraldehyde in 0.1 M phosphate buffer and post-fixed in osmium. Thin sections cut with an LKB ultratome were doubly stained with uranyl acetate (38) followed by lead citrate (42) and examined with either an RCA EMU-3D or Hitachi HU-11C electron microscope.

RESULTS

Specific granules were encountered in both the atrial and ventricular cardiocytes of the following nonmammalian vertebrate classes and representative species:

Class *Cyclostomata*:	Lamprey *(Petromyzon marinus)*
Class *Osteichthyes*:	Goldfish *(Carassius auratus)*
Class *Amphibia*:	Frog *(Rana pipiens)*; Toad *(Buffo marinus)*
Class *Reptilia*:	Turtle *(Pseudomis scripta* and *Pseudomis elegans)*
Class *Aves*:	Leghorn Chicken *(Gallus domesticus)*

The granules appeared in about the same numbers in both chambers. Moreover, each representative species showed a similar number of granules. Among these species, the structure of the granules in atrium and ventricle did not differ essentially (figs. 1 and 2). The specific granules in both atrial and ventricular cardiocytes appeared acid-phosphatase negative. Structures, evidently ly-

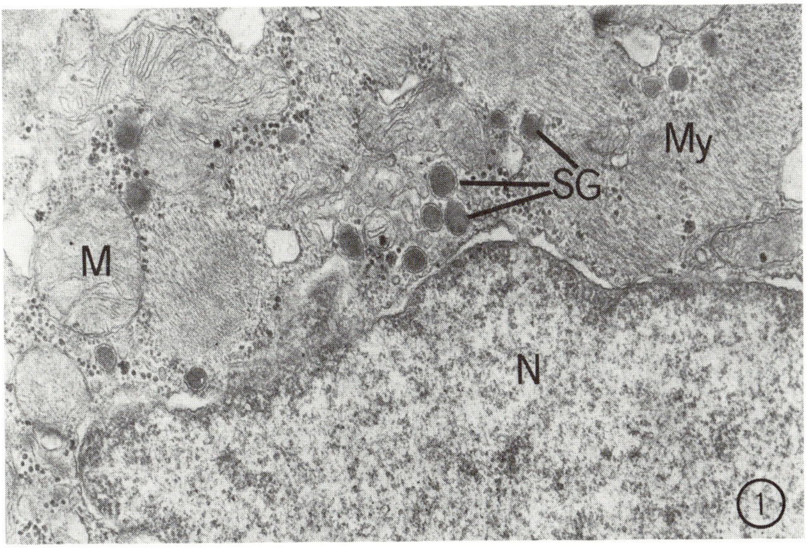

Fig. 1 Goldfish ventricle *(Carassius auratus)*. Nucleus (N) is seen partly surrounded by components of the sarcoplasm, containing mitochondria (M), myofilaments (My), and specific granules (SG). × 29,500.

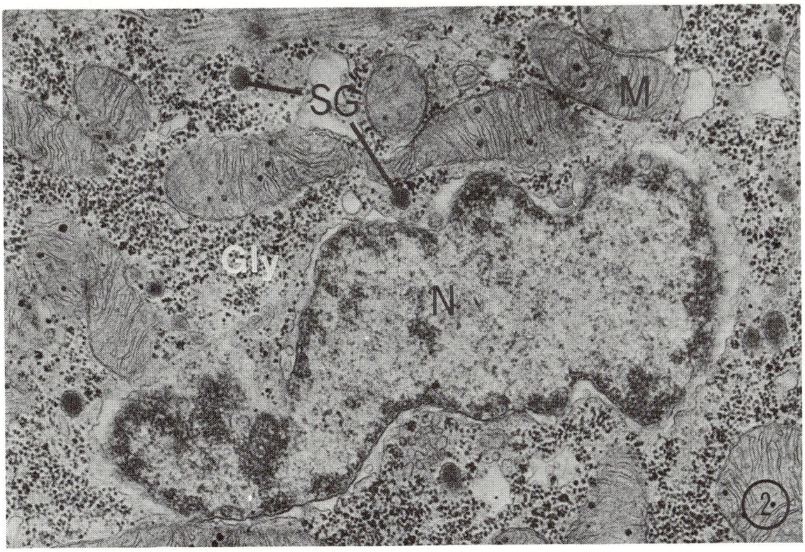

Fig. 2 Goldfish atrium *(Carassius auratus)*. In the sarcoplasm around the nucleus (N), abundant granules, presumably glycogen (Gly), can be seen. Among the glycogen particles and mitochondria (M), specific granules (SG) are evident. × 14,600.

sosomal, in these cardiocytes showed a positive reaction for acid-phosphatase activity, and were differentiated from specific granules on this basis (21, 22). Besides these main observations, several other interesting morphologic findings related to specific granules in the nonmammalian vertebrate have recently been reviewed elsewhere (1).

In the mammalian vertebrate, however, the most striking finding was the paucity of these specific granules in ventricular cardiocytes (figs. 3–6), although atrial cardiocytes in mammals contained abundant specific granules (fig. 7).

Class *Mammalia*: Sprague-Dawley rats *(Rattus norvegicus)*
Hooded rats *(Rattus norvegicus)*
New Zealand albino rabbits *(Oryctolagus cuniculus)*
Virginia opossum *(Didelphis virginiana)*
Guinea pig *(Cavia cobaya)*
Cat *(Felis catus)*
Dog *(Canis familiaris)*
Steer *(Bos taurus)*
Man *(Homo sapiens)*

These specific granules in mammals range from about 100–450 mμ in diameter and generally contain homogeneously electron-dense cores. Counting

Fig. 3 Rat ventricular cardiocyte *(Rattus norvegicus)*. The nucleus is associated with very little Golgi complex. Most important is the absence of specific granules in the paranuclear and sarcomeric zones. Absence of specific granules in mammalian ventricular cardiocytes is the more common finding. × 13,100.

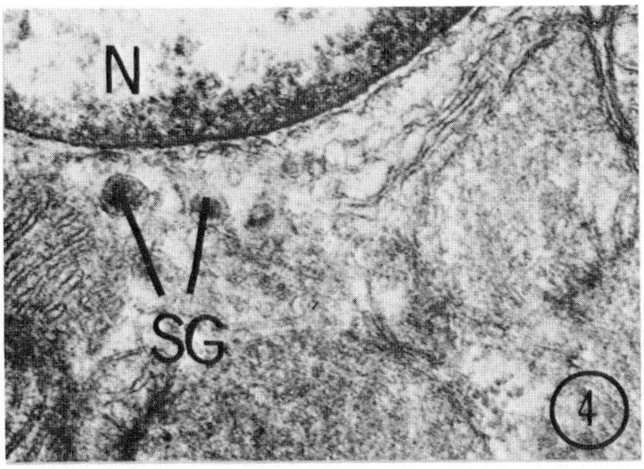

Fig. 4 Rat ventricular cardiocyte, showing a portion of the nucleus (N) and paranuclear zone. This zone contains a small Golgi complex near which two specific granules (SG) may be clearly seen. These ventricular specific granules morphologically resemble specific granules found near the cisternae of the atrial Golgi complex. × 46,000.

techniques showed fewer specific granules in left atrium of rat than in right, not only in normal animals but even with total increases in granules after adrenal regeneration hypertension (23). In either atrium, the granules appeared most concentrated in numbers in paranuclear zones. Substantial amounts of these granules occurred in sarcomeric zones of atrial cardiocytes. At higher magnifications, a considerable number of very small granules are found, particularly among the sarcomeres (fig. 7).

The findings in other mammalian species with respect to specific granules largely apply to specific granules in human atrial cardiocytes, in that the granules are concentrated in the paranuclear zones, appear membrane-bound, and have homogeneous cores. They were readily distinguished from human C-granules (lipofuscin), which appeared in greater numbers than usual, probably because

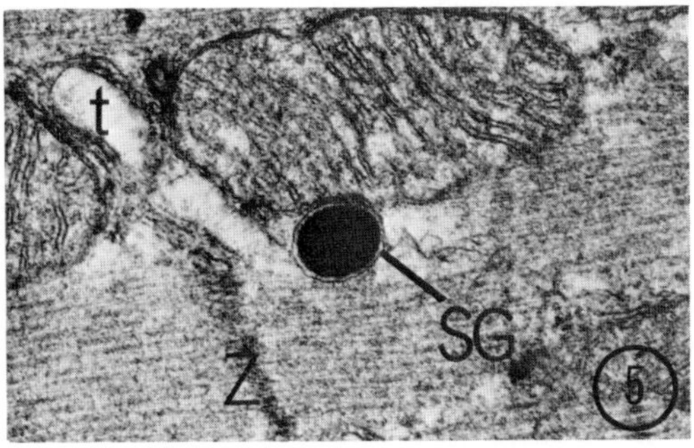

Fig. 5 Rat ventricular cardiocyte showing a specific granule (SG) with homogeneously electron-dense core retracted slightly from its limiting membrane, revealing a uniform clear space around the core. A triad with a transverse tubule (t) is present at the level of the Z line (Z). × 40,000.

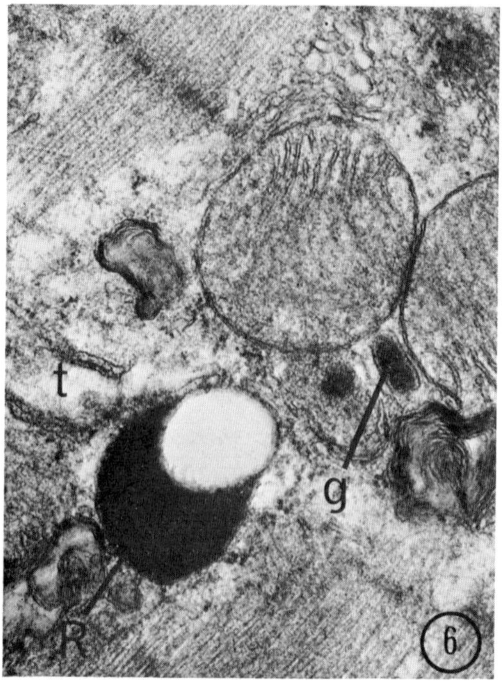

Fig. 6 Ventricular cardiocyte in rat, showing several granular structures. A residual body (R) containing a clear vacuole shows a well-defined limiting membrane and electron-dense core. There is a smaller elongated granule (g) resembling some specific granules seen in frog and toad. Transverse tubule (t). It may be difficult to differentiate ventricular specific granules on the basis of structure alone from images produced by tangential sections of residual bodies in thin sections. × 40,000.

human atrial tissues were obtained from senescent or pathologic hearts and at autopsy (figs. 8 and 9). It should be stressed that, in human autopsy material, the appearance of specific granules did not change with the general processes of post-mortem autolysis, which seemed to involve many other cellular structures (fig. 8). The C-granules in human biopsy (fig. 9) appeared as large, oblate, granular bodies composed of dense and irregular particulate matter. Frequently, these bodies contained clear vacuoles or vacuoles with moderately electron-dense, probably lipoid, material. Occasionally, within or abutting the C-granules, there were profiles closely resembling those of specific granules (fig. 9). Although in our material it was not possible to determine whether indeed phagocytosis accounted for this position, it is quite possible that lysosomal mechanisms contribute to the turnover of granule populations.

It is of more than passing interest, especially with regard to experiments designed for periods of several weeks, to mention that the number of specific

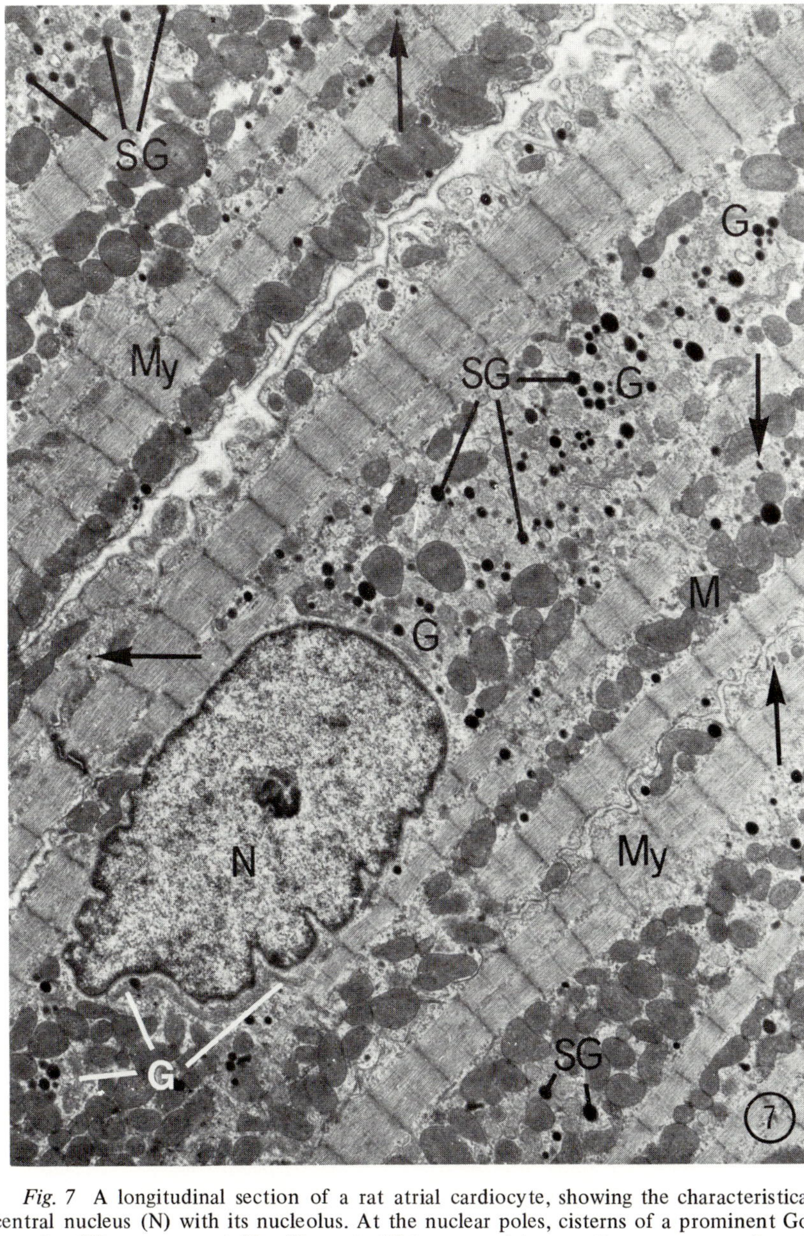

Fig. 7 A longitudinal section of a rat atrial cardiocyte, showing the characteristically central nucleus (N) with its nucleolus. At the nuclear poles, cisterns of a prominent Golgi complex (G) are apparent. Myofilaments (My), arranged in repeating sarcomeres, form the lateral boundaries of the conical paranuclear zones, which contain many mitochondria (M), much of the Golgi complex, and the greater concentration of specific granules (SG). Two populations of granules are evident with respect to size. The specific granules (SG) represent the medium-to-large variety. The smaller variety of specific granules (arrows), while not strikingly evident at this magnification, nevertheless forms a substantial proportion of the intermyofibrillar granule population. × 6,300.

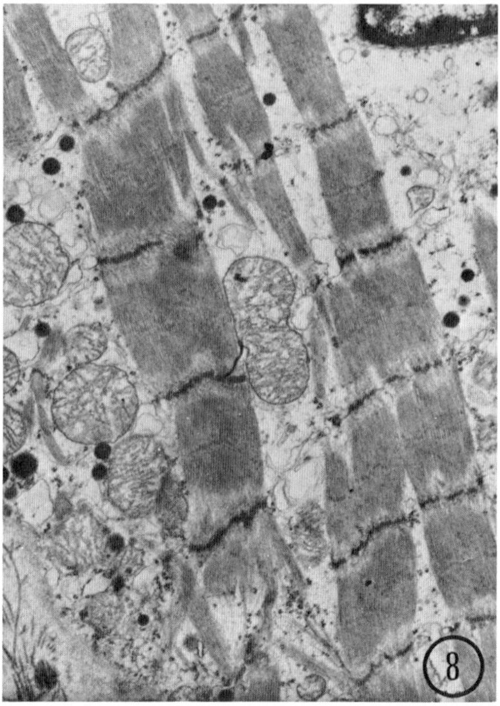

Fig. 8 Atrial cardiocyte from an autopsy specimen of man. In the upper right corner, a portion of the nucleus is evident and, in the lower left, some collagen. The mitochondria appear swollen, but the sarcomeres are still organized. Many specific granules can be seen among these intracellular components. × 14,000.

granules increased sharply with age in young adult rats and gradually thereafter (9).

DISCUSSION

The structure-function relationships in ventricle have been extensively investigated in recent years (3, 10, 25, 32, 34, 36, 37). Less numerous have been the investigations on the presence of granular components in cardiac muscle cells (cardiocytes). Since 1956, when Kisch (15) pointed out the presence of "microbodies" in electron microscopic sections of atrial muscle of the guinea pig, it has been known that, in mammals, a distinct group of membrane-bound granules less than 500 mμ in diameter could be found in abundant numbers in atrial cardiocytes, whereas in mammalian ventricle much fewer granules, resembling atrial granules, had been encountered. In nonmammalian vertebrates, however, these granules are abundantly present, both in ventricular and atrial cardiocytes (40, 41).

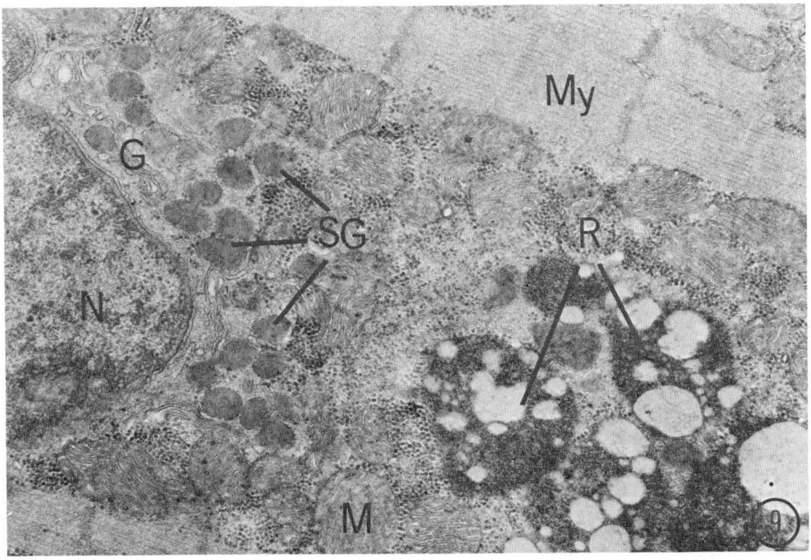

Fig. 9 Human auricular cardiocyte from biopsy. The paranuclear area is shown to contain a Golgi complex (G), numerous specific granules (SG), large and vacuolar residual bodies (R). Between the two residual bodies (C-granules) labeled, a specific granule appears in close apposition to one of the residual bodies. Mitochondria (M). The myofilaments (My) and nucleus (N) form two of the boundaries of the paranuclear zone. × 21,500.

Many investigators have called attention to the resemblance of the specific granules to secretory granules in endocrine glands (1, 12, 30). The possible secretory function of specific granules is supported indirectly by their association with the Golgi complex (14, 20, 24), their increase in number after stimulation by certain pharmacologic compounds (27, 29, 39), and the morphology (5) and topographic distribution (1) of the granules within the cardiocytes.

With improved techniques, the pale granules seen in endocrine cells have been considered to represent either precursors of the granules with darker or condensed cores, or as possible end products of a secretory process, or both. At present, however, the precise function of pale granules is still controversial. This may be due in part to possible oversimplification of interpretation of events observed in connection with stimulated, developing, or pathologic sequences in endocrine cells. Of great significance, however, is the close similarity of pale granules in most endocrine cells. In the rat, a number of pale specific granules, rarely encountered in right atrial cardiocytes, were shown in cardiocytes of the left atrium (5). It has been hypothesized that pale granules in atrial cardiocytes represent specific granules, in which electron-dense material may be absent from the cores (5, 6). A similar phenomenon has been reported in cortisone-treated rabbit pancreas by Bencosme and Martinez-Palomo (2), and in xylitol-treated chick pancreas by Machino (18), who showed that breakdown of the electron-dense cores of B-granules resulted in paler granular cores. The presence of pale specific granules normally observed in left atrium of rats as well as their relative abundance in the right atrial cardiocytes of rats treated with plasmocid (4) has lent further support to a possible storage and secretory role for specific granules (5, 6).

Granulophagia in α cells of pancreas and in anterior pituitary cells is not an uncommon finding (28, 33). Thus, apparent granulophagia in atrial residual bodies may represent a similar phenomenon of intracellular maintenance of a constant granular population. The occasional finding of specific granules within residual bodies, therefore, does not lessen the distinction between these two classes of granular bodies.

Despite the large number of attempts by many investigators to induce alterations by means of chemical agents and pathophysiologic manipulations, unequivocal changes in specific granule size, distribution, or number have rarely been demonstrated (1). Since, in instances where compounds have been reported to be particularly active in altering endocrine granules, the effects have usually been dramatic, one is compelled to conclude that, in the case of atrium, the proper chemical agent has not yet been applied. On the other hand, the converse could be argued, namely, that substantial biochemical and physiologic activity is actually taking place, although the morphology as evaluated by present methodology is not very much altered. A suitable example of this situation is seen in β cells, where short periods of stimulation by high concentrations of glucose, arginine, cortisone, glucagon, or xanthines will induce considerable insulin release *without* correlated morphologic findings during the early periods of activity studied. Some compounds, however, when acting for longer periods of time, such as cortisone after 48 h, or tolbutaminde after 5 h, will degranulate β cells and the changes of insulin secretory-activity after degranulation can be correlated.

Generally, endocrine cells show fewer granules in younger animals than in older ones. A similar situation obtains in the case of the atrial specific granules (9).

If indeed the secretory function of specific granules should be established, the intriguing question of its target organ will become the object of future investigations. Although it is possible that the ventricle is a target organ (1), in the case of endocrine organs, fairly distant targets have nevertheless proved to be the rule. Classical endocrinologic experimental design of organ ablation does not seem feasible as the primary approach to this problem.

On a morphologic basis, specific granules can be distinguished from lysosomal or lipofuscin bodies in most instances. This separation on a morphologic basis has given momentum to investigations on their possible functions. In humans, specific granules occur also as a distinct class of membrane-bound spheroids that, on most thin sections, can be distinguished from lipofuscin granules or residual bodies. One reason for this is the overlap in size and morphology between the larger specific granules and grazing sections through C-granules. In many instances, but not all, the non-homogeneity of core textures helps to identify C-granules.

It has been believed that mammalian ventricle may contain no specific granules (1). The present methods have not allowed us to exclude the possibility that mammalian ventricle has some specific granules (figs. 4-6), since in ventricular cardiocytes a few granules can be found that have enough resemblance to justify further study to establish their relationship to the atrial specific granules. Moreover, the presence of many such granules in ventricle of nonmammalian vertebrates suggests to us that the mammalian ventricle may indeed contain some specific granules, possibly as a phylogenetic atavism.

SUMMARY

Specific granules in human and nonhuman vertebrate cardiocytes were studied in relation to their species distribution, resemblance to C-granules, and to secretory granules in endocrine cells. Human atrial specific granules resembled those of other mammals, but ventricular cardiocytes contained a few granules similar to the atrial ones. By contrast, in nonmammalian vertebrates, specific granules appeared abundant in ventricular as well as in atrial cardiocytes. The specific granules were usually morphologically distinct from C-granules (lipofuscin). In human and nonhuman mammalian cardiocytes, specific granules in many respects resembled secretory granules in certain endocrine cells, and it is suggested that the specific granules are related to a secretory function in myocardium.

ACKNOWLEDGMENT

Figures 1, 2, 4–9 are adapted and reprinted with the kind permission of Methods and Achievements in Experimental Pathology, Functional Morphology of the Heart (S. Karger AG, Basel) 1971. The authors wish to thank Mrs. Eva Losinski for her technical assistance.

REFERENCES

1. BENCOSME, S. A. and BERGER, J. M.: Specific granules in mammalian and nonmammalian vertebrate cardiocytes, in BAJUSZ, E. and JASMIN, G. Functional Morphology of the Heart, pp. 173–213 (S. Karger AG, Basel 1971).
2. BENCOSME, S. A. and MARTINEZ-PALOMO, A.: Formation of secretory granules in pancreatic islet B cells of cortisone-treated rabbits. Lab. Invest. *18:* 746–756 (1968).
3. BENCOSME, S. A.; TRILLO, A.; ALANIS, J. and BENITEZ, D.: Correlative ultrastructural and electrophysiological study of the Purkinje system of the heart. J. Electrocardiology *2:* 27–38 (1969).
4. BERGER, J. M. and BENCOSME, S. A.: Divergence in patterns of atrial and ventricular cardiocyte degeneration: studies with plasmocid. J. Mol. Cell. Cardiol. *2:* 41–49 (1971).
5. BERGER, J. M. and RONA, G.: Fine structure of extra-nodal transitional cardiocytes in rat left atrium. J. Mol. Cell. Cardiol. *2:* 181–185 (1971).
6. BERGER, J. M. and RONA, G.: Functional and fine structural heterogeneity of atrial cardiocytes, in BAJUSZ, E. and JASMIN, G. Functional Morphology of the Heart pp. 540–590 (S. Karger AG, Basel 1971).
7. BJÖKERUD, S.: The isolation of lipofuscin granules from bovine cardiac muscle, with

observations on the properties of the isolated granules on the light and electron microscopic levels. J. Ultrastr. Res. *5* (Suppl.): 1–49 (1963).

8. BOMPIANI, G. D.; ROUILLER, C. et HATT, P. Y.: Le tissu de conduction du coeur chez le rat. Etude au microscope électronique. I. Le tronc commun du Faisceau de His et les cellules claires de l'oreillette droite. Arch. Mal. Coeur Vaiss. *52:* 1257–1274 (1959).
9. CHANG, W. W. and BENCOSME, S. A.: Quantitative electron microscopic analysis of the specific granule population of rat atrium. Canad. J. Physiol. Parmacol. *47:* 483–485 (1969).
10. FAWCETT, D. W. and McNUTT, N. S.: The ultrastructure of the cat myocardium. I. Ventricular papillary muscle. J. Cell Biol. *42:* 1–45 (1969).
11. FORSSMANN, W. G. and GIRARDIER, L.: A study of the T-system in rat heart. J. Cell Biol. *44:* 1–19 (1970).
12. HIBBS, R. G. and FERRANS, V. J.: An ultrastructural and histochemical study of rat atrial myocardium. Amer. J. Anat. *124:* 251–280 (1969).
13. HIBBS, R. G.; FERRANS, V. J.; WALSH, J. J. and BURCH, G. E.: Electron microscopic observations on lysosomes and related cytoplasmic components of normal and pathological cardiac muscle. Anat. Rec. *153:* 173–185 (1965).
14. JAMIESON, J. D. and PALADE, G. E.: Specific granules in atrial muscle cells. J. Cell Biol. *23:* 151–172 (1964).
15. KISCH, B.: Electron microscopy of the atrium of the heart; guinea pig. Exp. Med. Surg. *14:* 99–112 (1956).
16. KISCH, B.: A significant electron microscopic difference between the atria and the ventricles of the mammalian heart. Exp. Med. Surg. *21:* 193–221 (1963).
17. LINDNER, E.: Die submikroskopische Morphologie des Herzmuskels. Z. Zellforsch. *45:* 702–746 (1957).
18. MACHINO, M.: The stimulation of avian pancreatic islet B cells by xylitol. Anat. Rec. *166:* 342 (1970).
19. MALKOFF, D. B. and STREHLER, B. L.: The ultrastructure of isolated and in situ human cardiac age pigment. J. Cell Biol. *16:* 611–616 (1963).
20. MANASEK, F. J.: The appearance of granules in the Golgi complex of embryonic cardiac myocytes. J. Cell Biol. *43:* 605-610 (1969).
21. MARTINEZ-PALOMO, A.: The fine structure of the heart muscle cells with special reference to specific granules. Master of Science thesis. Queen's University, Kingston, Ontario (1966).
22, MARTINEZ-PALOMO, A. and BENCOSME, S. A.: Electron microscopic observations on myocardial specific granules and residual bodies in vertebrates. Anat. Rec. *154:* 473 (1966).
23. MARTINEZ-PALOMO, A. and BENCOSME, S. A.: Changes in the content of myocardial specific granules in adrenal regeneration hypertension. An electron microscopic study. Fed. Proc. *25:* 476 (1966).
24. McNUTT, N. S. and FAWCETT, D. W.: Ultrastructure of the cat myocardium. II. Atrial muscle. J. Cell Biol. *42:* 46–67 (1969).
25. MEESEN, H.: Ultrastructure of the myocardium. Its significance in myocardial disease. Amer. J. Cardiol. *22:* 319–327 (1968).
26. MILLER, F.: Acid phosphatase localization in renal protein absorption droplets, in BREESE, 5th International Congress for Electron Microscopy, Philadelphia, vol. 2., Q-2 (Academic Press, New York, 1962).
27. OKAMOTO, H.: An electron microscopic study of the specific granules in the atrial muscle cell upon the administration of agents affecting autonomic nerves. Arch. histol. jap. 467–478 (1969).

28. ORCI, L.; JUNOD, A.; PICTET, R. and ROUILLER, C.: Granulolysis in A cells of endocrine pancreas in spontaneous and experimental diabetes in animals. J. Cell Biol. *38:* 462–466 (1968).

29. OTSUKA, N.; OKAMOTO, H. and TOMISAWA, M.: Electron and fluorescence microscopic study of specific granules in rat atrial muscle cells. Arch. histol. jap. *30:* 367–374 (1969).

30. PALADE, G. E.: Secretory granules in the atrial myocardium. Anat. Rec. *139:* 262 (1961).

31. POCHE, R.: Elektronenmikroskopische Untersuchungen des Lipofuscin im Herzmuskel des Menschen. Zbl. allg. Path. path. Anat. *96:* 395 (1957).

32. SELYE, H.: The pluricausal cardiopathies. Ann. N.Y. Acad. Sci. *156:* 195–206 (1969).

33. SMITH, R. E. and FARQUHAR, M. G.: Lysosome function in the regulation of the secretory process in cells of the anterior pituitary gland. J. Cell Biol. *22:* 319–000 (1966).

34. SOMMER, J. R. and JOHNSON, E. A.: A comparative study of Purkinje fibers and ventricular fibers. J. Cell Biol. *36:* 497–526 (1968).

35. SOMMER, J. R. and JOHNSON, E. A.: Cardiac muscle. A comparative ultrastructural study with special reference to frog and chicken hearts. Z. Zellforsch. *98:* 437–468 (1969).

36. SONNENBLICK, E. H.; SPIRO, D. and COTRELL, T. S.: Fine structural changes in heart muscle in relation to the length-tension curve. Proc. Nat. Acad. Sci. *49:* 193–200 (1963).

37. STALEY, N. A. and BENSON, E. S.: The ultrastructure of frog ventricular cardiac muscle and its relationship to mechanisms of excitation-contraction coupling. J. Cell Biol. *38:* 99–114 (1968).

38. STEMPAK, J. G. and WARD, R. T.: An improved staining method for electron microscopy. J. Cell Biol. *22:* 697–701 (1964).

39. TOMISAWA, M.: Atrial specific granules in various mammals. Arch. histol. jap. *30:* 449–465 (1969).

40. TRILLO, A. and BENCOSME, S. A.: La ultraestructura de las celulas musculáres cardíacas con referencia especial a los gránulos específicos. I. El miocardio de la rana. Arch. Inst. Cardiol. Méx. *35:* 803–817 (1965).

41. TRILLO, A.; MARTINEZ-PALOMO, A. and BENCOSME, S. A.: Estudio ultraestructural del músculo cardíaco en relación a los gránulas específicos. II. Estudio comparativo. Arch. Inst. Cardiol. Méx. *36:* 45–67 (1966).

42. VENABLE, J. H. and COGGESHALL, R.: A simplified lead citrate stain for use in electron microscopy. J. Cell Biol. *25:* 407–408 (1965).

43. WHEAT, M. W., Jr.: Myocardial lysosomes in heart disease. Circulation *178* (Suppl. III): 178 (1964).

Mailing address: Sergio A. Bencosme, M.D., Department of Pathology, Richardson Laboratories, Queen's University, Kingston, Ontario (Canada).

Department of Pathology, Pathology Institute, McGill University, Montreal;
Warner-Lambert Research Institute of Canada, Sheridan Park, Ontario;
Department of Pathology, University of Toronto; and
Department of Pathology, Queens's University,
Kingston, Canada

RELATIONSHIP OF ATRIAL CATECHOLAMINES TO CYTOCHEMISTRY AND FINE STRUCTURE OF ATRIAL SPECIFIC GRANULES

J. M. BERGER, J. C. SOSA-LUCERO, F. A. DE LA IGLESIA,
G. LUMB, and S. A. BENCOSME

INTRODUCTION

Since 1956 (15) abundant granular structures have been described in vertebrate atrial muscle cells (cardiocytes) (2,5,7,8,12,13,15,16,19–21,24–27, 30–33) (fig. 1). Although structures resembling atrial granules have been reported in mammalian ventricle (27, 32, 33), it is now established that mammalian ventricular cardiocytes contain only very few such structures (2). By contrast, these granules are abundant in both the atrial and ventricular cardiocytes of non-mammalian vertebrates (2, 17, 23, 33).

The presence of catecholamine-containing granules within "specific chromaffin cells" in cyclostome hearts, morphologically similar to granules within cardiocytes (6, 7, 17, 25), fostered the concept that heart muscle cells had contractile and possible amine storage functions as well. The morphologic similarity of the granules in specific chromaffin cells to the granules in the cardiocytes probably led to adoption of the term "specific granules" for the granules within cardiocytes.

The "specific granules" (A-granules) (5) in mammalian atrial cardiocytes measure about 100–450 mμ and apparently comprise several classes. The A-granules, 100–450 mμ in diameter, appear most concentrated in the sarcoplasm at the paranuclear poles (4, 18), and D-granules, about 60–100 mμ in diameter, appear mainly in the sarcomeric zones (4, 5). Both A- and D-granules appear as membrane-bound circular profiles and contain very electron-dense cores retracted slightly from the inner lamella of the limiting membrane so as to show a narrow clear space. They are considered different here because of their size and distribution (figs. 2–6). It has been shown that both these types of granules may have elongated appearances (2). A unique feature of D-granules is their morphologic resemblance to amine-storing granules in heart axons (4). "B-granule" has been used to designate the pale specific granule (4, 5), and "C-granule" refers to lipofuscin granules, residual bodies, or lysosomal structures (4, 5). "Specific granules" is used loosely, here, to refer to A-, B-, or D-granules.

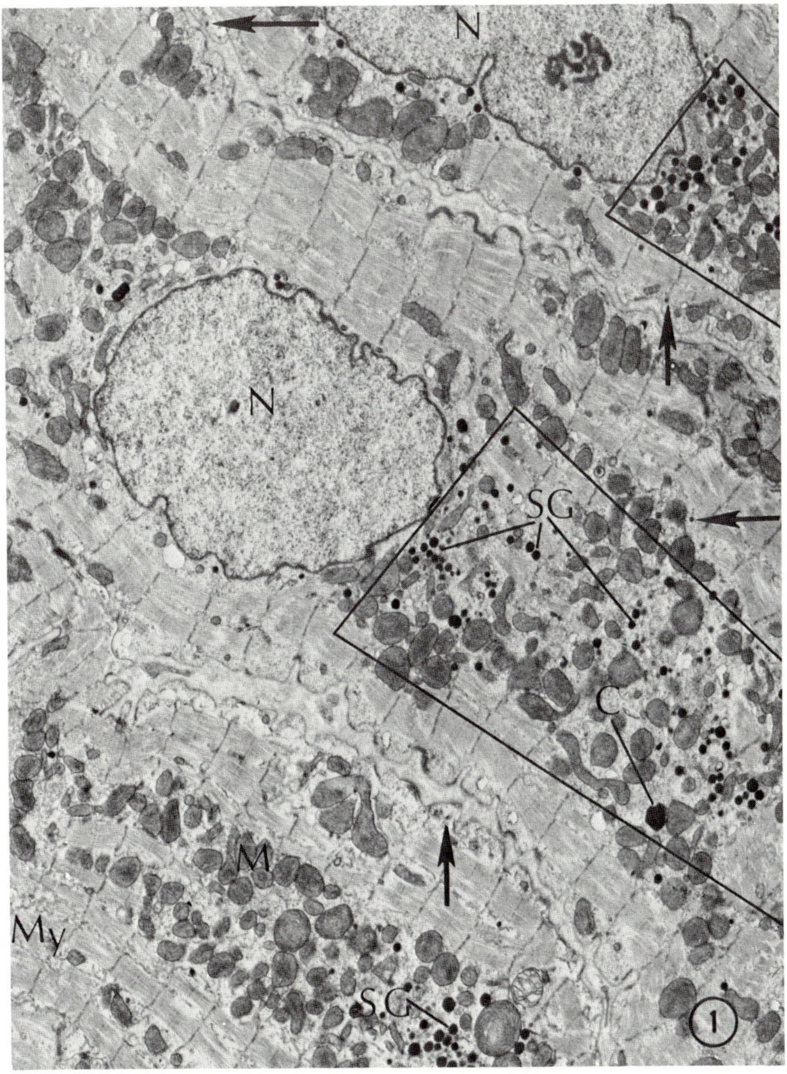

Fig. 1 Portions of several atrial cardiocytes in longitudinal array. Nucleus (N). In the upper cardiocyte nucleus, a nucleolus may be seen. In the sarcoplasm adjacent to the nuclear poles, mitochondria (M) and electron-dense specific granules, termed A-granules, (SG) appear concentrated in large numbers. The sarcomeres (My) fill most of the sarcoplasm. In one paranuclear zone, a portion of a residual body (C), or C-granule, is shown. A substantial population of very small specific granules, termed D-granules in the text, can still be appreciated at this magnification (arrows). ×6,200.

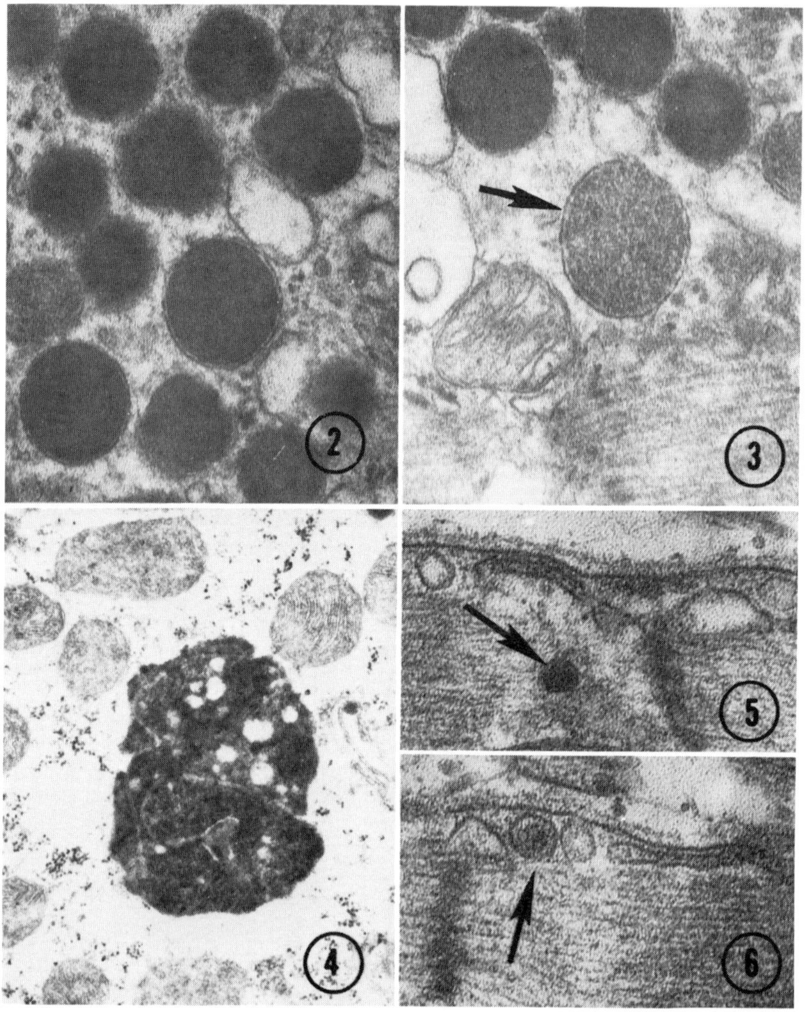

Fig. 2–6 (2) A-granules. Specific granule with homogeneously electron-dense core retracted slightly from the inner lamella of the limiting membrane so as to show a relatively uniform clear space. × 45,000. (3) B-granule (arrow). Specific granule similar in certain respects to the A-granule in figure 2, but containing a pale fibrogranular core. Some A-granules are shown close by. ×45,000. (4) C-granule. This large, irregularly shaped body appears to have a core filled with non-homogeneously distributed electron-dense material, some of which appears finely particulate. There are also several vacuoles, which may have contained lipoid material, in the core of this residual body. ×8,000. (5) D-granule (arrow) appears near a peripheral coupling. Very often, these granules may be found in the subsarcolemmal regions. This D-granule shows an electron-dense core. ×55,000 (6) D-granule (arrow) with a less dense core than the one in figure 5. In this case, the unit membrane and the uniform clear space are evident. ×65,000.

A certain amount of evidence has been interpeted as suggesting that the specific granules are involved in amine storage or binding (2, 23, 27, 30) and, recently, that they may be markedly altered, increased, or decreased in number after the administration of catecholamines and related compounds (24, 26, 31).

The objects of the present studies were: to determine the subcellular profile of norephinephrine distribution in rat atrium as opposed to ventricle; to locate the subcellular fractions containing specific granules, *i.e.*, A-, B-, or D-granules, in atrial homogenates; and to characterize the granules by a cytochemical technique that, at the ultrastructural level, has proved useful to differentiate norepinephrine- from epinephrine-containing granules in the adrenal medulla of the rat.

MATERIALS AND METHODS

Atria from rats were excised simultaneously with an equivalent weight of left ventricle and treated separately through all procedures. They were minced and homogenized in 0.25 M sucrose. Nuclear fractions and debris were removed after slow-speed centrifugation. Postnuclear supernatants were layered over discontinuous sucrose density gradients ranging from 1.50 M at the bottom to 0.90 M at the top. The tubes were centrifuged for 12 h at 17,500 ×G at 4°C. Tube contents were scanned through a spectrophotometer at a fixed wave length of 260 mμ. Fractions obtained from the gradient material were analysed for protein (10, 11) and catecholamines (1). The 25 fractions from each gradient tube were pooled to provide a sufficient amount for catecholamine determinations, as Zone I (fractions 1–15), Zone II (fractions 16–20), Zone III (fractions 21–25).

Verification of fraction contents was achieved after obtaining 100,000 ×G pellets, which were fixed with glutaraldehyde, post-fixed in osmium, dehydrated, and embedded in Epon. For cytochemical studies, selected areas of atria were fixed in 5% glutaraldehyde, buffered to pH 7.4. Tissues were dehydrated and embedded in Epon. Ultra-thin sections were treated with silver methenamine (3, 9). Certain of these silver-treated sections were counterstained with uranyl acetate or lead citrate, or both.

RESULTS

Figure 7 shows the analysis of fractions that were pooled into three zones. In both atrial and ventricular absorbance profiles, three maximal peaks were found, but of different zone distribution.

Electron microscopic examination revealed that Zone I of atrium contained mainly mitochondria (fig. 7, bottom). Specific granules and large vesicles were found at various density levels in Zone II. This latter zone also contained small vesicles with either pale or dense cores, elongated electron-dense vesicles, and elongated clear vesicles. Zone III contained small vesicles with either clear or

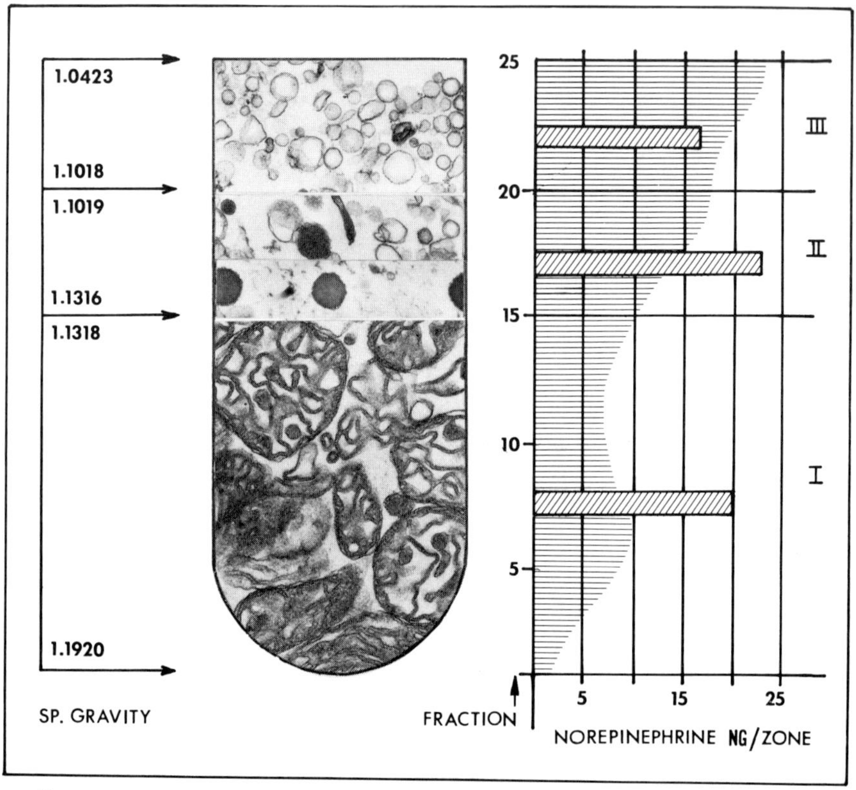

Fig. 7 Subcellular distribution of organelles after density gradient centrifugation of homogenates of rat atria. Zone I contains mitochondria, while the small and larger specific granules and fusiform structures are located between sp. gr. 1.1316 and 1.1019 (Zone II). The upper fraction is constituted mainly by vesicles about 150 mμ in diameter. At right, the horizontal shaded area shows the optical density of the gradient during fractionation. Each peak of absorbance has been correlated with the organelles shown in the photographs at center. The horizontal bars represent total norepinephrine content of each zone. Electron micrographs from lead- and uranyl-stained material. ×26,800.

dense cores. The granule fraction was localized between the lower density limit of 1.1316 and higher density limit of 1.1019, which corresponds to 1.02 and 0.80 M sucrose concentrations, respectively.

The results of catecholamine determinations are given also in figure 7 as total content in the three zones of fractions. Zone II of atrium evidenced the highest norepinephrine content. This finding indicates the presence of a significant amount of norepinephrine-binding material. The corresponding Zone II of ventricle contained comparatively very little norepinephrine.

After silver methenamine, the cores of the A-granules in ultra-thin sections of atrial cardiocytes were strikingly unstained (fig. 8). This negativity has been

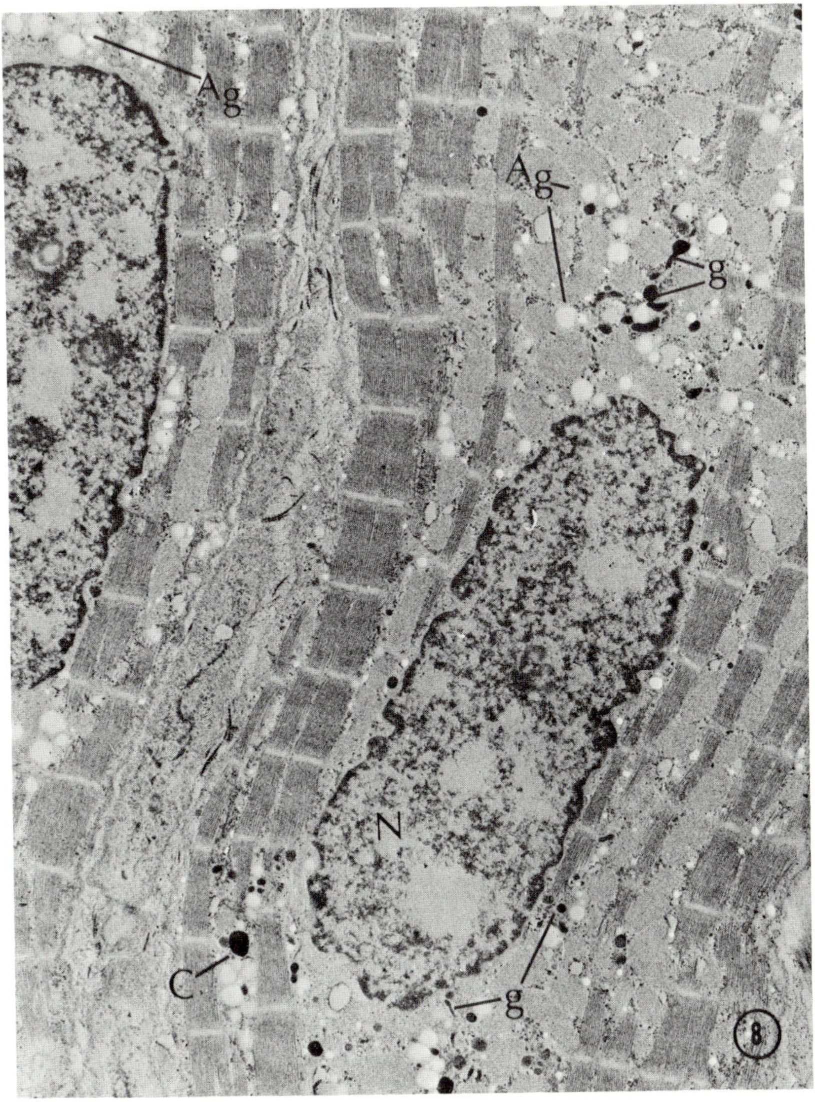

Fig. 8 Portions of two rat atrial cardiocytes treated with silver methenamine (9). Nucleus (N). In the paranuclear zones, the specific granules, termed A-granules (Ag), appear strikingly argentafugic. Many very small granules (g) are seen to be silver-positive. Some other silver-positive granules show elongated profiles; the latter granules are clearly distinguished from the circular profiles of A-granules. The larger silver-positive granules (C) are probably portions of residual bodies or C-granules. ×7,700.

termed "argentafugacity" (3–5), in view of the fact that, despite nonspecific background precipitation of silver on nearly all sarcoplasmic structures, the cores of specific granules appeared strikingly devoid of silver. The possibility that the argentafugic granule cores had been removed by the staining technique prior to silver impregnation was ruled out by counterstaining the section studied with uranyl acetate and lead citrate. After counterstaining, however, the previously very electron-lucent cores of specific granules were found to be electron dense (fig. 9). These cytochemical results showed that cores of specific granules did not all stain similarly to catecholamine granules in adrenal medulla.

A large number of granular structures, some small enough to resemble the D-granules (4) and others elongated, fusiform, or irregularly shaped, did acquire a silver positivity, however. The latter silver-positive granular structures appeared in both the paranuclear and sarcomeric zones and represented a substantial proportion of the total number of granular structures seen (fig. 8). Lipofuscin granules or residual bodies were readily distinguished from specific granules after silver methenamine, since the much larger cores of lipofuscin or residual bodies were silver-positive and appeared very electron dense.

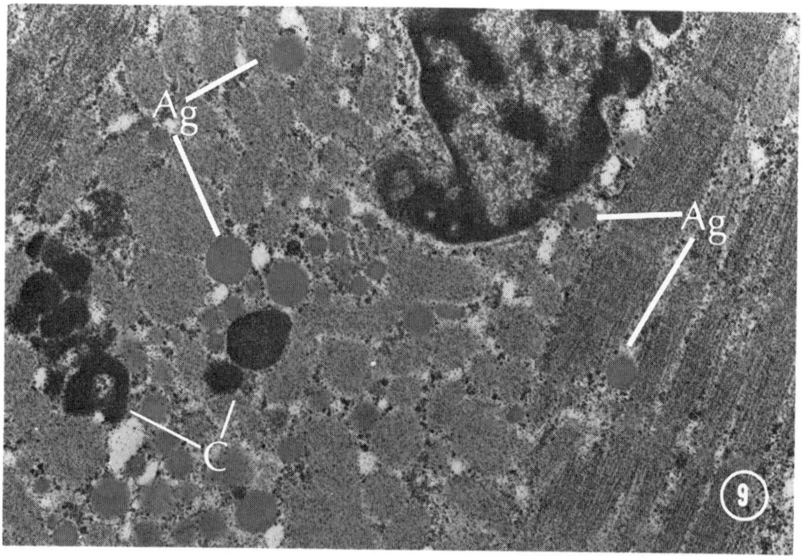

Fig. 9 Rat atrial cardiocyte that had previously been treated with silver methenamine similarly to the cells in figure 8, to demonstrate the silver-negative A-granules (Ag), and the silver-positive residual bodies (C). After silver methenamine, this section was doubly counterstained with uranyl acetate and lead citrate to show that the A-granules appear moderately electron-dense after counterstain. The residual bodies have retained their prominent electron density and may still be distinguished easily from specific granules. ×16,400.

DISCUSSION

Mammalian atrial cardiac muscle cells (cardiocytes) differ from their ventricular counterparts in several ways. One prominent difference is the presence of numerous specific granules mostly concentrated in the zones adjacent to the nuclear poles, but abundant in sarcomeric zones as well. The atrial cardiocytes are generally smaller than the ventricular ones, contain fewer tubules, and appear to be surrounded by considerably more collagen than do ventricular working cardiocytes. Both the small total size of atrium and the amount of collagen were found by us and by others (14) to be major limiting factors in obtaining homogenates easily; and in order to achieve sufficient material for centrifugation, a great many animals had to be used for each experimental run. The use of mammals with a larger atrial mass offers no advantage, since the total number of granules per cardiocyte decreases markedly with the increase in animal size (2, 13). With respect to these considerations, rats were satisfactory subjects.

The catecholamine assays and granule distribution in the subcellular fractions from atria differ strikingly from those from ventricle. Several publications on the subcellular distribution of norepinephrine in the rat heart (22, 28, 29) contain electron micrographs of pellets from fractions stated to be rich in norepinephrine. In these studies, whole heart homogenates were used, so that the subcellular distribution of catecholamines in atrium was considerably influenced by dilution with ventricular material. Although specific granules did not appear in these electron micrographs (22, 28, 29), smaller granules with electron-dense cores, morphologically very similar to the D-granules (4), could be seen.

Zone II of atrium, which contains large and small granules, fusiform and large vesicles, accounted for about 30% of the total norepinephrine content of atrial homogenates, whereas the comparable fraction from ventricle contained about 7.5 times less norepinephrine on a protein unit basis. The amount of norepinephrine found in Zone II shows that there may exist in atrial fractions a norepinephrine-binding material that does not appear to be present in comparable fractions from ventricle. The presence of granular structures in Zone II of atrium, but not in Zone II of ventricle, further suggests that the granules and the norepinephrine content of these fractions may be related (30).

While it is appreciated that after uranyl and lead and after silver treatment the axonal granules are morphologically similar to the D-granules within cardiocytes, it is not possible to be certain which granule is observed in the pellets. For this reason, it was necessary to know whether *in situ* correlations between cardiocyte granules and the norepinephrine content could be made. It is evident from figure 8 that a substantial population of granules in the cardiocytes themselves appeared silver positive. The technique employed is known to be able to distinguish silver-positive norepinephrine granules from epinephrine granules in adrenal medullary cells (9).

The silver method shows that a number of small, round to fusiform, but often elongated granular structures, readily distinguished from silver-negative specific granules and silver-positive residual bodies, constitute a portion of the granular structures from cardiocytes that may be encountered in sedimented zones of atrial homogenates. The A-granules (4, 5), showing circular profiles 100–450 mμ in diameter, were nearly always silver negative. These granules, we feel, do not represent all the granules storing the high catecholamine content of the atrial zones, although these silver-negative granules may be found in such zones. On the basis of cytochemistry, the silver-positive granules constitute more likely candidates for the structures responsible for the norepinephrine-binding protein in atrial homogenates. On the basis of the silver stain, moreover, it appears that a population of small or elongated atrial granules differ in their staining characteristics from specific granules, but in view of their number and morphology they do not appear to be synonymous with residual bodies. While the precise function of these small to elongated silver-positive granules will require further study, their appearance in glutaraldehyde-fixed, osmium post-fixed, uranyl- and lead-stained sections has already been documented in rat atrial cardiocytes (2). After the latter procedures, a number of small to elongated granules appeared very electron-dense, as did the cores of the larger circular specific granules (2). It was concluded that a substantial population of silver-positive granules in the cardiocytes may correspond to granular forms in zones of atrial fractions characterized as containing a high content of catecholamines.

In our studies, separate considerations of atrial fractions showed a distribution of catecholamines that differed substantially from those reported by other investigators using whole heart. In addition, we were able to find, in the catecholamine-rich zones, specific granules of the 100–450 mμ diameter variety, as well as much smaller granules. Although further work will be needed to determine to what extent small granules come from axons or from cardiocytes, their silver-positivity poses a challenging question concerning the origin of certain populations of amine storage granules in atrium.

SUMMARY

Studies undertaken in rats revealed a strikingly higher content of norepinephrine in heavy microsomal fractions of atrial homogenates than in comparable fractions from ventricle. In pellets from atrial fractions with high norepinephrine content, granules were identified by electron microscopy. In comparable ventricular fractions with low norepinephrine content, such granules were not found. A silver method for adrenal norepinephrine- storing granules was applied to ultra-thin sections of atrial cardiac muscle cells. Most specific granules were silver negative *in situ,* but a substantial number of granular structures in the muscle cells appeared silver positive. It is suggested that the silver-positive granules in atrium, both from nerve axons and cardiocytes, may account for the catecholamine binding in atrial fractions.

ACKNOWLEDGMENTS

Acknowledgments are gratefully made to Mrs. E. Losinski for her technical assistance; to *Laboratory Investigation* for permission to reproduce figures, from reference 30, as figure 7 and to reproduce excerpts from the text; to Methods and Achievements in Experimental Pathology for permission to reproduce figures 1, 8, and 9; and to *J. Mol. Cell. Cardiol.* for permission to reproduce figures 2–6.

REFERENCES

1. ANTON, A. H. and SAYRE, D. F.: A study of the factors affecting the aluminum oxide-trihydroxyindole procedure for the analysis of catecholamines. J. Pharmacol. Exp. Ther. *138:* 360–375 (1962).
2. BENCOSME, S. A. and BERGER, J. M.: Specific granules in mammalian and non-mammalian vertebrate cardiocytes, in BAJUSZ, E., and JASMIN, G. Functional Morphology of the Heart, pp. 173–213 (S. Karger AG Basel 1971).
3. BERGER, J. M. and BENCOSME, S. A.: Fine structural cytochemistry of granules in atrial cardiocytes. J. Mol. Cell. Cardiol., *in press.*
4. BERGER, J. M. and RONA, G.: The functional and fine structural heterogeneity of atrial cardiocytes, in BAJUSZ, E., and JASMIN, G. Functional Morphology of the Heart, pp. 540–590 (S. Karger AG Basel 1971).
5. BERGER, J. M. and RONA, G.: Fine structure of extra-nodal transitional cardiocytes in rat left atrium. J. Mol. Cell. Cardiol. *2:* 181–185 (1971).
6. BLOOM, G. D.: The fine structure of cyclostome cardiac muscle cells. Z. Zellforsch. *57:* 213–239 (1962).
7. BLOOM, G.; OSTLUND, E.; von EULER, U. S.; LISHAJKO, F.; RITZEN, M. and ADAMS-RAY, J.: Studies on catecholamine-containing granules of specific cells in cyclostome hearts. Acta physiol. scand. *53:* (Suppl. 185) 1–34 (1961).
8. BOMPIANI, G. D.; ROUILLER, Ch. et HATT, P. Y.: Le tissu de conduction du coeur chez le rat. Etude au microscope électronique. I. Le tronc commun du faisceau de His et les cellules claires de l'oreillette droite. Arch. Mal. Coeur *52:* 1257–1274 (1959).
9. CHANG, W. W. and BENCOSME, S. A.: Selective staining of secretory granules of adrenal medullary cells by silver methenamine: a light and electron microscopic study. Canad. J. Physiol. Pharmacol. *46:* 745–747 (1968).
10. CLELAND, K. W. and SLATER, E. C.: Respiratory granules of heart muscle. Biochem. J. *53:* 548 (1953).
11. GORNALL, A.; BARDAWILL, C. and DAVID, M.: Determination of serum proteins by means of the biuret reaction. J. Biol. Chem. *177:* 751–766 (1949).
12. HIBBS, R. G. and FERRANS, V. J.: An ultrastructural and histochemical study of rat atrial myocardium. Amer. J. Anat. *124:* 251–270 (1969).
13. JAMIESON, J. D. and PALADE, G. E.: Specific granules in atrial muscle cells. J. Cell Biol. *23:* 151–172 (1964).
14. JOHNSSON, G. and SACHS, C.: Subcellular distribution of ^{3}H-noradrenaline in adrenergic nerves of mouse atrium -- effect of reserpine, monoamine oxidase, and tyrosine hydroxylase inhibition. Acta Physiol. Scan. *77:* 344–357 (1969).
15. KISCH, B.: Electron microscopy of the atrium of the heart; guinea pig. Exp. Med. Surg. *14:* 99–112 (1956).
16. KISCH, B.: Further studies about the microbodies in the atria of the heart. Exp. Med. Surg. *23:* 1–12 (1965).
17. LEAK, L. V.: Electron microscopy of cardiac tissue in a primitive vertebrate *Myxine glutinosa.* J. Morph. *128:* 131–157 (1969).

18. MARTINEZ-PALOMO, A.: The fine structure of the heart muscle cells with special reference to specific granules. Master of Science Thesis. Queen's University, Kingston, Ontario (1966).
19. MARTINEZ-PALOMO, A. and BENCOSME, S. A.: Electron microscopic observations on myocardial specific granules and residual bodies in vertebrates. Anat. Rec. *154:* 473 (1966).
20. MARTINEZ-PALOMO, A. and BENCOSME, S. A.: Changes in the content of myocardial specific granules in adrenal regeneration hypertension. An electron microscopic study. Fed. Proc. *25:* 476 (1966).
21. McNUTT, N. S. and FAWCETT, D. W.: The ultrastructure of the cat myocardium. II. Atrial muscle. J. Cell. Biol. *42:* 46–67 (1969).
22. MICHAELSON, I. A.; TAYLOR, P. W., Jr.; RICHARDSON, K. C. and TITUS, E. O.: Uptake and metabolism of dl-norepinephrine by subcellular particles of rat heart. J. Pharmacol. Exp. Ther. *160:* 277–291 (1968).
23. NAYLER, W. G. and MERRILLEES, N. C.: Some observations on the fine structure and metabolic activity of normal and glycerinated ventricular muscle of the toad. J. Cell Biol. *22:* 533–550 (1964).
24. OKAMOTO, H.: An electron microscopic study of the specific granules in the atrial muscle cell upon the administration of agents affecting autonomic nerves. Arch. Histol. Jap. *30:* 467–478 (1969).
25. OSTLUND, E.; BLOOM, G.; ADAMS-RAY, J.; RITZEN, M.; SIEGMAN, M.; NORDENSTAM, H.; LISHAJKO, F. and von EULER, U. S.: Storage and release of catecholamines, and the occurrence of a specific submicroscopic granulation in hearts of cyclostomes. Nature (London) *188:* 324–325 (1960).
26. OTSUKA, N.; OKAMOTO, H. and TOMISAWA, M.: Electron and fluorescence microscopic study of specific granules in rat atrial muscle cells. Arch. Histol. Jap. *30:* 367–374 (1969).
27. PALADE, G. E.: Secretory granules in the atrial myocardium. Anat. Rec. *139:* 262 (1961).
28. POTTER, L. T. and AXELROD, J.: Subcellular localization of catecholamines in tissues of the rat. J. Pharmacol. Exp. Ther. *142:* 291–298 (1963).
29. ROTH, R. H.; STJARNE, L; BLOOM, F. E. and GIARMAN, J. J.: Light and heavy norepinephrine storage particles in the rat heart and in bovine splenic nerve. J. Pharmacol. Exp. Ther. *162:* 203–212 (1968).
30. SOSA-LUCERO, J. C.; DE LA IGLESIA, F. A.; LUMB, G.; BERGER, J. M. and BENCOSME, S. A.: Subcellular distribution of catecholamines and specific granules in rat heart. Lab. Invest. *21:* 19–26 (1969).
31. TOMISAWA, M.: Atrial specific granules in various mammals. Arch. Histol. Jap. *30:* 449–465 (1969).
32. TRILLO, A. and BENCOSME, S. A.: La ultraestructura de las celulas musculáres cardíacas con referencia especial a los gránulos específicos. I. El miocardio de la rana. Arch. Inst. Cardiol. Méx. *35:* 803–817 (1965).
33. TRILLO, A.; MARTINEZ-PALOMO, A. and BENCOSME, S. A.: Estudio ultraestructural del músculo cardíaco en relación a los gránulos específicos. II. Estudio comparativo. Arch. Inst. Cardiol. Méx. *36:* 45–57 (1966).

Mailing address: Warner-Lambert Research Institute, Sheridan Park, Clarkson, Ontario (Canada).

Section of Pathology and the Clinic of Surgery,
National Heart and Lung Institute,
National Institutes of Health, Bethesda, Maryland, USA

CARDIAC PRESERVATION: A MORPHOLOGIC STUDY

V. J. FERRANS, L. M. BUJA, S. LEVITSKY,
W. H. WILLIAMS, C. L. McINTOSH, AND W. C. ROBERTS

The methods thus far proposed for extended preservation of the heart have led to less than optimal results in terms of the duration of the period of preservation, the maintenance of the structural integrity of the myocardium, and the ability of preserved hearts to resume proper function. Histologic and electron-microscopic studies have shown that extra- and intracellular edema, dilatation of the sarcoplasmic reticulum, loss of glycogen, and mitochondrial alterations are the main anatomic changes in hearts preserved by *in vitro* perfusion (3, 4, 33). However, detailed studies of correlations between functional and structural changes have not been made in preserved hearts.

This communication reviews our morphologic studies (8, 12) on canine hearts preserved for up to 24 h by the methods described by Levitsky and his associates (24) and summarizes structural-functional correlations based on studies of cardiac function performed on these hearts.

MATERIALS AND METHODS

Twenty-three mongrel dogs, each weighing 18–24 kg, were used. The animals were free of microfilariae, and had normal blood erythrocyte, leukocyte and platelet counts, and erythrocyte sedimentation rates. Each animal was anesthetized with 30 mg/kg of pentobarbital. A thoracotomy was made and the heart was cooled to 28°C by infusing the coronary arteries with an iced balanced electrolyte solution (Normasol, Abbott Laboratories) until clear fluid issued from the coronary sinus. The heart was then excised, irrigated with the cooled electrolyte solution, and subjected to one of the procedures listed below.

Preserved Hearts (20 dogs)

Each heart was transferred directly to a silicone rubber, air-free preservation chamber[1] where the aortic root was perfused at 8–10°C with a 0.4 m^2 Kolobow membrane oxygenator (23), which also functioned as a pulsatile pump.

[1] Constructed by Messrs. Charles H. Haynes and Charles J. Burda, Medical Products Division, Dow Corning Corporation, Midland, Michigan 48640.

Fourteen hearts were perfused with filtered plasma: 1 heart was perfused for 3 h, 2 for 14 h, 5 for 18 h, and 6 for 24 h. For the preparation of filtered plasma, canine whole blood was collected in plastic bags with A.C.D. solution (Plasmapheresis double bag packs, Fenwal Laboratories, Morton Grove, Ill.). The plasma was then separated by centrifugation, placed in satellite plastic bags, and stored in the cold for not more than 7 days before freezing, rewarming and filtration, according to the method of Belzer *et al.* (2). The osmolarity of filtered plasma was 256 milliosmols.

The remaining 6 hearts were perfused for 18 h with filtered plasma to which dextran (molecular weight 60,000–90,000) was added to a final concentration of 4 g/100 ml. The osmolarity of this solution was 316 milliosmols.

At the end of the period of preservation, the hearts were removed from the chamber, mounted on a perfusion stand, and rewarmed by perfusion with warm, oxygenated blood from a support animal. Cardiac function was then measured by an isovolumetric balloon technique (25).

Control Hearts (3 dogs)

Control hearts were rewarmed immediately after cooling to 28°C, and mounted on the perfusion stand for functional studies, which were completed within 30 min of excision.

Myocardial biopsies

Biopsies of the right ventricular wall were obtained from each dog before cooling and excision of the heart (control biopsies) and 30 min after rewarming. Part of each biopsy was fixed with buffered 10% formalin and embedded in paraffin. Sections were stained with hematoxylin and eosin, phosphotungstic acid-hematoxylin, and the periodic acid-Schiff method with and without prior digestion of glycogen by amylase.

The remaining portion of each biopsy was fixed with cold 3% glutaraldehyde in 0.1 M phosphate buffer, pH 7.2, and washed with several changes of 5% sucrose in 0.1 M phosphate buffer, pH 7.2. The tissue was diced into 0.5 mm cubes, which were postfixed for 2 h with cold 1% osmium tetroxide in Millonig's phosphate buffer (27), dehydrated in a graded series of alcohols and propylene oxide, and embedded in Maraglas according to Freeman and Spurlock (15). Semifine sections (1 μ thick) of the Maraglas-embedded tissues were stained with alkaline toluidine blue and examined with the light microscope. Ultrathin sections were stained with lead citrate and uranyl acetate and studied with an RCA EMU 3G electron microscope.

RESULTS

Light microscopy

Light microscopic examination was useful in determining the presence and extent of edema and extravasation of blood (table I). Study of the semifine sections also helped in evaluating areas of artifactual tissue damage produced during fixation and preparation of the tissues for electron mircoscopy.

Table I

Histology and Ultrastructure of Control and Preserved Hearts

Heart no.	Pres. time	E	H	G	MF dam.	Swell. SR	M dam.	MC Ang.	dp/dt
Control hearts*									
1	0	0	1	0	0	1	1	1	1359
2	0	0	0	0	0	2	1	1	–
3	0	1	1	0	0	1	1	0	1306
Hearts preserved with filtered plasma									
4	3	0	0	0	0	1	1	2	1109
5	14	2	1	0	0	0	0	1	976
6	14	2	0	0	0	0	0	1	1276
7	18	3	3	2	2	3	2	0	†
8	18	2	2	0	0	2	2	1	842
9	18	1	0	0	0	1	1	3	2131
10	18	2	1	0	0	1	1	2	1624
11	18	1	0	0	0	0	1	1	1432
12	24	1	0	0	0	0	0	2	1000
13	24	2	0	0	0	1	2	0	951
14	24	2	0	1	0	1	3	1	†
15	24	1	1	0	0	2	2	0	579
16	24	2	2	0	0	0	0	1	1398
17	24	2	2	0	0	2	1	1	571
Hearts preserved with filtered plasma-dextran									
18	18	0	0	0	0	0	0	1‡	783
19	18	0	0	0	0	1	2	1‡	1213
20	18	0	0	0	0	0	0	0‡	876
21	18	0	0	0	0	0	1	1‡	1462
22	18	0	0	0	0	1	0	1	1404
23	18	0	0	0	0	0	0	1	928

Grading of changes: 0 = absent, 1 = minimal and focal, 2 = moderate, 3 = severe and extensive. Abbreviations: E = extracellular edema, H = hemorrhage, G = glycogen loss, MF dam. = myofibrillar damage, Swell. SR = swelling of sarcoplasmic reticulum, M dam. = mitochondrial damage, MC Ang. = angulations of mitochondrial cristae.

* In 9 control hearts, 7 of which were not studied morphologically, dp/dt values were 1310 = 109.

† These 2 hearts could not be defibrillated.

‡ These 4 hearts also showed stacked cristae and intramitochondrial concentric lamellar bodies.

Electron microscopy

Control biopsies

Electron microscopic examination of control biopsies showed the general characteristics of normal canine myocardium described by Bahr and Jennings (1). A review of these features is presented below, along with a description of artifacts and variations of normal that were encountered frequently and that sometimes gave rise to questions of interpretation.

The muscle cells were closely packed. Small interstitial spaces contained capillaries, neural elements, small bundles of collagen fibrils, and various types of connective tissue cells, *i.e.*, pericytes, fibroblasts, and occasional histiocytes. The muscle cells contained single, centrally located nuclei, surrounded by myofibrils and mitochondria (fig. 1). The inner and outer nuclear membranes were closely apposed and showed indentations, the extent of which was related to the state of contraction of the cell. The nuclear chromatin was distributed throughout the nucleus, with slight tendency to margination. A single nucleolus was found in the center of the nucleus (fig. 1). The myofibrils were composed of thick and thin filaments arranged in the characteristic A, I, and Z bands of striated muscle cells (fig. 2). They extended around the sides of the nucleus, leaving conical spaces at each nuclear pole. These spaces contained the Golgi apparatus, lysosomes, lipofuscin granules, numerous glycogen particles, and clusters of mitochondria (fig. 3). In some areas, the myofibrils showed severe degrees of contractions, with marked shortening of sarcomeres and contraction bands (fig. 4). These changes were most pronounced in peripheral areas of the biopsies (fig. 5) and were often accompanied by dislocation of the myofilaments, severe deformation of the nuclei, dilatation of the sarcoplasmic reticulum, mitochondrial damage, and displacement of the nucleus and mitochondria towards areas of less severely contracted myofibrils. Some muscle cells were ruptured, and extruded mitochondria and glycogen granules were present in the extracellular spaces.

The mitochondria contained many closely packed cristae, formed by numerous infoldings of the inner mitochondrial membranes, and a few electron-dense granules. The structure of the cristae showed considerable variation. In many mitochondria they were straight and parallel, while in others they followed a zig-zag course with periodic angulations (fig. 6) similar to those described as "energized-twisted" configurations (11, 19, 30); in a few mitochondria, they formed polygonal profiles that completely surrounded small areas of mitochondrial matrix. Areas of mitochondrial damage, with swelling of the mitochondria and fragmentation or loss of the cristae, were frequently associated with contraction bands (fig. 5). A few isolated damaged mitochondria were found in cells that had no contraction bands or other evidence of damage. Elongated mitochondria that extended for more than one sarcomere length, and

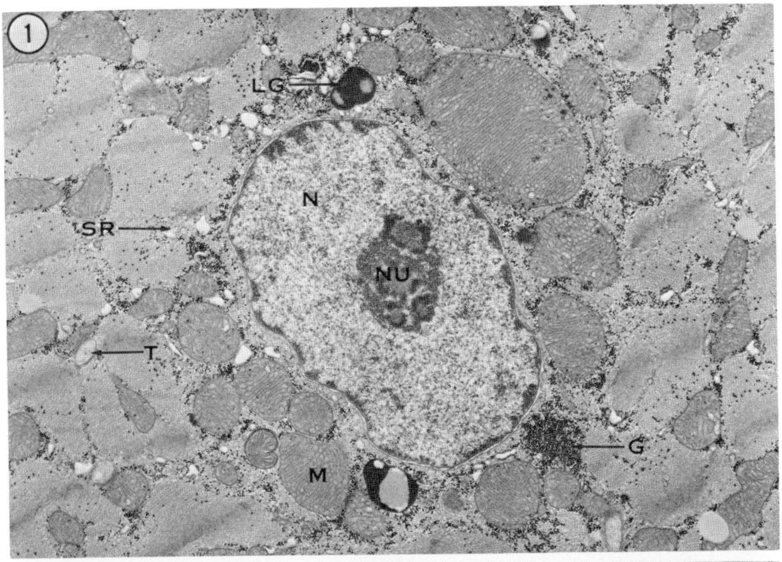

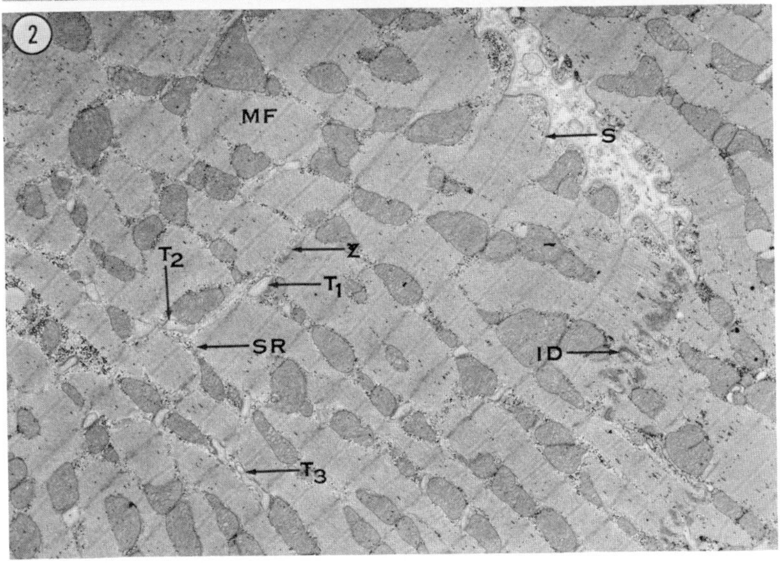

Fig. 1 Control biopsy. Central area of a cardiac muscle cell showing nucleus (N), nucleolus (NU), mitochondria (M), sarcoplasmic reticulum (SR), T tubules (T), lipofuscin granules (LG), and glycogen particles (G). ×7,250.

Fig. 2 Control biopsy. Longitudinal section through portions of two muscle cells showing scalloping of the sarcolemma (S), an intercalated disc (ID), contracted myofibrils (MF), the sarcoplasmic reticulum (SR), and the T tubules. The T tubules generally are located transversely (T_1) at the levels of the Z lines (Z), but some course obliquely (T_2) or longitudinally (T_3) between adjacent Z lines. × 5,200.

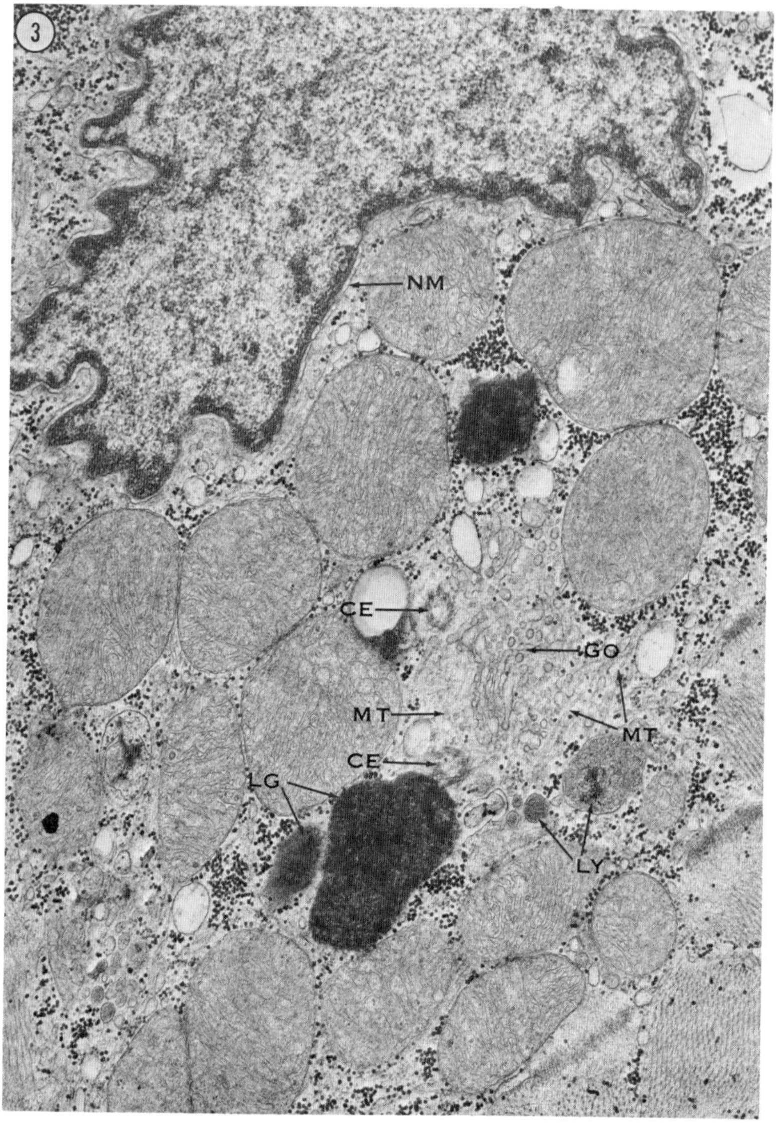

Fig. 3 Perinuclear area of a muscle cell from a control biopsy, showing indentations in the nuclear membrane (NM), clusters of mitochondria, several lysosomes (LY), the Golgi apparatus (GO), two centrioles (CE), lipofuscin granules (LG), and microtubules (MT). × 21,200.

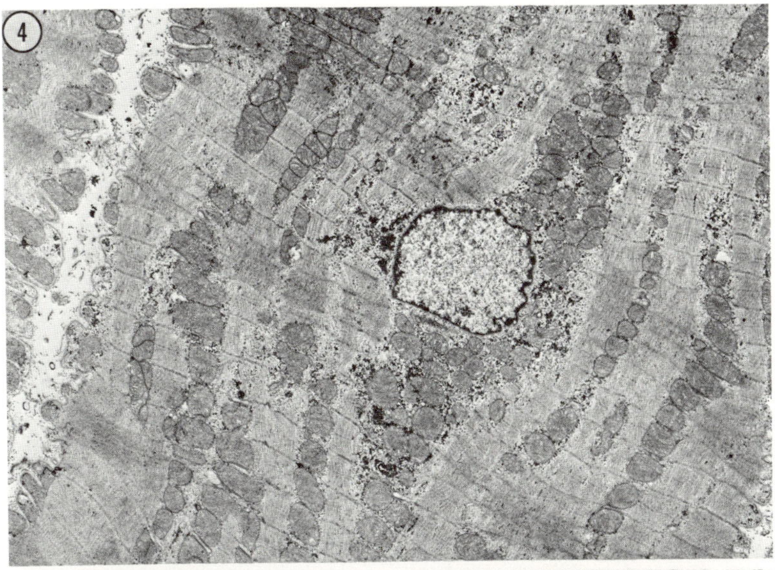

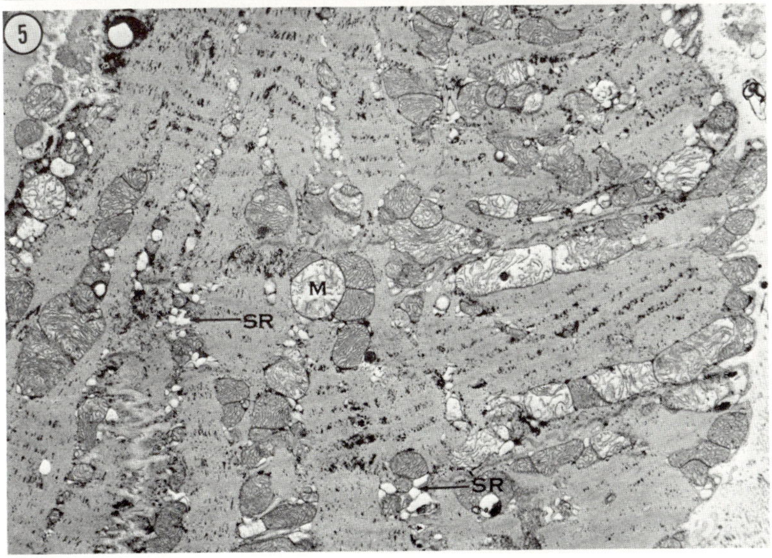

Fig. 4 Low power view of peripheral area of a control biopsy. Contraction bands are evident in the muscle cells. × 5,800.

Fig. 5 Peripheral area of control biopsy, showing severe contraction bands, damaged mitochondria (M) and dilated sarcoplasmic reticulum (SR). × 5,800.

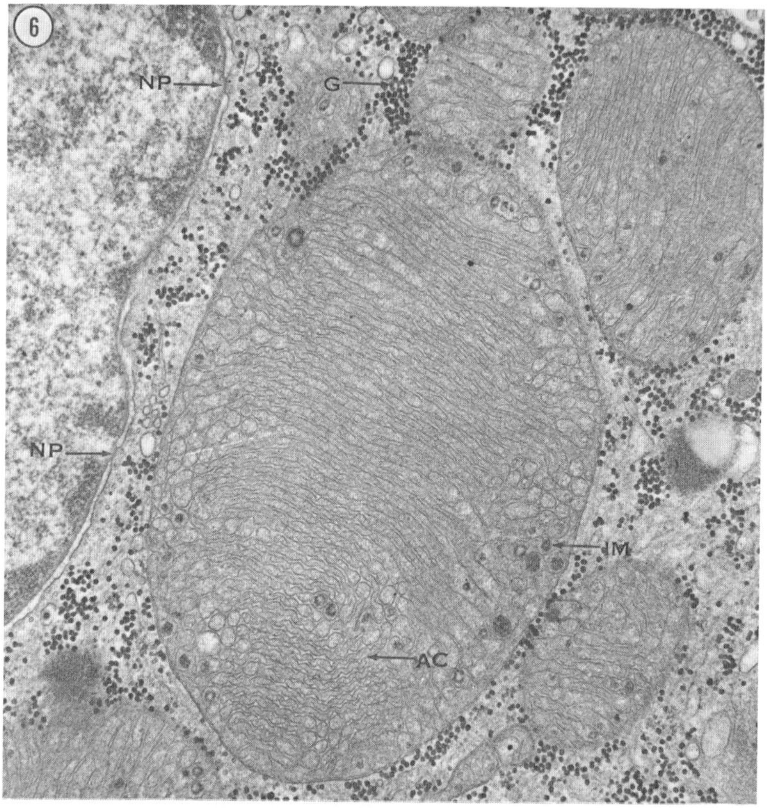

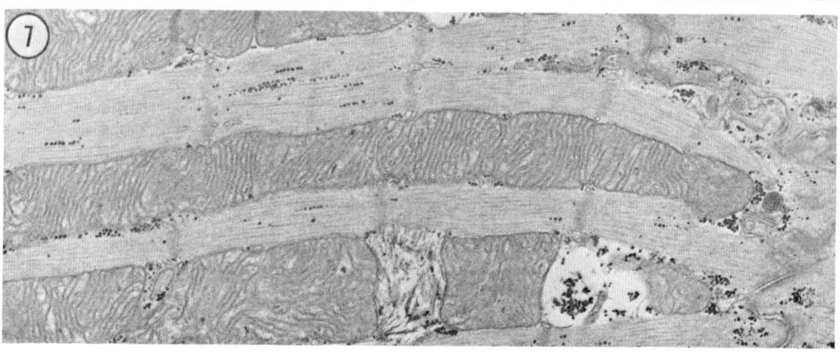

Fig. 6 High power view of mitochondria from control biopsy, showing angulations in the cristae (AC). (NP) = nuclear pore; (G) = glycogen; (IM) = intramitochondrial granules. × 26,340.

Fig. 7 Area of control biopsy, showing a greatly elongated mitochondrion that extends for the length of 4 sarcomeres. × 13,600.

contained cristae oriented in more than one plane, were also found at times in otherwise normal cells (fig. 7).

The Golgi apparatus was small and inconspicuous, and consisted of a few vesicles and groups of flattened cisterns located in the perinuclear area (fig. 3). The number of lysosomes and lipofuscin granules varied considerably from one cell to another. The lysosomes were small (0.5–2 μ in diameter) and composed of a homogeneous or finely granular matrix surrounded by single membranes (fig. 3). The lipofuscin granules consisted of aggregates of coarse, osmiophilic particles in which irregularly arranged lamellae were sometimes discernible. Some lipofuscin granules were surrounded by single membranes; others were free in the sarcoplasm.

The sarcolemma was formed by the plasma membrane and the external lamina (glycocalyx, fuzzy coat, basement membrane) of the muscle cells (fig. 2). The degree of scalloping of the sarcolemma was related to the state of contraction of the cell. Sarcolemmal invaginations were consistently present at the level of the Z bands. As in other mammalian hearts, these tubular invaginations extended deeply into the cell, forming the transverse tubular system (T system) (figs. 2, 8, and 9). The luminal surfaces of the T tubules were invested by an extension of the external lamina of the sarcolemma, with which it was continuous. This investment, the transverse orientation of the T tubules, their relationship to the Z bands, and their larger size have served as criteria to differentiate the tubules of the T system from those of the sarcoplasmic reticulum. However, recent histochemical studies on rat myocardium have shown that the T system possesses longitudinally and diagonally oriented branches, many of which lack an external lamina and are similar in size to the tubules of the sarcoplasmic reticulum (9, 10, 13, 14). Although no histochemical techniques were used in this study, these branches of the T system were identified occasionally in our material (fig. 9). Nevertheless, unequivocal differentiation between the tubules of the sarcoplasmic reticulum and the smaller branches of the T system often proved impossible.

The sarcoplasmic reticulum (fig. 8) was composed of a complex meshwork of branching and anastomosing tubules (longitudinal elements) that surrounded the myofibrils and formed specialized dilatations (terminal cisterns or transverse elements) at the levels of the Z bands. These terminal cisterns were closely apposed to the T tubules, with which they formed dyads and triads, thus providing points of close approximation of the sarcoplasmic reticulum to the extracellular environment. Similar specialized dilatations (subsarcolemmal cisterns) were present immediately adjacent to the sarcolemma. The intercalated discs (fig. 2), formed by the close apposition of the plasma membranes of adjacent cells, were similar in structure to those of other mammalian hearts (7). Most cells were apposed end-to-end, but side-to-side apposition was occasionally

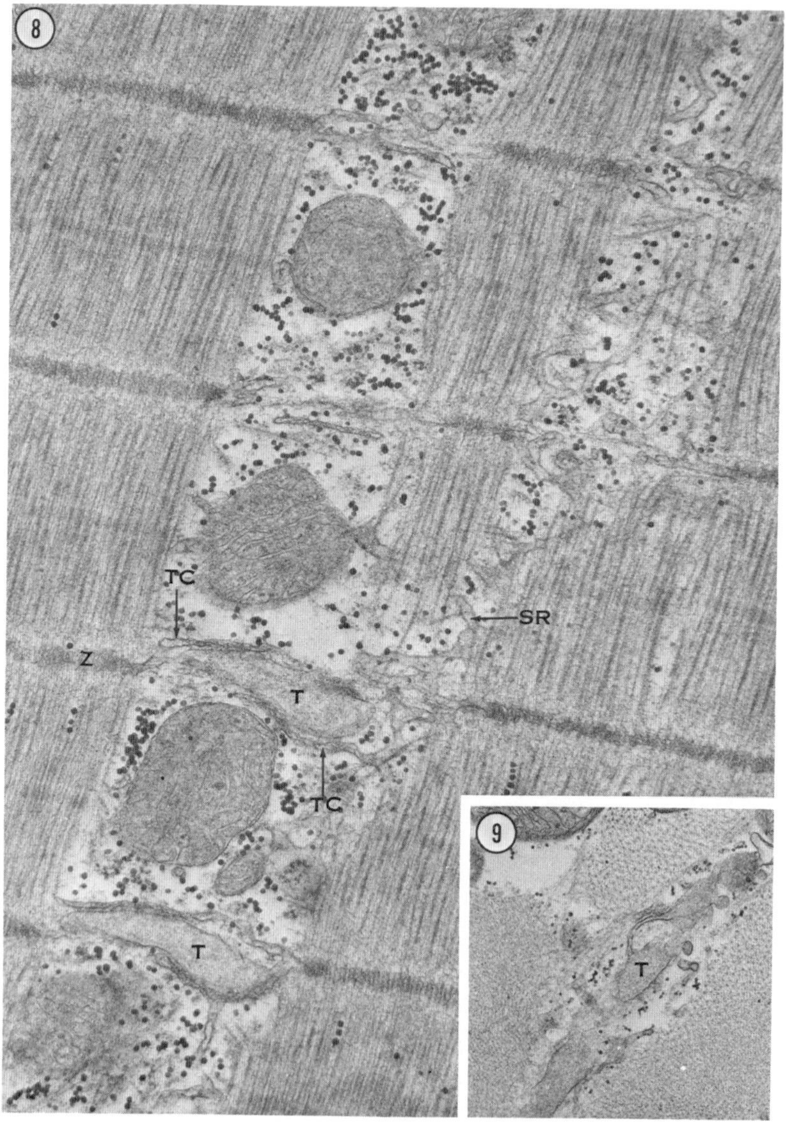

Fig. 8 Longitudinal section through muscle cell of a control biopsy, showing the branching and the anastomosing tubules of the sarcoplasmic reticulum (SR), and their relationship to the T tubules (T), with which the terminal cisterns (TC) of the sarcoplasmic reticulum form triads at the level of the Z bands (Z). × 33,740.

Fig. 9 Control biopsy, showing area of branching of a T tubule (T) and marked narrowing of its lumen at the point of branching. × 21,200.

observed. Bizarre distortions of the intercalated discs were often present in extremely contracted muscle cells.

Rough-surfaced sarcoplasmic reticulum, microtubules and centrioles (fig. 3) were rarely observed. The size of the glycogen granules (fig. 6), measuring from 250 to 350 Å in diameter, differentiated them from ribosomes, which measured 150 Å in diameter and were present only in very small numbers. The glycogen granules were located in the perinuclear areas, interfibrillary spaces, and on each side of the Z bands. Glycogen granules were often concentrated on one side of the cell, in an example of "glycogen flight," which is related to displacement of glycogen during the process of penetration of the fixative into the tissue. Glycogen granules were occasionally found within the lumen of tubules of the sarcoplasmic reticulum and T system. Localization here was probably also artifactual. Thus, the arrangement of glycogen particles observed in ultrathin sections of cardiac muscle cells, particularly after fixation by immersion, does not reflect exactly its *in vivo* distribution in the cell.

Biopsies taken after cardiac preservation

The electron-microscopic and dp/dt data are summarized in table I. Individual variations in morphology and function were observed in each group of preserved hearts. The morphologic findings consisted primarily of interstitial edema, mitochondrial damage, various other types of changes in mitochondrial structure, and dilatation of the sarcoplasmic reticulum.

Interstitial edema. Some degree of interstitial edema occurred in each heart preserved with filtered plasma for 14 h or longer. The weight of these hearts was increased by up to 5% at the end of the preservation period. No interstitial edema and no increase in cardiac weight were observed in hearts preserved with the plasma-dextran mixture.

By electron microscopy, interstitial edema was characterized by widened intercellular spaces. In some areas, the intercellular spaces appeared optically empty (fig. 10); in others, they contained an amorphous material similar to that within the capillaries. Edema of the heart was also readily recognizable on gross examination.

Intracellular edema. Light microscopy was helpful only in detecting severe intracellular edema, which was manifested by separation of myofibrils and a watery-appearing sarcoplasm. These changes were obvious by electron microscopy. They were observed in only one heart in which an abnormally high perfusion pressure developed during the course of preservation for 18 h with filtered plasma (fig. 10). This heart, which also showed extensive capillary damage (fig. 11) and extravasation of erythrocytes, could not be defibrillated. Ultrastructural changes in capillaries of other preserved hearts were observed infrequently and consisted of a decrease in thickness of capillary walls and, less often, of rarefaction of the cytoplasm of endothelial cells (fig. 12).

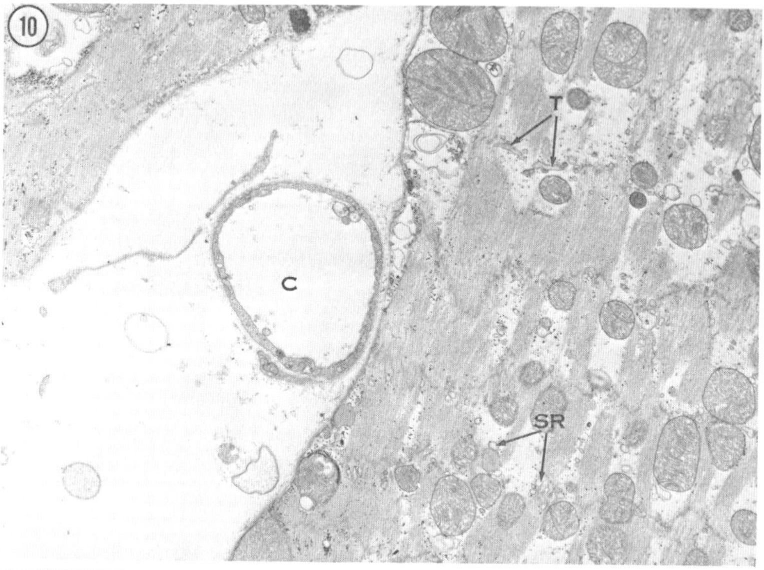

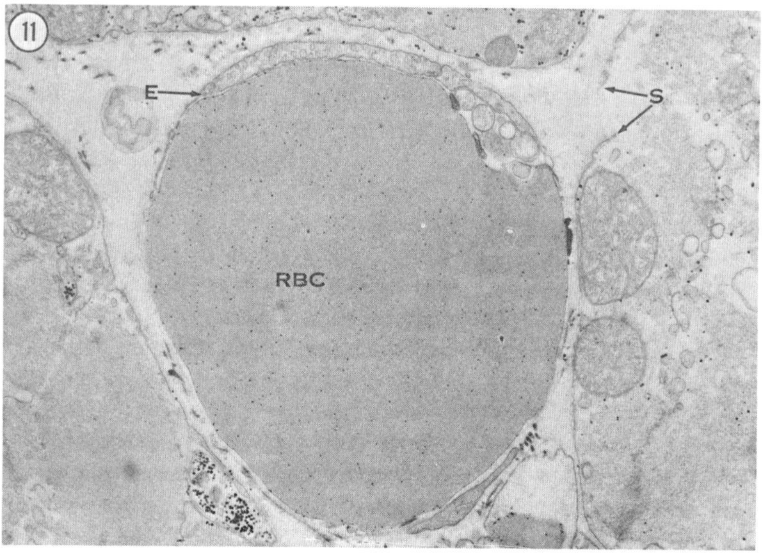

Fig. 10 Area of marked inter- and intracellular edema in heart No. 7, which could not be defibrillated after perfusion with filtered plasma for 18 h. The interstitial space is widened; the capillary (C) is thin-walled and dilated; the myofibrils are widely separated and partially lysed; the sarcoplasmic reticulum (SR) is vesiculated and fragmented; some T tubules (T) are recognizable. × 5,650.

Fig. 11 Capillary damage in same heart shown in figure 10. The capillary endothelium (E) surrounding the red blood cell (RBC) is severely disrupted, and the sarcolemma (S) of an edematous muscle cell is fragmented. × 13,360.

Less severe degrees of intracellular edema were recognized only by electron microscopy and were manifested by swelling of the mitochondria and dilatation of the sarcoplasmic reticulum. It should be emphasized, however, that these two alterations are not necessarily consequences of extracellular edema, since they may result from a variety of other causes.

Mitochondrial alterations. Changes observed in the mitochondria of preserved hearts were mitochondrial damage, stacking of cristae with formation of intramitochondrial concentric lamellar bodies, and angulations of the mitochondrial cristae. As shown in figures 13-20, considerable variation occurred in the extent to which individual mitochondria in a cell showed these changes.

Mitochondrial damage (figs. 13 and 14) was defined as consisting of mitochondrial swelling and loss of cristae. Minimal, focal mitochondrial damage was found in each non-preserved control heart. Of 14 hearts perfused with filtered plasma, this change was absent in 5, minimal in 5, moderate in 3, and severe in 1. The heart that showed severe, selective mitochondrial damage could not be defibrillated after preservation. In the 6 hearts perfused with filtered plasma-dextran, mitochondrial damage was more focal in distribution and was considered moderate in 1, minimal in 1, and practically absent in the other 4. However, the mitochondria in 4 of the 6 hearts preserved with plasma and dextran showed a different type of alteration (figs. 15, 17, and 18), which was observed very uncommonly in hearts perfused with filtered plasma. This change consisted of intramitochondrial concentric lamellae (fig. 17) that were strongly osmiophilic, showed a periodicity of 55 Å and were associated with stacks of closely apposed cristae in the central areas of the mitochondria (fig. 18). Mitochondria that showed this change were seldom swollen; in fact, they usually appeared smaller than normal and their outer membranes were often wrinkled. Concentric lamellae similar to those present in mitochondria were seen occasionally within vesicles that could have been derived either from mitochondria or from dilated elements of sarcoplasmic reticulum.

The data on hearts preserved with filtered plasma suggested a correlation between marked degrees of angulation of the mitochondrial cristae and high dp/dt values. In hearts preserved for 18 h, the dp/dt values obtained in the group perfused with filtered plasma-dextran (1111 ± 248) were not as high as in the group perfused with filtered plasma (1286 ± 547). In hearts perfused with filtered plasma-dextran, the angulations of the cristae were not prominent in biopsies taken after preservation and occurred more frequently in control biopsies.

In both groups of preserved hearts, the electron-dense intramitochondrial granules were present in all areas except in those with mitochondrial damage. Mitochondrial fusion was observed to the same extent in control biopsies and in biopsies taken at the end of the period of preservation.

Swelling of the sarcoplasmic reticulum. Swelling of the sarcoplasmic reticulum

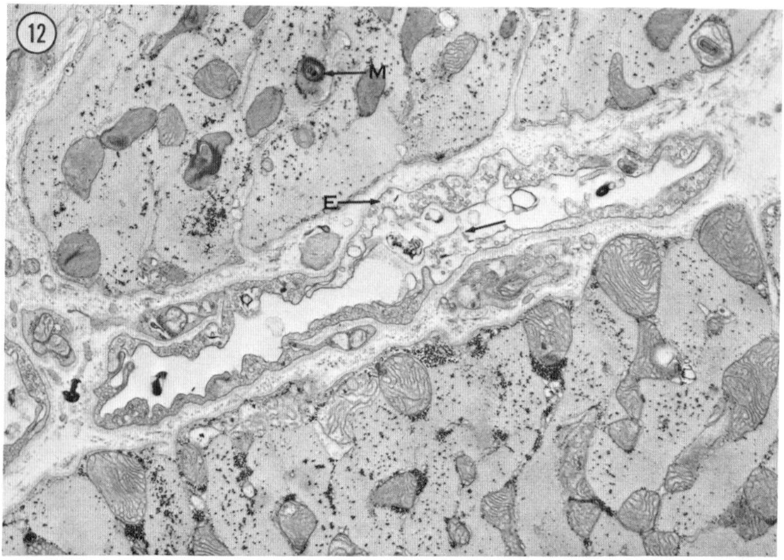

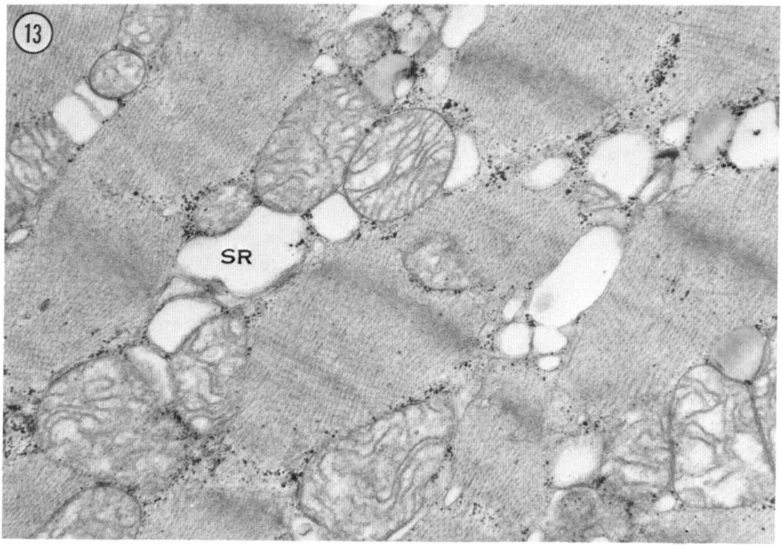

Fig. 12 Area of focal damage in capillary from heart No. 19 after preservation for 18 h with filtered plasma-dextran. The capillary is composed of several endothelial cells, one of which (E) shows cytoplasmic rarefaction and fragmentation (arrows) of its plasma membrane. Concentric lamellae are present on some mitochondria (M) and glycogen is abundant in the muscle cells. × 7,640.

Fig. 13 Marked swelling of the sarcoplasmic reticulum (SR) and moderate mitochondrial damage in muscle cell from heart No. 15, which functioned poorly after 24 h of preservation with filtered plasma. × 18,800.

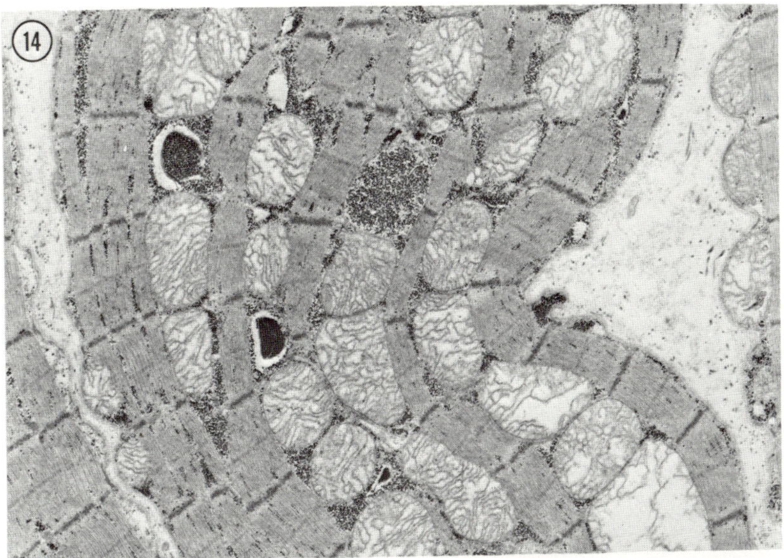

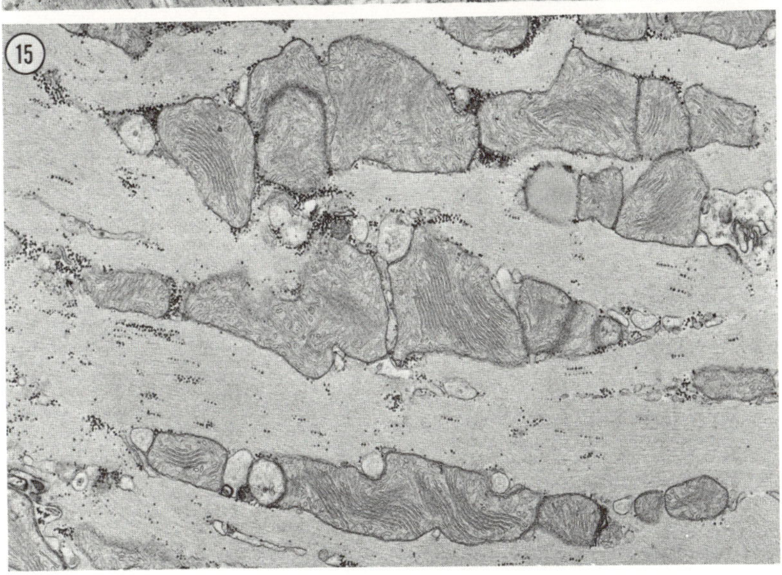

Fig. 14 Severe, diffuse mitochondrial damage in muscle cell from heart No. 14, which could not be defibrillated after preservation for 24 h with filtered plasma. The myofibrils are normal and glycogen granules are abundant. × 9,130.

Fig 15 Mitochondria in muscle cell from heart No. 19 after perfusion for 18 h with filtered plasma-dextran. The cristae form stacks in the central areas. × 13,240.

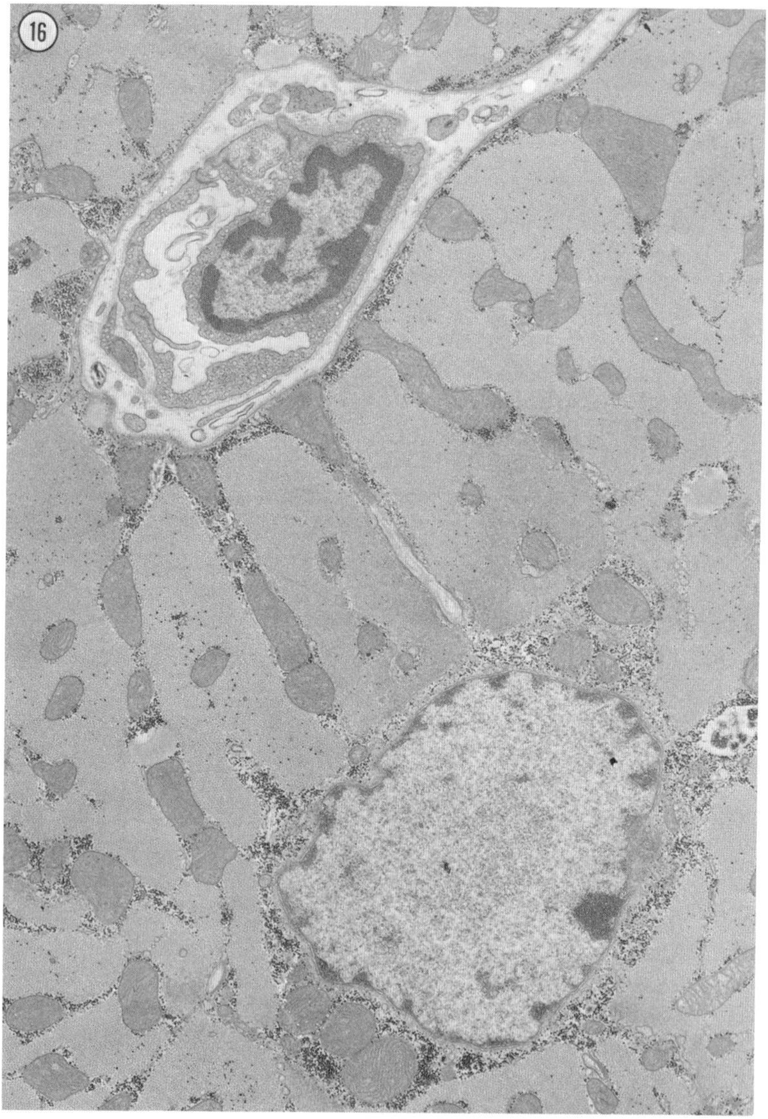

Fig. 16 Completely normal ultrastructure in capillary and two adjacent muscle cells from heart No. 23 after perfusion for 18 h with filtered plasma-dextran. × 7,325.

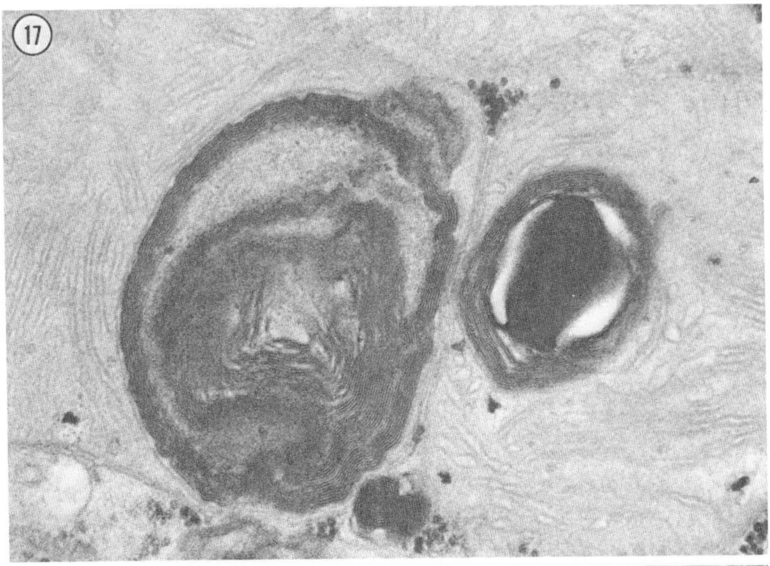

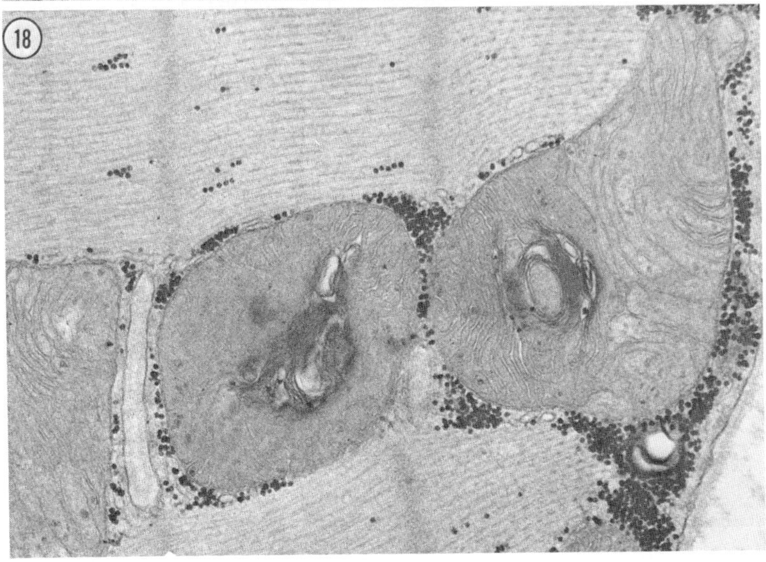

Fig. 17 Intramitochondrial concentric lamellae in heart No. 20, perfused for 18 h with filtered plasma-dextran. × 60,880.

Fig. 18 View of mitochondria from same heart seen in figure 17, showing the derivation of concentric lamellae from mitochondrial cristae. × 21,830.

was characterized (figs. 13, 19, and 20) by dilatation of the tubules and cisterns of which this compartment is composed. These structures, which normally appear as narrow channels, formed spherical or elongated vesicles scattered throughout the cell. Thus, the swollen sarcoplasmic reticulum occupied a considerably greater than normal fraction of the total cell volume. As shown in table I, swelling of the sarcoplasmic reticulum was present (severe in 1, moderate in 3, and minimal in 5) in 9 of 14 hearts preserved with filtered plasma. This change was observed to a lesser extent in 2 of 6 hearts preserved with the plasma-dextran solution; the sarcoplasmic reticulum appeared normal in the other 4 hearts.

Changes in other structures of the muscle cells. Considerable amounts of glycogen were present in all preserved hearts, except in one that could not be defibrillated at the end of the preservation period. This was the only heart (dog no. 7) in which the myofibrils and intercalated discs were damaged. In all other hearts (fig. 10), the myofibrils and discs appeared normal. No accumulation of lipid droplets or changes in Golgi apparatus or lysosomes were observed in preserved hearts.

Nuclear alterations were uncommon in preserved hearts; when present, they consisted of clumping of chromatin and of separation of the outer and inner nuclear membranes. This separation left a widened space between the nuclear membranes that was continuous with the lumen of the sarcoplasmic reticulum. Marked dilatation of the sarcoplasmic reticulum was present in cells that showed this change.

The sarcolemma was altered in a few hearts preserved without dextran, in which the external lamina was frayed in areas of marked interstitial edema. Dilatation of the T tubules was observed occasionally in preserved hearts.

DISCUSSION

Several studies have shown that the main morphologic changes that develop in myocardium after elective cardiac arrest are mitochondrial damage, swelling of the sarcoplasmic reticulum, clumping of nuclear chromatin, loss of glycogen, and progressive lysis of myofibrils (5, 22, 26). These changes are qualitatively similar to those of postmortem autolysis (20), but progress faster. Protection against these changes is provided by hypothermia, and particularly by combined hypothermia and coronary perfusion with oxygenated nutrient solutions (28, 34).

Limited data are available on the morphology of preserved hearts. Wedell *et al.* (35) showed that mitochondrial ultrastructure and oxidative phosphorylation remained largely normal in hearts preserved without perfusion for 16 h at 5°C. Taheri *et al.* (33) preserved hearts up to 24 h by perfusion at 4°C, with and without hyperbaric oxygenation. They observed marked inter- and intracellular

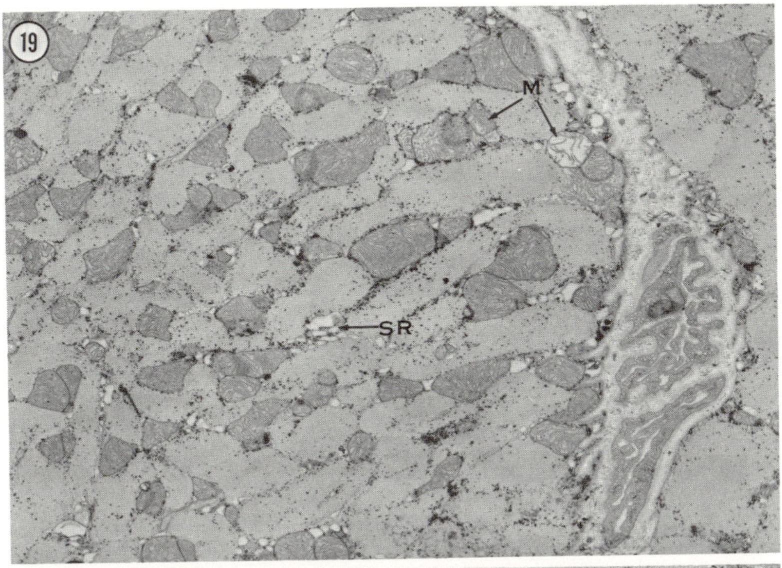

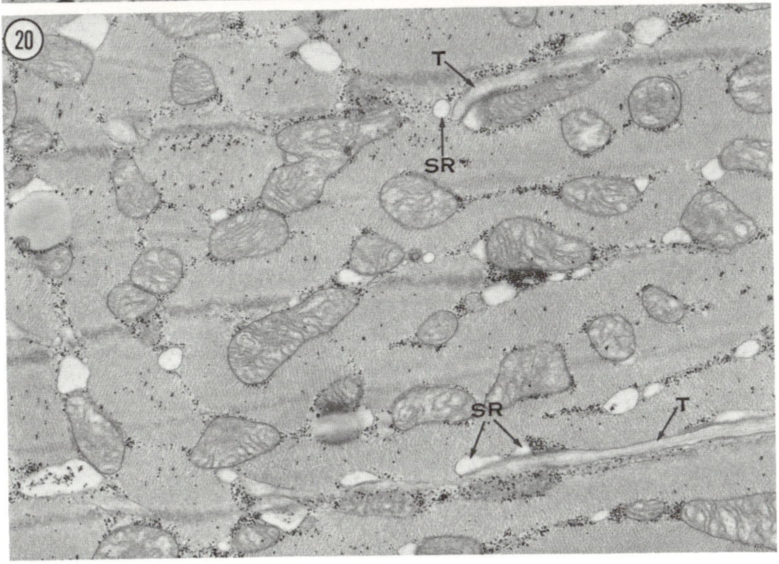

Fig. 19 Minimal dilatation of the sarcoplasmic reticulum (SR) in heart No. 22 after preservation for 18 h with filtered plasma-dextran; 2 mitochondria (M) show damage, while other mitochondria are normal. × 6,060.

Fig. 20 Minimal mitochondrial damage and moderate swelling of the sarcoplasmic reticulum in heart No. 9 after preservation for 18 h with filtered plasma. Dilated elements of the sarcoplasmic reticulum (SR) are immediately adjacent to the t tubules (T). × 10,040.

edema, myofibrillar damage, and severe mitochondrial degeneration; the use of hyperbaric oxygenation reduced slightly the mitochondrial changes. Alterations similar to those described by Taheri *et al.* were found by Paulussen *et al.* (28) in hearts preserved by perfusion for 11 h at 5°C with an oxygenated solution of electrolytes, glucose, and procaine. The studies of Blackstone *et al.* (3) also showed that severe edema developed in hearts perfused *in situ* under mild hypothermia (28°C) with an oxygenated solution of low molecular weight dextran. Edema was also found in hearts perfused with whole blood under conditions of graded hypoxia, both *in situ* and in the isolated state, for up to 90 min at 37°C. Blackstone *et al.* concluded that hypoxia, glycogen loss, and the lack of removal of toxic metabolites from the perfusate were the main reasons for the ultrastructural abnormalities. Brown *et al.* (4) studied isolated canine hearts perfused up to 150 min at 160 mm Hg and 37°C. They observed congestion, extravasation of blood, and reduced ventricular compliance. These changes were not observed in hearts perfused at 100 mm Hg, from which Brown *et al.* concluded that myocardial injury was produced by the high perfusion pressure.

The ultrastructural changes seen in our dogs were less severe than those observed in previous studies of preserved hearts, and consisted of extra- and intracellular edema, alterations in mitochondrial morphology, and swelling of sarcoplasmic reticulum. Changes in the myofibrils were associated with severe intracellular edema and were observed only in one (no. 7) of two hearts that could not be defibrillated at the end of the period of preservation. Significant depletion of glycogen was found only in these two hearts (no. 7 and 14).

Considerable individual variation was observed in the extent of ultrastructural changes in preserved hearts. These changes were more marked in hearts perfused with filtered plasma, in which some degree of edema was always present, and in which the greatest range in cardiac function was observed. Fewer morphologic variations were present in hearts perfused with filtered plasma-dextran, in which studies of cardiac function gave more uniform results. In both groups, ultrastructure was well preserved when cardiac function was good; but when cardiac function was poor, ultrastructural changes in mitochondria and sarcoplasmic reticulum or both were prominent.

Changes related to edema. Factors important in the pathogenesis of the edema found in preserved hearts are: direct damage by high perfusion pressure; hypoxia; changes in osmolarity; cold-induced injury; and metabolic and toxic factors. The capability of a high perfusion pressure to produce damage to capillaries was demonstrated in one heart perfused with filtered plasma, in which an unexplained elevation of perfusion pressure occurred during preservation. High perfusion pressure can be excluded as a cause of edema in the other hearts, in which the perfusion pressure was approximately 25 mm Hg. Possibly, hypoxia occurred during cardiac excision and, subsequently, as the result of regional underperfusion. This possibility is unlikely, since high pO_2 levels were maintained in the perfusate. Furthermore, large amounts of glycogen were present in most muscle cells at the end of the period of preservation, a finding indicative of little ischemia and anaerobic glycolysis during preservation. Direct capillary damage may have been produced by release of vasoactive amines from neural elements at the time of nerve transection during cardiac excision, as suggested by Sutherland and Young (32). This is not surprising in view of the extensive innervation of the heart, the critical nature of the control of cardiac activity by the nervous system, and since various cardiac lesions can be produced by neurogenic stimuli (6, 22a). Cold-induced injury and metabolic factors, such as the accumulation of toxic metabolites in the perfusate, may have played an additional role in the development of edema. The possible contribution of these factors to an increase in capillary permeability is uncertain. This study clearly demonstrates, however, that edema did not develop in hearts perfused with filtered plasma-dextran. The osmolarity of the perfusion fluid, therefore, is a critical factor in controlling edema.

Mitochondrial changes. Although mitochondrial structure was generally well maintained, several types of mitochondrial changes were observed in some areas of the preserved hearts. Mitochondrial damage, with swelling and loss of the cristae, was more prominent in hearts perfused with filtered plasma than in hearts perfused with filtered plasma-dextran. These changes may have resulted from ischemia. That these changes were less pronounced in hearts perfused with filtered plasma-dextran suggests that the higher osmolarity and higher oncotic pressure of this solution counteracts the tendency for mitochondria to swell. Focal mitochondrial damage, however, always less extensive than after preservation, was observed at times in the control biopsies. Therefore, minor degrees of mitochondrial damage may have resulted from the progression of changes induced by anesthesia and the operative procedure, or from the process of mitochondrial turnover.

Other changes in mitochondrial morphology observed in both groups of preserved hearts consisted of modifications of the ususal pattern of arrangement of the mitochondrial cristae and of changes in density of the intramitochondrial matrix. Particularly prominent among these changes was the formation of "energized-twisted" configurations in the mitochondrial cristae, a phenomenon that was frequently observed in the hearts perfused with filtered plasma. In hearts preserved with filtered plasma-dextran, this change was much less prominent; in fact, it was much more pronounced in the control biopsies than in those taken at the end of the period of preservation. Current research indicates that these various types of alterations in mitochondrial structures are associated either with changes in the functional state of mitochondria (mechanochemical structural transformations), or with osmotic alterations brought about by the energized accumulation of ions into the water phase of the inner mitochondrial compartment (16–19, 29, 30). Exact determination of the nature of these changes is dependent upon various biochemical measurements of uptake of ions, energy metabolism, and electron transport in isolated mitochondria. Such studies will be necessary to evaluate the significance of these changes in preserved hearts, particularly since the work of Stoner and Sirak (31) has raised questions about the influence of fixation in the development of morphologic changes of this type.

A change possibly brought about by osmotic alterations of the type described by Hackenbrock and Caplan (17) was observed in the mitochondria of some of the hearts perfused with filtered plasma-dextran. This change consisted of stacking of the cristae and the formation of intramitochondrial concentric lamellar bodies with a well-defined periodicity. Our observations suggest that in these mitochondria the cristae first came into close apposition, their structure then became partially disorganized and, finally, their lipid constituents were gradually rearranged into concentric lamellae. Hackenbrock and Caplan have described similar changes under conditions in which calcium ions (up to 190 nanomoles of Ca per mg of mitochondrial protein) were taken up by isolated mitochondria in the absence of phosphate. That these bodies were seldom present in hearts perfused with filtered plasma also indicated their relationship to osmotic alterations.

Swelling of the sarcoplasmic reticulum. Swelling of the sarcoplasmic reticulum of cardiac muscle has been observed in a variety of circumstances and is considered a nonspecific change (21). The sarcoplasmic reticulum is believed to contain the calcium-binding sites for contraction and relaxation of cardiac muscle (23a). Narrowly regulated concentrations of calcium and sodium appear necessary for proper functioning of this system. It remains to be determined what changes in concentration and binding of calcium and sodium occur in the sarcoplasmic reticulum, as this compartment increases several times in volume as its tubules and cisterns dilate. The importance of the integrity of the sarcoplasmic reticulum was shown by the fact that cardiac function was poor in all preserved hearts in which swelling of the sarcoplasmic reticulum was either moderate or severe. Such swelling was much less marked in hearts perfused with filtered plasma-dextran, which again reflects the effect of this solution in decreasing both intra- and extracellular edema.

SUMMARY

Morphologic and functional studies were performed on control (3) and preserved (20) canine hearts, cardiac biopsies being obtained before cardiac excision and after functional evaluation: 14 hearts were preserved (1 for 3 h, 2 for 14, 5 for 18, and 6 for 24 h) with filtered plasma (256 milliosmols); 6 hearts were preserved for 18 h with filtered plasma with added dextran (316 milliosmols).

Individual variations in morphology and function occurred within each group of preserved hearts; however, morphologic and functional data correlated in each heart. Of the

15 hearts preserved with filtered plasma, 2 could not be defibrillated: 1 showed severe edema and damage to capillaries and myofibrils and the other had extensive mitochondrial damage. The 13 other hearts preserved with filtered plasma showed variable degrees of extracellular edema, mitochondrial damage, and swelling of the sarcoplasmic reticulum. Poor function was observed in hearts in which these changes were prominent; the other hearts showed good function and minimal morphologic changes. Good function, no interstitial edema, and fewer ultrastructural alterations were observed in hearts preserved with filtered plasma-dextran.

It is concluded that adequate preservation of canine hearts can be attained for up to 24 h. The use of a perfusate with a high osmolarity reduces cardiac edema and improves structural preservation.

REFERENCES

1. BAHR, G. F. and JENNINGS, R. B.: Ultrastructure of normal and asphyxic myocardium of the dog. Lab. Invest. *10:* 548–581 (1961).
2. BELZER, F. O.; ASHBY, B. S.; HUANG, J. S. and DUNPHY, J. E.: Etiology of rising perfusion pressure in isolated organ perfusion. Ann. Surg. *168:* 382–396 (1968).
3. BLACKSTONE, E. H.; EVANS, R. H.; ECKNER, F. A. O.; DRAKE, A. and MOULDER, P. V.: Perfusion-induced myocardial injury. J. Thor. Cardiov. Surg. *56:* 689–698 (1968).
4. BROWN, A. H.; BRAIMBRIDGE, M. V.; NILES, N. R.; GERBODE, F. and AGUILAR, M. J.: The effect of excessively high perfusion pressures on the histology, histochemistry, birefringence and function of the myocardium. J. Thor. Cardiov. Surg. *58:* 655–663 (1968).
5. BURDETTE, W. J. and ASHFORD, T. P.: Response of myocardial fine structure to cardiac arrest and hypothermia. Ann. Surg. *158:* 513–525 (1963).
6. CONNOR, R. C. R.: Myocardial damage secondary to brain lesions. Amer. Heart J. *78:* 145–148 (1969).
7. FAWCETT, D. W. and McNUTT, M. S.: The ultrastructure of the myocardium. I. Ventricular papillary muscle. J. Cell. Biol. *42:* 1–45 (1969)
8. FERRANS, V. J.; BUJA, L. M.; LEVITSKY, S. and ROBERTS, W. C.: Effects of hyperosmotic perfusate on ultrastructure and function of the isolated canine heart. Lab. Investig. *24:* 265–272 (1971).
9. FERRANS, V. J.; HIBBS, R. G.; CIPRIANO, P. R. and BUJA, L. M.: Nucleoside monophosphatase activity in ventricular myocardium of normal and norepinephrine-treated rats. Anat. Rec. *163:* 301 (1969).
10. FERRANS, V. J.; HIBBS, R. G.; CIPRIANO, P. R. and BUJA, L. M.: Histochemical and electron microscopic studies of norepinephrine-induced myocardial necrosis. This Book, pp. 495–525.
11. FERRANS, V. J., HIBBS, R. G.; WEILY, H. S.; WEILBAECHER, D. G.; WALSH, J. J. and BURCH, G. E.: A histochemical and electron microscopic study of epinephrine-induced myocardial necrosis. J. Molec. Cell. Cardiol. *1:* 11–22 (1970).
12. FERRANS, V. J.; LEVITSKY, S.; BUJA, L. M.; WILLIAMS, W. H.; McINTOSH, C. L. and ROBERTS, W. C. Histologic and ultrastructural studies on preserved hearts. Circulation *41:* (Suppl. II) 104–109 (1970).
13. FORSSMANN, W. G. and GIRARDIER, L.: A study of the T system in rat heart. J. Cell. Biol. *44:* 1–19 (1970).
14. FORSSMANN, W. G.; MALTER, A.; DALDRUP, J. and GIRARDIER, L.: A study of the T system in the heart muscle by means of horseradish peroxidase tracing. J. Cell Biol. *39:* 45A–46A (1968).
15. FREEMAN, J. and SPURLOCK, B.: A new epoxy embedment for electron microscopy. J. Cell. Biol. *13:* 437–443 (1962).

16. GREEN, D. E.; ASAI, J.; HARRIS, R. A. and PENNISTON, J. T.: Conformational basis of energy transformations in membrane systems. III. Configurational changes in the mitochondrial inner membrane induced by changes in functional states. Arch. Biochem. Biophys. *125:* 684–705 (1968).

17. HACKENBROCK, C. R. and CAPLAN, A. I.: Ion-induced ultrastructural transformations in isolated mitochondria. The energized uptake of calcium. J. Cell Biol. *42:* 221–234 (1969).

18. HARRIS, R. A.; PENNISTON, J. T.; ASAI, J. and GREEN, D. E.: The conformational basis of energy conservation in membrane systems. II. Correlation between conformational change and functional states. Proc. Natl. Acad. Sci. U.S. *59:* 830–837 (1968).

19. HARRIS, R. A.; WILLIAMS, C. H.; CALDWELL, M.; GREEN, D. E. and VALDIVIA, E.: Energized configurations of heart mitochondria in situ. Science *165:* 700–702 (1969).

20. HIBBS, R. G. and BLACK, W. C.: Electron microscopy of post-mortem changes in the rat myocardium. Anat. Rec. *147:* 261–272 (1963)

21. HIBBS, R. G.; FERRANS, V. J.; BLACK, W. C.; WEILBAECHER, D. G.; WALSH, J. J. and BURCH, G. E.: Alcoholic cardiomyopathy: An electron microscopic study. Amer. Heart J. *69:* 766–779 (1965).

22. HÖLSCHER, B.; JUST, O. H. and MERKER, H. J.: Studies by electron microscope on various forms of induced cardiac arrest in dog and rabbit. Surgery *49:* 492–499 (1961).

22a. KLOUDA, M. and BRYNJOLFSSON, G.: Cardiotoxic effects of electrical stimulation of the stellate ganglia. Ann. New York Acad. Sci. *156:* 271–280 (1969).

23. KOLOBOW, T. and BOWMAN, R. L.: Construction and evaluation of an alveolar membrane artificial heart-lung. Trans. Amer. Soc. Artif. Int. Organs *9:* 238–243 (1963).

23a. LANGER, G. A.: The ionic basis for control of myocardial contractility. Progr. Cardiovasc. Dis., *9:* 194–203 (1966).

24. LEVITSKY, S.; WILLIAMS, W. H.; DETMER, D. E.; McINTOSH, C. L. and MORROW, A. G.: Extended preservation of the excised heart: Functional evaluations. Circulation *40:* (Suppl. III) 132 (1969).

25. LEVITSKY, S.; WILLIAMS, W. H.; DETMER, D. E.; McINTOSH, C. L. and MORROW, A. G.: A functional evaluation of the preserved heart. J. Thor. Cardiov. Surg. *60:* 625–634 (1970).

26. MILLER, D. R.; RASMUSSEN, P. and KLIONSKY, B.: Reversibility of morphologic changes following elective cardiac arrest. Ann. Sur. *159:* 208–218 (1964).

27. MILLONIG, G.: Advantage of a phosphate buffer for OsO_4 solution in fixation. J. Appl. Physics *32:* 1367 (1961).

28. PAULUSSEN, F.; HUBNER, G.; GREBE, D. und BRETSCHNEIDER, H. J.: Die Feinstruktur des Herzmuskels während einer Ischämie mit Senkung des Energiebedarfes durch spezielle Kardioplegie. Klin. Wschr. *46:* 165–171 (1968).

29. PENNISTON, J. T.; HARRIS, R. A.; ASAI, J. and GREEN, D. E.: The conformational basis of energy transformations in membrane systems. I. Conformational changes in mitochondria. Proc. Natl. Acad. Sci. U. S. *59:* 624–631 (1968).

30. SORDAHL, L. A.; BLAILOCK, Z. R. KRAFT, G. H. and SCHWARTZ, A.: The possible relationship between ultrastructure and biochemical state of heart mitochondria. Arch. Biochem. Biophys. *132:* 404–414 (1969).

31. STONER, C. D. and SIRAK, H. D.: Passive induction of the "energized-twisted" conformational state in bovine heart mitochondria. Biochem. Biophys. Res. Comm. *35:* 59–66 (1969).

32. SUTHERLAND, T. M. and YOUNG, D. A. B.: Increased permeability of the capillaries of the rat heart to plasma albumin with asphyxiation and with perfusion. J. Physiol. *183:* 112–122 (1966).

33. TAHERI, S. A.; MASER, M. D. and STOESSER, F. G.: Canine heart storage under hyperbaric and hypothermic conditions. Angiology *19:* 573–580 (1968).

34. WEBB, W. R.; SUGG, W. L. and ECKER, R. O.: Heart preservation and transplantation: experimental and clinical studies. Amer. J. Cardiol. *22:* 820–825 (1968).

35. WEDELL, J.; MERKER, H. J. und NEUBERT, D.: Mitochondrienstruktur und Atmungskettenphosphorylierung im Herzmuskel nach vollständiger Kreislaufunterbrechung. Virchows Arch. path. Anat. *338:* 355–370. (1965).

Mailing address: Victor J. Ferrans, M.D., Ph. D., Section of Pathology, National Heart and Lung Institute, National Institutes of Health, Bethesda, Maryland 20014 (USA).

Pathology Institute, McGill University, Montreal, Quebec, Canada

ULTRASTRUCTURAL STUDIES ON MYOCARDIAL FIBRIN DEPOSITION IN EXPERIMENTAL HYPERTENSION [1]

I. HÜTTNER,[2] G. RONA, D. THEODOSIS, and R. H. MORE

Focal myocardial lesions have been described in different types of experimental hypertension models (1, 3, 5, 13, 15–19, 22). Although these alterations are predominantly perivascular in location and they are associated with vascular lesions, only a few studies have made significant contributions with regard to the pathogenetic importance of plasma insudation (16). Since fibrin can be regarded as a natural tracer of plasma proteins, fine structural identification of fibrin in vascular and myocardial lesions supports the role of plasma insudation in the development of these alterations. Electron microscopic studies in experimental malignant hypertension have revealed polymerized fibrin in coronary artery injury (6). In recent investigations, we observed fibrin deposition in the vicinity of the involved vessels and also within cardiac muscle cells (9, 10).

The present work was designed to study the pathogenesis of this lesion and to trace the pathways of plasma protein insudation. For tracing the physiological pathways of plasma constituents through the vascular endothelium, horseradish peroxidase, a protein of relatively low molecular weight, and ferritin, a protein comparable in size to most plasma proteins, were used as probe molecules. A third tracer, colloidal carbon particles, was also applied for detection of discontinuity of the arterial endothelium.

MATERIALS AND METHODS

The experiments were carried out in a total of 98 adult male rats of the Sprague-Dawley strain weighing 300-400 g. Arterial hypertension was produced by unilateral renal ischemia with complete ligation of the aorta between the origin of the renal arteries (17) or by bilateral compression of the renal cortex with a rubber capsule (13). Rats found to be hypertensive were sacrificed at various intervals ranging from 2 to 26 days following the operation.

Tracer experiments were carried out on the 1st and 2nd postoperative day, parallel with control rats. Animals were divided into three groups (each

[1] Supported by MRC Grants No. MT-3635 and MA-3683.

[2] Research Associate of the Hungarian Academy of Sciences at the 2nd Department of Pathology, Medical University of Budapest, Hungary; Research Fellow of the Canadian Arthritis and Rheumatism Society at the Department of Pathology, McGill University, Montreal, during the course of this study.

containing 12 hypertensive and 12 control rats) for peroxidase, ferritin, and colloidal carbon studies. Each group of rats received, by injection into a femoral vein under ether anesthesia, either 8 mg/100 g body weight of peroxidase in 0.5 ml/100 g saline, 100 mg/100 g of ferritin in 1.0 ml/100 g saline, or 0.1 ml/100 g colloidal carbon. The heart was directly perfused (2) with a mixture of aldehyde fixatives (12), 1, 6, and 30 min after peroxidase injection and 6 and 30 min after ferritin and colloidal carbon injection. Tissues selected for fine structural studies from the right and left ventricle were further fixed, sectioned at 30-50 μ with a Smith-Farquhar tissue sectioner, and subsequently incubated for the detection of peroxidase in the medium of Graham and Karnovsky (4) before being postfixed in OsO_4. Specimens taken from the heart of ferritin or colloidal carbon injected rats, as well as tissues from the ventricular myocardium for nontracer fine structural studies, were processed similarly except for sectioning with the Smith-Farquhar apparatus and incubation in the peroxidase medium. Following either procedure, tissue blocks were then dehydrated in graded ethanols and embedded in Epon (14). Sections 1μ in thickness were stained with toluidine blue for selection of areas for fine structural studies. Unstained sections for tracer studies and sections stained with ethanolic uranyl acetate and lead citrate (20) were examined with a Phillips EM 300 microscope at 60 KV.

RESULTS

The systolic blood pressure in the hypertensive animals of both experimental groups increased as early as the 1st postoperative day, and in several animals rose to 200 mm Hg by the end of the 1st postoperative week. At the time of the tracer experiment, on the 1st and 2nd day after operation, the average value of the hypertensive group was 150–170 mm Hg, while the blood pressure of controls ranged from 120–140 mm Hg. Gross, histological, and ultrastructural changes of the myocardium were comparable in the two hypertensive groups. The vascular and myocardial lesions appeared later and were less severe in the left ventricle than in the right one. For this reason, electron microscopic studies were carried out only on the right ventricle.

The detailed fine structural morphology of myocardial fibrin deposition was the subject of our previous studies (10). Therefore, in the description that follows, we summarize only the two distinct patterns of fibrin deposition within cardiac muscle cells:

Membrane-bound fibrin appeared in cardiocytes within sacs that, as shown by serial sections, communicated with the extracellular space (fig. 1). One part of these fibrin-filled sacs displayed a topographic relationship to intercellular junctions. Some of them became visible as casts of the extracellular space at lateral cell junctions, whereas others appeared to correspond to dilated portions of the nonspecialized contact area of the intercalated discs. Fibrin became

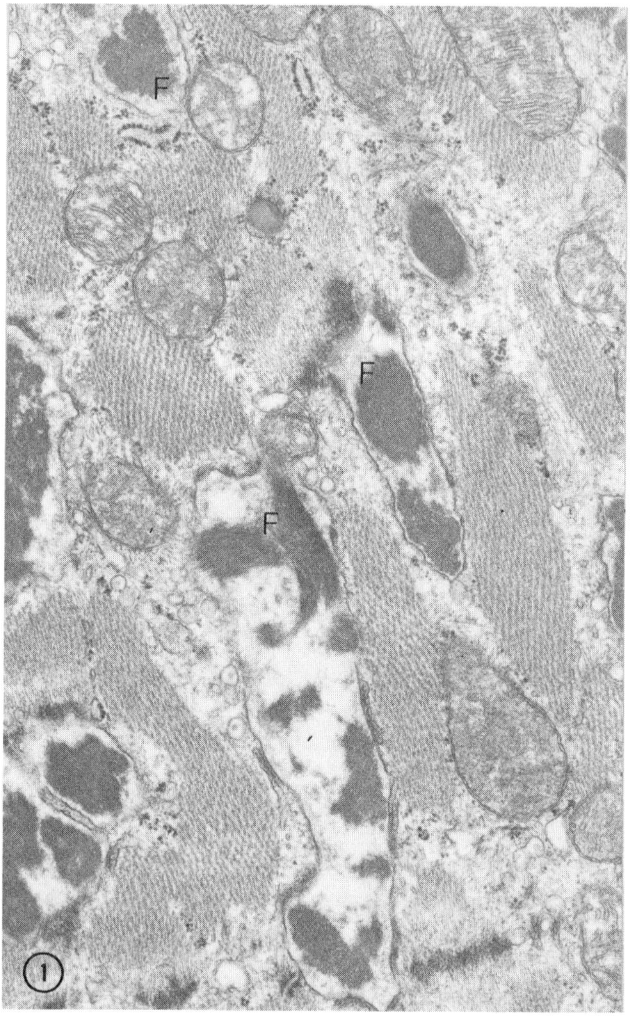

Fig. 1 Surviving periarterial cardiac muscle cell with preserved basic fine structure, from a hypertensive rat 16 days after operation. Fibrin (F) appears within membrane-bound sacs deriving from sarcolemmal invaginations. Uranyl acetate and lead citrate staining, ×26,100.

polymerized also within tubules of the T system. Other fibrin-filled sacs represented profiles of irregular sarcolemmal folds of varying geometry. Fibrin-containing sacs might also have developed from a fibrin-infiltrated portion of a cardiocyte that became surrounded and isolated from the viable remainder of the cell by a membrane. Whatever the mechanism responsible for the formation of membrane-bound fibrin, cardiac muscle cells exhibiting this change survived the initiating injury.

Whereas membrane-bound fibrin indicated reversible cardiocyte damage, intrasarcoplasmic fibrin polymerization occurred in severely altered cells (fig. 2). Destruction of sarcoplasmic reticulum, organelles, and contractile elements of cardiocytes observed in association with this type of fibrin deposition impeded the regeneration of such cells. The irreversible damage to these cells was reflected by the fact that they were replaced by proliferating connective tissue.

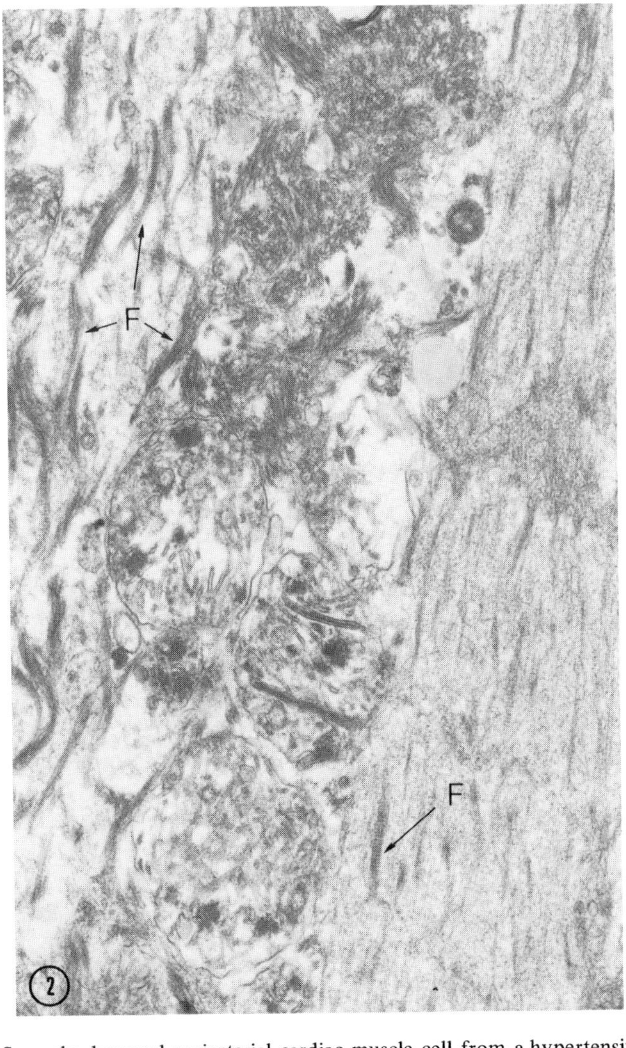

Fig. 2 Severely damaged periarterial cardiac muscle cell from a hypertensive rat on the 2nd post operative day. Polymerized fibrin (F) appears in a true intrasarcoplasmic location, intermingled with dissociated myofilaments. Some of the fibrin fibers exhibit characteristic ~200Å axial periodicity (arrows). Uranyl acetate and lead citrate staining, ×26,100.

Fine structural evidence of increased endothelial permeability. In hypertensive animals, 24 h after operation, fibrin deposition had not yet appeared in arterial walls or in the myocardium. In this stage, electron microscopy did not reveal endothelial discontinuity in arterial or in capillary endothelium. Morphologically, the interendothelial junctions appeared normal. In the coronary arteries, the endothelial cells generally appeared to be larger in size and contained considerable amounts of rough endoplasmic reticulum and many ribosomes. The smooth muscle cells of the media were hypertrophic and the adventitial fibroblasts showed swelling and proliferation. Activated fibroblasts were also seen in the myocardial interstitium.

Tracer studies in this early stage of experimental hypertension revealed an increased passage of peroxidase and ferritin molecules from blood to interstitium, without detectable gap formation. Colloidal carbon particles did not escape from the circulation in hypertensive animals at this stage. At 6 and 30 min after injection, the distribution and intensity of peroxidase reaction product in arterial walls and in the myocardial interstitium was similar in hypertensive and control animals. At 1 min after injection, however, the distribution of the tracer exhibited distinct differences between the two animal groups. In the controls, only low density reaction product was present in the subendothelial layers of arteries and in the pericapillary spaces as compared with the high intensity reaction found in hypertensive animals. Both in control and hypertensive rats, peroxidase reaction product appeared in the interendothelial junctions. These results suggest that, although the portal of entry remains the same, there is an increased passage of small molecules, such as peroxidase, through the cell junctions in both coronary arteries and capillaries in the early stage of the experimental hypertension model. At both 6 and 30 min after intravenous injection, there was a strikingly increased number of ferritin molecules in the myocardial interstitium as well as in the subendothelial layers of arteries as compared with those of the controls. In the endothelial cells of arteries and of some of the capillaries, ferritin molecules appeared in increased numbers within plasmalemmal vesicles. Ferritin molecules were not encountered in endothelial cell junctions, indicating that large molecules such as ferritin may enter the arterial wall and pericapillary interstitium in increased amounts by vesicular transport.

In the case of substantial fibrin deposition in the myocardium in hypertensive animals from the 2nd postoperative day, electron microscopy revealed endothelial discontinuity in the involved arteries, varying in extent from opening of endothelial cell junctions, only detectable by serial sections, to loss of one to several endothelial cells. In the latter case, the denuded intima was often covered by platelets.

In hypertensive animals presenting such vascular lesions, interendothelial gaps appeared to be the main passageway of tracer molecules. Colloidal carbon particles were detectable between separated endothelial cells on the basement membrane of capillaries, and focally within the arterial wall. There was a strikingly increased number of ferritin molecules within the arterial walls and in the myocardial interstitium as compared with those of controls. Ferritin molecules were seen in opened endothelial junctions. Similarly, in hypertensive animals, peroxidase reaction product appeared to be more intensive in the extracellular compartment of the myocardium.

In cardiac muscle cells, it was possible to detect cell membrane damage by applying the small peroxidase molecule before fibrin deposition occurred in the sarcoplasm (fig. 3). The location of ferritin tracer, a molecule comparable in size to most plasma proteins, followed the pattern of fibrin deposition within damaged cardiac muscle cells. In cardiocytes with membrane-bound fibrin deposition, ferritin molecules appeared within these basically extracellular sacs. In cardiocytes with intrasarcoplasmic fibrin deposition, a great number of ferritin molecules appeared within the sarcoplasm a few minutes after intravenous injection (fig. 4).

COMMENT

The sequence of fine structural changes in the coronary arteries as well as observations obtained by tracer molecules seem to confirm the primary role of an increased endothelial permeability in the development of hypertensive vascular and myocardial injury. The results suggest that, in the early stage of experimental hypertension, the increased passage of plasma proteins through the endothelium is related to an enhanced physiological transport rather than to gap formation, a mechanism previously described by us in hypertensive aortic endothelium (7, 8). Molecules similar in size to peroxidase pass increasingly through cell junctions, whereas molecules similar in size to ferritin, such as most plasma proteins, are transported in increased amounts by plasmalemmal vesicles. In advanced hypertensive lesion of the coronary arteries, interendothelial gap formation may represent the main portal of entry for plasma proteins, in agreement with the observations of other authors (21). These suggestions are in accord with previous light microscopic observation that revealed accumulation of colloidal iron in the vessel wall before the vessels leaked colloidal carbon particles (11). In capillary endothelium, the portals of increased protein passage appeared to be comparable.

Increased vascular permeability evidently leads to insudation of plasma with a high protein content into the perivascular cardiac muscle, where the increased concentration of fibrinogen and, possibly, factors released from injured cells may be responsible for the polymerization of fibrinogen to fibrin. In surviving cardiocytes, the cell membrane is not permeable to plasma proteins, and fibrin polymerization occurs within membrane-bound invagination of the extra-cellular

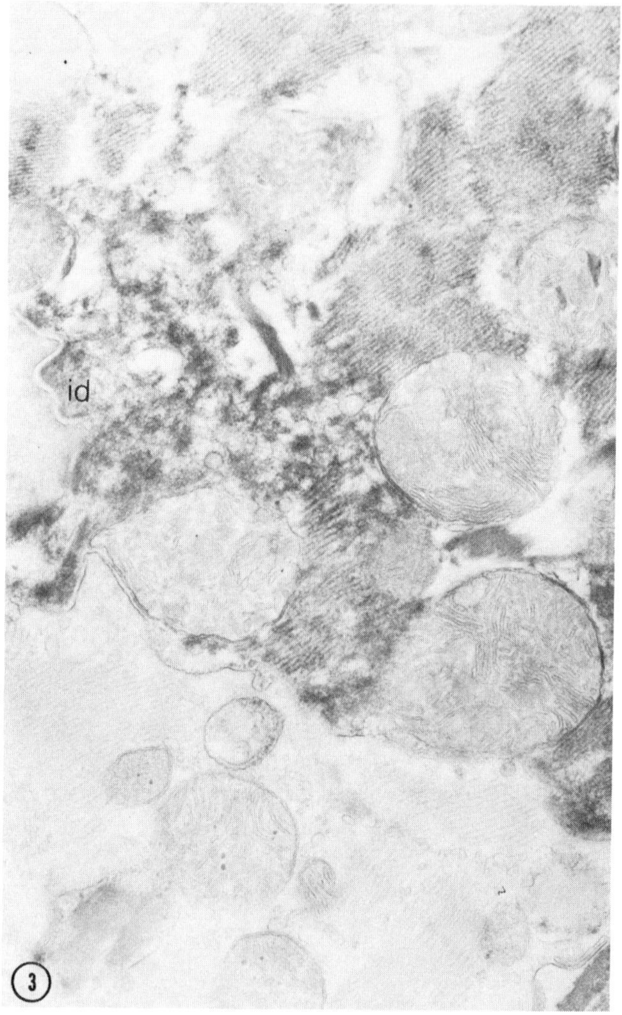

Fig. 3 Two cardiac muscle cells separated by intercalated disc (id), from a hypertensive rat on the 2nd postoperative day, 1 min after intravenous peroxidase injection. The cell membrane of the cardiocyte on the right is permeable to peroxidase. Note electron-dense reaction product in the sarcoplasm. Unstained section, × 26,100.

space. In cardiocytes with cell-membrane damage, plasma proteins enter the sarcoplasm, where fibrin polymerizes intracellularly.

This study afforded information concerning the passageways of protein molecules through coronary arterial and capillary endothelium in experimental hypertension. However, the primary causes of the increased protein passage and the structural changes of vascular endothelium remain unknown. Similarly, the

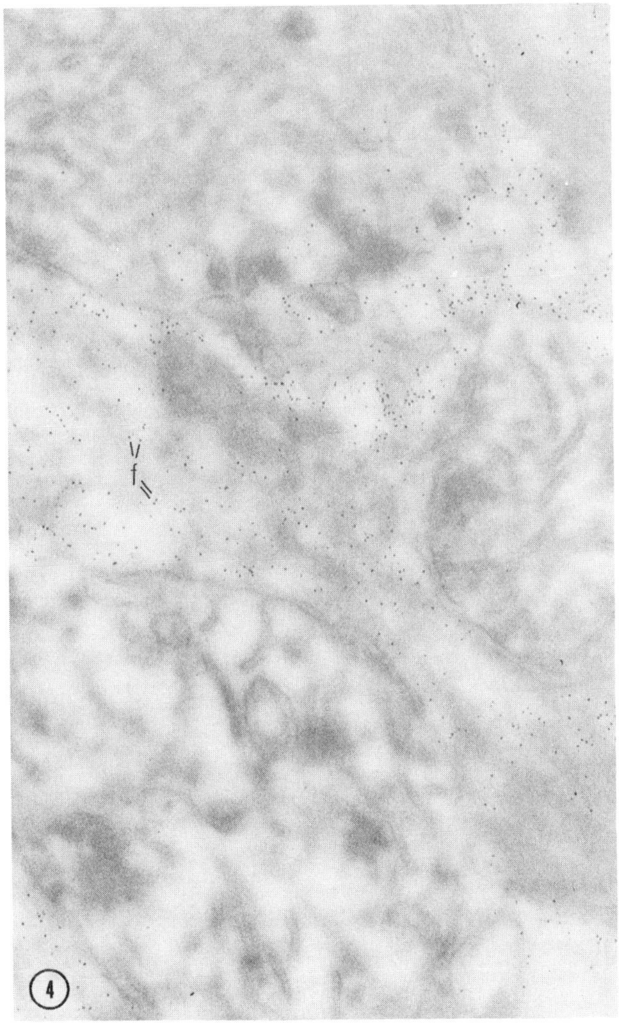

Fig. 4 Detail of a severely damaged cardiac muscle cell from a hypertensive rat on the 2nd postoperative day, 6 min after intravenous ferritin injection. Note ferritin molecules (f) throughout the sarcoplasm. Unstained section, ×77,200.

present work does not furnish direct information as to which factor is primarily responsible for the initial cardiac muscle injury. Nevertheless, it is conceivable that an increased vascular permeability plays a primary role and the subsequent interstitial accumulation of plasma materials may contribute to the development of cardiac muscle lesions. This proposed mechanism is supported by tracer studies on the coronary vasculature during the early stages of experimental hypertension, which demonstrated both increased arterial and capillary endothelial permeability and increased amounts of tracer molecules in the myocardial interstitium before cardiac-muscle damage becomes manifest. The subsequent cell damage, with release of chemical mediators, may further increase endothelial permeability, thus resulting in a vicious circle. This latter mechanism may be responsible for the capillary endothelial discontinuity in the areas involved.

SUMMARY

Earlier, fine structural studies on the myocardium of hypertensive rats revealed two distinct patterns of fibrin deposition within cardiac muscle cells in the vicinity of vascular lesions. Fibrin appeared within sarcolemmal invaginations in surviving cardiocytes and in true intrasarcoplasmic location in severely damaged cardiac muscle cells. The present tracer studies on the coronary vasculature suggest an increased endothelial permeability as the primary pathogenetic factor in the development of fibrinous cardiac muscle cell lesions. An increased passage of peroxidase and ferritin molecules was observed in the myocardial interstitium, prior to the development of cardiac muscle lesions. In cardiocytes, the peroxidase tracer revealed the presence of membrane damage before intrasarcoplasmic fibrin deposition occurred. In damaged cardiac muscle cells, ferritin, a tracer comparable in size to most plasma proteins, followed the pattern of fibrin deposition.

REFERENCES

1. CHAPPEL, C. I.: RONA, G. and CAHILL, J.: Studies on the pathogenesis of adrenal-regeneration hypertension in rat. Arch. Path. (Chicago) *65:* 636–645 (1958).
2. FORSSMANN, W. G.: A method for *in vivo* diffusion tracer studies combining perfusion fixation with intravenous tracer injection. Histochemie *20:* 277–286 (1969).
3. GORÁCZ, G. und KONYÁR, E.: Histologische Veränderungen im Herzen bei Hypertonie, pp. 707-714, in Gottsegen Acta Secundi Conventus Medicinae Internae Hungarici (Academiai Nyomda, Budapest 1963).
4. GRAHAM, R. C. and KARNOVSKY, M. J.: The early stages of absorption of injected horseradish peroxidase in the proximal tubules of mouse kidney: Ultrastructural cytochemistry by a new technique. J. Histochem. Cytochem. *14:* 291–302 (1966).
5. HALL, C. E. and HALL, O.: Macromolecular hypertension and associated pathologic changes resulting from treatment with polyvinyl alcohol. Amer. J. Path. *41:* 247–257 (1962).
6. HÜTTNER, I.; JELLINEK, H. and KERÉNYI, T.: Fibrin formations in vascular fibrinoid change in experimental hypertension. An electron microscopic study. Exp. Molec. Path. *9:* 309–321 (1968).
7. HÜTTNER, I. and MORE, R. H.: Ultrastructural evidence for specific endothelial permeability in experimental hypertension (Abstract). Fed. Proc. *29:* 499 Abs. (1970).
8. HÜTTNER, I.; MORE, R. H. and RONA, G.: Fine structural evidence of specific mechanism for increased endothelial permeability in experimental hypertension. Amer. J. Path. *61:* 395–412 (1970).
9. HÜTTNER, I.; RONA, G. and MORE, R. H.: Fibrin deposition within cardiac muscle cells in experimental malignant hypertension – An electron microscopic study (Abstract). Amer. J. Path. *59:* (1970).
10. HÜTTNER, I.; RONA, G. and MORE, R. H.: Fibrin deposition within cardiac muscle cells in experimental malignant hypertension – An electron microscopic study (Chicago) *91:* 19–28 (1971).

11. JELLINEK, H.; NAGY, Z.; HÜTTNER, I.; BÁLINT, A.; KOCZÉ, A. and KERÉNYI, T.: Investigations of the permeability changes of the vascular wall in experimental malignant hypertension by means of a colloidal iron preparation. Brit. J. Exp. Path. *50:* 13–16 (1969).
12. KARNOVSKY, M. J.: A formaldehyde-glutaraldehyde fixative of high osmolality for use in electron microscopy (Abstract). J. Cell. Biol. *27:* 137A–138A (1965).
13. LÖRINCZ, J. and GORÁCZ, G.: A new method of inducing experimental hypertension in the rat. Acta Physiol. Acad. Sci. Hung. *5:* 489–494 (1954).
14. LUFT, J. H.: Improvements in epoxy resin embedding methods. J. Cell Biol. *9:* 409–414 (1961).
15. MOLTENI, A.; BROWNIE, A. C. and SKELTON, R. F.: Production of hypertensive vascular disease in the rat by methyltestosterone. Lab. Invest. *21:* 129–137 (1969).
16. MOORE, R. D.: SCHOENBERG, M. D. and KOLETSKY, S.: Cardiac lesions in experimental hypertension. Arch. Path. (Chicago) *75:* 28–44 (1963).
17. ROJO-ORTEGA, J. M. and GENEST, J.: A method for production of experimental hypertension in rats. Can. J. Physiol. Pharmacol. *46:* 883–885 (1968).
18. SALGADO, E. and SELYE, H.: The production of hypertension, nephrosclerosis and cardiac lesions by methylandrostenediol treatment in the rat. Endocrinology *55:* 550–560 (1954).
19. SELYE, H. and STONE, H.: Pathogenesis of the cardiovascular and renal changes which usually accompany malignant hypertension. J. Urol. *56:* 399–419 (1946).
20. VENABLE, J. H. and COGGLESHALL, R.: A simplified lead citrate stain for use in electron microscopy. J. Cell. Biol. *25:* 407–408 (1965).
21. WIENER, J.; LATTES, R. G.; MELTZER, B. G. and SPIRO, D.: The cellular pathology of experimental hypertension. IV. Evidence for increased vascular permeability. Amer. J. Path. *54:* 187–207 (1969).
22. WILSON, C. and BYRON, F. B.: Renal changes in malignant hypertension. Lancet *ii:* 136–139 (1939).

Mailing address: G. Rona, M. D., Pathology Institute, McGill University, 3775 University Ave., Montreal 110, Que. (Canada).

From the Departments of Pathology and Medical Ecology,
Hadassah University Hospital and the Hebrew University–Hadassah
Medical School, Jerusalem, Israel

EXPERIMENTAL MYOCARDITIS: AN AUTOIMMUNE DISEASE [1]

A. LAUFER, A. M. DAVIES, I. GERY, and I. FRIEDMAN

Since the morphological description of the myocardial changes found in acute rheumatic fever, many investigators have tried to reproduce these changes experimentally in an attempt to explain the pathogenesis of the human disease. The well-known relationship between streptococcal infection and rheumatic fever stimulated numerous authors to perform different experiments in which either injections of heat-killed hemolytic streptococci and emulsions of homologous heart, subcutaneous or intramyocardial injections of streptococci and other microorganisms, or streptococcal enzymes or other immunization techniques were used to produce these lesions (12, 14, 21).

The myocardial lesions obtained consisted of muscle myofiber degeneration or necrosis and an inflammatory reaction. These lesions were either focal and granulomatous, or diffuse and interstitial in nature. The cellular infiltrate consisted mainly of mononuclear cells, *i.e.*, lymphocytes and histiocytes, giant cells and occasional polymorphonuclear leucocytes. In general, most of the lesions were microscopic and sparse, but at times they were extensive in nature. Great variations between different animal species and in animals of the same species were observed.

Animals immunized with homologous or heterologous heart extracts or homogenates with admixtures of Freund's adjuvant showed circulating anti-heart antibodies (4). These antibodies have been detected by different methods including precipitation, flocculation, complement fixation, tanned hemagglutination, and immunofluorescence (13).

The similarity between the histological lesions obtained and such a wide variety of different biologically noxious agents, and the limited extent of myocardial reaction made us consider the following hypothesis as an explanation for the phenomenon. The various noxious agents, acting in different ways, damage the myofiber and evoke a common sequence of events, which in turn leads to a histologically identical pattern of myocardial damage that includes myofiber degeneration and necrosis and an inflammatory reaction. The primary myofibrillar lesion is probably a change in permeability leading to electrolyte

[1] Supported by Grant HE-05739 from the National Institutes of Health, United States Public Health Service, and a grant from the Common Research Fund of the Hebrew University-Hadassah Medical School.

imbalance and the end target is lysosomal damage. The lysosomal damage is 'triggered' off by initial mitochondrial damage and both are eventually affected by, and in turn influence, changes in cell-wall permeability. These changes result in the extension of the lesion to adjacent cells by the action of lysosomal enzymes, and are instrumental in the production of altered cellular components that may be a source of new antigens. This antigenic stimulus serves to sensitize immunologically competent cells and these in turn may cause damage to the new cross-reacting antigens or normal myocardial cells.

Previous studies performed in our laboratory (5, 6) on the effect of different concentrations of lysosomal labilizers such as vitamin A acid and alcohol, chlorpromazine, and dimethylsulfoxide on rat heart and kidney cell cultures, showed that the damage induced primarily concerned the cell membrane, producing altered permeability, and not directly the lysosome. When isolated lysosomes from liver cells, polymorphonuclear leucocytes, and monocytes were added to cultured rat heart cells no damage was produced, even if the pH of the medium was lowered from 7.2 to 6.2.

Negative results obtained with lysosomal labilizers and exogenous lysosomes *in vitro,* however, do not negate the possibility that lysosomal damage may occur *in vivo* and produce myofiber damage. It is possible, under conditions where excessive production and accumulation of lactic acid occurs, that the intracellular acidosis may influence lysosomal enzyme activity, and the release of cardiac lysosomal enzymes may be another mechanism of cardiac damage.

Monayenkov and Repin (20) studied autoimmune processes during intermittent high altitude hypoxia. They maintained white rats in an altitude chamber daily for 5 h at a reduced barometric pressure corresponding to an altitude of 7,500-7,700 m above sea level for 36 days. Titers of anti-heart antibodies measured by the latex agglutination method were 1 : 90 at 14 days, 1 : 72 at 36 days, and 1 : 64 at 51 days. These authors concluded that intermittent hypoxia altered the antigenic properties of cardiac tissue proteins and provided a stimulus for an autoimmune reaction. In addition, different workers have shown that cardiac stress secondary to chronic hypoxia in rabbits kept at high altitudes caused some aggravation of immunopathological lesions in the myocardium of animals that had been sensitized to heterologous heart antigen (28). The production of circulating antibodies and sensitized lymphocytes to these newly formed antigens may perpetuate the process, even after removal of the primary cause (3).

Preliminary electron microscopical studies performed in guinea pigs immunized with homologous and heterologous heart homogenate and Freund's adjuvant showed organelle changes that were focal and affected the myocardial cells randomly (16). Areas of subsarcolemmal edema were frequently observed. Crowding of mitochondria was also present and, in some, mitochondrial granules were lacking. Dilatation of sarcoplasmic reticulum and T tubules was a frequent and striking finding. In a more recent study on Lewis rat hearts immunized with rabbit heart and Freund's adjuvant, similar changes were also evident, and some mitochondria showed matrix condensation while in others there was disappearance of cristae. Partial or complete vacuolization was also seen, with absence of mitochondrial granules (fig. 1). Myelin figures, lipofuscin, and lipid droplets were also present (fig. 2) and, occasionally, the tubular system was dilated (fig. 3). Similar electron microscopical changes, although not specific, have been described in animals by other workers, in response to streptolysin O (25), anoxia (23), toxic agents (1), and electrolyte imbalance in mammalian myocardium perfused with varying concentrations of sodium, calcium, and chloride (17).

These findings provide further support for the concept that different biological noxious stimuli lead to similar alterations at a subcellular level. It is possible that these changes are related to cell wall damage, causing electrolyte imbalance and disturbed function of the damaged organelles. These myocardial changes may induce an antigenic protein change that is followed by an immunological reaction.

Anti-heart antibodies have been detected by the tanned hemagglutination, precipitation, flocculation, complement fixation, and immunofluorescent techniques in animals immunized with homologous or heterologous heart extract or homogenates, together with complete Freund's adjuvant, endotoxin and killed streptococci (14). In this type of experimental myocarditis, as with other experimental autoimmune diseases such as allergic encephalomyelitis, thyroiditis and orchitis, no correlation was found between the presence of specific anti-heart antibodies and pathological lesions, even if the titer of the circulating anti-heart antibodies was high during the experiment (4, 13, 16). Lewis rats, which have been shown to be more susceptible than other strains to a number of autoimmune diseases, are also more susceptible to the development of immune myocarditis than are the outbred strain of Hebrew University rats, when immunized with rabbit and rat heart extracts plus the addition of Freund's adjuvant and pertussis vaccine. As in previous studies, no correlation between the presence of cardiac lesions and anti-heart antibodies was found in this study (9). Furthermore, passive immunization with homologous or heterologous sera failed to produce myocardial lesions *in vivo* (15). Using heart tissue cultures, cytotoxicity was obtained *in vitro*, but only with heterologous serum (19).

As in other immune diseases, the role of anti-heart antibodies in relation to the pathogenesis of experimental immune myocarditis is controversial.

Kaplan's immunofluorescent studies (12) showed the presence of bound gamma globulin in myocardial lesions induced experimentally by injections of heterologous heart and Freund's adjuvant in rabbits. However, it is difficult to produce binding of gamma globulin to myocardial fibers under normal conditions; and whether the bound gamma globulin found in these experimental conditions was identical to the globulin of the specific circulating anti-heart

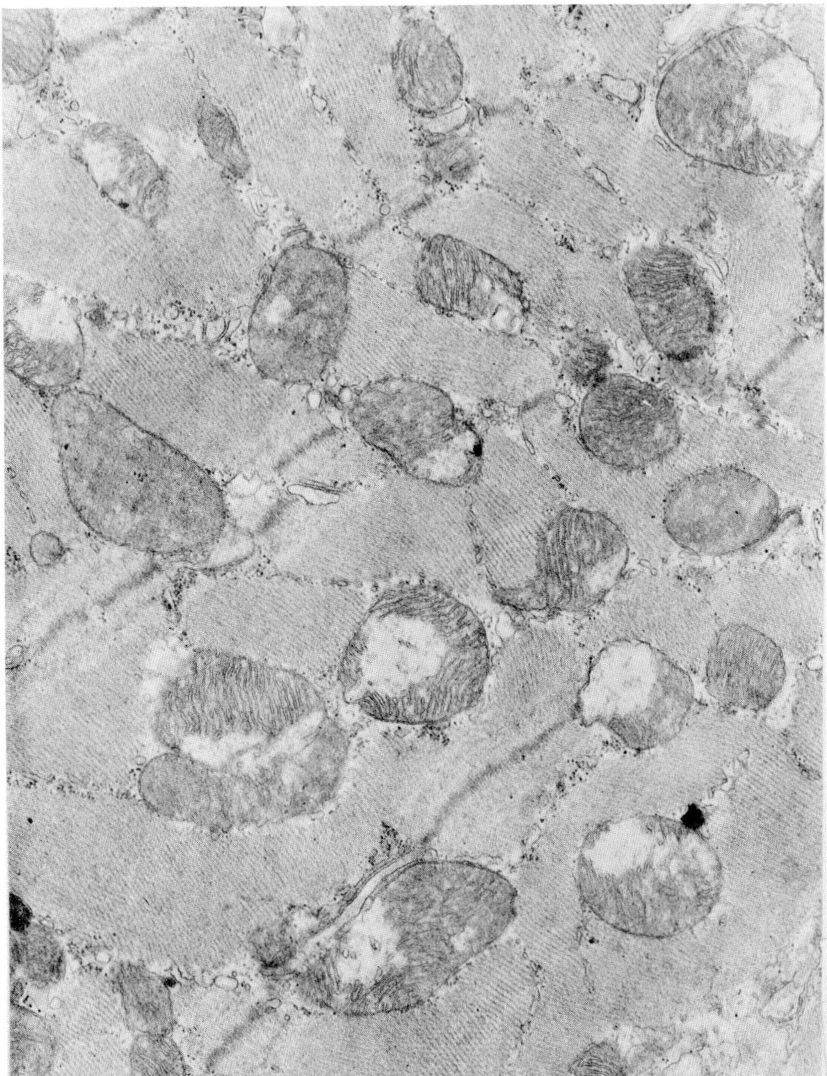

Figs. 1–3 Lewis rat heart immunized with rabbit heart and Freund's adjuvants:

Fig. 1 Mitochondrial vacuolization and area of condensation of matrix. × 18,620.

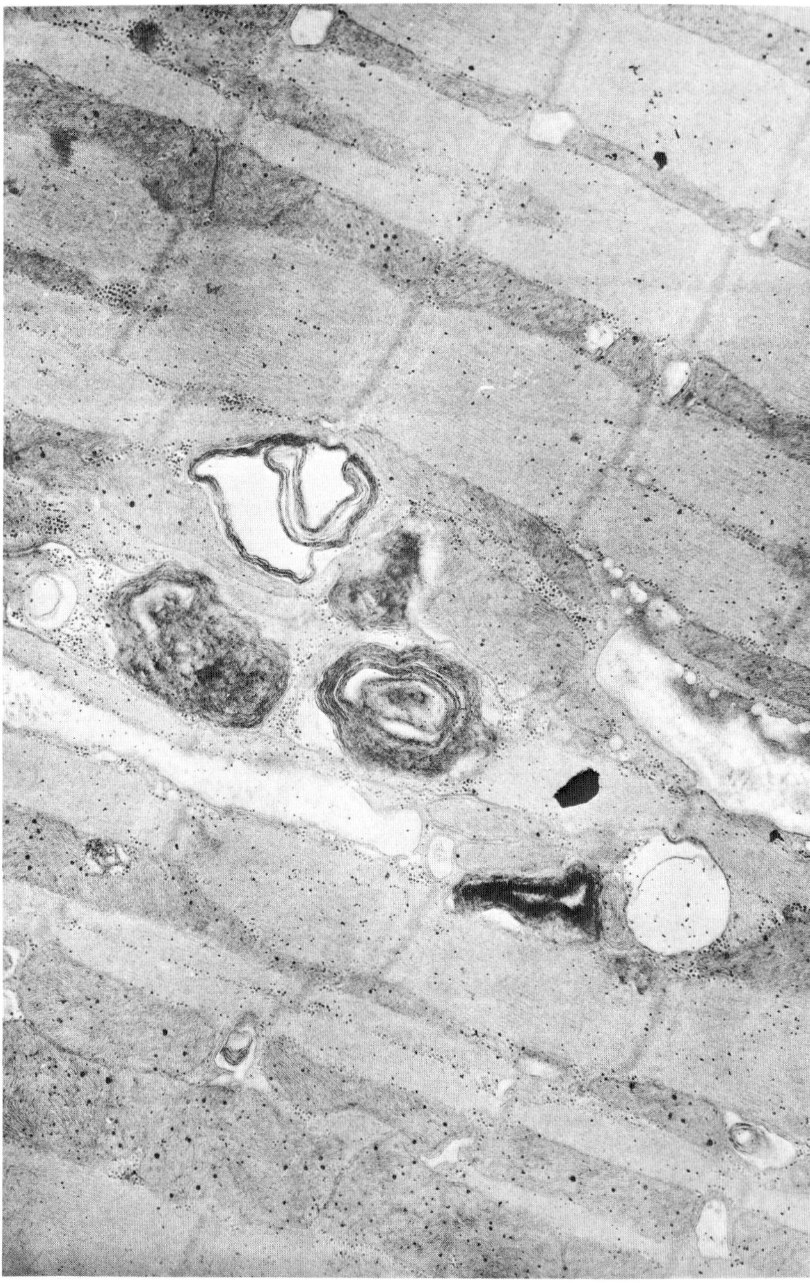

Fig. 2 Note myelin figures. × 15,970.

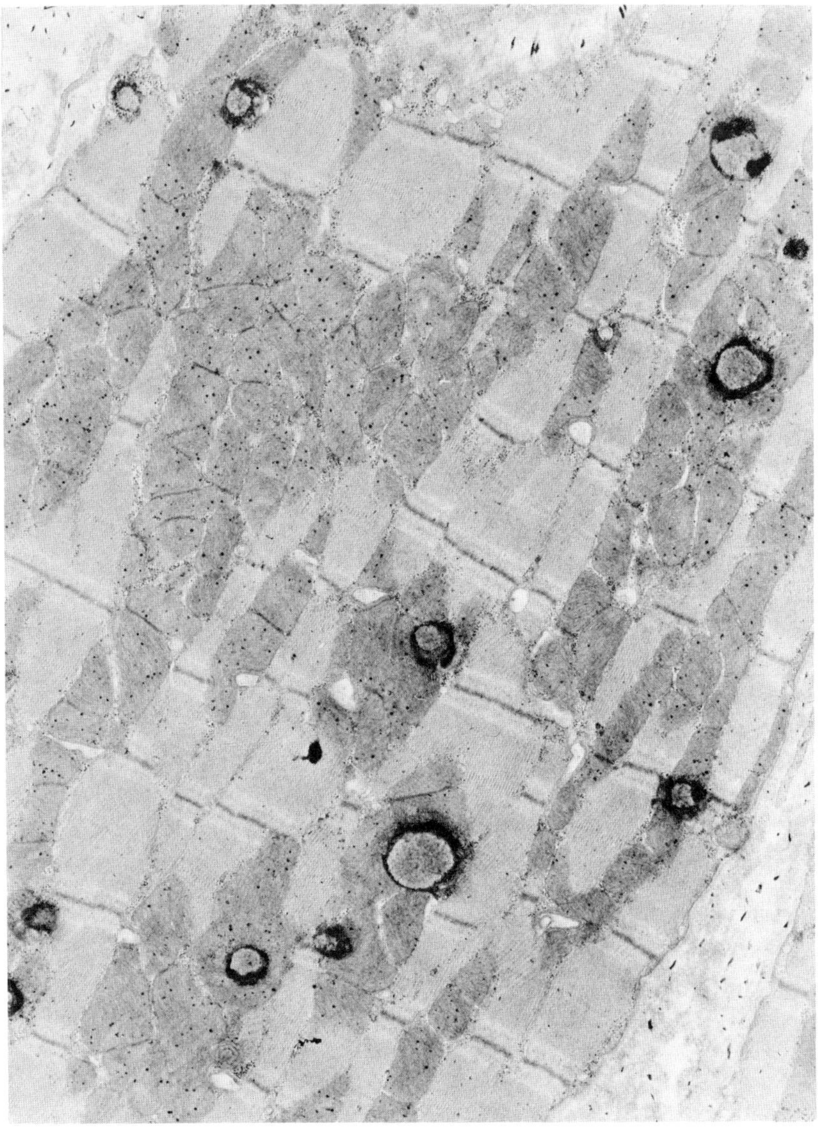

Fig. 3 Mitochondrial accumulation and lipid droplets. × 9,500.

antibodies remains to be proven. In order to enable specific antibodies to reach their specific antigenic destination, altered cellular permeability or cell barrier damage is required. Doniach (24) showed that passive transfer of heterologous anti-thyroid serum can induce thyroiditis in animals treated previously with Freund's adjuvant or proteolytic enzymes, or by irradiation of the thyroid. Rats pre-immunized against homologous or heterologous heart extracts and Freund's adjuvant developed circulating anti-heart antibodies when treated with a single injection of isoproterenol, and showed a marked extension of the myocardial lesions as compared with those resulting from a single injection of isoproterenol in non-immunized rats (10). The elution of anti-heart antibodies from the myocardium of transplanted hearts (26), as well as the induction of experimental glomerulonephritis (18), allergic thyroiditis (22), and encephalomyelitis (11) by injection of specific antibodies, points to the possible role of anti-heart antibodies in the production of experimental myocarditis.

The results of the work performed by our group demonstrate the importance of sensitized lymphocytes in the production of experimental immune myocarditis (7, 8). Lymphocytes sensitized to heart antigens and to cross-reacting antigen of *Streptococcus haemolyticus* are cytotoxic *in vitro* (fig. 4) and, *in vivo,* their transfer to inbred normal rats causes focal myocarditis in these animals. The results of our *in vitro* study has demonstrated that lymphocytes from rats sensitized to xenogenic (rabbit) heart and allogenic (rat) heart, in the ratio of

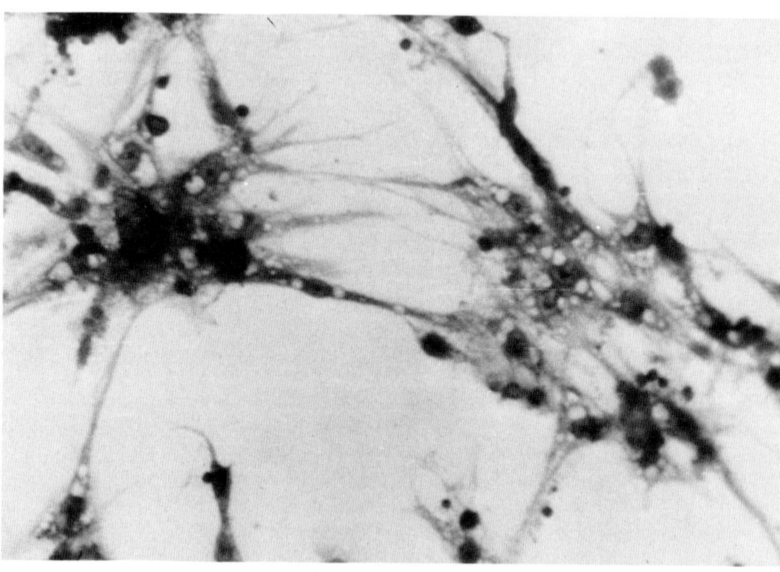

Fig. 4 Rat heart-cell culture 48 h after addition of lymphocytes sensitized to rabbit heart. Note cell vacuolization, shrinkage and destruction. Some lymphocytes are attached to the cells. Giemsa stain, × 640.

40 : 1, are cytotoxic to rat heart tissue cultures. The degree of cytotoxicity and specificity of the reaction observed grossly and microscopically was more marked with lymphocytes sensitized to xenogenic (rabbit) heart. Cells obtained from lymph nodes of rats sensitized to rabbit heart were cytotoxic to 78% of rat heart cell cultures, and lymphocytes sensitized to rat heart produced a similar effect in 65% of the cell cultures. The mean grading was 2.1 or 1.95 for lymphocytes sensitized to rabbit or rat heart as compared to the mean grading of 0.3–0.4 of lymphocytes sensitized to rabbit liver or adjuvant alone. A high degree of cytotoxicity (82%) was obtained when rat heart cultures were exposed to lymphocytes sensitized to group A streptococcus as compared to 18% when exposed to lymphocytes sensitized to *Streptococcus viridans.* Consequently, the role played by lymphocytes sensitized to group A streptococci may be of importance in the development of rheumatic fever and further studies in this direction may be of great value in the understanding of the pathogenesis of rheumatic fever.

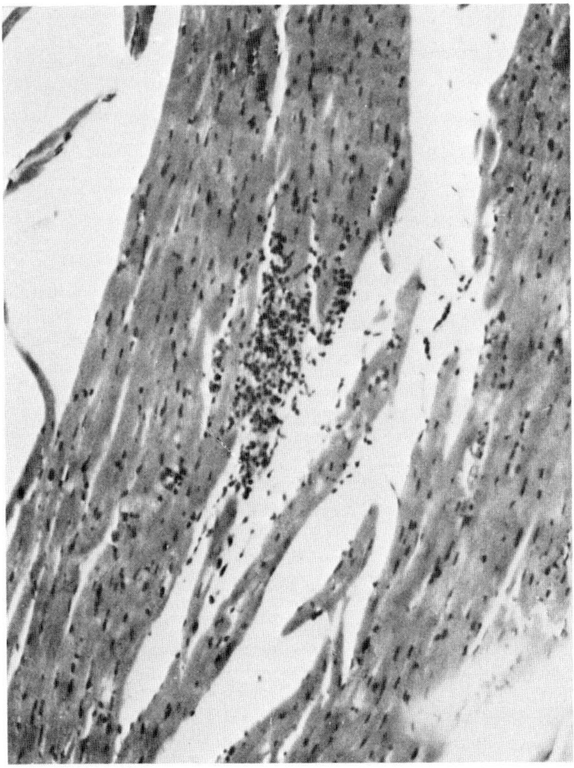

Fig. 5 Focus of myocarditis in Lewis rat heart 7 days after passive transfer of lymphocytes from donor Lewis rats immunized with rabbit heart. H & E, × 150.

Direct contact between sensitized lymphocytes and the heart cells is required in order to obtain cytotoxicity. This has been well demonstrated by the inhibition of their cytotoxic effect obtained by the agglutination of lymphocytes with anti-lymphocytic serum. *In vivo* studies confirmed the *in vitro* results: 13 of the 14 normal Lewis rats receiving lymphocytes from donor Lewis rats immunized with rabbit heart developed mild to moderate myocarditis (fig. 5); the myocardium of the donor rats rarely showed myocardial lesions.

Both the *in vitro* and *in vivo* results tend to support the hypothesis that one of the important mechanisms in the production of immune myocarditis may be mediated through sensitized lymphocytes.

The mechanism by which lymphocytes recognize antigens and cause tissue injury is obscure. It has been shown that lymphocytes stimulated specifically by antigens and nonspecifically by phytohemagglutinin release substances that damage cell cultures. The release of these substances from lymphocytes sensitized to heart extracts, in contact with specific antigens or heart cultures *in vitro*, may explain the cytotoxic effect obtained (27). It is also feasible to consider the possibility of a synergistic effect of sensitized lymphocytes and specific antibodies. Recent studies on experimental allergic orchitis tend to support the existence of such a synergistic effect (2). *In vitro* and *in vivo* studies on experimental myocarditis being performed at present in our laboratory may provide additional information on the possibility of the synergistic effect as a pathogenic mechanism involved in this process.

SUMMARY

Circulating specific antibodies to heart antigens have been demonstrated in the sera of animals immunized with extracts of xenogenic or allogenic heart and complete Freund's adjuvant. No correlation between the antibody titer and the presence of cardiac lesions was found. Passive immunization with specific anti-heart sera did not produce myocardial lesions; however, rats preimmunized against heart extracts showed a marked increase in the severity and extent of the myocardial lesions produced by a standard injection of isoproterenol. This would suggest that the specific antibodies become cytotoxic only after the cell wall is broken down.

In vitro, lymphocytes sensitized to heart antigens proved to be cytopathogenic to rat heart cultures. The transfer to normal syngenic rats of lymphocytes sensitized to heart antigens produced focal myocarditis in these animals. Lymphocytes sensitized to group A streptococci were cytotoxic to 82% of rat heart cell cultures, whereas lymphocytes sensitized to *Streptococcus viridans* produced cytotoxic changes in only 8% of the cultured cells. The close histological similarity between myocardial lesions induced immunologically and those obtained via a number of biologically different agents suggests that different stimuli may trigger off a common sequence of events to produce these lesions. The initial myofiber lesion is probably one of increased cell membrane permeability, leading to electrolyte imbalance and, eventually, lysosomal damage. The results of *in vitro* studies on the effect of lysosomal labilizers and isolated lysosomes on heart tissue cultures, as well as electron microscopical studies of the myocardium of animals immunized with heart extracts and complete Freund's adjuvant, tend to support the hypothesis that the primary effect is on the cell membrane. *In vivo*, it is possible that lysosomal damage occurs and may play a role in the production of myofiber damage at a later stage.

REFERENCES

1. ALEXANDER, C. S.: Idiopathic heart disease. I. Analysis of 100 cases, with special reference to chronic alcoholism. II. Electron microscopic examination of myocardial biopsy specimens in alcoholic heart disease. Amer. J. Med. *41:* 213–234 (1966).
2. BROWN, P. S. and GLYNN, L. E.: The early lesion of experimental allergic orchitis in guinea-pigs: an immunological correlation. J. Path. *98:* 277–282 (1969).
3. BURCH, P. S. and GLYNN, L. E.: The early lesion of experimental allergic orchitis in guinea-pigs: an immunological correlation. J. Path. *98:* 277–282 (1969).
4. DAVIES, A. M.;LAUFER, A.; GERY, I. and ROSENMANN, E.: Organ specificity of the heart. Circulating antibodies and immuno-pathological lesions in experimental animals. Arch. Path. *78:* 369–376 (1964).
5. FRIEDMAN, I.; LAUFER, A. and DAVIES, A. M.: Studies on lysosomes in rat heart cell cultures. I. The effect of lysosomal labilizers. Brit. J. Exp. Path. *50:* 213–218 (1969).
6. FRIEDMAN, I.; LAUFER, A. and DAVIES, A. M.: Studies on lysosomes in rat heart cell cultures. II. The effect of exogenous lysosomes. Experientia *25:* 1092–1093 (1969).
7. FRIEDMAN, I.; LAUFER, A.; RON, N. and DAVIES, A. M.: Experimental myocarditis: in vitro and in vivo studies of lymphocytes sensitized to heart extracts and group A streptococci. Immunology, in press (1971).
8. FRIEDMAN, I.; RON, N.; LAUFER, A. and DAVIES, A. M.: Experimental myocarditis: The effect of sensitized lymphocytes in vitro and in vivo (Abstract). Israel J. Med. Sci. *5:* 448 (1969).
9. FRIEDMAN, I.; RON, N.; LAUFER, A. and DAVIES, A. M.: Experimental myocarditis: enhancement by the use of pertussis vaccine in Lewis rats. Experientia, in press (1970).
10. GAZENFELD, E.; ROSENMANN, E.; DAVIES, A. M. and LAUFER, A.: Isoproterenol-induced myocardial lesions in the immunized and non-immunized rat. II. Histo-pathological lesions after specific immunization. Immunology *10:* 193–198 (1966).
11. JANKOVIC, B. D.; DRASKOCI, M. and JANJIC, M.: Passive transfer of 'allergic' encephalomyelitis with antibrain serum injected into the lateral ventricle of the brain. Nature *207:* 428–429 (1965).
12. KAPLAN, M. H. and CRAIG, J. M.: Immunologic studies of heart tissue. VI. Cardiac lesions in rabbits associated with autoantibodies to heart induced by immunization with heterologous heart. J. Immunol. *90:* 725–733 (1963).
13. KAPLAN, M. H. and FRENGLEY, J. D.: Autoimmunity to the heart in cardiac disease. Current concepts of the relation of autoimmunity to rheumatic fever, postcardiotomy and postinfarction syndromes and cardiomyopathies. Amer. J. Cardiol. *24:* 459–473 (1969).
14. LAUFER, A. and DAVIES, A. M.: Allergic granulomatous myocarditis, In methods and achievements in experimental myocarditis; vol. III (Karger, Basel 1966).
15. LAUFER, A. and DAVIES, A. M.: Experimental granulomatous myocarditis: genesis and immunological aspects. Ann. N. Y. Acad. Sci. *156:* 91–104 (1969).
16. LAUFER, A.; DAVIES, A. M. and BEN-ISHAY, Z.: Myocarditis. Accad. med. dei Lincei, Roma, in press (1970).
17. LEGATO, M. J.; SPIRO, D. and LANGER, G. A.: Ultrastructural alterations in mammalian myocardium by variation in perfusate ionic composition. J. Cell Biol. *37:* 1–12 (1968).
18. LERNER, R. A. and DIXON, F. J.: Transfer of ovine experimental allergic glomerulonephritis (EAG) with serum. J. Exp. Med. *124:* 431–442 (1966).

19. MARCUS, Z. A.; DAVIES, A. M. and GINSBURG, I.: Oxygen stable hemolysin of group A streptococci. V. Effect on rat heart and kidney cells grown in tissue culture. Exp. molec. Path. *5:* 93–107 (1966).

20. MONAYENKOV, A. M. and REPIN, Y. M.: Relationship between cardiac hypertrophy and autoimmunological processes during prolonged high altitude hypoxia. In: physiology and pathology of blood circulation, (translated), Foreign Technol. Division, FTD-TT-63-295/1 and 2, AFSC, Wright Patterson AFB, Ohio, pp. 109–111 (1963).

21. MURPHY, G. E.: The characteristic rheumatic lesions of striated and of non-striated or smooth muscle cells of the heart. Genesis of the lesions known as Aschoff bodies and those myogenic components known as Aschoff cells or as Anitschkow cells or myocytes. Medicine *42:* 73–89 (1963).

22. NAKAMURA, R. M. and WEIGLE, W. O.: Transfer of experimental autoimmune thyroiditis by serum from thyroidectomized donors. J. Exp. Med. *130:* 263–283 (1969).

23. ONISHI, S.: Die Feinstruktur des Herzmuskels nach Aderlass bei der Ratte. Beitr. Path. Anat. *136:* 96–132 (1967).

24. ROITT, I. M.; JONES, H. E. H. and DONIACH, D.: Mechanisms of tissue damage in human and experimental autoimmune thyroiditis. In: Proc. 2nd. Int. Symp. Immunopath. (Benno Schwabe, Basel, 1962).

25. WALDMAN, G.: Submikroskopische Herzmuskelveranderungen bei Ratten und Meerschweinchen nach Streptolysin-O-Gaben. Acta Biol. Med. German. *15:* 788–809 (1965).

26. WILLIAMS, G. M.; DePLANQUE, B.; LOWER, R. and HUME, D.: Antibodies and human transplant rejection. Ann. Surg. *170:* 603 (1969).

27. WILLIAMS, T. W. and GRANGER, G. A.: Lymphocyte in vitro cytotoxicity: Mechanism of lymphotoxin-induced target cell destruction. J. Immunol. *102:* 911–918 (1969).

28. ZEFT, H. J.; MARABLE, I. W.; CASEY, H. W. and GLENN, W. G.: The effect of high altitude hypoxia on antiheart antibodies and experimental myocarditis. Arch. Path. *82:* 434–442 (1966).

Mailing address: Dr. A. Laufer, Department of Pathology, Hebrew University-Hadassah Medical School, Jerusalem (Israel).

Myocardial Infarction

Institute of Morbid Anatomy, School of Medicine,
University of Milan, Milan, Italy

HUMAN MYOCARDIAL INFARCTION: CORONAROGENIC OR NONCORONAROGENIC COAGULATION NECROSIS?[1]

G. BAROLDI

Coronarogenic or non-coronarogenic or both? The question arises from the critical review of the different vascular theories proposed for the human cardiac infarct. The term "coronarogenic" implies a primary mechanical impairment at some point of the arterial coronary tree, which leads to an interference of the blood supply and is responsible for a myocardial coagulation necrosis. In other words an *acute primary* obstructive lesion of a vessel, lasting for a certain period of time to induce the irreversible myocellular damage typical of the cardiac infarct in man.

The following two main categories of cardiac necrosis have been distinguished (44): a) "Coronarogenic necrosis", mainly due to atherosclerosis as the most predisposing factor, on which a functional overstrain of the heart muscle, a supervening coronary spasm, or a supervening thrombosis may act; b) "Non-coronarogenic necrosis", secondary to anemia, anoxia, mechanical stress, shock (four groups that are akin to coronarogenic necrosis), metabolic processes, hormonal action, toxic conditions, allergic-infectious state. In the first case, the necrosis may be micro- or macrofocal, while in the second one the lesion is microfocal.

This useful distinction, however, groups together patterns that differ from a morphologic and, likely, from a pathogenetic point of view. In fact, the use of the generic term "necrosis" neglects the various types of myocellular death. Morphologically speaking, the myocytolysis and the necrosis secondary to inflammatory processes behave in a different way from the infarctual coagulation necrosis, even if all of them share a common final fibrous repair. Also, the distinction between macrofocal and microfocal lesions is misleading, since each one of the above-mentioned types of necrosis may be micro- or macrofocal. We had the opportunity to examine some cases in which an acute extensive myocytolysis (7) was the only cause of death found (fig. 1). Myocytolysis in particular is, as yet, a little-known, often overlooked pattern, frequently associated with coagulation necrosis, but even observed as a unique myocardial lesion in different pathologic conditions (13, 50, 52).

[1] Supported by the National Institute of Health, United States Postdoctoral Research Fellowship Grant, Award No. FF–485.

a

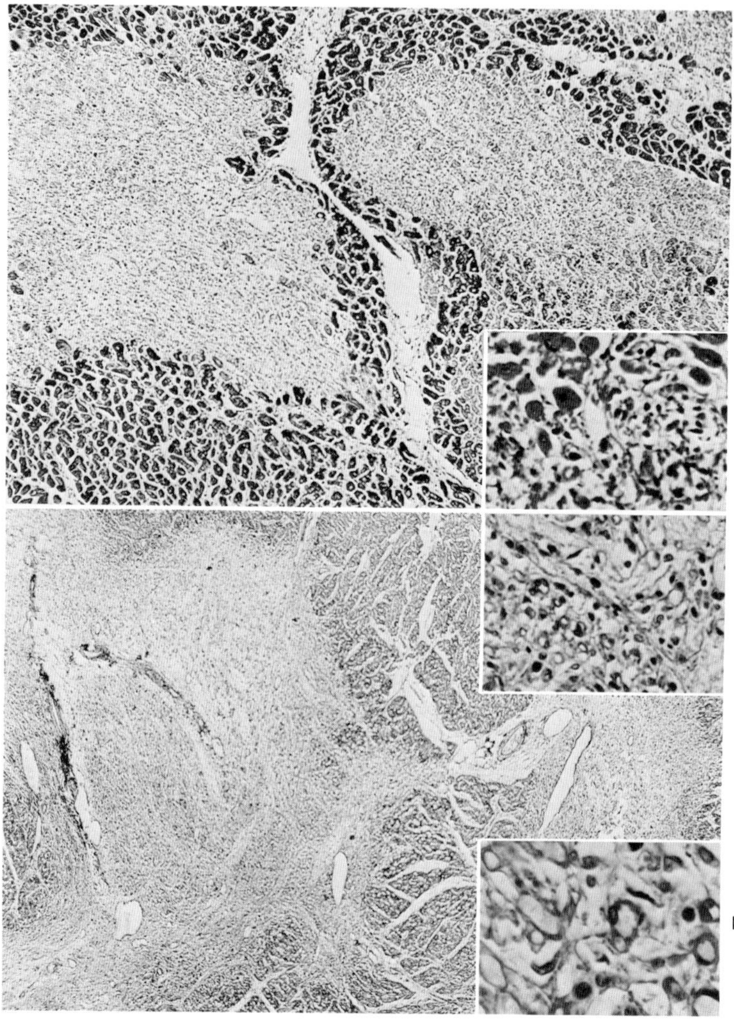

a′

b″

b

b′

Fig. 1 a, a′ Large, confluent foci of severe disruption and lysis of the myocells without cellular reaction (few lymphocytes and pigment-loaded macrophages) as main lesion in a 59-year-old man dying suddenly (during moderate physical activity; short, mild precordial pain 4 days before death) with rupture of the cardiac wall and hemopericardium. Severe stenoses of the middle tracts of both the anterior descending branch and the circumflex branch of the left coronary artery. Heart weight 500 g. (a, PTAH, ×95; a′, Movat, ×285).

b Different stages of extensive, focal-confluent myocytolysis, showing lysis of the myocells, alveolar pattern (b′) and early fibrous repair (b″) in both ventricles of the heart (65 g) as a unique finding in a 6-month-old baby (5,940 g) without previous symptoms and signs. Rapid death (within 35 h) with general malaise, muscular hypotonia, bradycardia. (b, Movat, × 60; b′ and b″, PAS-HE, × 360).

For a discussion on the pathogenesis of the human cardiac infarct–interpreted as a coagulation necrosis–a synthesis of the main proposed theories could be of some help (17, 19–22, 24, 26, 29, 33, 35, 37, 40–43, 47–49, 53, 54, 58), according to the following distinction.

I. *Acute coronary insufficiency due to mechanical arterial occlusion; coronarogenic infarct*: a) At the level of the subepicardial or extramural arteries or branches: thrombosis, embolization, rupture of atheroma, intimal hemorrhage, dissecting aneurysm, intimal edema, spasm. b) At the level of the intramyocardial or intramural arterial branches: thrombosis, platelet thrombosis or aggregation, erythrocyte aggregation, settling of blood cells, embolization, capillary damage, spasm. Most of these causes may ensue in previously diseased vessels and may act synergically.

II. *Acute coronary insufficiency due to extracoronary causes; non-coronarogenic ischemic infarct*: Overstrain of the heart muscle, acute massive hemorrhage, shock, "steal syndrome," subendocardial ebb-flow mechanism, intramural vascular compression, Hb/O_2 abnormal affinity. These causes, in general, ought to act on the background of a chronic hypoxia due to different conditions, such as chronic obstructive coronary diseases, severe cardiac hypertrophy, cardiac malformations, stenosing valvulopathies, chronic anemia, asthma, and other similar, clinicopathologic patterns that are considered responsible for a chronic absolute or relative coronary insufficiency. Obviously, when the association is with a chronic obstructive coronary disease the ischemic infarct could be better defined as partially coronarogenic.

III. *Acute primary myocardial coagulation necrosis* due to a neurohormonal disorder leading to a derangement of the ionic equilibrium or dysionism (48) of the myocell.

IV. *Pluricausal theory,* according to which the coagulation necrosis could be the result of the association of coronarogenic and non-coronarogenic dysmetabolic factors.

The number of the proposed theories is proportional to the uncertainty as regards the pathogenesis of the human cardiac infarct. At present, none of them has been conclusively documented in man and even the leading cause, *i.e.,* coronary thrombosis, is open to discussion.

Most of our information derives from experimental models, which, however, may easily mislead us in our interpretation of the human pathology. The experimental occlusion of a normal coronary artery in animals could be an example of this, since an equivalent pathogenetic mechanism can be proposed only exceptionally in man. In fact, about 90% of the human infarcts occur in the presence of a long-lasting, severe obstructive arterio-atherosclerosis, generally involving more than one extramural artery or branch; the eventual acute occlusion takes place in a severely stenosed vessel (3). A discussion on the previously defined "coronarogenic infarct" must start from this well-documented basis in order to evaluate the significance of an eventual acute occlusion in these conditions–conditions in which an essential role is played by the arterial anastomotic vessels. Let me first summarize very briefly the following pertinent data obtained in a study on the human collaterals (9). 1) In any area of the normal heart, countless communications exist between the coronary arteries (intercoronary anastomoses) and the branches of the same coronary artery (homocoronary anastomoses), the diameter of which ranges from less than 20 to 350 μ. 2) This collateral reserve is capable of enlarging in direct proportion to the degree and the extension of the obstructive lesions. The number of the anastomoses showing an impressive increase in diameter (up to 2,000 μ) is maximum in cases with multiple occlusions and/or stenoses. 3) The compensatory function of these vessels is reflected in man by the high frequency of the cases with one or more occlusions (44%) or with one or more severe (lumen reduction of more than 60-70%) stenosis (88%) without an infarct. 4) The enlargement of the anastomotic system is related to the vascular stenosing-occlusive damage of the parent vessels, but not to the myocardial status. In other words, the same degree of enlargement has been found in hearts with the same coronary lesions but with different myocardial involvement, namely, with or without infarction. 5) The behavior of the collaterals is independent of age, sex, and the type of death ("clinical" or "sudden" death) (fig. 2). From these data, we may postulate that when a human myocardial infarct occurs, in a small percentage of the cases there is no proof of a manifest coronarogenic origin, while in most of the cases it ensues in the presence of an already functioning collateral circulation that, for a long time, was able to compensate the previous vascular lesions. Apart from the questionable assumption that the normal collaterals are not functional (6, 9), we have to conclude that in human infarction the coronary arteries, as a result of a chronically pre-existing obstructive disease, are not, even physiologically speaking, end arteries. On that basis, what can be the meaning of a supervening acute occlusion of any type? The retrograde collateral flow distally to the severe stenosis will minimize or nullify the efficacy of this acute occlusion in stopping the proximal arterial flow. This mechanism is supported, not only by the high frequency of the human cases with occlusion and/or stenosis without infarct, but even by an experiment in dogs in which the occlusion performed at the level of a severe stenosis, established a few days before, was not followed by infarct or fibrillation or by coronary vasoconstrictive reflexes (32).

One may wonder if the particular hemodynamic situation at the site of the severe stenosis could not explain why and when a thrombosis in the residual lumen may occur. The decrease of the proximal flow, proportional to the degree and the length of the stenosis, is paralleled by an increase of the distal backflow

a b
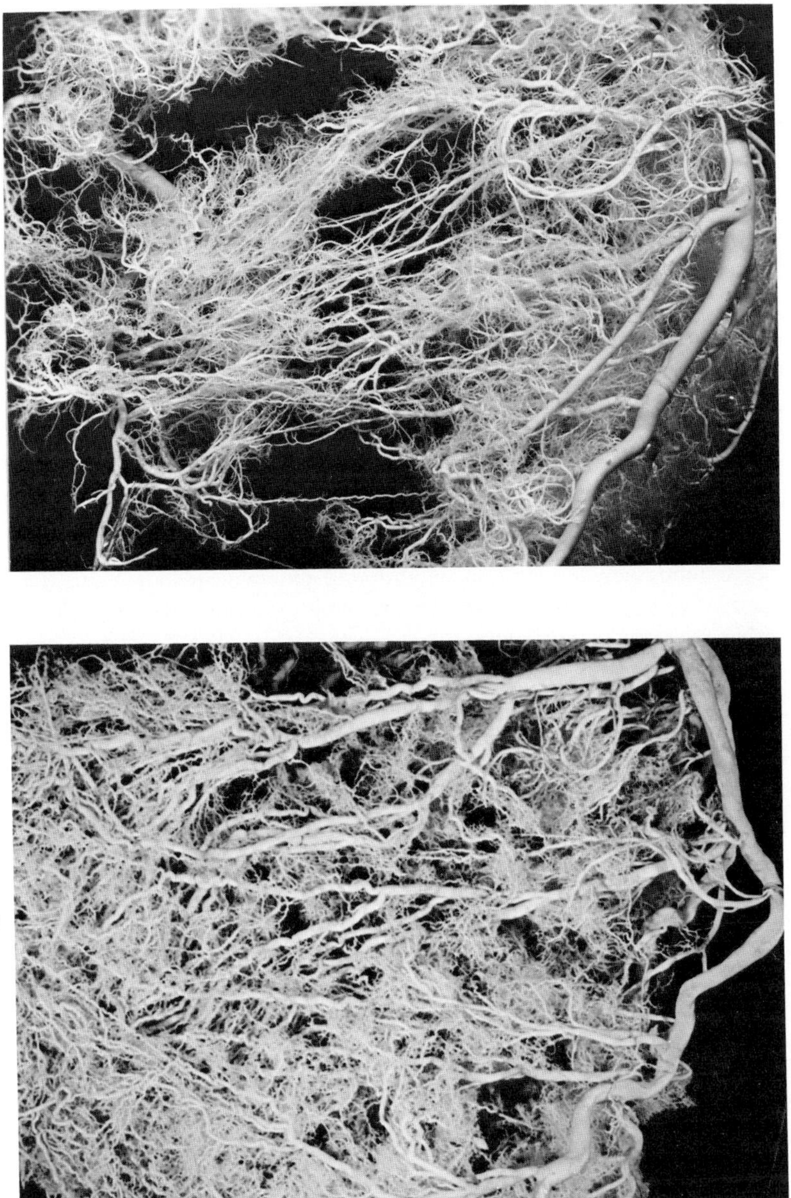

Fig. 2 Comparison between the normal aspect (a) and apparently increased in number and enlarged septal anastomoses in a case of right coronary occlusion (b).

through the anastomoses that is capable of restoring the blood pressure below the stenosis. The consequent reduction and slowing of blood flow within the stenosis is a favorable condition for the establishment of a thrombus, the probability of which will be enhanced by any cause that will increase the peripheral resistance in the vascular area of the stenosed vessel. In fact, this will aggravate its already critical hemodynamic equilibrium as, for example, in cases of a massive coagulation necrosis. In this instance, already in the very early stage, there occur many functional and structural changes (paradoxic systolic lengthening of the cardiac wall, acidosis, interstitial edema, secondary thrombosis of the intramural vessels, etc., resulting in an "avascular area" (2, 7)) capable of blocking the intramural flow in the damaged area. This block implies a further increase of the blood pressure in the vascular tract distal to the stenosis, with a further increase of the hindrance of the flow within the stenosis. It sounds, therefore, logical to theorize that the thrombus in the human myocardial infarct may be a secondary event (4, 9), particularly if we keep in mind the increased thromboplastic activity of the blood following a necrosis (16, 27, 38) and that the atherosclerotic walls are less protected against the thrombosis (45, 46). This fits well with previous observations that the frequency of the thrombus increases with the extension of the myocardial lesion (3, 11, 15, 28) and with the increase of the survival time (55), and that the age of the thrombus may not correlate with the age of the myocardial coagulation necrosis (3, 10). The claim that the secondary formation of a thrombus is contradicted by the finding that the thrombus is always separated from the closest margin of the associated infarct (11) does not take into consideration the peripheral redistribution of the blood flow around the necrotic zone, the true hemodynamic impairment being limited to or predominant at the stenosis level. In the plastic-cast study of the coronary arteries (9), we were always able to inject the distal portion of an occluded vessel, showing the persistence of the anastomotic channels around the avascular area, which is typical of a coagulation necrosis (fig. 3). Since the blood flow distribution is chiefly dependent on the pressure gradients, it is likely that the blood flow, when prevented from running down in an area with an increased peripheral resistance, will escape in the adjacent vessels with lower resistance. Such a flow redistribution in addition to the well-preserved integrity of the vascular wall distally to the stenosis explains why the thrombus is confined to the area of the stenosis or above it.

The lack of an occlusive function of the thrombus and its secondary formation lessens the sense of the controversy on the frequency of the coronary thrombosis in the human cardiac infarct. For the sake of argument, however, it must be recalled that figures in the literature vary over a wide range as far as this frequency is concerned (from 7 to 96%) (10-12, 14, 15, 18, 23, 31, 36, 56). The discrepancy has been attributed to a fault of technique and to a different definition of the cardiac infarct or to a lysis of the thrombus (11). In this

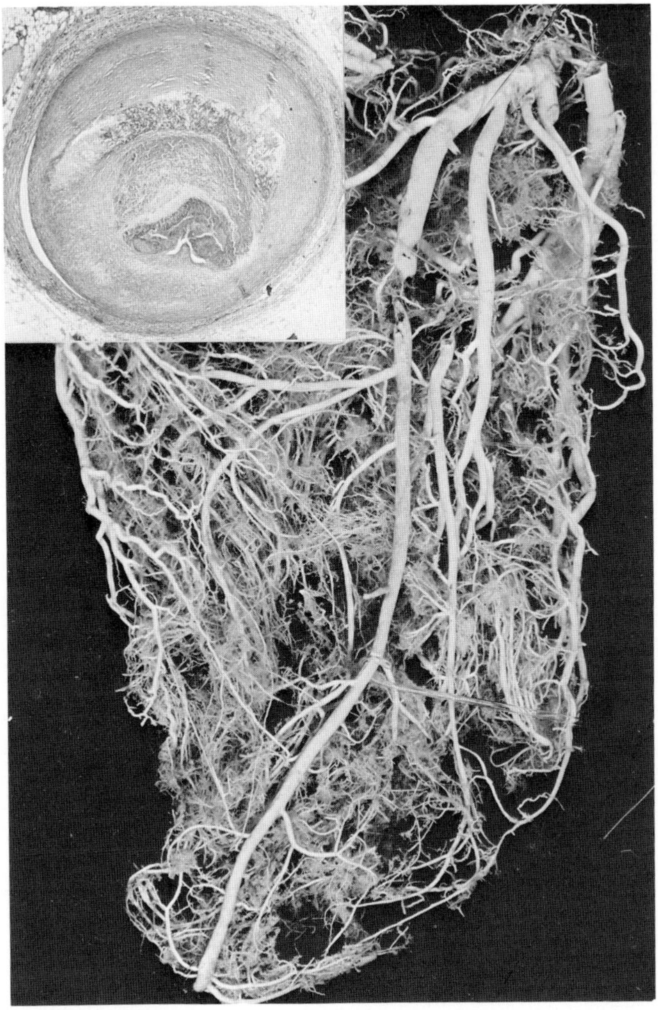

Fig. 3 Plastic cast of coronary arteries, showing an occlusion of the anterior descending branch and of its secondary branch with extensive apical-septal posterior "avascular" area corresponding to an acute infarct. The origin of the distal portion of the main occluded vessel, well injected through collaterals deriving from all the adjacent branches, shows a severe reduction of its diameter. The histologic counterpart of such an occlusion can be exemplified in the inset, which demonstrates the case of an extremely severe stenosis occluded by a recent thrombus (HE, × 14). In this condition, the increased collateral circulation induced by the severe stenosis exists prior to the thrombus formation and may nullify its functional occlusiveness. The thrombus could result from the increased peripheral resistance in the avascular area plus thromboplastic local and general factors. The lack of thrombosis in the whole distal segment of the occluded vessel is possibly due to the flow redistribution around the infarcted area (see text).

last-mentioned critical review, only the studies carried out by longitudinal incision of the decalcified coronary arteries and in which the concept of infarct was limited to a necrosis at least 2.5 cm in greatest diameter were accepted. Selection of the cases according to these parameters gave a thrombus frequency ranging between 80 and 90% and the conclusion was that the thrombus is the main cause of the cardiac infarct. Two considerations, however, seem justifiable. First, that by longitudinal incision of the vessels we lose the possibility of evaluating the previous lumen stenosis and, therefore, the occlusive efficacy of the thrombus. Second, we cannot limit the concept of a disease to the size of its lesion.

For its definition, the pathogenesis of a disease has to be unique, independently of the extension of the morphologic changes, or we have to distinguish different diseases having a diverse pathogenesis, even if the only morphologic difference is circumscribed to the extension of a similar lesion. The myocardial infarct starts as spotty damage limited to a single cell or a group of cells, and it may then become focal, focal-confluent, or massive; at any stage of its evolution, death may occur, mainly owing to electrical instability of the myocardium. At present, because of the pathogenetic uncertainty, there is no justification to consider the small and the extensive coagulation necrosis as patterns of different diseases, particularly if we interpret the thrombosis as a non-causal secondary phenomenon, more frequent in the presence of extensive alteration in relation to the above-mentioned hemodynamic and thromboplastic factors.

Intravitam or postmortem lysis of the thrombus has been frequently assumed, in explanation of the absence of the thrombus in cases of infarct without occlusion at autopsy. In normal dogs, the induced coronary thrombosis, followed by an infarct, did not show any lytic change of the thrombus before 10 days; on the 17th day of survival the lumen was seen to be completely occluded in 66.6% of the cases, and partially occluded (75% of lumen reduction) in 26.6%, with the remaining 6.6% showing a 25% reduction of the lumen (57). In our material (3), the absence of an acute occlusion was 53% in cases of acute-recent infarct, 53% in cases of sudden and unexpected "coronary" death, and 54% in cases of sudden but not unexpected "coronary" death. If we limit our figures only to the cases with early coagulation necrosis or to the sudden death cases with an apparently normal myocardium – that is to say, with a very short survival time of much less than 10 days–the absence of an acute occlusion is 56.7, 52.8, and 56.4%, respectively. Moreover, we must emphasize that the previous experiment in the healthy animals differs from those in the athero-sclerotic animals in which the reduced fibrinolytic activity was unable to resolve the thrombi following an intracoronary injection of thrombin (45, 46). From these data, it seems unlikely that the disappearance of a thrombus may explain the negative autoptic finding, particularly in human coronary arterio-atherosclerosis in which a reduced fibrinolytic and anticoagulant activity has been reported (46). The postmortem lysis of the thrombus with its total dissolution despite the lack of any washing-out effect does not seem equally easily acceptable.

Until now, we have been referring to the whole-blood thrombus. However, we must consider the role of the evanescent, reversible platelet thrombus not found at autopsy if the organ fixation is delayed (43). Its role is worthy of discussion, not as a possible cause of occlusion at the level of a severe stenosis of an extramural artery or branch already compensated by collaterals, but because the occurrence of transient platelet thromboemboli in the microcirculation has been suggested as a possible pathogenetic mechanism in sudden "coronary" death and in myocardial infarction (29, 43). This interpretation has been derived from the results obtained by the intracoronary and intraventricular infusion of adenosine diphosphate in swine, resulting in most of the animals in transient circulatory collapse, electrocardiographic ischemic changes, transient fall in the circulating platelet count, endothelial damage, and formation of platelet aggregates in the microcirculation, the latter being more prominent during or immediately after the infusion than in the later stages; in the animals allowed to live for 2 h or more, gross as well as light- and electron-microscopic evidence of myocardial infarcts was found. Animals made thrombocytopenic or with platelets made refractory to ADP infusion did not show evidence of severe circulatory changes or myocardial infarcts. From this finding, it has been assumed that a primary myocardial injury due to a direct ADP action on myocardial cells was unlikely and the conclusion was drawn that the formation of platelet aggregates caused by ADP released by different sources (injured tissue, red blood cells, or platelet themselves) in the microcirculation could produce myocardial infarcts (29). This experimental model is very provocative and the hypothesis is worthy of investigation, particularly if we keep in mind that the platelet adhesiveness can be increased by many other substances and that, in ischemic heart disease, an augmented adhesive index has been demonstrated (27). However, some points need further elucidation before this experimental pathogenetic mechanism can be transferred to human pathology, since an identical lesion can be reproduced by different pathogeneses. The same myocardial coagulation necrosis seems obtainable experimentally by acute coronary occlusion or catecholamines, by a hormone-salts combination, or ADP platelet aggregation, apparently without the possibility of demonstrating a common denominator. The point is to understand what really is acting in man, without preconceived ideas and without being influenced by the experimental results, even if we pay the maximum attention to them. We know that, as a consequence of myocardial coagulation necrosis, hyaline thrombi usually occur within and around the damaged zone and that the intramural vessels in the center of the necrotic area undergo necrotic wall changes until they disappear, while at the periphery, in the reactive zone, they show an obliterative arteritic-like process (2, 9)— a vascular pattern never found in normal areas of infarcted hearts or of hearts with coronary arterio-atherosclerosis. If, in the latter, we admit the possibility of recurrent showers of platelet thromboemboli with endothelial injury, the observation of a wall reaction, namely a vasculitis, in the absence of myocardial coagulation necrosis should be a usual finding, and this is not so. The "ADP infusion theory" implies a primary source, such as injured tissues or red blood cells, or the platelets themselves. In most of the cases of infarct or sudden "coronary" death, a pre-existing injured tissue capable of releasing the requested amount of ADP is not easily found. The distortion of red blood cells in the sclerotic vessel as suggested, does not explain why the infarct occurs only in some atherosclerotic subjects; furthermore, a selective coronary infusion of ADP, as in the experiment cited, seems highly unlikely. Therefore, we may expect an infusion of the total coronary tree with disseminated infarcts everywhere in the myocardium, which is not the usual pattern in man, the infarct being limited, in general, to one well-defined area, no matter if the lesion is massive, focal-confluent, or focal. On the other hand, there exist some diseases in human pathology that appear as good "experimental models" in man for the evaluation of many related "microcirculatory disturbances" theories, including red blood cell aggregation (sludging), the settling of blood cell masses, and the disseminate intravascular coagulation syndrome. The first is thrombotic thrombocytopenic purpura (Moschcowitz disease), a concurrence of many hypoxiating and microvascular factors, including the "ADP infusion" with showers of platelet thromboemboli. In fact, an extremely severe, long-lasting hemolytic anemia starting in previously healthy people, severe thrombocytopenia, and hemorrhagic and neurologic disorders are all associated with an equally extremely severe impairment of the small arterial intramural vessel showing an obstructive wall damage frequently combined with platelet thrombi. In only 10.2% of 39 cases examined (8) were we able to demonstrate rare, microfocal foci of an early coagulation necrosis and, therefore, a terminal event not related to the vascular lesions (fig. 4). A review of the cases reported in the literature as well as of our own cases did not evidence true clinical and electrocardiographic signs of cardiac ischemia. Another example is sickle cell anemia, in which blood sludging has been demonstrated *in vivo* (34) and in which a vicious cycle of sickling and plugging of the red blood cells is accepted. In not one instance out of 53 cases studied were foci of acute-recent coagulation necrosis detected (5). Furthermore, the same absence of acute necrotic changes was noted in human subjects dying after prolonged traumatic and/or hemorrhagic shock. In only one case out of 56 was an acute myocardial infarct present (5). These examples make us cautious in accepting the microcirculatory disturbances and the acute insufficiency *per se* as a primary cause of the human cardiac infarct. The steal syndrome and the ebb-flow mechanisms, as well as spasm, the last one of which has not been convincingly proved to occur

a c b d

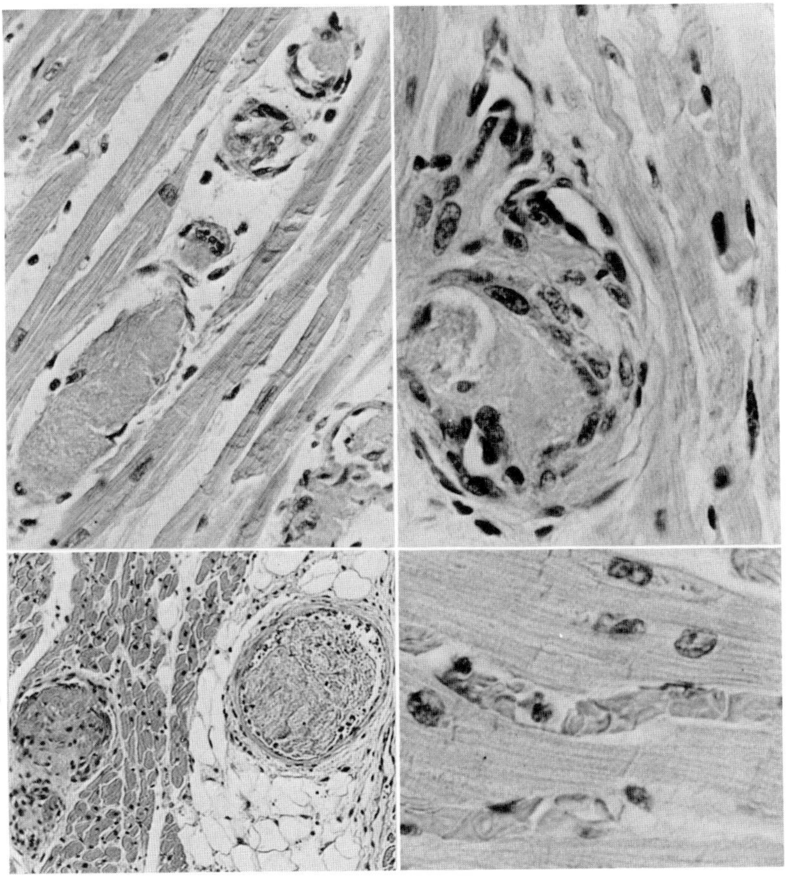

Fig. 4 Normal myocardium associated with severe obstructive impairment with platelet thrombi of the intramyocardial small arterial vessels, in cases of thrombotic thrombocytopenic purpura. (a, HE, × 450; b, HE, × 170; c, HE, × 440). Absence of myocardial damage in a case of sickle cell anemia with "erythrostasis" (d, HE, × 490).

in man and is not supported by experimental data (30–39), seem to be contradicted by the richness of the collaterals present in any area and, as far as the spasm is concerned, by their structural make-up, which is mainly capillary in type (9). On the other hand, the same caution is necessary in evaluating the so-called chronic ischemic predisposing factors, whether coronarogenic or not, particularly if we keep in mind, for example, the rarity of an infarct in cases with bilateral occlusion or severe stenosis of the coronary ostia secondary to an aortitis (51) (in general associated with severe cardiac hypertrophy) or in instances of aortic or mitral stenosis. In that sense, the same relatively low frequency of an infarct in the cases with severe coronary arterio-atherosclerosis has to be emphasized. If it is true that about 90% of the human infarcts are associated with a severe obstructive coronarosclerosis, it is also true that most of the coronary arteriosclerotic subjects do not show an infarct or die suddenly. This lack of correlation has also been documented clinically (25). Why, then, may or may not the same degree of coronary disease showing the same pattern of a previously long-lasting collateral compensation be associated with an infarct? Because of the presence of functioning anastomoses, the role of the coronary arterio-atherosclerosis *per se* is questionable, whether as a primary cause or as a predisposing factor. If, as is probably the case, a link exists between these two lesions, it only acts, however, in a particular group of coronary arterio-atherosclerotic subjects and at a certain moment, for some as yet not well-established reasons that are not necessarily definable as coronarogenic or ischemic. We need more information as regards the eventual primary metabolic disorders and their relationship with the dynamics of the cardiac wall as well as concerning the relationship between the latter and the intramural circulation.

SUMMARY

The coronarogenic and non-coronarogenic main pathogenetic theories of the human cardiac infarct are briefly reviewed and summarized. The importance of the coronary collateral in understanding the natural history of the lesion is outlined. In most of the cases of infarct with a coronary thrombus (about 50% of our material), the latter takes place in severely stenosed extramural vessels and therefore already compensated by functioning anastomoses. Under these conditions, the retrograde collateral flow makes the acute occlusion ineffective and the possibility of a secondary formation of the thrombus is suggested. The impairment of the intramural branches by stenosing processes and by platelet- and red blood cell-aggregation seems to have little if any significance, as is shown in the cases of thrombotic thrombocytopenic purpura and sickle cell anemia.

REFERENCES

1. ALBERTINI, A.; BRUNCK, H. J. und PAPERNITZKI, A.: Die Coronarosklerose in der schweizerischen Bevölkerung. Eine statistische Erhebung an Hand der Sektionsfälle eines Jahres (1.3. 1955–20.2. 1956). Bull. Schweiz. Akad. med. Wissensch. *13:* 17–37 (1957).
2. BAROLDI, G.: Histopathological study of the intramural artery vessels in relation to the pathology of extramural coronary arteries and myocardial damage. Cardiologia *41:* 364–380 (1962).
3. BAROLDI, G.: Acute coronary occlusion as a cause of myocardial infarct and sudden coronary heart death. Amer. J. Cardiol. *16:* 859–880 (1965).
4. BAROLDI, G.: Lack of correlation between coronary thrombosis and myocardial infarction or sudden "coronary" heart death. Ann. N.Y. Acad. Sci. *156:* 504–525 (1969).
5. BAROLDI, G.: High resistance of the human myocardium to shock and red blood cell aggregation (sludge). Cardiologia *54:* 271–277 (1969).
6. BAROLDI, G.: Functional morphology of the anastomotic circulation in human cardiac pathology. In: BAJUSZ, E. and JASMIN, G.: Functional Morphology of the Heart. Methods and Achievements in Experimental Pathology, pp. 438–473 (Karger, Basel, 1971).

7. BAROLDI, G. and FALZI, G.: Miocitolisi diffusa. Cuore e circolazione *50:* 365–373 (1966).
8. BAROLDI, G. and MANION, W. C.: Microcirculatory disturbances and human myocardial infarction. Amer. Heart J. *74:* 171–178 (1967).
9. BAROLDI, G. and SCOMAZZONI, G.: Coronary circulation in the normal and the pathologic heart. (Government Printing Office, A.F.I.P., American Registry of Pathology, Washington 20305, U.S.A., 1967).
10. BRANWOOD, A. W. and MONTGOMERY, G. L.: Observations on the morbid anatomy of coronary artery disease. Scot. Med. J. *1:* 367–375 (1956).
11. CHAPMAN, L.: Relationship of recent coronary artery occlusion and acute myocardial infarction. J. Mt. Sinai Hosp. *35:* 149–154 (1968).
12. CLAWSON, B. J.: Coronary sclerosis. An analysis of nine hundred twenty-eight cases. Amer. Heart J. *17:* 387–400 (1939).
13. CONNOR, R. C. R.: Focal myocytolysis and fuchsinophilic degeneration of the myocardium of patients dying with various brain lesions. Ann. N.Y. Acad. Sci. *156:* 261–270 (1969).
14. EDWARDS, J. E.: What is myocardial infarction? Circulation *40:* (suppl. IV): 5–12 (1969).
15. EHRLICH, J. C. and SHINOHARA, Y.: Low incidence of coronary thrombosis in myocardial infarction. A restudy by serial block technique. Arch. Path. *78:* 432–445 (1964).
16. EISENBERG, S.: Blood viscosity and fibrinogen concentration following cerebral infarction. Circulation *33–34* (Suppl. II): 10–14 (1966).
17. ELIOT, R. S. and MIZUKAMI, H.: Oxygen affinity of hemoglobin in persons with acute myocardial infarction and in smokers. Circulation *34:* 331–336 (1966).
18. ELIOT, R. S. and STREIFF, R.: The infrequency of coronary thrombosis causing death after recovery from myocardial infarction and the effect of anticoagulants. Circulation *39–40*, (Suppl. III): 75 (1969).
19. FERRANS, V. J.: HIBBS, R. G.; WALSH, J. J. and GURCH, G. E.: Histochemical and electron microscopical studies on the cardiac necroses produced by sympathomimetic agents. Ann. N.Y. Acad. Sci. *156:* 309–332 (1969).
20. FULTON, W. F. M.: The dynamic factor in enlargement of coronary arterial anastomoses and paradoxical changes in the subendocardial plexus. Brit. Heart J. *26:* 39–50 (1964).
21. HANDFORTH, C. P.: Isoproterenol-induced myocardial infarction in animals. Arch. Path. *73:* 161–165 (1962).
22. HARDAWAY, R. M.; McKAY, D. G. and HOLLOWELL, O. W.: Vascular spasm and disseminated intravascular coagulation. Influence of the phenomena one on the other. Arch. Surg. *83:* 173–180 (1961).
23. HARLAND, W. A. and HOLBURN, A. M.: Coronary thrombosis and myocardial infarction. Lancet *ii:* 1158–1160 (1966).
24. HAUSS, W. H.; GERLACH, U.; JUNGE-HULSING, G.; THEMANN, H. and WIRTH, W.: Studies on the "nonspecific mesenchymal reaction" and the "transit zone" in myocardial lesions and atherosclerosis. Ann. N.Y. Acad. Sci. *156:* 207–218 (1969).
25. HELFANT, R. H.; KEMP, H. G. and GORLIN, R.: The interrelation between extent of coronary artery disease, presence of collaterals, ventriculographic abnormalities and hemodynamics. Amer. J. Cardiol. *25:* 102 (1970).
26. HIRSCH, S.: Le rôle des petites branches coronaires dans la pathogénèse de l'infarctus myocardique. Acta med. Scand. *138:* 449–456 (1950).
27. HORLICK, L.: The adesive index, a measure of platelet stickiness. Its response to myocardial injury and to anticoagulant treatment. Intern. Symposium "Anticoagulants and fibrinolysins" pp. 56–57 (Toronto 1961).

28. HORT, W.: Mikroskopische Beobachtungen an menschlichen Infarktherzen. Virch. Arch. path. Anat. *345:* 61–70 (1968).

29. JORGENSEN, L.; ROWSELL, H. C.; HOVIG, T.; GLYNN, M. F. and MUSTARD, J. F.: Adenosine diphosphate-induced platelet aggregation and myocardial infarction in swine. Lab. Invest. *17:* 616–644 (1967).

30. JOYCE, E. E. and GREGG, D. E.: Coronary artery occlusion in the intact unanesthetized dog: intercoronary reflexes. Amer. J. Physiol. *213:* 64–70 (1967).

31. KAGAN, A.; LIVSIC, A. M.; STERNY, N. and VIHERT, A. M.: Coronary-artery thrombosis and the acute attack of coronary heart-disease. Lancet *ii:* 1199–1202 (1968).

32. KHOURI, E. N.; GREGG, D. E. and LOWENSOHN, H. S.: Flow in the major branches of the left coronary artery during experimental coronary insufficiency in the unanesthetized dog. Circulat. Res. *23:* 99–109 (1968).

33. KNISELY, M. H.: The settling of blood cell masses to the lower sides of vessels during life and the probable significance. H. Ford Symposium: The etiology of myocardial infarction. pp. 557–585 (Little, Brown & Co., Boston, 1963).

34. KNISELY, M. H.; BLOCH, E. H.; ELIOT, T. S. and WARNER, L.: Sludged blood. Science *106:* 431–440 (1947).

35. KORB, G. and TOTOVIC, V.: Electron microscopical studies on experimental ischemic lesions of the heart. Ann. N.Y. Acad. Sci. *156:* 48–60 (1969).

36. KURLAND, G. S.; WEINGARTEN, Ch. and PITT, B.: The relation between the location of coronary occlusions and the occurrence of shock in acute myocardial infarction. Circulation *31:* 646–650 (1965).

37. LAPLANE, R.; PAUTRAT, J. et GRAVELEAU, D.: L'aptitude reflexogène de la paroi aortique: son rôle dans l'infarction du myocarde. Presse Méd. *64:* 766–769 (1956).

38. LEVINE, H. D. and WELLS, R. E.: Blood sludging observed at the bedside: flexibility and limitations of examination by ophthalmoscopy. Circulation *32:* Suppl. II: 136 (1065).

39. LEVY, M. N.; CUTARELLI, R.; ZIESKE, H., Jr. and NELSON, G.: Effect of acute coronary artery ligation upon blood flow to the normal myocardium. Am. J. Cardiol. *13:* 48–50 (1964).

40. LONG, D. N.; MYER, M. V.; BROWN, E. B. and LILLEHEI, C. W.: Myocardial necrosis and electrocardiographic changes related to microcirculatory abnormalities. Amer. J. Cardiol. *10:* 695–705 (1962).

41. MASTER, A. M.; JARFE, H. L.; DACK, S. and GRISMAN, A.: Coronary occlusion, coronary insufficiency and angina pectoris. Amer. Heart J. *27:* 803–816 (1944).

42. MEESSEN, H.: Ueber den plotzlichen Herztod bei Frühsklerose und Frühthrombose der Koronararterien bei Männern unter 45 Jahren. Z. Kreislaufforsch. *36:* 185–201 (1944).

43. MUSTARD, J. F. and PACKHAM, M. A.: Platelet function and myocardial infarction. Circulation *40:* (Suppl. IV): 20–30 (1969).

44. MYASNIKOV, A. L.: Myocardial necroses of coronary and noncoronary genesis. Amer. J. Cardiol. *13:* 435–440 (1964).

45. MYASNIKOV, A. D.; CHAZOV, E. I.; KOSHEVNIKOVA, T. L. and NIKOLAEVA, L. F.: Some new data on the occurrence of coronary thrombosis in conjunction with atherosclerosis. J. Atheroscl. Res. *1:* 401–412 (1961).

46. MYASNIKOV, A. L.; KIPSCHIDZE, N. N. and TCHAZOV, E. I.: The experimental induction of myocardial infarction. Amer. Heart J. *61:* 76–80 (1961).

47. PATERSON, J. C.: Coronary artery intramural hemorrhage in the etiology of myocardial infarction. H. Ford Symposium "The etiology of myocardial infarction" pp. 247–254 (Little, Brown & Co., Boston, 1963).

48. RAAB, W.: Myocardial electrolyte derangement: crucial feature of pluricausal, so-called coronary, heart hisease (Dysionic cardiopathy). Ann. N.Y. Acad. Sci. *147:* 627–686 (1969).

49. RONA, G.; KAHN, D. S. and CHAPPEL, C. I.: Studies on infarct-like myocardial necrosis produced by isoproterenol: a review. Symposium. Rev. Can. Biol. *22:* 241–255 (1963).

50. SARAM, M.: Ueber die azelluläre Entstehung von Narben dei Durchblutungsstörungen im Herzmuskel. Beitr. Path. Anat. *118:* 275–291 (1957).

51. SCHARFMAN, W. B.; WALLACH, J. B. and ANGRIST, A.: Myocardial infarction due to syphilitic coronary ostial stenosis. Amer. Heart J. *40:* 603–613 (1950).

52. SCHLESINGER, M. J. and REINER, L.: Focal myocytolysis of the heart. Amer. J. Path. *31:* 443–459 (1955).

53. SELYE, H.: The chemical prevention of cardiac necrosis. (Ronald Press Co., New York 1958).

54. SHIMAMOTO, T.: Microcirculatory aspects of atherogenesis, thrombogenesis and antiatherosclerotics. Amer. Heart J. *76:* 105–113 (1968).

55. SPAIN, D. M. and BRADESS, V. A.: Frequency of coronary thrombi as related to duration of survival from onset of actue fatal episodes of myocardial ischemia. Circulation *22:* 816 (1960).

56. THOMAS, W. A.; HARTROFT, W. S. and O'NEAL, R. M.: Myocardial infarction in man and experimental animals. Arch. Path. *69:* 104–109 (1960).

57. WEISSE, A. B.; LEHAN, P. H.; ETTINGER, P. O.; MOSCHOS, C. B. and REGAN, T. J.: The fate of experimentally induced coronary artery thrombosis. Amer. J. Cardiol. *23:* 229–237 (1969).

58. ZINCK, K. H.: Sondervorrichtungen an Kranzgefässen und ihre Beziehung zu Coronarinfarkt und miliaren Nekrosen. Virch. Arch. path. Anat. *305:* 288–297 (1940).

Mailing address: Giorgio Baroldi, M.D., Istituto di Anatomia e Istologia Patologica, Facolta di Medicina, Universita di Milano, via Francesco Sforza 38, Milano (Italy).

Warner-Lambert Research Institute of Canada,
Sheridan Park, and Department of Pathology,
University of Toronto, Toronto, Ontario, Canada

ULTRASTRUCTURAL MYOCARDIAL CHANGES FOLLOWING EXPERIMENTAL NARROWING OF THE CORONARY ARTERIES

F. A. DE LA IGLESIA and G. LUMB

The occlusion of a coronary vessel either by ligation (8) or by hygroscopic plastic constrictors (15) induces myocardial morphologic changes that have been well recognized.

In an attempt to reproduce the condition occurring in humans (17), experimental gradual narrowing was employed in our studies to assess the cellular changes that might result in this condition. Previous light microscopic studies were conducted with a view to determining the early changes that occur before overt myocardial necrosis (21). Recently, we have undertaken studies designed to extend those observations, by investigating the presence of changes that may develop in early stages of coronary narrowing before complete occlusion.

Pigs were subjected to the surgical procedure previously described (20). The thorax was opened through the third right intercostal space under anesthesia and a wide pericardial incision was made. The pericardium was not sutured after the procedures. The initial segment of the anterior descending branch of the left coronary artery was dissected and the constrictors were applied to it 5 mm from the ostium of the artery. Plexiglass constrictors were used for controlled narrowing in this study. Flow probes were applied distally to the constrictor, to perform coronary blood flow measurements.

In preparations lasting longer than 6 h, the chest was closed without drainage. Electrocardiograms were monitored during the operation and daily, when applicable. Contrast medium perfusion (19) and gross and histologic examinations were used to evaluate the degree of narrowing of the coronary arteries at the area of constriction.

For electron microscopic studies, samples of myocardium were taken from an area in the left ventricle, arbitrarily chosen but always in the proximity of the narrowed vessel. Sampling was made usually within 30 to 60 min of constrictor removal after post-constrictor flow measurements. Portions of tissue were excised, fixed, dehydrated, and embedded in Epon. Ultrathin sections were stained with uranium and lead salts.

After application of the constrictor, a significant reduction of flow was obtained and this was monitored throughout the constriction period. In the acute preparations, there was a prolonged lag after removal of the constriction before normal flow conditions were regained. In the chronic preparations, the diminished flow produced by the constriction returned to original values shortly after constrictor removal. A significant variation in the distribution of the alterations was found early in the experiment and not all cardiac muscle fibers were affected.

Changes in mitochondria constituted a prominent feature after 1 h of narrowing. Interfibrillar mitochondria were within normal limits, while mitochondria in paranuclear areas evidenced significant changes in volume. Mitochondria showed matrix condensation and concentric cristae (fig. 1). Mitochondrial dense granules were changed. Other findings were the occurrence of small subsarcolemmal vacuoles. However, such observations were sparse. These changes were probably of incipient nature.

After 2 h, pynocytotic activity was observed in the sarcolemma and vascular endothelium. Sarcoplasmic reticulum dilatations and increased interfibrillar sarcoplasmic space resulted. Mitochondria were not swollen as seen at 1 h and their matrices were lighter, with disarranged cristae. Glycogen was preserved and a few small vacuoles appeared, engulfing glycogen particles. No significant changes were seen in the fibrils.

Marked changes were observed after 5 h of narrowing. Mitochondria seemed to revert to the appearance observed at 1 h. In extremely enlarged mitochondria, other changes were seen, including matrix rarefaction with areas of heterogeneous density. Concentric cristae were less frequent (fig. 2). Large vacuoles appeared and they varied in size and content, containing glycogen and proteinaceous material or a few mitochondria (fig. 3). Vacuoles were paranuclear or subsarcolemmal. Fibrils were altered with disarranged myofilaments.

At 48 h of narrowing, the changes were severe and diffuse. The vacuoles, either paranuclear or subsarcolemmal, were also observed and sometimes reached the size of the nucleus. Their contents and size were varied. These membrane-bound vacuoles contained glycogen, membranes, mitochondria, and other unidentified material. Dissociation of myofilaments was present and the changes in mitochondria were similar to those found at 5 h.

Cellular changes found after 72 h of narrowing did not differ significantly from those observed at 48 h. The large vacuoles were also present, but no substantial increase in number was found. The subsarcolemmal vacuoles were more prominent and there was fragmentation of myofilaments near the intercalated disc. The sarcoplasmic reticulum was not altered except for the few dilated lamellae already mentioned above. Mitochondria showed concentricity of cristae and clear matrix.

In our studies, detectable myocardial changes occurred within 1 h of coronary narrowing and a 20% reduction of the coronary flow.

The nature of the ultrastructural ischemic myocardial changes following

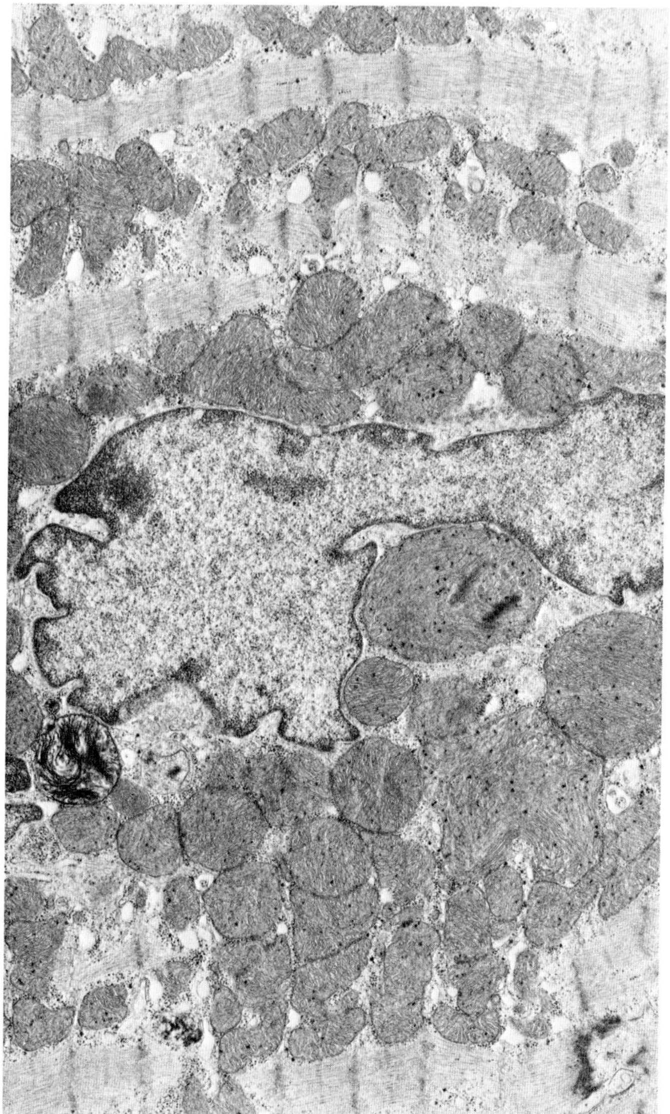

Fig. 1 1 h of narrowing. Electronmicrograph of a left venticular myocardial cell showing mitochondrial enlargement, condensation of the matrix, and parallel cristae. Lead and uranium stain, ×8,400.

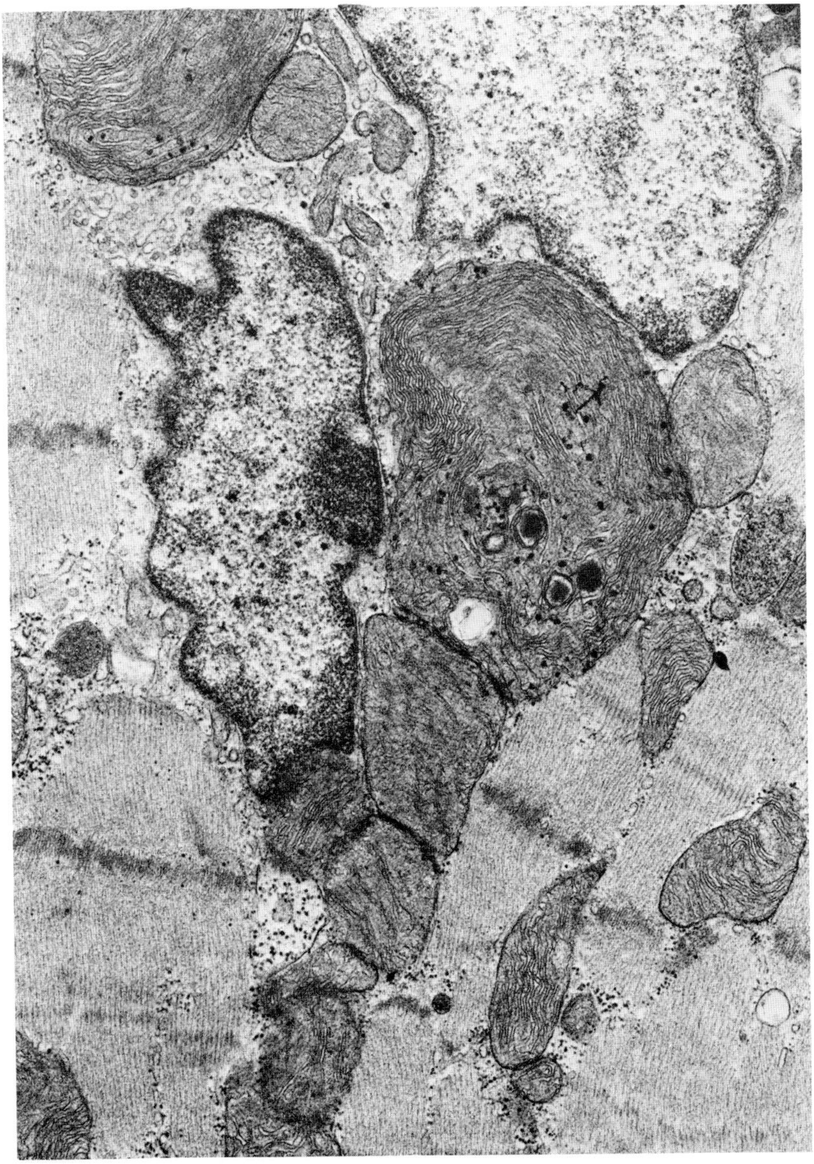

Fig. 2 5 h of narrowing. Marked mitochondrial enlargement with matrical alterations. Lead and uranium stain, ×19,000.

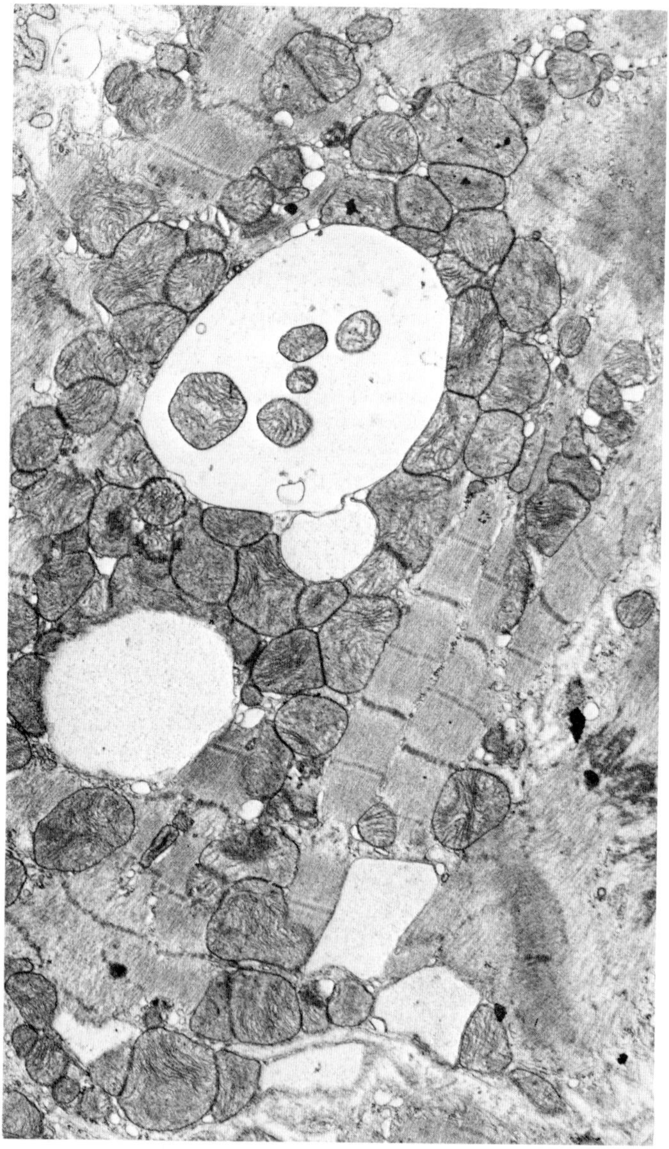

Fig. 3 5 h of narrowing. Electron microscopic aspect of the large vacuoles appearing in myocardial cells at this period. Mitochondria appear engulfed (center top). Lead and uranium stain, ×7,560.

occlusion in several species is well documented (2–5, 10, 12, 14). However, subcellular changes in experimental narrowing have not been defined. A detailed light microscopic study was carried out together with histochemical assays (21). There were obvious cellular changes evidenced by 48 and 72 h. The changes at 72 h resulted from occlusion, which was produced gradually. Since we were able to demonstrate changes within 48 h of narrowing, the morphologic alterations seen were considered to be the results of a partially occluded vessel.

A striking resemblance exists between the already known ischemic alterations and the lesions we have described. This suggests that the initiation of cellular changes takes place as soon as reduction of the blood supply occurs. Also, the changes due to narrowing and occlusion appear to be qualitatively similar. Degenerative changes within a 20 min period of ischemia were induced after the ligation of the descending branch of the left coronary (6). The alterations were time-related, as seen in our studies, and only slight differences exist with our findings.The mitochondrial changes observed in our studies are difficult to interpret. Decreased density of the matrix appears with ischemia (3, 4, 11). Dark matrices probably represent a "condensed" conformational state (7). It is probable that the volume changes taking place are due to electrolyte shifts induced by a derangement of the local circulation.

We have not detected the reported changes in intramitochondrial granules (4, 6, 10, 12, 14). Damaged mitochondria display abnormal granules, owing to alterations in their cationic balance (22, 23). Others have suggested that the small granules seen in canine heart mitochondria could be the site where the abnormal granules originate (9, 13).

It has been observed that membranes show changes that would represent serious alteration of vital functions. In our studies, the most noticeable changes were the occurrence of large vacuoles engulfing other organelles. We have not established the origin of these, and more studies are underway to determine their nature. There are no comparable descriptions in the literature nor has any significance been attached to these vacuoles. They appear to be intracellular and we could not determine whether extracellular connections existed. The mechanism by which the sarcoplasmic components are engulfed in these large vacuoles is also not known. That disproportionate dilatation of the sarcoplasmic reticulum would be the source of the vacuoles is unlikely, since the degree of dilatation detected was not of extreme proportions.

It is possible that some of these changes are reversible (1), although other alterations might be irreversible (6). Irreversible changes occurring within 72 h with complete occlusion and reversible alterations within 24 h have been found (16). The reversibility of the changes described here will be a subject for further investigations, since our studies were aimed at establishing the induction of lesions following narrowing of the coronary vessels.

SUMMARY

In myocardial infarction, well-known morphologic tissue changes have been established in the past few years. There exists a wealth of documentation on the changes in the myocardium following the occlusion of a coronary vessel. However, little or no attention has been paid to the possibility that subcellular myocardial changes might occur with certain degrees of narrowing of the coronary vessels. In the investigations described here, consideration was given to this possibility by studying the myocardium of pigs after a relative narrowing was produced in the anterior descending branch of the left coronary artery. The appearance of changes was correlated with time and the degree of impairment of coronary blood flow. The results show that early myocardial changes are detectable within a short period after the onset of partial occlusion. These changes were progressive, with a definite time relationship. The changes described in this report bear some resemblance to those found after complete occlusion. It would appear that the difference is related to the intensity of the changes and not to the type of subcellular organelle affected. In addition, important alterations occurred in the membranes, including the formation of large vacuoles. This could reflect that the electrolyte balance is an important factor in the pathogenesis of the changes that occur during coronary artery narrowing. Final characterization of these cytoplasmic vacuoles, not described in previous electron microscopic studies, awaits the results of further investigations in our laboratories.

REFERENCES

1. AFIFI, A. K.: BERGMAN, R. A. and HARVEY, J. C.: Steroid myopathy. Hopkins Med. J. *123:* 158–174 (1968).
2. BAHR, G. F. and JENNINGS, R. B.: Ultrastructure of normal and asphyxic myocardium of the dog. Lab. Invest. *10:* 548–571 (1961).
3. BRYANT, R. E.; THOMAS, W. A. and O'NEAL, R. M.: An electron microscopic study of myocardial ischemia in the rat. Circulat. Res. *6:* 699–709 (1958).
4. BURDETTE, W. J. and ASHFORD, T. P.: Structural changes in human myocardium following hypoxia. J. Thorac. Cardiovasc. Surg. *50:* 210–220 (1965).
5. CAULFIELD, J. and KLIONSKY, B.: Myocardial ischemia and early infarction: an electron microscopic study. Amer. J. Path. *35:* 489–523 (1959).
6. DENKER, M. W.; BERGMAN, R. A. and NACHLAS, M. M.: Ultrastructural changes in myocardium during experimental ischemia. Hopkins Med. J. *124:* 311–330 (1969).
7. HACKENBROCK, C. R.: Ultrastructural basis for metabolically linked mechanical activity in mitochondria. I. Reversible ultrastructural change with changes in metabolic steady state in isolated liver mitochondria. J. Cell Biol. *30:* 269–297 (1966).
8. HARRIS, A. S.: Delayed development of ventricular ectopic rhythms following experimental coronary occlusion. Circulation *1:* 1318–1328 (1950).
9. HERDSON, P. B.; KALTENBACH, J. P. and JENNINGS, R. B.: Fine structural and biochemical changes in dog myocardium during autolysis. Amer. J. Path. *57:* 539–557 (1969).
10. HERDSON, P. B.; SOMMERS, H. M. and JENNINGS, R. B.: A comparative study of the fine structure of normal and ischemic dog myocardium with special reference to early changes following temporary occlusion of a coronary artery. Amer. J. Path. *46:* 367–386 (1965).
11. HOELSCHER, B.; JUST, O. H. and MERKER, H. J.: Studies by electron microscope on various forms of cardiac arrest in dog and rabbit. Surgery *49:* 492–499 (1961).
12. JENNINGS, R. B.; BAUM, J. H. and HERDSON, P. B.: Fine structural changes in myocardial ischemic injury. Arch. Path. *79:* 135–143 (1965).
13. JENNINGS, R. B.; HERDSON, P. B. and SOMMERS, H. M.: Structural and functional abnormalities in mitochondria isolated from ischemic dog myocardium. Lab. Invest. *20:* 548–557 (1969).

14. LEV, M.; BIRKNER, M. L.; FAROUK, S. J.; GRANA, L. and LARK, R. G.: Extracorporeal circulation effect on dog's heart. Ultrastructure study. Arch. Path. *79:* 257–269 (1965).

15. LITVAK, J.; SIDERIDES, L. E.; and VINEBERG, A. M.: The experimental production of coronary artery insufficiency and occlusion. Amer. Heart J. *53:* 505–518 (1957).

16. LUMB, G. D.: Experimentally incuded cardiac failure in swine: pathological changes. In: Swine in Biomedical Research (Frayin Printing Co., Seattle, Washington 1966).

17. LUMB, G. D. and HARDY, L. B.: Collateral circulation and survival related to gradual occlusion of the right coronary artery in the pig. Circulation *27:* 717–721 (1963).

18. LUMB, G. D. and HARDY, L. B.: Collaterals and coronary artery narrowing: I. The effect of coronary artery narrowing on collateral channels in swine. Lab. Invest. *13:* 1530–1540 (1964).

19. LUMB, G. D. and HARDY, L. B.: Technique for dissection and perfusion of heart. Arch. Path. *77:* 233–238 (1964).

20. LUMB, G. D.; SINGLETARY, H. P. and HARDY, L. B.: Collateral circulation following experimental gradual narrowing of the coronary arteries. Angiology *13:* 463–465 (1962).

21. MALININ, T. I.; STOKES, J. R.; HARDY, L. B. and LUMB, G. D.: Experimental coronary artery narrowing in swine. I. Morphologic and histochemical myocardial changes. Hopkins Med. J. *122:* 102–111 (1968).

22. SALADINO, A. J.; BENTLEY, P. J. and TRUMP, B. F.: Ion movements in cell injury. Effect of Amphotericin B on the ultrastructure and function of the epithelial cells of toad bladder. Amer. J. Path. *54:* 421–466 (1969).

23. TRUMP, B. F.; GOLDBLATT, P. J. and STOWELL, R. E.: Studies on necrosis of mouse liver in vitro. Ultrastructural alterations in the mitochondria of hepatic parenchymal cells. Lab. Invest. *14:* 343–371 (1965).

Mailing address: Felix A. de la Iglesia, M.D., The Warner-Lambert Research Institute of Canada, Sheridan Park, Clarkson, Ontario (Canada).

Pathological Institute, McGill University,
Montreal, Quebec, Canada

STUDIES ON THE MECHANISM OF INCREASED MYOCARDIAL RESISTANCE [1]

G. RONA and J. DUSEK[2]

Our previous investigations indicated that the extent of myocardial necrosis is influenced by many variables in addition to the intensity of the stimulus applied (14). Various procedures that cause necrosis, *e.g.*, temporary and permanent coronary ligation (16), injection of sympathomimetic amines (3, 6) and plasmocid (1), intoxication by cobalt (7), and experimental renal hypertension (9), also result in some differences in the involvement of subcellular organelles of the cardiac muscle cells in the alterations of steady state preceding frank necrosis. This variation of the reaction pattern of the myocardium should be considered whenever protective studies are designed.

The etiology and possible mechanisms of myocardial protection are reviewed in table I. In our Department, the isoproterenol model has been used to investigate the myocardial reaction to injury (10, 13, 15). As we suggested in our first studies in 1959, this beta adrenergic catecholamine, by affecting oxygen and energy utilization of the cardiac muscle, produces necrobiotic myocardial alterations. One of the factors that seems to considerably reduce myocardial sensitivity to isoproterenol is a previous myocardial damage. In 1960, Selye *et al.* (17) showed that, in rats, a preparatory procedure (left coronary-artery ligation), while in itself causing myocardial necrosis, appears to protect the rest of the myocardium against a subsequent challenging insult (toxic dose of isoproterenol one week later). This finding served as a basis for the concept of increased myocardial resistance and we have now investigated this phenomenon in more detail. In these studies of myocardial resistance, we were seeking factors that could eventually protect the myocardium from irreversible damage. The investigations are still under way; but, to date, they have led to the following conclusions.

1) The induced myocardial resistance was independent of the localization of the preparatory necrosis in the heart. Preparatory myocardial necrosis following either right or left coronary-artery ligation afforded a statistically significant increase in the resistance against the challenging, necrotizing effect of isoproterenol. It appeared to be preferable in future experiments to use right coronary-artery ligation as a preparatory procedure, since, in the rat, it produces

---

[1] Supported by Medical Research Council of Canada Grant MT-3635.

[2] Research Fellow of the Medical Research Council of Canada.

Table I

The Etiology and Pathogenesis of Myocardial Protection and Resistance

ETIOLOGY

GENERAL AND ENVIRONMENTAL FACTORS

species
breed
age
body weight
pregnancy
temperature deviations
muscular exercise
blood letting
starvation
forced restraint
dietary effects: ions
fat, carbohydrates, protein

CHANGES INHERENT TO THE MYOCARDIUM

preparatory procedures
myocardial hypertrophy

HUMORAL AGENTS (DRUGS)

thyroid hormones
insulin
adrenal cortical hormones
sex hormones
propranolol
iproveratril
adrenalectomy

MYOCARDIAL RESISTANCE and PROTECTION

MECHANISMS

CIRCULATORY FACTORS

a. Systemic
Systemic circulatory adaptation
Humoral factors in the blood

b. Local
Increased coronary blood supply

LOCAL TISSUE FACTORS

accumulation of the metabolic substrate
changes in enzymatic activities
shift towards anaerobic metabolism

right ventricular infarct and, hence, the severity of the subsequent left ventricular necrosis caused by the challenging dose of isoproterenol can be readily assessed.

2) The resistance was time dependent. It reached its peak between 48 and 96 h, then gradually waned, and was statistically insignificant after 4 weeks.

3) Persistence of necrotic muscle was not required for the increase in resistance; therefore, the postulation that a humoral substance continuously liberated from the necrotic muscle cells was responsible for the increase in resistance could not be substantiated.

4) Hypertrophy of the cardiac muscle cells in the well-developed steady state did not seem a likely explanation for the increased resistance.

The question was then raised whether the increased myocardial resistance could be correlated with some of the metabolic and morphological changes occurring in the left ventricular myocardium remote from the preparatory infarct (4). Biochemical studies showed no significant change of tissue electrolytes,with the exception of a drop in myocardial potassium concentration after coronary artery ligation (fig. 1). There was a temporary postoperative accumulation of glycogen (fig. 2) and increased nucleic acid, particularly RNA, content (fig. 3) of the cardiac muscle. Whereas the glycogen accumulation appeared to be related to the postoperative fasting, the increase in RNA content correlated roughly with the postoperative increase in myocardial resistance.

Enzyme histochemical studies indicated in some animals an increase in the activity of glucose-6-phosphate dehydrogenase and, occasionally, of phosphorylase, which suggested an increased importance of the glycolytic pathway and hexose monophosphate shunt (5). In their investigations of the metabolic activity of necrosis and of the myocardium remote from the infarct, Gudbjar-

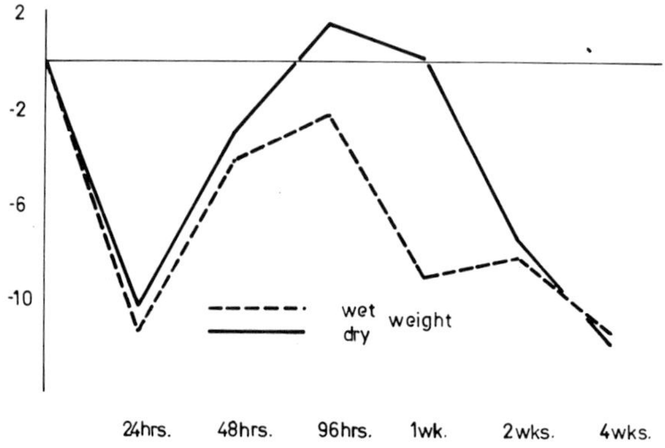

Fig. 1 Potassium: Concentration of potassium in the left ventricular myocardium. The values are percentually related to the concentration found in controls (100%), calculated for both wet and dry weight of the tissue. The decreases at 24 h, 2 weeks, and 4 weeks are significant compared with the controls (values for wet weight of the tissue).

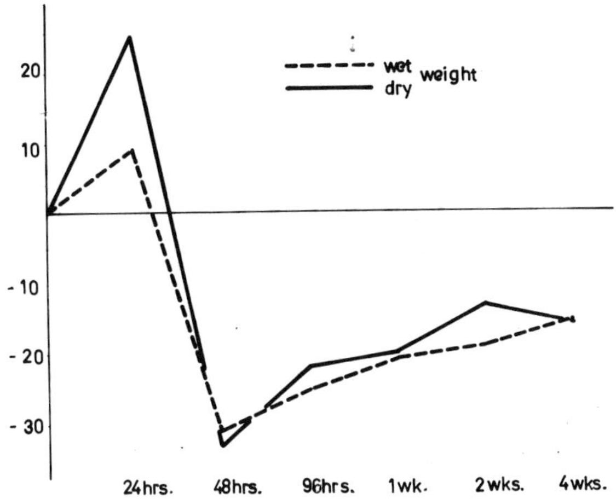

Fig. 2 Glycogen: Concentration of glycogen in the left ventricular myocardium. Note increase 24 h after the right coronary-artery ligation, followed by a sharp decrease to values significantly below the level of the control group.

nason *et al.* (8) found a more than 3-fold increase in the activity of glucose-6-phosphate dehydrogenase in the noninfarcted myocardium. With the use of biochemical methods alone, however, it was not possible to differentiate the degree to which the stimulated cardiac muscle cells or the cells participating in the repair process contributed to this increase. Our enzyme histochemical study showed that the activity of glucose-6-phosphate dehydrogenase in cardiac muscle cells sometimes even exceeded that of the connective tissue cells. Thus, it is conceivable that increased myocardial glucose-6-phosphate dehydrogenase and

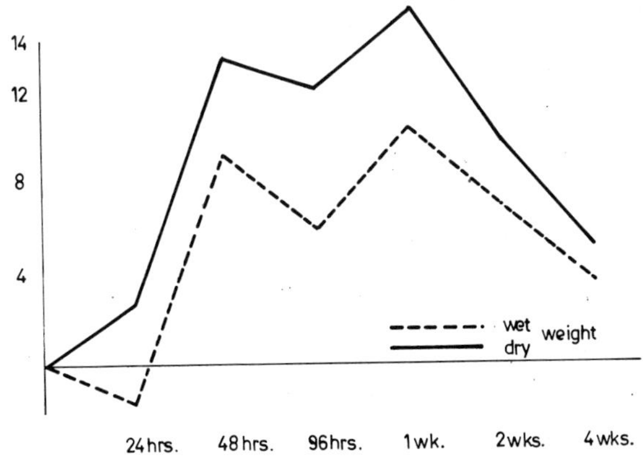

Fig. 3 RNA: Concentration of RNA in left ventricle of rats, following right coronary-artery ligation.

phosphorylase activity might contribute to the energy requirements of the stimulated cardiac muscle cells and facilitate the synthesis of ribonucleic acid; both processes are essential for cellular repair (11).

Electron microscopic observations on the protected remote left ventricular myocardium similarly revealed certain features that suggested an increased metabolic activity of the cardiac muscle cells. Changes were most apparent from 24 to 96 h, corresponding to the increasing myocardial resistance. There was swelling of the cardiac muscle cells with accumulation of fat droplets (fig. 4) and intracellular fluid, accumulation of ribosomes, and Golgi hyperplasia (fig. 5). Mitochondrial changes were characterized by an increase in the staining and a peculiar fingerprint-like arrangement of the cristae (fig. 6). Since similar morphologic changes of the mitochondria have been reported in rats after swimming exercise and following anoxia (12, 18), this change may be regarded as an indication and morphologic equivalent of increased metabolic mitochondrial activity (2).

From the results of our studies, it appears that myocardial necrosis produced by the preparatory procedures activates the metabolism of the remote

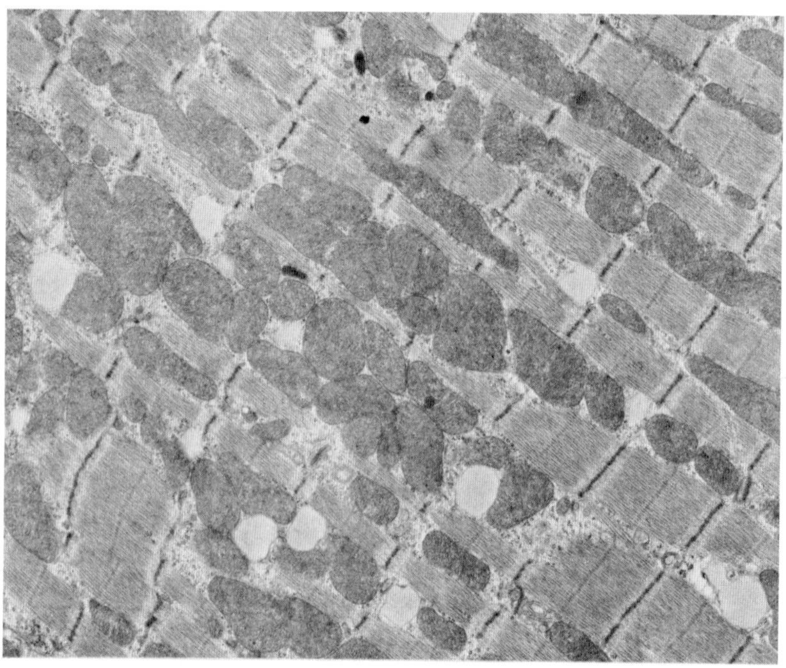

Fig. 4 Subendocardial layer of the left ventricular myocardium, 24 h after right coronary-artery ligation. Note many fat vacuoles. The number of small granules (possible glycogen) seems to be increased. Some of the mitochondria reveal clearing of the matrix. ×12,500.

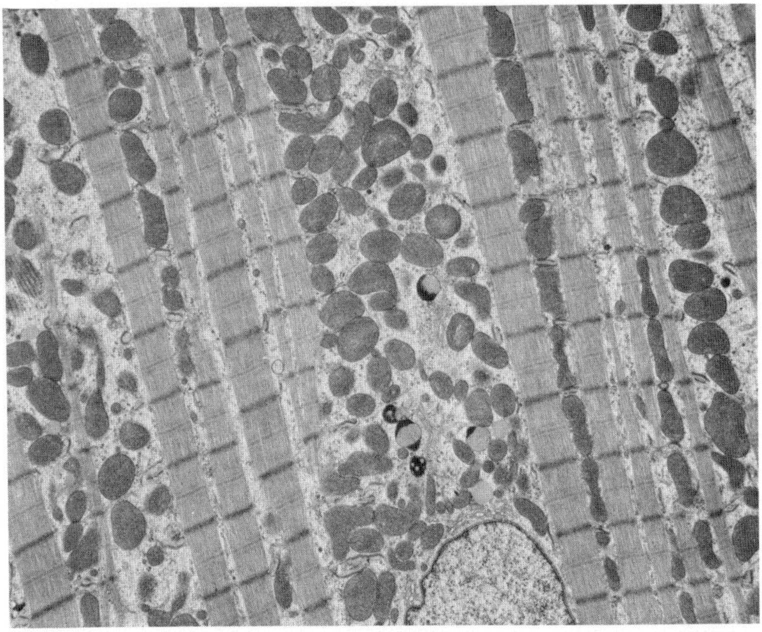

Fig. 5 Left ventricular myocardium, subendocardial layer, 4 days after right coronary-artery ligation. Note some swelling of the cardiac muscle cells, leading to separation of myofibrils and clearing of the perinuclear space. The perinuclear area contains several Golgi zones and a few dense bodies, probably lysosomes. Some of the mitochondria show circular arrangement of cristae. × 7,800.

myocardium, following which the cardiac muscle cells acquire an enhanced adaptation to withstand an exaggerated demand (challenging dose of isoproterenol) and an increased nonspecific resistance to noxious stimuli. Thus, our experimental results demonstrate the importance of myocardial versus coronary factors in the pathogenesis, development, and outcome of the so-called anoxic myocardial disease and may have some important bearing on its prevention and treatment.

SUMMARY

Our investigations on various experimental models indicated that the severity of myocardial necrosis is influenced by many variables. The isoproterenol model has been used to study the myocardial reaction to injury. It was seen that a previous myocardial injury (preparatory procedure) produced either by right or left coronary ligation or by isoproterenol is followed by less necrotic effects after a further dose of isoproterenol (challenging insult). Biochemical and histochemical studies of the remote myocardium did not show time correlation between the intensity of myocardial resistance and the tissue concentration of glycogen and potassium. The RNA curve, however, closely followed the postoperative changes of myocardial resistance. Electron microscopic studies similarly showed an apparent activation of proteosynthesis that could be correlated with the postoperative increase of myocardial resistance.

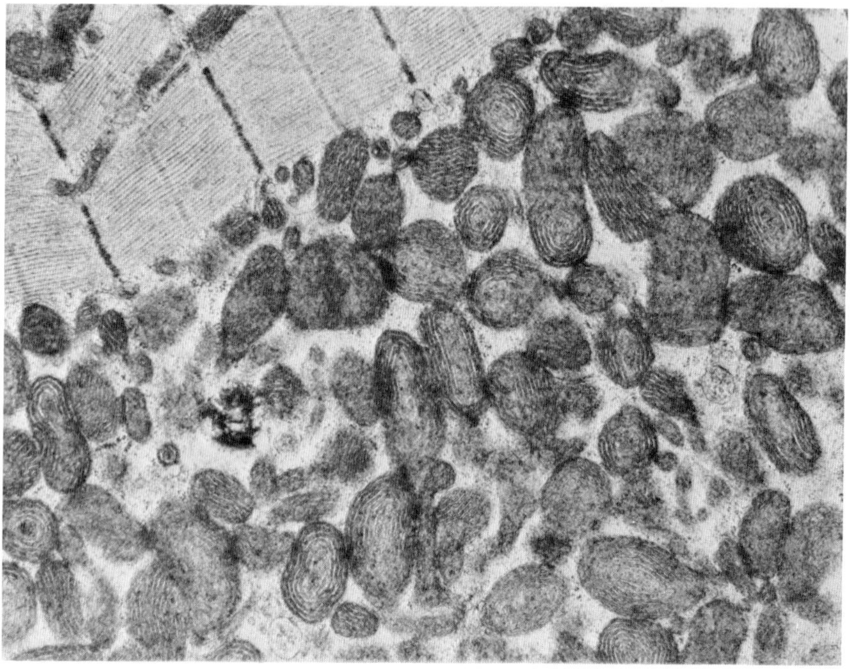

Fig. 6 Subendocardial layer of the left ventricular myocardium, 24 h after right coronary ligation. Mitochondrial cristae stain heavily and many mitochondria exhibit a circular, fingerprint-like arrangement. ×12,500.

It appears that myocardial necrosis produced by a preparatory procedure activates the metabolism of the remote myocardium, following which cardiac muscle acquires an enhanced adaptation to an exaggerated demand (challenging dose of isoproterenol) and an increased nonspecific resistance to various noxious stimuli.

REFERENCES

1. BERGER, J. M.; RONA, G. and LAMM, S.: Relation of reversible to irreversible toxic degeneration in myocardium (Abstract). Amer. J. Path. *59:* 73a–74b (1970).
2. BOZNER, A. und MEESEN, H.: Die Feinstruktur des Herzmuskels der Ratte nach einmaligen und wiederholten Schwimmtraining. Virchows Arch. Abt. B Zell Path., *3:* 284–269 (1969).
3. CSAPO, Z.; DUSEK, J. and RONA, G.: Early alterations of cardiac muscle cells in isoproterenol necrosis. Arch. Path. (in press).
4. DUSEK, J.: Experimental studies on myocardial protection. M. Sc. Thesis, McGill University, Montreal, 1969.
5. DUSEK. J.: Myocardial reaction to experimental ischemic injury in the rat. Ph. D. Thesis, McGill University, 1970.
6. FERRANS, V. J.; HIBBS, R. G.; WALSH, J. J. and BURCH, G. E.: Histochemical and electron microscopic studies on the cardiac necroses produced by sympathomimetic agents. Ann. N.Y. Acad. Sci. *156:* 309–332 (1969).
7. GRICE, H. C.; GOODMAN, T.; MUNRO, I. C.; WIBERG, G. S. and MORRISON, A. B.:

Myocardial toxicity of cobalt in the rat. Ann. N.Y. Acad. Sci. 189–194 (1969).

8. GUDBJARNASON, S.; BRAASCH, W.; COWAN, C. and BING, R. J.: Metabolism of infarcted heart muscle during tissue repair. Amer. J. Cardiol. *22:* 360–369 (1968).
9. HÜTTNER, I.; RONA, G. and MORE, R. H.: Fibrin deposition within cardiac muscle cells in experimental hypertension–An electron microscopic study. Arch. Path. (Chicago) *91:* 19–28 (1971).
10. KAHN, D. S.; RONA, G. and CHAPPEL, C. I.: Isoproterenol-induced cardiac necrosis. Ann. N.Y. Acad. Sci. *156:* 285–293 (1969).
11. MEERSON, F. Z.: The myocardium in hyperfunction hypertrophy and heart failure. Circulat. Res. 25 (Suppl. II): 1–163 (1969).
12. ONISHI, S.: Die Feinstruktur des Herzmuskels nach Aderlass bei der Ratte. Beitr. Path. Anat. *136:* 96–132 (1967).
13. RONA, G.; CHAPPEL, C. I.; BALAZS, T. and GAUDRY, R.: An infarct-like myocardial lesion and other toxic manifestations produced by isoproterenol in the rat. Arch. Path. *67:* 443–455 (1959).
14. RONA, G.; CHAPPEL, C. I. and KAHN, D. S.: The significance of factors modifying the development of isoproterenol-induced myocardial necrosis. Amer. Heart J. *66:* 389–395 (1963).
15. RONA, G. and KAHN, D. S.: Experimental studies on the healing of cardiac necrosis. Ann. N.Y. Acad. Sci. *156:* 177–188 (1969).
16. SELYE, H.: The Pluricausal Cardiopathies (C. C Thomas Publ., Springfield, Ill. 1961).
17. SELYE, H., VEILLEUX, R., GRASSO, S.: Protection by coronary ligation against isoproterenol-induced myocardial necrosis. Proc. Soc. Exp. Biol. Med. *104:* 343–345 (1960).
18. SULKIN, N. M. and SULKIN, D. F.: Age differences in response to chronic hypoxia on the fine structure of cardial muscle and autonomic ganglion cells. J. Gerontol. *22:* 485–501 (1967).

Mailing address: G. Rona, M.D., Pathological Institute, McGill University, 3775 University Avenue, Montreal 110, Que. (Canada).

Pathological Institute, McGill University,
Montreal, Quebec, Canada

SIGNIFICANCE OF MORPHOLOGICAL CHANGES AT THE PERIPHERY OF AN EXPERIMENTAL MYOCARDIAL INFARCT

J. DUSEK

Morphological studies of a myocardial infarct usually deal with the central area of complete or almost complete necrosis. In only a few reports have the periphery of the necrosis and the marginal area of the surviving myocardium been studied. As regards the periphery of the necrosis, it has been shown that its development resembles the morphological evolution of disseminated cardiac necrosis of diverse etiology (1, 6, 7). Some of the cardiac muscle cells of this zone apparently suffer more or less severe anoxic damage, but survive mainly because of an improving collateral circulation. The present communication describes findings in the surviving myocardial fibers at the periphery of the infarct and in the marginal zone.

The study was performed on Sprague-Dawley rats in which the left or right coronary artery was ligated. The animals were sacrificed at various intervals from 24 h up to 4 weeks after the operation and the margin of the infarct was studied by light and electron microscopy. Several histochemical methods were applied for light microscopy: demonstration of glycogen, nucleic acids, neutral fat, and potassium. The following histoenzymological methods were also applied: phosphorylase, succinate dehydrogenase, beta hydroxybutyrate dehydrogenase, lactate dehydrogenase, and beta naphthylamidase at acid pH (believed to demonstrate cellular cathepsins (8)). Perfusion fixation with glutaraldehyde-paraformaldehyde mixture was used for electron microscopy.

Histochemical methods showed alterative changes in the early postoperative phase, namely, focal depletion of glycogen with a decrease in activity of phosphorylase in the surviving fibers; further abnormalities in the formation of formazan were revealed by methods used for the demonstration of dehydrogenases, employing nitro blue tetrazolium as a hydrogen ion acceptor (fig. 1). Instead of diffuse staining of cellular structures, the formazan formed coarse granules. This finding was believed to result from the increased permeability of the cellular membranes (2, 5). Most of the granular areas disappeared by 96 h postoperatively. Slightly increased activity of beta naphthylamidase indicated an increase in the cellular catabolism.

Electron microscopy in the acute phase showed an increase in the number of fat droplets, swelling of the whole cells, and mitochondrial swelling with clearing of the matrix. Some mitochondria exhibited increased staining and circular

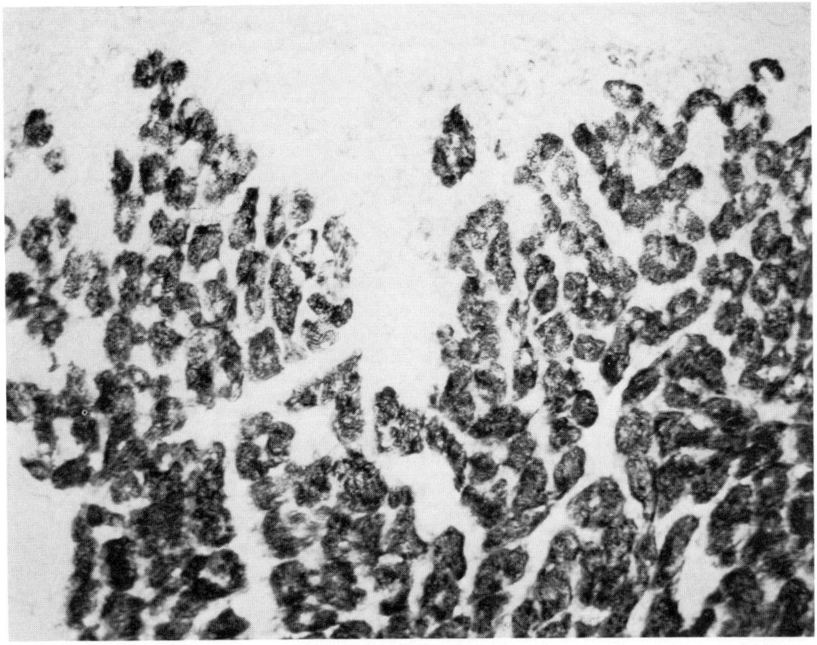

Fig. 1 Abnormal granular pattern of the tetrazolium formazan at the periphery of a myocardial infarct (top and center of the photograph). Rat 767, 96 h after ligation of the right coronary artery. Incubation for succinate dehydrogenase with nitro blue tetrazolium. ×125.

arrangement of cristae; others contained increased numbers of dark granules or electron-dense lines (fig. 2).

The later development of the marginal area (from 1 to 4 weeks) was characterized by fibrosis of the newly formed connective tissue with a new reduction of the blood supply to the surviving marginal muscle cells. Some of the cardiac muscle cells in this area were markedly hypertrophic; the degree of hypertrophy exceeded that of cardiac muscle cells undergoing compensatory hypertrophy in areas more remote from the infarct. The reason for this locally exaggerated hypertrophy was probably the rigidity of connective tissue surrounding the fibers. In addition to the hypertrophic cardiac muscle cells, thinner pale fibers appeared, apparently as a result of anoxic atrophy.

Histochemical methods revealed the following: coarse granulation of formazan was mostly absent at this time interval; but another abnormality of the staining pattern appeared mainly in the atrophic fibers, *i.e.*, more or less diffuse staining of the cytoplasm of the whole fibers, more pronounced with the use of some methods, *e.g.*, incubation for lactate dehydrogenase (fig. 3), glucose-6-phosphate dehydrogenase than with others, *e.g*, succinate dehydrogenase, beta

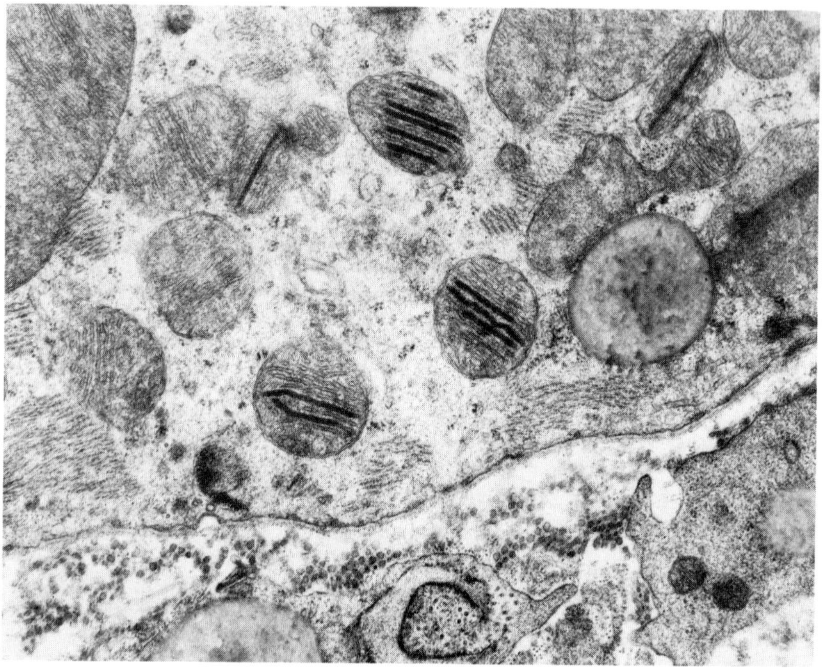

Fig. 2 Oblique section through a cardiac muscle cell at the periphery of a 1-week-old left ventricular infarct. Myofibrils have partly disappeared; some of the mitochondria show linear densities involving their cristae. Fat droplet appears near the right margin. Collagen fibers surround the muscle cell at the bottom. Rat 1341, 7 days after left coronary-artery ligation. ×20,000.

hydroxybutyrate dehydrogenase, Although permeability changes with abnormal localization of the reaction product could not be excluded, the difference was interpreted as an increase in extramitochondrial enzymatic activity. The fibers showed generally increased staining for phosphorylase, glucose-6-phosphate dehydrogenase, and beta naphthylamidase, as well as increased pyroninophilia, demonstrating higher cellular content of RNA (fig. 4). These changes suggested increased importance of glycolysis and of the hexose monophosphate shunt and increased turnover of cellular proteins, in accordance with the biochemical results of others (3, 4).

The electron microscopic pattern at this stage was variable. Some of the fibers showed simple thickening with preservation of the normal ultrastructure. Other thickened fibers contained areas of accumulated abnormal mitochondria with dark granules and linear densities, prominent tubules of the sarcoplasmic reticulum, increased number of membranes of rough endoplasmic reticulum, and numerous particles probably representing ribosomes. Irregular accumulation of thin (fig. 5) and thick (fig. 6) filaments could be seen, as well as disperse

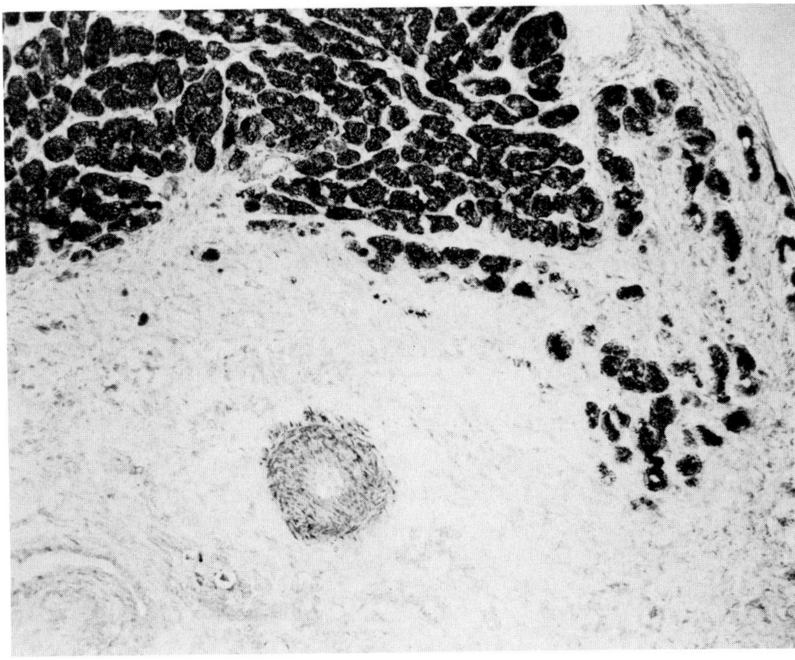

Fig. 3 Margin of a right ventricular infarct 7 days old. Note dark and diffuse staining of some of the marginal fibers (right) surrounded by granulation tissue, as compared with the more remote muscle fibers. Granulation tissue (bottom half) with an artery show only mild enzymatic activity. Rat 1602, incubation for lactate dehydrogenase with nitro blue tetrazolium. ×100.

accumulation of electron-dense Z-band-like material under the sarcolemma and changes in intercalated discs. Atrophic fibers generally showed a marked reduction of myofibrils and mitochondria (fig. 7), signs of cellular autophagy, and the presence of a granular material in the sarcoplasm. Compression of the cells by fibrotizing connective tissue led to bizarre distortion of cardiac muscle cells.

The changes described seem to indicate that, to some extent, cardiac muscle cells adjust their metabolism in accordance with the changing local conditions and that the increase in protein synthesis with a possible regeneration of the inner structure of the cells takes place (9). Proliferation of the connective tissue with progressive reduction of blood supply, however, markedly reduces the possibility of cellular repair and causes delayed damage to the marginal cells of the infarct. The description shows that there is a certain dynamic aspect to the evolution of healing of a myocardial infarct and that, during the scarring, further damage and eventual death of the marginal cells occurs. We do not know how long this process continues (apparently, after some time, an equilibrium may be

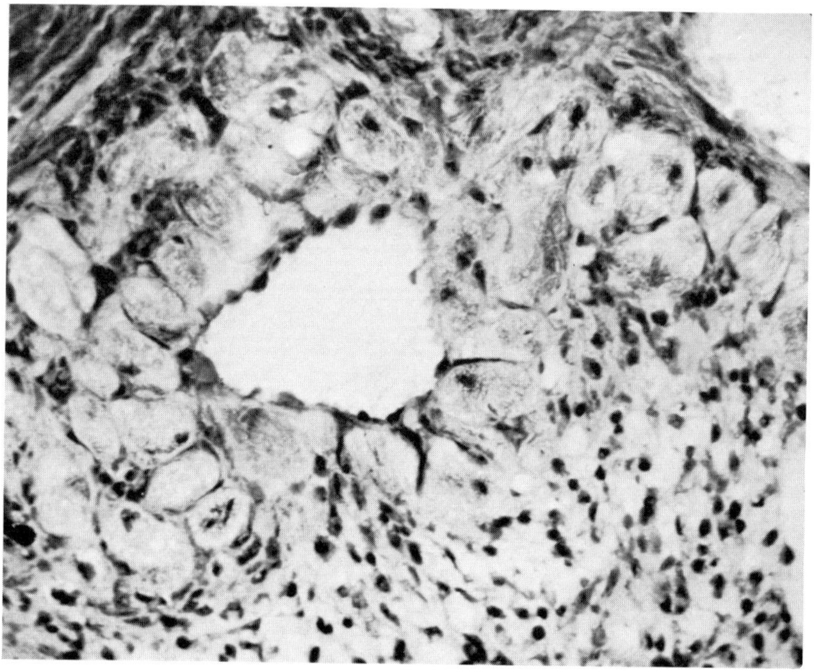

Fig. 4 Granular increase in pyroninophilia of the surviving myocardial fibers around a subepicardial vein. Granulation tissue (left upper corner) shows marked staining. Rat 2005, periphery of an infarct, 1 week following left coronary-artery ligation. Methyl green-pyronin stain. ×350.

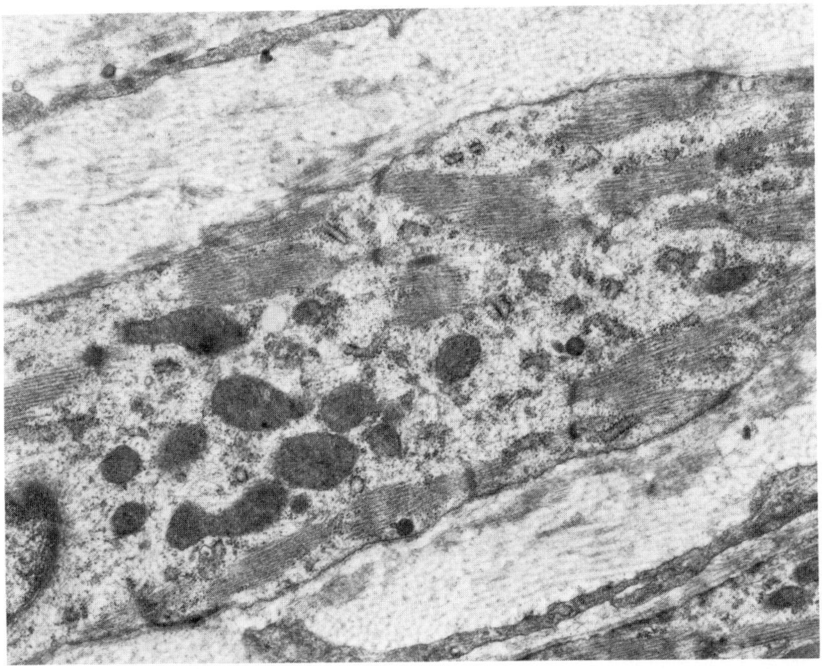

Fig. 5 Right ventricular infarct, 1 month old. The surviving cardiac muscle cell contains an irregular network of fine filaments and an abundant rough endoplasmic reticulum. Rat 1436. ×13,000.

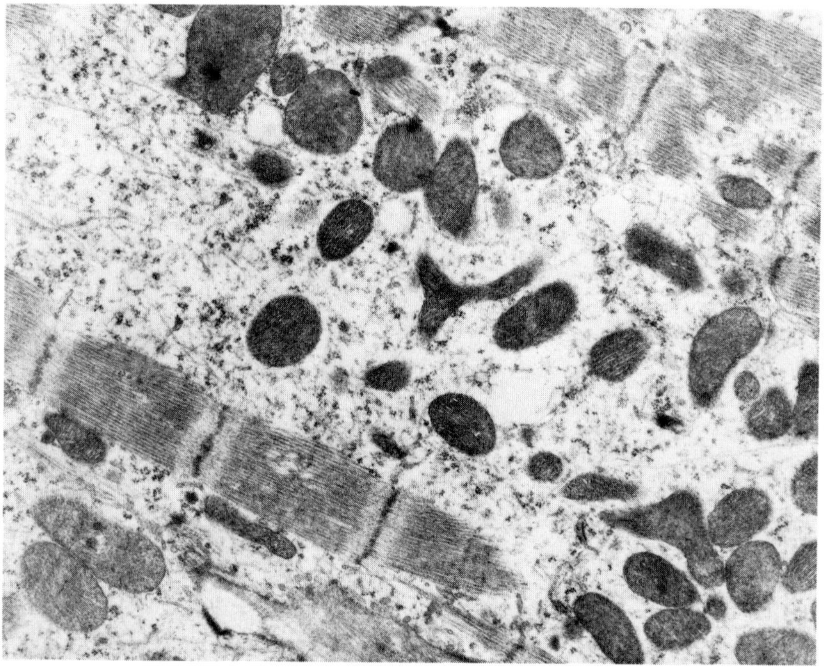

Fig. 6 Subsarcolemmal part of a cardiac muscle cell surviving at the periphery of a 1-month-old infarct. Note increase in rough endoplasmic reticulum and scattered thick filaments that seem to be connected with the filaments of an A band (left upper corner). Rat 1435, right coronary-artery ligation. ×13,000.

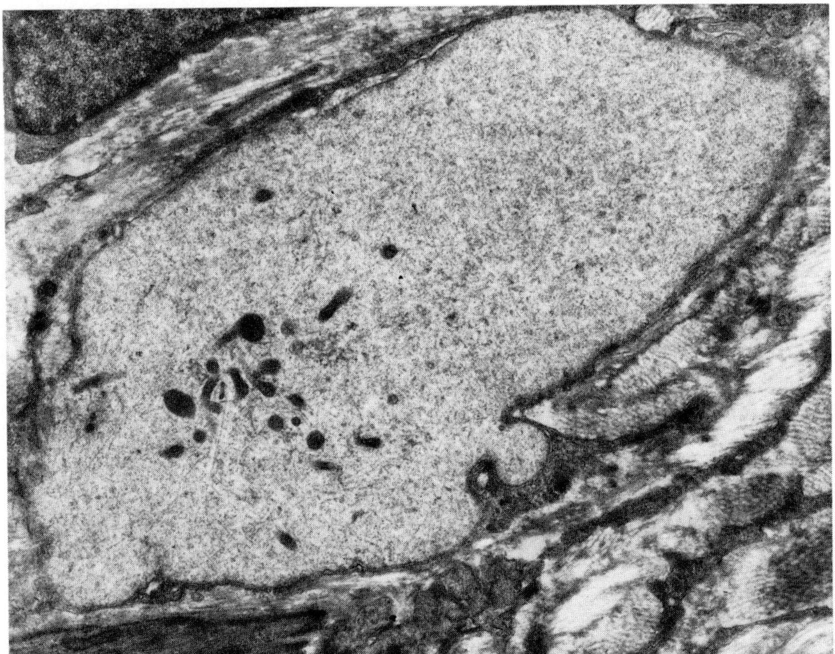

Fig. 7 Periphery of an atrophic cardiac muscle cell at the periphery of a 1-month-old infarct. The sarcoplasm contains granular material, several dark mitochondria, and a few myofilaments (left). Dense collagen fibrils surround the deformed cell. Rat 1346, 1 month after right coronary-artery ligation. × 8,000.

reached). It is possible that the events at the margin of the infarct influence the surrounding myocardium. If so, this would explain some cases of arrhythmia that occur during the healing of a myocardial infarct.

SUMMARY

The development of changes in cardiac muscle cells at the periphery of a myocardial infarct was studied in Sprague-Dawley rats. The muscle cells showed early damage after the coronary ligation; this was followed by signs of regeneration of the intracellular structures. Delayed anoxia connected with a decrease of the blood supply in the post-infarction scar led to secondary degenerative changes in the cardiac muscle cells. The biological and clinical significance of the findings is discussed.

REFERENCES

1. COX, J.; McLAUGHLIN, V. W.; FLOWERS, N. C. and HORAN, L. G.: The ischemic zone surrounding acute myocardial infarction. Its morphology as detected by dehydrogenase staining. Amer. Heart J. *76:* 650–659 (1968).
2. FERRANS, V. J.; HIBBS, R. G.; WALSH, J. J. and BURCH, G. F.: Histochemical and electron microscopical studies on the cardiac necroses produced by sympathomimetic agents. Ann. N.Y. Acad. Sci. *156:* vol. 1 309–332 (1969).
3. GUDBJARNASON, S.; COWAN, C. and BING, R. J.: Increase in hexosemonophosphate shunt activity during tissue repair. Life Sci. *6:* 1093–1097 (1967).
4. GUDBJARNASON, S.; BRAASCH, W.; COWAN, C. and BING, R. J.: Metabolism of infarcted heart muscle during tissue repair. Amer. J. Cardiol. *22:* 360–369 (1968).
5. HECHT, A.: Fermenthistochemische Frühveränderungen beim experimentellen Herzinfarkt. Virchows Arch. path. Anat. *337:* 414–424 (1964).
6. KORB, G. und TOTOVIC, V.: Elektronenmikroskopische Untersuchungen über Frühveränderungen im Zentrum und in die Peripherie experimenteller Herzinfarkte. Virchows Arch. path. Anat. *342:* 85–96 (1967).
7. KORB, G. and TOTOVIC, V.: Electron microscopical studies on experimental ischemic lesions of the heart. Ann. N.Y. Acad. Sci. *156:* 48–60 (1969).
8. NIEMI, M. and SYLVEN, B.: The naphthylamidase reaction as a diagnostic tool for the demonstration of cellular injury and autophagy. Histochemie *18:* 40–47 (1969).
9. RING, P. A.: Myocardial regeneration in experimental ischemic lesions of the heart. J. Path. Bact. *62:* 21–27 (1950).

Mailing address: J. Dusek, M.D., Pathological Institute, McGill University, Faculty of Medicine, 3775 University Avenue, Montreal 110, Que. (Canada).

Departments of Medicine and Biochemistry, Wayne State University, School of Medicine, Detroit, Michigan, USA, and Science Institute, University of Iceland, Reykjavik, Iceland

METABOLIC CHANGES IN INFARCTED AND NON-INFARCTED MYOCARDIUM DURING THE POSTINFARCTION PERIOD [1]

S. GUDBJARNASON, K. G. RAVENS, and P. MATHES

The purpose of this study was to examine biochemical changes in noninfarcted myocardium following coronary artery occlusion. The effect of anabolic hormones and a protein-free diet upon myocardial metabolism and tissue repair during the postinfarction period was also examined.

Myocardial infarction was produced in anesthetized dogs by ligation of several branches of the anterior descending and circumflex coronary arteries (1). The animals were restudied at varying intervals, from 1 day to 10 weeks following infarction. After hemodynamic measurements were made, myocardial biopsy samples were taken for determination of metabolites and high-energy phosphates.

The results illustrate a significant diminution in ATP and creatine phosphate (CP) content of noninfarcted muscle following myocardial infarction. The ATP levels fell from 5.73 μmole/g in normal animals to 2.63 μmole/g 24 h after infarction (table I). The ATP content remained significantly reduced during the first 10 days of the postinfarction period, but 10 days after infarction the ATP levels had increased again to 4.65 μmole/g. In sham-operated animals, the myocardial ATP level was not significantly altered (table I).

The creatine phosphate (CP) level in noninfarcted muscle fell from 9.18 μmole/g to 4.87 μmole/g 24 h after infarction. The CP level subsequently increased and reached a level of 6.51 μmole/g on the 10th day. The CP level of sham-operated animals showed a small but significant decrease from 9.18 μmole/g to 7.54 μmole/g 24 h after the surgical procedure. This decrease in CP may, in part, be due to the postoperative stress; but since neither ATP nor the other metabolites changed significantly at this time, no further study of sham-operated animals at a later period was considered necessary.

The decreased tissue levels of ATP and CP in noninfarcted muscle were not accompanied by significant alterations in ADP, lactate, or pyruvate levels of the muscle (table I). The lactate showed some variation in tissue levels 1 to 2 days after infarction (table I), but this increase was not statistically significant.

[1] Supported by United States Public Health Service Career Development Award 5-KO3-HE-31271 (SG), American Medical Association Education and Research Foundation, and Michigan Heart Association.

Table I

Tissue Levels of High Energy Phosphates and Metabolites in Noninfarcted Cardiac Muscle before and after Coronary Artery Occlusion

Time after infarction	ATP μmole/g	CP μmole/g	ADP μmole/g	Lactate μmole/g	Pyruvate μmole/g
Control	5.73	9.18	0.50	2.00	0.26
N = 13	±0.21	±0.25	±0.09	±.21	±.05
Sham					
operated	5.45	7.54*	0.65	1.93	0.37
(1 Day)	±0.31	±0.24	±0.04	±0.26	±0.05
N = 5					
1 Day	2.63*	4.87*	0.61	2.68	0.35
N = 8	±0.36	±0.48	±0.09	±0.41	±0.07
2 Days	3.48*	5.97*	0.65	3.03	0.36
N = 18	±0.32	±0.80	±0.07	±0.42	±0.04
4 Days	3.12*	5.26*	0.70	2.02	0.34
N = 5	±0.56	±0.13	±0.10	±0.38	±0.12
10 Days	4.65†	6.51*	0.67	2.20	0.30
N = 10	±0.37	±0.64	±0.11	±0.38	±0.07

* $p < 0.01$.
† $p < 0.02$.

Determinations of water and hydroxyproline content of noninfarcted muscle did not indicate any changes in these parameters, excluding edema or fibrosis as the factors responsible for the diminution in high-energy phosphate levels.

A correlation appears to exist between the diminution in ATP content of noninfarcted muscle and the impairment of left ventricular function following myocardial infarction (fig. 1). The stroke work (SW) and the rate of left ventricular pressure rise (dp/dt) were significantly diminished 1 day after infarction and increased slowly after that (fig. 1).

The impairment in left ventricular function was accompanied by the diminution in high-energy phosphate levels and followed by a significant reduction in norepinephrine (NE) content of the noninfarcted muscle (3). The infarcted tissue lost its NE-content entirely after 48 h (fig. 2), and a rapid decline in the NE-content of the functioning myocardium can be noted. The endogenous level of NE fell from a control value of 0.99 ± 0.02 μg to 0.35 ± 0.01 μg NE/g in the noninfarcted portion of the left ventricle ($p<0.001$) 10 days after infarction. The NE-levels rose gradually after the 2nd week and reached control values again 6 weeks after infarction. The decrease in NE-stores did not correlate with changes in energy-rich phosphates or with changes in myocardial function. The changes in NE-stores could not be related to ischemia, edema, or fibrosis of noninfarcted muscle, and the uptake, accumulation or subcellular distribution of NE-7-C^{14} in noninfarcted muscle was not affected by the coronary artery occlusion.

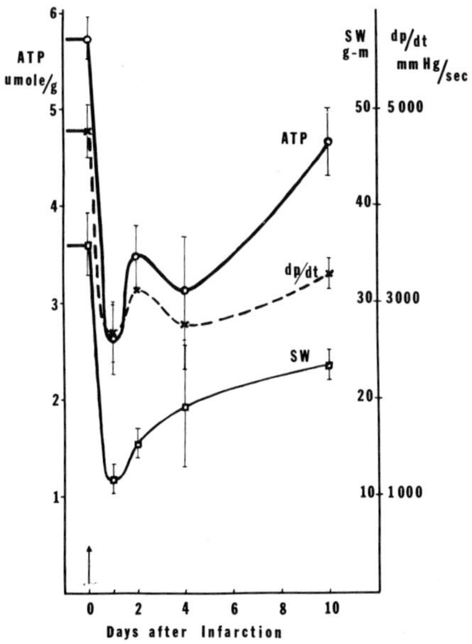

Fig. 1 Changes in stroke work (SW), rate of left ventricular pressure rise (dp/dt), and ATP content of noninfarcted muscle during the postinfarction period.

An increase in NE-release due to the increased activity of the sympathetic nervous system following myocardial infarction appears to be the major factor responsible for the temporary reduction in myocardial NE-stores. Following completion of compensatory processes and restoration of myocardial function, the activity of the sympathetic nervous system returns to normal and subsequently the NE-stores are restored.

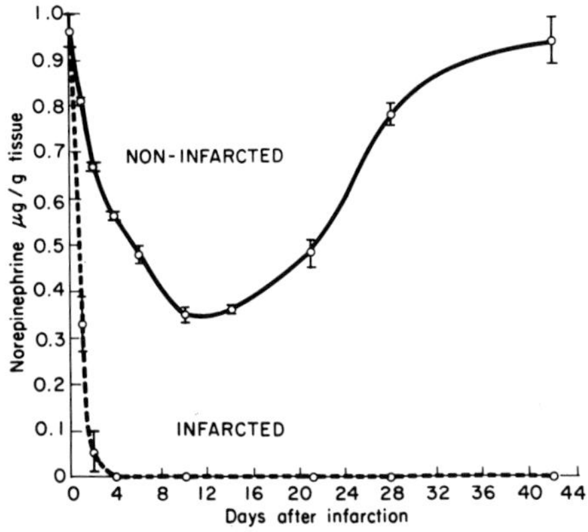

Fig. 2 Changes in tissue content of norepinephrine in infarcted and noninfarcted heart muscle.

Attempts were made to influence the rate of tissue repair and muscle recovery following injury and the results indicated that it is possible to accelerate tissue repair and enhance compensatory processes in noninfarcted muscle by treatment with anabolic steroids. A protein-free diet results, on the other hand, in delayed healing, impaired development of compensatory hypertrophy in noninfarcted muscle, and markedly increased mortality rate.

The results illustrate that treatment with anabolic steroids (Norbolethone, 2 mg/kg per day) significantly increased the ATP and CP levels of noninfarcted muscle 10 days and 3 weeks after myocardial infarction. The ATP level (fig. 3) increased from 4.65 μmole/g in untreated animals to 5.88 μmole/g in steroid-treated animals, 10 days after myocardial infarction.

The protein-free diet significantly reduced the ATP levels of infarcted and noninfarcted muscle. The ATP level decreased to 3.10 μmole/g in noninfarcted tissue and to 1.3 μmole/g in infarcted tissue. Treatment with insulin (0.5 unit NPH/kg per day) was without effect. The CP level (fig. 4) increased from 6.51 μmole/g to 7.52 μmole/g in steroid-treated animals on the 10th day, but in animals on a protein-free diet the CP level fell to 4.45μmole/g. Insulin treatment increased the CP level to 7.6 μmole/g.

This decrease in high-energy phosphate levels of noninfarcted muscle of animals on the protein-free diet was accompanied by a significant increase in mortality rate, from 10% in control animals to 60% in animals on a protein-free diet.

The connective tissue formation during tissue repair was examined by determining the changes in hydroxproline and proline content of necrotic tissue in treated and untreated animals.

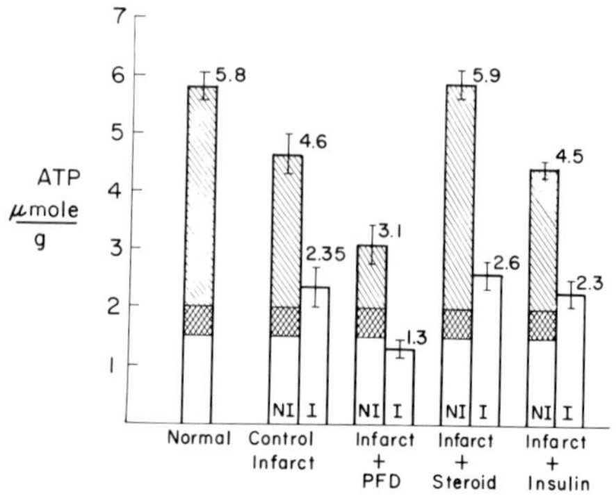

Fig. 3 Changes in tissue content of ATP in infarcted (I) and noninfarcted (NI) heart muscle 10 days after myocardial infarction. Normal = normal cardiac muscle without injury. Control Infarct = animals with myocardial infarction, normal diet without anabolic hormones. Infarct + PFD = animals on a protein-free diet. Infarct + Steroid = animals treated with Norbolethone, 2 mg/kg per day. Infarct + Insulin = animals treated with NPH Insulin, 0.5 unit/kg per day.

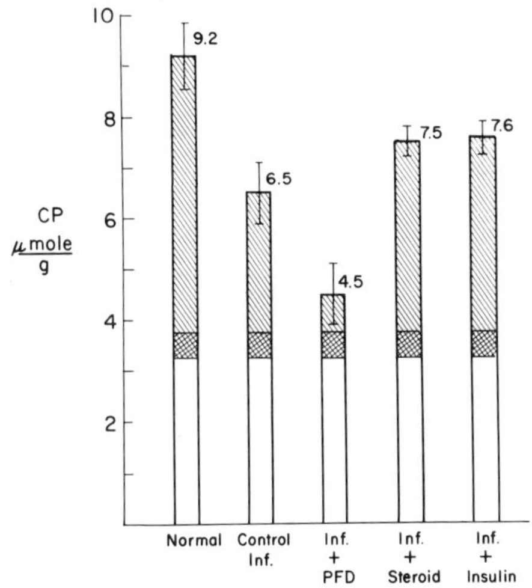

Fig. 4 Changes in tissue content of creatine phosphate (CP) in noninfarcted heart muscle 10 days after myocardial infarction. Groups are identical to those in **figure 3**.

The hydroxyproline content of normal left ventricular muscle was 4.02 ± 0.10 mg/g muscle (dry weight), with a proline/hydroxyproline ratio of 10.3 (fig. 5). Following myocardial infarction, the hydroxyproline content, reflecting the amount of collagen in necrotic muscle, decreased slightly at first but increased on the 4th day and rose in a logarithmic fashion until the 10th day, at which time the hydroxyproline content was 17.61 ± 1.45 mg/g, with a proline/hydroxy-

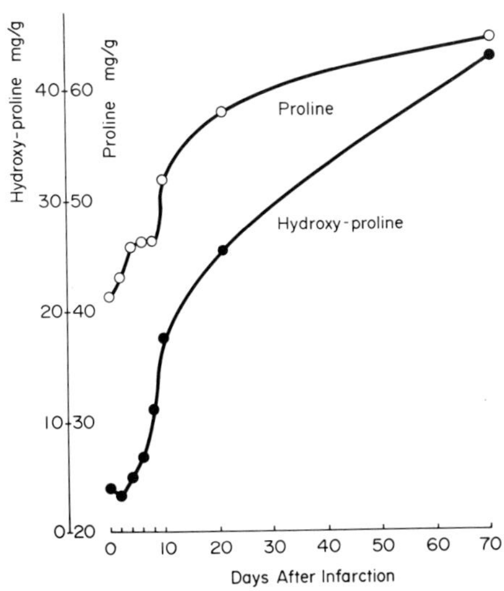

Fig. 5 Changes in proline and hydroxyproline content of infarcted heart muscle.

proline ratio of 3.3. At 3 weeks after infarction, the tissue content was 25.6 ± 2.85 mg/g and, at 10 weeks, the scar contained 43.14 ± 2.31 mg/g hydroxyproline, with a proline/hydroxyproline ratio of 1.58. The proline content of infarcted muscle increased rapidly at first, but then remained relatively constant from the 4th to the 8th day. Subsequently, it once again increased, rapidly for 2 weeks and then more gradually after that (fig. 5).

The rapid rise in hydroxyproline content during the first 10 days after infarction illustrates the fibroblastic proliferation during this period and the rapid synthesis of collagen. The subsequent period of slower increase in hydroxyproline content reflects, not only collagen formation, but also the increase in density of the developing scar tissue.

The discontinuous increase in proline content of infarcted muscle probably reflects the balance between degradation of necrotic muscle and synthesis of new scar tissue (fig. 6). The proline content failed to increase significantly from the 4th to the 8th day after infarction. This period represents the time of maximum lysosomal activity, as illustrated here in the change in activity of soluble or nonparticle bound β-glucuronidase and acid phosphatase (fig. 6). When the activity of these hydrolytic enzymes decreases again the proline content increases, reflecting a greater rate of scar tissue formation compared with the rate of tissue degradation.

Treatment with anabolic steroids increases the rate of protein synthesis (2), but appears to reduce scar formation in the necrotic myocardium. Cross-sections of the infarcted tissue, containing a longitudinal section of the papillary muscle and additional myocardial tissue inside the previously ischemic area, were analyzed for proline and hydroxyproline content.

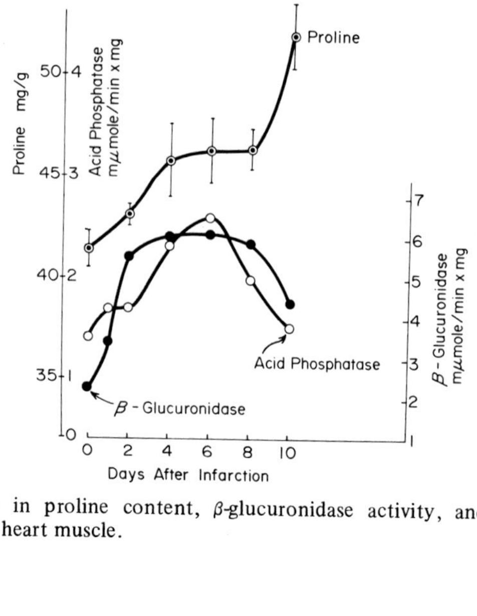

Fig. 6 Changes in proline content, β-glucuronidase activity, and acid phosphatase activity in infarcted heart muscle.

Control samples from infarcted tissue of untreated animals had a hydroxyproline content of 20.9 mg/g 3 weeks after infarction. Samples from animals treated with anabolic steroids (2 mg/kg per day) contained only 15.5 mg/g of hydroxyproline. In animals treated with anabolic agents, the infarcted area appears smaller than in untreated animals. Table II illustrates attempts to determine the effect of the anabolic steroid upon the size of the scar. The first column illustrates the relative weight of the damaged tissue or developing scar tissue, *i.e.*,% of the left ventricle that is infarcted, in control and treated animals. An attempt to determine the change in scar size from the decrease in hydroxyproline and collagen content of treated animals is illustrated in the next two columns. The cross-section of the infarcted area showed an increase of 17.3 mg of hydroxyproline per g of tissue in control animals, indicating that 80.2% of the tissue was new, developing scar tissue. In steroid-treated animals, the increase in hydroxyproline content is 11.6 mg/g, indicating that 53.9% of the tissue is scar tissue, a decrease of 26.3% from the controls.

Several mechanisms might be postulated to explain the apparent decrease in collagen formation in steroid-treated animals, such as: 1) inhibition of collagen synthesis, 2) stabilization of lysosmal membranes and diminished muscle destruction by lysosomal enzymes at the periphery of the infarct, and 3) enhanced muscle recovery and muscle protein formation. At the present time, we are attemtpting to examine the various possibilities.

These results indicate that anabolic agents may increase the rate of compensatory processes in noninfarcted muscle and reduce scar formation and permanent injury following myocardial infarction, whereas a protein-deficient diet may have the opposite effect.

Table II

Effect of Anabolic Steroid upon Scar Formation 3 Weeks after Myocardial Infarction

		Increase in hydroxyproline content of infarcted muscle		
	Size of infarcted area (scar) % of left ventricle	Hydroxyproline mg/g	% Scar of "infarcted" tissue	Proline / Hydroxyproline Ratio
Control, untreated animals N = 8	20.6% ±1.2	+17.31	80.2%	2.7
Steroid-treated animals N = 8	15.6% ±0.6	+11.64	53.9%	3.3
Change	−24.3% Decrease in scar size	−32.8% Decrease in collagen content	−26.3% Decrease in scar size	

SUMMARY

Following acute myocardial infarction in dogs, there is a significant, reversible diminution in tissue levels of ATP, creatine phosphate, and norepinephrine in noninfarcted heart muscle.The diminution in ATP and CP levels of the noninfarcted muscle is accompanied by a significant impairment in myocardial function as reflected in the decrease in rate of pressure rise (dp/dt) or stroke work. Treatment with anabolic steroids significantly increases the ATP level of noninfarcted muscle, whereas a protein-free diet results in a significant diminution in myocardial ATP level and increased mortality. Scar formation in cardiac muscle is markedly reduced in animals treated with the anabolic steroid, and collagen formation is reduced by 26%. These results suggest that diet and anabolic hormones may play an important role during tissue repair and muscle recovery, following acute myocardial infarction.

REFERENCES

1. GUDBJARNASON, S.; BRAASCH, W.; COWAN, C. and BING, R. J.: Metabolism of infarcted heart muscle during tissue repair. Amer. J. Cardiol. *22:* 360–369 (1968).
2. GUDBJARNASON, S.; FENTON, J.; WOLF, P. and BING, R. J.: Stimulation of reparative processes following experimental myocardial infarction. Arch. Int. Med. *118:* 33–40 (1966).
3. MATHES, P. and GUDBJARNASON, S.: Changes in norepinephrine stores in the canine heart following experimental myocardial infarction. Amer. Heart J. *81:* 211–219 (1971).

Mailing address: Sigmundur Gudbjarnason, Ph.D., Science Institute, University of Iceland, Reykjavik (Iceland).

IInd Institute of Medical Pathology and Institute of Morbid Anatomy,
University of Milan, Milan, Italy

STUDY OF INFARCTED MYOCARDIAL TISSUE WITH FLUORESCEINATED ANTI-HAPTOGLOBIN AND ANTI-OROSOMUCOID (ALPHA$_1$-ACID GLYCOPROTEIN) SERA

A. AGOSTONI, G. BAROLDI, P. C. BINAGHI,
R. STABILINI, and F. RADICE

Changes in serum concentration of "acute phase reactive proteins" have been demonstrated to occur in stress situations as a consequence of an increase of hepatic synthesis (2, 4, 5). A similar increase of these proteins, mostly haptoglobin and orosomucoid, has been demonstrated in the sera of patients with myocardial infarction (1). The serial changes of these serum proteins seem to be related to the clinical evolution: in fact, the normalization of their levels seems to indicate that the infarcted area is healed, whereas high levels indicate either that the repair is still in process or that other specific pathological conditions (infections, trauma) are acting.

In order to elucidate the meaning of these changes, we have examined the possibility that the acute phase reactive proteins might be fixed by the injured tissue. The present study was undertaken by a direct immunofluorescent technique on necrotic areas of hearts of patients who had died from a recent myocardial infarction.

MATERIAL AND METHODS

Antisera

Specific FITC-conjugated antisera against haptoglobin (Hp), orosomucoid, and beta$_{1C/A}$ globulin from rabbits were used. These antisera were supplied by Behringwerke (Marburg/Lahn) in lyophilized form to be made up with distilled water when employed.

The conjugated antisera were diluted 1 : 5 in order to reduce the excessive concentration of the dye present in the original conjugate, avoiding as much as possible the occurrence of a non-specific staining.

For the blocking test controls, undiluted specific antisera against Hp, orosomucoid, and beta$_{1C/A}$-globulin from rabbits (Behringwerke) were used.

Tissues

The hearts of 5 patients who had died from recent myocardial infarction (less than 15 days from the onset of the symptoms and the signs) were selected. The samples of myocardium were obtained at necropsy, performed 24 h after death. Each sample was divided into 2 portions, the first for routine histological examination and the second for the immunofluorescence technique. The latter was sliced in small blocks, wrapped in aluminium foil, and quickly frozen by immersion in isopentane cooled at $-70°C$ with solid CO_2; the frozen blocks were then preserved at $-20°C$ in small sealed polystyrene boxes.

Immunofluorescence

The 7 μ tissue sections, obtained by a cryostat, were fixed for 10 min in cold acetone, and then stained with the fluorescent antisera as indicated by Mellors (3).

All the controls customary in immunofluorescence were included, and the staining was not considered specific unless it could be completely (or almost completely) blocked by prior treatment of the tissue with the corresponding unconjugated specific antiserum.

Finally, the stained sections were viewed in a Leitz Orthoplan microscope equipped with an HBO-200W mercury lamp (Osram-Germany), using BG38/4mm and UG1/2mm as primary filters and Leitz Wetzlar K 430 as secondary filter. Dark background Leitz dry condenser was used. The photographic records were obtained with a Kodak Ektachrome High Speed daylight 23 DIN film.

RESULTS AND DISCUSSION

An increase of fluorescence was present in all the infarcted areas treated with FITC-conjugated anti-Hp and antiorosomucoid sera in respect of the blocked controls. The same was not observed when an FITC-conjugated antiserum against $beta_{1C/A}$-globulin was used.

Evaluation of the data obtained by the immunofluorescence technique is still a difficult task, particularly because of the well-known affinity of the necrotic tissue for the fluorescent dye. This implies that we are limited to a comparison in evaluating the intensity of the fluorescence between specifically and non-specifically stained tissue.

The sole aim of the present preliminary attempt was to find out whether a histopathological counterpart of the clinical finding of an increase of the reactive proteins in the cardiac infarct could be established. The results obtained seem to indicate that haptoglobin and orosomucoid are fixed to the necrotic myocells.

In contrast to this, a negative finding was recorded with the use of FITC-conjugated anti-$beta_{1C/A}$-globulin serum. This is in agreement with the previously reported clinical data in which the serum increase of the $beta_{1C/A}$-globulin was delayed and much less pronounced than the other reactive proteins

(1). That led us to exclude this protein from the group of so-called reactive proteins.

The logical consequence of these first results could be an extensive study, in both human and experimental material, in order to grade and to time the fixation phenomenon of these reactive proteins in infarcted areas in relation to other clinical parameters (shock, liver damage, therapy, etc.).

SUMMARY

In order to elucidate the meaning of the increase of haptoglobin and orosomucoid observed in the serum of patients with recent myocardial infarction, the possibility that these proteins might be fixed by the injured tissue was examined by a direct immunofluorescent technique. The results obtained provide evidence that haptoglobin and orosomucoid are present in recently injured tissue.

ACKNOWLEDGMENT

We would like to express our gratitude to Prof. G. Chiappino for his help and advice.

REFERENCES

1. AGOSTONI, A.; VERGANI, C.; STABILINI, R.; MARASINI, B.; ARCIDIACONO, R.; SBAFFIA, A. and BINAGHI, P. C.: Immunochemical quantitation of acute phase reactive proteins in myocardial infarction. Amer. Heart J. *80:* 313–318 (1970).
2. DOBRYSZYCKA, W.; ZEINEH, R.; EBROON, E. and KUKRAL, J. C.: Metabolism of plasma proteins in injury states. II. Incorporation of (^{14}C) glucosamine and (^{35}S) methionine into human and canine haptoglobin. Clin. Sci. *36:* 231–240 (1969).
3. MELLORS, R. C.: Analytical Cytology, pp. 1-67 (McGraw-Hill Book Co. New York) (1959).
4. OWEN, J. A.: Effect of injury on plasma proteins. Advanc. Clin. Chem. *9:* 1–41 (1967).
5. WERNER, M. and COHNEN, G.: Changes in serum proteins in the immediate postoperative period. Clin. Sci. *36:* 173–184 (1969).

Mailing address: A. Agostoni, M.D., Instituto di Patologia Speciale Medica II, Pad. Beretta Est, Via F. Sforza 35, 20122 Milano (Italy).

Department of Pharmacology and Therapeutics, and Pathology
Institute, Department of Pathology, McGill University, Montreal, Canada

CARDIOVASCULAR AND CARDIAC FINE STRUCTURAL EFFECTS OF THE NONEQUILIBRIUM ALPHA-BLOCKING AGENT, EEDQ

K. I. MELVILLE and J. M. BERGER

For several decades, experimental pharmacologists have measured responses of tissues *in vivo* and *in vitro* to characterize the effects of pharmacologic agents. This interest has been especially active in the case of cardiovascular pharmacology. Despite the large amount of pharmaco-physiologic observations, virtually no correlated ultrastructural studies on the appearance of cardiac muscle cells (cardiocytes), affected by such compounds, have been pursued. Part of the reason for this lack of correlation has been the difficulty posed by the lack of specificity or sustained pharmacologic effect of compounds studied, since it was considered doubtful that the morphology would appear altered simultaneously with the peak activity of the compound or that the effect on a specific tissue could be separated sufficiently well from other effects to make such a study realistic. A compound with a relatively prolonged activity and a somewhat intriguing profile of effects was considered suitable for a pilot study into the ultrastructural correlates of drug action.

The compound, N-ethoxycarbonyl-2 ethoxyl-1, 2 dihydroquinoline, or EEDQ (fig. 1), is one of the new series of "nonequilibrium" (irreversible) types of alpha adrenergic receptor blocking agents, first reported by Belleau *et al.* (3). It was also shown by Belleau and Malek (2) and Belleau (1) that EEDQ leads to inactivation of enzyme hydrolases and enhances peptide and amide syntheses *in vitro*. In recent studies by Melville *et al.* (2), it was observed that EEDQ induced a rapid (in 2 to 3 min) alpha adrenergic receptor blockade (epinephrine reversal). This effect was prolonged and not associated with evidence of beta adrenergic blockade, *i.e.*, tachycardia induced by exogenous epinephrine, norepinephrine or isoproterenol was not antagonized. Indeed, the alpha-blockade induced by EEDQ was associated with increasing degrees of sinus tachycardia, which could be prevented by prior injection of the beta blocker, propranolol (2). While it is not yet clear what role reflexes may play in the observed tachycardia, it appears that EEDQ does have potent alpha-blocking properties (1-3, 11-13).

Figure 1 shows the comparative chemical structures of EEDQ and the alpha-blocking agent phenoxybenzamine, as well as the beta-blocking agent propranolol. It is of interest to note that, in spite of the basic quinoline ring in

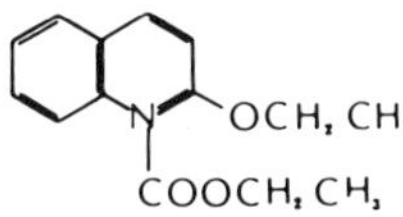

EEDQ (BC-681)

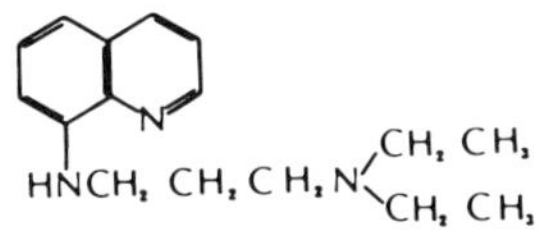

PLASMOCID (SN-3115)

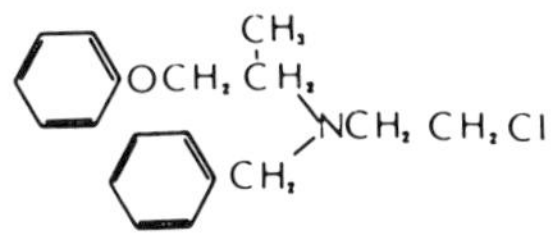

PHENOXYBENZAMINE

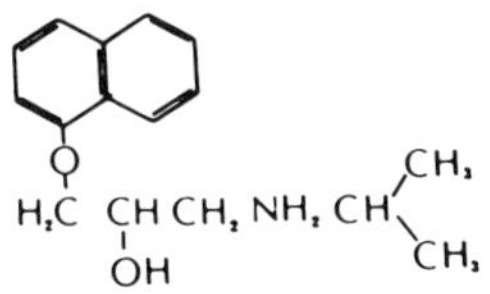

PROPRANOLOL

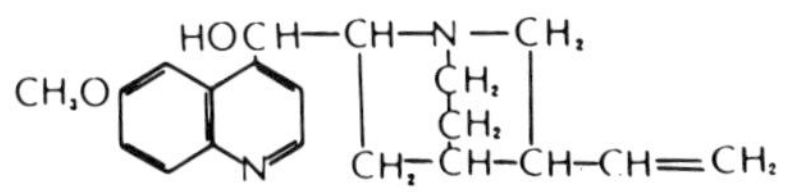

QUINIDINE

Fig. 1 EEDQ and phenoxybenzamine are shown as representative of alpha adrenergic blockers. Plasmocid, recently characterized[1] as showing beta adrenergic blocking properties, is compared with propranolol, a known beta blocker. The similarity in both the presence of a quinoline ring nucleus and di-ethyl terminal groups on plasmocid and EEDQ seems worthy of note. Plasmocid was originally synthesized as an antimalarial with a quinoline nucleus like that in quinidine (shown also). Like quinidine, plasmocid appears to have a greater affinity for atrium than for ventricle at lower doses. [1]Berger and Melville, Abstracts of the III Annual Meeting of the International Study Group for Research in Cardiac Metabolism, Stowe, Vermont, June 29–July 1, 1970.

its structure, EEDQ could not be shown to lead to any direct "quinidine-like" myocardial depressant action (12). Moreover, repeated i.v. injection of increasing doses of EEDQ (fig. 2, a and b) was followed by increasing depressor responses, but concomitantly, increasing degrees of sinus tachycardia (11-13). EEDQ was observed to enhance rather than antagonize digitalis arrhythmias (13). In view of the above observations, it was of some interest to investigate the effects of EEDQ on the fine structure of the myocardium.

Light and electron microscopic investigations in this pilot study were carried out on a total of 10 tomcats weighing 2.5–3.5 kg and fasted for 12 h. The animals received chloralose anaesthesia (50 mg/kg) either alone or concomitantly with EEDQ (0.5 mg/kg, i.v.). During the anaesthesia, cats were monitored for blood pressure and ECG changes and were maintained by a Harvard respirator. The animals were sacrificed at periods of 1 and 5 h after induction of anaesthesia, or 1–5 h after administration of anaesthesia and EEDQ.

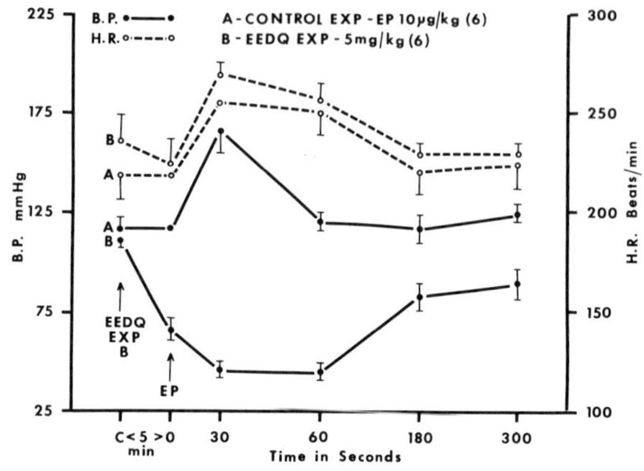

Fig. 2a Graph showing the changes in both the mean heart rate (H.R.) and blood pressure (B.P.), following epinephrine (EP) in doses of 10 μg/kg i.v. One series of experiments (A) records the responses following EP alone. The other set (B) represents the responses when EEDQ was injected 5 min before the epinephrine. Experiment B shows "epinephrine reversal" in the blood pressure response, characteristic of an alpha adrenergic blocker. The number of experiments appears in brackets, and the vertical lines show ± S.E. of the means. (Modified from Melville *et al.* (12)).

Hearts were perfused *in situ* by buffered 1% followed by 3% glutaraldehyde. Areas in atrium and ventricle were selected and excised. These were then post-fixed in buffered osmium tetroxide solution, dehydrated, embedded in Epon (9), and cured for 48 h at 60°C. Toluidine blue stained (19) thick sections of 1 μ were studied for orientation and appearance of cardiocytes. Ultrathin sections 600–800 Å in thickness, as judged from silver to light-gold interference colours, were cut with glass knives on an LKB I ultratome. Sections were contrasted with lead citrate (20) and uranyl acetate (17), examined, and photographed in a Hitachi HU-11B electron microscope.

FINDINGS

Atrial and ventricular working cardiocytes of the control, chloralose-treated cats showed a central nucleus and surrounding myofibrils. In the proximity of the nucleus, myofibrils diverged and coursed around, so as to create a fusiform, filament-free paranuclear zone. Within the atrial paranuclear zones, there appeared large numbers of mitochondria, many atrial specific granules (A-granules) 150–450 mμ in diameter with homogeneously electron-dense cores, a prominent Golgi complex composed of flattened and sacular cisterns, as well as small vesicles and tubules, a moderate amount of glycogen granules, and frequently microtubules. In ventricular cardiocytes, paranuclear zones showed a small Golgi complex and did not contain A-granules. The ventricular cardiocytes did show moderate amounts of glycogen particles, a T-system, well-organized

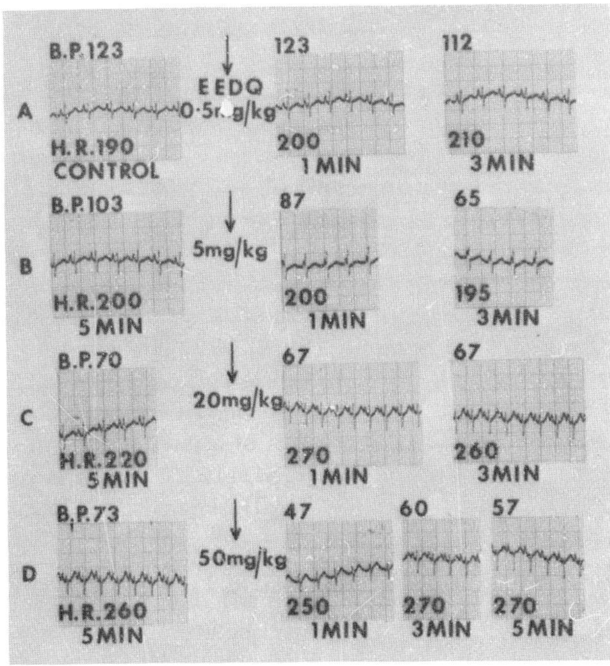

b

Fig. 2b Recordings of cat ECG (lead II), with mean blood pressure (B.P.) above and mean heart rate (H.R.) below each strip. Cardiac responses following successive superimposed i.v. injections of EEDQ are seen. The doses used were 0.5 (A), 5 (B), 20 (C), and 50 (D) mg/kg. With increasing doses, there developed increasing degrees of hypotension, associated with moderately increasing sinus tachycardia. At doses up to 20 mg/kg, however, a dose within the toxic range, only a transient (30–60 sec) disappearance of the P-wave with development of a junctional type of rhythm occurred. No evidence of myocardial depression was found at these doses. (Modified from Melville *et al.* (12)).

sarcomeres, and a relatively electron-lucent sarcoplasmic background (fig. 3).

The most striking change encountered was the appearance, within 1 h after EEDQ administration, of large amounts of granular particles 200–300 Å in diameter, but occasionally smaller, filling the sarcoplasm of the paranuclear poles and intermyofibrillar spaces. The granules appeared to be glycogen of the beta variety (figs. 4–6), Rough sarcoplasmic reticulum in test sections was judged to be about the same as that in control animals.

Within some atrial cardiocytes, the nucleus showed clusters of rosette configurations of glycogen of the alpha variety as membrane-bound inclusions. As a rule, intranuclear glycogen of the alpha type of configuration was found only within nuclei of atrial working cardiocytes (fig. 7). The beta glycogen, however, could be seen in the sarcoplasm of both atrial and ventricular working

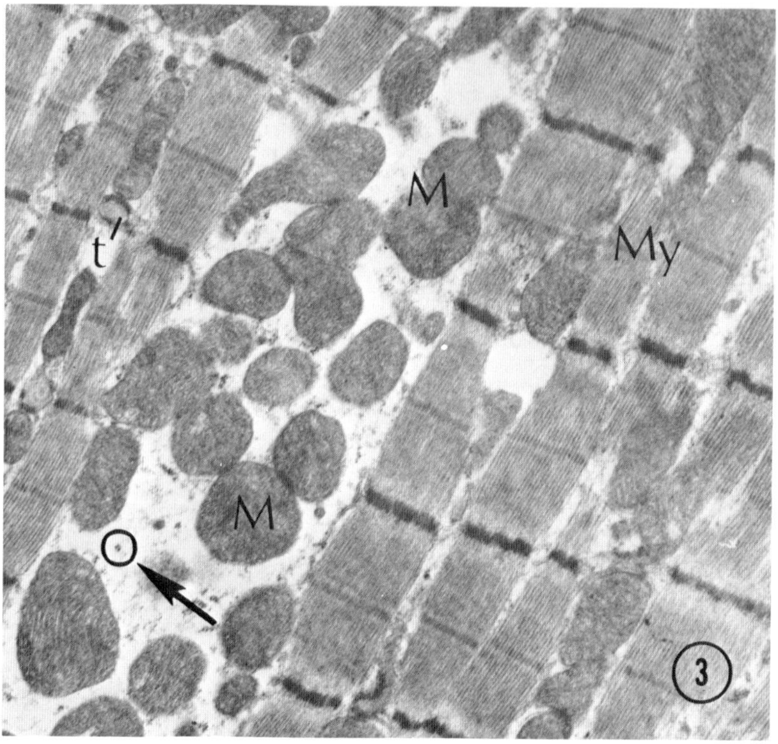

Fig. 3 Portion of a paranuclear zone of a typical cat ventricular cardiocyte several hours after chloralose anaesthesia. In the paranuclear zone, many mitochondria (M) but small amounts of glycogen particles (arrow and circle) may be observed. The background cytoplasm of this zone appears relatively electron-lucent after chloralose alone. On either side of the paranuclear zone, numerous sarcomeres (My) are found, showing a typical banding pattern and organization. Close to the Z lines of some sarcomeres, the transverse tubule (t) can be identified. Atrial cardiocytes likewise contained little glycogen in the chloralose controls. ×17,500.

cardiocytes. The beta glycogen appeared often closely associated in groups, as has been described to be characteristic for young adult cats (7, 10). In most cases, however, it was distributed in a diffusely uniform field of single particles.

In many atrial paranuclear areas, C-granules (residual bodies) 800–1200 mμ in their long axes were prominent also at 1 h after EEDQ. The variable frequency of such C-granules, seen in controls as well, did not appear at this time to be a prominent feature of the cardiocyte reaction to EEDQ.

It is important to note that the Golgi complex in ventricular working cardiocytes, in contrast to that in atrium, is ordinarily meagre. Its moderate enlargement, therefore, within 1 h after EEDQ administration in test tomcats appeared remarkable.

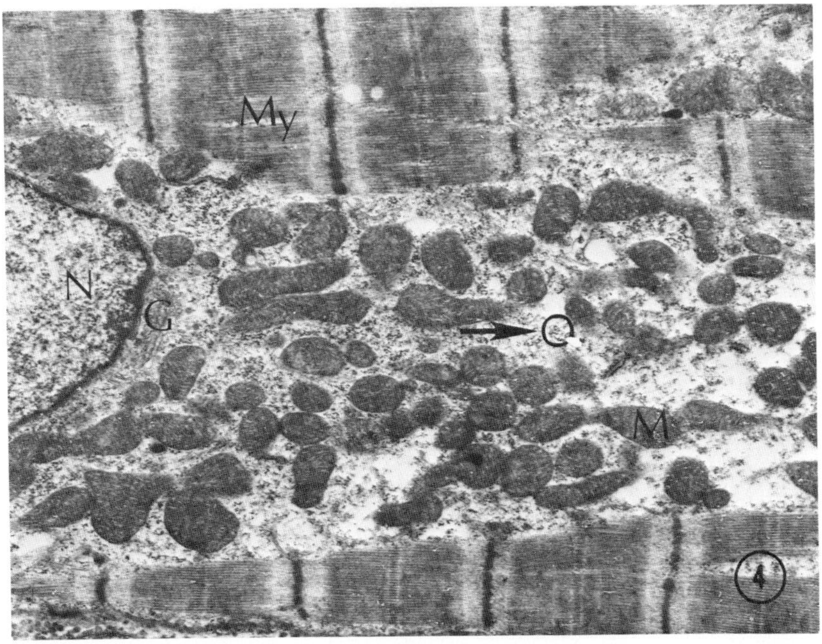

Fig. 4 Nucleus (N) and paranuclear zone of a cat ventricular cardiocyte 1h after EEDQ administration. Between the numerous mitochondria (M) in the paranuclear zone, considerable numbers of electron-dense particulate matter, apparently glycogen particles (arrow and circle), have already accumulated. There appears only a moderate infiltration of these particles into the adjacent sarcomeric zones (My). In association with this finding, the Golgi complex (G) appears only moderately enlarged. ×15,200.

At 5 h after administration of EEDQ (figs. 8 and 9), the amount of glycogen appeared to be about as much as that seen after 1 h. The number of C-granules, however, appeared greater after 5 h than at 1 h. Many of the C-granules at 5 h showed within their cores granular particles very similar to the beta glycogen particles in the sarcoplasm, and it was presumed that, to some extent, phagocytosis accounted for this appearance. In many atrial cardiocytes, giant residual bodies 1.5–2.5 μ, composed of coalescent C-granules, could be seen in the paranuclear areas (fig. 8). In such cases, the cores of each of the component C-granules of these giant residual bodies showed particles resembling the glycogen in the neighbouring sarcoplasm. The abundance of C-granules in ventricle and the presence of macrobodies of apparently coalescent C-granules in atrium were the main new features found at 5 h.

COMMENT

The appearance of beta glycogen particles has been described and is well known to occur in heart muscle cells (cardiocytes) (5, 7, 10, 16). Such particles

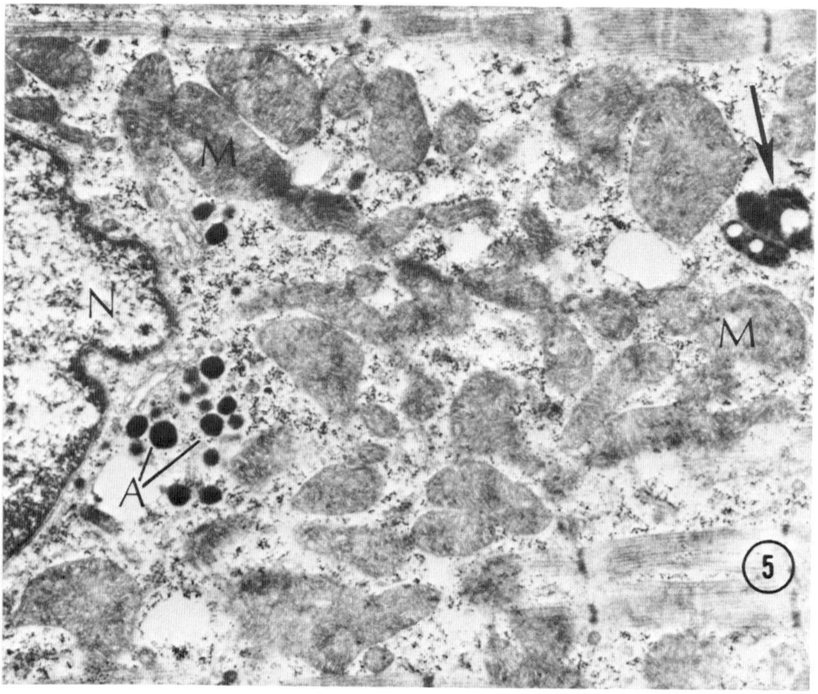

Fig. 5 Nucleus (N) and paranuclear zone of a cat atrial cardiocyte 1 h after EEDQ. Atrial specific granules (A) are located in the region of the Golgi complex and appear similar to those seen in control atrium. Considerable amounts of glycogen particles, similar in distribution to those encountered in ventricle, appear between the Golgi and the mitochondria (M). A C-granule, or residual body, is shown (arrow). ×19,650.

may be as large as 300 Å but, when smaller than 200 Å, cannot easily be distinguished from free ribonucleoprotein (RNP) particles. In our material, the abundance of particles in the range of 250–300 Å and their prominent contrasting after lead alone, strongly suggested that these particles were probably not RNP. Counterstaining with uranyl, after lead, did not appear to increase the total number of these particles. The appearance of polysomes of RNP has been reported to show a linearity that, although multiangular with respect to any straight line drawn through its midpoint, can be easily distinguished from the geometry of aggregates of beta glycogen particles. The latter tend to be aggregated in configurations suggesting oblate circles or polygons and to show in a single aggregation both a greater number of particles than RNP polysomes and a greater interparticulate distance between discrete particulate granules. In this way, RNP particles and polysomes may often be distinguished from the beta glycogen particles.

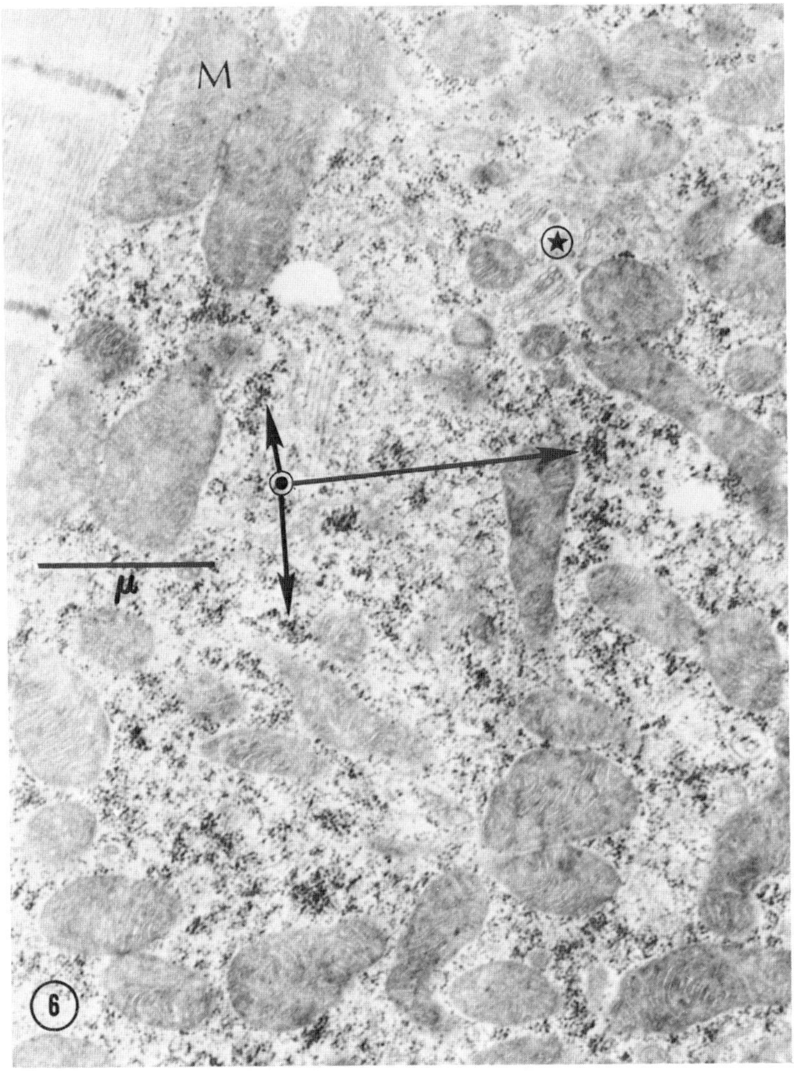

Fig. 6 Higher magnification of part of an atrial paranuclear zone at 5 h after EEDQ. Between the mitochondria (M), accumulations of glycogen are apparent (arrows emanating from circle). Some cisterns of the Golgi complex (star) are present.

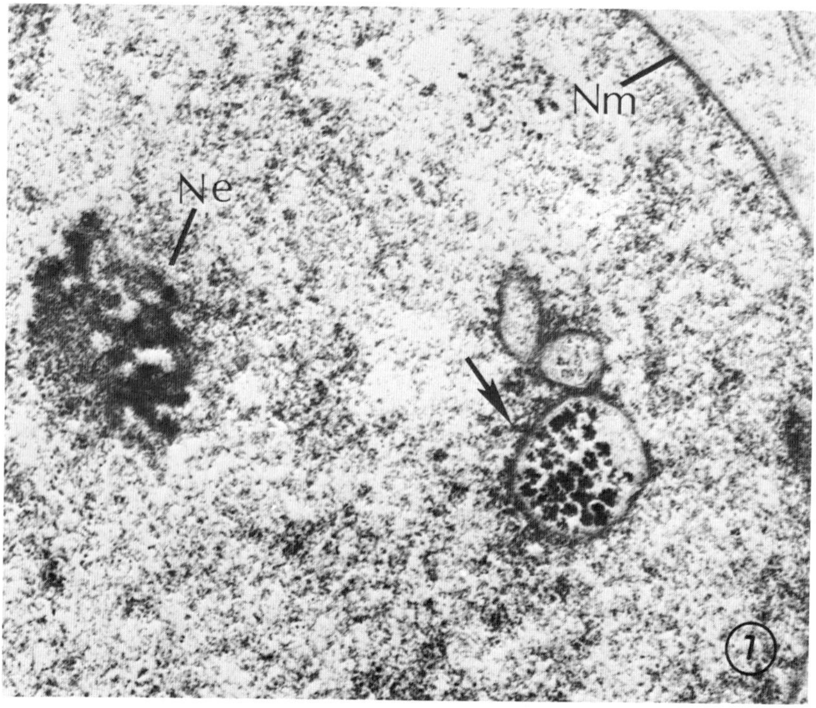

Fig. 7 Portion of a nucleus from atrial cardiocyte after EEDQ, showing a segment of the nuclear membrane (Nm) and the nucleolonema (Ne). The arrow points to an inclusion containing rosettes of electron-dense particulate granules that closely resemble alpha-glycogen particles, such as is seen in hepatocytes. ×24,800.

The abundance of C-granules (residual bodies) after 5 h suggests that the cardiocyte reacts to the marked increase of glycogen by producing phagocytic organelles to reduce the intracellular content of this substance. This conclusion is supported both by the presence within C-granules of granular particles similar to the beta glycogen in the sarcoplasm, as well as by the moderately enlarged ventricular Golgi complex, which is considered to be an important organelle in the production of C-granules. The significance of coalescent C-granules appearing as giant residual bodies is less clear, however, since such bodies have not been described in untreated cats and were never encountered in controls. The normal appearance of myofibrils and mitochondria agrees with the pharmacologic finding of no "quinidine-like" action of EEDQ (4, 11, 13). The findings of particles that were apparently glycogen suggested a possible biochemical mechanism related to the action of EEDQ.

It is now recognized that the action of epinephrine and norepinephrine upon the heart is associated with activation of the adenyl cyclase system of enzymes

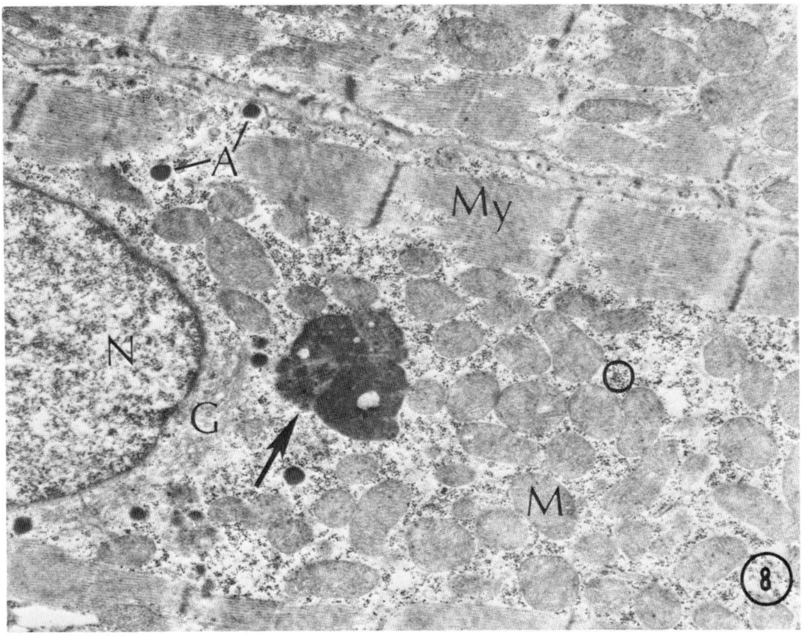

Fig. 8 Nucleus (N) and paranuclear zone of a cat atrial cardiocyte 5 h after EEDQ administration. The Golgi complex (G) seems large, but its greater size than that of the ventricular Golgi is typical. Atrial specific granules (A) appear essentially similar to those in controls, as do the mitochondria (M) and sarcomeres (My). The amount and distribution of the fine particulate glycogen granules (circle) appears unchanged from that seen 1 h after EEDQ, but the C-granules, which ordinarily lie separated from one another, now appear to have coalesced with apposed limiting membranes into a macrobody (arrow). The cores of the macrobody appear to contain particulate material resembling the glycogen particles in the sarcoplasm. ×15,200.

(6, 8, 14, 18). Considerable evidence at present seems to support the hypothesis that cyclic AMP acts as secondary messenger for epinephrine and digitalis in mediating the positive inotropic effect of these compounds on myocardial cells (8, 14, 18).

Although the precise significance of the action of cyclic AMP has yet to be elucidated, it appears to be collaterally associated with the positive inotropic effect produced by epinephrine on the heart (15). Thus, the stimulation of cyclic AMP appears to result both in an inotropic effect and glycogenolysis. For this reason, a block at or before the level of cyclic AMP would be expected to interfere with both effects. The finding of "irreversible" block to alpha adrenergic responses, together with the accumulation during this period of glycogen particles, suggests the intriguing conclusion that the mechanism that accounted for the alpha blockade may have resulted in inhibition of glycogen

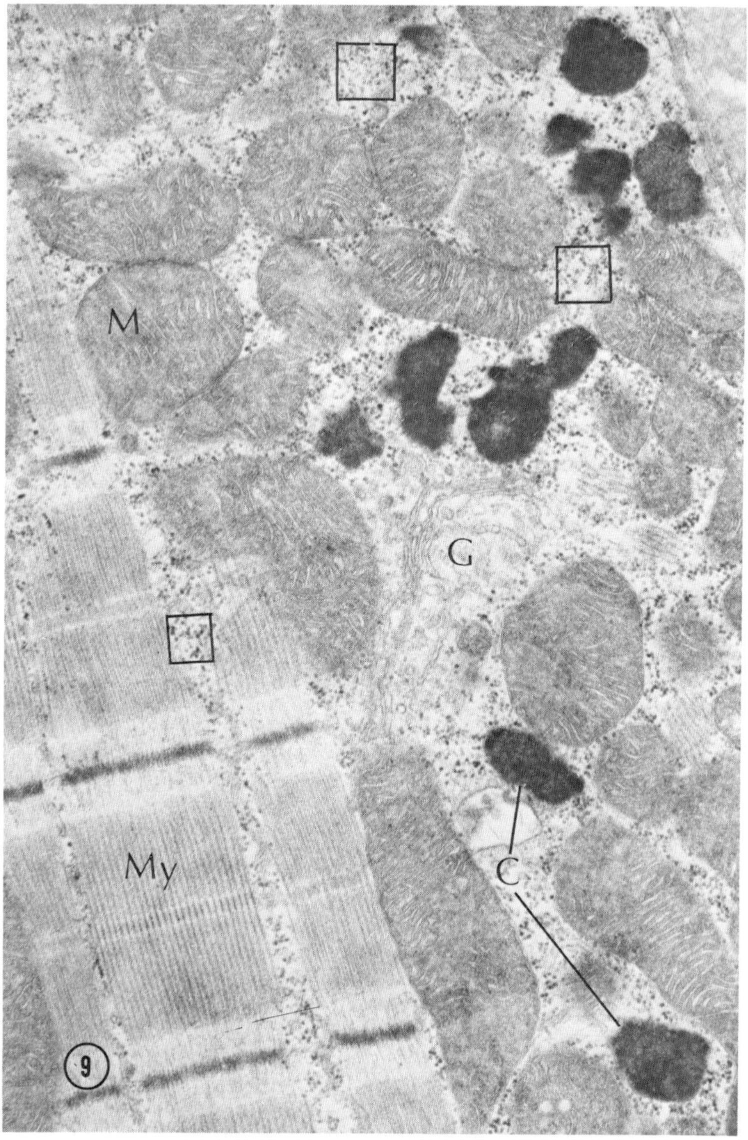

Fig. 9 Part of the paranuclear zone of a typical cat ventricular cardiocyte 5 h after EEDQ administration. The Golgi complex (G) still appears moderately enlarged. Despite the cardiac adrenergic blockade at this time, the sarcomeres (My) and mitochondria (M) appear morphologically unaffected. The main findings were the great abundance of beta glycogen particles (enclosed in squares), similar in amount to that seen at 1 h, and the markedly increased number of C-granules (C) since 1 h. × 25,800.

breakdown. While further work will be necessary to characterize the interrelationship of these events after EEDQ, the present results support the hypothesis that the abundance of glycogen particles in these cats was linked to the action of a nonequilibrium alpha adrenergic block in the heart.

SUMMARY

EEDQ (N-ethoxycarbonyl-2 ethoxyl-1, 2 dihydroquinoline) has been characterized as exerting an "irreversible" cardiac alpha adrenergic blockade without "quinidine-like" effects. A fine structural study correlating the drug action with cardiac morphology was undertaken on chloralosed cats. The prominent change in cardiocytes at 1 and 5 h after EEDQ was manifested as marked glycogen accumulation. The unchanged appearance of myofilaments and mitochondria was correlated with the pharmacodynamic findings of no "quinidine-like" action. At 5 h, a marked increase in C-granules was also seen. These results support the hypothesis that the action of EEDQ also resulted in a biochemical block of the adenyl cyclase system in the heart.

REFERENCES

1. BELLEAU, B.; DITULLIO, V. and GODIN, D.: The mechanism of the irreversible adrenergic blockade by N-carbethoxy-dihydroquinolines. Model studies with serine hydrolases. Biochem. Pharmacol. *18:* 1039–1044 (1969).
2. BELLEAU, B. and MALEK, G.: A new convenient reagent for peptide syntheses. J. Amer. Chem. Soc. *90:* 1651–1952 (1968).
3. BELLEAU, B.; MARTEL, R.; LACASSE, G.; MENARD, M.; WEINBERG, N. L. and PERRON, Y. G.: N-carboxylic acid esters of 1, 2- and 1,4-dihydroquinolines. A new class of irreversible inactivators of the catecholamine alpha receptors and potent central nervous system depressants. J. Amer. Chem. Soc. *90:* 823 (1968).
4. BERGER, J. M. and MELVILLE, K. I.: Effects of an irreversible alpha-receptor blocker on myocardial ultrastructure. 3rd Ann. Symp. Canad. Ass. for Research in Toxicology, Toronto, 1969.
5. BERGER, J. M. and RONA, G.: Functional and fine structural heterogeneity of atrial cardiocytes, in BAJUSZ and JASMIN, Methods and Achievements in Experimental Pathology, Vol. 5, pp. 540–590 (Karger, Basel, 1971).
6. ENTMAN, M. L.; LEVEY, G. S. and EPSTEIN, S. E.: Mechanism of action of epinephrine and glucagon on the canine heart. Evidence for increase in sarcotubular calcium stores mediated by cyclic 3′, 5′,-AMP. Circulat. Res. *15:* 429–438 (1969).
7. FAWCETT, D. W. and McNUTT, N. S.: The ultrastructure of the cat myocardium. I. Ventricular papillary muscle. J. Cell. Biol. *42:* 1–45 (1969).
8. LEVEY, G. S. and EPSTEIN, S. E.: Activation of adenyl cyclase by glucagon in cat and human heart. Circulat. Res. *24:* 151–156 (1969).
9. LUFT, J. H.: Improvements in epoxy resin embedding methods. J. Cell Biol. *9:* 409–414 (1961).
10. McNUTT, N. S. and FAWCETT, D. W.: The ultrastructure of the cat myocardium II. Atrial muscle. J. Cell Biol. *42:* 46–47 (1969).
11. MELVILLE, K. I.: Alpha adrenergic-blocking and antiarrhythmic actions of N-ethoxycarbonyl-2-ethoxyl-1, 2-dihydroquinoline (EEDQ). pp. 274–275, Proc. 4th International Congress on Pharmacology, Basel, July, 1969.
12. MELVILLE, K. I.; EU, H. H.; BERGER, J. M. and DOOKHOO, A.: Effects of N-ethoxycarbonyl-2 ethoxy-1,2-dihydroquinoline (EEDQ) in experimental chloroform-epinephrine and digitalis dysrhythmias. J. Pharmacol. Exp. Ther. *175:* 147–156 (1970).

13. MELVILLE, K. I.; EU, H. Y. and DOOKHOO, A.: Cardiovascular changes following combined EEDQ and propranolol and effects on digitalis arrhythmias. Can. J. Physiol. Pharmacol. (in press).
14. ROBISON, G. A.; BUTCHER, R. W. and SUTHERLAND, E. W.: Cyclic AMP. Annu. Rev. Biochem. *37:* 149–174 (1968).
15. SHANFELD, J.; FRAZER, A. and HESS, M. E.: Dissociation of the increased formation of cardiac adenosine 3′,5′-monophosphate from the positive inotropic effect of norepinephrine. J. Pharmacol. Exp. Ther. *169:* 315–320 (1969).
16. SOMMER, J. R. and JOHNSON, E. A.: A comparative study of Purkinje fibers and ventricular fibers. J. Cell Biol. *36:* 497–526 (1968).
17. STEMPAK, J. G. and WARD, R. T.: An improved staining method for electron microscopy. J. Cell Biol. *22:* 697–701 (1964).
18. SUTHERLAND, E. W. and ROBISON, G. A.: Metabolic effects of catecholamines. Pharmacol. Rev. *18:* 145 (1966).
19. TRUMP, B. F.; SMUCKLER, E. A. and BENDITT, E. P.: A method for staining epoxy sections for light microscopy. J. Ultrastruct. Res. *5:* 343–348 (1961).
20. VENABLE, J. H. and COGGESHALL, R.: A simplified lead citrate stain for use in electron microscopy. J. Cell Biol. *25:* 407–408 (1965).

Mailing address: Dr. K. I. Melville, Department of Pharmacology and Therapeutics, McGill University, Montreal, Quebec (Canada).

May Institute for Medical Research of the Jewish Hospital, and
Departments of Medicine and Pathology, University of
Cincinnati, College of Medicine, Cincinnati, Ohio, USA

PATHOPHYSIOLOGIC CHANGES IN ARTERIOSCLEROTIC AND NONARTERIOSCLEROTIC RATS, FOLLOWING ISOPROTERENOL-INDUCED MYOCARDIAL INFARCTION[1]

B. C. WEXLER, J. T. JUDD, R. F. LUTMER, and J. SAROFF

The potent beta-adrenergic stimulating agent, isoproterenol, has been used to produce myocardial infarction experimentally (4, 11). The myocardium is stimulated so intensely that the coronary arteries are unable to supply adequate quantities of blood, a state of relative ischemia or anoxia ensues, and myocardial necrosis results. The area of myocardium infarcted can be regulated by the dose of isoproterenol employed (4, 11).

Wexler and his associates (6–8, 14), have found that repeatedly bred male and female rats will develop arteriosclerosis spontaneously. The incidence and severity of the arteriosclerosis is commensurate with the number and frequency of litters raised by the female breeder and the number and frequency of litters sired by the male breeder. Although the female breeder rats develop severe, grossly-visible arteriosclerosis, they survive considerably longer than the males. Paradoxically, although the male breeder rats develop relatively mild arterial lesions, *i.e.*, microscopic aortic lesions, they often succumb prematurely owing to myocardial infarction.

For the past several years we have been inducing myocardial infarction, with isoproterenol, in virgin rats free of arterial disease and comparing their response with the response of breeder rats having pre-existing arteriosclerosis (1–3, 9–13). A summation of our most cogent findings is described here.

MATERIALS AND METHODS

All of the animals used in these studies were Sprague-Dawley rats. Virgin male and female rats, comparable in age to the breeder rats, were used as subjects having no arteriosclerosis. Female breeder rats that had given birth to and suckled four to five litters were used as subjects with pre-existing, grossly-visible

[1] Supported by the Southwestern Ohio Heart Assn., the National Heart Institute (HE-13, 222, HE-8868 and HE-K3-1734-Res. Career Awd. to B.C.W.), and (The National Institutes of Health) FR-5633 (U.S. Public Health Service).

arteriosclerosis. Male breeder rats that had sired five to six litters of rats in close succession were used as subjects with pre-existing microscopic aortic sclerosis.

Isoproterenol was administered subcutaneously on 2 successive days. Non-arteriosclerotic virgin rats received 50 mg/100 g body weight and the arteriosclerotic breeder rats were given 25 mg/100 g. The Sprague-Dawley rat is very resistant to the myocardial-infarct-inducing effects of isoproterenol (5, 11). Despite the disparity in dose, the area of myocardium infarcted is equal in the virgin and breeder rats with the regimen of injections described above. Because breeder rats are consistently much more obese than their virgin counterparts, the total dose received by the heavier breeder rats approaches that of the virgin rats (10–13).

All of the animals were sacrificed by decapitation to avoid the stress of anesthesia. At autopsy, pertinent organs were trimmed carefully and weighed. The severity of the hydrothorax was measured by syringe aspiration of thoracic fluid. Blood from the severed neck vessels was collected, centrifuged under refrigeration, and stored in the deep freeze until time for analysis. Serum creatine phosphokinase (CPK), transaminases (SGOT and SGPT), lactic dehydrogenase (LDH), glucose, triglycerides, free fatty acids, cholesterol, blood urea nitrogen (B.U.N.), and other factors were analyzed by means of an Auto-analyzer (Technicon), utilizing methods prescribed for the Auto-analyzer. Adrenocorticosteroid determinations and connective tissue biochemistry, *e.g.*, collagen and mucopolysaccharides, as well as histochemical and histopathologic procedures and statistical methods, were as detailed in previous publications (1–3, 10–13).

RESULTS

General observations

Within minutes of receiving an injection of isoproterenol, the animals become prostrate and stuporous. In addition to marked tachycardia, the animals manifest dyspnea, epistaxis, and hydrothorax. Within 4 h after the first injection, the animals recover and move about normally. Following the second injection, the animals once again display tachycardia, dyspnea, prostration, and exacerbation of their congestive heart failure. By the 3rd day, the survivors manifest severe myocardial infarction (figs. 1 and 2). The animals with pre-existing arteriosclerosis do not exhibit the signs of severe myocardial shock shown by the nonarteriosclerotic subjects, although their myocardium is permeated with the same degree of confluent, through-and-through necrosis. Male rats appear to be much more sensitive to the infarct-inducing effects of isoproterenol than do female subjects, *i.e.*, male rats succumb in significantly greater numbers than females and show definitely less ability to affect subsequent myocardial repair.

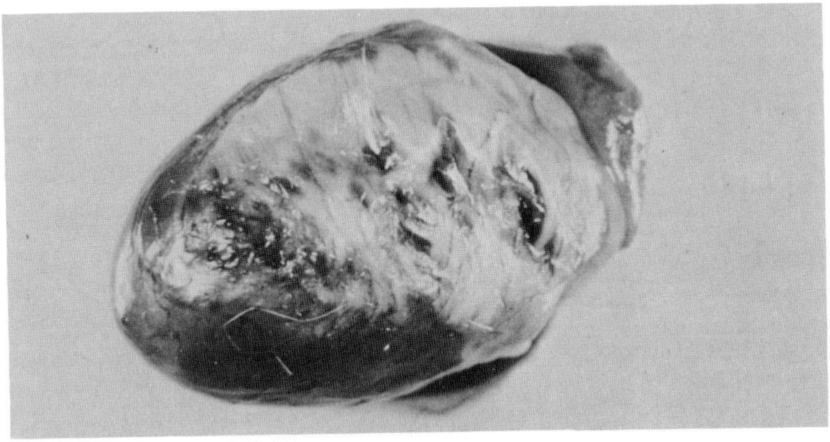

Fig. 1 Massive myocardial infarct (white area) produced in a nonarteriosclerotic male Sprague-Dawley rat.

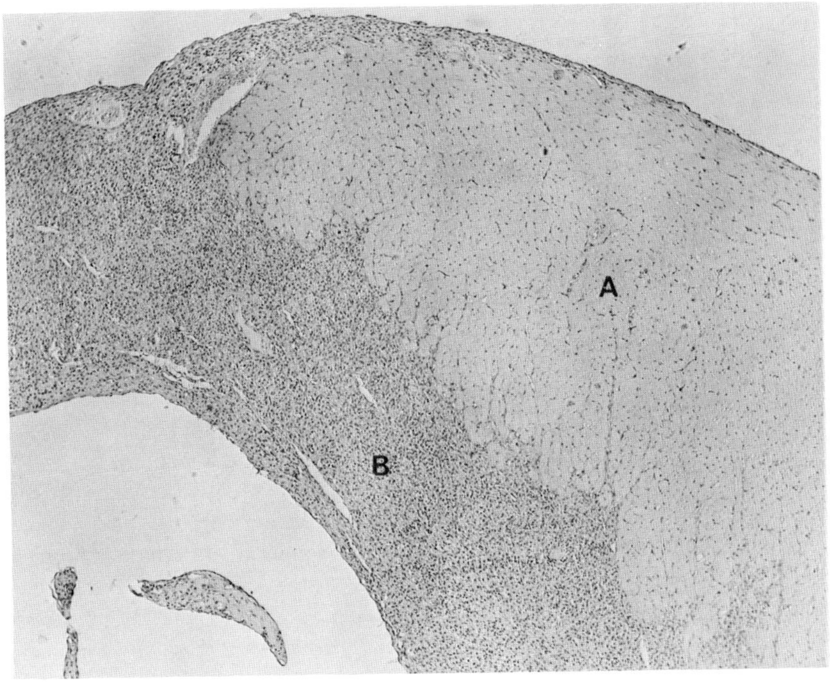

Fig. 2 Cross section of the apex and left ventricle of the heart of an arteriosclerotic male breeder 5–6 h after a 2nd injection of isoproterenol. There is extensive confluent necrosis throughout the apex and into the left ventricle; the light grey area in the photo (A) is viable myocardium; all of the granular area (B) is necrotic myocardium showing massive cellular infiltration and complete destruction of all muscle fibers in this area of infarction. H&E, ×60.

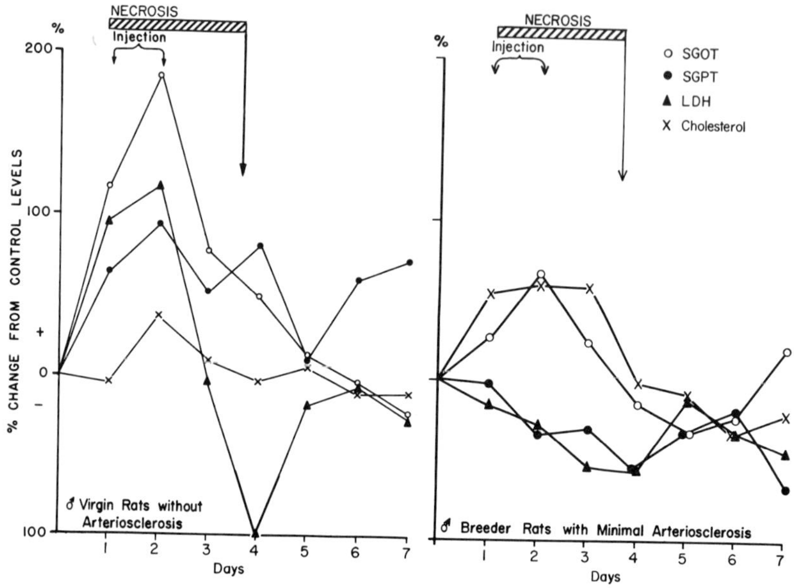

Fig. 3 Fluctuations in serum SGOT, SGPT, LDH, and cholesterol following the induction of cardiac necrosis by two injections of isoproterenol (arrows). Statistical treatment of the data by the analysis of variance method showed that the fluctuations within each parameter measured, *i.e.*, SGOT, SGPT, LDH, and cholesterol, were significant ($p = 0.01 - 0.05$) in the time period covered.

Serum enzyme changes

Although serum transaminase (SGOT and SGPT) and LDH levels rise and fall dramatically during the acute onset and repair of myocardial infarction in these rats (fig. 3), serum CPK levels are, by far, the most sensitive and dynamic index of myocardial cell death (fig. 4) (9, 11, 13).

Adrenal steroid changes

During the acute onset of myocardial necrosis, there is a pronounced increase in aldosterone production, concomitant with anuria, the hydrothorax or congestive heart failure (fig. 5)(12). When myocardial necrosis reaches a zenith (3rd day post-isoproterenol), the animals begin to diurese, the chest clears of fluid, and the aldosterone production returns to normal (fig. 5). During this same period, glucocorticoid production, *e.g.*, corticosterone, falls considerably (fig. 6), commensurate with prostration, shock or low blood pressure, and death (12). Histopathologically, the adrenal glands and thymuses of these animals demonstrate intense cortical lipid depletion and thymic involution, respectively, which are indicative of active adrenal steroid production and secretion.

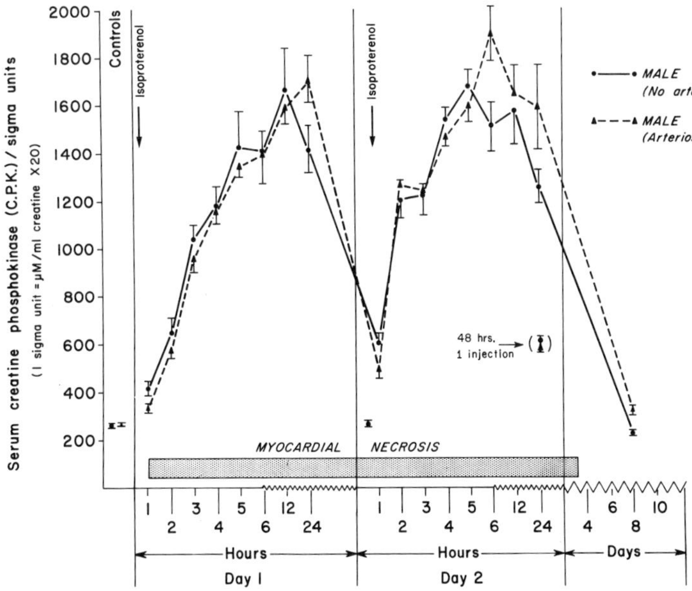

Fig. 4 Changes in serum CPK levels (mean ± standard error) of arteriosclerotic and nonarteriosclerotic male Sprague-Dawley rats at various time intervals after two injections of isoproterenol spaced 24 h apart. (CPK levels shown separately, 48 h after the first injection of isoproterenol, are included to demonstrate the drop in CPK, in time, if only one injection of isoproterenol is given.) Myocardial necrosis is an ongoing event and becomes maximal between the 2nd and 3rd day. Myocardial repair is rapid and occurs between the 4th and 10th day.

Lipid changes

During the acute stages of myocardial necrosis, peripheral adipose tissue sites, *e.g.*, periadrenal and mesenteric fat, undergo intense dissolution, the blood becomes lipemic, and the liver exhibits marked fatty infiltration (fig. 7). Serum triglycerides and free fatty acids are greatly elevated, but cholesterol levels show little or no change (10). Chemical analyses of hepatic lipids demonstrate that the intense fatty infiltration of the liver is largely due to great accumulations of triglyceride (fig. 8)(1). Again, when myocardial necrosis reaches its zenith (day 3) and myocardial repair begins, the fatty liver condition begins to recede, the hyperlipemia clears, and the various serum lipid levels are restored to normal.

Glucose

Marked hyperglycemia also accompanies the hyperlipemia during the acute cardiac necrosis phase, followed by normal levels of blood glucose during the repair phase (10).

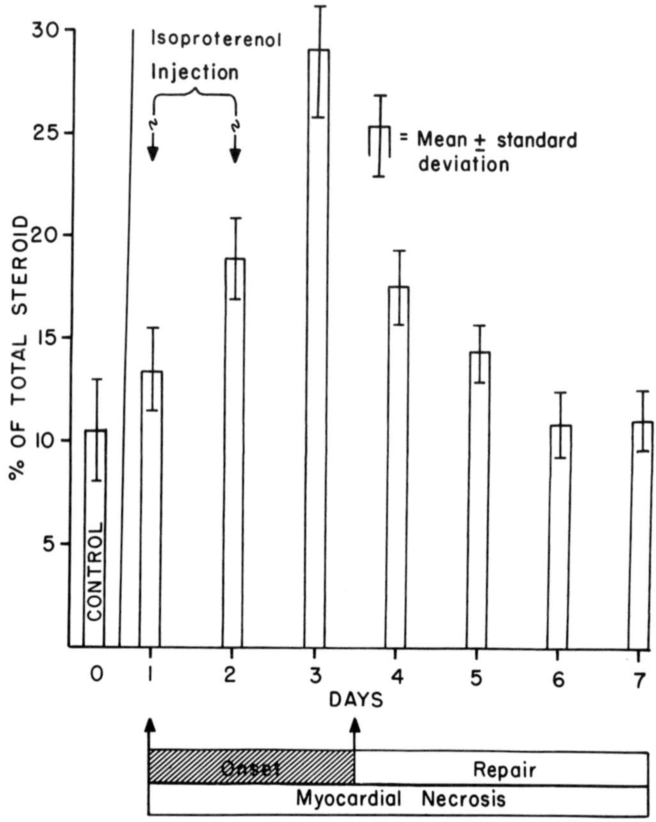

Fig. 5 Aldosterone production *(in vitro)* by adrenals from arteriosclerotic female Sprague-Dawley rats during and after the induction of myocardial infarction.

Biochemical and histochemical connective tissue changes

Shortly after the appearance of myocardial necrosis (first injection of isoproterenol), large myocardial interfascicular edematous spaces appear. These spaces are accentuated following the second dose of isoproterenol and, shortly thereafter, become filled with a mucoid material that appears to be mucopolysaccharide when tested histochemically (2). Chemical tests demonstrate that the material is rich in hexosamine (2). The concentration of chondroitin sulfate increases in conjunction with beginning fibrilogenesis or myocardial repair (3).

Arteriosclerotic vs. nonarteriosclerotic subjects

Although the connective tissue repair process appears to be similar in both the arteriosclerotic and nonarteriosclerotic subjects, it is quite apparent that the

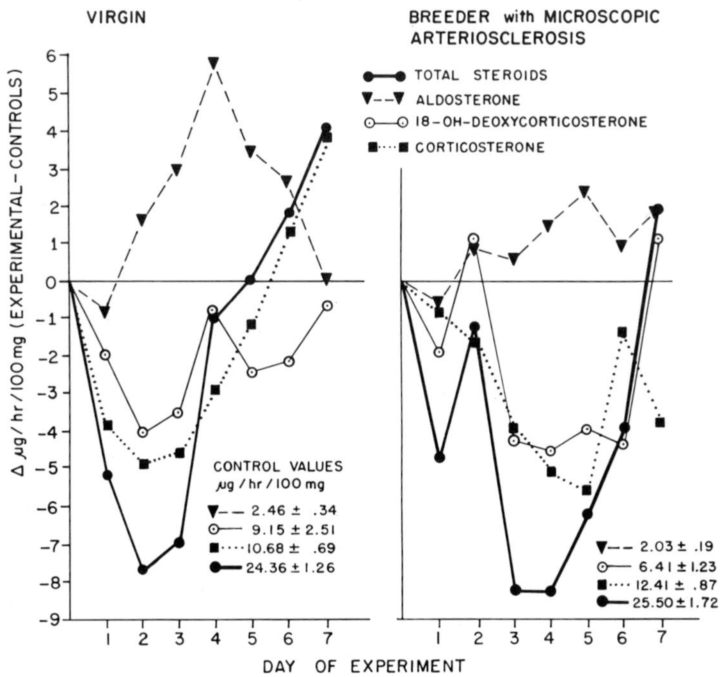

Fig. 6 Changes in the *in vitro* production of steroids by adrenal glands of arteriosclerotic and nonarteriosclerotic male Sprague-Dawley rats during the onset and repair of myocardial infarction (control values: µg/hr/100 mg).

pathophysiologic changes, *e.g.*, enzymes, lipids, glucose, adrenal steroids, etc., are much less dramatic in those subjects with pre-existing arteriosclerosis when challenged with the myocardial infarct-inducing effects of isoproterenol (2, 3, 10–13).

DISCUSSION

These series of investigations demonstrate that the potent beta-adrenergic stimulating agent, isoproterenol, can induce a "myocardial infarct syndrome" that mimics the changes observed in man during the acute stages of a heart attack. Serum enzymes such as CPK, SGOT, SGPT, and LDH all rise and fall commensurate with the onset and repair of myocardial injury and the severity of the myocardial infarction. Mobilization of lipid, hyperlipidemia, fatty metamorphosis of the liver, and hyperglycemia all attest to the intense stimulation of body metabolic defense mechanisms. We interpret the profound shock and prostration in these animals as due to the temporary alterations in adrenal steroidogenesis, *i.e.*, reduced glucocorticoid synthesis concomitant with in-

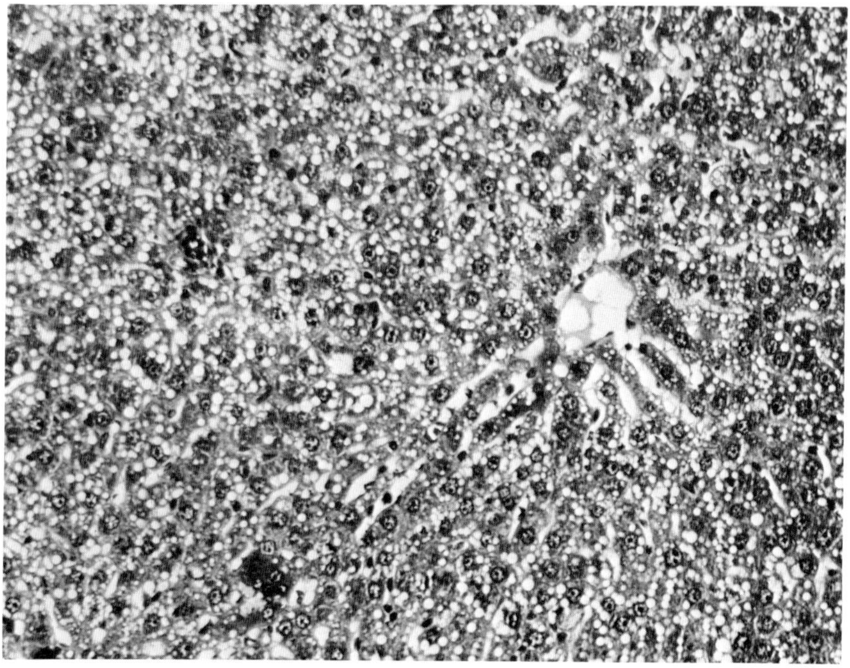

Fig. 7 Liver of a nonarteriosclerotic male rat 4 h after a second (s.c.) injection of isoproterenol. The multitude of droplets demonstrate the intense fatty infiltration of the liver that occurs coincidental with the development of acute myocardial necrosis. This condition of hepatic fatty metamorphosis clears rapidly coincidental with myocardial repair. H&E, ×100.

creased aldosterone production. The temporarily increased aldosterone production is, we believe, correlated with the temporary development of congestive heart failure.

Although definitive information is lacking, we are stimulated to investigate the possibility that subjects with pre-existing arteriosclerosis respond less dramatically to their drug-induced myocardial infarction because their pre-existing arteriosclerosis has affected a greater collateralization of coronary arteries than in their nonarteriosclerotic counterparts. Also intriguing is the fact that male rats are much more susceptible to myocardial infarction than female rats.

SUMMARY

Male and female, arteriosclerotic and nonarteriosclerotic rats were given large doses of isoproterenol to induce massive myocardial infarction. During the acute stages of myocardial necrosis, the animals manifested evidence of shock, increase in serum enzymes (CPK, SGOT, SGPT, and LDH), hyperlipidemia, hyperglycemia, increased aldosterone

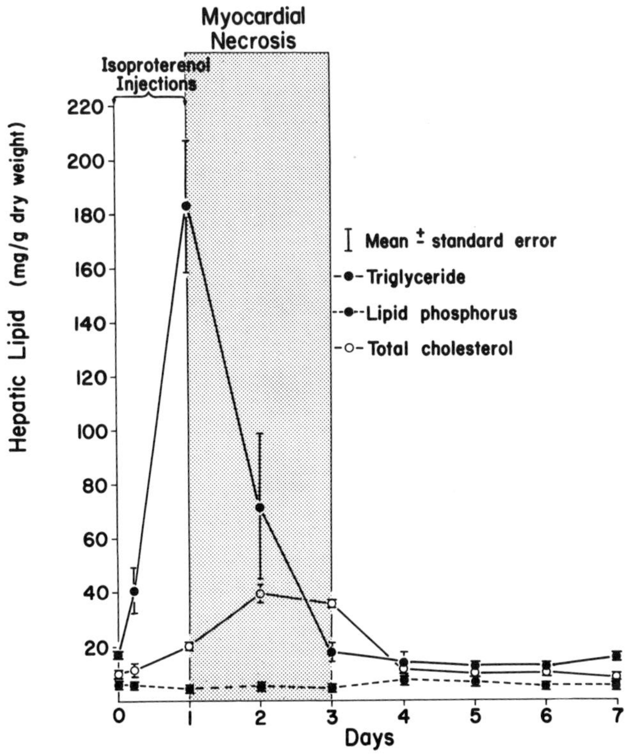

Fig. 8 Dynamic changes in hepatic triglyceride, cholesterol, and phospholipid, following the injection of isoproterenol. Each point represents the average of the analyses of five livers. The shaded area indicates the time period when maximum myocardial necrosis is observed histologically.

production, reduced corticosterone production, and congestive heart failure. During myocardial repair, this abnormal metabolism became normal, the signs of shock and congestive heart failure disappeared, and ground substance changes demonstrated that scar tissue formation had begun. Subjects with pre-existing arteriosclerosis appeared to recover more efficiently than those with no pre-existing arteriosclerosis.

REFERENCES

1. JUDD, J. T. and WEXLER, B. C.: Acute changes in liver lipids during myocardial infarction induced by isoproterenol. Proc. Soc. Exper. Biol. Med. *130:* 1302–1305 (1969).
2. JUDD, J. T. and WEXLER, B. C.: Myocardial connective tissue metabolism in response to injury: histological and chemical studies of mucopolysaccharide and collagen in rat hearts after isoproterenol-induced infarction. Circulat. Res. *25:* 201–214 (1969).
3. JUDD, J. T. and WEXLER, B. C.: Myocardial connective tissue metabolism in response to injury: II. Investigation of the mucopolysaccharides involved in isoproterenol-induced necrosis and repair in rat hearts. Circulat. Res. *26:* 101–109 (1970).

4. RONA, G.; CHAPPEL, C. I.; BALAZS, T. and GAUDRY, R.: An infarct-like myocardial lesion and other toxic manifestations produced by isoproterenol in the rat. Arch. Pathol. *67:* 443–455 (1959).

5. RONA, G.; CHAPPEL, C. I.; BALAZS, T. and GAUDRY, R.: The effect of breed, age and sex on myocardial necrosis produced by isoproterenol in the rat. J. Genontol. *14:* 169–173 (1959).

6. WEXLER, B. C.: Spontaneous arteriosclerosis in repeatedly bred male and female rats. J. Atheroscler. Res. *4:* 57–80 (1964).

7. WEXLER, B. C.: Spontaneous coronary arteriosclerosis in repeatedly bred male and female rats. Circulat. Res. *14:* 32–43 (1964).

8. WEXLER, B. C.: Spontaneous arteriosclerosis of the mesenteric, renal and peripheral arteries of repeatedly bred rats. Circulat. Res. *15:* 485–496 (1964).

9. WEXLER, B. C.: Serum creatine phosphokinase activity following isoproterenol-induced myocardial infarction in male and female rats with and without arteriosclerosis. Amer. Heart J. *79:* 69–79 (1970).

10. WEXLER, B. C.; JUDD, J. T. and KITTINGER, G. W.: Myocardial necrosis induced by isoproterenol in rats: Changes in serum protein, lipoprotein, lipids and glucose during active necrosis and repair in arteriosclerotic and nonarteriosclerotic animals. Angiology *19:* 665–682 (1968).

11. WEXLER, B. C. and KITTINGER, G. W.: Myocardial necrosis in rats: serum enzymes, adrenal steroid and histopathological alterations. Circulat. Res. *13:* 159–171 (1963).

12. WEXLER, B. C. and KITTINGER, G. W.: Steroids produced in vitro by adrenal glands of normal and arteriosclerotic rats during and after drug-induced myocardial necrosis. Circulat. Res. *16:* 322–331 (1965).

13. WEXLER, B. C.; KITTINGER, G. W. and JUDD, J. T.: Responses to drug-induced myocardial necrosis in rats with various degrees of arteriosclerosis. Circulat. Res. *20:* 78–87 (1967).

14. WEXLER, B. C. and TRUE, C. W.: Carotid and cerebral arteriosclerosis in the rat. Circulat. Res. *12:* 659–666 (1963).

Mailing address: Dr. B.C. Wexler, May Institute for Medical Research of the Jewish Hospital, 421 Ridgeway Avenue, Cincinnati, Ohio 45229 (USA).

Department of Pathological Physiology and 2nd Surgical Clinic,
School of Medicine, J.E. Purkinje University,
Brno, Czechoslovakia

INFLUENCE OF MECHANICAL ASSISTING SYSTEMS ON THE ACUTE FAILING HEART

J. VAŠKŮ, B. BEDNAŘÍK, and E. URBÁNEK

In our experimental laboratory, we have concerned ourselves for more than three years with the problem of mechanical assistance of the failing heart and have developed several types of pumps that we tried out on the healthy hearts of normotonic dogs (1–4). Of late, we have performed a series of experiments using a bypass pump in acute heart failure induced in experimental animals by ligation of ramus circumflexus of the left coronary artery. The evaluation of the experimental results and our first experience with the clinical application of the pump form the subject of this communication.

MATERIALS AND METHODS

Description of pumping system

The mechanical assisting device consists of a pump and an external power and controlling system. The pump has a volume of 42 ml (fig. 1) and features a pneumatic drive; the output may be asynchronous or maintain synchronization by initiating a stroke from each ECG signal, or from every other or every third signal.

The experiments were carried out on 23 dogs weighing 15–28 kg, anesthetized with an oxygen-ether mixture. Left-sided thoracotomy was performed through the 5th intercostal space. A terylene prosthesis was sutured to the descending aorta, while a Tygon tube was introduced into the left atrium through the left heart auricle. After administration of heparin at a dose of 2 mg per kg body weight, the pump was attached to both connectors and placed on the chest of the experimental animal. Acute heart failure was produced through ligation of ramus circumflexus. During the experiment, the pressures in the left and right ventricles and in the left atrium were recorded. The blood flow through the pulmonary artery and the pump outlet was measured with the aid of a Nycotron electromagnetic flow meter. All pressures, flow rates, and electrocardiograms were recorded by means of an Elema eight-channel mingograph. The mean pressure values, the metabolic cardiac work, and the peripheral vascular resistance were calculated with the aid of a Meda analogue computer and printed by an automatic digital central unit. The pump was not started until 15 min after ligation of ramus circumflexus, while the hemodynamic evaluation was

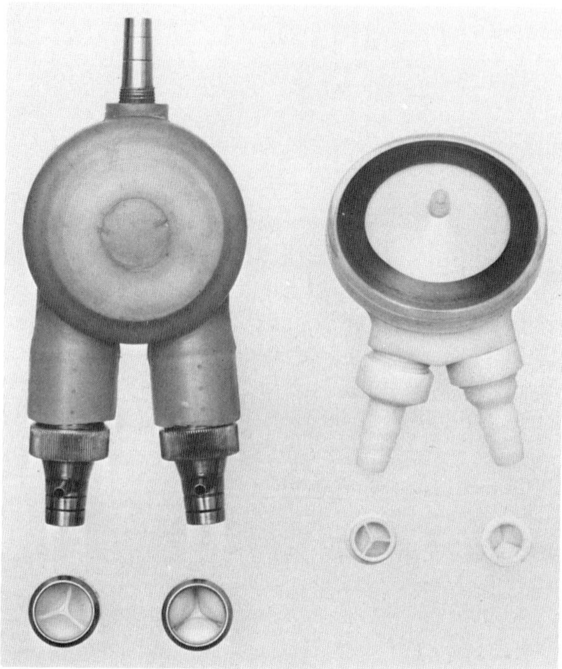

Fig. 1 Bypass pump. Left, the pump used in the animal experiments; right, the pump used for the clinical case.

always performed after another 15 min of pump operation. If after stopping the pump no improvement of the hemodynamic conditions occurred, the pump was restarted, usually for another 15 minutes, but for no longer than 60 min. After another 30 min following the termination of bypass, the final hemodynamic evaluation was performed.

RESULTS

Out of 23 experiments carried out on dogs, 10 animals were not evaluated. Bypass was introduced in 4 dogs for 45 min and in 6 dogs for 60 min. The mean duration of the pump's run was 48 min.

Hemodynamic changes

Left atrial pressure

In 10 dogs, the left atrial pressure decreased after termination of bypass to the original preligation value (77%). In 3 dogs, it diminished after termination of bypass (LHBP), as compared with the postligation values, but remained higher than 10 mm Hg (23%). (Fig. 2, table I.)

Aortic pressure

In 7 dogs, the systolic pressure, after termination of LHBP, reached the initial value recorded prior to ligation (54%). In 6 dogs, it remained lower than before ligation (46%), and was higher than after ligation in 3 dogs. (Fig. 2, table I.)

Left ventricular pressure

In 6 dogs, the end diastolic pressure in the left ventricle returned to the preligation value (46%). In another 6 dogs, it diminished in comparison with the postligation values by an average of 11 mm Hg, but remained higher than before ligation. In 1 dog, it remained unchanged (8%). (Fig. 3, table I.)

Right ventricular pressure

The right ventricular pressure, like the venous pressure, did not undergo any essential changes during the experiment. (Fig. 3, table I.)

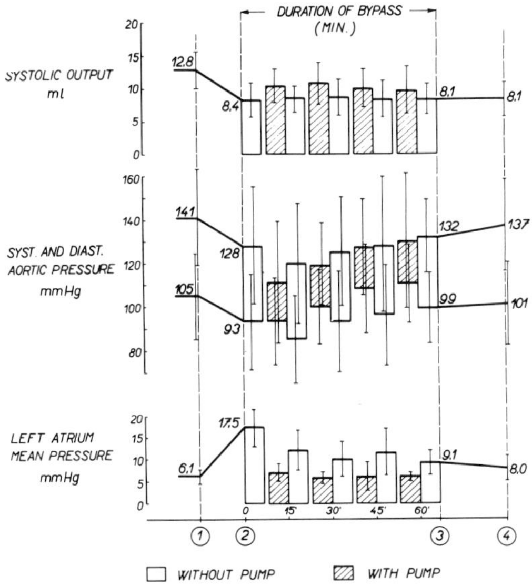

Fig. 2 Stroke volume, aortic pressure, and mean pressure in left atrium (1) before ligation of r. circumflexus, (2) after ligation, (2–3) during bypass, and (4) 30 min after termination of LHBP.

Table I

Hemodynamic Changes

	Pressures in mmHg										
	Aorta			Left ventricle		Left atrium	Right ventricle	Pulmonary blood flow	Heart rate	Systolic output	PVR
	Syst.	Diast.	Mean	Diast.	Mean	Mean	Syst.	ml/min	per min	ml	mmHg·min/l
Before ligation	141	105	120	5.5	57	6.1	54	1986	158	12.8	63
	±22	±20	±18	±2.6	± 9	±1.4	±11	±446	±23	±2.8	±16
Start of bypass	128	93	107	20.0	58	17.5	56	1437	172	8.4	81
	±27	±22	±24	±5.7	±13	±4.2	±14	±537	±29	±2.6	±25
End of bypass	132	99	110	10.5	58	9.1	50	1302	161	8.1	94
	±17	±16	±16	±3.7	± 8	±3.1	±11	±432	±25	±2.3	±29
30 min after bypass	137	101	112	10.4	60	8.0	51	1373	170	8.1	92
	±21	±19	±18	±5.6	±13	±2.7	±11	±536	±29	±2.6	±32

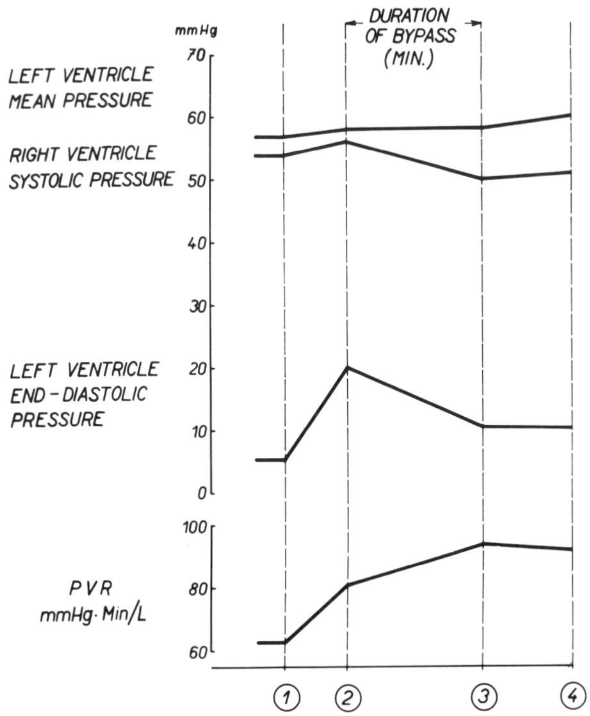

Fig. 3 Pressures in ventricles and peripheral vascular resistance (1) before ligation of r. circumflexus, (2) after ligation, (2–3) during LHBP, and (4) min after termination of LHBP.

Cardiac output, systolic stroke volume, and heart rate

In 8 dogs, the cardiac output was the same after termination of LHBP (62%), while in 5 dogs it rose in comparison with the postligation values (38%). The heart rate increased after ligation on the average by 9%; during LHBP, it decreased in all experimental animals; and after termination of LHBP it was, on the average, 7% higher than prior to ligation. (Fig. 4, table I.)

In no dog did the systolic stroke volume reach the preligation values after termination of the pump's run. In comparison with the postligation values, it rose by an average of 50% in 3 dogs (23%), remained the same in 7 dogs (54%), while it continued to decrease as compared with the postligation values in 3 dogs (23%).

Peripheral vascular resistance

The peripheral vascular resistance increased after ligation by 29%; 30 min following termination of LHBP it was 46% higher than before ligation. (Fig. 3, table I.)

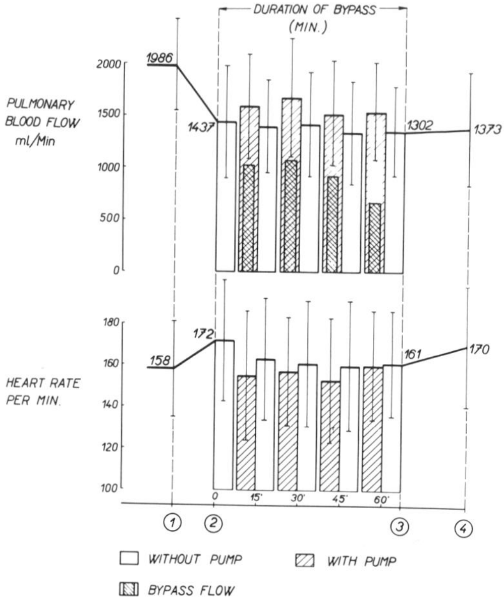

Fig. 4 Cardiac output and heart rate (1) before ligation of r. circumflexus, (2) after ligation, (2–3) during LHBP, and (4) 30 min after termination of LHBP.

Bypass flow

On the average, 63% of the total cardiac output was pumped from the left atrium into the descending aorta. (Fig. 4.)

Left ventricle efficacy

The efficacy of the left ventricle and the metabolic work of the left ventricle were determined through integration of the area under the ventricular pressure curve with the aid of an analogue computer (fig. 5). After ligation, the efficacy of the left ventricle dropped on the average to 82% of the preligation value. In the course of LHBP, the pump reduced the metabolic work of the left ventricle by 5%; 30 min after termination of LHBP, the efficacy was 94% of the preligation value.

Hematological changes

A summarized survey of the hematological changes is presented in figure 6.

CLINICAL APPLICATION

On December 8th, 1969, a 7-year-old girl in whom Fallot's tetralogy had been diagnosed on the basis of an angiocardiographic examination was admitted to

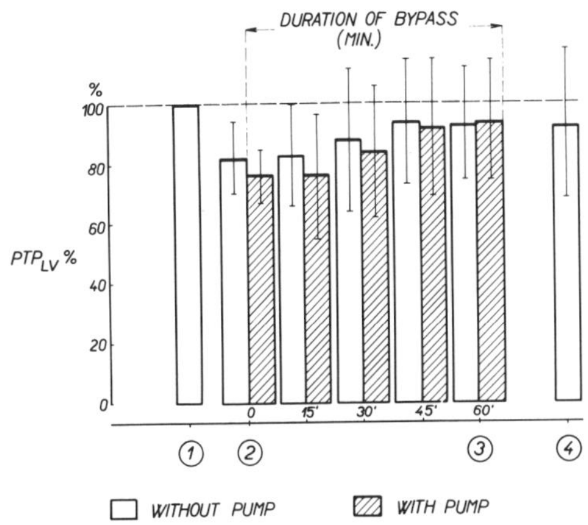

Fig. 5 Efficacy of left ventricle (1) before ligation of r. circumflexus, (2) after ligation, (2–3) during LHBP, and (4) 30 min after termination of LHBP.

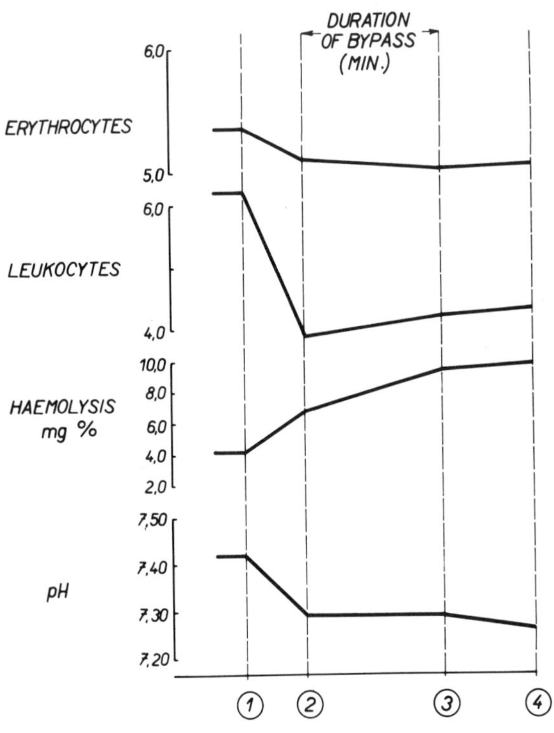

Fig. 6 Hematological values (1) before ligation of r. circumflexus, (2) after ligation, (2–3) during LHBP, and (4) 30 min after termination of LHBP.

our cardiovascular department. Following appropriate cardiologic preparation, surgery was performed on December 12th, 1969. After stoppage of bypass, the blood pressure maintained itself at 80 mm Hg. During completion of the operation, heart arrest occurred twice and pulmonary edema set in. Through massage and defibrillation, we succeeded in restoring the heart's activity, but the pulmonary edema continued to increase and could not be controlled by conventional means. It was therefore decided to connect the bypass pump (fig. 1). A prosthesis was stitched to the ascending aorta and a Tygon tube was introduced into the left atrium. Both connectors were run through the thorax wall and joined to the pump. The thoracotomic wound was closed in anatomical layers and the pump was fixed to the child's thorax (fig. 7). Figures 8 and 9 show the course and evaluation of bypass.

After the pump had been in operation for 4 h, it was decided to change the asynchronous bypass to a synchronous one; but the temporary stoppage of the pump led to new fibrillation of the ventricles and further attempts to restore the heart's activity were without success. Prior to connection of the pump, a mean left atrial pressure of 38 mm Hg was recorded. After starting the bypass system, the left atrial pressure progressively decreased to a mean value of 6.9 mm Hg.

The pump was controlled by pneumatic pulses of 250 mm Hg pressure and 40 mm Hg negative pressure. It pumped 800–2,496 ml of blood per min from the left atrium, which represents 26 to 83% of the total cardiac output.

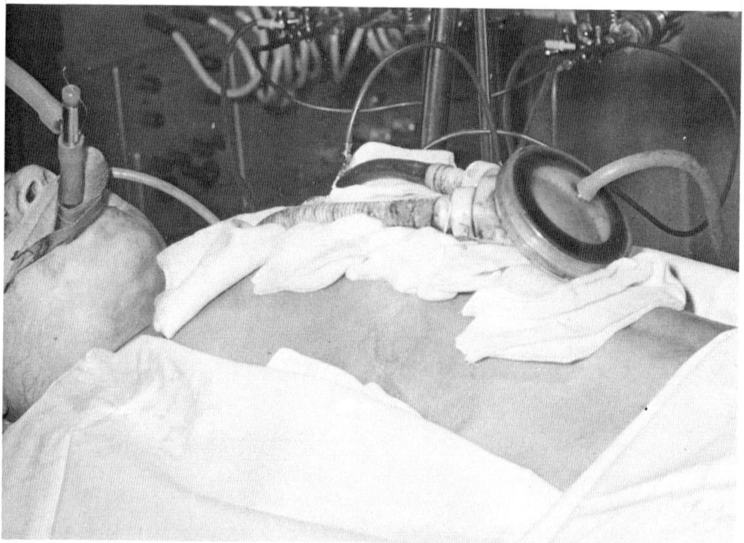

Fig. 7 Pump attached to connectors is placed in location and fixed to the patient's chest wall.

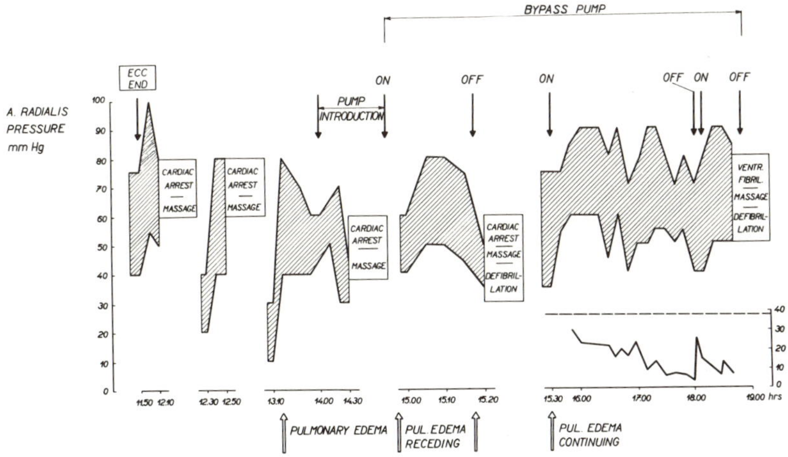

Fig. 8 Graphic presentation of pressure courses in a. radialis and in left atrium after termination of ECG and during bypass.

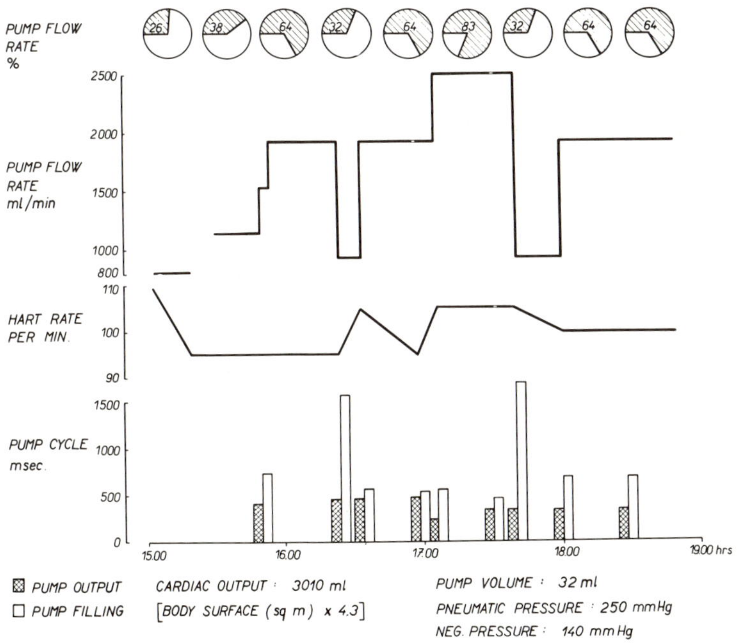

Fig. 9 Graphic presentation of quantity of blood delivered by the pump during LHBP from the left atrium to the aorta, presentation of heart rate time, correlation between pump filling and delivery.

DISCUSSION

Our experimental results with left heart bypass in acutely induced heart failure show that this bypass may favorably influence disturbed cardiovascular hemodynamics. During the pump's run, the cardiac work diminished, the left atrial and left ventricular end-diastolic pressure decreased, while the cardiac and systolic outputs increased

The fact that after termination of bypass the cardiac output did not remain at the level recorded during the pump's run is, in our opinion, connected with the methods used in the experiments, where various factors, primarily surgical trauma, maintained a shock condition, even though the work of the heart muscle improved.

Clinical application of mechanical assistance to a failing heart has been tried several times, but with more or less unsatisfactory results (5–7). We believe that, besides a sometimes inadequate indication, the cause lies mainly in the fact that the assisting system is customarily only applied to moribund patients in whom all other means of resuscitation have failed. This was also the case with our patient where, after several episodes of heart arrest accompanied by pulmonary edema, the hope of restoring a normal heart activity was minimum. The postmortem showed the cause of left heart failure to be marked left ventricular hypoplasia. The ionogram displayed a rather large disturbance of potassium metabolism in the analyzed heart. Only the postmortem revealed that the barrier later maintaining the pulmonary edema was, rather, intrapulmonary. In the lungs were found signs of pulmonary edema, with the presence of edematous fluid in the alveoli. The lung tissue exhibited large focal hemorrhages and the pulmonary capillaries showed signs of venostasis. No conclusion can be drawn from our first clinical application of bypass. It had some shortcomings that we carefully analyzed; nevertheless, certain positive aspects encourage us to use this method again in suitably and responsibly chosen cases.

SUMMARY

In 23 dogs, acute heart failure was induced through ligation of ramus circumflexus of the left coronary artery. In 10 of the dogs, fibrillation of the ventricles occurred within 15 min following ligation or shortly after. A bypass pump was connected to the remaining 13 dogs for an average period of 48 min, after a standard model of heart failure had been accomplished.

The left atrial pressure, which rose on the average by 188% after ligation, normalized in 77% of the experimental animals, while it diminished in 23%. The end-diastolic pressure in the left ventricle, which increased by 265% after ligation, stabilized after termination of bypass at the original value in 46% of the cases, while it decreased in 46% of the experimental animals and remained unchanged in the remaining 8%. The cardiac output remained 31% lower after termination of bypass while the stroke volume was 37% higher as compared with the preligation values. The efficacy of the left ventricle, which diminished after ligation by 12%, reached 94% of this value following termination of bypass. Hemolysis rose as a result of the pump's activity from an average of 4.2 mg% to a value of 9.7 mg%.

The bypass pump was used clinically for a failing heart after radical surgery for Fallot's tetralogy. It was kept in operation for a period of 4 h; subsequently, as a result of the technical design of the assisting system, irreversible fibrillation of the ventricles occurred when the pump was disconnected.

REFERENCES

1. BEDNARIK, B.; BOLF, J.; VASKU, J.; URBANEK, E.; PERESTY, S.; DOLEZEL, S. und ANDREOVSKY, F.: Kreislaufpumpe zur Entlastung des linken Herzens. Wiederbeleb. Organers. Intensivmed. *6:* 1–14 (1969).
2. BEDNARIK, B.; NAVRATIL, J.; TOMECEK, J.; PERESTY, S. and MUSIL, F.: A new type of pump for the support of failing left ventricle. Bratisl. Lek. Listy *50:* 200–207 (1968).
3. BEDNARIK, B.; NAVRATIL, J.; TOMECEK, J.; PERESTÝ, S. and MUSIL, F.: Synchronized blood pump for the support of the activity of the left ventricle. Bratisl. Lek. Listy *50:* 338–345 (1968).
4. BEDNARIK, B.; VASKU, J.; URBANEK, E.; PERESTY, S.; BOLF, J.; ANDREOVSKY, F. und KLIMES, F.: Mechanische Herzunterstützung: Experimentelle Auswertung des linksseitigen Bypasses und der Gegenpulsation. Wiederbeleb. Organers. Intensivmed. *6:* 28–37 (1969).
5. DEBAKEY, M.E.; LIOTTA, D. and HALL, C. W.: Left heart bypass using an implantable blood pump. Reprinted from: Mechanical Devices to Assist the Failing Heart, Chapter 18. National Academy of Sciences-National Research Council, Washington, D.C., 1966.
6. DENNIS, C.; HALLP, D. P.; MORENO, J. R. and SENNING, Å.: Clinical use of a cannula for LHBP without thoracotomy. Amer. Surg. *156:* 623–632 (1962).
7. KANTROWITZ, A.; AKUTSU, T.; CHAPTAL, P. A.; KRAKAUER, J.; KANTROWITZ, A. and JONES, R. T.: A clinical experience with an implanted mechanical auxiliary ventricle. J. Amer. Med. Assoc. *197:* 525–531 (1966).

Mailing address: Prof. Dr. J. Vasku, Institute of Pathophysiology, Purkinje University, Komenskeho nam. 2, Brno (Czechoslovakia).

Institute of Experimental Surgery, Slovak Academy of Sciences, Bratislava, Czechoslovakia, and Department of Physiology, University of Manitoba, Faculty of Medicine, Winnipeg, Canada

INFLUENCE OF ASSISTED CIRCULATION ON CARDIAC METABOLISM

A. ZIEGELHOFFER, K. SISKA, V. HOLEC, M. FEDELESOVA, AND I. FROLLO

The main objective of mechanical circulatory assist is the restoration of normal cardiac function by reducing cardiac work while maintaining systemic circulation. At present, counter pulsation (balloon pumping) and bypass are frequently used techniques to support the infarcted myocardium (4, 15). Recent estimates (2, 3) indicate that counter pulsation and bypass are capable of reducing mechanical load up to 29% and 66%, respectively.

The present investigation was designed to study the metabolism of the heart under the influence of supportive bypass.

METHOD

The experimental animals were anesthetized with pentobarbital (30 mg/kg body weight, Spofa) and received heparin (3 mg/g, Spofa), glucose (400 ml of 5% glucose) and gilurytmal (2 ml, Giulini, Ludvigshafen).

Infarction of the myocardium was produced by clamping the ventral intraventricular branch of the left coronary artery in adult mongrel dogs. The development of infarction was then assessed by the following criteria: a) increase in the blood pressure in left atrium to double the initial value, or at least 12 mm of mercury; and b) elevation of the S-T segment on the ECG

Supportive bypass was applied for 60 min (fig. 1). Using this technique, the oxygenated blood from lungs returns to the left atrium, is bypassed to a reservoir, and then pumped hydraulically into the aorta through the left subclavian artery. The blood pressure was recorded from the aorta, left atrium, and left ventricle. The coronary sinus was cannulated for collection of blood samples. The tissue samples were taken from the area above the infarcted myocardium and quickly frozen by a Wollenberger clamp precooled in liquid nitrogen. Levels of lactate, pyruvate, glucose in the blood, and high energy phosphate compounds such as creatine phosphate (PCr), ATP, ADP, AMP and glycogen (GLY) in the ventricle were determined by enzymatic methods. Uptake of substrates was calculated according to the formula:

$$\% \frac{A - CS}{A} \times 100$$

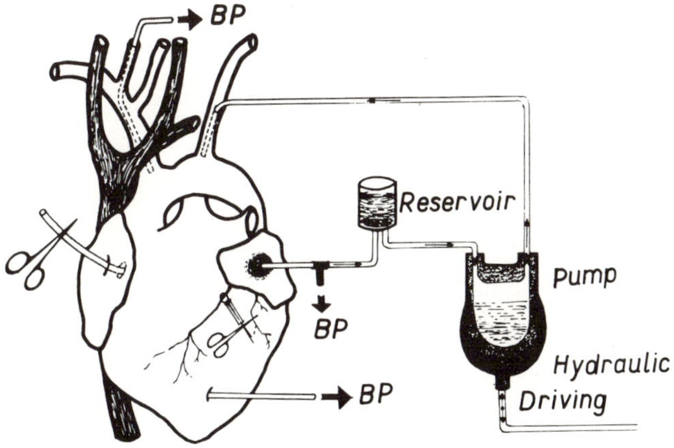

Fig. 1 The blood from the left atrium is bypassed to the reservoir and, by hydraulic means, is pumped into the aorta through the left coronary artery. Blood pressure (BP) is recorded from the aorta, left atrium, and left ventricle.

lactate excess according to Huckabee (5), and redox potential, Eh and Δ Eh, according to Gudbjarnason (4). For spectrophotometric determinations, a Beckman DU spectrophotometer and an oxygen saturation Elema cuvette oximeter were used.

RESULTS

Table I shows the oxygen, lactate, pyruvate, and glucose content in blood samples obtained from the aorta and the coronary sinus prior to clamping of the artery, following the development of infarction, and after assisted circulation.

Myocardial infarction resulted in a reduction of the oxygen saturation in the arterial and venous blood, together with a marked increase in the oxygen uptake by the myocardium. Uptake of lactate and pyruvate assumed a negative balance, correlating well with the lower oxygen saturation of venous samples; however, glucose uptake by the heart did not change appreciably.

Assisted circulation led to only a partial restoration of oxygen uptake, although the absolute values of oxygen saturation in arterial and venous blood suggested a shift from anaerobic to aerobic status of the myocardium. This was further indicated by a positive uptake of lactate and pyruvate by the myocardium. However, lactate and pyruvate levels in the arterial blood were higher during assisted circulation. A similar increase in lactate/pyruvate ratios across the heart was afforded by extracorporeal circulation, and this may be due to a high rate of ketone body utilization.

The increase of glucose utilization in the myocardium appears to constitute evidence of its increased transport through the membrane. The redox potential across the heart, and excess lactate produced from the ratios between lactate and

Table I

Influence of Assisted Circulation on Cardiac Metabolism

	Oxygen saturation (%)			Lactate (mmole/l)			Pyruvate (mmole/l)			Glucose (mmole/l)		
	A*	CS†	%	A	CS	%	A	CS	%	A	CS	%
Control	99.80	77.00	22.86	2.55	1.61	36.90	0.117	0.096	17.95	10.62	10.18	4.14
n = 10	±0.15	±1.00		±0.15	±0.09		±0.002	±0.001		±1.02	±0.90	
Infarction	90.80	61.10	32.10	3.59	4.16	−15.90	0.097	0.146	−50.90	10.40	10.10	2.99
n = 10	±2.10	±5.10		±0.21	±0.23		±0.001	±0.003		±1.00	0.95	
Assisted circulation	98.50	69.50	29.45	5.14	4.88	5.06	0.189	0.144	23.80	14.06	11.89	15.50
n = 10	±0.50	±2.10		±0.26	±0.24		±0.005	±0.003		±1.05	1.09	

* A = aorta.

† CS = coronary sinus.

pyruvate in blood entering and leaving the myocardium are shown in figure 2. Myocardial anoxia is associated with the formation of excess lactate and a negative redox potential across the heart. Assisted circulation decreased the negative redox potential; however, the elevated lactate levels persisted.

Table II shows the content of the high energy phosphate compounds and glycogen in the control, noninfarcted, left ventricular muscle. Without assisted circulation, in the same experimental design, the levels of PCr, ATP, and glycogen were decreased, and ADP was increased, while there was no appreciable difference in the total adenine nucleotide concentrations. With assisted circulation, there was a total restoration in the level of creatine phosphate and a partial restoration in the levels of ATP and glycogen. However, ADP was lower and AMP higher than in the controls.

DISCUSSION

Vasku *et al.* (15) described the changes that occur in Ca^{++}, Mg^{++}, K^{+}, and Na^{+} metabolism in tissue when assisted circulation is applied to the experimentally injured heart, distinguishing these, at the same time, from changes originating from experimental myocardial infarction. Similarly, their findings pointed to an imbalance of ional homeostasis associated with changes in active transport of cations, and to an altered function of cell membranes. Since part of these changes were of hypoxic origin and, moreover, the model of experimental infarction used leads to similar energy-linked changes, we devoted particular attention, in the present study, to investigating the energetic and carbohydrate metabolism of the heart.

At 15-min intervals after induction of myocardial infarction by clamping of the coronary artery, the metabolic status of the myocardium was characterized by diminution of oxygen saturation in both arterial and venous blood and a

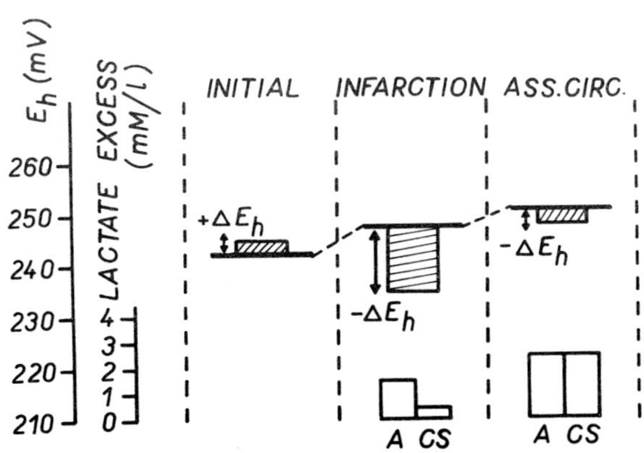

Fig. 2 Eh and Δ Eh redox potential; A = aorta; CS = coronary aorta

Table II

Concentrations of High Energy Phosphate Compounds in Left Ventricle*

	PCr	ATP	ADP	AMP	Σ ADN	GLY
Control hearts in situ n = 10	9.10 ± 1.95	5.58 ± 0.53	1.25 ± 0.15	0.31 ± 0.06	7.14 ± 0.35	49.90 ± 5.52
Infarction with assisted circulation n = 4	9.10 ± 1.90	4.75 ± 0.41	0.70 ± 0.19	0.51 ± 0.08	5.96 ± 0.29	43.60 ± 4.05
Infarction without assisted circulation n = 4	7.10 ± 0.78	3.45 ± 0.29	1.55 ± 0.27	0.66 ± 0.10	5.66 ± 0.25	37.90 ± 3.10

* All the values are expressed as μmole/g wet wt of heart ± S.E.

resultant, growing demand for oxygen in the tissues (table I). This was manifested as a marked increase of oxygen uptake by the myocardium. Uptake of lactate and pyruvate from coronary blood assumed a negative balance. There occurred formation of "excess lactate" and a negative redox potential throughout the heart (fig. 2) (6, 8–10, 13).

Establishment of assisted circulation in the heart led, in fact, to a partial correction of increased oxygen uptake; but the elevated absolute values of oxygen saturation in both arterial and venous blood bore evidence of a gradual shift from anaerobiosis to aerobic metabolism (table I). This conclusion is also corroborated by a balanced (formerly negative) equilibrium of lactate and an improved, though not normal, pyruvate balance in the heart. The great increase of glucose utilization in the myocardium appears to reflect its increased transport through the membrane, possibly brought about by an elevated serum insulin level in shock or anoxia (9).

The observed shift from anaerobiosis to aerobic metabolism was also confirmed by a decrease of the negative redox potential through the heart (fig. 2), in spite of the fact that the elevated arterial redox potential and lactate excess persisted. The latter may be due to the assisted circulation itself (7).

Table II gives a survey of the state of energetic reserves of the uninfarcted myocardium. The results shown refer to the state after 60 min of supportive bypass, rounding off the resultant metabolic picture in the uninfarcted myocardium. The ATP level in tissue approximated the lower normal limit; the increase in AMP level did not, however, balance the decrease of the total content of adenine nucleotides. This decrease can be explained by their breakdown after the AMP step in the phase of shock following infarction, prior to the establishment of assisted circulation; it also reflects the development of small foci of infarction in the areas of the myocardium above the site at which the left coronary artery was ligated (1).

In one particular group of hearts where, although the same experimental model was employed no assisted circulation was established, in all instances there occurred an irreversible dilatation and cardiac arrest, with typical catabolic metabolism. Thus, the findings presented constitute evidence that assisted circulation does have an influence on the metabolism of the heart, also.

Cessation of heart contraction may not be due only to the reduction in the ATP stores of the myocardium, but may well be due to impairment in the energy transfer process. Recent evidence from various laboratories has indicated that the contraction-supporting pool of ATP or other high energy compounds constitutes a better index of correlation between the metabolic and functional state of the myocardium than does the total content of high energy phosphate stores. A precise understanding of the compartmentalization of such pools and their functional participation is necessary before any definite conclusions regarding the influence of energy metabolism on the cardiac function can be made.

Assisted circulation is regulated by impulses taken from the ECG; furthermore, the ECG becomes normalized within a few minutes of its application. In view of the lagging resynthesis of adenine nucleotides following anaerobiosis (12) and the normal creatine phosphate and glycogen levels observed, which is in keeping with the more positive redox potential through the heart, it can be said that assisted circulation exerts beneficial effects in restoring the metabolic status and cardiac function in this experimental design.

SUMMARY

The influence of assisted circulation was tested on the model of the failing dog heart during shock induced by experimental infarction. Specimens of blood were taken from the femoral artery and coronary sinus as well as from the noninfarcted areas of the left ventricle. Myocardial anoxia in our experimental design resulted in a reduction of the oxygen saturation in the arterial and venous blood, together with a marked increase in the oxygen uptake by the myocardium and an increased formation of excess lactate and negative redox potential across the heart. At the same time, the levels of PCr, ATP, and glycogen were decreased and ADP increased in the left ventricle. Assisted circulation of 60 min duration led to total restoration of the level of creatine phosphate and partial restoration of ATP and glycogen in the left ventricle, in keeping with the more positive redox potential throughout the heart.

REFERENCES

1. ANDREEV, S. V. and CECULIN, J. S.: Some notes about the reactivity of the coronary system. (In Russian). Izd. Medicina, Moskva. *39:* 372 (1965).
2. BEDNARIK, B.; NAVRATIL, J.; PERESTY, S.; TOMECEK, J.; VASKU, J.; URBANEK, E.; DOLEZEL, S.; HOFFMANN, K. and ANDREOVSKY, F.: Supportive blood circulation by means of a synchronous blood pump connected with the ascending aorta. (In Russian). Bratisl. lekar. Listy *50:* 474–480 (1968).
3. BEDNARIK, B.; VASKU, J.; URBANEK. E.; PERESTY, S.; ANDREOVSKY, F.; TOMECEK, J.; SOLEZEL, S. and BEDNAR, O.: Prispevek k hodnoceni ucinnosti podpurneho krevniho cerpadla napojeneho na hrudnu aortu. Bratisl. lekar. Listy. (in press) (1969).
4. GUDBJARNASON, S.; HAYDEN, R. O.; WENDT, V. E.; STOCK, T. B. and BING, R. J.: Oxidation reduction in heart muscle: Theoretical and clinical considerations. Circulation *24:* 937–945 (1962).
5. HUCKABEE, W. W.: Relationships of pyruvate and lactate during anaerobic metabolism. I. Effects of infusion of pyruvate or glucose and of hyperventilation. J. Clin. Invest. *37:* 244–254 (1958).
6. McDOWAL, D. G.: Biochemistry of hypoxia; current concepts. II. Biochemical derangements associated with hypoxia and their measurement. Brit. J. Anaest. *41:* 251–256 (1969).
7. MOFFITT, E. A.; ROSEVEAR, J. W. and McGOON, D. C.: Myocardial metabolism in open-heart surgery: Arterial levels of metabolites, electrolytes, oxygenation and acid-base balance. Anaest. Analg. *48:* 633–641 (1969).
8. OLSON, R. E.: "Excess lactate" and anaerobiosis. Ann. Int. Med. *59:* 960–963 (1963).
9. OPIE, L. H.: Metabolism of the heart in health and disease. Part I. Am. Heart J. *76:* 683–698 (1968).
10. SCHEUER, J.: Myocardial metabolism in cardiac hypoxia. Amer. J. Cardiol. *19:* 385–394 (1967).

11. SCHEUER, J. and STEZOSKI, S. W.: Disturbance between the electrocardiogram and ATP levels in the isolated rat heart. Am. J. Med. Sci. *257:* 218–227 (1969).
12. SISKA, K.; FEDELESOVA, M.; HOLEC, V.; ZIEGELHOFFER, A.; STADELMANN, G.; DJOULFAIAN, O. and KUHN, J.: Metabolic changes after electrically induced cardiac fibrillation as a substitute for artificial asystole. J. Cardiovasc. Surg. *10:* 386–394 (1969).
13. SISKA, K.; HOLEC, V.; FEDELESOVA, M. und ZIEGELHOFFER, A.: Einfluss von Persantin auf das vermehrt drucbelastete Hundreherz. Z. Kreislaufforschg. *57:* 119–127 (1968).
14. SOROFF, H. S.; GIRON, F.; RUIZ, U.; BIRTWELL, W. C.; HIRSCH, L. J. and DETERLING, R. A.: Physiological support of heart action. New England J. Med. *280:* 693–704 (1969).
15. VASKU, J.; BEDNARIK, B.; URBANEK, E. and DOLEZEL, S.: Hemodynamic effects of the auxiliary ventricle and left heart bypass with subsequent metabolic and morphologic changes in the myocardium. 1st Annual Meeting of the International Study Group for Research in Cardiac Metabolism. Dubrovnik 14–18 September, 1968.

Mailing address: Dr. M. Fedelesova, Department of Physiology, University of Manitoba, Faculty of Medicine, 770 Bannatyne Ave., Winnipeg 3, Manitoba (Canada).

Degenerative and Toxic Cardiomyopathies

Section of Pathology, National Heart and Lung Institute,
National Institutes of Health, Bethesda, Maryland, and the
Departments of Medicine and Anatomy, Tulane University
School of Medicine, New Orleans, Louisiana, USA

HISTOCHEMICAL AND ELECTRON MICROSCOPIC STUDIES OF NOREPINEPHRINE-INDUCED MYOCARDIAL NECROSIS IN RATS [1]

V. J. FERRANS, R. G. HIBBS, P. R. CIPRIANO,
AND L. M. BUJA

Focal myocardial damage has been produced by large amounts of circulating norepinephrine in patients with pheochromocytomas (25, 36, 37, 44, 46, 47) and after treatment with l-norepinephrine bitartrate (44), and in experimental animals given overdoses of this agent (10, 11, 16, 22, 29, 30, 32, 39, 41, 44, 45, 46, 47). Although several studies have been made of the fine structure of the lesions induced by norepinephrine (11, 16, 22), the nature and pathogenesis of the changes observed in the sarcoplasmic reticulum (SR) and transverse tubular systems (T system) of the muscle cells remain unclear. Investigation of these problems has been made possible by the development of techniques of electron microscopic histochemistry, which have shown that, in glutaraldehyde-fixed rat myocardium, nucleoside triphosphatase is localized in the terminal cisterns of the SR; nucleoside diphosphatase, in the Golgi apparatus and throughout the SR; nucleoside monophosphatase, in the intercalated discs and T system; and acid phosphatase in the lysosomes (12, 14, 15, 40). These observations and the recent studies of Forssman and his associates (17, 18) have led to the formulation of new concepts concerning the distribution of the T system in cardiac muscle cells, and have pointed out the need for using histochemical criteria to distinguish elements of the T system from those of the SR. Therefore, electron microscopic and histochemical techniques were employed in the present studies to further define some of the characteristics of norepinephrine-induced myocardial necrosis.

MATERIALS AND METHODS

Fifty-two adult male rats, weighing 200–250 g each, were used: 40 animals received one subcutaneous injection of 2 mg/kg of l-norepinephrine bitartrate, and were killed at 1/2, 1, 2, 4, 6, 12, 16, 24, and 72 h after injection; 12 animals received two doses, given either 24 or 48 h apart. Tissue samples from each heart

[1] Supported in part by Research Grants HE-06100 and HE-12075, and Career Development Award 5-K3-GM-15-281 from the United States Public Health Service.

were processed immediately for histologic, histochemical, and electron microscopic studies.

Light microscopic studies

Paraffin sections of formalin-fixed tissue were used for histologic studies and for carbohydrate stains. For histologic study, the sections were stained with hematoxylin-eosin and iron hematoxylin. The following stains were used for carbohydrate substances: periodic acid-Schiff, with and without prior digestion with amylase; Alcian blue (pH 2.5); Alcian blue-periodic acid-Schiff, and toluidine blue (43). Cryostat sections postfixed in 95% ethanol were stained for the fluorescence microscopy of ribunucleic acid (RNA) and desoxyribonucleic acid (DNA) (6). Some preparations were pretreated with ribonuclease to ascertain the specificity of the staining for RNA.

Electron microscopic studies

For electron microscopy, 0.5 mm cubes of tissue were taken from subendocardial areas of the left ventricle, fixed for 2 h with cold 3% glutaraldehyde in 0.1 M phosphate buffer, pH 7.4, and washed with several changes of 5% sucrose in 0.1 M phosphate buffer, pH 7.4. The tissues were then postfixed for 2 h with cold 1% OsO_4 in 0.1 M phosphate buffer, pH 7.4, and embedded in Maraglas (18a). Ultrathin sections were stained with uranyl acetate and lead citrate and examined with the electron microscope.

Electron microscopic-histochemical studies

The experiments on the electron microscopic localization of nucleoside phosphatases and esterase were performed on tissues from 4 animals: 2 animals were sacrificed 16 h after a single dose of norepinephrine; the other 2 were killed 20 h after two doses, given 24 h apart. These choices were made after electron microscopic study showed that the greatest variety of morphologic changes was present at these times.

For these studies, the beating heart was perfused for 10 min through a polyethylene tube placed in the ascending aorta, with a cold solution of 3% glutaraldehyde, 3% dextran (60,000 to 80,000 molecular weight) and 3% glucose, pH 7.2 (40). Tissue samples were then selected and placed in the fixative for an additional 50 min, after which they were washed for 1 h in the cold with a solution of 3% dextran and 3% glucose. This technique of fixation was necessary for these experiments, since preliminary data showed that reproducible results could not be obtained in tissued fixed only by immersion.

After fixation, tissue blocks were frozen rapidly and sectioned in a cryostat at a thickness of 50μ. For the experiments on phosphatases, the sections were collected in cold 0.1 M Tris-maleate buffer with 5% sucrose, pH 7.2, and then transferred to the appropriate incubating medium. Sections for esterase experiments were collected in cold 7.5% sucrose.

The following incubation media were used: the adenosine triphosphatase medium of Wachstein and Meisel (48) as modified by Rostgaard and Behnke (40); the nucleoside diphosphatase medium of Novikoff and Goldfischer (35), using uridine-5-diphosphate (UDP) or inosine-5-diphosphate (IDP) as substrates; the thiamine pyrophosphatase medium of Novikoff and Goldfischer (35); the nucleoside monophosphatase medium of Wachstein and Meisel (48) as modified by Rostgaard and Behnke (40); the acid phosphatase medium of Barka and Anderson (5) using 0.00625 M cytidine-5-monophosphate (CMP) as the substrate, and the thiolacetic acid esterase medium of Karnovsky and Hug (24). These media were prepared, filtered, and brought to the incubating temperature immediately before use.

After incubation in the phosphatase media, sections were washed with four changes of a solution of 0.1 M Tris-maleate buffer with 5% sucrose, pH 7.2. Sections incubated in the thiolacetic acid medium were washed with four changes of 7.5% sucrose. After washing, the sections were postfixed for 1 h with cold 1% buffered osmium tetroxide, dehydrated, and embedded in Maraglas. Control sections were incubated in substrate-free media and then processed simultaneously with sections incubated in the complete histochemical media and with nonincubated tissue samples from the same hearts. For purposes of comparison, tissues from normal animals were processed simultaneously with those of the experimental animals.

Ultrathin sections of the Maraglas-embedded 50μ sections were examined by electron microscopy with and without prior staining with uranyl acetate and lead citrate. Such staining of the ultrathin sections did not alter the pattern of histochemical localization and did not interfere with the recognition of deposits of reaction product.

RESULTS

Light microscopic observations

Small, focal lesions, characterized by increased eosinophilia and loss of the striations of the muscle cells, appeared in the subendocardial areas of the left ventricular wall 8 h after the administration of norepinephrine. The inflammatory cell infiltrate, which consisted mostly of mononuclear cells (histiocytes and occasional Anitschkow cells), was first evident at 12–16 h (fig. 1). The damaged muscle cells stained intensely with iron hematoxylin and contained a diastase-resistant, non-metachromatic, PAS-positive material that did not stain with Alcian blue, either at pH 1.0 or at pH 2.5. A metachromatic material was present in the extracellular areas of edema and inflammation; this material often accumulated on the surface of the muscle cells (fig. 2). The lesions became larger and the inflammatory cell infiltrate increased in severity until 48–72 h following the administration of norepinephrine, when the lesions began to heal by fibroblastic proliferation and subsequent scarring.

Sections of normal heart stained for the fluorescence microscopy of nucleic acids showed very little red fluorescence, indicative of RNA, in the muscle fibers; an intense red fluorescence was seen in the cytoplasm of connective tissue cells. In norepinephrine-treated animals, a definite increase in red fluorescence (fig. 3) was observed in the cytoplasm of many of the muscle fibers throughout the myocardium. The cytoplasm of the inflammatory cells showed a very intense red fluorescence. These cells appeared spindle or stellate-shaped, with several cytoplasmic processes that extended between the muscle fibers (fig. 4). All red fluorescence was abolished by treatment of the sections with ribonuclease prior to the histochemical staining.

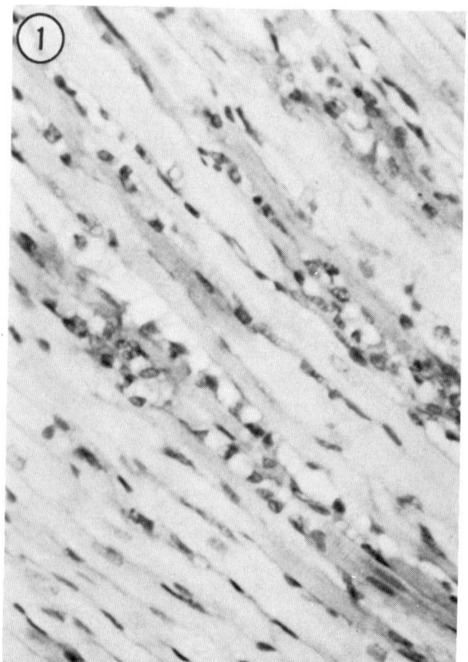

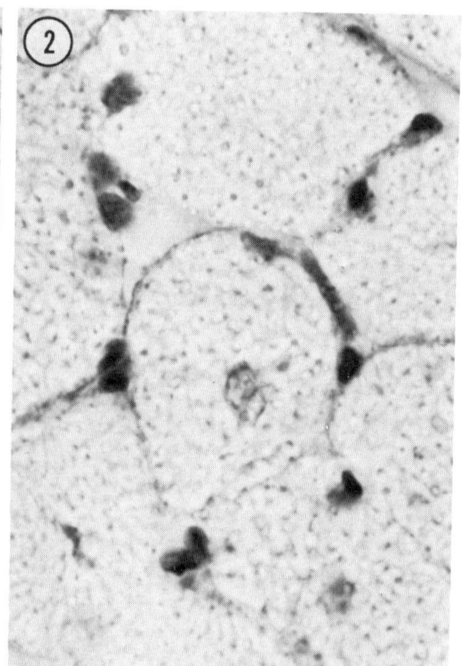

Fig. 1 Rat myocardium 16 h after administration of norepinephrine. Stained by the periodic acid-Schiff and hematoxylin method. Note the mononuclear cellular infiltrate and the positive PAS reaction in the cytoplasm of some of the muscle cells. ×100.

Fig. 2 Rat myocardium 24 h after norepinephrine. Stained with toluidine blue to show metachromasia in interstitial spaces and on surfaces of muscle cells. ×1,000.

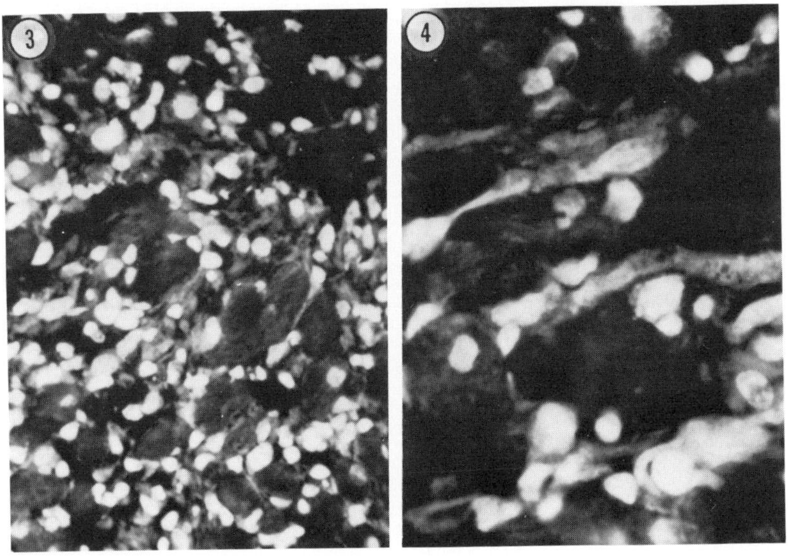

Figs. 3 and 4 Rat myocardium 24 h after norepinephrine. Sections stained for the fluorescence microscopy of nucleic acids.

Fig. 3 Area of inflammatory infiltrate, showing fluorescence in the inflammatory cells and in the cytoplasm of muscle cells. ×250.

Fig. 4 High power view of area of inflammation. The inflammatory cells are elongated and show intense cytoplasmic fluorescence. ×830.

Electron microscopic observations

Electron microscopic examination of the ventricular myocardium of control animals (fig. 5) showed the usual features described previously (13).

Areas of intra- and extracellular edema were observed 30 min after administration of norepinephrine. An amorphous or finely fibrillar material was present in areas of extracellular edema (fig. 6). Most mitochondria of the muscle cells appeared normal, except in some cells in which there was focal loss of the cristae. Needle-like intramitochondrial deposits of an electron-dense material, presumed to represent calcium salts, were observed occasionally (fig. 7). Elements of the T system, identified by the presence of an external lamina on their luminal surfaces, often showed moderate, diffuse swelling. The sarcoplasmic reticulum was normal in some cells. Other cells, particularly those that showed mitochondrial damage, contained numerous dilated vesicles. These vesicles were especially prominent between the myofibrils and under the sarcolemma. The fact that no normal elements of either the SR or T system could be identified with certainty in some of these areas (fig. 8) indicates that both of these cellular components gave rise to the vesicles at this stage. Glycogen

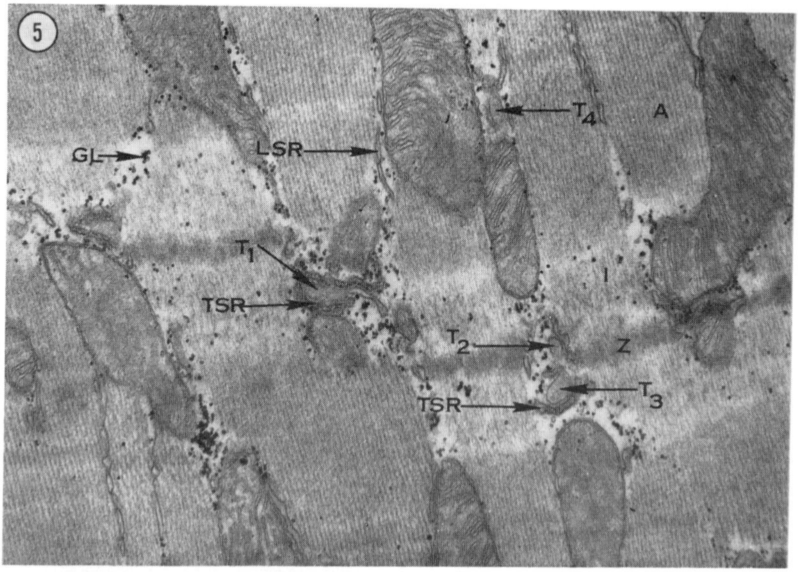

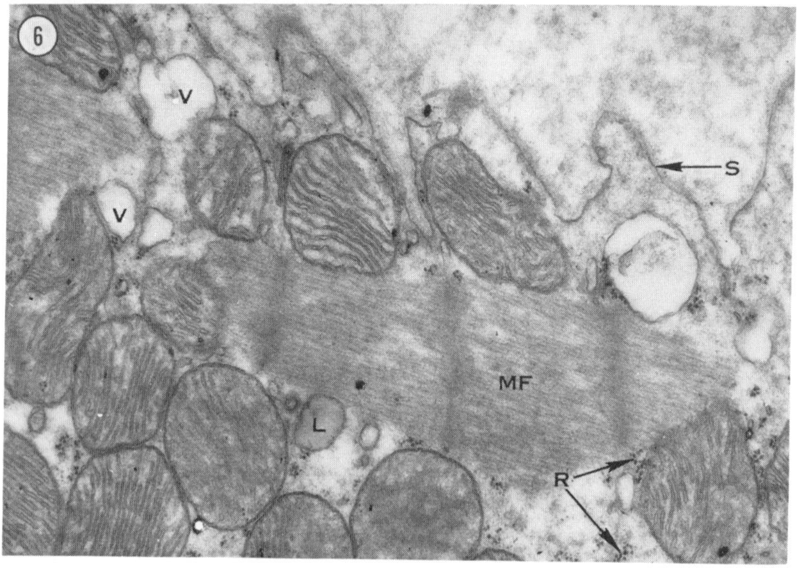

Fig. 5 Ventricular myocardium of normal rat. Longitudinal section through part of a muscle cell showing myofibrils, mitochondria, sarcoplasmic reticulum and T system. The myofibrils, which are relaxed, are composed of A bands (A) and I bands (I), which are bisected by Z bands. The sarcoplasmic reticulum has longitudinal (LSR) and transverse (TSR) elements. The transverse elements form dyads and triads with T tubules (T_1). Two branches of a T tubule (T_2 and T_3) are seen at the level of a Z band; another T tubule (T_4) courses at the level of an A band. Glycogen granules (GL) are scattered throughout the interfibrillary spaces. ×28,200.

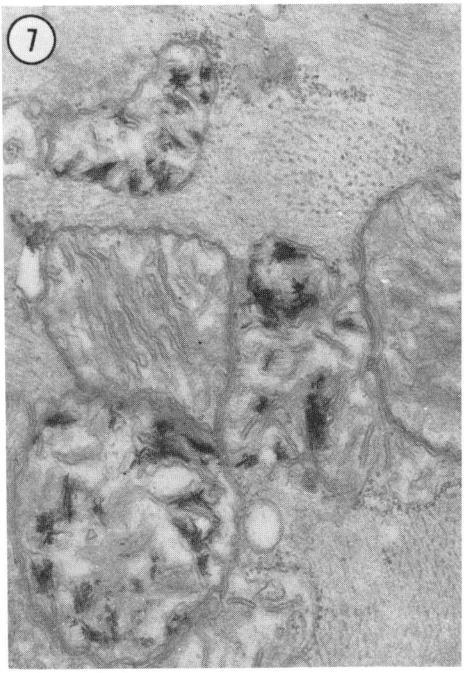

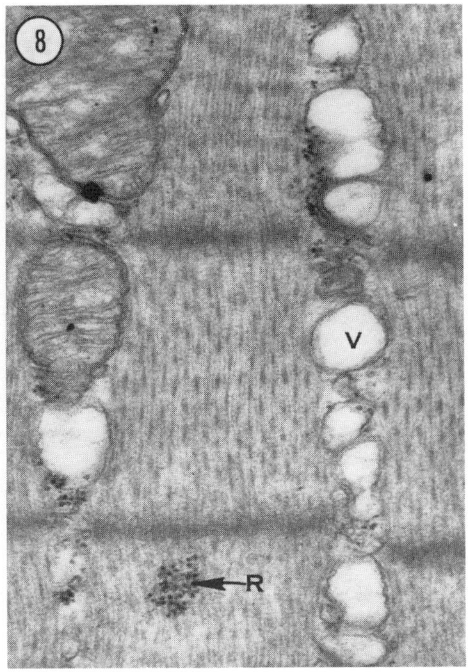

Fig. 7 Thirty minutes after norepinephrine. Portion of a muscle cell showing focal loss of mitochondrial cristae and deposits of electron-dense material, presumably calcium salts, within mitochondria. ×44,000.

Fig. 8 Thirty minutes after norepinephrine. Longitudinal section through part of a muscle cell, showing dilated vesicles (V) between myofibrils. Definite distinction between SR and T elements cannot be made. Several clusters of ribosomes (R) are evident. ×40,500.

granules were almost completely absent from the muscle cells. There was a marked increase in the number of free ribosomes, which measured about 150 Å in diameter and were scattered throughout the interfibrillary spaces and in the perinuclear areas (figs. 6, 8, and 9). Increased numbers of granules morphologically similar to the ribosomes were observed within the nuclei (fig. 9). Many cells contained increased numbers of lipid droplets. The myofibrils often were strongly contracted, with marked shortening of the sarcomeres (figs. 6, 8, and 10). In some areas, these changes were accompanied by focal destruction of the myofilaments (fig. 10). In other areas, the myofibrils only showed an increase in

---

Fig. 6 Thirty minutes after norepinephrine. Longitudinal section through a muscle cell that shows marked edema and strongly contracted myofibrils (MF). Amorphous deposits are present in areas of intra- and extracellular edema. Dilated vesicles (V) are present near the sarcolemma (S). Numerous ribosomes (R) are scattered throughout the sarcoplasm. Mitochondria show minimal alterations. (L)- lipid droplet. ×25,400.

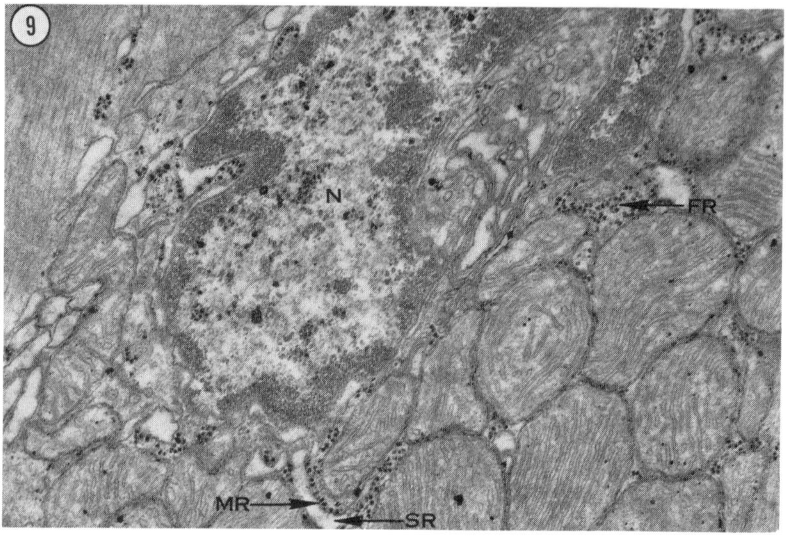

Fig. 9 Thirty minutes after norepinephrine. The perinuclear area of a muscle cell shows numerous free (FR) and membrane bound (MR) ribosomes. Particles similar to the ribosomes are present in the nucleus (N). The mitochondria are normal. Some elements of the sarcoplasmic reticulum (SR), identified by their relationship to the ribosomes, appear dilated. ×31,500.

the width and electron density of the Z bands (figs. 6 and 8). The intercalated discs and the sarcolemma were normal.

At 1 h after the injection of norepinephrine, an increase in the intercellular edema was observed in some areas (fig. 11). Myelin figures began to develop in close association with the mitochondria, and glycogen granules began to reappear in some cells. Nucleoli were prominent in some cells. Other structures of the muscle cells showed no change from their appearance at 30 min.

At 2, 4, and 6 h, evidence of mitochondrial fusion was seen in many areas. In other areas, the T system was dilated and its lumen showed an increase in electron density. Elements of the sarcoplasmic reticulum were recognizable as normal in many areas. The myofibrils, ribosomes, glycogen, lipid droplets, and the areas of subsarcolemmal damage appeared the same as at 1 h.

At 12 h following the administration of norepinephrine, the areas of myofibrillar damage were more clearly evident. In some areas, the myofilaments were still disorganized (fig. 12); in others they were completely lysed (fig. 13). There was no correlation between myofibrillar damage and mitochondrial alterations; mitochondria showed only minimal changes in many cells in which the myofilaments were severely damaged. Areas of mitochondrial fusion were still present in some cells; in others, the mitochondria showed swelling and focal loss of their cristae. The muscle cells often contained numerous lipid droplets

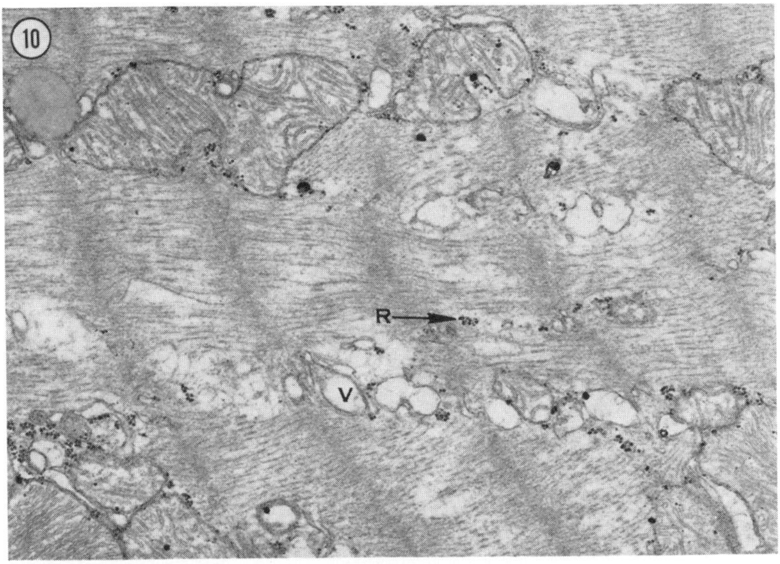

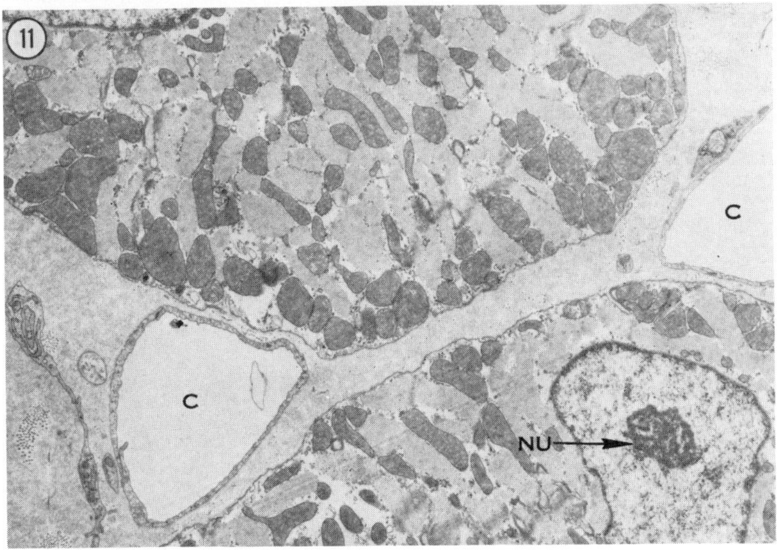

Fig. 10 Thirty minutes after norepinephrine. The sarcomeres of the muscle cell are markedly shortened; focal disorientation and lysis of the myofilaments is present; and glycogen is absent. The elements of the SR and T system, which are not recognizable as such, appear to have transformed into the dilated vesicles (V) present throughout this area. Ribosomes (R) are found in moderate numbers. ×19,500.

Fig. 11 One hour after norepinephrine. Area of intra- and extra- cellular edema. A large amount of amorphous material is present between the muscle cells. (C)-capillary; (NU)-nucleolus. ×5,150.

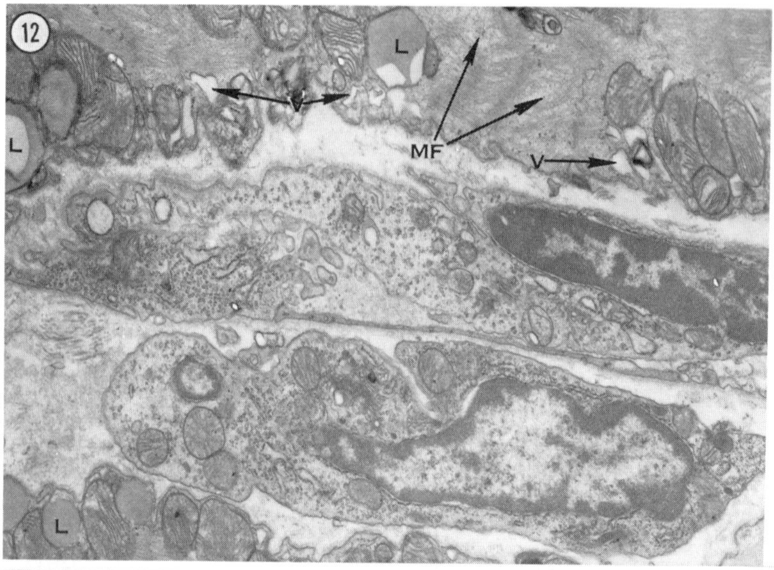

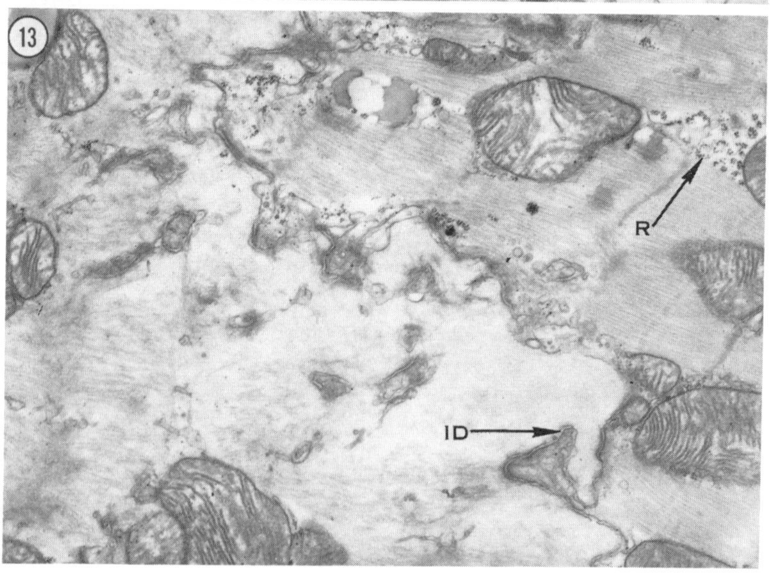

Fig. 12 Twelve hours after norepinephrine. View of two interstitial cells located between two muscle cells, one of which contains several lipid droplets (L), damaged myofilaments (MF) and subsarcolemmal vesicles (V). ×9,200.

Fig. 13 Twelve hours after norepinephrine. Section through part of two muscle cells separated by an intercalated disc (ID). One cell shows extensive disruption of myofilaments; the other contains several clusters of ribosomes (R). The mitochondria in both cells show minimal, focal loss of cristae. ×11,800.

(fig. 12). Large amounts of glycogen were present in some cells, but not in those that showed mitochondrial damage. Large numbers of ribosomes (fig. 13) were found in the interfibrillary spaces and perinuclear areas of most cells, particularly in cells that contained little or no glycogen. The capillary endothelial cells and Schwann cells also contained large numbers of ribosomes. At this time, and in subsequent stages, poorly differentiated, elongated and irregularly shaped inflammatory cells were present in the interstitial space (fig. 12) and adjacent to the capillaries (fig. 14). These cells had single nuclei, with dense clumps of chromatin adjacent to the nuclear membrane, and moderate numbers of mitochondria, which contained loosely packed cristae. The Golgi apparatus of these cells was large and prominent; it consisted of several stacks of cisterns and numerous vesicles. The most distinctive feature of these cells was the presence of very large numbers of ribosomes, which were either free in the cytoplasm or attached to the membranes of the endoplasmic reticulum. At this stage, the majority of these cells contained mostly free ribosomes, and had a poorly developed rough-surfaced endoplasmic reticulum that formed cisterns filled with a material of low electron density; they contained few lysosomes and showed little evidence of phagocytosis.

At *16* and *24 h*, the previously described areas of damage to the contractile elements had progressed to form characteristic lesions in which the alterations ranged from clumping and loss of the banding pattern to complete lysis of the myofilaments. These lesions now occupied larger areas of the cells or, in some cases, entire cells. Characteristically, these cells contained numerous ribosomes, but little or no glycogen. Accumulation of lipid droplets was variable. Other cells appeared normal, except for the presence of large numbers of free ribosomes. The areas in which the myofilaments were either damaged or strongly contracted often contained optically empty, vesicular structures (fig. 15), which could not be definitely identified on a morphologic basis as components of either the SR or the T system. These vesicles were often found just under the sarcolemma. However, in many of these areas nondilated elements of the SR could be recognized; ribosomes were often attached to the outer surfaces of their membranes. As reported below, study by electron microscopic histochemistry demonstrated that these vesicles formed part of the T system. The inflammatory cells formed clusters in the interstitial spaces, particularly near blood vessels. By now, three types of these cells could be distinguished. As described in detail above, cells of the first type (fig. 14) were most prominent at 12 h; we consider these cells to be "activated" perivascular mesenchymal cells. Cells of the second type resembled fibroblasts and contained well-developed, dilated cisterns of rough-surfaced endoplasmic reticulum (fig. 16). The third type of cells contained few rough-surfaced cisterns of endoplasmic reticulum, but had numerous lysosomes and showed evidence of intense phagocytic activity (fig. 17); these cells were considered to be histocytes or macrophages. Cells were often found

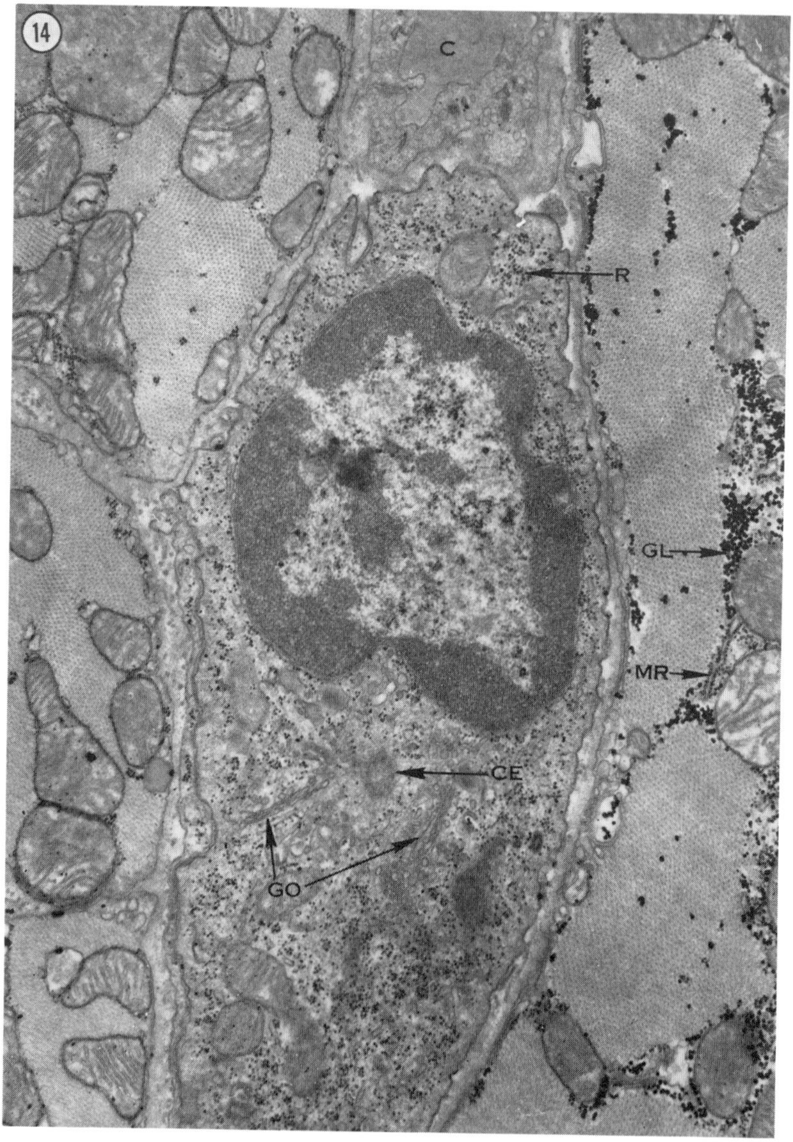

Fig. 14 Sixteen hours after norepinephrine. View of a perivascular cell that has a prominent Golgi apparatus (GO), numerous free ribosomes (R), and a centriole (CE). Large numbers of glycogen granules (GL) and a few membrane-bound ribosomes (MR) are present in one muscle cell. The other two muscle cells shown (upper and lower left) contain a few ribosomes, but no glycogen. (C)-capillary. ×16,800.

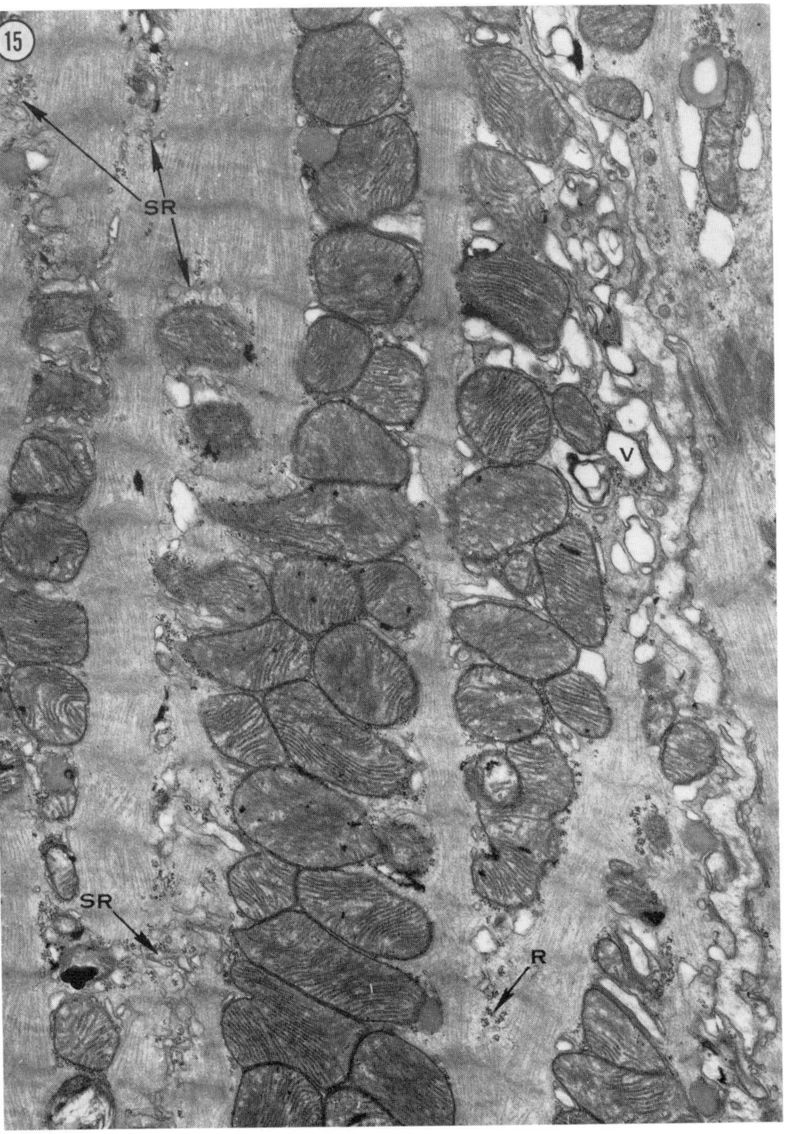

Fig. 15 Twenty-four hours after norepinephrine. Longitudinal section through damaged muscle cell that contains numerous subsarcolemmal vesicles (V), highly contracted myofibrils, and scattered ribosomes (R). The mitochondria and sarcoplasmic reticulum (SR) appear normal. ×12,850.

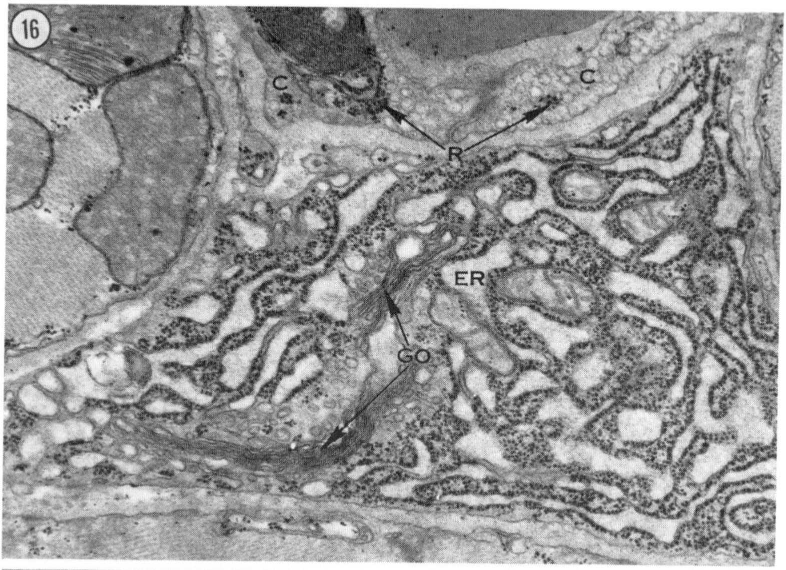

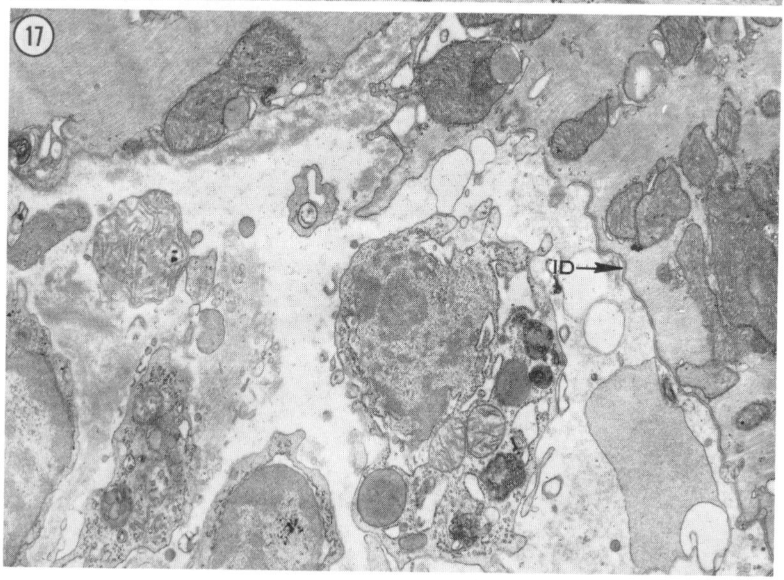

Fig. 16 Sixteen hours after norepinephrine. View of a fibroblast-like cell with numerous dilated cisterns of endoplasmic reticulum (ER) and a large Golgi apparatus (GO). Ribosomes (R) are present in both capillary endothelial cells (C). ×19,500.

Fig. 17 Twenty-four hours after norepinephrine. View of several macrophages invading a necrotic muscle cell in the vicinity of an intercalated disc (ID). Remnants of mitochondria and contractile elements are still recognizable in this cell. ×13,700.

that appeared intermediate in structure between these three types of cells. At 24 h, the macrophages contained greater quantities of phagocytized material than at 16 h; these cells were often seen invading necrotic muscle cells (fig. 17).

In animals killed *1 and 2 days following the second injection* of norepiphrine, the changes observed resembled those seen at 16 to 24 h after a single injection (fig. 18). Increased amounts of free ribosomes were seen in the heart muscle cells of animals sacrificed as late as 12 days after the administration of norepinephrine.

Electron microscopic-histochemical observations

Since the pattern of histochemical localization was determined by the type of substrate used, the histochemical data on control and experimental animals are presented accordingly. The same patterns of localization were found in the 4 norepinephrine-treated animals. The results in control animals were identical to those described in previous reports (12, 14, 40). Histochemical control preparations showed no selective deposition of lead particles in the tissue.

Adenosine triphosphate (ATP)

Following incubation with ATP, reaction product in normal ventricular muscle cells was localized in the terminal cisterns and subsarcolemmal cisterns of the sarcoplasmic reticulum. The T system and the longitudinal elements of the sarcoplasmic reticulum showed no reaction. This pattern of reaction was unchanged in the norepinephrine-treated rats (fig. 19). Reaction product was also present on the outer surfaces of the plasma membranes of the inflammatory cells.

Thiamine pyrophosphate (TPP)

After incubation with TPP, the reaction product in normal rat ventricle was localized in some of the cisterns of the Golgi apparatus, which was small and inconspicuous. After prolonged incubation, the reaction product was also found in the terminal cisterns of the sarcoplasmic reticulum. In norepinephrine-treated animals, TPP-incubated preparations showed a striking increase in the size and reactivity of the Golgi apparatus of the muscle cells (fig. 20). The Golgi apparatus of the inflammatory cells also showed a very intense reaction (fig. 21). Other structures of the cardiac muscle cells and inflammatory cells were unreactive.

Nucleoside diphosphates

In preparations of normal rat ventricular muscle incubated with either IDP, UDP, or ADP, the reaction product was localized in some of the cisterns and

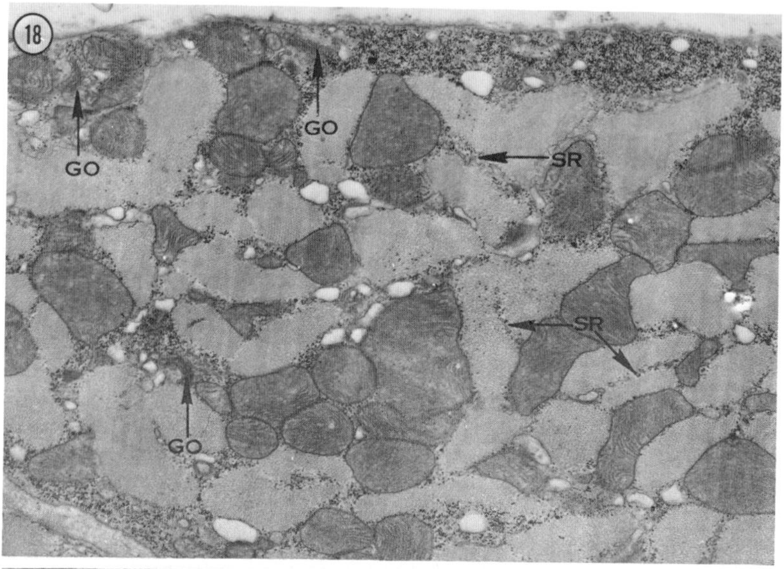

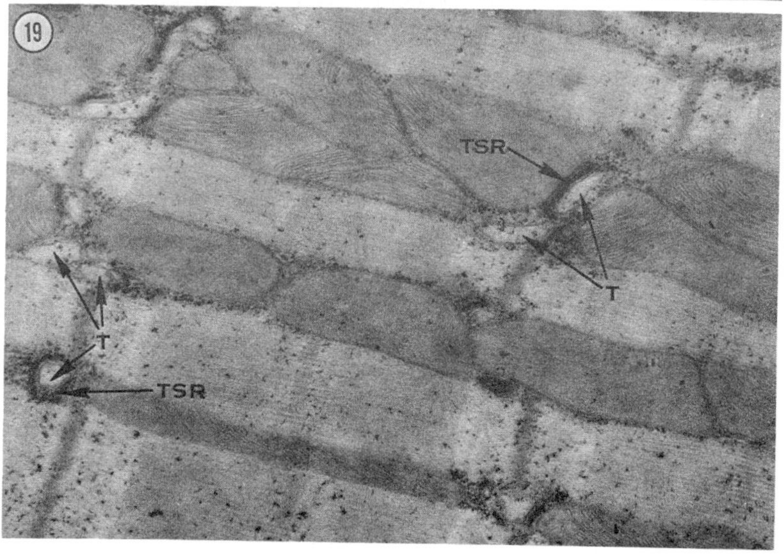

Fig. 18 Forty-eight hours after two doses of norepinephrine given 24 h apart. Cross-section through a muscle cell showing marked vacuolization. Normal elements of sarcoplasmic reticulum (SR) are present in some areas. The Golgi complex is greatly enlarged, with several stacks of Golgi cisterns (GO) scattered through the cell. Compare with figure 27. ×8,700.

Fig. 19 Twenty hours after two doses of norepinephrine given 24 h apart. Preparation incubated with ATP to demonstrate nucleoside triphosphatase activity. Reaction products is present in the transverse elements (TSR) of the sarcoplasmic reticulum. The T tubles (T) do not contain reaction product. ×31,550.

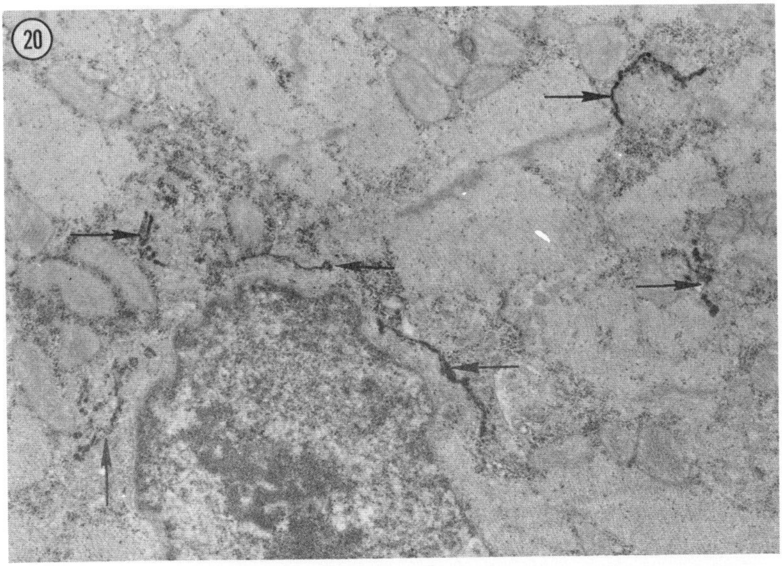

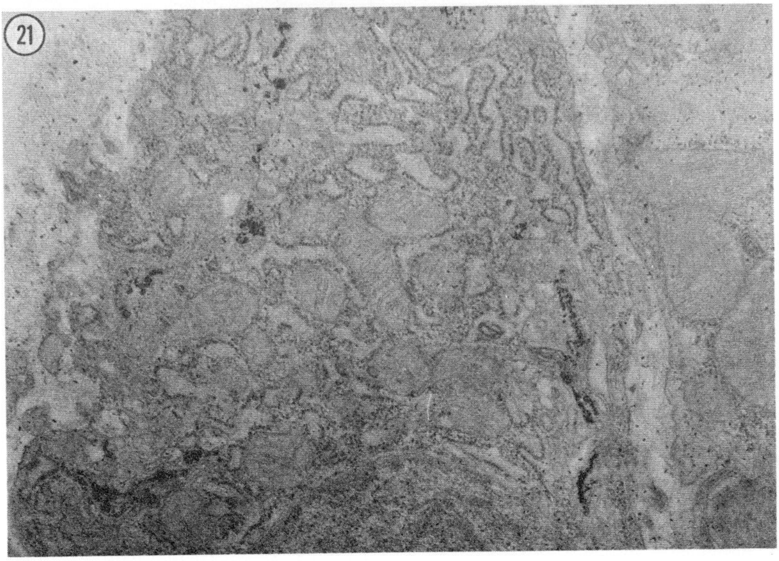

Figs. 20 and 21 Thiamine pyrophosphatase preparations of rat myocardium after two doses of norepinephrine given 24 h apart.

Fig. 20 Reaction product is localized in the Golgi apparatus, which is enlarged considerably and consists of several scattered groups of cisterns (arrows) and associated vesicles. ×15,000.

Fig. 21 Portion of fibroblast-like cell with deposits of reaction product in the Golgi cisterns. ×26,500.

subsarcolemmal cisterns of the sarcoplasmic reticulum; a much weaker reaction was observed in the longitudinal elements of the sarcoplasmic reticulum. As in the case of the TPP method, the nucleoside diphosphatase techniques also showed a marked increase in the size of the Golgi apparatus of the cardiac muscle cells of norepinephrine-treated rats.

Adenosine monophosphate (AMP)

In normal rat ventricular myocardium, the deposits of reaction product observed after incubation with AMP were localized in the lumina of the T tubules, in the interspace between the two opposed plasma membranes of the intercalated discs, and in some areas of the sarcolemma (figs. 22–24). A reaction was also observed in longitudinally and diagonally oriented T tubules that coursed between the myofibrils, and extended even to the proximity of the nuclear membrane (fig. 24). After the administration of norepinephrine (figs. 25–29), the branching of the T system increased, with the formation of extensive, longitudinally and diagonally oriented branches, which extended from the level of a Z band across the length of one or more sarcomeres (figs. 26 and 29). Most of the dilated vesicular structures observed in the late stages of norepinephrine toxicity were part of the T system, rather than of the sarcoplasmic reticulum (figs. 27 and 28). The reaction in the intercalated discs appeared unchanged from that seen in normal animals; however, the number of reactive T elements in the proximity of the intercalated discs appeared increased in norepinephrine-treated animals.

Cytidine monophosphate (CMP)

Following incubation of normal rat ventricle with CMP at pH 5.0, the reaction product was found only in the lysosomes, which are few in number and small in size. In norepinephrine-treated animals, the number and size of the lysosomes in the muscle cells was considerably greater than in normal animals (figs. 30–32). In these animals, reaction product was often observed in some of the cisterns of the Golgi apparatus (fig. 31), and continuity between the Golgi apparatus and the lysosomes was demonstrated in some instances (fig. 32). The inflammatory cells also showed heavy deposits of reaction product in the lysosomes (fig. 33) and, in some instances, in the Golgi cisterns (fig. 34).

Thiolacetic acid

In normal rat ventricular myocardium, the deposition of reaction product observed after incubation with thiolacetic acid was confined to the lysosomes and the mitochondria of the muscle cells and connective tissue cells. The localization in the mitochondria of muscle cells was similar in normal and norepinephrine-treated animals. The activity in lysosomes of inflammatory cells (fig. 35) and muscle cells was similar to that seen in acid phosphatase

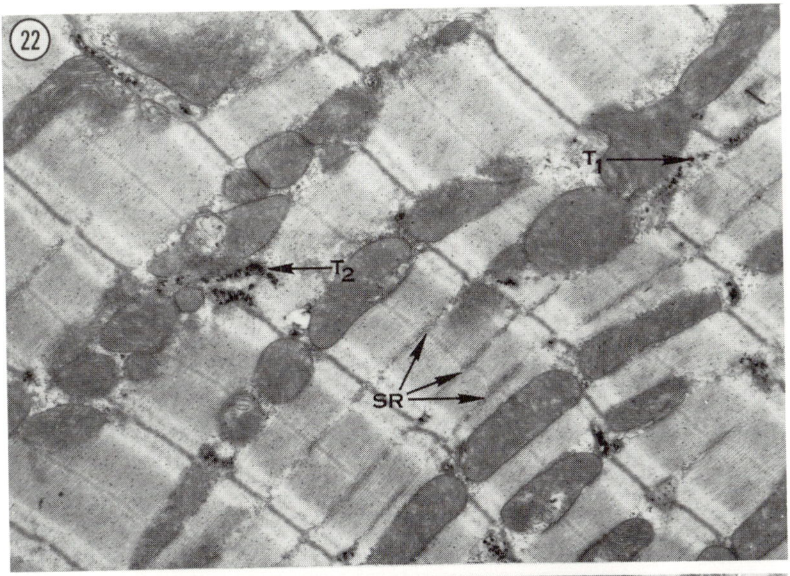

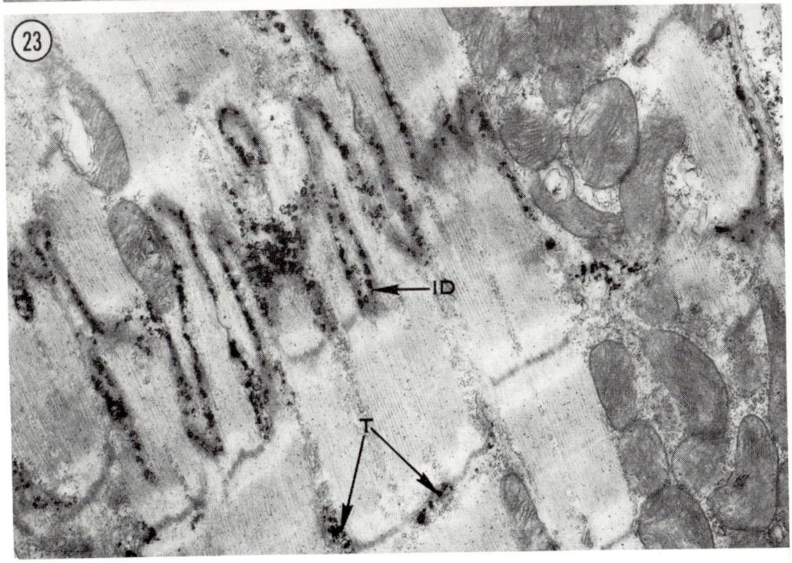

Figs. 22 and 23 Sections of normal rat myocardium incubated with adenosine monophosphate for demonstration of nucleoside monophosphatase activity.

Fig. 22 Reaction product is localized in the T system. Some elements of the T system course longitudinally (T_1) and obliquely (T_2) along the myofibrils. The sarcoplasmic reticulum (SR) shows no reaction. ×11,800.

Fig. 23 Reaction product is localized in the intercalated disc (ID) and adjacent elements of the T system (T). ×13,450.

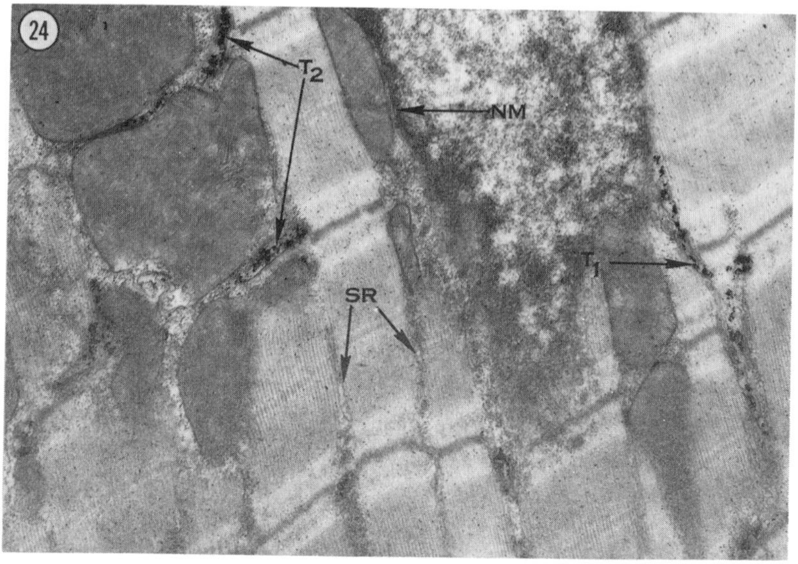

Fig. 24 Nucleoside monophosphatase preparation from normal rat myocardium, showing the perinuclear area of a muscle cell, with reaction product present in small, longitudinally oriented (T_1), and larger, transversely oriented T tubules (T_2). The sarcoplasmic reticulum (SR) and the nuclear membrane (NM) show no reaction. ×15,000.

Fig. 25 Nucleoside monophosphatase preparation from rat myocardium after two doses of norepinephrine given 24 h apart. Reaction product is present in transversely (T_1) and longitudinally oriented (T_2) T tubules; the latter extend through the length of one sarcomere. ×11,800.

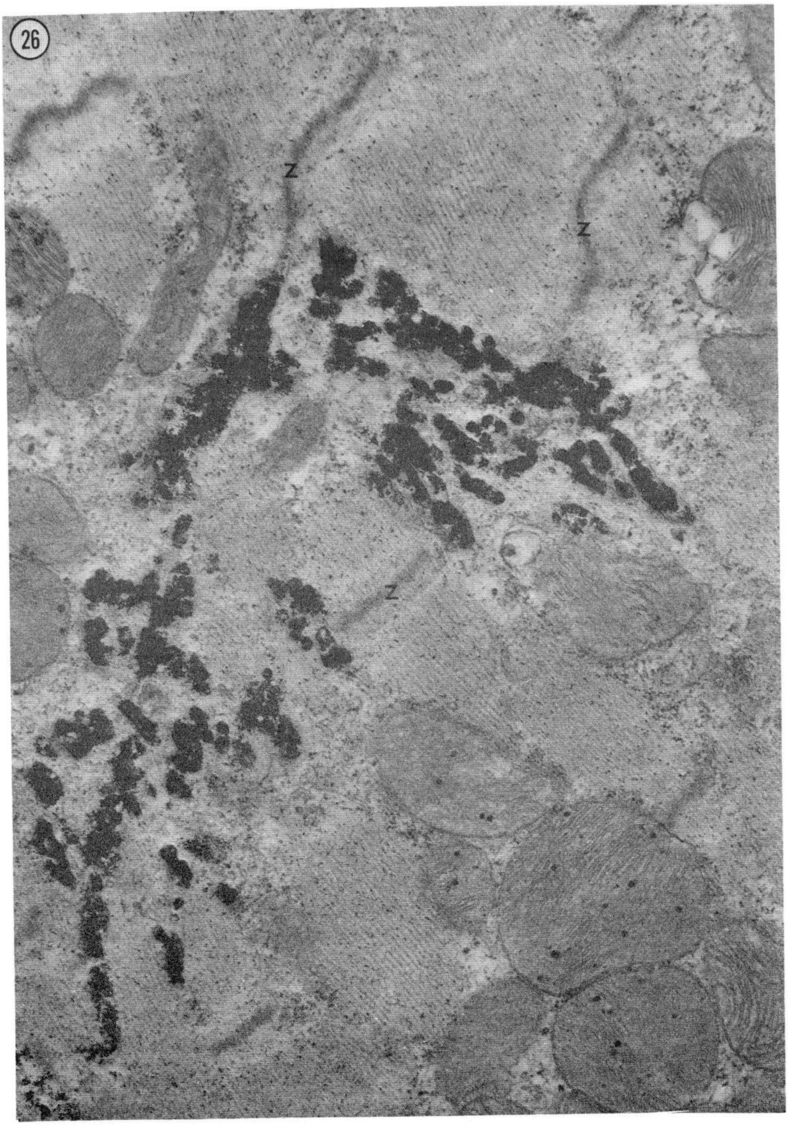

Fig. 26 Nucleoside monophosphatase preparation from same heart shown in figure 21. Section through a portion of a muscle cell showing reaction product in T tubules that branch extensively between adjacent Z bands (Z). ×29,200.

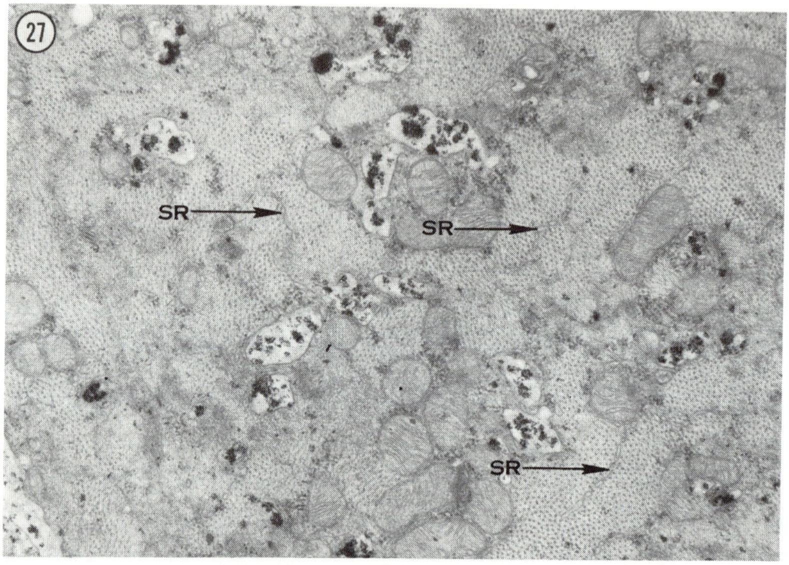

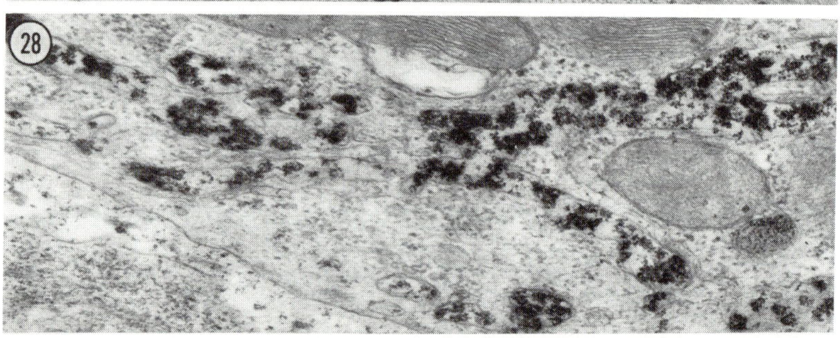

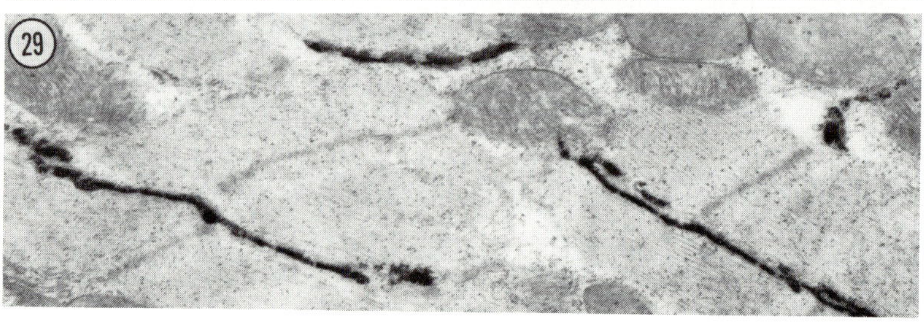

Figs. 27–29 Nucleoside monophosphatase preparations from myocardium of rat given two doses of norepinephrine 24 h apart.

Fig. 27 Cross section through a muscle cell that shows marked vesiculation. Deposits of reaction product are localized in the vesicles, indicating that they represent areas of the T system. Normal elements of sarcoplasmic reticulum (SR) are present in some areas. Compare with figure 18. ×26,500.

Fig. 28 Subsarcolemmal area of a muscle cell with deposits of reaction product in the T system, which shows vesiculation and extensive branching. Compare with figure 15. ×31,550.

Fig. 29 Portion of a muscle cell, showing reaction product in elongated, diagonally and longitudinally oriented T tubules, some of which are branching. ×15,750.

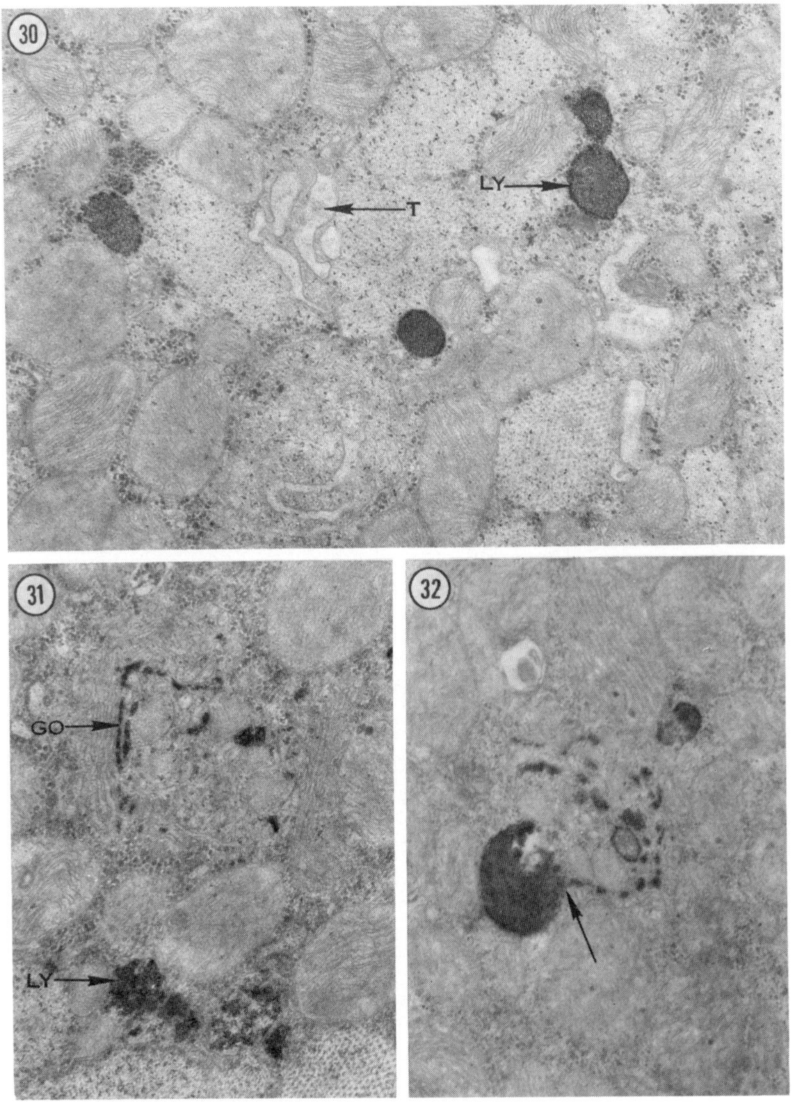

Figs. 30–32 Rat myocardium 20 h after two doses of norepinephrine, given 24 h apart. Preparations incubated with cytidine-5-monophosphate at pH 5.0 to demonstrate acid phosphatase activity.

Fig. 30 Reaction product is localized in lysosomes (LY). The T tubules (T) are unreactive. ×28,850.

Fig. 31 Acid phosphatase activity is localized in lysosomes (LY) and in some Golgi cisterns (GO). × 15,000.

Fig. 32 Acid phosphatase activity is present in Golgi cisterns and lysosomes. One lysosome (arrow) shows continuity with a Golgi cistern. ×28,600.

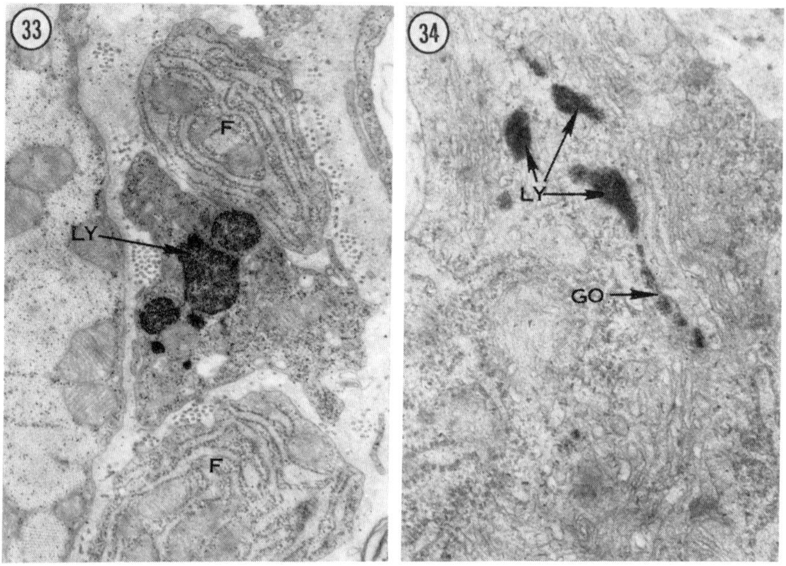

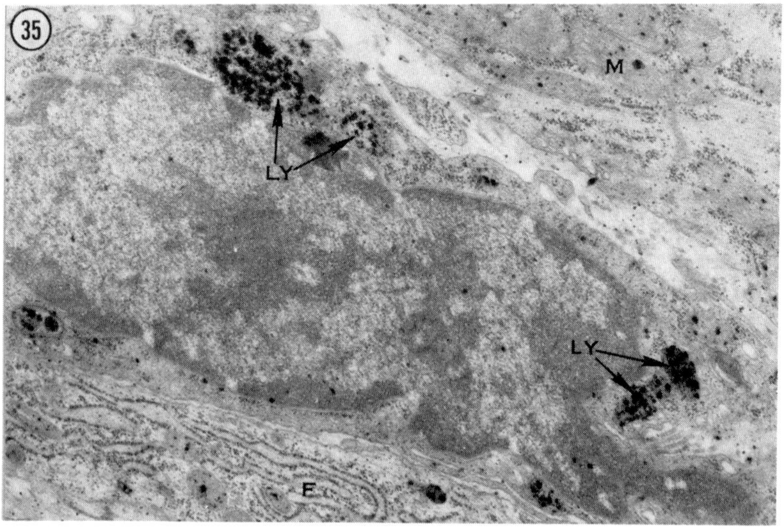

Figs. 33 and 34 Acid phosphatase activity in myocardium of rat given two doses of norepinephrine 24 h apart.

Fig. 33 Reaction product is present in several large lysosomes (LY) in a macrophage; 2 adjacent fibroblasts (F) show no reaction. ×16,800.

Fig. 34 Acid phosphatase activity in same specimen and preparation as in figure 33. Reaction product is present in the Golgi cisterns (GO) and developing lysosomes (LY) of a macrophage. ×37,850.

Fig. 35 Rat myocardium 20 h after two doses of norepinephrine given 24 h apart. Preparation incubated with thiolacetic acid for the demonstration of esterase activity. The lysosomes (LY) in the macrophage contain large clumps of reaction product; small, scattered deposits are present in the mitochondria (M) of cardiac muscle cells. The fibroblast (F) shows very little reaction. ×16,600.

preparations; however, no reaction product was found in the Golgi apparatus in the esterase preparations.

DISCUSSION

The alterations observed in rat myocardium after administration of large doses of norepinephrine consisted of marked contraction of myofibrils and damage to myofilaments; dilatation of the tubules of the sarcoplasmic reticulum and T system; loss of glycogen; marked increase in the number of ribosomes; mitochondrial damage; hypertrophy of the Golgi apparatus; increase in the number of lipid droplets and lysosomes, and infiltration of mononuclear cells in areas of muscle cell damage.

Alterations in contractile elements

Contraction bands such as those found in this study have been described in the hearts of patients with angina pectoris (8), at the periphery of experimentally induced myocardial infarcts (27, 28), in various ischemic lesions (20, 42), and after the administration of isoproterenol (13, 26). They are not found, however, in the center of areas of infarction (27) or in the myocardium of asphyxiated animals (1); in both of these instances the myofibrils show relaxation rather than contraction. Although the norepinephrine-induced contraction bands were associated with damaged myofilaments, the number of subsequently damaged myofibrils was relatively small when compared with the number of myofibrils that showed contraction bands during earlier stages. Thus, contraction bands are at least a partially reversible phenomenon, and other factors must influence the progression of this change to lysis of the myofilaments. David *et al.* (11) postulated that the myofibrils are the primary site of the damage produced by norepinephrine. There may be a relationship between the positive inotropic effect of norepinephrine and the development of contraction bands.

Dilatation of the T system and sarcoplasmic reticulum

In normal myocardium, the elements of the T system are usually identified by their continuity with the sarcolemma, by their orientation in relation to the Z lines of the myofibrils, by their size, which is usually larger than that of the elements of the sarcoplasmic reticulum, and by the presence in their lumen of an external lamina that is continuous with that of the sarcolemma. However, recent histochemical studies (17, 18) have shown that the T system has longitudinally and diagonally oriented branches from which the external lamina may be absent. Since the size and appearance of some of these branches is similar to that of the elements of the sarcoplasmic reticulum, Forssmann and Girardier have concluded that the T system may be defined only in operational terms, *i.e.*, by means of the histochemical demonstration of the distribution of protein tracers such as peroxidase. As an alternative to the use of tracers, we employed the adenosine monophosphatase technique, which also demonstrates the T system and intercalated discs of cardiac muscle cells with a high degree of specificity (14, 40). Our data using this technique were complemented by the results of the adenosine triphosphatase and nucleoside diphosphatase techniques for the staining of the sarcoplasmic reticulum. Unfortunately, the histochemical demonstration of the longitudinal elements of the SR often proved difficult because of their low reactivity with nucleoside diphosphates. Improvements in fixation techniques may help solve this problem.

The work of Forssmann and Girardier and the findings of this study indicate the need for a reevaluation of previous observations concerning the nature of the optically empty, dilated vesicles commonly found as a nonspecific abnormality in damaged cardiac muscle cells. While it has been widely assumed that these vesicles represent dilated elements of the SR, this study shows that, at least in the case of norepinephrine-induced myocardial necrosis, these vesicles are derived from the T system as well as from the SR. Swelling of the SR appeared to be a much more rapidly reversible phenomenon than that of the T system. Most of the vesicles observed in the later stages of norepinephrine toxicity (16 h or longer) showed the histochemical characteristics of the T system. It is evident that morphologic criteria are often inadequate to distinguish the tubules of the SR from the smaller, branching tubules of the T system in both normal and damaged myocardial cells, and that such a distinction must be made on the basis of histochemical data. Additional studies will be necessary to investigate the time course of histochemical changes in the SR and T system in norepinephrine-induced myocardial necrosis.

However, the evidence presented in this study suggests that in some instances irreversible cell damage is mediated through marked swelling of the T system in subsarcolemmal areas, which leads eventually to rupture of the sarcolemma. The T system, which is continuous with the sarcolemma, often showed significant changes since the early stages of cell damage, while the sarcolemma itself showed no abnormalities until much later. These observations suggest that some of the alterations produced by injury in the bioelectric properties of the muscle cell membrane may reflect changes in the T system rather than in the nonspecialized areas of the sarcolemma.

Alterations in glycogen and ribosomes

As reviewed previously (21), loss of glycogen from cardiac muscle cells is a nonspecific change that occurs in a variety of types of cell injury. We observed marked depletion of glycogen after the administration of norepinephrine, but rarely found the "glycogenosomes" described by David *et al.* (11). These workers also observed many small particles, which they assumed to be glycogen, in the cytoplasm of muscle cells. We believe that these particles are ribosomes rather than glycogen. This interpretation is based on the size and electron density of both the free and membrane-bound particles, on their similarity to ribosomes present in connective tissue cells, and on the indirect evidence provided by preparations stained for the fluorescence microscopy of nucleic acids.

The fact that the number of ribosomes in cardiac muscle cells was increased as early as 30 min after administration of norepinephrine suggests that this agent has a stimulatory effect on the synthesis of ribonucleic acid in cardiac muscle cells. This increase in RNA may be related to an increase in protein synthesis, directed to the formation of new contractile elements. This is suggested by the fact that large amounts of RNA are present in embryonic cardiac muscle (31) and in muscle cells that are undergoing hypertrophy (34). Proliferation of ribosomes also was evident in the inflammatory cells, in capillary endothelial cells, and even in Schwann cells, which implies a more generalized phenomenon. Isoproterenol also produces a marked increase in myocardial RNA and in the incorporation of amino acids into cardiac muscle protein. Of interest is the fact that this agent does not increase the incorporation of amino acids into skeletal muscle protein.

The mechanism by which catecholamines produce these changes has not been investigated in cardiac muscle. The possibility cannot be excluded that these are secondary phenomena due to hypoxia, which is believed (33) to provide the critical stimulus to cardiac hypertrophy. On the other hand, it is also possible that this effect of hypoxia is mediated through the release of catecholamines. The work of Barka (2, 3, 4) on the proliferation of salivary gland tissue produced by isoproterenol in the rat has demonstrated conclusively that catecholamines can exert a profound effect on the metabolism of DNA, RNA, and protein. Barka showed that isoproterenol produced a marked rise in the synthesis of DNA, RNA, and protein in the acinar cells of the salivary gland. These changes were preceded by a marked rise in DNA polymerase and thymidine kinase, and were inhibited by pretreatment with actinomycin-D.

There is a need for additional biochemical study of these problems in cardiac muscle, especially since decreased concentrations of norepinephrine have been observed in the ventricular myocardium of humans and dogs with cardiac failure (7). It has been suggested that these depleted stores of norepinephrine in myocardium result in inadequate adrenergic support to the failing heart. Our observations suggest that stimulation of the synthesis of RNA and protein may be one way by which the adrenergic nervous system influences cardiac function.

Mitochondrial alterations

The lack of correlation between mitochondrial and myofibrillar damage suggests that these two alterations are mediated through different mechanisms.

The relatively mild changes produced by norepinephrine in mitochondria contrast sharply to the more severe damage in hypoxia (9), infarction (27, 42), and after administration of isoproterenol (13, 19, 26). These mitochondrial changes also differ from those caused by the administration of epinephrine (16a), which results in marked angulations of the cristae. The reasons for the different effects of norepinephrine, epinephrine, and isoproterenol on the mitochondria of cardiac muscle cells remain unclear. It is possible that such differences reflect the relative potency of these agents in stimulating oxidative metabolism. The data available, however, suggest that these various alterations may reflect qualitatively different pharmacologic effects of these drugs.

Changes in the Golgi apparatus and lysosomes

The histochemical studies with the nucleoside diphosphatase and thiamine pyrophosphatase methods provided reliable techniques for the identification of the Golgi apparatus in cardiac muscle cells and in inflammatory cells. The hypertrophy of the Golgi zone observed in this study may be related to the increase in the metabolic activity of the muscle cells.

The role of the Golgi apparatus in the formation of lysosomes in cardiac muscle was demonstrated by its positive acid phosphatase reaction in norepinephrine-treated animals, and by the continuity observed in some instances between acid phosphatase-reactive Golgi cisterns and lysosomes.

Comparison of the two histochemical techniques used for the demonstration of lysosomes, *i.e.*, thiolacetic acid esterase and acid phosphatase, showed that the latter gave better localization of the reaction product in lysosomes, and did not result in mitochondrial staining as did the esterase method. This mitochondrial staining appears to be due to another type of esterase, the nature of which has not been completely elucidated.

Inflammatory cells

The inflammatory exudate found in the norepinephrine-induced lesions was formed by mononuclear cells, rather than by the ususal polymorphonuclear leukocytic response observed in areas of myocardial infarction. The origin of the mononuclear inflammatory cells could not be determined from the methods used in this study. However, the electron microscopic evidence suggests that at least some of these cells may have been derived in situ from perivascular mesenchymal cells. This interpretation is based on the fact that in the early stages of the lesions many of these cells, which normally are poorly differentiated, showed increasing numbers of free ribosomes; as the lesions progressed, cells were found that appeared intermediate in structure between the poorly differentiated cells and the macrophages and fibroblasts, which were observed in later stages of the lesions. Rona and Kahn (38) noted that the myocardial lesions produced by catecholamines healed much faster than those caused by myocardial infarction. This rapid healing may be related to the presence of large amounts of RNA in both the muscle fibers and the inflammatory cells, which can result in a very high rate of protein synthesis.

SUMMARY

Cardiac lesions produced in rats by administration of 2 mg/kg of l-norepinephrine consisted of focal myocardial necrosis with infiltration by mononuclear inflammatory cells. Three main types of inflammatory cells were observed: poorly differentiated cells with numerous free ribosomes, few lysosomes, and few cisterns of rough-surfaced endoplasmic reticulum; macrophages with numerous esterase and acid phosphatase-positive lysosomes; and fibroblasts, with numerous cisterns of rough-surfaced endoplasmic reticulum and few lysosomes. The main alterations in muscle fibers were: marked contraction of myofibrils and damage to myofilaments; dilatation of the T system, with a pronounced increase in its branching; swelling of the sarcoplasmic reticulum; focal mitochondrial damage with swelling and loss of the cristae; increase in the size of the Golgi apparatus and in the number of ribosomes and lysosomes, and loss of glycogen. Studies on the electron microscopic histochemistry of various nucleoside phosphatases showed that these enzymes can be used as cytologic markers to follow changes in the T system, sarcoplasmic reticulum, Golgi apparatus, and lysosomes. The cardiotoxic effects of norepinephrine are exerted primarily on the myofibrils and T system. This drug may have a stimulatory effect on the synthesis of ribonucleic acid in the rat heart.

REFERENCES

1. BAHR, G. F. and JENNINGS, R. B.: Ultrastructure of normal and asphyxic myocardium of the dog. Lab. Invest. *10:* 548–571 (1961).
2. BARKA, T.: Stimulation of DNA synthesis by isoproterenol in the salivary gland. Exp. Cell. Res. *39:* 355–364 (1965).
3. BARKA, T.: A model of induced cell proliferation: the effect of isoproterenol on salivary glands. J. Cell Biol. *31:* 9A (1966).
4. BARKA, T.: Further studies on the stimulation of DNA synthesis in the salivary gland by isoproterenol. J. Cell Biol. *35:* 9A–10A (1967).
5. BARKA, T. and ANDERSON, P. J.: Histochemistry, Theory, Practice and Bibliography, p. 212 (Hoebner, New York, 1963).
6. BERTALANFFY, L., MASIN, M. and MASIN, F.: A new and rapid method for diagnosis of vaginal and cervical cancer by fluorescence microscopy. Cancer *11:* 873–887 (1958).
7. BRAUNWALD, E.; ROSS, J. and SONNENBLICK, E. H.: Mechanisms of Contraction of the Normal and Failing Heart, Ch. 9, p. 157–163 (Little, Brown and Company, Boston 1968).
8. BÜCHNER, F.: Die Rolle des Herzmuskels bei der Angina pectoris. Beitr. path. Anat. *89:* 644–667 (1932).
9. BÜCHNER, F. und ONISHI, S.: Die akute hypoxische Herzinsuffizienz an der Ratte. Verh. Dtsch. Ges. Path. *51:* 139–144 (1967).
10. CHAPPEL, C. I.; RONA, G.; BALAZS, T. and GAUDRY, R.: Comparison of cardiotoxic actions of certain sympathomimetic amines. Canad. J. Biochem. Physiol. *37:* 35–42 (1959).
11. DAVID, R.; HECHT, A. und UERLINGS, I.: Noradrenalinbedingte Feinstrukturveränderungen des Herzmuskels der Ratte. Beitr. path. Anat. *137:* 1–18 (1968).
12. ESSNER, E.; NOVIKOFF, A. B. and QUINTANA, M.: Nucleoside phosphatase activity in rat cardiac muscle. J. Cell Biol. *25:* 201–215 (1965).

13. FERRANS, V. J.; HIBBS, R. G.; BLACK, W. C. and WEILBAECHER, D. G.: Isoproterenol-induced myocardial necrosis: a histochemical and electron microscopic study. Amer. Heart J. *68:* 71–90 (1964).

14. FERRANS, V. J.; HIBBS, R. G. and BUJA, L. M.: Nucleoside phosphatase activity in atrial and ventricular myocardium of the rat: a light and electron microscopic study. Amer. J. Anat. *125:* 47–86 (1969).

15. FERRANS, V. J.; HIBBS, R. G.; CIPRIANO, P. R. and BUJA, L. M.: Nucleoside monophosphatase activity in ventricular myocardium of normal and norepinephrine-treated rats. Anat. Rec. *163:* 301 (1969).

16. FERRANS, V. J.; HIBBS, R. G.; WALSH, J. J. and BURCH, G. E.: Histochemical and electron microscopic studies on the cardiac necroses produced by sympathomimetic agents. Ann. New York Acad. Sci. *156:* 309–332 (1969).

16a. FERRANS, V. J.; HIBBS, R. G.; WEILY, H. S.; WEILBAECHER, D. G., WALSH, J. J. and BURCH, G. E.: A histochemical and electron microscopic study of epinephrine-induced myocardial necrosis. J. Molec. Cell. Cardiol. *1:* 11–22 (1970).

17. FORSSMAN, W. G. and GIRARDIER, L.: A study of the T system in rat heart. J. Cell Biol. *44:* 1–19 (1970).

18. FORSSMANN, W. G.; MATTER, A.; DALDRUP, J. and GIRARDIER, L.: A study of the T system in the heart muscle by means of horseradish peroxidase tracing. J. Cell Biol. *39:* 45A–46A (1968).

18a. FREEMAN, J. A., and SPURLOCK, B. O.: A new epoxy embedment for electron microscopy. J. Cell. Biol. *13:* 437–442 (1962).

19. HAUSAMEN, T. U. und POCHE, R.: Elekronenmikroskopische Untersuchungen über die Wirkung von Alupent[R] auf die Ultrastruktur des Herzmuskels der Ratte. Virchows Arch. path. Anat. *339:* 225–233 (1965).

20. HERDSON, P. B.; SOMMERS, H. M. and JENNINGS, R. B.: A comparative study of the fine structure of normal and ischemic dog myocardium with special reference to early changes following temporary occlusion of a coronary artery. Amer. J. Path. *46:* 367–386 (1965).

21. HIBBS, R. G.; FERRANS, V.J.; BLACK, W. C.; WEILBAECHER, D. G.; WALSH, J. J. and BURCH, G. E.: Alcholic cardiomyopathy. An electron microscopic study. Amer. Heart J. *69:* 766–779 (1965).

22. HOAK, J. C.; WARNER, E. D. and CONNOR, W. E.: New concept of levarterenol-induced acute myocardial ischemic injury. Arch. Path. *87:* 332–338 (1969).

23. JENNINGS, R. B.; BAUM, J. H. and HERDSON, P. B.: Fine structural changes in myocardial ischemic injury. Arch. Path. *79:* 135–143 (1965).

24. KARNOVSKY, M. J. and HUG, K.: The nature of the M band enzyme in rat ventricular muscle. J. Cell Biol. *19:* 255–260 (1963).

25. KLINE, I. K.: Myocardial alterations associated with pheochromocytomas. Amer. J. Path. *38:* 539–551 (1961).

26. KORB, G. H.: Elektronenmikroskopische Untersuchungen zur Aludrin (Isoproterenolsulfat)-Schädigung des Herzmuskels. Virchows Arch. path. Anat. *339:* 136–150 (1965).

27. KORB, G. H. und TOTOVIC, V.: Electronenmikroskopische Untersuchungen über Frünveränderungen im Zentrum und in der Peripherie experimenteller Herzinfarkte. Virchows Arch. path. Anat. *342:* 85–96 (1967).

28. KORB, G. H. und TOTOVIC, V.: Elektronenmikroskopische Veränderungen im Bereich sog. hyaliner Querbänder in Herzmuskelzellen. Verh. Dtsch. Ges. Path. *51:* 145–146 (1967).

29. MALING, H. M. and HIGHMAN, B.: Exaggerated ventricular arrhythmias and myocardial fatty changes after large doses of norepinephrine and epinephrine in unanesthetized dogs. Amer. J. Physiol. *194:* 590–596 (1958).

30. MALING, H. M.; HIGHMAN, B. and THOMPSON, E. C.: Some similar effects after large doses of catecholamines and myocardial infarction in dogs. Amer. J. Cardiol. *8:* 628–633 (1960).
31. MANASEK, F. J.: Histogenesis of the embryonic myocardium. Amer. J. Cardiol. *25:* 149–168 (1970).
32. NAHAS, G. G.; BRUNSON, J. G.; KING, W. M. and CALVERT, H. M.: Functional and morphologic changes in heart lung preparations following administration of adrenal hormones. Amer. J. Path. *34:* 717–729 (1958).
33. NORMAN, T. D.: The pathogenesis of cardiac hypertrophy. Progr. Cardiovasc. Dis. *4:* 439–463 (1962).
34. NOVI, A. M.: Beitrag zur Feinstruktur des Herzmuskels bei experimenteller Herzhypertrophie. Beitr. path. Anat. *137:* 19–50 (1968).
35. NOVIKOFF, A. B. and GOLDFISCHER, S.: Nucleoside diphosphatase activity in the Golgi apparatus and its usefulness for cytological studies. Proc. Nat. Acad. Sci. *47:* 802–810 (1961).
36. RAAB, W: Pathogenic significance of adrenalin and related substances in heart muscle. Exper. Med. Surg. *1:* 188–225 (1943).
37. RAAB, W.: Key position of catecholamines in functional and degenerative cardiovascular pathology. Amer. J. Cardiol. *5:* 571–578 (1960).
38. RONA, G. and KAHN, D. S.: The healing of cardiac necrosis as reflected by experimental studies, in: BAJUSZ, E. and JASMIN, G., Methods and Achievements in Experimental Pathology (Vol. 3, p. 200–249) (Year Book Medical Publishers, Inc., Chicago) (1967).
39. ROSENBLUM, I.; WOHL, A. and STEIN, A. A.: Studies in cardiac necrosis I. Production of cardiac lesions with sympathomimetic amines. Toxicol. Appl. Pharmacol. *7:* 1–8 (1965).
40. ROSTGAARD, J. and BEHNKE, O.: Fine structure localization of adenine nucleoside phosphatase activity in the sarcoplasmic reticulum and the T system of rat myocardium. J. Ultrastruct. Res. *12:* 579–591 (1965).
41. SCHENK, E. A. and MOSS, A. J.: Cardiovascular effects of sustained norepinephrine infusions. II Morphology. Circulat. Res. *18:* 605–615 (1966).
42. SOMMERS, H. M. and JENNINGS, R. B.: Experimental acute myocardial infarction. Lab. Invest. *13:* 1491–1503 (1964).
43. SPICER, S. S.; HORN, R. G. and LEPPI, T. J.: Histochemistry of connective tissue mucopolysaccharides, in: WAGNER, B. M. and SMITH, D. E., The Connective Tissue Internat. Acad. Path. Monograph No. 7, (Williams and Wilkins, Co., Baltimore 1966).
44. SZAKACS, J. E. and CANNON, A.: L-norepinephrine myocarditis. Amer. J. Clin. Path. *30:* 425–430 (1958).
45. SZAKACS, J. E.; DIMMETTE, R. M. and COWART, E. C.: Pathologic implication of the catecholamines, epinephrine, and norepinephrine. U.S. Armed Forces Medical J. *10:* 908–985 (1959).
46. SZAKACS, J. E. and MEHLMAN, B.: Pathologic changes induced by 1-norepinephrine. Quantitative aspects. Amer. J. Cardiol. *5:* 619–627 (1960).
47. VAN VLIET, P. D.; BURCHELL, H. B. and TITUS, J. L.: Focal myocarditis associated with pheochromocytoma. New Eng. J. Med. *274:* 1102–1108 (1966).
48. WACHSTEIN, M. and MEISEL, E.: Histochemistry of hepatic phosphatase at a physiological pH with special reference to the demonstration of the canaliculi. Amer. J. Clin. Path. *27:* 13–23 (1957).

Mailing address: Victor J. Ferrans, M.D., Ph.D., Section of Pathology, National Heart and Lung Institute, National Institutes of Health, Bethesda, Maryland 20014 (USA).

Department of Pharmacology, New York Medical College,
Flower & Fifth Avenue Hospitals, New York, N.Y., USA

HEALING OF MYOCARDIAL NECROSIS CAUSED BY SYMPATHOMIMETIC AMINES [1]

D. LEHR

In his pioneering observations on the origin and prevention of degenerative heart disease, Raab consistently emphasized that adrenergic amines, in excessive amounts, whether produced endogenously (as in pheochromocytoma) or administered extrinsically, can elicit hypoxiating disproportions between supply and demand for oxygen in the myocardium (20, 21). As a consequence, one may encounter electrocardiographic alterations, arrhythmias, necrotizing structural changes and, eventually, death. This important concept has only gradually received the wide acceptance it rightfully deserves. In recent years, however, Raab's ideas experienced substantial impetus from a general upsurge of interest in the adrenergic agonists and their specific antagonists, and especially from the introduction of potent *beta*-adrenergic blocking agents, which were shown to possess cardioprotective properties of a high order. These fortunate developments finally triggered an ever-increasing stream of studies from all over the world, on the effects of sympathomimetic amines and their antagonists upon the heart.

In earlier work reported from this laboratory (15) on the cardiotoxic effects of adrenergic amines, it was shown that the disseminated necrosis of the heart that develops in the albino rat within hours after the injection of such amines in excessive dosage (fig. 1) was consistently preceded or accompanied by ionic shifts in the myocardium. Rapid loss of cardiac magnesium (Mg) was associated with accumulation of calcium (Ca). A rise in sodium (Na) and drop in inorganic phosphate (P) occurred within 24 h, whereas the potassium (K) concentration remained, initially, remarkably constant. The changes of the myocardial electrolyte content and the concomitant morphologic alterations could be readily obviated by pretreatment with the beta-adrenergic blocking agent propranolol in the case of isoproterenol, and the alpha-adrenergic blocking agent tolazoline in the case of phenylephrine, whereas the prevention of epinephrine-induced myocardial changes necessitated combined alpha- and beta- adrenergic blockade.

The present study focuses upon the *healing* of myocardial necrosis induced by the same three sympathomimetic amines. It will be demonstrated that

[1] Supported by Grant HE-00890-18 and 19 from the National Institutes of Health, National Heart Institute, U.S. Public Health Service.

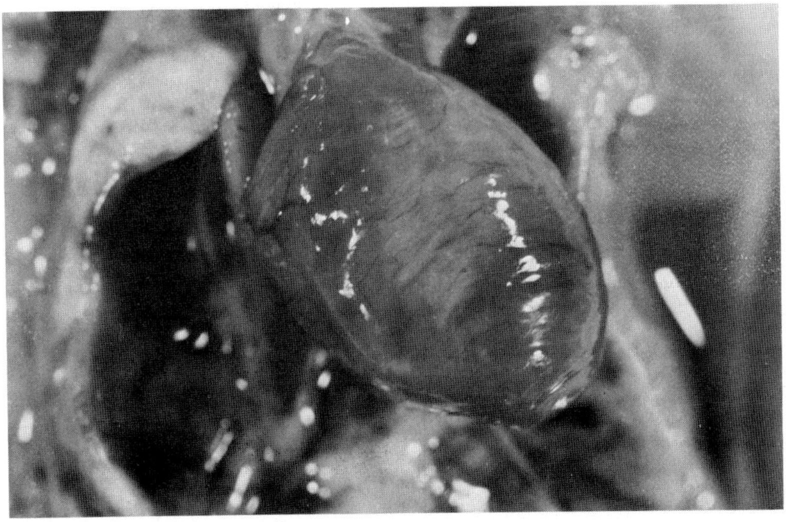

Fig. 1 Rat heart *in situ*, 24 h after epinephrine (4 mg/kg body weight, s.c. The heart has been twisted to the left to bring the right ventricle into full view. A large area, especially in the center of the right ventricle, consists of numerous light-grey patches and streaks, clearly visible to the naked eye, which consitute necrotic muscle bundles. The areas of necrosis are distinctly offset against the darker shading of the surrounding intact myocardium. A lesion of this type was classified as severe.

successive stages of structural repair are correlated with certain characteristic alterations of the cardiac electrolyte content and that the extent of protracted electrolyte shifts is roughly proportional to the amount of young scar tissue formed in the myocardium.

MATERIALS AND PROCEDURES

A total of 310 adult male albino rats of the Sprague-Dawley strain, weighing between 300 and 350 g, was employed. The procedures have been outlined earlier in detail (18) and will be only briefly summarized here.

Aqueous solutions of the hydrochloride salts of the three adrenergic amines were administered as a single subcutaneous injection. The dosages, per kg body weight, were: epinephrine, 4 mg; phenylephrine, 15 mg; and isoproterenol, 5.25 mg and 85 mg. While the mortality was low even for the high dose of isoproterenol (about 10%), it was in the neighborhood of 30 – 40% for phenylephrine, and reached 50 – 60% for epinephrine. All surviving rats were killed at predetermined intervals by decapitation under light ether anesthesia.

Tissue electrolytes were determined in heart, diaphragm, stomach wall, and kidney. To that end, pieces of these organs adjacent to areas utilized for histologic evaluation were homogenized and prepared for determination of individual samples by automated chemical analysis. All values were expressed as mEq/kg wet tissue. An arbitrary scale of 1 plus to 4 plus, indicating mild to severe damage or fibrosis, was employed for histologic evaluation of structural injury or repair in conventional stains.

The progression of the healing process was followed in groups of albino rats under the light and electron microscope and correlated with determinations of the cardiac electrolyte content (Ca, Mg, P, Na and K) at 1, 2, 4, 6, and 13 days following the production of the cardiac injury.

Control values for the electrolyte content of rat connective tissue were obtained from pieces of fascia lata and carefully shaved skin. Serum samples derived from terminal neck blood were likewise utilized for determination of the same five electrolytes.

RESULTS

Structural and Ultrastructural changes

In the interest of conciseness, microphotographs and electron microscopic photographs of pathologic anatomic alterations have been limited to the evolvement of full myocardial injury and the end-stage of healing under our experimental conditions (13th day of repair). Moreover, illustrations of phenylephrine-induced changes were omitted because of their close resemblance to those depicting epinephrine injury.

Under the light microscope, cardiac muscle fibres in affected areas showed, within 6 to 12 h, segmental loss of cross striation and homogenization, as well as granular or fatty degeneration, often running band-like across several muscle fibres. Many of these fibres, or segments of fibres, were eosinophilic and PAS-positive and had either shrunken or pale-staining nuclei. In others, the nuclei were no longer demonstrable.

Within 24 h, fragmentation of numerous necrotic muscle fibres and appearance of mononuclear inflammatory cells and a few polymorphonuclear leucocytes could be observed. After 24 h, the apparent height of injury, the accumulation of large mononuclear cells in muscular debris and between muscle fibres was prominent and there was considerable accumulation of neutral fat in the wall of both the left and right ventricle in the case of epinephrine and phenylephrine, whereas with isoproterenol, fatty infiltration was limited primarily to the sub-endocardial layer of the left ventricle (figs. 2–4).

After 48 h, many of the affected fibres had disappeared by lysis, although in some, the sarcolemmic outline had remained intact.

By the 4th day, almost all necrotic debris had been removed, so that some foci containing only empty sarcolemmic sheets of dissolved fibres resembled fatty tissue. Most of these foci contained or were surrounded by surviving fragments of muscle fibres, often still showing cross striation and normal nuclei.

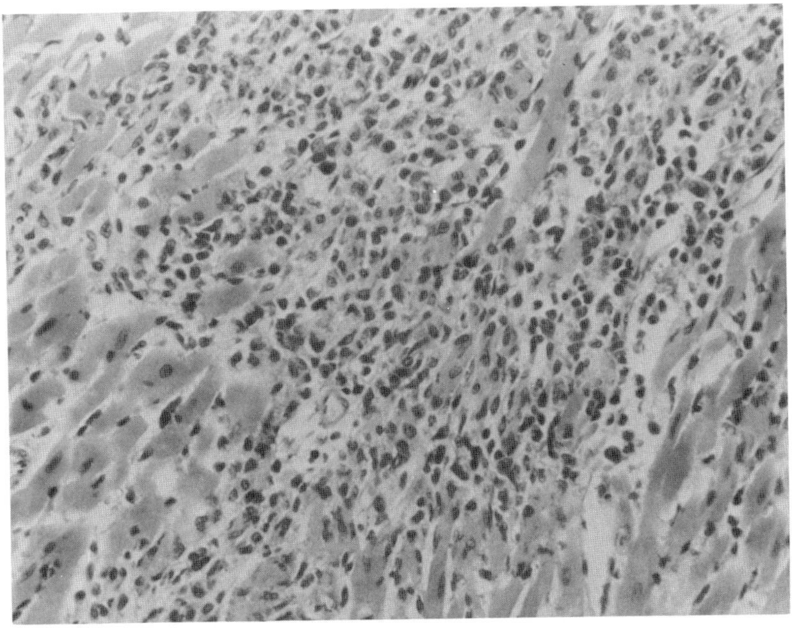

Fig. 2 Rat myocardium, 24 h after isoproterenol (5.25 mg/kg body weight, s.c.). Typical focus of necrosis containing muscular debris and disintegrating muscle fibres, some of which stain PAS-positive. Note that the cross striations have disappeared, giving the sarcoplasma a homogeneous appearance. Many nuclei of the dying muscle cells are pale-staining or pyknotic. The area of necrosis is invaded by mostly mononuclear inflammatory cells and some polymorphonuclear leucoytes. PAS stain; medium power.

By the 6th day following initiation of the injury, granulation tissue had replaced all foci of necrosis and the mononuclear cells had been joined by fibroblasts and sprouting capillaries.

Finally, on the 13th day, fibroblasts and fibrocytes imbedded in collagen fibres dominated the picture, the healed foci giving the appearance of young connective tissue still containing considerable numbers of round cells. In the case of epinephrine, in particular, the early interstitial infiltration with inflammatory cells had led to extensive fibroblastic proliferation. The eventual result was diffuse fibrosis, characterized by the thickened sheets of connective tissue, interspersed between apparently healthy muscle bundles. This phenomenon was distinct from the large focal, often aneurysmic, scars that had formed as a replacement of areas of muscular necrosis in the same hearts (figs. 5–8).

For more precise correlation with early ionic shifts in the myocardium, the *evolvement* of the injury was studied also under the electron microscope.

Early changes, preceding overt histologic evidence of injury, occurring within hours, and frequently restricted to part of a single myocyte, consisted of swelling and vacuolization of mitochondria with distortion and disintegration of

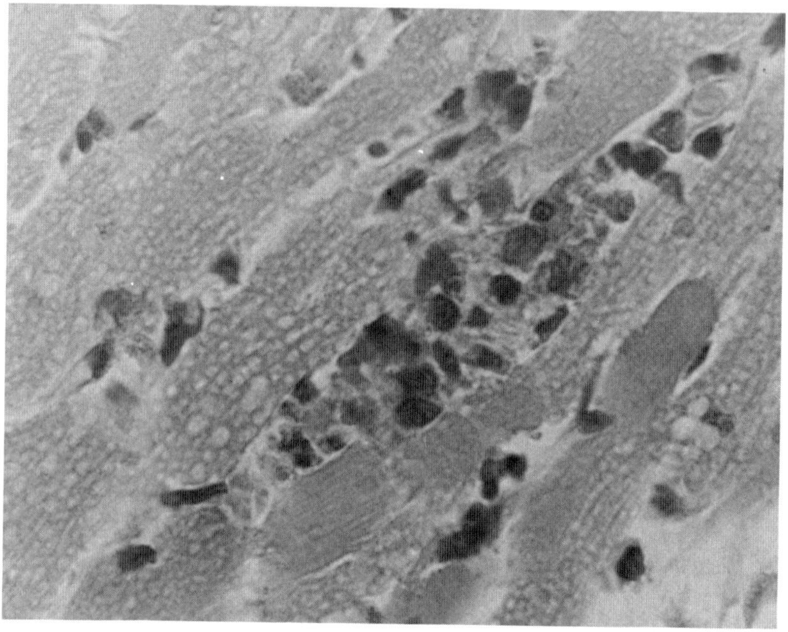

Fig. 3 Rat myocardium, 24 h after epinephrine (4 mg/kg body weight, s.c.). Peculiar honeycomb appearance of myocardial cells due to extraction of heavy accumulation of neutral fat globules. Note that some muscle cells are free of fat and still show cross striation. In the center of the photograph, typical interstitial proliferation of mononuclear inflammatory cells. PAS stain; high power.

cristae, enlargement of the sarcotubules, and margination of nuclear chromatin. This was followed by focal fraying, rupture, and homogenization of myofilaments. There was a striking increase in the number and size of fat droplets, especially in the vicinity of mitochondria. In confirmation of light-microscopic findings, this accumulation of fat was found to be more pronounced and to occur earlier with the two amines possessing an alpha-adrenergic component of action, and led, within about 12 h, through confluence of droplets, to the formation of large globules of fat. This stage was further characterized by the appearance of numerous myelin figures of varying electron density, and by total disruption of the myofilament pattern in severely affected areas. The fragmentation ultimately led to loss of myofibrils, with the formation of intracellular spaces containing clusters of translocated mitochondria and dilated components of the sarcoplasmic reticulum (figs. 9–14).

These observations are in full agreement with the detailed and thorough description of damage and healing of isoproterenol lesions provided by Rona and his associates (23, 24), and of the histologic and electronmicroscopic pictures of isoproterenol and epinephrine injury presented by Ferrans *et al.* (3).

Fig. 4 Same heart as depicted in figure 3. The heavy accumulation of neutral fat is clearly demonstrated. Near the top and bottom of the photograph, a fat-loaded myocyte is seen immediately adjacent to another one that is essentially devoid of fatty infiltration (arrows). Frozen section; Sudan IV stain; high power.

Mention should also be made here of the considerable similarity of ultrastructural alterations noted in experimental Mg depletion of albino rats. The changes were detected after only 5 days of dietary Mg depletion and preceded overt histologic evidence of myocardial necrosis by the same length of time (4, 5). It was this close resemblance of changes seen under the electron microscope, and the consistent early occurrence of Mg loss from the injured heart, that led us to postulate that the initial ionic disequilibrium might serve as a common biochemical denominator in the pathogenesis of myocardial necrosis from various causes. The demonstrated effectiveness of Mg salts in the prevention of experimental myocardial necrosis, induced by two entirely unrelated procedures (16), would seem to provide supportive evidence of this concept.

As regards the other organs examined, the kidneys of phenylephrine-injected rats had a peculiar mottled, pale greyish-yellow appearance, and showed parenchymatous degeneration. Epinephrine caused similar, but less pronounced changes and some nephrocalcinosis, whereas no substantial alterations were noted with isoproterenol. A more detailed discussion of the changes is beyond the scope of this paper. No injury was seen in the diaphragm and smooth muscle

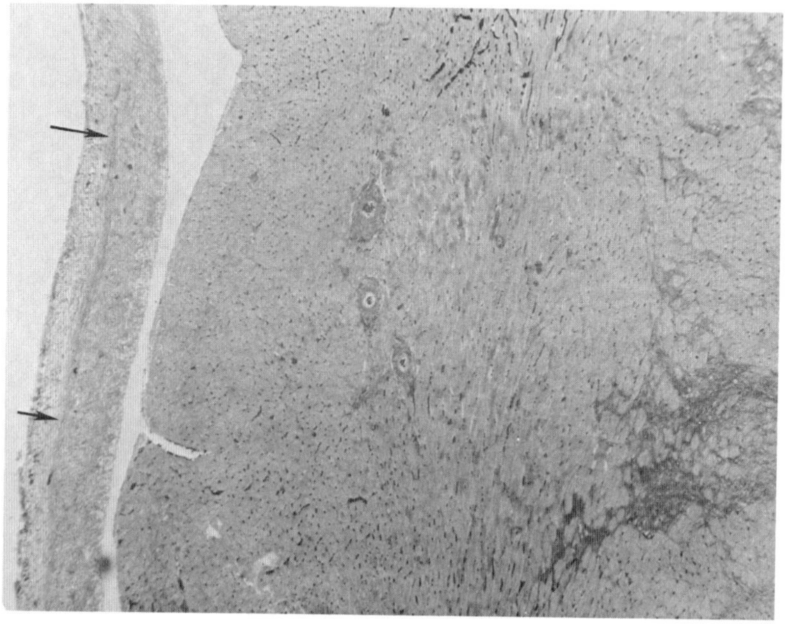

Fig. 5 Rat myocardium, 13th day of repair of isoproterenol-induced injury (85 mg/kg body weight, s.c). A long subepicardial scar (between arrows) can be seen in the thinned-out wall of the right ventricle, which also contained a large aneurysm, partly illustrated in figure 6. Note epicardial thickening over the area of the scar. Heavy focal scarring is apparent near the endocardial layer of the left ventricle (darker shading). Mallory stain; medium power.

of the stomach wall and there were also no deviations of their electrolyte content from the norm.

Alterations of the Myocardial Electrolyte Content

In the determination of the myocardial electrolyte content at predetermined intervals during the various stages of structural repair, it was found that except for a return of the raised Ca level into the normal range by the 4th day, the early alterations of the cardiac electrolyte concentrations did not disappear. In fact, depletion of K, which usually became manifest only 3 days after initiation of injury, was noted as an additional disturbance. At about the 4th to the 6th day of healing, the alterations of the cardiac electrolyte content became stabilized and, thereafter, persisted generally with little change for the remainder of the 13 day observation period. Since the foci of necrosis had been invariably replaced by young scars within 2 weeks, the protracted abnormalities of the electrolyte content were ascribed to the presence of fibroblastic proliferation. In hearts exhibiting extensive scar formation, the electrolyte changes consisted in

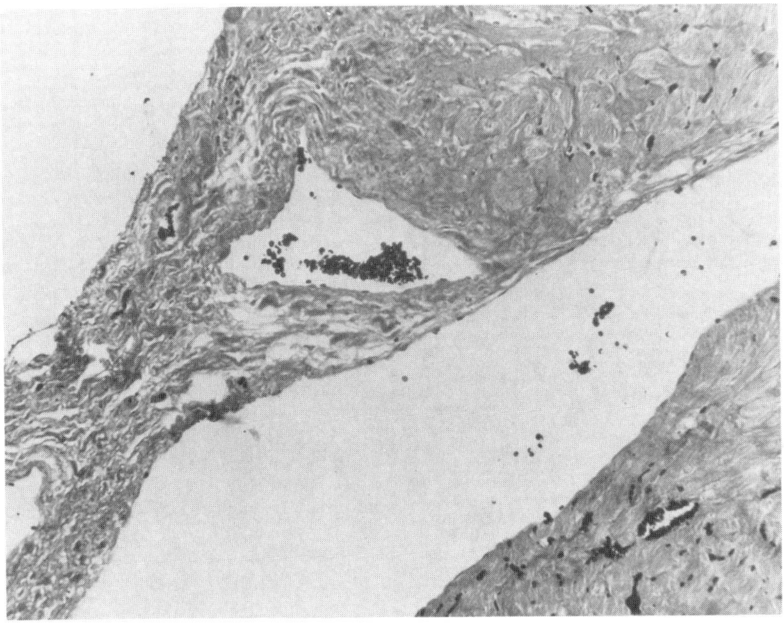

Fig. 6 Same heart illustrated in figure 5, photographed near insertion of right auricle at septum. The wall of the right ventricle contains a large vein and consists in this area of scar tissue only (part of aneurysm). Near the top of the photograph can be seen residual muscle bundles that have lost their normal architecture and are encased in thick layers of fibrous tissue. At the right bottom is part of the left ventricle. Mallory stain; high power.

substantial lowering of the content of Mg, P and K, and considerable elevation of Na.

The described sequence of events is exemplified in table I. It will be noted that a high incidence of predominantly severe lesions was obtained with epinephrine, and that the characteristic changes of the electrolyte pattern are in evidence 24 h after the injection of a highly cardiotoxic dose of this catecholamine, that is, at a time when disseminated cardiac necrosis was fully developed. The foci of necrosis were, incidentally, clearly visible to the naked eye in every instance.

It should be emphasized that there is, as yet, no loss of K, while all other electrolyte concentrations show significant changes from the norm. Rats permitted to survive for longer periods (2 to 13 days) exhibit a similar high incidence of severe injury although, at certain stages of healing, the lesions may become less conspicuous to the naked eye. While the P and Mg loss remain highly significant at all periods of observation, the early Ca accumulation is sharply reduced or transiently eliminated. It can be seen further that during the progression of healing there is a significant drop of the K content, which occurs

Table I

Changes of Cardiac Electrolyte Content in Myocardial Necrosis and Fibrosis Induced by Epinephrine

Group	Days after injection	No. of rats	Myocardial damage: Gross appearance Mild	Severe	Total incidence	Myocardial electrolytes mean mEq/kg ± S.E. (wet weight): Ca	P	Mg	Na	K
Control	–	42	–	–	–	2.8 ±0.1	57.6 ±0.6	18.7 ±0.2	39.8 ±0.5	72.0 ±0.6
Epinephrine 4 mg/kg s.c.	1	25	5	20	100%	9.1^c ±1.3	53.1^c ±0.9	16.5^c ±0.3	50.4^c ±1.0	74.2^a ±0.8
	2	12	0	10	83%	5.2^b ±1.2	44.4^c ±1.6	14.9^c ±0.5	47.1^c ±1.8	71.2 ±1.5
	4	11	3	5	82%	2.8 ±0.2	51.7^b ±1.6	16.6^c ±0.3	41.9^a ±1.2	69.2^a ±0.9
	6	13	3	10	85%	2.8 ±0.2	52.5^b ±0.9	16.6^c ±0.2	44.6^c ±1.1	69.9^a ±0.6
	13	10	2	7	90%	4.1^c ±0.2	45.3^c ±2.1	16.6^b ±0.6	42.8^a ±1.2	64.3^c ±1.1

$^a p < 0.05$.
$^b p < 0.01$.
$^c p < 0.001$.

simultaneously with a reduction of the Na accumulation on the 4th day. This phenomenon coincides in time with the disappearance of necrotic muscular debris. It is of particular interest that 13 days after production of the injury, when healing is complete and young scars, in most instances readily recognizable with the naked eye, have replaced all foci of necrosis, the concentration of P and K show highly significant deviations. In this particular experiment, the Ca content is again increased.

Closely similar findings were encountered with phenylephrine, a pure alpha-adrenergic amine (table II), except that grossly visible lesions were less conspicuous, especially in later stages of repair. (Note in particular the 6th and 13th day of healing.)

There is no need to recount in detail what obviously represents a recapitulation of ionic shifts just described for epinephrine-induced necrosis, including the secondary drop of the K content on the 4th day of repair and the concurrent reduction of Na accumulation, except to point out that the terminal deviations of the electrolyte content on the 13th day of healing were less

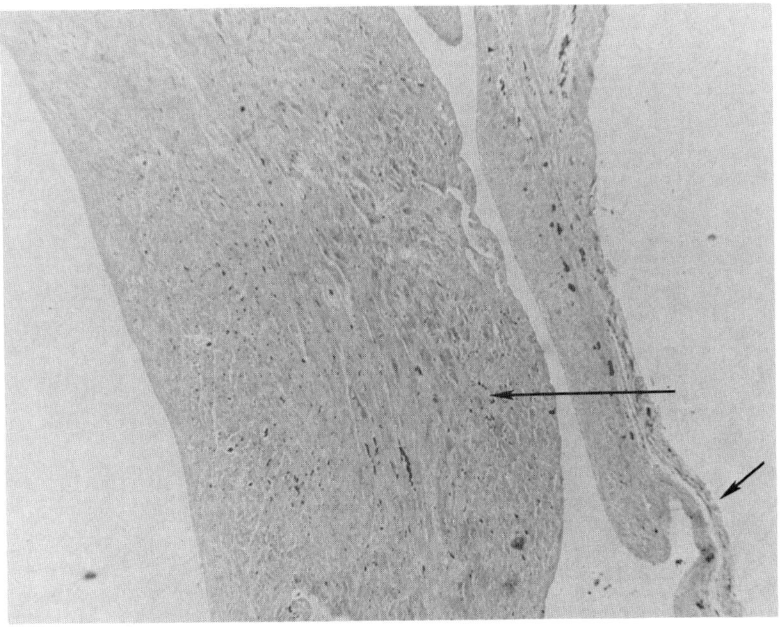

Fig. 7 Rat myocardium, 13th day of repair of epinephrine-induced injury (4 mg/kg body weight, s.c.). Aneurysm formation in the wall of the right ventricle is indicated by the arrow on top. A diffuse bluish tinge was seen in the wall of both the right and left ventricles, in the original observation under the microscope, which unfortunately is lost in this black and white illustration. This bluish color was due to extensive interstitial fibroblastic proliferation and also to the formation of numerous tiny scars as substitution of dead muscle fibres. This is seen in figure 8, which focuses upon the approximate area indicated by the long arrow in the center of the photograph. Mallory stain; low power.

pronounced with phenylephrine. This appeared to correlate well with the far less pronounced scarring found upon histologic examination.

In order to investigate the relationship between extent of fibrosis and degree of electrolyte changes further, the third amine, isoproterenol, was used in two widely divergent dosages. As first shown by Rona and his associates (22), the small dose (5.25 mg/kg body weight) produces only small foci of disseminated necrosis that result in fine, diffuse fibrosis upon healing. The large dose (85 mg/kg body weight) causes extensive infarctoid-type necrosis, usually involving, among other areas, the entire tip of the right and left ventricle. Hence, it leads to massive scarring.

Table III depicts the data obtained with the small dose of isoproterenol. While the initial electrolyte changes (24 h) are typical and as severe as with epinephrine and phenylephrine, except for a less pronounced rise in Ca content, there is considerable normalization of the electrolyte pattern on the 4th day of repair. And yet, there is again indication of K depletion and reduction of Na

Table II

Changes of Cardiac Electrolyte Content in Myocardial Necrosis and Fibrosis Induced by Phenylephrine

Group	Days after injection	No. of rats	Myocardial damage: Gross appearance: Mild	Severe	Total incidence	Myocardial electrolytes mean mEq/kg ± S.E. (wet weight): Ca	P	Mg	Na	K
Control	–	42	–	–	–	2.8 ±0.1	57.6 ±0.6	18.7 ±0.2	39.8 ±0.5	72.0 ±0.6
Phenylephrine 15 mg/kg s.c.	1	19	6	8	74%	10.4[c] ±2.7	54.5[a] ±1.5	17.3[c] ±0.4	46.4[c] ±2.2	73.5 ±1.5
	2	10	3	4	70%	5.3[b] ±1.5	52.1[b] ±1.4	17.9 ±0.4	43.0[b] ±0.7	73.5 ±1.9
	4	16	3	2	31%	3.0 ±0.1	50.8[c] ±1.4	16.8[c] ±0.3	41.0 ±0.6	67.9[b] ±1.1
	6	14	0	2	14%	3.5[a] ±0.2	48.6[c] ±0.9	16.8[c] ±0.3	42.3[a] ±1.0	65.5[c] ±1.0
	13	11	2	0	17%	3.0 ±0.2	54.6[a] ±1.5	18.1 ±0.5	43.1[a] ±1.2	69.8 ±1.8

[a] $p < 0.05$.
[b] $p < 0.01$.
[c] $p < 0.001$.

excess at that time. Terminally, when only small scars are visible to the naked eye in 25% of the animals, the Mg and P content are in the normal range, but there are residual deviations of the other electrolytes.

With the large dose of isoproterenol (table IV), the highly significant ionic shifts at the 1-day interval encompassed all five electrolytes tested. Thus, K loss occurred much earlier than in any of the previously described experiments. In other aspects, however, the changes are fully comparable with those seen after epinephrine and phenylephrine. As noted with these agents on the 4th day, the Na level transiently returns to the normal range with a concurrent downward movement of the K concentration and, except for Ca, the disturbance of the electrolyte pattern is still severe on the 13th day. It should be noted that these changes are associated with a 100% incidence of grossly visible large scars.

These findings suggest that the long-lasting alterations of the myocardial electrolyte content encountered in our experiments may have been caused by dilution of the muscle mass with scar tissue and that the degree of protracted deviations of electrolytes may be directly proportional to the extent of fibrosis. Strong additional support for this view was obtained from a comprehensive and

Table III

Changes of Cardiac Electrolyte Content in Myocardial Necrosis and Fibrosis Induced by Isoproterenol

Group	Days after injection	No. of rats	Myocardial damage: Gross appearance Mild	Severe	Total incidence	Myocardial electrolytes mean mEq/kg ± S.E. (wet weight): Ca	P	Mg	Na	K
Control	–	42	–	–	–	2.8 ±0.1	57.6 ±0.6	18.7 ±0.2	39.8 ±0.5	72.0 ±0.6
Isoproterenol 5.25 mg/kg s.c.	1	26	10	12	71%	4.2[c] ±0.3	48.4[c] ±1.3	15.9[c] ±0.2	50.8[c] ±1.0	78.9 ±1.2
	2	11	4	1	45%	4.1[c] ±0.2	60.1[a] ±0.9	17.2[b] ±0.3	43.1[b] ±0.7	72.6 ±0.7
	4	11	2	0	18%	2.8 ±0.2	55.1[a] ±1.3	18.1 ±0.4	41.5[a] ±0.7	68.8 ±1.1
	6	12	3	2	42%	3.5[a] ±0.4	55.2[a] ±1.0	19.0 ±0.5	43.3[b] ±1.3	71.4 ±1.4
	13	16	1	3	25%	3.4[a] ±0.2	56.1 ±1.2	18.4 ±0.3	42.7[b] ±0.6	68.9[b] ±0.8

[a] $p < 0.05$.
[b] $p < 0.01$.
[c] $p < 0.001$.

detailed microscopic review of the hearts of all animals employed in this investigation. In favor of this assumption, furthermore, is the fact that connective tissue is low in Mg, K and P, and high in Na. The actual figures for the electrolyte content of rat connective tissue as contrasted to normal myocardium are shown in table V. It should be kept in mind that, in connective tissue, the minerals are predominantly in the ground substance and thus extracellular. The mineral content of connective tissue corresponds, therefore, more closely to that of the plasma, and is very different from a typical cellular tissue such as myocardium.

In view of the fact that the normal P content of myocardium and connective tissue shows the greatest difference (ratio 1:4), changes of P values would have to be considered the most reliable index for judging the percentage of scar tissue in the myocardium. Next in accuracy are the values for Mg and K (approximate ratio 1:3), while Na, with a 2:1 ratio, would be the least accurate for that purpose.

Table V depicts, in addition, the increasing deviations of the myocardial electrolyte content when rat connective tissue is added to healthy cardiac tissue

Table IV

Changes of Cardiac Electrolyte Content in Myocardial Necrosis and Fibrosis Induced by Isoproterenol

Group	Days after injection	No. of rats	Myocardial damage: Gross appearance: Mild	Severe	Total incidence	Myocardial electrolytes mean mEq/kg ± S.E. (wet weight): Ca	P	Mg	Na	K
Control	–	42	–	–	–	2.8 ±0.1	57.6 ±0.6	18.7 ±0.2	39.8 ±0.5	72.0 ±0.6
Isoproterenol 85 mg/kg s.c.	1	9	0	9	100%	7.3[c] ±1.3	47.8[c] ±1.1	14.9[c] ±0.3	56.4[c] ±1.5	63.8[c] ±1.3
	2	11	0	9	82%	3.6[b] ±0.2	48.1[c] ±2.3	14.9[c] ±0.3	47.4[c] ±1.6	68.5[a] ±1.7
	4	10	2	3	50%	1.4[c] ±0.2	52.1[b] ±2.3	16.9[b] ±0.6	40.0 ±2.7	62.7[c] ±2.9
	6	9	2	5	88%	3.0 ±0.2	51.0[b] ±1.5	16.4[c] ±0.4	44.3[b] ±1.1	63.6[c] ±0.9
	13	10	3	7	100%	3.0 ±0.1	50.2[c] ±1.4	17.0[b] ±0.4	43.4[b] ±1.6	62.1[c] ±1.5

[a] $p < 0.05$.
[b] $p < 0.01$.
[c] $p < 0.001$.

in increasing proportions. It can be seen that there is good agreement between the values found and those calculated. If the values for P, Mg, and K (framed in heavy black in this table) are placed in juxtaposition to values for the same electrolytes obtained on the 13th day of healing, as seen in the preceding four tables, one can, at a glance, obtain a rough estimate of the percentage of scarring (table VI). This is indicated in the right-hand column. Note the striking similarity of the epinephrine pattern with that of the 22% admixture of connective tissue and the far less marked electrolyte deviation in the phenylephrine and low-dosage isoproterenol group, whereas, with the large dose of isoproterenol, the epinephrine pattern is approached again. More accurate figures can be read off from the slope of the straight lines prepared from the admixture figures in the top section of the table. The values depicted in the column "calculated" represent averages of the three electrolytes read off from figure 15 and are in good agreement with estimates obtained from thorough histologic examination, including serial sections of entire hearts.

The procedure thus provides a rapid, objective, and fairly accurate estimate of the extent of myocardial fibrosis. It would appear to be more precise and less

Table V

Changes of Myocardial Electrolyte Content by Admixture of Connective Tissue

Group	Myocardial electrolytes mean mEq/kg ± S.E. (wet weight)			
	P	Mg	K	Na
Normal Myocardium (M)	52.1 ±1.8	19.9 ±0.5	70.7 ±1.3	36.4 ±0.3
Connective Tissue (CT)	13.3 ±0.3	6.5 ±0.1	22.3 ±1.6	61.4 ±1.4
CT/M Ratio	1:4	1:3	1:3	2:1
H + CT (9.2%) Found	49.3 ±1.2	18.4 ±0.1	66.0 ±1.0	39.9 ±1.1
Calculated	48.5	18.7	66.3	38.7
H + CT (22%) Found	44.8 ±1.1	16.1 ±0.2	58.9 ±1.3	42.4 ±0.9
Calculated	43.6	17.0	60.1	41.9
H + CT (33%) Found	35.1 ±0.8	14.6 ±0.2	50.7 ±1.4	43.7 ±0.9
Calculated	39.3	15.5	54.7	44.7

time-consuming than routine histologic review of stained sections of affected hearts and the classification of mild, moderate, and severe fibrosis, or a 1 to 4 plus scale of cicatrization.

DISCUSSION

Detailed studies on the repair of cardiac injury induced by sympathomimetic amines are available thus far for isoproterenol only (11, 25), although the development of biochemical and structural alterations in the myocardium, produced by adrenergic agonists, have been investigated extensively by many workers (for Refs., see among others: 1, 2, 8, 18, 20, 25–29).

Variation in dosage employed may account for some differences in the extent and also the nature of injury encountered by some authors. With regard to isoproterenol, reference was made earlier to the observation of Rona and his associates (22) that low dosage causes finely disseminated necrosis, whereas high dosage results in large, infarctoid-type foci of necrosis, often leading to ventricular aneurysm formation. In the case of epinephrine, Ferrans *et al.* (3),

Table VI

Estimate of Percentage of Amine-Induced Myocardial Fibrosis on the Basis of the Extent of Deviation of the Electrolyte Content

Group	Myocardial electrolytes Mean mEq/kg ± S.E. (wet weight)			Percentage CT	
	P	Mg	K	Routh estimate	Calculated
H + CT (9.2%)	48.5	18.7	66.3		
H + CT (22%)	43.6	17.0	60.1		
H ± CT (33%)	39.3	15.5	54.7		
Control	57.6 ±0.6	18.7 ±0.2	72.0 ±0.6		
Epinephrine, 4 mg/kg (13th day)	45.3 ±2.1	16.6 ±0.6	64.3 ±1.1	22%	21%
Phenylephrine, 15 mg/kg (13th day)	54.6 ±1.5	18.1 ±0.5	69.8 ±1.8	< 9%	5.5%
Isoproterenol, 5.25 mg/kg (13th day)	56.1 ±1.2	18.4 ±0.3	68.9 ±0.8	<9%	4%
Isoproterenol, 85 mg/kg (13th day)	50.2 ±1.4	17.0 ±0.4	62.1 ±1.5	> 9% < 22%	17%

using 1 mg/kg body weight, found only minor heart lesions, if compared with the injury produced by a large isoproterenol dose, although interstitial edema and inflammation were relatively more prominent than the alterations of the muscle fibres. In the present study, using 4 mg/kg body weight of epinephrine, we have shown that the cardiac injury was closely similar to and as severe as that following a massive dose of isoproterenol. It manifested the additional feature of heavy infiltration of the ventricular myocardium with neutral fat. Since this latter phenomenon also occurred with phenylephrine and a third alpha-adrenergic agonist, metaraminol (Lehr, unpublished data), it is assumed that the alpha-adrenergic activity, in particular, may be responsible for the heavy fat accumulation, perhaps by inducing coronary vasoconstriction or by increasing peripheral resistance. Either phenomenon would tend to augment myocardial oxygen deficiency and thus favor neutral fat aggregation as described for the hypoxic myocardium (14, 19).

It would seem, therefore, that the difference in cardiotoxicity of alpha-and beta-adrenergic amines may encompass, in part, true differences in pathogenesis. Rosenblum *et al.* (26) have suggested three types of pathogenetic mechanisms for the toxic effects of sympathomimetic amines, namely: 1)metabolic changes;

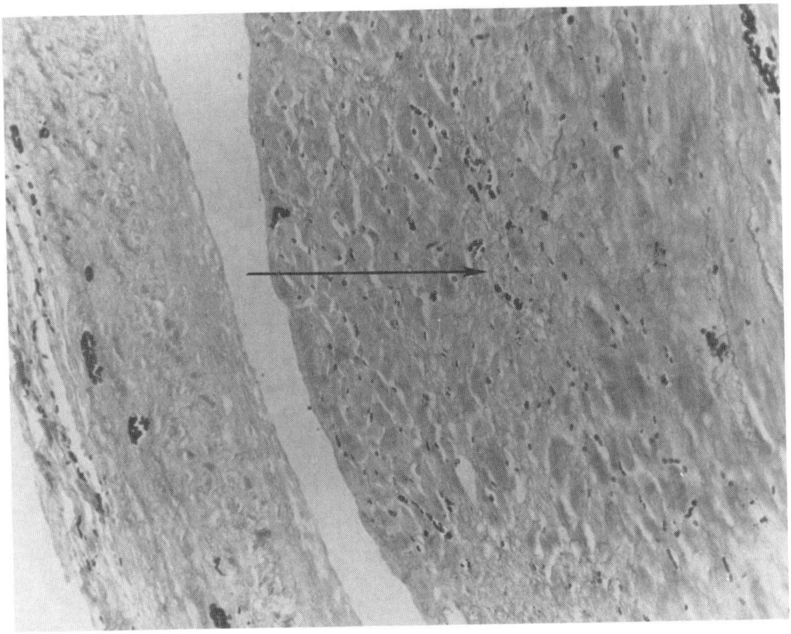

Fig. 8 Same section as figure 7, under high power. The head of the arrow is in the identical area as that of figure 7. The extensive diffuse fibrous proliferation is now clearly apparent.

2)hemodynamic disturbances; and 3)myogenic alterations related to a reduction in myocardial contractile force. It is the view of these authors that a variety of sympathomimetic agents cause primarily interstitial myocarditis, whereas only isoproterenol induces severe myocardial necrosis. While I cannot subscribe to this concept, on the basis of the severe cardiac necrosis obtained with epinephrine and phenylephrine in this laboratory, I agree with them and with Ferrans *et al.* (3) that the severity of the inflammatory response is not always directly related to the degree of myocardial necrosis. In other words, sympathomimetic amines may exert separate and distinct influences on muscle fibres and on other cellular components of the heart. The extensive diffuse interstitial fibrosis, over and above the scars of repaired foci of necrosis, observed in the present epinephrine experiments may serve as a case in point. It is not surprising, therefore, that the total percentage of fibrosis induced by the 4 mg dose of epinephrine exceeded that of the 85 mg dose of isoproterenol.

In view of the fact that scars shrink with age (diminution of the ground substance), and thus lead to a reduction of the space occupied by fibrous tissue in relation to the remaining muscle mass, comparative studies of factors influencing scar formation are best carried out in earlier phases of repair, as done

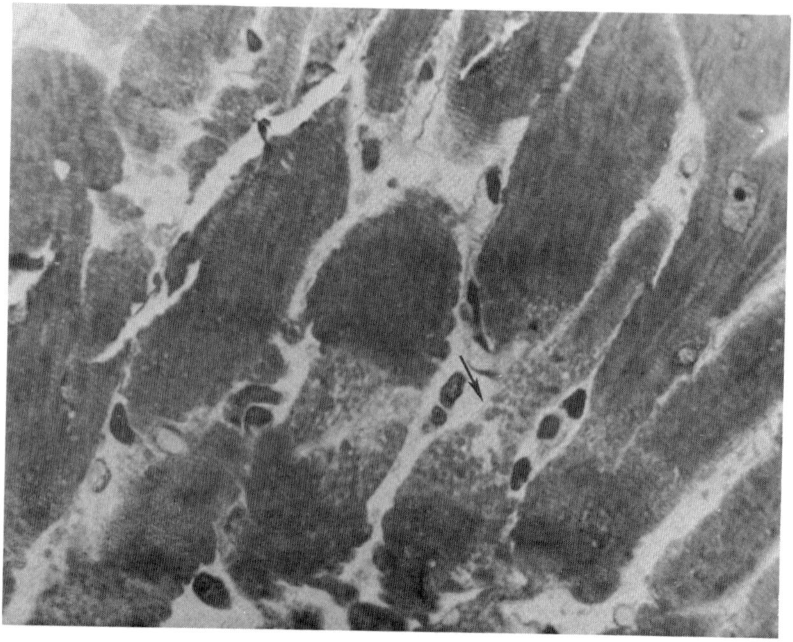

Fig. 9 Rat myocardium, 12 h after epinephrine (4 mg/kg body weight, s.c.). In the lower center of the photograph, the degeneration of two single muscle cells separated by intercalated disks from intact cells is clearly apparent. There is considerable interstitial edema and some cellular proliferation. Toluidine blue stain; high power.

in the present instance, because of a more favorable ratio of connective tissue over myocardium.

The young scars studied in the present instance could be considered as *loose* connective tissue, which means they are water rich and contain relatively small amounts of insoluble, highly aggregated colloid. Structurally, loose connective tissue consists of a matrix composed of fibres and separated by a plastic fluid. The materials localized in the interstices of the connective tissue are referred to as ground substance. They consist, in the main, of water, colloids, and electrolytes. The ground substance is in constant association with circulating fluids. In studies of the interaction of Na, K, and other ions with connective

---

Fig. 10 Rat myocardium, 12 h after epinephrine (4 mg/kg body weight, s.c.). Composite of several E.M. photographs from approximate area indicated in figure 9. Upper half of the picture shows a fairly intact myocyte. The nucleus (left, top) shows margination of chromatin. The sarcomeres are contracted, the sarcotubules are dilated, and many mitochondria display partial cristae-lysis. Numerous fat globules are in evidence. The lower half of the photograph illustrates extensive cystic degeneration of an adjacent myocyte, starting at the intercalated disk, with disintegration and homogenization of myofilaments. Within the debris of myofilaments are seen a cluster of enlarged, electron-dense mitochondria and fat droplets.

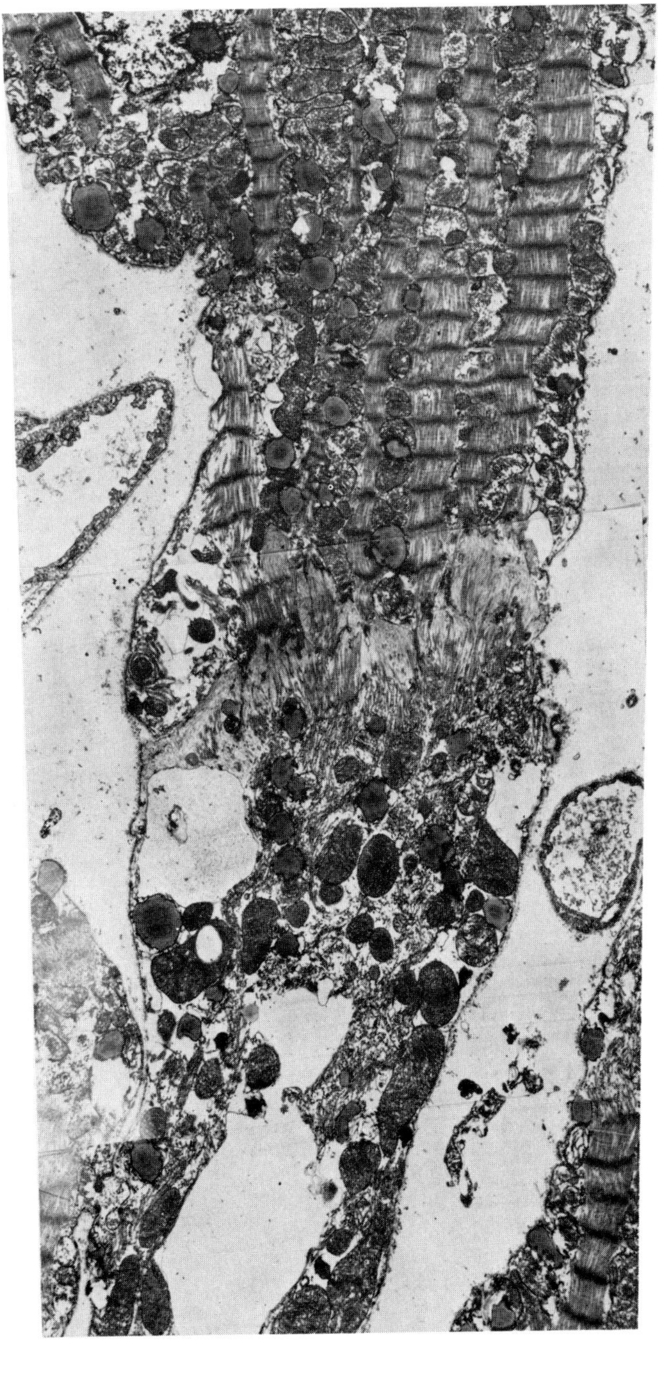

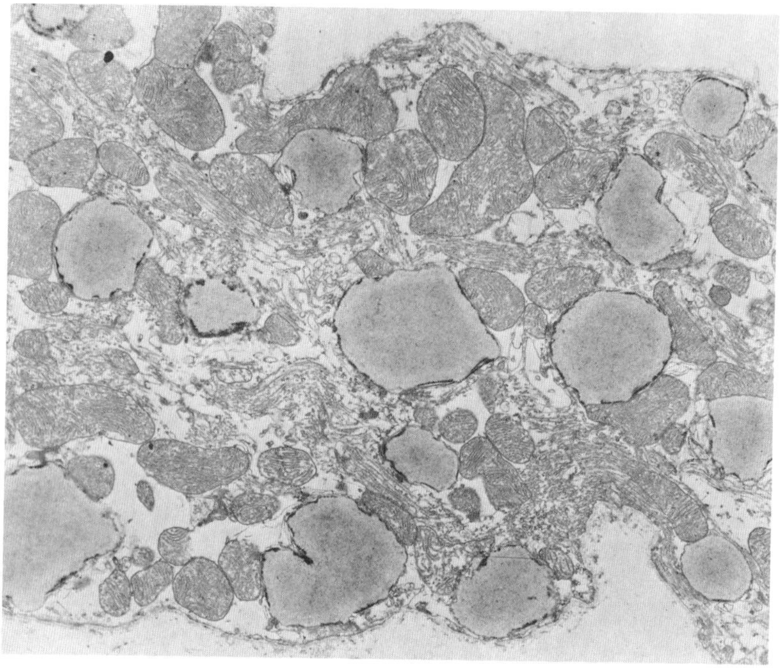

Fig. 11 Rat myocardium, 12 h after epinephrine (4 mg/kg body weight, s.c.). There is considerable intracellular edema. The sarcoplasmic reticulum is dilated and the myofilaments are disintegrated. The mitochondria are substantially intact but display considerable variations in size; some of the very large ones have assumed odd shapes. There is an abundance of large fat globules

tissue (10), it was found that homeostasis ultimately depends on the invariance of blood electrolyte concentrations and that K and Ca concentrations approach those of the blood. Also, there is a strong tendency for K to replace Na from the colloidal phase. Thus, in colloid-rich connective tissue (cartilage), K concentrations may be more than ten times that of the blood.

Old scars represent *firm* connective tissue, consisting mostly of aggregated colloid (collagen fibres) and little ground substance. They are thus smaller and usually contain fewer electrolytes, except for Ca phosphate in the case of calcification.

It should be kept in mind that the injury induced in myocardial cells by adrenergic amines is due to *temporary* ischemia that may be survived by many cells. In agreement with Rona and Kahn (23), we found that those cells that die are usually completely removed within 3 to 5 days and replaced by an organization process, probably because of the patency of the coronary arteries, since the same process may take weeks if myocardial necrosis is produced by ligation of coronary arteries (12, 13). The speedy removal of Na-loaded muscular

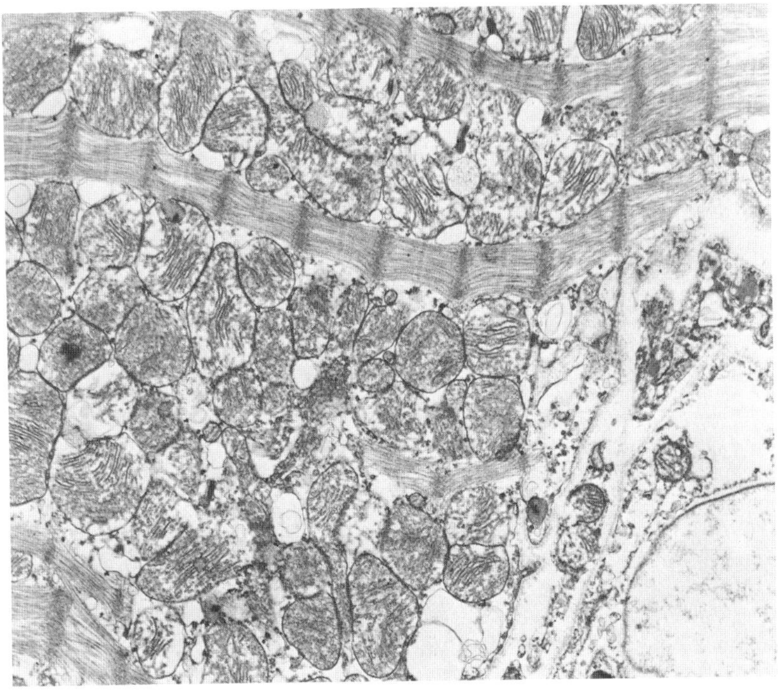

Fig. 12 Rat myocardium, 12 h after isoproterenol (5.25 mg/kg body weight, s.c.). As in figure 11, there is intracellular edema, dilatation of the sarcotubular system, variation in size of mitochondria, and disarray of the normal intracellular structure. In striking contrast to the epinephrine picture, there is complete absence of fat globules and the mitochondria exhibit considerable marginal cristae-lysis.

debris could explain the normalization of the myocardial Na content on the 4th day, invariably seen in all our experiments. The simultaneous decline of the K content might be due to an observed decrease, at that time, in mononuclear inflammatory cells that are rich in K. By the same token, the presence of these cells in massive concentration in earlier phases of the injury may account for the lack of initial K decline. In infarctoid-type necrosis caused by the large dose of isoproterenol, the massive destruction of myocardial cells might outweigh the K-balancing effect of infiltration with inflammatory cells.

Studies of temporary coronary ligation in the dog and rat, with restoration of blood flow, revealed a more rapid development and progression of the lesions than after permanent vascular occlusion (6, 7). Herdson and his associates suggested that tissue electrolyte disturbances develop more rapidly with restored flow. It is of interest that extensive mitochondrial calcification is usually not seen after permanent coronary ligation but is readily apparent after temporary occlusion. Jennings *et al.* (9) point out that mitochondrial damage occurs early in

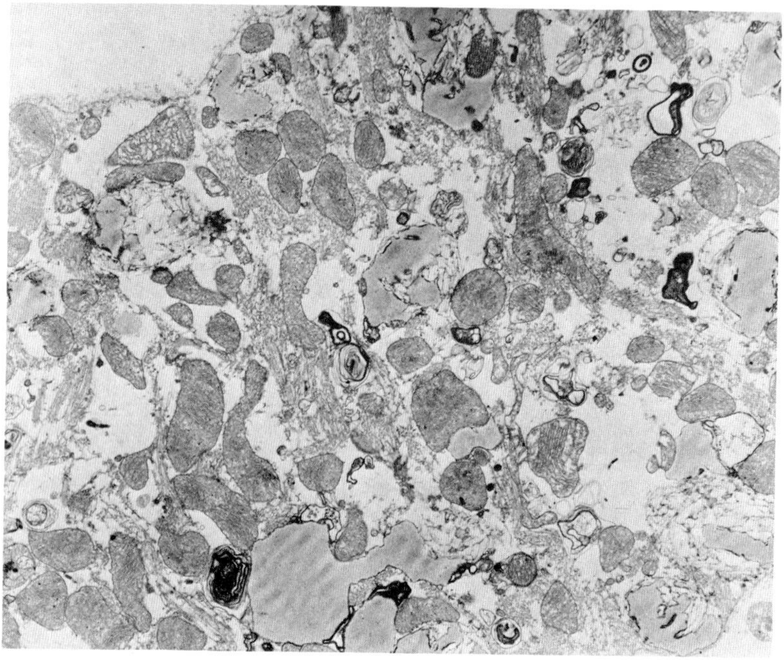

Fig. 13 Rat myocardium, 24 h after epinephrine (4 mg/kg body weight, s.c.). There is extensive intracellular disorganization. Only remnants of the disrupted myofilaments are still in evidence. This loss of myofibrils has resulted in the formation of empty spaces containing clusters of translocated mitochondria of varying shape and size with their cristae still largely intact. Some dilated components of the sarcoplasmic reticulum are still present. The fat globules apparently grow through confluence (see lower border of photograph). Numerous myelin figures of varying shape and electron density can be seen.

cells subjected to ischemic injury because mitochondria contain a very complex, highly organized multienzyme system. Among the many possible causes of deficient function are spatial disorganization or inactivation of mitochondrial enzymes, loss of essential cofactors, and maldistribution of essential anions or cations within the mitochondrion.

Since temporary oxygen deficiency may cause reversible injury of myocardial cells, or slow death, it is conceivable that the cell membrane may retain its semipermeable qualities much longer than in irreversible coronary occlusion. Thus, at a time when energy-dependent active transport may be seriously disturbed (*e.g.*, Na pump), equilbration of ions dependent upon passive diffusion only may continue unhampered and actually contribute to more rapid distortions of the myocardial electrolyte content.

It is of considerable interest that the myocardial Ca content showed highly significant increases 24 h after the injection of epinephrine and phenylephrine, whereas this phenomenon did not occur with the low dose of isoproterenol and

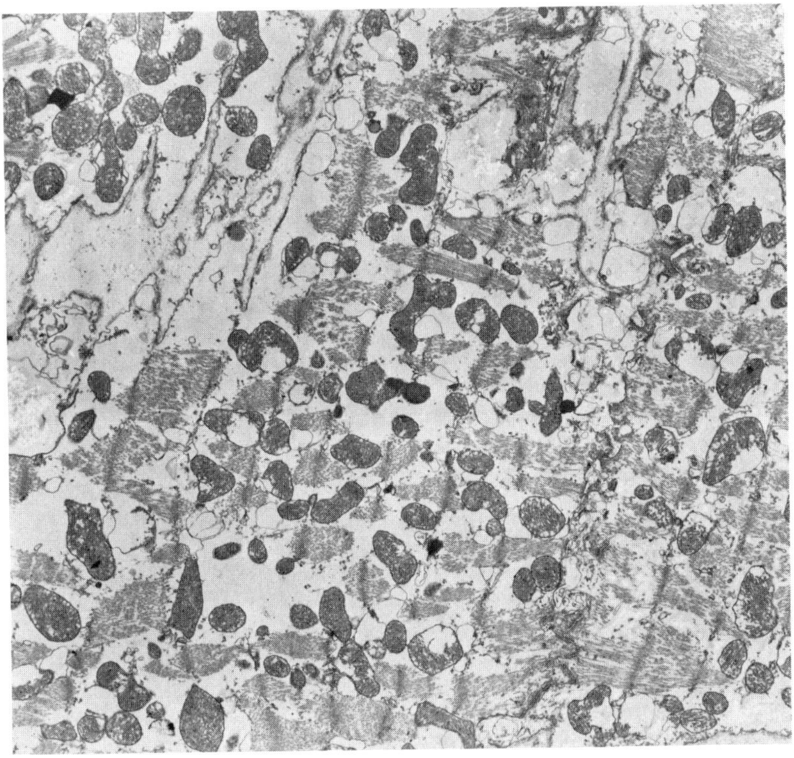

Fig. 14 Rat myocardium, 24 h after isoproterenol (5.25 mg/kg body weight, s.c.). The picture is in many ways similar to the preceding one showing epinephrine-induced injury (fig. 13). As in figure 13, disruption and disappearance of myofibrils with the formation of empty spaces, the dilatation of the sarcotubular system, and the variation in size and shape of mitochondria can be seen. However, many of these mitochondria show extensive cristae-lysis. In striking contrast to the epinephrine-induced injury, no fat globules and no myelin figures are in evidence despite the severity of the intracellular disorganization.

was less pronounced even with the high dose, although the latter induced massive cardiac injury. It is our present view that the transient substantial accumulation of Ca in the myocardium, which regularly vanished within 4 days, is related primarily to the effects of the alpha component of adrenergic activity, since we found earlier that it is also conspicuous with metaraminol and that it can be prevented by pretreatment with the alpha-adrenergic blocking agents tolazoline and dibozane, but not with the beta-adrenergic antagonist propranolol (15).

In conclusion, it can be said that the determination of changes in the myocardial electrolyte content provides an objective and fairly accurate estimate of the extent of cardiac fibrosis following the production of standardized myocardial injury. It is thus expected to aid in the assessment of noxious and

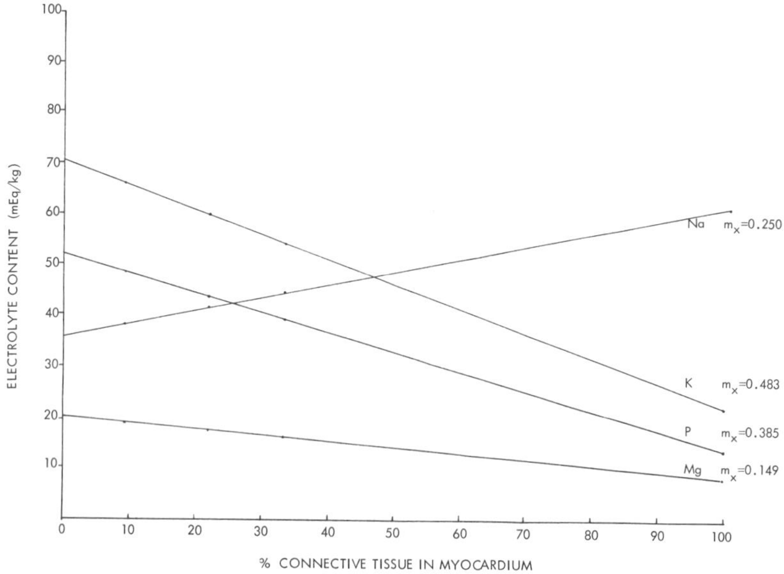

Fig. 15 Changes of electrolyte content of normal myocardium by admixture of connective tissue in increasing concentrations. (mx = slope of the straight lines.)

beneficial influences upon myocardial repair by many factors, including electrolytes and drugs, diet, sex, age, and physical activity. There is little doubt that there is a great need for additional information in this vital area.

SUMMARY

A chemical method for the objective and fairly accurate assessment of the degree of experimental myocardial fibrosis is described. The method is based on significant differences in electrolyte compositon of healthy myocardium and connective tissue, thus permitting estimation of the extent of myocardial fibrosis on the basis of the degree of deviation of the myocardial electrolytes from the norm. The applicability of the method was demonstrated in studies of standardized myocardial fibrosis induced by three adrenergic amines in which the extent of scarring observed with the naked eye and under the light microscope showed good correlation with the expected degree of ionic shifts of the fibrotic myocardium.

REFERENCES

1. BAJUSZ, E.: Experimental pathology and histochemistry of heart muscle. In: E. BAJUSZ AND G. JASMIN, Methods and achievements in experimental pathology, Vol. 2, pp. 172–223 (Karger, Basel New York 1967).
2. FERRANS, V. J.; HIBBS, R. G.; BLACK, W. C. and WEILBAECHER, D. G.: Isoproterenol induced myocardial necrosis. A histochemical and electron microscopic study. Amer. Heart J. *68:* 71–90 (1964).
3. FERRANS, V. J.; HIBBS, R. G.; WALSH, J. J. and BURCH, G. E.: Histochemical and electron microscopical studies on the cardiac necrosis produced by sympathomemetic agents. Ann. N.Y. Acad. Sci. *156:* 309–332 (1969).

4. HEGGTVEIT, H. A.: The cardiomyopathy of Mg-deficiency. In: BAJUSZ, E., Electrolytes and Cardiovascular Diseases, Vol. 1, pp. 204–220 (Karger, Basel New York 1965).
5. HEGGTVEIT, H. A.; HERMAN, L. and MISHRA, R. K.: Cardiac necrosis and calcification in experimental magnesium deficiency. A light and electron microscopic study. Amer. J. Path. *45:* 757–782 (1964).
6. HERDSON, P. B.; SOMMERS, H. B. and JENNINGS, R. B.: A comparative study of the fine structure of normal and ischemic dog myocardium with special reference to early changes following temporary occlusion of a coronary artery. Amer. J. Path. *46:* 367–386 (1965).
7. HORT, W. und DACANALIS, S.: Untersuchungen an Kattenherzen nach temporärer Unterbindung der linken Kranzarterie. In: Virchows Archiv für Pathologische Anatome und Physiologie und für Klinische Medizin. *339:* 61–71, ed. H. Hamperl and E. Venlinger (1965).
8. JASMIN, G.: Factors influencing the production of cardiomyopathies by methoxamine and metaraminol. Ann. N.Y. Acad. Sci. *156:* 333–343 (1969).
9. JENNINGS, R. B.: SOMMERS, H. M.; HERDSON, P. B. and KALTENBACH, J. P.: Ischemic injury of myocardium. Ann. N.Y. Acad. Sci. *156:* 61–78 (1969).
10. JOSEPH, N. R.; ENGEL, M. B. and CATCHPOLE, H. R.: Homeostasis of connective tissue. II. Potassium-sodium equilibrium. Amer. Med. Assoc. Arch. Path. *58:* 40–58 (1954).
11. KAHN, D. S.; RONA, G. and CHAPPEL, C. I.: Isoproterenol-induced cardiac necrosis. Ann. N.Y. Acad. Sci. *156:* 285–293 (1969).
12. KASNER, H. T. and DWYER, J. E.: Studies in infarction. IV. Experimental bland infarction of the myocardium, myocardial regeneration and cicatrization. J. Med. Res. *34:* 21–39 (1916).
13. KAUFMAN, N.; GAVAN, T. L. and HILLS, R. W.: Experimental myocardial infarction in the rat. Arch. Path. (Chicago) *67:* 482–488 (1959).
14. KORB, G. and TOTOVIC, V.: Electron microscopical studies on experimental ischemic lesions of the heart. Ann. N.Y. Acad. Sci. *156:* 48–60 (1969).
15. LEHR, D.: Tissue electrolyte alteration in disseminated myocardial necrosis. Ann. N.Y. Acad. Sci. *156:* 344–378 (1969).
16. LEHR, D.; CHAU, R. and KAPLAN, J.: Prevention of experimental myocardial necrosis by electrolyte solutions. This Book, pp. 684–698.
17. LEHR, D.; KRUKOWSKI, M. and CHAU, R.: Acute myocardial injury produced by sympathomimetic amines. Israel J. Med. Sci. *5:* 519–524 (1969).
18. LEHR, D.; KRUKOWSKI, M. and COLON, R.: Correlation of myocardial and renal necrosis with tissue electrolyte changes. J. Amer. Med. Assoc. *197:* 105–112 (1966).
19. POCHE, R.: Metrastructure of heart muscle under pathological conditions. Ann. N.Y. Acad. Sci. *156:* 34–47 (1969).
20. RAAB, W.: Hormonal and neurogenic cardiovascular disorders. (Williams and Wilkins Co., Baltimore, Md. 1953).
21. RAAB, W.: Myocardial electrolyte derangement: Crucial feature of pluricausal, so-called coronary, heart disease. Ann. N.Y. Acad. Sci. *147:* 627–686 (1969).
22. RONA, G.; CHAPPEL, C. I.; BALAZS, T. and GAUDRY, R.: An infarct-like myocardial lesion and other toxic manifestations produced by isoproterenol in the rat. Arch. Path. *67:* 443–455 (1959).
23. RONA, G. and KAHN, D. S.: Experimental studies on the healing of cardiac necrosis. Ann. N.Y. Acad. Sci. *156:* 177–188 (1969).
24. RONA G.; KAHN, D. S. and CHAPPEL, C. I.: Study on the healing of cardiac necrosis in the rat. Amer. J. Path. *39:* 473–485 (1961).

25. RONA, G.; KAHN, D. S. and CHAPPEL, C. I.: Studies on infarct-like myocardial necrosis produced by isoproterenol. A review. Rev. Canad. Biol. *22:* 241–255 (1963).
26. ROSENBLUM, I.; WOHL, A. and STEIN, A. A.: Studies in cardiac necrosis. I. Production of cardiac lesions with sympathomimetic amines. Toxic. Appl. Pharmacol. *7:* 1–8 (1965).
27. SELYE, H.: The pluricausal cardiopathies, pp. 438 (Charles C Thomas, Springfield, Illinois 1961).
28. WENZEL, D. G.: Drug-induced cardiomyopathies. J. Pharmacol. Sci. *56:* 1209–1224 (1967).
29. WENZEL, D. G. and CHAU, R. Y. P.: Effect of adrenergic blocking agents on reduction of myocardial aspartate aminotransferase activity by sympathomimetics. Toxic. Appl. Pharmacol. *9:* 514 (1966).

Mailing address: David Lehr, M.D., Department of Pharmacology, New York Medical College, Flower and Fifth Avenue Hospitals, Fifth Avenue at 106th Street, New York, N.Y. 10029 (USA).

Institut de Médecine et de Chirurgie expérimentales,
Université de Montréal, Montreal, Canada

ULTRASTRUCTURAL STUDIES OF VARIOUS FORMS OF EXPERIMENTALLY INDUCED CARDIAC NECROSES [1]

K. KOVÁCS, C. GARDELL, AND J.A. BLASCHECK

There are several procedures whereby disseminated focal myocardial necrosis can be elicited in animals. These experimental cardiopathies in many respects simulate various pathologic processes in the human heart. By studying the morphologic and biochemical aspects of the lesions during various phases of their development and the factors that influence their severity, useful information can be obtained on the basic mechanisms underlying their pathogenesis.

Three experimental cardiopathies were selected for electron microscopic studies in order to analyze similarities and differences from the early to the advanced stages of their development: 1) the "Electrolyte-Steroid Cardiopathy with Necrosis" (ESCN), which can easily be elicited by the concurrent administration of 9a-fluorocortisol, disodium phosphate, and corn oil; 2) the digitoxin cardiopathy, which develops in rats treated conjointly with digitoxin, disodium phosphate, and corn oil; and 3) the uremic cardiopathy, which is induced by bilateral nephrectomy.

MATERIALS AND METHODS

Female Sprague-Dawley rats were used throughout these studies. In the first experiment, the rats were given 9a-fluorocortisol acetate (750 μg microcrystals in 0.2 ml water), subcutaneously, once daily; disodium phosphate (1 mM in 1 ml water), orally through a soft rubber catheter, twice daily; and corn oil (1 ml), orally twice daily. In the second experiment, the rats received digitoxin (0.4 mg in 1 ml water); disodium phosphate (1 mM in 1 ml water); and corn oil (1 ml); all three substances were given orally, twice daily. In the third experiment, the rats were bilaterally nephrectomized. The animals were sacrificed (in groups of 3), without anesthesia, by mechanical destruction of the medulla oblongata according to the following schedule: 1, 2, 3, and 4 days after the commencement of treatment in the first experiment; 1, 2, and 3 days in the second experiment; and 1, 2, and 3 days after nephrectomy in the third experiment. In each series, 3–6 untreated rats were used as controls.

[1] Supported by grants from le Ministère de la Santé, Québec, the Medical Research Council of Canada (Block Term Grant MT-1829), and the Quebec Heart Foundation, Montreal.

Material for electron microscopy was obtained from the apical part of the still-beating heart; small pieces were placed in a 2.5% glutaraldehyde solution, postfixed in osmium tetroxide, dehydrated, embedded in Epon resin, and examined under a Carl Zeiss EM9 A electron microscope. Some of the findings were described previously (8, 9, 21).

RESULTS

Conjoint treatment with 9α-fluorocortisol, disodium phosphate, and corn oil

Electron microscopic examination revealed marked abnormalities 1 to 2 days after the commencement of treatment. However, the lesions were focal and considerable variations were seen in different fields of the same myocardium. One of the earliest changes was the separation and fragmentation of the myofibrils, resulting in the formation of intermyofibrillary spaces that later increased in size and number. In the early stage, contraction bands were observed, consisting of markedly compressed sarcomeres. Compared with the controls, the Z-bands were broad, irregular and more electron-dense at first, and broken and discontinuous later. Three to four days after commencement of treatment, the myofibrils in the myocytes undergoing necrosis were replaced by structureless homogeneous masses embedded in edema fluid and intermingled with loose fibrin masses (fig. 1).

The mitochondria appeared to be slightly swollen, with the cristae dilated and often disrupted. The density of the matrix decreased progressively. The double limiting membranes remained well preserved, but at the sites of more advanced lesions they ruptured, transforming the mitochondria into a homogeneous mass. Round electron-dense bodies, small at first but becoming larger with the progression of the lesions, were located on or between the cristae. They had coalesced in some places and occupied a considerable part of the mitochondrial structure. In the later phases, large clusters of mitochondria were seen in a state of total disorganization. The mitochondrial changes were more pronounced where cellular damage was intense; but abnormalities were also observed in areas that appeared, under the electron microscope, to be otherwise well preserved. The severity of mitochondrial and cellular lesions was not always related and, in some cases, normal mitochondria were recognized in cells showing signs of marked injury. In cells undergoing necrosis, the quantity of glycogen granules was strikingly reduced. There was a considerable accumulation of lipid droplets among the mitochondria. The nuclei did not show marked abnormalities and remained intact even in those cells where degenerative alterations were already evident. The coronary vessels and the capillaries of the myocardium seemed to be intact. Thrombus formation and vascular occlusions were not seen. The endothelial cells were occasionally swollen. Edema fluid and actively phagocytic

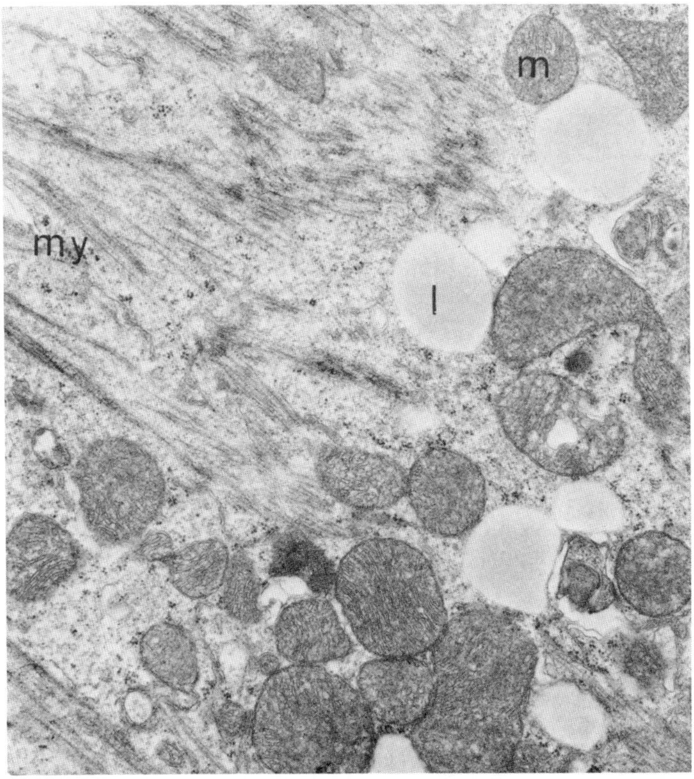

Fig. 1 Apical myocardium of a rat given 9a-fluorocortisol, Na_2HPO_4, and corn oil, 4 days after commencement of the treatment. The myofibrils (my) are fragmented and embedded in edema fluid. Many myofibrils have disappeared. The mitochondria (m) seem to be intact, a few showing swelling and vacuoles. There is an accumulation of lipid droplets (l). × 16,200.

mononuclear cells accumulated in the interstitium between the injured myocytes.

Conjoint treatment with digitoxin, disodium phosphate and corn oil

Under the electron microscope, changes were already seen 1 day after the commencement of treatment. Sarcolemmic ruptures appeared quite frequently. The intercalated disks were stretched and contraction bands were present. The mitochondria appeared to be swollen and showed partial homogenization of their internal structures. Vacuolization and disruption of the cristae followed, often accompanied by electron-dense deposits in the vacuoles. The sarcoplasmic reticulum was dilated. There was marked edema of the muscle fibers, with obliteration of the normal banded pattern, the Z-bands being the only ones remaining and even these were broader and denser than normal and often segmented and disrupted. In the more advanced lesions, the myofilaments had disintegrated completely and become a homogeneous mass containing damaged mitochondria with large electron-dense inclusions (fig. 2). Nuclear changes in the early phases were not conspicuous and the amount of glycogen granules did not seem to be significantly decreased. Striking changes occurred in the capillary endothelium. The intercellular terminal bars had opened and several cells were markedly swollen and light in density. Pinocytotic vesicles were very numerous and seemed to contain a light amorphous substance. The perivascular spaces were enlarged and filled by a proteinaceous material. Occasionally, the capillary endothelium was infiltrated with lipid droplets. In other places, the nuclei of the endothelial cells protruded into the lumen of the vessels, thus almost completely obstructing them. Phagocytic macrophages, other mononuclear cells and myocytic debris, polymorphonuclear leucocytes, and extravasated erythrocytes were frequently seen in the dilated interstitial spaces.

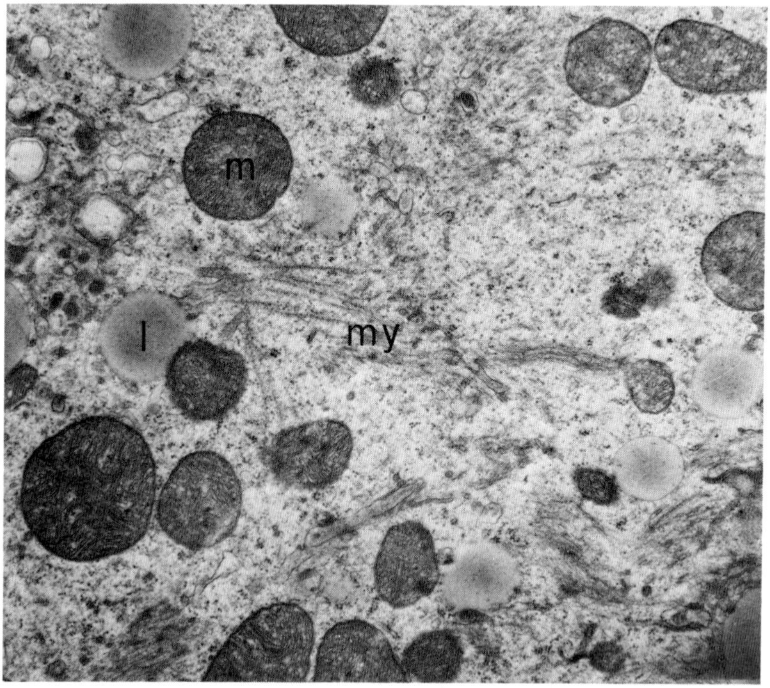

Fig. 2 Apical myocardium of a rat given digitoxin, Na_2HPO_4, and corn oil, 3 days after commencement of the treatment. The changes are similar to those shown in figure 1. The myofibrils (my) are fragmented and embedded in edema fluid. The mitochondria (m) are swollen. Lipid droplets (l) are also seen. × 14,000.

Bilateral nephrectomy

Electron microscopy showed that changes were already present 1 day after the removal of both kidneys. A slight dilation of the sarcoplasmic reticulum and swelling and homogenization of the mitochondria were observed, with a decrease in some places in the number of glycogen granules. The interstitial spaces were broader and contained edema fluid. The capillaries showed some early changes, i.e., a slight edema of the perivascular space, while the endothelium exhibited evaginations of its cytoplasm. Pinocytotic vesicles increased in number and reached a large size. Sometimes, the intercellular junctions of the endothelium were open, and coagulated proteinaceous material was apparent in the so-called transit zone. Two days after bilateral nephrectomy, necrotic foci were recognizable. In their periphery, the myocytes displayed rather widespread contraction bands. An accumulation of lipid droplets was also evident. The sarcoplasmic reticulum and the tubular system were somewhat dilated and the swelling and homogenization of the mitochondria were more extensive. The mitochondria became vacuolated, with extensive deposition of electron-dense bodies in, or on, damaged cristae. The vascular changes progressed and numerous inflammatory cells, principally polymorphonuclear neutrophil leucocytes, were present in the interstitial spaces. Three days after removal of both kidneys, the lesions became more extensive and severe (fig. 3). Cellular infiltration was more widespread and numerous leucocytes were seen. Phagocytic macrophages were also observed, with ingested debris in their cytoplasms. In the mitocondria, the deposition of electron-dense bodies continued. The areas of advanced necrosis included unrecognizable debris, forming a more or less homogeneous mass intermingled with extensive deposits of fibrin easily identified by their specific periodicity. The walls of the damaged capillaries often contained electron-dense bodies, and the mitochondria of the endothelial cells were vacuolated and dense.

DISCUSSION

Our investigations revealed that the fine structural changes in the myocardial cells were indistinguishable from one another at the sites of advanced necrosis in the three cardiopathies studied. This was not surprising, since cellular death is necessarily associated with a loss of the characteristic signs that could have thrown some light on the sequence of events leading to complete destruction of the myocytes. Thus, in this phase, only nonspecific damage was expected, regardless of any differences in the causative factors. However, in the early prenecrotic period, the lesions differed from one another, suggesting a dissimilarity in the processes leading to necrosis.

On the basis of electron microscopic studies, it is difficult to determine which of the various fine structural changes is the most important factor in the

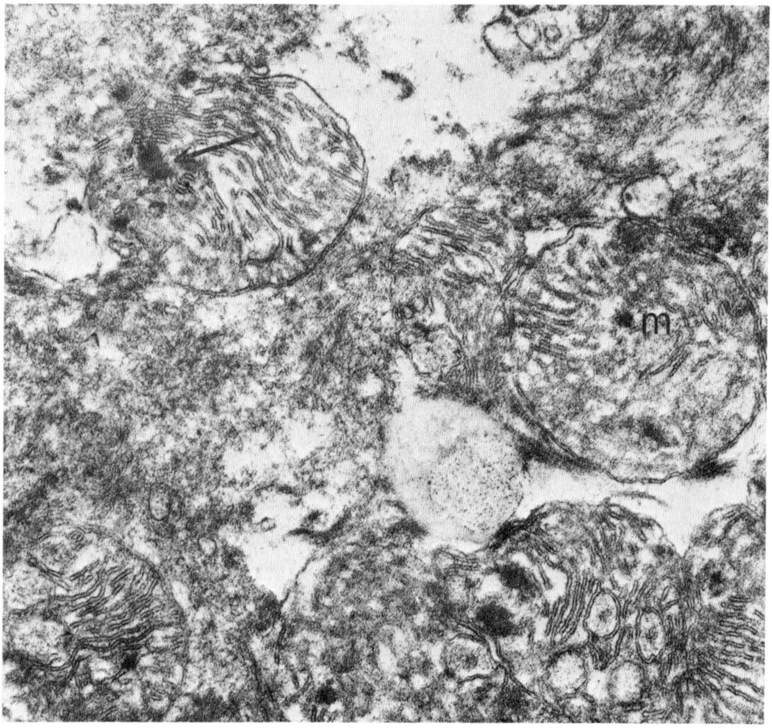

Fig. 3 Apical myocardium of a rat, 3 days after bilateral nephrectomy. The mitochondria (m) are swollen and vacuolated. The mitochondrial cristae are broken. Electron-dense bodies (arrow) are seen. The myofibrils have disappeared. ×46,200.

induction of the lesion and it is rarely possible to distinguish satisfactorily between primary cause and secondary result. Structural alterations are rather late manifestations of a series of characteristic metabolic processes and several biochemical changes are not readily detectable under the electron microscope. Interpretation of electron microscopic findings is somewhat similar to the assembling of a jigsaw puzzle, and a certain amount of creative imagination is therefore needed to trace dynamically the pathogenesis from a large number of electron microscopic pictures that represent only static portraits of the actual structural state of the cells. Despite these reservations, it is always a challenge to recreate what occurred in the cells, and to discover the factors responsible for cellular injury.

In rats treated concurrently with 9α-fluorocortisol, disodium phosphate, and corn oil, various abnormalities occurred simultaneously during the development of myocardial necrosis. We have failed to find convincing evidence to support the assumption that injury to the mitochondria is the primary event. For example, in cells undergoing necrosis, marked damage of the myofibrils was frequently observed while many mitochondria still remained intact. D'Agostino (7) suggested that the presence of electron-dense bodies in the mitochondria, representing calcium deposits and thus, in the functional sense, indicating uncoupling of oxidative phosphorylation, might be the primary factor in the production of phosphate-steroid myocardial necrosis. The present findings, however, showed that mitochondrial damage, including the appearance of electron-dense bodies, is only one of the events occurring in the course of cellular injury.

If, in any organ, a circumscribed area of the parenchyma undergoes necrosis, cellular death is due either to ischemia or to some direct toxic action primarily affecting cellular metabolism. The present electron microscopic studies indicate that ischemia cannot play a major role in the pathogenesis of the phosphate-steroid cardiopathy. Data reported in the literature show that, in cardiac ischemia induced by ligature of the coronary arteries, the most striking changes were a sudden decrease in the amount of glycogen granules, marked margination of nuclear chromatin, and mitochondrial abnormalities (5, 15, 17, 18). In our experiments, however, these lesions were prominent only in the later phases, when necrosis was well advanced. The role of ischemia in eliciting myocardial necrosis can also be excluded on the basis that we could not observe any alterations in the coronary vessels or obstruction of the vascular lumina. Nevertheless, the presence of marked intracellular edema and simultaneous myofibrillary damage seems to indicate the existence of increased membrane permeability, resulting in or arising from an intracellular electrolyte imbalance i.e., a change in intracellular osmolarity. While the importance of the mitochondrial damage in the process of necrosis cannot be neglected, it should be emphasized that similar mitochondrial abnormalities, including the appearance of electron-dense bodies, are also seen in various other pathologic conditions that affect the myocardium (6, 11, 14–17, 23, 31). It seems reasonable, therefore, to assume that the mitochondrial damage observed in our experiments is the nonspecific ultrastructural equivalent of severe cellular injury unrelated to any one particular causative factor such as phosphate-steroid administration or potassium deficiency.

As far as the pathogenesis is concerned, it has been assumed that potassium depletion plays an important role in the induction of phosphate-steroid myocardial necrosis (24, 28). This view is supported by the observation that both potassium chloride and amiloride, a potassium-sparing diuretic agent, exert a preventive action (28, 29). However, similar myocardial lesions occur in rats maintained on a magnesium-deficient diet, which indicates that the role of potassium is not exclusive, particularly as magnesium chloride also inhibits the myocardial changes caused by corticoids and phosphates (28). Potassium depletion would therefore appear to be only one of the factors in the complex pathogenesis of myocardial necrosis.

It has long been known that overdosage of cardiac glycosides can lead to the development of structural changes in the myocardium of man as well as of laboratory animals (3, 28). Recently, Selye and his associates (30) observed that conjoint administration of digitoxin, disodium phosphate, and corn oil results in focal myocardial necrosis in rats. Our electron microscopic findings confirmed their results. The vascular changes are very marked from the start and they provide a perfect picture of the hypoxic endothelial lesions as reported by several authors (4, 12, 20, 26). It seems reasonable to assume, therefore, that myocardial circulation is disturbed in the digitoxin-treated rats and the resulting hypoxia plays an important role in the genesis of myocardial necrosis. Digitoxin given in toxic doses may also directly damage the capillary endothelium. The endothelial swelling and the opening of the intercellular gaps brought about by these primary actions could then lead to a further degree of hypoxia. Indeed, there is edema of the perivascular space of the transit zone, as described by Hauss *et al.* (13), which further decreases the necessary metabolic exchanges between the blood capillaries and the muscle cells.

Another factor worthy of consideration is the inhibition of membrane ATP-ase by digitoxin overdosage, leading to a loss of potassium and a simultaneous accumulation of sodium in the myocardial cells (2). We do not assume that potassium depletion *per se* would be primarily responsible for the development of necrosis, but the possibility exists that intracellular ionic imbalance may sensitize the cardiac cells to other noxious influences acting simultaneously. The injurious effect of hypoxia might be aggravated by the disturbance of intracellular electrolyte equilibrium, i.e., intracellular depletion of potassium.

In agreement with earlier observations (22, 28), the present findings clearly showed that removal of both kidneys was accompanied in the rat by disseminated focal necrosis as well as calcification of the myocardium. According to Selye (28), this disease model belongs to the group of calcifying cardiopathies and can easily be differentiated from the necrotizing and hyalinizing cardiopathies because, in the former, not only the myocytes but also the blood vessels of the heart are involved in the calcification.

It is known that, following bilateral nephrectomy, parathyroid hormone accumulates in the blood and tissues (1), largely through increased release due to a marked imbalance of electrolyte metabolism associated with the cessation of renal excretory function (19, 25, 32). It was also shown that the severity and extent of the myocardial lesions following removal of both kidneys is greatly reduced by previous parathyroidectomy (22, 28). These findings clearly indicate that parathyroid hormone plays a determining role in the genesis of the uremic cardiopathy. The induction, with parathyroid extracts, of cardiac lesions structurally similar to those occurring after bilateral nephrectomy, further emphasized the causative role of parathyroid hormone in the pathogenesis of the uremic cardiopathy (28).

It is generally accepted that parathyroid hormone excess exerts a direct toxic action on cells. By increasing the permeability of cellular and subcellular membranes, the hormone facilitates phosphate transfer from the extracellular fluid, thus accelerating mitochondrial phosphate uptake (27). We assume that phosphate and calcium transport across cellular and subcellular membranes is also increased in the myocardium of renoprival rats, since massive mineral deposition occurs in the myofibers following removal of both kidneys. Nevertheless, the question remains whether or not the deposition of calcium salts is the primary cause of myocardial necrosis. In rats suffering from sulfonamide-induced obstructive nephropathy, Lehr (22) found that the areas of cardiac necrosis were those that later became calcified, thus suggesting that there is always a primary injury of cellular metabolism followed by a secondary "dystrophic" calcification. On the other hand, in their electron microscopic study of the myocardium of dihydrotachysterol-treated rats, Gerlach and Themann (10) found that the mitochondria calcify at a very early stage, and assumed that this process had some etiologic importance in eliciting necrosis of the myocytes. On the basis of our investigations, it would appear that, in the case of the uremic cardiopathy, cellular death cannot be attributed solely to the intramitochondrial deposition of calcium salts since, in the hearts of renoprival rats, we found severely damaged myocytes that showed no signs of mineralization even under the electron microscope. Therefore, it seems reasonable to suppose that some primary biochemical changes not readily detectable with the electron microscope alter cellular function and, probably, membrane permeability. Calcium deposition follows these changes and contributes to the deterioration of intracellular metabolism, thus accelerating the progression of the lesions until cellular death ultimately ensues.

While emphasizing the primary importance of parathyroid hormone excess in the pathogenesis of the uremic cardiopathy, consideration should also be given to the participation of extraparathyroidal mechanisms.

Previous observations and our present findings suggest that vascular changes may also play a role in myocardial injury following removal of both kidneys.It was convincingly demonstrated that the coronary vessels are severely damaged in renoprival rats (22, 28); the necrosis with calcification that occurs in the muscle fibers of the larger vessels presumably impairs coronary circulation, resulting in hypoxia of the myocytes. Circulatory disturbances due to vascular injury may also afford an explanation for the peculiar localization of the myocardial lesions in the perivascular areas. It could be supposed that there is an escape of parathyroid hormone and phosphate in large amounts through the damaged vascular walls and that those myocytes that are more intensely exposed to the deleterious action of these compounds will die and calcify. Perhaps this is why only isolated myofibers scattered in the vicinity of the damaged vessels undergo necrosis and calcification. Myocardial hypoxia may also sensitize the myocytes to the actions of parathyroid hormone. However, further experiments are needed to verify these suppositions. Factors other than the action of parathyroid hormones and circulatory disturbances might also play a role in the pathogenesis of the uremic cardiopathy. It is well known that, following the removal of both kidneys, various toxic metabolites accumulate in the body fluids, resulting in a marked electrolyte imbalance. There is also a sudden rise of blood pressure (22). All these alterations might directly damage the myocytes or sensitize the heart to parathyroid hormone action or to hypoxia. The significance of these factors cannot be ignored, since the myocardium is not completely normal in renoprival rats previously subjected to parathyroidectomy.

SUMMARY

Administration of 9a-fluorocortisol, disodium phosphate, and corn oil, or of digitoxin, disodium phosphate, and corn oil, as well as bilateral nephrectomy, results in disseminated patches of necrosis in the myocardia of rats. The lesions were studied at various times under the electron microscope and the fine structural alterations observed are described in this work. It was found that similar changes occurred at the sites of advanced necrosis in all three cardiopathies; but in the early prenecrotic period, the lesions differed from one another, thus indicating that the sequence of events leading to cellular death is not the same in each instance.

ACKNOWLEDGMENTS

The authors wish to thank the Upjohn Co. for the fluorocortisol used in these experiments.

REFERENCES

1. BERSON, S. A. and YALOW, R. S.: Parathyroid hormone in plasma in adenomatous hyperparathyroidism uremia and bronchogenic carcinoma. Science *154:* 907–909 (1966).
2. BONTING, S. L.; CARARAGGIO, L. L. and HAWKINS, N. M.: Studies on sodium-potassium-activated adenosin triphosphatase. IV. Correlation with cation transport sensitive to cardiac glycosides. Arch. Biochem. *98:* 413–419 (1962).
3. BÜCHNER, F.: Herzmuskelnekrosen durch hohe Dosen von Digitalis-glykosiden. Naunyn-Schmiedeberg's Arch. exp. Pathol. Pharmak. *176:* 59–64 (1934).
4. BÜCHNER, F. und ONISHI, S.: Fruhstadien der akuten hypoxischen Veranderung des Herzmuskels im elektronenmikroskopischen Bild und ihre Bedeutung fur die akute hypoxische Herzinsuffizienz. Beitr. pathol. Anat. *135:* 153–182 (1967).
5. CAULFIELD, J. and KLIONSKY, B.: Myocardial ischemia and early infarction: an electron microscopic study. Amer. J. Path. *35:* 489–523 (1959).
6. D'AGOSTINO, A. N.: An electronmicroscopic study of skeletal and cardiac muscle of the rat poisoned by plasmocid. Lab. Invest. *12:* 1060–1071 (1963).
7. D'AGOSTINO, A. N.: An electron microscopic study of cardiac necrosis produced by 9α-fluorocortisol and sodium phosphate. Amer. J. Path. *45:* 633–644 (1964).
8. GARDELL, C.; BLASCHECK, J. A. and KOVACS, K.: Digitoxin cardiopathy in the rat. A light and electron-microscopic study. J cell. molec. cardiol. *1:* 175–182 (1970).
9. GARDELL, C.; HORVATH, E.; SOMOGYI, A.; BLASCHECK, J. A. and KOVACS, K.: Myocardial changes in rats following bilateral nephrectomy. A histochemical and electron microscopic study. Beitr. path. Anat. *140:* 166–181 (1970)

10. GERLACH, U. und THEMANN, H.: Elektronenmikroskopische Untersuchung der metastatischen Calcifizierung. Klin. Wchnschr. *43:* 1262–1268 (1965).

11. HALL, J. L. and SMITH, E. B.: Cobalt heart disease. An electron microscopic and histochemical study in the rabbit. A.M.A. Arch. Path. *86:* 403–412 (1968).

12. HAUSAMEN, T. V. und POCHE, R.: Die Ultrastruktur des Herzmuskels der Ratte nach einmaligen und wiederholten Unterdruckversuchen. Virch. Arch. path. Anat. *339:* 212–224 (1965).

13. HAUSS, W. H.; GERLACH, U.; JUNGE-HÜLSING, G.; THEMANN, H. and WIRTH, W.: Studies on the "nonspecific mesenchymal reaction" and the "transit zone" in myocardial lesions and atherosclerosis. Ann. N.Y. Acad. Sci. *156:* 207–218 (1969).

14. HEGGTVEIT, H. A.; HERMAN, L. and MISHRA, R. K.: Cardiac necrosis and calcification in experimental magnesium deficiency. A light and electron microscopic study. Amer. J. Path. *45:* 757–782 (1964).

15. HERDSON, P. B.; SOMMERS, H. M. and JENNINGS, R. B.: A comparative study of the fine structure of normal and ischemic dog myocardium with special reference to early changes following temporary occlusion of a coronary artery. Amer. J. Path. *46:* 367–386 (1965).

16. HIBBS, R. G.; FERRANS, V. J.; BLACK, W. C.; WEILBAECHER, D. G.; WALSH, J. J. and BURCH, G. E.: Alcoholic cardiomyopathy. An electron microscopic study. Amer. Heart J. *69:* 766–779 (1965).

17. JENNINGS, R. B.; BAUM, J. H. and HERDSON, P. B.: Fine structural changes in myocardial ischemic injury. A.M.A. Arch. Path. *79:* 135–143 (1965).

18. JENNINGS, R. B.; SOMMERS, H. M.; HERDSON, P. B. and KALTENBACH, J. P.: Ischemic injury of myocardium. Ann. N.Y. Acad. Sci. *156:* 61–78 (1969).

19. KATZ, A. I.; HAMPERS, C. L. and MERRILL, J. P.: Secondary hyperparathyroidism and renal osteodystrophy in chronic renal failure. Analysis of 195 patients with observations on the effect of chronic dialysis, kidney transplantation and subtotal parathyroidectomy. Medicine (Baltimore) *48:* 333–374 (1969).

20. KORB, G. and TOTOVIC, V.: Electron microscopic studies on experimental ischemic lesions of the heart. Ann. N.Y. Acad. Sci. *156:* 48–60 (1969).

21. KOVACS, K.; BLASCHECK, J. A. and GARDELL, C.: Electron microscopic study of myocardial changes produced in rats by the concurrent administration of 9α-fluorocortisol, Na_2HPO_4 and corn oil. Cardiologia (Basel) *54:* 16–29 (1969).

22. LEHR, D.: Causative relationships of parathyroid hormone to renogenic and renoprival cardiovascular disease. Ann. N.Y. Acad. Sci. *72:* 901–969 (1959).

23. MARUFFO, C. A.: Fine structural study of myocardial changes induced by isoproterenol in Rhesus monkeys (Macaca mulatta). Amer. J. Path. *50:* 27–37 (1967).

24. NICKERSON, M.; KARR, G. W. and DRESEL, P. E.: Pathogenesis of "electrolyte-steroid-cardiopathy". Circulat. Res. *9:* 209–217 (1961).

25. PARFITT, A. M.: Soft-tissue calcification in uremia. A.M.A. Arch intern. Med. *124:* 544–556 (1969).

26. POCHE, R.: Ultrastructure of heart muscle under pathological conditions. Ann. N.Y. Acad. Sci. *156:* 34–47 (1969).

27. RASMUSSEN, H.; SALLIS, J.; FANG, M.; DELUCA, H. F. and YOUNG, R.: Parathyroid hormone and anion uptake in isolated mitochondria. Endocrinology *74:* 388–394 (1964).

28. SELYE, H.: The Pluricausal Cardiopathies. Springfield, Ill.: Charles C Thomas Publ., pp. 438 (1961).

29. SELYE, H.: Prevention of cardiac necrosis by amiloride. J. Amer. Med. Ass. *206:* 103–104 (1968).

30. SELYE, H.; KRAJNY, M. and SAVOIE, L.: Digitoxin poisoning: prevention by spironolactone. Science *164:* 842–843 (1969).
31. SUSIN, M. and HERDSON, P. B.: Fine structural changes in rat myocardium induced by thyroxine and by magnesium deficiency. A.M.A. Arch. Path. *83:* 86–98 (1967).
32. WILLS, M. R.: Biochemical consequences of chronic renal failure: a review. J. Clin. Path. *21:* 541–554 (1968).

Mailing address: Dr. K. Kovács, St. Michael's Hospital, Department of Pathology, 30 Bond Street, Toronto 2, Ontario (Canada).

Department of Pharmacology, Faculty of Medicine,
University of Ottawa, Ottawa, Canada

ROLE OF THE AUTONOMIC NERVOUS SYSTEM IN DIGITALIS INTOXICATION [1]

H. L. GARVEY

Increasing evidence has suggested that a relationship exists between the autonomic nervous system and the cardiovascular actions of cardiac glycosides (5, 7, 9, 26). However, whether or not these agents act directly on the autonomic nervous system (3, 16) or exert their effects through a modification of the functions of either the central nervous system (2, 4, 12) or cardiovascular reflex mechanisms (15, 17, 24) is still controversial.

Most of the observations suggesting that the autonomic nervous system might be involved in the actions of cardiac glycosides are indirect (7, 9, 26). Since these cardiovascular observations are the resultant effects of several intrinsic cardiac and reflex variables (19), they are more meaningful when correlated with a study of impulse traffic in autonomic nervous pathways involved in cardiovascular regulation. Thus, this study was undertaken in order to correlate changes in postganglionic cardiac sympathetic and parasympathetic efferent nerve activity with cardiovascular responses during progressive ouabain intoxication.

METHODS

Cats of either sex, weighing 2.5–4.5 kg, were anesthetized with chloralose (90–100 mg/kg, i.p.), artificially respired, and maintained at a constant body temperature 37 ± 2°C. The chest was opened by a sternal splitting incision and the right stellate ganglion exposed. Cardiac sympathetic and parasympathetic postganglionic fibres were identified by observations of the resultant alterations in heart rate and force of cardiac contraction following brief periods of electrical stimulation. Twigs of these nerves were placed on bipolar, stainless-steel recording electrodes and the preparation immersed in mineral oil. The traffic of nerve impulses was amplified by using 2 Tektronics (model 122) preamplifiers and then displayed on 2 oscilloscopes to one of which was coupled a kymographic camera (Grass C4F). These impulses were also audibly reproduced by the simultaneous use of audio-amplifiers. To ensure that efferent nerve activity was being monitored, in some experiments the nerves were sectioned distal to the recording electrodes.

[1] Supported by the Canadian Heart Foundation.

Modifications of the activity of the cardiac autonomic nervous system were produced by the following experimental procedures:

a) Carotid sinus nerve stimulation. Both carotid sinus nerves were desheathed and placed on stimulating electrodes. Stimulation was effected by 10 sec trains of impulses (100 C.P.S., 2 msec duration, ½ mA) at 5 min intervals.

b) Carotid sinus denervation. Both sinuses and the reflexogenic areas of the aortic arch were denervated ("debuffered") and infiltrated with a local anesthetic (xylocaine 0.5%).

c) Interruption of nervous pathways. Postganglionic cardiac sympathetic and parasympathetic nerves were sectioned near the stellate ganglia to interrupt efferent nervous pathways. To produce "total cardiac denervation" both the stellate ganglia and sections of the vagosympathetic trunk were removed, followed by denervation of the carotid sinus and the aortic arch.

d) Carotid artery stenosis. The artery was ligated together with a 14-gauge hypodermic needle, followed by removal of the needle. This procedure resulted in a 15–25% reduction in the lumen of the artery. Stenosis was produced in both the intact and "debuffered" animal.

Concurrent changes in femoral arterial blood pressure and electrocardiograms (Lead II) were recorded. Ouabain was administered by intravenous infusion (2 μg/kg/min) until ventricular fibrillation ensued.

Statistical evaluation of the data is based on the methods of Steel and Torrie (25) and results were considered significant if they exceeded the 95% probability level.

RESULTS

Influence of ouabain administration

Both cardiac sympathetic and parasympathetic postganglionic nerve activity was influenced by ouabain administration, as is illustrated in figure 1. Control sympathetic nerve activity was of 2 types: a regular or synchronous spontaneous activity that could be correlated with either the systolic blood pressure or the QRS complex of the ECG, and irregular or asynchronous bursts of activity with a frequency of 5–20/sec. These asnychronous outbursts were of greater amplitude than the spontaneous synchronous activity. Parasympathetic nerve activity was similarly of two types, but these were of lesser amplitude and asynchronous firing was less frequent.

Ouabain administration induced an initial suppression of asynchronous activity in the sympathetic nerve within 2–5 min. Complete inhibition of both nerves then ensued (35–50 min). Prior to the onset of ouabain intoxication

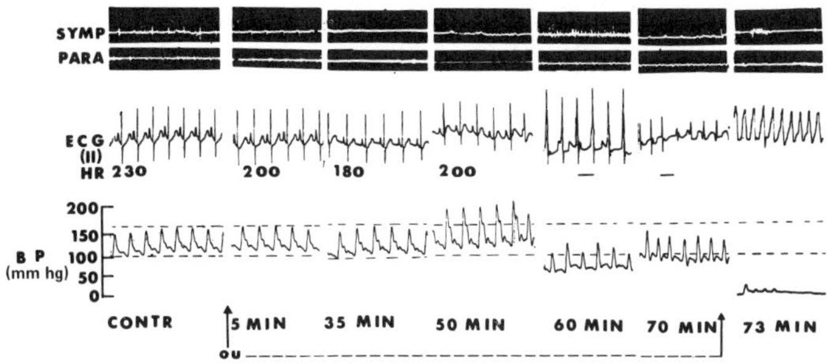

Fig. 1 Effect on cardiac sympathetic (SYMP) and parasympathetic (PARA) nerve activity, ECG, heart rate (HR) and blood pressure of progressive ouabain (ou) intoxication. Broken lines indicate the highest levels of end systolic and end diastolic control blood pressures.

(ectopic beats and A-V dissociation), increased asynchronous sympathetic nerve activity was observed. This was associated with a return of parasympathetic nerve firing. Sympathetic activity decreased as multifocal arrhythmias and terminal ventricular fibrillation ensued, but occasional bursts of activity could still be observed for periods of up to 5 min after all apparent cardiac electrical activity had disappeared.

As is seen in figure 1, no pronounced alterations in heart rate, blood pressure, or the ECG occurred until a large cumulative dose of ouabain (100 μg/kg at 50 min) was administered. The onset of ectopic beats was associated with irregular blood pressure changes. Terminal ventricular fibrillation ensued within 73 min.

Table I summarizes the results of progressive ouabain intoxication in these animals. The onset of toxic ECG changes (A-V dissociation and ectopic beats) was observed within 41.2 ± 5.0 min at a cumulative dose of 82.5 ± 10 μg/kg.

Table I

Cumulative Effects of Ouabain Administration in Control Cats

Exp. no.	Onset of toxic ECG effects (min)	Cumulative dose (μg/kg)	Onset of vent. fibrillation (min)	Cumulative dose (μg/kg)
1	60	120	70	140
2	33	66	55	112
3	35	70	47	94
4	40	80	73	146
5	60	120	70	140
6	30	60	45	90
7	31	62	51	102
Mean	41.2	82.5	58.8	117
± S.E.	5.0	10.0	4.5	9.0

Death by ventricular fibrillation occurred in all animals within 58.8 ± 4.5 min following a total cumulative dose of 117 ± 9.0 μg/kg.

Influence of carotid sinus nerve stimulation

In these experiments, the carotid sinus nerves were stimulated at 5 min intervals during ouabain administration, as illustrated in figure 2. Carotid sinus nerve stimulation (before ouabain administration) induced a suppression of parasympathetic nerve activity and also decreased firing in the sympathetic nerve. As observed in control experiments, ouabain administration induced an inhibition of spontaneous activity in both nerves. However, when the stimulus was repeated after ouabain infusion (5 min), an immediate outburst of asynchronous sympathetic and parasympathetic nerve activity ensued and persisted for several minutes. Within 10 min, sympathetic nerve activity decreased resulting in phasic bursts with corresponding parasympathetic inhibition. Stimulation after this period resulted in a return of increased sympathetic and parasympathetic nerve activity. The intense sympathetic outbursts preceded an early onset of ventricular arrhythmias. When ventricular fibrillation ensued, nerve activity decreased and was replaced by phasic firing of the sympathetic nerve with corresponding parasympathetic inhibition.

Associated changes in ECG, heart rate, and blood pressure were preceded by alterations in cardiac nerve activity. An enhanced depressor response to carotid sinus nerve stimulation that became progressively more intense as the cumulative dose of ouabain increased is also seen.

Table II summarizes the effects of progressive ouabain intoxication in these animals. Toxic ECG manifestations occurred within 14.0 ± 2.2 min (34% of the time observed in unstimulated control animals, table I). Ventricular fibrillation ensued within 22 ± 2.5 min (39.7% of the time observed in unstimulated control animals). Statistical comparisons of these values indicated that the mean time of onset of both toxic effects and ventricular fibrillation as well as the cumulative

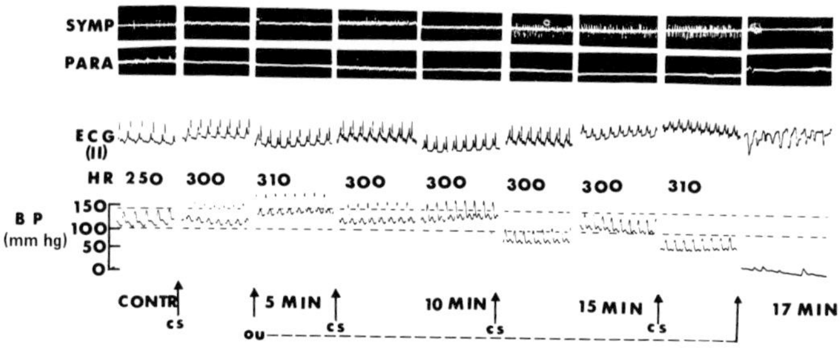

Fig. 2 Effect of carotid sinus nerve stimulation (CS) during progressive ouabain intoxication. Records were taken after cessation of the stimulus.

Table II

Cumulative Effects of Ouabain Administration in Cats Subjected to Carotid Sinus Nerve Stimulation

Exp. no.	Onset of toxic ECG effects (min)	Cumulative dose (μg/kg)	Onset of vent. fibrillation (min)	Cumulative dose (μg/kg)
1	12	24	20	40
2	12	24	14	28
3	15	30	27	54
4	14	28	21	42
5	17	34	28	56
Mean	14.0	28	22	44
± S.E.	2.2	2.0	2.5	1.2
% of value in Table I	34	33.7	39.7	37.6
$p<$ Table I *vs.* Table II	0.01	0.01	0.01	0.01

doses of ouabain necessary to induce these effects were all significantly reduced ($p < 0.01$) following repeated carotid sinus nerve stimulation.

A comparison of concurrent maximal changes in blood pressure and heart rate at 5 min intervals in both groups of animals is graphically summarized in figure 3. Control blood pressure and heart rate were both higher in animals with exposed carotid sinus nerves. Control carotid sinus nerve stimulation in these

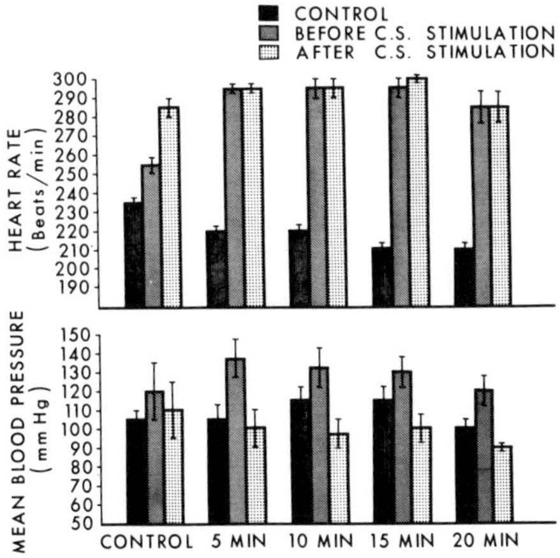

Fig. 3 Concurrent maximal changes in heart rate and blood pressure at 5 min intervals in control unstimulated animals (first column) and in animals with exposed carotid sinus nerves before (second column) and after (third column) nerve stimulation during ouabain administration.

animals (before ouabain administration) induced a slight decrease in blood pressure but a marked increase in heart rate (285 ± 5 compared with 255 ± 4 beats/min). Following ouabain administration for 5 min, a decrease in heart rate but no associated alterations in blood pressure were observed in the unstimulated control group of animals. In contrast, both blood pressure and heart rate increased during this period in the stimulated animals. Carotid sinus nerve stimulation at this time induced a dramatic fall in blood pressure but no concurrent changes in heart rate. This pattern of response to nerve stimulation was readily repeatable at 5 min intervals until terminal arrhythmias ensued in these animals.

Influence of carotid artery stenosis

Carotid artery stenosis induced marked alterations in the spontaneous activity of both sympathetic and parasympathetic postganglionic cardiac nerves, as illustrated in figure 4. The control pattern of activity in both nerves was succeeded by marked asynchronous sympathetic and parasympathetic outbursts following stenosis. Ouabain administration induced an inhibition of parasympathetic nerve activity with a concurrent reduction in sympathetic outbursts within 2 min. There was an associated increase in both systolic and diastolic blood pressure but no alterations in heart rate or ECG. Sympathetic activity increased progressively (4 min), while parasympathetic nerve activity remained inhibited. Pulse pressure also increased markedly. Sympathetic nerve activity became quite intense within 8 min, while parasympathetic outbursts also became pronounced. Arrhythmias preceded by intense persistent activity in both nerves ensued within 12 min and ceased abruptly as ventricular fibrillation ensued within 13 min.

In contrast, as is seen in figure 5, no alterations in either sympathetic or parasympathetic nerve activity were observed following either carotid artery stenosis or ouabain administration, when the animals were "debuffered" before

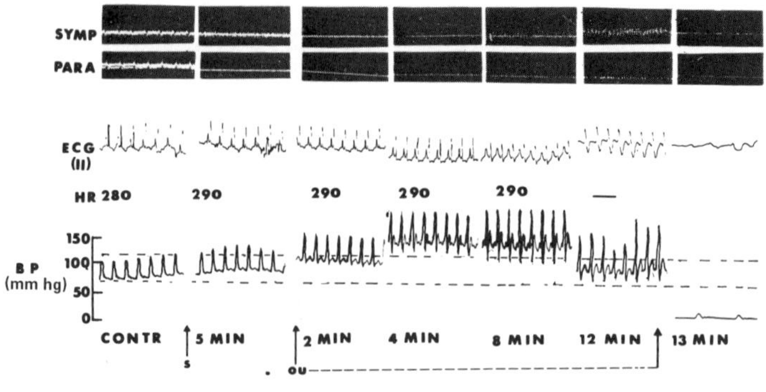

Fig. 4 Effects of carotid artery stenosis (S) upon progressive ouabain intoxication.

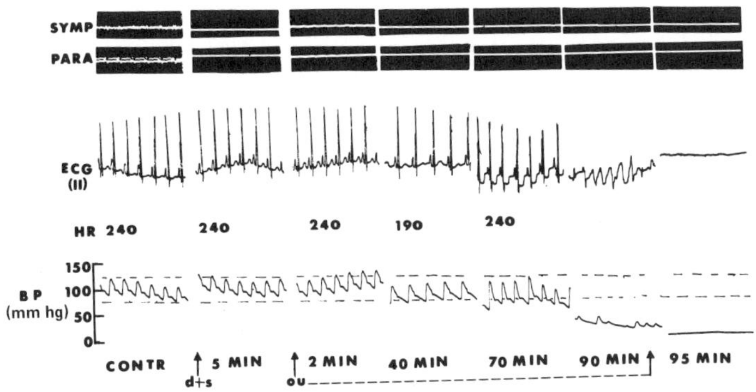

Fig. 5 Influence of denervation of the buffer areas (d) upon progressive ouabain intoxication, following carotid artery stenosis (s).

these procedures. The control level of activity in both nerves decreased after denervation of the buffer areas and was not influenced by ouabain administration. In addition, the marked pressor responses observed in animals with intact buffer nerves were absent in these "debuffered" animals. Terminal ventricular fibrillation occurred within 95 min (*cf.* 13 min in fig. 4).

Comparative changes in pulsatile blood pressure and heart rate in both intact and "debuffered" cats in response to ouabain administration are summarized in figure 6. In animals with intact nerves, carotid artery stenosis induced an increase in both systolic and diastolic pressures to a maximum of 210/130 from the control value of 107/75 mm Hg within 4 min. There were no associated changes in heart rate during this interval. Thereafter, blood pressure declined as arrhythmias ensued and ventricular fibrillation with a precipitous fall in blood pressure occurred within 13 min.

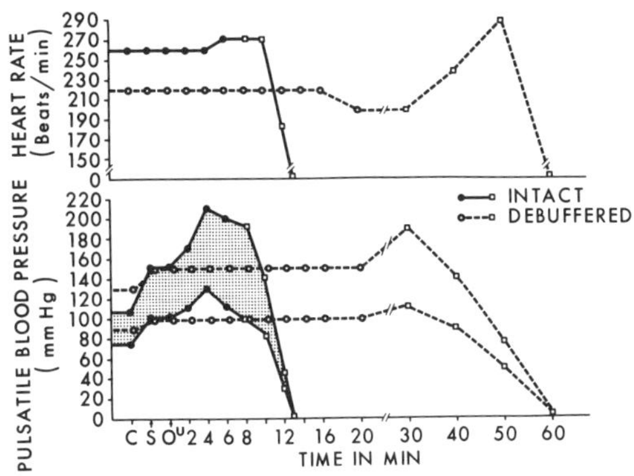

Fig. 6 Comparative effects of ouabain administration in "debuffered" animals and in animals with intact nervous system. □ denotes the onset of arrhythmias. The carotid arteries were stenosed in both groups of animals.

In "debuffered" animals, there was a slight increase in blood pressure following carotid artery stenosis. However, neither blood pressure nor heart rate was influenced by initial ouabain administration. The onset of intoxication was associated with variations in both heart rate and blood pressure. These changes were succeeded by a progressive fall in blood pressure as multifocal arrhythmias and ventricular fibrillation ensued.

Changes in blood pressure and heart rate induced by ouabain administration following section of the cardiac postganglionic nerves as well as "total cardiac denervation" were identical to those observed in "debuffered" animals.

DISCUSSION

The carotid sinus exerts a significant role in the regulation of vital cardiovascular functions. It is believed that afferent impulses initiated by changes in pulsatile pressure are relayed to the medullary vasomotor and cardiac centres. Efferent impulses from these areas will produce adjustments in the cardiovascular system (13, 18). In the present investigations, phasic sympathetic and parasympathetic cardiac postganglionic nerve activity was observed. This activity persisted following section of these cardiac nerves distal to the recording electrodes, but were reduced or disappeared following denervation of the buffer areas or cardiac denervation. It would appear that these were efferent impulses that exerted a tonic influence on the heart.

Ouabain administration caused an initial reduction or inhibition of cardiac nerve activity. However, when a toxic dose was administered, increased nerve activity that preceded the onset of ventricular arrhythmias ensued. Gillis (8), using decerebrated cats, concluded that the effects of ouabain upon cardiac sympathetic nerve activity was dependent on the dose of ouabain as well as upon the time after administration; inhibition was observed with low doses and increased activity with toxic doses. While no correlation between the patterns of neural activity and concurrent changes in blood pressure or heart rate could be demonstrated in the present investigation, it was consistently observed that the development of cardiac arrhythmias was preceded (within 3–5 min) by increased sympathetic nerve activity.

Periodic excitation of the carotid sinus nerves following ouabain administration induced increased and persistent spontaneous sympathetic and parasympathetic nerve activity. In addition, an enhanced depressor response to nerve stimulation as well as early ouabain intoxication at very low cumulative dose levels were observed in these animals. It is believed that cardiac glycosides sensitize the carotid sinus reflex mechanisms (14, 17). This has been of clinical importance, as carotid sinus stimulation or massage has been advocated as a diagnostic aid in determining the state of digitalization (1, 10). The increased neural activity in response to excitation noted in ouabain-pretreated animals may then constitute the cardiac autonomic expression of increased carotid sinus sensitivity and activity. It would also appear that, in the presence of increased cardiac autonomic nerve activity, fatal cardiac arrhythmias can be precipitated by low doses of ouabain. This observation is supported by clinical reports of fatal cardiac arrhythmias following carotid sinus massage in digitalized patients (10, 21, 22).

Greenwood and Dupler (10) suggested that carotid sinus stimulation or massage should be employed with caution in the presence of such circulatory impairments as atherosclerosis of the carotid arteries and cerebrovascular diseases. Experiments conducted on animals with stenosed carotid arteries but intact nervous system suggested that cardiac autonomic neural activity is increased by procedures that interfere with the normal function of the carotid sinus. It is also evident that the cardiotoxic action of ouabain is increased by such procedures. Section of the buffer nerves as well as "denervation" reduced or abolished the ouabain-induced changes in neural activity and also significantly increased the cumulative dose of the drug that induced ventricular fibrillation. It would therefore appear that increased baroreceptor activity is reflected as an increase in vagal activity in areas supplied by postganglionic cardiovagal nerve endings, *i.e.* -the sinus node, atria, and A-V node. The concurrently increased cardiac sympathetic activity observed might then contribute to the early onset of arrhythmias, in view of the widespread distribution of sympathetic nerve fibres to the ventricles (23) and the postulated permissive role of catecholamines in the actions of cardiac glycosides (6, 11, 20, 27).

Finally, it was observed that the inotropic action of ouabain was evident shortly after administration of the drug, when cardiac neural activity was increased. Moreover, this inotropic action was also significantly greater than in control, "debuffered", or "denervated" animals. The positive inotropic action of cardiac glycosides is readily demonstrable in the decompensated heart. This effect is not readily observed in the normal heart and circulation, in view of the glycoside-induced cardiovascular adjustments that are known to occur; consequently, several investigators have been led to postulate that cardiac glycosides do not act on the normal cardiovascular system. Whether or not increased cardiac sympathetic and parasympathetic efferent activity occurs in the presence of cardiac decompensation and how the activity of these nerves is influenced by cardiac glycosides in this abnormal cardiovascular state remains to be elucidated.

SUMMARY

The influence of progressive ouabain intoxication upon cardiac sympathetic and parasympathetic nerve activity was studied in chloralose-anesthetized open-chest cats. Measurement of concurrent changes in postganglionic nerve activity, arterial pressure, and the ECG (Lead II) revealed that ouabain (2 μg/kg/min) induced an initial inhibition of both sympathetic and parasympathetic nerve activity without positive inotropic action. The onset of intoxication was preceded by increased neural activity, followed by multifocal arrhythmias and terminal ventricular fibrillation. Carotid sinus nerve stimulation at 5 min intervals increased the positive inotropic action of ouabain and induced persistent nerve activity that was followed by early ventricular fibrillation. Increased neural activity induced by carotid artery stenosis was initially supressed by ouabain, but increased prior to the early onset of ventricular fibrillation. The positive inotropic action of ouabain was also enhanced. Alterations in baroreceptor activity induced by section of the buffer nerves or cardiac denervation prevented these ouabain-induced changes and significantly delayed the onset of ventricular fibrillation. It is concluded that ouabain exerts its cardiac action partly through a modification of the activity of the cardiac autonomic nervous system.

ACKNOWLEDGMENTS

The author is grateful to Messrs. H. Van den Bergen and J. Verboon, and to Mrs. A. Nunn, for their excellent technical assistance.

REFERENCES

1. ALEXANDER, S. and PING, W. C.: Fatal ventricular fibrillation during carotid sinus stimulation. Amer. J. Cardiol. *18:* 289–291 (1966).
2. AQUINO-CIAS, J.; HARMONY, T.; GUMA, E.; TORO, A.; ANIEROS, S.; HERNANDEZ, N. and FERNANDEZ, G.: Effect of strophantine and ouabain on cortical after-discharges. Arch. Int. Pharmacodyn. *168:* 352–365 (1967).
3. BATTERMAN, R. C. and GUTNER, L. B.: Hitherto undescribed neurological manifestations of digitalis toxicity. Amer. Heart J. *36:* 582–586 (1948).
4. BORISON, H. L. and WANG, S. C.: Locus of the central emetic action of cardiac glycosides. Proc. Soc. Exp. Biol. Med. *76:* 335–338 (1951).
5. BRONK, D. W.; FERGUSON, L. K. and SOLANDT, D. Y.: Inhibition of cardiac accelerator impulse by the carotid sinus. Proc. Soc. Exp. Biol. Med. *31:* 579–580 (1934).
6. DAGGET, W. M. and WEISFELDT, M. L.: Influence of the sympathetic nervous system on the response of the normal heart to digitalis. Amer. J. Cardiol. *16:* 394–405 (1965).
7. ERLIJ, D. and MENDEZ, R.: The modification of digitalis intoxication by excluding adrenergic influences on the heart. J. Pharmacol. Exp. Therap. *144:* 97–103 (1964).
8. GILLIS, R. A.: Cardiac sympathetic nerve activity: changes induced by ouabain and propranolol. Science *166:* 508–510 (1969).
9. GILLIS, R. A. and STANDAERT, F. G.: Effect of carotid sinus and vagal nerve section on some cardiovascular responses induced by ouabain. Pharmacologist *10:* 220 (1968).
10. GREENWOOD, R. J. and DUPLER, D. A.: Death following carotid sinus pressure. J.A.M.A. *181:* 605–609 (1962).
11. HAN, J.; DEJALON, P. G. and MOE, G. K.: Adrenergic effects on ventricular vulnerability. Circulat. Res. *14:* 516–524 (1964).
12. HATCHER, R. A. and EGGLESTON, G.: The emetic action of the digitalis bodies. J. Pharmacol. Exp. Therap. *4:* 113–134 (1912).
13. HERING, H. E.: Compression of the carotid artery. München. Med. Wchschr. *70:* 1287–1290 (1923).
14. HEYMANS, C.; BOUCKAERT, J. J. et REGNIERS, P.: Sur le mecanisme reflexe de la bradycardie provoquee par les digitaliques. C.R. Seanc. Soc. Biol. *110:* 572–574 (1932).
15. JARISCH, A. und RICHTER, H.: Der Bezold Effekt: Eine vergessene Kreislaufreaktion. Klin. Wchnshr. *18:* 185–187 (1939).
16. KONZETT, H. and ROTHLIN, E.: Effects of cardioactive glycosides on a sympathetic ganglion. Arch. Int. Pharmacodyn. *89:* 343–352 (1952).
17. LOWN, B. and LEVINE, S. A.: The carotid sinus: Clinical value of its stimulation. Circul. *23:* 766–789 (1961).

18. McCUBBIN, J. W.; GREEN, J. H. and PAGE, I. H.: Baroreceptor function in chronic renal hypertension. Circulat. Res. *4:* 205–210 (1956).

19. McLAIN, P. L.: Effects of cardiac glycosides on spontaneous efferent activity in vagus and sympathetic nerves of cats. Int. J. Neuropharmac. *8:* 379–387 (1969).

20. MENDEZ, C.; ACEVES, J. and MENDEZ, J.: The anti-adrenergic action of digitalis on the refractory period of the A-V transmission system. J. Pharm. Exp. Therap. *131:* 199–204 (1961).

21. NICHOL, A. D. and STRAUSS, H.: Effects of digitalis, urginin, congestive heart failure and atropine on hyperactive carotid sinus. Amer. Heart J. *25:* 746–759 (1943).

22. PORUS, R. L. and MARCUS, F. I.: Ventricular fibrillation during carotid sinus stimulation. New Eng. J. Med. *268:* 1338–1342 (1963).

23. RANDALL, W. C.; SZENTIVANYI, M.; PACE, J. B.; WECHSLER, J. S. and KAYE, M. P.: Patterns of sympathetic nerve projections into the canine heart. Circulat. Res. *22:* 315–323 (1968).

24. SLEIGHT, P.; LALL, A. and MUERS, M.: Reflex cardiovascular effects of epicardial stimulation by acetylstrophanthidin in dogs. Circulat. Res. *25:* 705–711 (1969).

25. STEEL, R. D. and TORRIE, J. H.: Principles and Procedures of Statistics (McGraw-Hill, New York 1960).

26. STICKNEY, J. L. and LUCCHESI, B. R.: The effect of sympatholytic agents on the cardiovascular responses produced by the injection of acetylstrophanthidin into the cerebral ventricles. Europ. J. Pharm. *6:* 1–7 (1969).

27. TANZ, R. D.: The action of ouabain on cardiac muscle treated with reserpine and dichloroisoproterenol. J. Pharm. Exp. Therap. *144:* 205–213 (1964).

Mailing address: Dr. H. Lloyd Garvey, Department of Pharmacology, Faculty of Medicine, University of Ottawa, Ottawa (Canada).

School of Veterinary Medicine, University of Pennsylvania,
Philadelphia, Pa., USA, and Institute of Veterinary Pathology,
University of Bern, Bern, Switzerland

OBSERVATIONS ON NATURALLY OCCURRING MYOCARDIAL FIBROSIS AND NECROSIS IN DOGS [1]

D. K. DETWEILER, D. F. PATTERSON, and H. LUGINBÜHL

INTRODUCTION

Focal areas of myocardial necrosis and fibrosis are the characteristic lesions found in the hearts of dogs suffering from chronic hemodynamic cardiac overload. These lesions are invariably found in the hearts of dogs that have developed clinical signs of cardiac disability associated with such diverse diseases as acquired mitral insufficiency, various congenital cardiac malformations, and heartworm heart disease (cor pulmonale). Typically, intramural coronary arteriosclerosis is observed to accompany these myocardial alterations. The arteriosclerotic lesions in histological sections are often found in the absence of myocardial changes detectable by light microscopy. On the other hand, myocardial necrosis and fibrosis did not occur in our material in the absence of the vascular lesions. When the associated heart condition involves primarily the left heart (*e.g.*, acquired mitral insufficiency), the vascular and myocardial lesions are found most commonly and are most extensive in the papillary muscles of the left ventricle. Correspondingly, in lesions involving primarily the right heart (*e.g.*, congenital pulmonic stenosis), the papillary muscles of the right ventricle are sites of predilection. The clinical signs accompanying these changes include enlargement of one or more cardiac chambers, various ectopic arrhythmias, and atrioventricular and intraventricular conduction disturbances. The term "chronic myocardial disease" has been used to designate cases exhibiting these clinical signs in the absence of evidence of acute myocarditis. While this designation is something of a clinical catch-all, these lesions are practically always found in dogs suffering from chronic cardiac disability accompanied by the clinical changes mentioned.

MATERIALS AND METHODS

During the past 22 years, some 15,000 dogs have been the subject of clinical examinations of the cardiovascular system; 4,831 were given a screening examina-

[1] Supported by Grant HE 4885, National Heart Institute, National Institutes of Health.

tion and individuals with detectable heart disease were studied further in an epidemiological investigation; and some 2,600 dogs with various types of heart disease were identified, of which approximately 1,500 came to necropsy. Over 290 cases of various types of congenital heart disease have been found in this material and a series of 140 dogs with congestive heart failure have been studied (1, 3–5, 8–10).

RESULTS

In dogs with acquired heart disease, the most common lesions found at necropsy are: atrioventricular valvular fibrosis, intramural coronary arteriosclerosis, and focal areas of myocardial necrosis and fibrosis (1, 2, 8, 10). This pathological triad is the usual post-mortem finding in dogs that have developed clinical signs of heart disease.

Valvular fibrosis is designated "chronic valvular disease" when clinical diagnosis is possible. This is the most frequent cardiac lesion of dogs. Insufficiency of the mitral and less often of the tricuspid valve occurs, while stenosis of the atrioventricular valves is rare. Semilunar valve involvement may occur, but is seldom sufficient to produce clinical evidence of disease. The usual clinical signs of chronic valvular disease are a grade 3 or louder pansystolic murmur in the mitral or tricuspid area, left atrial and left ventricular enlargement, cough, dyspnea, pulmonary rales, and roentgenographic evidence of pulmonary congestion. The valvular murmurs alone are often detected fairly early in life and are tolerated for some years before other evidence of cardiac disease supervenes (3). When this does occur and the animal comes to post-mortem, the discrete lesions of myocardial necrosis and fibrosis and intramural coronary arteriosclerosis are usually present. This combination of lesions can be inferred from the clinical signs and has been designated in our clinic "chronic myocardial disease" to distinguish it from inflammatory, degenerative, and neoplastic myocardial lesions that produce similar clinical signs. Clinically, the presence of these chronic myocardial and coronary arterial lesions is considered likely in middle-aged and older dogs with mitral insufficiency, left ventricular enlargement, and ventricular ectopic arrhythmias, complete atrioventricular block, or intraventricular conduction disturbances. Similar lesions are also found in dogs with congenital subaortic stenosis, congenital pulmonic stenosis, and in hearts of animals that have developed congestive heart failure from *Dirofilaria immitis* infestation (heartworm heart disease) or from any type of congenital cardiac malformation.

On the basis of these clinical signs, chronic myocardial disease was diagnosed in 3.4% of 4,831 dogs screened for the presence of heart disease. In the age groups 0–1, 1–4, 5–8, 9–12, and over 13 years, the prevalences increased as follows: 0.7, 1.5, 3.7, 9.0, and 14.9%, respectively (4). From clinicopathological correlation, it is clear that a number of other myocardial lesions (*e.g.*, myocarditis, thrombo-embolic infarction, etc.) produce similar clinical signs and that the myocardial lesions may be present in the absence of clinical manifestations.

The fibrotic and necrotic lesions themselves are discrete, may involve only a few myocardial fibers or include an area several millimeters in diameter. Large fibrotic areas resulting from the coalescence of smaller focal lesions are seldom found and the percentage of normal myocardial tissue replaced by these microscopic lesions appears insufficient to account alone for the loss of contractile strength in dogs with chronic congestive heart failure. Microscopically, degeneration and necrosis of myocardial cells, resorption, and reparative fibrosis are seen.

Tissue sections of the myocardium were studied in 19 dogs with fibrous subaortic stenosis and 15 dogs with isolated pulmonary stenosis (6). Subaortic stenosis was accompanied by extensive alterations of the intramural coronary arteries within the left ventricle in every dog, but neither the intramural arteries of the right ventricle nor the extramural coronary arteries were altered pathologically. Numerous areas of myocardial necrosis and fibrosis were found in the left ventricular myocardium in association with these intramural arteriosclerotic lesions, especially in the anterior papillary muscle. Neither of these lesions were found in the myocardium of the right ventricle.

Hearts with congenital pulmonary stenosis and intact ventricular septa from 15 dogs were studied. Neither myocardial fibrosis nor lesions involving the intramural arteries were found in 10 of these 15 hearts. In 4, an occasional intimal thickening was observed in intramural arteries of the right ventricle but myocardial fibrosis was absent. The intramural coronary arterial lesions were similar to those seen in the dogs with subaortic stenosis, but their frequency of occurrence and severity were less. In one dog with severe pulmonary stenosis and a right ventricular peak systolic pressure (190 mm Hg) that greatly exceeded aortic peak pressure (128 mm Hg), there were microscopic areas of myocardial fibrosis in the right ventricle and changes in the intramural coronary arteries similar to those seen in the left ventricles of dogs with subaortic stenosis.

DISCUSSION

On the basis of these observations, any explantation given for the intramural coronary arterial lesions and the myocardial lesions described can only be speculative. The lesions themselves cannot be said to be specific, either with respect to the species involved or the underlying cardiac conditions with which they are associated. But that they are reactions to injury associated with hemodynamic cardiac overload and relative or absolute myocardial hypoxia seems probable. Experimental work with both chickens and swine, in which the animals were grouped so as to produce psycho-social interactions of differing intensities, demonstrated that the incidence of such cardiac lesions can be increased experimentally in the absence of underlying heart disease (11). A feature that characterized the pathological changes described here in dogs, those produced experimentally in swine and chickens, and those occurring spontaneously in zoo animals is the consistent occurrence of intramural coronary arteriosclerosis associated with myocardial necrosis and fibrosis. At present, it seems most reasonable to consider these lesions secondary reactions to a variety of conditions that produce a sufficiently marked disparity between tissue nutritive needs and supply.

SUMMARY

Focal lesions of myocardial necrosis and fibrosis of unknown etiology and pathogenesis are found in necropsied dogs, most frequently associated with chronic valvular fibrosis and intramural coronary arteriosclerosis. This triad of pathological changes is the characteristic finding in dogs with acquired cardiovascular disease. Clinically, the triad is typically manifested by the occurrence of the following signs: systolic murmur of mitral insufficiency, left ventricular enlargement, left atrial enlargement, ventricular ectopic arrhythmias and conduction disturbances. The myocardial lesions are more extensive in dogs with chronic congestive heart failure. The prevalence of these clinical signs increases with aging. The mere presence of the myocardial lesions does not appear to be age related, but their extensiveness is positively correlated with the degree of cardiac disability.

REFERENCES

1. DETWEILER, D. K.; HUBBEN, K. and PATTERSON, D. F.: Survey of cardiovascular disease in dogs. Preliminary report on the first 1,000 dogs screened. Amer. J. Vet. Res. *21:* 329–359 (1960).
2. DETWEILER, D. K. and LUGINBÜHL, H.: Arteriosclerosis in animals. In: BREST, A. N. and MOYER, J. H.. Atherosclerotic Vascular Disease, pp. 50–76 (Appleton-Century-Crofts, New York 1967).
3. DETWEILER, D. K.; LUGINBÜHL, H.; BUCHANAN, J. W. and PATTERSON, D. F.: The natural history of acquired cardiac disability of the dog. Ann. N.Y. Acad. Sci. *147:* 318–329 (1968).
4. DETWEILER, D. K. and PATTERSON, D. F.: The prevalence and types of cardiovascular disease in dogs. Ann. N.Y. Acad. Sci. *127:* 481–516 (1965).
5. DETWEILER, D. K.; RATCLIFFE, H. L. and LUGINBÜHL, H.: The significance of naturally occurring coronary and cerebral arterial disease in animals. Ann. N.Y. Acad. Sci. *149:* 868–881 (1968).
6. FLICKINGER, G. and PATTERSON, D. F.: Coronary lesions associated with congenital subaortic stenosis in the dog. J. Path. Bact. *93:* 133–140 (1967).
7. JONES, T. C. and ZOOK, B. C.: Aging changes in the vascular system of animals. Ann. N.Y. Acad. Sci. *127:* 671–684 (1965).
8. LUGINBÜHL, H. and DETWEILER, D. K.: Cardiovascular lesions in dogs. Ann. N.Y. Acad. Sci. *127:* 517–540 (1965).
9. LUGINBÜHL, H.; DETWEILER, D. K. und RATCLIFFE, H. L.: Über die Bedeutung von arteriosklerotischen Veränderungen an Koronar- und Hirngefässen bei Tieren. Schweiz. Arch. Tierheilk. *110:* 180–189 (1968).

10. LUGINBÜHL, H.; JONES, J. E. T. and DETWEILER, D. K.: The morphology of spontaneous arteriosclerotic lesions in the dog.In: ROBERTS, J. C. and STRAUS, R., Comparative Arteriosclerosis, pp. 161–169 (Harper and Row, Publishers, Inc., Hoeber Medical Division, New York 1965).

11. RATCLIFFE, H. L.; LUGINBÜHL, H.; SCHNARR, W. R. and CHACKO, S. K.: Coronary arteriosclerosis in swine: evidence of a relation to behavior. J. Comp. and Physiol. Psychol. *68:* 385–392 (1969).

Mailing address: Dr. D.K. Detweiler, Comparative Cardiovascular Studies Unit, School of Veterinary Medicine, University of Pennsylvania, 3800 Spruce Street, Philadelphia, Pa. 19104 (USA).

McGill University Clinic of the Royal Victoria Hospital,
and the Pathological Institute, McGill University,
Montreal, Canada

ACUTE HYPERKALEMIC CARDIOMYOPATHY IN THE DOG [1]

H. P. OJIAMBO,[2] R. PETRIE, J. DUSEK,[3] and G. A. KLASSEN

Methyl salicylate (MS) is a known uncoupler of oxidative phosphorylation (2). The possible effects of this agent on myocardial performance and metabolism have not been studied extensively. The work reported in the present communication was prompted by the death of two patients following accidental ingestion of MS. Their electrocardiograms (fig. 1) showed evidence of hyperkalemia, which could not be reversed by the usual methods. At autopsy, the subendocardium of one patient showed fragmentation of myofibrils. Muscle necrosis of the lower extremities occurred in another patient who survived. On the basis of these findings, an experimental animal model was devised to study this toxic process.

METHODS

Healthy mongrel dogs with a known metabolic history and in the fasting state were lightly anaesthetized with pentobarbital sodium. Catheters were placed in the femoral artery and veins, coronary sinus, left ventricle, and right atrium, for sampling and pressure measurement. A standard lead II electrocardiogram was recorded. Following a control period, MS was administered intragastrically in the dosage of 700 mg/kg. Experimental procedures were carried out over a 4 to 5 h period, at which time the survivors were sacrificed. The results were compared with a control series.

Analytic Procedures

Blood gases, pH, PCO_2, and PO_2 were measured using an I.L. electrode. Oxygen saturation was measured by oximetry. Bicarbonate was calculated from the Sigaard-Anderson nomogram. Total plasma salicylate was determined by the method of Stevenson (11). Lactate was measured by the method of Horn and Bruns (3). Potassium was determined by flame photometry, using internal

---

[1] Supported by the Canadian Muscular Dystrophy Association and the Canadian Internal Agency for Development.

[2] CIDA Research Fellow sponsored under the McGill-Kenya Project.

[3] Medical Research Council of Canada Research Fellow.

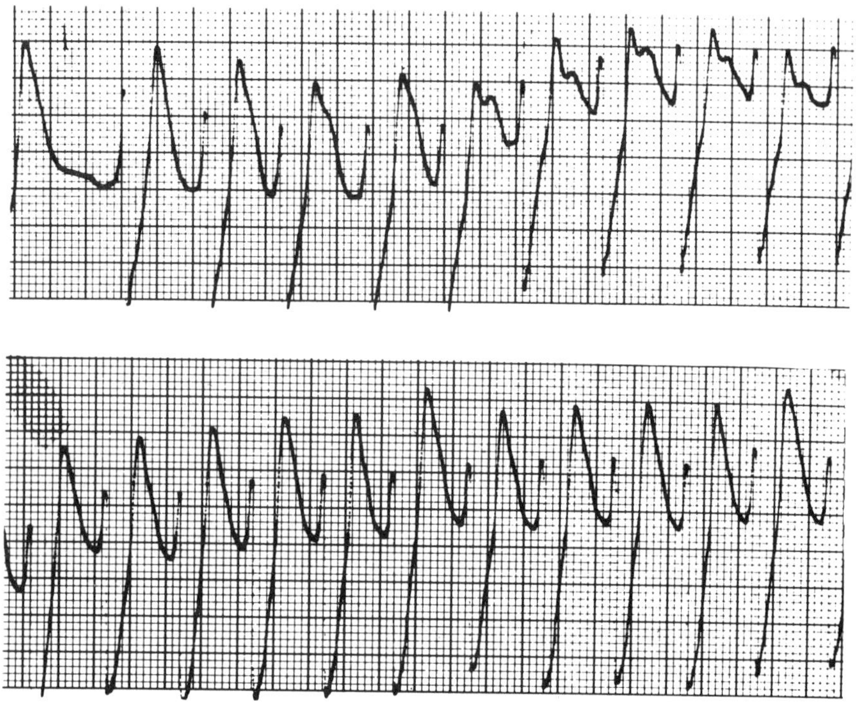

Fig. 1 Electrocardiogram of a patient who ingested methyl salicylate. Note widening of QRS and peaked T waves.

standards. Creatine phosphokinase was estimated by the method of Kuby *et al.* (6). Plasma inorganic phosphorus was measured by the Hurst (4) modification of the method of Kuttner and Cohen (7). Cardiac output was determined by the Fick method.

Histochemical methods

Portions of myocardium were quenched in isopentane, precooled in liquid nitrogen. Sections 10μ in thickness were cut in a cryostat and mounted on coverslips. For the visualization of glycogen, sections were fixed in cold Gendre's solution. Sections for enzymatic incubations were air-dried and those used for the demonstration of potassium were freeze-dried for 1 h before staining. Glycogen was detected by PAS stain. Control sections were pretreated with diastase. Succinic dehydrogenase, lactate dehydrogenase, and beta hydroxybutyrate dehydrogenase were determined using a standard method (9). Freeze-dried floating sections were stained for the detection of potassium by the method of Poppen *et al.* (10) and Amon (1).

Electronmicroscopy

Immediately after death or sacrifice, the hearts were perfused through a cannula inserted into the anterior descending or circumflex branch of the left coronary artery. Karnovsky fluid (5), modified by the addition of 5% sucrose, was used as perfusate. Tissue blocks were excised from the perfused areas of the myocardium. After an additional immersion fixation in glutaraldehyde-paraformaldehyde mixture, the blocks were post-fixed in 1% osmium tetroxide, dehydrated in graded ethanols, and embedded in Epon 812. Silver sections from selected areas of the myocardium were cut and stained with uranyl acetate and lead nitrate (12).

RESULTS

Total plasma salicylate

Arterial concentrations of plasma salicylate are shown in table I. A linear rise in the concentration of the total plasma salicylate was observed up to 4 h, after which time a plateau occurred.

Creatine phosphokinase

A rise in CPK was observed in the coronary sinus effluent, a mean arterial level of 12.5 sigma units being recorded after 2 h (normal < 4 units).

Hemodynamics

The cardiac output remained constant, falling only terminally. A slight fall in the mean blood pressure was observed after 2 h (fig. 2). There was a slight increase in heart rate of MS dogs as compared with controls (fig. 2), but this increase was not of statistical significance.

Oxygen consumption

An increase in oxygen consumption, which doubled at 4 h, was seen. This observation was associated with a widening of arteriovenous difference for oxygen (fig. 3).

Table I

Mean ± SEM Plasma Salicylate Levels as Measured in Arterial Blood throughout the Study

Plasma salicylate					
Time	1 h	2 h	3 h	4 h	5 h
Mean	12.2	19.6	29.9	41.3	37.6
SEM	±1.3	±2.7	±3.4	±6.6	±6.9

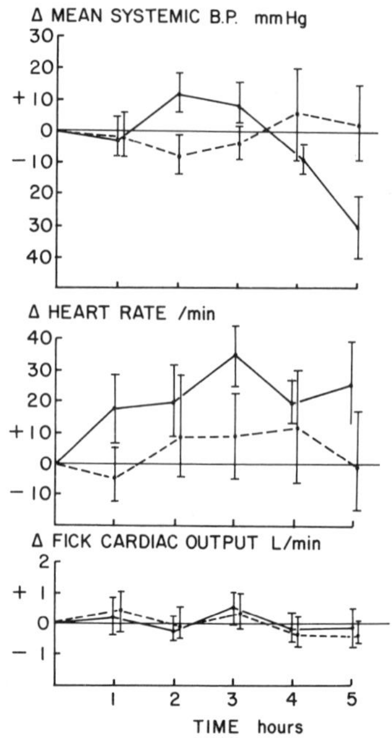

Fig. 2 Changes in blood pressure, heart rate, and cardiac output of MS-treated dogs (solid lines) and controls (interrupted). Points represent means; bars indicate ± SEM. Resting values were: blood pressure = 126 ± 4.8 mm Hg; heart rate = 175 ± 4/min; cardiac output = 2.2 ± 0.2L/min.

Acid-base changes

A period of respiratory alkalosis was initially observed, followed after the 3rd h by a metabolic acidosis.

Electrocardiogram

The observed changes were those of hyperkalemia. Tall slender peaked T waves developed 2 h after the intervention. The PR interval was prolonged. Supraventricular and ventricular tachycardia was observed. The changes then culminated in the development of bundle branch block, ventricular fibrillation and, terminally, cardiac standstill (fig. 4).

Metabolism

The results of these studies can be seen in table II.

Potassium

There was a gradual rise in the arterial concentration of potassium throughout the procedures. The arterial minus coronary sinus difference for potassium

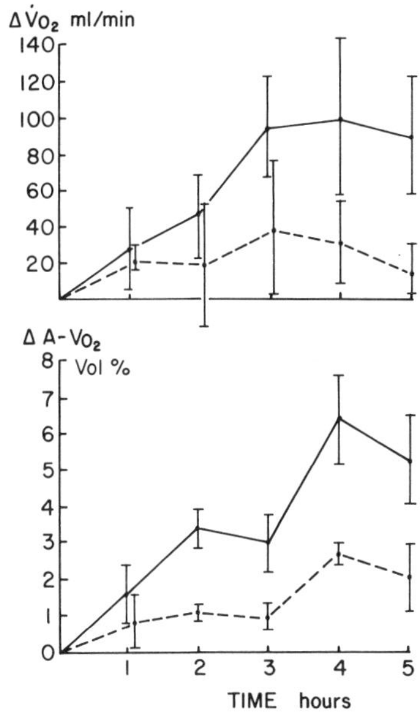

Fig. 3 Changes in oxygen consumption ($\Delta\ V_{O_2}$) in ml/min, and arterial minus venous oxygen difference (Δ A-V) in volumes %. Points represent means; bars indicate ± SEM. Resting values were: oxygen consumption 100 ± 9 ml/min; arterial-venous difference 4.1 ± 0.36 volumes %.

showed a net uptake of K^+ up to the 3rd h of study. Thereafter, a net efflux of K^+ was observed.

Lactate

Arterial lactate concentration increased during the procedure. A normal myocardial extraction of lactate was observed up to the 2 h period, after which time lactate output occurred.

Table II

Percent Extraction of Metabolites by the Myocardium*

	Control	2 h	4 h
PO_4	+0.9	−2.3	−3.1
Lactate	+8.9	−3.1	−55.8
K^+	+1.7	+4.8	−16.8

* Mean percent extraction by the myocardium of phosphate, lactate, and potassium at control, 2 h, and 4 h. (Arterial minus coronary sinus divided by arterial × 100.)

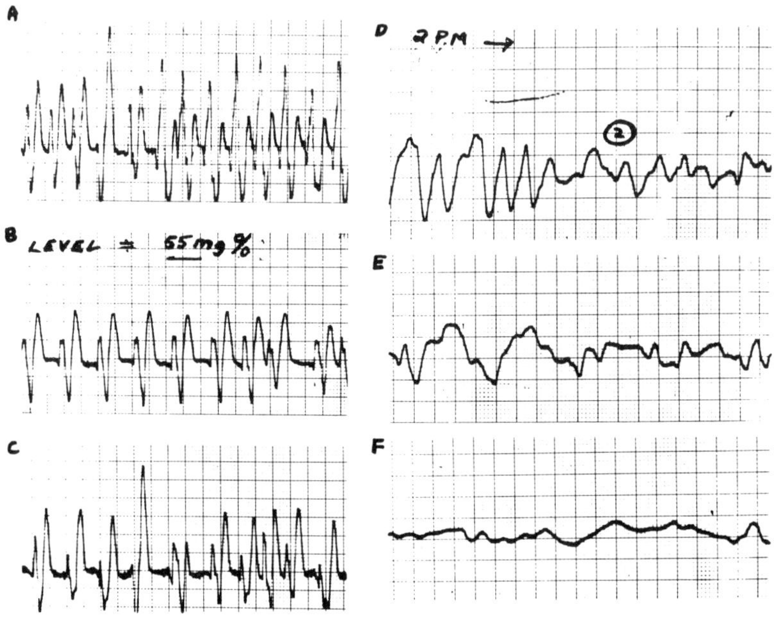

Fig. 4 Changes in the electrocardiogram of a dog receiving MS. Tracing A was taken at a serum salicylate of 45 mg%. Note peaked T waves and widening of QRS complex. Further tracings B to F indicate progression to final asystole.

Oxygen

Despite hyperventilation, only a slight increase in arterial PO_2 was observed. A widening of the arterial minus coronary sinus PO_2 was noted. The lowest coronary sinus PO_2 was recorded terminally.

Phosphate

An initial uptake of inorganic phosphate was observed, followed by a net phosphate efflux, beginning at 2 h (table II).

Morphology

No signs of microvascular occulsion were observed and the staining method for potassium revealed no diffuse or focal changes. There was swelling of cardiac muscle cells, which caused separation of myofibrils and bulging of the sarcolemma (fig. 5). Clear spaces were present, suggesting dilatation of sarcoplasmic reticulum (fig. 6). Mitochondrial swelling and focal areas of clearing of matrix were seen. In some mitochondria, vacuolation and bulging of the outer limiting membrane (fig. 7) was observed. The morphological changes were discrete and focal.

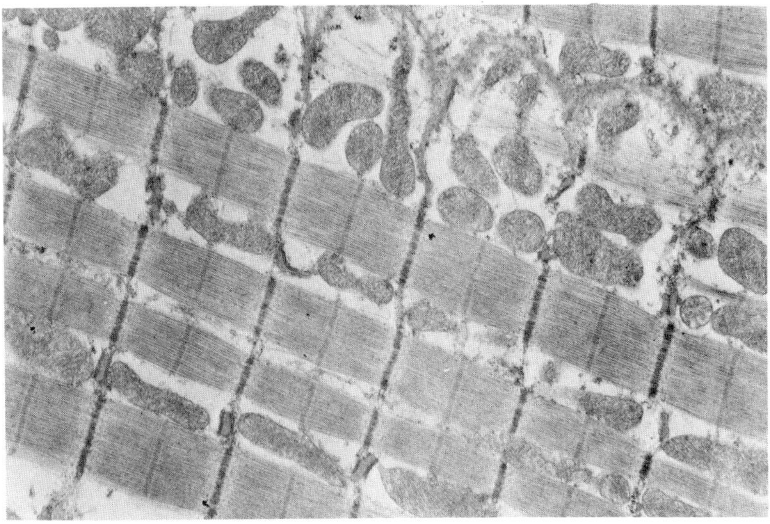

Fig. 5 Electron micrograph showing swelling of cardiac muscle cell, with separation of myofibrils and bulging of the sarcolemma between T-tubules. Mitochondria appear well preserved, with a slight clearing of the matrix. ×10,850.

Fig. 6 Section through a part of cardiac muscle cell showing moderate degree of swelling. Some of the clear spaces represent distended tubules of the sarcoplasmic reticulum. ×11,000.

Fig. 7 Photomicrograph showing the area of an intercalated disc, with swollen neighbouring mitochondria and clearing of their matrix. There is a bleb formed under the outer limiting membrane of one of the mitochondria in the lower left corner. The disc itself appears normal. ×10,350.

DISCUSSION

The observed metabolic changes suggest the following hypothesis for the mode of action of MS. By uncoupling oxidative phosphorylation, MS interferes with the production of high energy phosphate. There is accumulation of phosphate and phosphate nucleotides, compounds that are known to stimulate glycolysis by activating phosphofructokinase (8). Lactate production from the myocardium occurs as glycolysis is stimulated. The sodium pump requires ATP as well as the presence of Na^+/K^+ activated magnesium-dependent ATPase to maintain normal potassium influx. Decreased availability of ATP, resulting from uncoupling, leads to a decrease in K^+ influx; hence, efflux dominates and, in turn, extracellular concentrations of K^+ increase. Hyperkalemia further reduces the resting membrane potential, leading to arrythmias and, terminally, to a cardiac standstill. The loss of CPK is indicative of cardiac cell membrane damage, with a leak of this enzyme into the coronary sinus effluent.

CONCLUSION

In an acute myocardiopathy induced in dogs by administration of MS, haemodynamic changes were minimal, while profound metabolic derangements were observed. These derangements are probably due to a lack of ATP production. It is speculated that the morphological changes observed may be the result of a metabolic insufficiency. Derangements in energy metabolism could lead to imbibition of water by the cell and subcellular organelles. Further studies should be conducted to determine the possibility of a similar process occurring in man. It is thought that, on a chronic basis, this process could give rise to discrete myocardial scarring as is often seen in some forms of "idiopathic" cardiomyopathy.

SUMMARY

Dogs challenged with toxic doses of methyl salicylate were observed to develop metabolic manifestations of uncoupling of oxidative phosphorylation. Oxygen consumption increased two-fold; haemodynamic changes were minimal. Profound metabolic derangements were observed, with a net efflux of phosphate, lactate, and potassium from the myocardium. Terminally, marked hyperkalemia occurred. Morphologic abnormalities were also observed.

ACKNOWLEDGMENTS

The technical assistance of Mrs. Edie Summers was greatly appreciated.

REFERENCES

1. AMON, H.: Methodische Untersuchungen zum histochemischen Nachweis von Electrolyten II Die lichtmikroskopische Darstellung von K^+ in der Niere. Histochemie *14:* 251–259 (1968).
2. BRODY, T. M.: The action of sodium salicylate and related compounds on tissue metabolism in vitro. J. Pharmacol. Exp. Therap. *117:* 39–51 (1956).
3. HORN, H. D. und BRUNS, F. H.: Quantitative Bestimmung von L (+) - Milchsäure mit Milchsäuredehydrongenase. Biochim. biophys. Acta *21:* 378–380 (1956).
4. HURST, R. O.: The determination of nucleotide phosphorus with a stannous chloride-hydrazine sulphate reagent. Can. J. Biochem. *42:* 287–292 (1964).
5. KARNOVSKY, M. M.: A formaldehyde-glutaraldehyde fixative of high osmolarity for use in electromicroscopy. J. Cell. Biol. *27:* 137A (1965).
6. KUBY, S. A.; NODA, L. and LARDY, H. A.: Adenosine triphosphate-creatine transphorylase. J. Biol. Chem. *210:* 65–82 (1954).
7. KUTTNER, T. and COHEN, H. R.: Micro colorimetric studies; molybdic acid, stannous chloride reagent; micro estimation of phosphate and calcium in pus, plasma, and spinal fluid. J. Biol. Chem. *75:* 517–531 (1927).
8. NEWSHOLME, E. A. and RANDLE, P. J.: Regulation of glucose uptake by muscle. 6. Fructose 1, 6-diphosphatase activity of rat heart and rat diaphragm. Biochem. J. *83:* 387–392 (1962).
9. PEARSE, A. E.: Histochemistry: Theoretical and applied. 2nd ed. (J. & A. Churchill Ltd., London, 1961).
10. POPPEN, K. J.; GREEN, D. M. and WRENN, H. T.: Histochemical localization of potassium and glycogen. J. Histochem. *1:* 160–173 (1953).
11. STEVENSON, G. W.: Rapid ultraviolet spectrophotometric determination of salicylate in blood. J. Anal. Chem. *32:* 1522–1525 (1960).
12. VENABLE, J. H. and COGGESHALL, R.: A simplified lead citrate stain for use in electron microscopy. J. Cell. Biol. *25:* 407–408 (1965).

Mailing address: Dr. G. A. Klassen, Department of Cardiology, The Royal Victoria Hospital, Montreal 112, Que. (Canada).

Research Laboratories, Food and Drug Directorate, and
Department of Pathology, University of Ottawa,
Ottawa, Canada

CARDIOTOXICITY OF BROMINATED VEGETABLE OILS

I. C. MUNRO, S. HASNAIN, F. A. SALEM, T. GOODMAN,
H. C. GRICE, and H. A. HEGGTVEIT[1]

Brominated vegetable oils (BVO) have been used in North America for nearly forty years to adjust the density of essential flavoring oils used in the manufacture of certain citrus flavored beverages. Until recently, little information was available with respect to their toxicity. The studies reported herein summarize the results of some investigations on BVO toxicity recently conducted in our laboratory.

In initial toxicological studies, brominated cottonseed oil (BCO) was fed to young male rats, at 0, 0.5, or 2.5% of the diet (1), for a period of 80 days. During that time, body-weight gain and food consumption were recorded weekly. Following completion of the feeding period, the animals were killed and subjected to biochemical and pathological examination. The weight gain of animals fed BCO is shown in figure 1. Animals fed 2.5% BCO, but not those fed 0.5%, had reduced body-weight gains and this was accompanied by a reduction

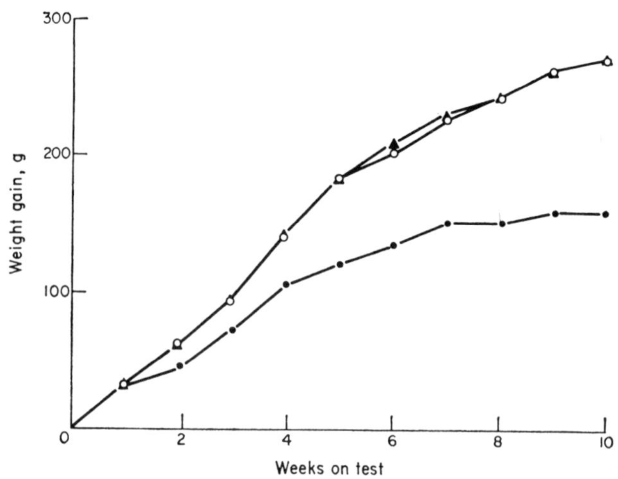

Fig. 1 Body-weight gain of male rats fed brominated cottonseed oil at dietary levels of 0 (▲), 0.5 (o), or 2.5% (•) for 80 days.

[1] Research Associate, Ontario Heart Foundation.

in the efficiency of food utilization. Animals fed 0.5 or 2.5% BCO had increased absolute and relative heart weights, while those fed 2.5% BCO had increased absolute and relative liver weights (table I). The increase in liver weight could be accounted for primarily by an accumulation of lipid in the liver, as measured by biochemical techniques (table II). The increase in liver-fat level was reflected morphologically by an accumulation of lipid material in the liver, when frozen sections of liver tissue from BCO-treated animals were stained with Oil Red O. Grossly, the hearts of most BCO-treated rats appeared normal; but on microscopic examination, the increased heart weight was reflected morphologically by fatty degeneration involving the entire myocardium (fig. 2). Interstitial edema, with separation, fraying and fragmentation of myocardial fibres was also noted. In addition, scattered foci of myocytolysis and overt foci of necrosis accompanied by an accumulation of reactive inflammatory cells were seen in the hearts of BCO-treated animals (figs. 3 and 4).

Since we were unable to determine the no-effect level for BCO cardiotoxicity from the first study, a second sub-acute study was undertaken in which the toxicity of four commercially available brominated oils was determined. In this study, brominated corn, sesame, cottonseed, or olive oil was fed to male weanling rats, at 0, 0.1, or 2.5% of the diet, for 105 days. The animals were then killed and subjected to pathological examination. The incidence of myocardial lesions is shown in table III. With the exception of those fed brominated olive oil, about 85% of rats fed 0.5% BVO had degenerative myocardial lesions. No cardiac changes were noted in animals fed 0.1% BVO, with the exception that 2 of 15 animals fed brominated sesame oil had minimal myocytolytic changes and some hearts showed mild fatty change. The livers of animals fed 0.5% BVO showed diffuse fatty change as previously noted and the thyroids showed varying degrees of microfollicular hyperplasia.

Since the principal pathological lesion was found in the heart, biochemical studies were undertaken to establish if BCO administration altered the metabolism of palmitic acid in the heart. Palmitic acid metabolism in the liver was also investigated. In these studies, male rats weighing approximately 250 g were given daily oral doses of BCO ranging from 160-2,500 mg per kg of body weight for 3 days. The heart was then removed, chilled, and homogenized. Aliquots of the homogenate were placed in a container with adequate amounts of radioactive substrate. The vessel was then sealed and the radioactive CO_2 liberated during the incubation period was measured using a scintillation counter. As indicated in table IV, the hearts of animals dosed with BCO showed a dose-related reduction in ability to metabolize palmitic acid. BCO treatment did not alter hepatic metabolism of palmitate. The reduction in cardiac utilization of palmitate was accompanied by a striking accumulation of lipid in the myocardium as can be seen in figure 5. In this electron micrograph, numerous lipid globules are arranged in rows between the myofibrils. The globules were closely associated with the mitochondria.

Table I

Mean Absolute and Relative Organ Weights of Male Rats Receiving Dietary Levels of 0–2.5% Brominated Cottonseed Oil for 80 Days

Dietary level (%)	No. of animals	Terminal body weight (g)	Organ weight: Liver	Heart	Kidney	Testes	Spleen
			Absolute organ weight (g)				
0	18	334 ± 10	10.19 ± 0.31	0.98 ± 0.02	2.00 ± 0.03	2.63 ± 0.21	0.53 ± 0.08
0.5	20	332 ± 6	9.97 ± 0.18	1.10 ± 0.03**	2.15 ± 0.03	2.69 ± 0.07	0.54 ± 0.01
2.5	9	235 ± 9	20.68 ± 0.95**	1.43 ± 0.14**	2.12 ± 0.07	2.11 ± 0.08*	0.47 ± 0.02
			Relative organ weight (g/100 g body weight)				
0			3.07 ± 0.01	0.302 ± 0.008	0.622 ± 0.030	0.885 ± 0.061	0.165 ± 0.007
0.5			3.01 ± 0.04	0.328 ± 0.008*	0.651 ± 0.140	0.816 ± 0.023	0.164 ± 0.007
2.5			8.76 ± 0.19**	0.658 ± 0.029*	0.915 ± 0.250*	0.898 ± 0.087	0.203 ± 0.043**

Results are given as the mean ± SE. Those marked with asterisks differ significantly from the control values: * $p < 0.05$; ** $p < 0.01$.

Table II

Fat Content of Livers of Male Rats Receiving Dietary Levels of 0–2.5% Brominated Cottonseed Oil for 80 Days

Dietary level (%)	No. of animals	Total fat (%, wet weight)
0	18	6.59 ± 0.05
0.5	20	8.17 ± 0.05**
2.5	9	10.96 ± 0.33**

Results are given as the mean ± SE and degrees of significance are indicated by asterisks: ** $p < 0.01$.

In an attempt to determine the reason for the reduced cardiac utilization of palmitate in BCO-treated animals, several co-factors necessary for optimal metabolism of fatty acids were added to the assay flasks of both treated and control animals. In this study, the animals were given doses of 400 mg of BCO per kg of body weight for 3 days; the results are shown in table V. As expected, addition of succinate depressed cardiac utilization of palmitate in both control and treated animals, since it probably competed with palmitate as a substrate. Addition of DL-carnitine and ATP significantly increased palmitate metabolism in both treated and control animsls, but it was not until CoA was added to the media that palmitate metabolism in treated animals returned to control values.

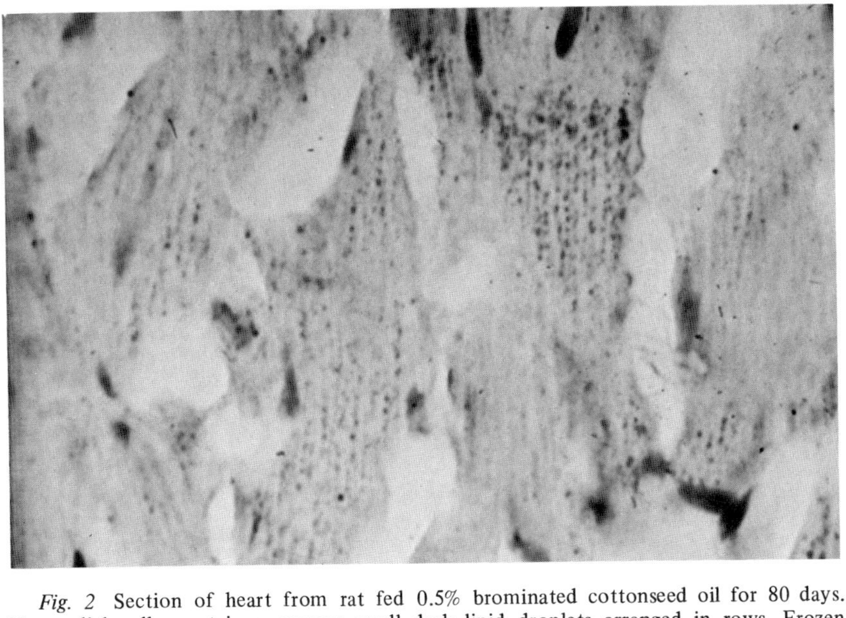

Fig. 2 Section of heart from rat fed 0.5% brominated cottonseed oil for 80 days. Myocardial cells contain numerous small dark lipid droplets arranged in rows. Frozen section, Oil Red O stain, ×500.

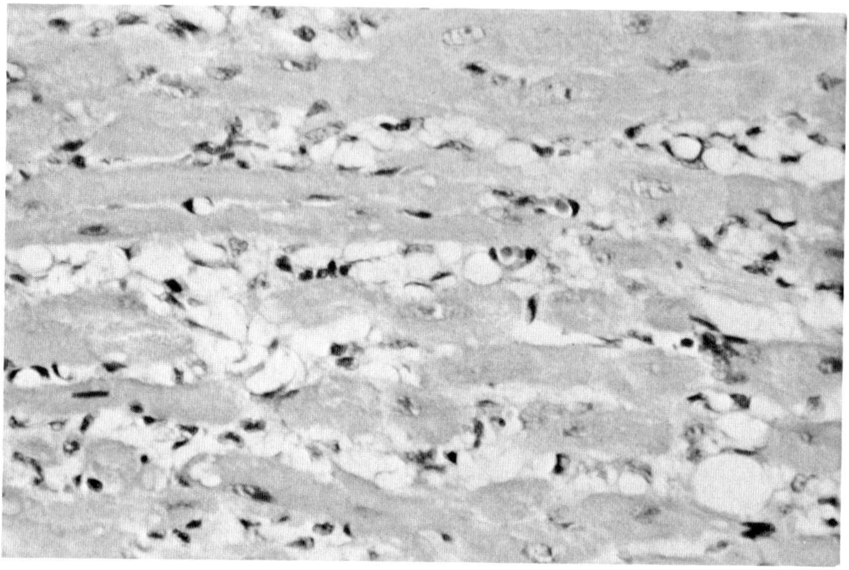

Fig. 3 Section of heart from rat fed 2.5% brominated cottonseed oil for 80 days, showing widespread myocytolysis with edematous separation, fragmentation and vacuolization of myofibers, and a few interstitial inflammatory cells. Hematoxylin, phloxine and saffron, ×250.

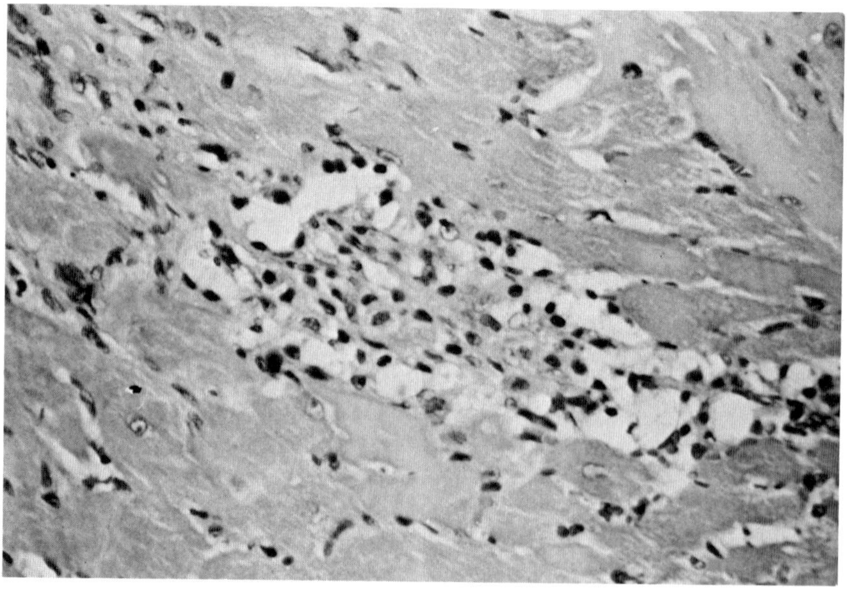

Fig. 4 Section of heart from rat fed 2.5% brominated cottonseed oil for 80 days, showing focal myocardial necrosis with reactive changes consisting primarily of histiocytic infiltration. Hematoxylin, phloxine and saffron, ×250.

Table III

Pathological Findings in Male Rats Fed Brominated Vegetable Oils

	Heart		Liver	Thyroid
	Degenerative lesions	Fat accumulation	Fat accumulation	Microfollicular hyperplasia
Controls	0/14	0/14	2/14	1/14
0.1% Brominated cottonseed oil	0/15	3/15	3/15	0/15
0.1% Brominated corn oil	0/13	3/13	5/13	1/13
0.1% Brominated olive oil	0/15	0/15	6/15	1/15
0.1% Brominated sesame oil	2/15	1/15	5/15	3/15
0.5% Brominated cottonseed oil	12/14	8/14	14/14	3/14
0.5% Brominated corn oil	11/15	7/15	14/15	7/15
0.5% Brominated olive oil	4/15	10/15	4/15	1/15
0.5% Brominated sesame oil	13/15	12/15	14/15	6/15

Table IV

Palmitate Metabolism in the Livers and Hearts of Male Rats Treated with Brominated Cottonseed Oil for 3 Days[a]

Treatment (mg/kg body weight)	DPM $^{14}CO_2$/h/mg Protein	
	Liver	Heart
0 (Control)	430 ± 55	407 ± 31
160 BCO	–	393 ± 29
400 BCO	418 ± 98	137 ± 16**
1,000 BCO	562 ± 81	117 ± 13**
2,500 BCO	486 ± 76	72 ± 5**

Results are given as mean ± SE. Those marked with asterisks differ significantly from the control values: ** $p < 0.01$.

[a] Each reaction flask contained 1 ml of homogenate (15–20 mg protein) and the following in μmole: phosphate buffer (pH 7.4), 50; palmitic acid (90,000 dpm) 1.3; $MgCl_2$, 2.0. Total volume 3.0 ml.

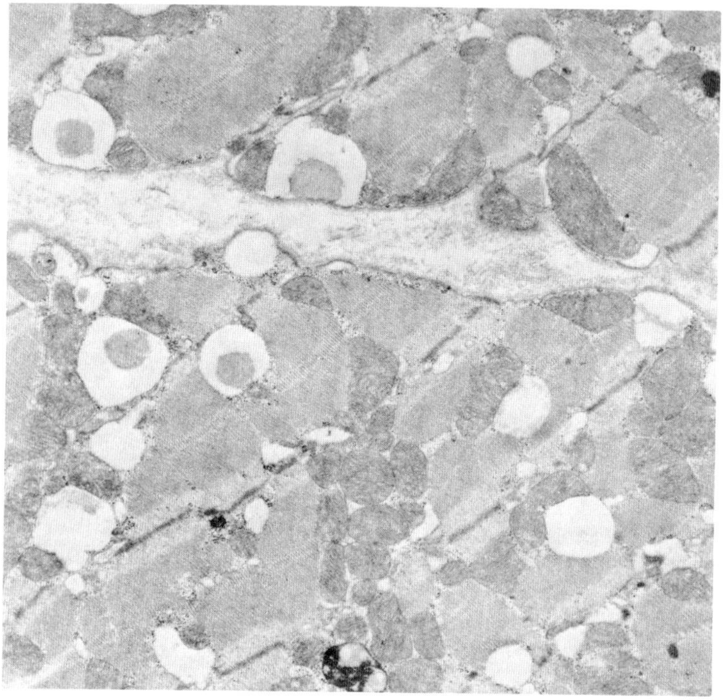

Fig. 5 Heart from rat receiving a very high oral dose of brominated vegetable oil for 3 days, showing fat accumulation within myocardial cells. The fat vacuoles are closely associated with mitochondria and, in some, the lipid contents have contracted to an osmiophilic central mass. A lysosome is seen at the bottom of the electron micrograph. ×11,400.

Addition of the other co-factors was without significant effect. It would appear from the results obtained to date that a combination of carnitine, ATP, and CoA is required to produce optimal palmitate utilization in the hearts of BCO-treated animals.

The results of these investigations indicated that BVO produce biochemical and pathological changes in the heart of rats. Using, in part, the data described herein, a no-effect level for BVO cardiotoxicity was established in the rat and regulatory action was taken to reduce the level of BVO permitted for use in foods in Canada.

SUMMARY

Studies conducted in these laboratories have indicated that rats fed brominated vegetable oils (cottonseed, corn, sesame, and olive), at 0.5% of the diet for 100 days, developed

Table V

Palmitate Metabolism in the Hearts of Male Rats Treated with Brominated Cottonseed Oil for 3 Days, with the Addition of Various Co-factors [a]

Additions to basic media	DPM $^{14}CO_2$/h/mg Protein	
	Control	Treated
Nil	299 ± 22	148 ± 9**
Succinate	261 ± 23	116 ± 8**
Succinate + Carnitine	524 ± 31	282 ± 27**
Succinate + Carnitine + ATP	900 ± 42	476 ± 27**
Succinate + Carnitine + ATP + CoA	1336 ± 128	1190 ± 108
Succinate + Carnitine + ATP + CoA + Cytochrome C	1387 ± 112	1292 ± 98
Succinate + Carnitine + ATP + CoA + Cytochrome C + NAD	1443 ± 190	1435 ± 138

Results are expressed as mean ± SE. Those marked with asterisks differ significantly from controls: ** $p < 0.01$.

[a] Media same as shown in table IV. Each reaction flask contained the following concentration of co-factor in μmole: succinate, 0.12; DL-carnitine, 1.0; ATP, 2.25; CoA, 0.13; cytochrome C, 0.048; NAD, 5.0.

cardiomyopathy and fatty degeneration of heart muscle. The myocardial lesions were characterized by prominent interstitial edema, myocytolysis, and focal necroses. Other changes included microfollicular hyperplasia of the thyroid gland and fatty metamorphosis in the liver. The hearts of rats dosed for only 3 days with 40–250 mg of brominated cottonseed oil per 100 g of body weight metabolized palmitic acid at about 35% of the rate of controls, but cardiac utilization of pyruvic acid was not impaired. Histological and electron microscopic observations revealed that the diminished cardiac utilization of palmitate was accompanied by intracellular accumulation of fat, but no necrotic changes were noted in the hearts of rats given large amounts of brominated cottonseed oil for only 3 days.

REFERENCE

1. MUNRO, I. C.; MIDDLETON, E. J. and GRICE, H. C.: Biochemical and pathological changes in rats fed brominated cottonseed oil for 80 days. Fd. Cosmet. Toxicol. *7:* 25–33 (1969).

Mailing address: Ian C. Munro, Ph.D., Food and Drug Directorate, Tunney's Pasture, Ottawa 3, Ontario (Canada).

Department of Physiology, Faculty of Medicine,
University of Ottawa, Ottawa, Canada

MECHANISM OF ETHANOL-INDUCED TRIGLYCERIDE ACCUMULATION IN THE RABBIT HEART [1, 2]

K. J. KAKO and T. KIKUCHI

Following ingestion of ethanol, the fat content of the heart increases (6, 13). This is so, even after a moderate dose of ethanol administered in a relatively short period. The aim of this study, therefore, was to find the mechanism responsible for the observed increase in myocardial triglyceride (TG). Rabbits (av. 2.5 kg) were infused with 74 μl ethanol/min for the first 12 min, then 37 μl/min for 102 min, making a total of 4.8 ml over the period of 2 h. The experiments and analyses were carried out 3 h after the completion of the infusion. This mode of administration was chosen, since Regan *et al.* (13) previously employed this dose schedule and were able to provide extensive hemodynamic and biochemical data.

Figure 1 shows the possible metabolic pathways whereby TG can be synthesized in heart muscle.

Firstly, an increase in plasma free fatty acids (FFA) results in an increase in myocardial TG, as FFA are readily taken up by the heart (4). In addition, an increase in plasma TG can also result in TG accumulation. Although the mechanism of TG uptake is not well understood (1), one of the possible pathways is hydrolysis of TG to fatty acids (FA) by lipoprotein lipase (LPL) situated at the cell membrane, and the subsequent entry of the resultant FA into the cell. If this is so, the increased heart uptake of plasma TG should be associated with the increased activity of heart LPL, increased plasma TG, or both.

The second hypothesis is that ethanol increases the rate of carbohydrate metabolism and/or change in redox states of heart muscle. This would inevitably increase tissue α-glycerophosphate (α-GP), one of the percursors of TG. Reduced tissue DPN provides a favorable environment for FA synthesis in the cell (14). Both of these factors would tend to increase heart TG.

Lastly, a block in oxidation of FA at a site as yet unknown may produce extra FA CoA, which is likely to result in an increased esterification. These three hypotheses were tested (11) and the results are summarized below.

---

[1] Supported by the Medical Research Council of Canada, Ontario Heart Foundation, J.P. Bickell Foundation, and Licensed Beverage Industries, Inc.

[2] Dedicated to Dr. Yuzuru Kako on the occasion of his 74th birthday.

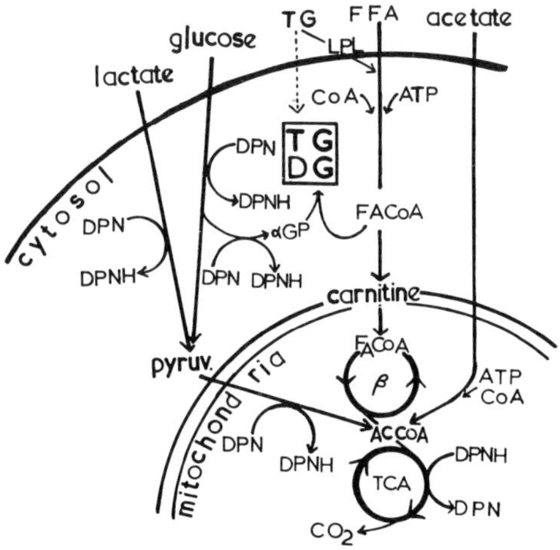

Fig. 1 Outline of the metabolic pathways related to fat accumulation in the heart.

Figure 2 illustrates the extent of the myocardial TG accumulation under our experimental conditions. Within 5 h after alcohol administration, the TG level increased to approximately 180% of the control, as shown by the first and second columns. There was no significant change in the plasma level of FFA and TG in the ethanol-infused rabbits (fig. 3). The plasma concentrations were

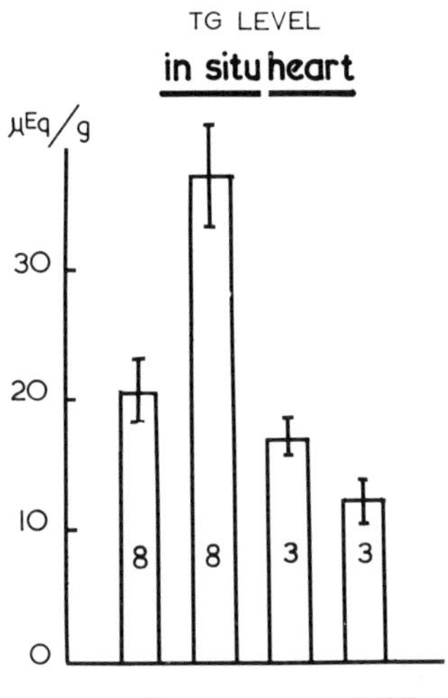

Fig. 2 Change in heart triglyceride (TG) under the influence of ethanol (EtOH). TG concentration is expressed as µEq/g wet weight of the left ventricle (ordinate). The mean, standard error (I), and number of experiments (figures in the columns) are shown. Norepinephrine (NE) was infused at a rate of 2–3 µg base/kg/min for 2 h and 20 mg of hydrocortisone succinate (COMP.F) was infused over 1 h.

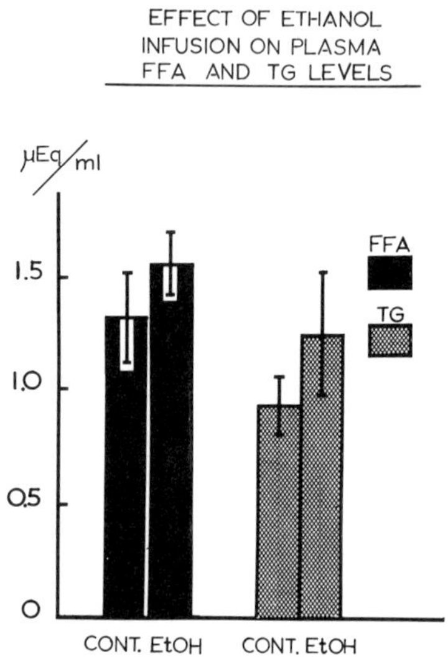

Fig. 3 Effect of ethanol (EtOH) infusion on plasma free fatty acid (FFA) and triglyceride (TG) levels. The concentration is expressed as μEq/ml of plasma (ordinate). The number of determinations was four in each case.

compared during the ethanol infusion, immediately following cessation, and 2 h after the infusion, giving the same results.

Figure 4 shows that no significant change in LPL activity (91 ± 22% of the control) occurs in heart muscle in response to ethanol. In this experiment, the activity was measured by using Ediol as the substrate (7). Although TG transport across the membrane is not solely controlled by these two factors, no further attempt was made to examine, quantitatively, the TG uptake by the heart. These results do suggest, however, that an increased penetration of FFA and TG as the cause of myocardial TG accumulation is unlikely.

Figure 5 shows the ratios of three pairs of substrates; the open columns represent the controls, and the shaded columns the ethanol-infused hearts. These substrate ratios indicate the states of oxidation and the reduction of the cell (2). A decreased ratio indicates oxidation and an increased ratio reduction. It may be concluded from these results that ethanol caused oxidation of the cellular oxidation-reduction state; in other words, tissue free DPNH decreased. This conclusion excludes our second possibility. At the same time, these results suggest that ethanol does not increase FA synthesis. The fact that mitochondrial pyridine nucleotides were more oxidized suggests that FA oxidation is actually depressed. These findings are in contrast to the action of ethanol on the liver (8, 15), where it produces a dramatic reduction of the redox state owing to the existence of the enzymes, alcohol and aldehyde dehydrogenases, which are not present in heart muscle.

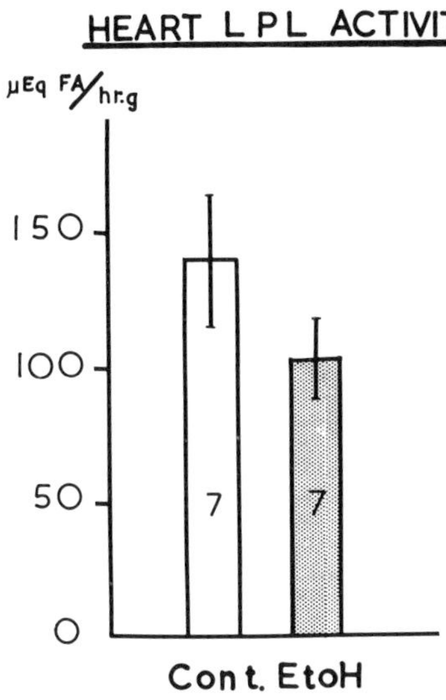

Fig. 4 Effect of ethanol (EtOH) infusion on heart lipoprotein lipase (LPL) activity, expressed as μEq fatty acids released per h per g wet weight (7). The standard error of the means and the number of experiments are shown.

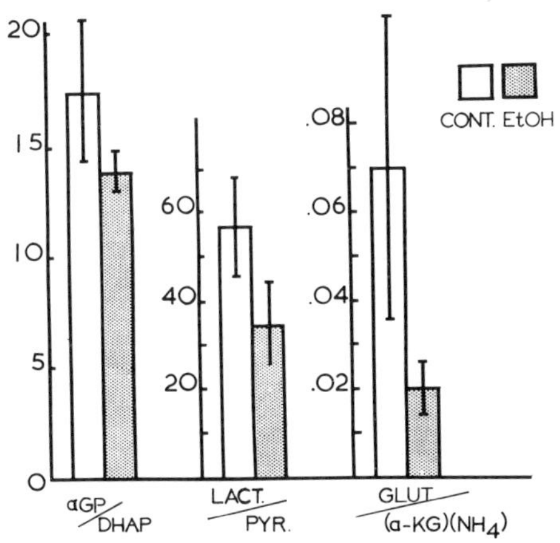

Fig. 5 Effect of ethanol (EtOH) on heart redox states. EtOH was infused for between 5 and 3 h before the heart was perfused with diluted (1 : 1) autologous blood for 30 min; the perfusion pressure was 60–80 mm Hg. The heart was freeze-quenched and the substrates determined by enzymatic-fluorometric methods. The ratio of concentrations of α-glycerophosphate (αGP) to dihydroxyacetone phosphate (DHAP), of lactate (LACT) to pyruvate (PYR), and of glutamate (GLUT) to the product of α-ketoglutarate (α-KG) and NH_4 are illustrated. The number of experiments was 9 in control and 5 in EtOH-treated hearts.

In the next experiment, the results of which are shown in figure 6, a decrease in FA oxidation was demonstrated directly. The effect of ethanol on the distribution of palmitate in heart homogenates is illustrated here. Homogenates prepared from the ethanol-infused animal were incubated in the presence of C-14 labelled potassium palmitate. Comparison between the results obtained in the controls (left) and those following ethanol-infusion (right) reveals that ethanol increased the esterification of FA to TG (270%) and decreased FA oxidation to CO_2 (70%). Since enhanced esterification of FA is observed in these heart homogenates in the absence of exogenous TG, increased membrane permeability to TG cannot be a major factor in the TG accumulation of heart muscle. The results also exclude the possibility that, under our experimental conditions, an increased level of acetate or lactate competitively inhibits the oxidation of the fatty acyl moiety in the myocardium, as neither substrate was present in the medium.

The results of FA oxidation and esterification studies were confirmed by

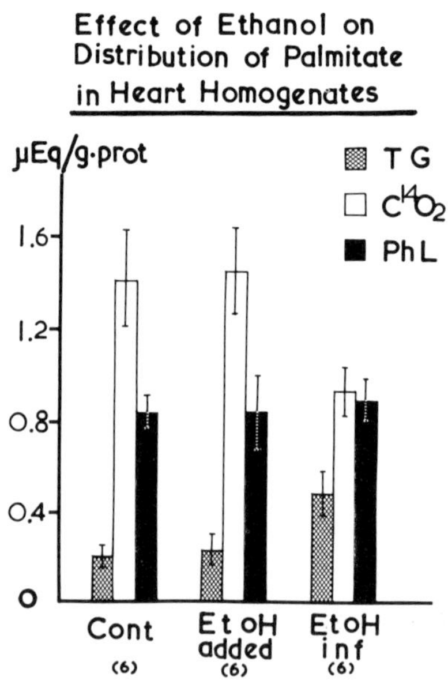

Fig. 6 Effect of ethanol (EtOH) on fatty acid oxidation and esterification by heart homogenates. Approximately 150 mg of 1 : 10 homogenates of rabbit left ventricle were incubated at 37°C for 30 min in 1.5 ml of medium containing 0.14 μEq(0.06 μc) of labelled potassium palmitate and Krebs phosphate buffer (pH 7.4); the concentration of ethanol added was 24 mM. $C^{14}O_2$ was absorbed on a filter paper soaked in 0.1 ml of 10% KOH after the pH of the medium was reduced to 4.5. The homogenates were treated with chloroform-methanol and the lipid extract was separated by thin-layer chromatography. Specific activities of the precursor in individual experiments were calculated by using the results of chemical and radio-chemical determinations of fatty acids in the incubation mixture; the results are expressed as μEq/g protein (ordinate).

parallel experiments in which rabbit hearts were perfused with 1 : 1 diluted autologous blood. Figure 7 shows the results of this series. Here again, we observed a large increase in the incorporation of FA carbon into the tissue TG fraction (174%), while no such increase was observed in any other lipid fraction. Although it is not shown here, ethanol decreased $C^{14}O_2$ production by half, supporting the results obtained by tissue homogenates. In order to localize the rate-limiting step, mitochondrial palmitoylcarnitine oxidation (9) was measured in a few experiments (fig. 8), and was found to be within the normal range. Similarly, lactate and pyruvate metabolism of the perfused rat heart was reported to be unchanged in the presence of a toxic concentration of ethanol (12). More sensitive methods (5) appear to be necessary to detect minute changes in mitochondria.

Preliminary observations indicated that two other factors control myocardial FA esterification, *i.e.*, the feeding condition of the animal and the rate of TG catabolism in the heart. Both heart TG and phospholipid fractions contain proportionally more unsaturated FA and less saturated FA in fasted than in fed rabbits (fig. 9)(10). The reason for this difference in FA compositon is unknown, since adipose tissue, the exclusive source of FA in the body during

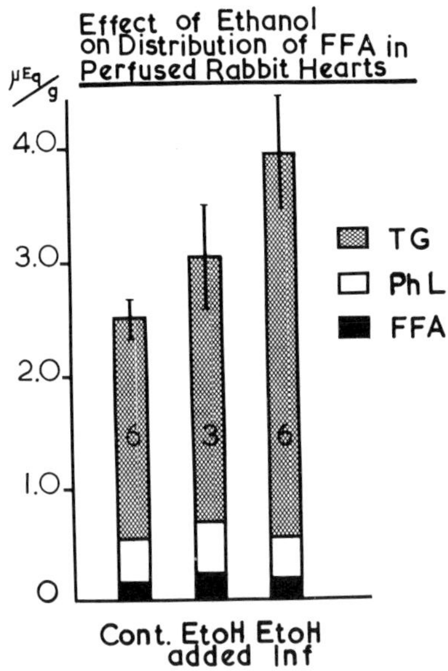

Fig. 7 Effect of ethanol on fatty acid esterification by perfused rabbit hearts. The experiment was carried out as outlined in the legend to figure 5. The concentration of ethanol added was 44.5 mM, and that of free fatty acids (FFA) was 1.0–1.3 μEq/ml. The specific activity of labelled FFA in the perfusate was 20–60 × 10^3 cpm/ μEq. Ordinate indicates μEq/g wet weight. The number of experiments and SEM are also shown.

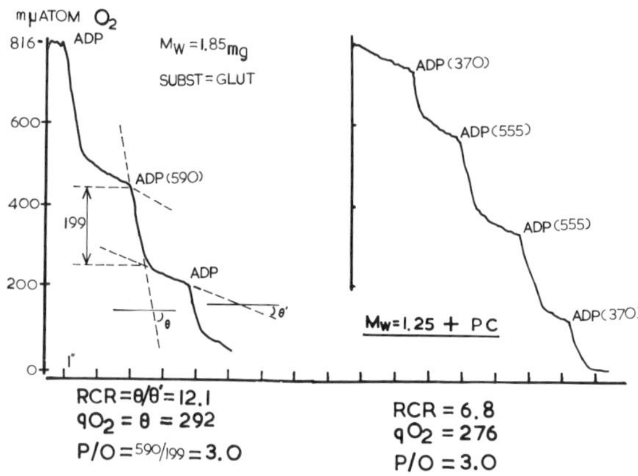

Fig. 8 The polarographic records of normal mitochondrial respiration. Heart mitochondria were prepared as previously reported (9). Mitochondria and either 17 μmole glutamate (left side) or 0.85 μmole palmitoylcarnitine and albumin (PC, right-side) were added to a reaction chamber containing 1.7 ml of 0.25 M sucrose, 10 mM Tris, and 8.5 mM K_2HPO_4 (final pH, 7.4), and the recording was started. ADP was added at the points indicated. The quantity of ADP is shown in parentheses in mμmole. Assays were carried out at 25°C. The ordinate illustrates oxygen content, and the abscissa, the time scale in min. RCR signifies the respiratory control ratio, qO2, the oxygen consumption in the presence of ADP, and P/O, the ADP (mμmole) to oxygen (mμatom) ratio. Mw denotes mitochondrial protein in mg. Palmitoylcarnitine was synthesized using DL-carnitine HCl and palmitoyl chloride, according to the method described by Brendel and Bressler (1a).

fasting, contains a relatively large proportion of palmitic acid. In contrast, ethanol administration results in a disproportionate increase in palmitic and a decrease in arachidonic acids in the myocardial neutral lipid of the fasted animal, suggesting that adipose tissue fat is the source of FA that accumulated in the heart under these conditions.

Although there is a paucity of information concerning the factors that control the rate of myocardial TG breakdown (3), it is possible that the rate of catabolism is greater in fasted than in fed animals. This was shown by our previous results of heart perfusion experiments, in which heart TG decreased from 18.6 to 6.1 μEq/g during a 2 h perfusion period in the fasted rabbit, while, in the fed rabbit, it was changed from 4.9 to 3.2 μEq/g (10). Furthermore, we found recently that ethanol infusion raised heart TG content to 124% of the control value in the fasted, in contrast to 173% in the fed animals, despite a lower plasma FFA concentration in the latter rabbits as compared with the former.

SUMMARY

We have found that the raise in myocardial triglyceride (TG) resulting from ethanol administration is not due to an increase in plasma free fatty acid (FA) and TG, or to a change in heart lipoprotein lipase activity. Glycolytic flux and the supply of α-glycero-

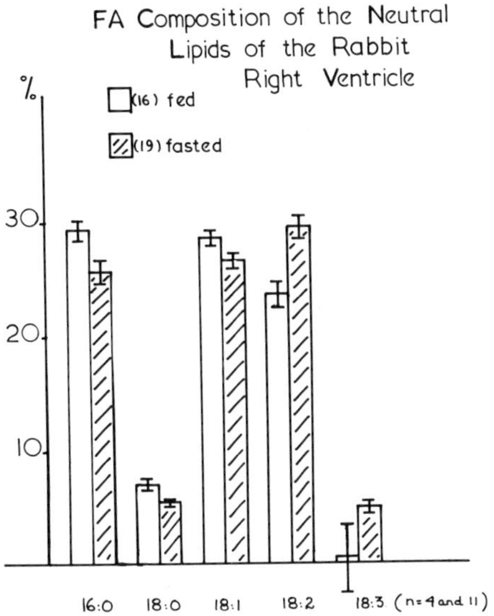

Fig. 9 Percentage of esterified fatty acids in the neutral lipid fraction of the heart of 48 h fasted and fed rabbits. Lipids were extracted (chloroform-methanol), separated (silicic acid), and methylated (boron trifluoride). Gas chromatography was carried out at 200°C, using a 9 ft × 1/8" column coated with 20% DEGS on Chromosorb W. Heptadecanoic acid was used as an internal standard. In this figure, the fatty acids are designated in terms of chain length: unsaturation, and the mean ± SEM as well as the number of observations are shown.

phosphate appear to be diminished, as reflected by oxidation of the cytoplasmic redox state. An increase in FA synthesis and competition by acetate are also both unlikely. Palmitoylcarnitine transferase activity does not appear to be a critical factor. We conclude from our results that the cause of TG accumulation is the suppression of FA oxidation at an, as yet, unknown step, followed by acceleration of esterification of the excess fatty acyl CoA, which becomes more available in the cytosol. The extent of the participation of α-glycerophosphate acyltransferase in ethanol-induced myocardial TG accumulation has still to be investigated.

REFERENCES

1. BING, R. J. Cardiac metabolism. Physiol. Rev. *45:* 171–213 (1965).

1a. BRENDEL, K. and BRESSLER, R.: Resolution of (±)-carnitine and the synthesis of acylcarnitines. Biochim. Biophys. Acta *137:*98–106 (1967).

2. BÜCHER, T. und KLINGENBERG, M.: Wege des Wasserstoffs in der lebendigen Organisation. Angew. Chem. *70:* 552–570 (1958).

3. CRASS, M. F., III and SHIPP, J. C.: Metabolism of exogenous and endogenous fatty acids in heart muscle. This Book, pp. 115–126.

4. EVANS, J. R.: Cellular transport of long chain fatty acids. Can. J. Biochem. *42:* 955–969 (1964).

5. FERRANS, V. J.; HIBBS, R. G.; WEILBAECHER, D. C.; BLACK, W. C.; WALSH, J. J. and BURCH, G. E.: Alcoholic cardiomyopathy. A histochemical study. Amer. Heart J. *69:* 748–765 (1965).

6. GUDBJARNASON, S. and PRASAD, A. S.: Cardiac metabolism in experimental alcoholism. *In* V. M. SARDESAI Biochemical and clinical aspects of alcohol metabolism, p. 266–272 (Charles Thomas, Springfield, 1969).
7. HUANG, C. C. and KAKO, K. J.: Mechanism of triglyceridemia in hypercholesterolemic rabbits. Circulat. Res. *26:* 771–782 (1970).
8. ISSELBACHER, K. J. and GREENBERGER, N. J.: Metabolic effects of alcohol on the liver. New Engl. J. Med. *270:* 351–356 and 402–410 (1964).
9. KAKO, K. J.: ADP-dependent palmitoylcarnitine sensitivity of mitochondria isolated from the perfused rabbit heart. Can. J. Biochem. *47:* 611–618 (1969).
10. KAKO, K. and DUBUC, M. J. G.: Changes in esterified fatty acids in the isolated, perfused rabbit heart. Can. J. Biochem. *46:* 1241–1246 (1968).
11. KIKUCHI, T. and KAKO, K. J.: Metabolic effect of ethanol on the rabbit heart. Circulat. Res. *26:* 625–634 (1970).
12. LOCHNER, A.; COWLEY, R. and BRINK, A. J.: Effect of ethanol on metabolism and function of perfused rat heart. Amer. Heart J. *78:* 770–780 (1969).
13. REGAN, T. J.; KOROXENIDIS, G.; MOSCHOS, C. B.; OLDEWURTEL, H. A.; LEHAN, P. H. and HELLEMS, H. K.: The acute metabolic and hemodynamic responses of the left ventricle to ethanol. J. Clin. Invest. *45:* 270–280 (1966).
14. WHEREAT, A. F.; HULL, F. E.; ORISHIMO, M. W. and RABINOWITZ, J. R.: The role of succinate in the regulation of fatty acid synthesis by heart mitochondria. J. Biol. Chem. *242:* 4013–4022 (1967).
15. WILLIAMSON, J. R.; SCHOLZ, R.; BROWNING, E. T.; THURMAN, R. G. and FUKAMI, M. H.: Metabolic effects of ethanol in perfused rat liver. J. Biol. Chem. *244:* 5044–5054 (1969).

Mailing address: Dr. K. J. Kako, Department of Physiology, University of Ottawa, Faculty of Medicine, Ottawa (Canada).

Institut de Médecine et de Chirurgie expérimentales,
Université de de Montréal, Montréal, Québec, Canada

INFLUENCE OF SPIRONOLACTONE AND OTHER STEROIDS ON THE PLASMA LEVEL OF DIGITOXIN [1]

B. SOLYMOSS, S. TOTH, S. VARGA, and M. KRAJNY

Pretreatment with spironolactone or certain other steroids (e. g., norbolethone, fluoxymesterone, oxandrolone, methylandrostenediol, ethylestrenol) prevents the convulsions and mortality induced by digitoxin in rats (18, 19). Gitalin, proscillaridin, digoxin, or digitalin intoxication is similarly suppressed by spironolactone or norbolethone (20). These observations initiated a systematic study of the resistance-increasing or "catatoxic" effect of various steroids (17) and, subsequently, it was shown that spironolactone, norbolethone, and ethylestrenol stimulate hepatic drug-metabolizing enzymes and enhance the degradation of various drugs (e. g., pentobarbital, hexobarbital, aminopyrine, bishydroxycoumarin) (22–24). It is probable that the production of hydroxylated polar metabolites by NADPH-dependent mixed-function oxidases is largely responsible for the prevention of various types of intoxication.

Digitoxin is metabolized through different pathways (fig. 1). Cardiac glycosides contain an unsaturated lactone ring, the reduction of which diminishes their biologic activity (25). These substances also possess a hydroxyl group at C_3, which can bind various sugar molecules to the steroid skeleton.

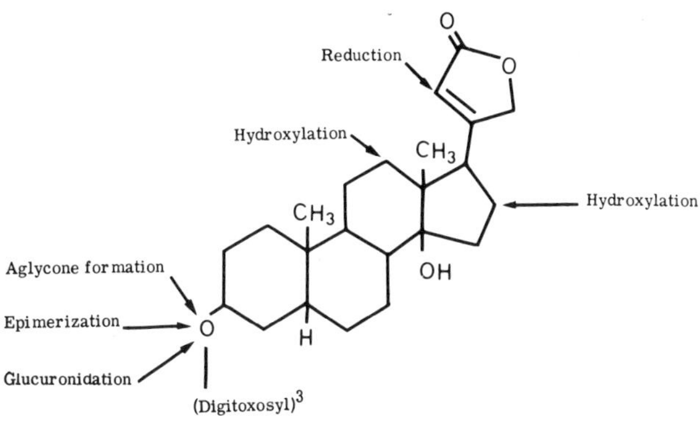

Fig. 1 Pathways of biotransformation of digitoxin.

[1] Supported by grants from le Ministère de la Santé de Québec and the Medical Research Council of Canada (Block Term Grant MT-1829).

These sugars inhibit the degradation of glycosides to less active derivatives (10, 14). Enzymatic fission of cardenolids, production of aglycones, and the subsequent epimerization of the free hydroxyl group at C_3 are important steps in the detoxication of digitalis compounds (14, 15). The enzymes of epimerization are mainly found in the liver and are apparently the same as those that catalyze similar processes in steroid hormones (14). Furthermore, hepatic 12- or 16β-hydroxylation of digitoxin leads to the production of digoxin or gitoxin; these compounds are not accumulated in target organs in the same way as digitoxin, but are excreted more easily in the urine and feces (3, 5, 14). The polarity and, consequently, the excretion of cardiac steroids is also influenced by glucuronidation (14). Some of these processes were shown to occur even *in vitro* if digitoxin is incubated with liver slices (10).

Drug-metabolizing enzymes and steroid hydroxylases are considered to be similar in many respects (9); hence, it can be supposed that the stimulation of these hepatic mixed-function oxidases by various steroids might influence the biotransformation of digitoxin, enhance the production of its polar metabolites and, in turn, decrease the plasma level of the drug. Therefore, we studied the influence of spironolactone, norbolethone, and ethylestrenol on the plasma level of digitoxin as well as the potential suppressive action on these processes of SKF 525-A, a well-known drug-metabolizing enzyme inhibitor (1, 6, 16), and cycloheximide, a protein synthesis inhibitor (12).

MATERIALS AND METHODS

Female ARS/Sprague-Dawley rats[2] with a mean initial body weight of 100 g (range 90–110 g), maintained *ad libitum* on Purina Laboratory Chow and tap water, were used throughout the experiments. The doses of the compounds administered are expressed per 100 g of body weight.

First experiment

Seventy rats were divided into 7 equal groups and treated as indicated in table I. Digitoxin was given at different dose levels as micronized suspension in 1 ml of H_2O by stomach tube. Blood samples were taken at different time intervals under light ether anesthesia by aortic puncture. The plasma was extracted by chloroform (13), the extract purified through acid alumina Ag-columns (7), and digitoxin was measured fluorometrically (8, 26) with a Hitachi-Perkins-Elmer MPF 2A recording spectrofluorometer.

Second experiment

One hundred and ten rats were divided into 11 equal groups and treated as indicated in table II. Groups 2–4, 6–8, and 11 were pretreated with

[2] Madison, Wisconsin, U.S.A.

Table I

Influence of Dose and Time on the Plasma Level of Digitoxin (Mean ± S.E.)

Group	Dose of digitoxin (mg)	Hours elapsed after p.o. digitoxin administration	Plasma level of digitoxin (μg %)
1	0.5	18	7.7 ± 2.0
2	1.0	18	14.4 ± 1.8[a]
3	2.0	18	33.4 ± 4.9[a]
4	1.0	3	56.1 ± 3.1
5	1.0	6	58.3 ± 3.3
6	1.0	18	14.8 ± 1.7[a]
7	1.0	24	10.0 ± 2.5

[a] Significant difference ($p < 0.05$) as compared with the preceding group. Each group consisted of 10 rats, treated as described in the text.

spironolactone, norbolethone, and ethylestrenol (10 mg in 1 ml of H_2O as micronized emulsion, p.o., twice daily) for 3 days. Cycloheximide (100 μg in 0.2 ml of physiological NaCl, s.c.) was also given to the animals of group 4, 1 h before each spironolactone treatment. Twenty-four hours before digitoxin administration, the animals of group 9 were partially hepatectomized by ablation of the left lateral and the median lobes, thus removing about 70% of

Table II

Effect of Spironolactone and Other Pretreatments on the Plasma Level of Digitoxin (Mean ± S.E.)

Group	Pretreatment	Hours elapsed after 1 mg digitoxin (p.o.)	Plasma level of digitoxin (μg %)	Convulsions (scale: 0 – +++)
1	None	6	56.6 ± 2.3	
2	Spironolactone	6	24.8 ± 1.4[a]	
3	Spironolactone + SKF 525-A	6	33.6 ± 3.6[b]	
4	Spironolactone + cycloheximide	6	35.0 ± 4.7[b]	
5	None	18	17.8 ± 2.3	++
6	Spironolactone	18	6.0 ± 0.7[c]	0
7	Norbolethone	18	7.1 ± 2.0[c]	0
8	Ethylestrenol	18	5.7 ± 0.9[c]	0
9	Partial hepatectomy	18	27.4 ± 3.2[c]	+++
10	SKF 525-A	18	20.3 ± 5.0	++
11	Spironolactone + SKF 525-A	18	14.3 ± 0.9[d]	±

[a,c] Significantly different ($p < 0.05$) from corresponding controls (groups 1 and 5, respectively).

[b] Significantly different ($p < 0.05$) from group 2.

[d] Significantly different ($p < 0.05$) from group 6.

Each group consisted of 10 rats, treated as described in the text.

hepatic tissue (27). On the 4th day at 9 a.m., the animals of groups 3, 10, and 11 were injected with SKF 525-A (5 mg in 0.5 ml of H_2O, i.p.); 1 h later, all animals received a test dose of digitoxin (1 mg in 1 ml of H_2O as micronized suspension, p.o.), and blood samples were taken for digitoxin determination, as described above, 6 h later in groups 1–4 and 18 h later in groups 5–11. The convulsive motor disturbances in groups 5–11 were assessed in terms of an arbitrary scale of 0–+++ (18).

The differences in the plasma digitoxin level of the different groups were statistically evaluated by Student's *t*-test (21).

RESULTS

Table I shows that the plasma level of digitoxin is dose related: the administration of increasing amounts of cardiac steroid results in increasing plasma levels of the drug, as measured 18 h later. Furthermore, approximately the same amount of digitoxin is found in the plasma 3 or 6 h after its oral administration, but there is a significant decrease 18 h later. Thereafter, elimination is slow (*cf.* groups 6 and 7).

Table II shows that 6 h after digitoxin administration, its plasma concentration is lower in spironolactone-pretreated animals (group 2) than in controls (group 1). The results in groups 3 and 4 demonstrate that SKF 525–A or cycloheximide partially suppresses the enhanced decrease of the plasma digitoxin level induced by spironolactone. The results in groups 6, 7, and 8 as compared with the controls (group 5) indicate that norbolethone and ethylestrenol share the decreasing action of spironolactone on plasma digitoxin concentration, as measured 18 h after the test dose of the drug. On the other hand, the plasma digitoxin content is significantly higher in partially hepatectomized animals (group 9) than in controls. Moreover, while SKF 525-A *per se* does not significantly influence the plasma digitoxin level (group 10), this microsomal inhibitor also suppresses the action of spironolactone (group 11) at the 18 h time interval.

Furthermore, as can be seen in table II, severe motor disturbances were observed in control animals (group 5) 18 h after digitoxin administration; while spironolactone, norbolethone and ethylestrenol pretreatment completely prevented these disturbances (groups 6–8), partial hepatectomy increased their severity (group 9). SKF 525-A slightly suppressed the preventive effect of spironolactone against digitoxin-induced convulsions (group 11).

DISCUSSION

Cardiac glycosides can be quantitatively measured by biologic (2), colorimetric (13), fluorometric (8, 26), enzymatic (4), or isotopic (11) techniques.

The fluorometric method is applicable if relatively high amounts of cardiac glycosides are given. However, this approach is not suitable for a quantitative assessment, following therapeutic doses of cardenolids. In order to diminish the influence of interfering compounds, plasma extracts were purified through acid alumina Ag-4 columns (7). With the use of this method, a good correlation was found between the plasma level of digitoxin and the dose administered, as is indicated in table I.

After partial hepatectomy, the administration of digitoxin results in a higher plasma level of the drug and more severe convulsions than in control animals (table II). Spironolactone, norbolethone and ethylestrenol, on the other hand, decrease the plasma digitoxin concentration and prevent the development of convulsions otherwise elicited by the cardiac steroid in the rat.

It is noteworthy that the decrease in plasma digitoxin content, following spironolactone pretreatment, is suppressed by SKF 525-A and cycloheximide. It was shown earlier that SKF 525-A prevents the metabolic transformation of various compounds *in vivo* (1); in liver slices or homogenates, it inhibits the enzyme systems that play an important role in drug metabolism (6). The compound affects the orderly arrangement of the endoplasmic reticulum in hepatocytes and, presumably, thereby influences a large number of metabolic pathways and microsomal drug-metabolizing processes (16). Cycloheximide blocks enzyme production by inhibiting the translation of mRNA information into protein synthesis, since it suppresses the transfer of amino acids from soluble RNA to polypeptide chains in liver extracts (12). Consequently, our findings are consistent with the view that induction of hepatic microsomal enzymes is at least partly responsible for the influence of spironolactone pretreatment on the plasma digitoxin content and on convulsions. Besides other alterations, the biotransformation of digitoxin into its hydroxylated derivatives, digoxin and gitoxin, might be enhanced by mixed function oxidases and these metabolites are more rapidly excreted in urine or feces than is the original compound (3, 5, 14).

In liver microsomal + supernatant fraction, spironolactone, norbolethone, and ethylestrenol increase the NADPH-dependent enzymatic degradation of pentobarbital, hexobarbital, aminopyrine or bishydroxycoumarin, with a corresponding decrease in the blood concentration of these drugs (22–24). Direct evidence of the degradation of digitoxin by liver homogenates or hepatic microsomes is not yet available. All attempts, including our own, were unsuccessful in this respect, probably because of the strong binding of the drug to proteins (10), or owing to other factors that might prevent extraction into the organic solvents. However, the observed action of SKF 525-A and cycloheximide indirectly supports the view that the effect of various steroids on the disappearance of digitoxin from plasma is mediated, at least partly, through the stimulation of drug-metabolizing enzymes in hepatic microsomes.

SUMMARY

In spironolactone-, norbolethone- or ethylestrenol-pretreated rats, the plasma-clearance of digitoxin is enhanced in proportion to the decreasing severity of the drug-induced convulsions. Partial hepatectomy reduces digitoxin elimination. The action of spironolactone is suppressed by SKF 525-A and cycloheximide.

Spironolactone, norbolethone, and ethylestrenol were previously shown to increase NADPH-dependent mixed-function oxidation in hepatic microsomal + supernatant fraction; hence, the present findings are consistent with the view that these steroids probably enhance the production of polar digitoxin metabolites, thus facilitating their excretion. Since spironolactone and digitoxin are sometimes given conjointly, these results might also have clinical implications.

REFERENCES

1. AXELROD, J.; REICHENTHAL, J. and BRODIE, B. B.: Mechanism of the potentiating action of beta-diethylaminoethyl diphenylpropylacetate. J. Pharmacol. Exp. Ther. *112:* 49–54 (1954).
2. BINE, R.; FRIEDMAN, M.; BYERS, S. O. and BLAND, C.: The deposition of digitoxin in the tissues of the rat after parenteral injection. Circulation *4:* 105–107 (1951).
3. BROWN, B. T.; WRIGHT, S. E. and OKITA, G. T.: C_{12}–Hydroxylation of digitoxin. Nature *180:* 607–608 (1957).
4. BURNETT, G. H. and CONKLIN, R. L.: The enzymatic assay of plasma digitoxin levels. J. Lab. Clin. Med. *71:* 1040–1044 (1968).
5. CLOETTA, M.: Preparation and chemical composition of active substances of digitalis leaves: their pharmacologic-therapeutic properties. Arch. Exp. Path. Pharmak. *112:* 261–342 (1926).
6. COOPER, J. R.; AXELROD, J. and BRODIE, B. B.: Inhibitory effects of beta--diethylaminoethyl diphenylpropylacetate on a variety of drug metabolic pathways in vitro. J. Pharmacol. Exp. Ther. *112:* 55–63 (1954).
7. JELIFFE, R. W. and STEPHENSON, R. G.: A fluorometric determination of myocardial digoxin at autopsy, with identification of digitalis leaf, digitoxin and gitoxin. Amer. J. Clin. Path. *51:* 347–357 (1969).
8. JENSEN, K. B.: Fluorometric determination of gitoxingenin. Acta Pharmacol. Toxicol. *8:* 101–109 (1952).
9. KUNTZMAN, R.; JACOBSON, M.; SCHEIDMEN, K. and CONNEY, A. H.: Similarities between oxidative drug-metabolizing enzymes and steroid hydroxylases in liver microsomes. J. Pharmacol. Exp. Ther. *146:* 280–285 (1964).
10. LAUTERBACH, F. und REPKE, K.: Die fermentative Abspaltung von d-digitoxose, d-cymarose und 1-thevetose aus Herzglykosiden durch Leberschnitte. Naunyn-Schmiedeberg's Arch. Exp. Path. Pharmak. *239:* 196–218 (1960).
11. LULLMAN, H. and VAN ZWIETEN, P. A.: The kinetic behaviour of cardiac glycosides in vivo, measured by isotope techniques. J. Pharm. Pharmacol. *21:* 1–8 (1969).
12. MIS, H. L. and LUBIN, M.: Cycloheximide: aspects of inhibition of protein synthesis in mammalian cells. Science *146:* 1474–1476 (1964).
13. REPKE, K.: Die chemische Bestimmung von Digitoxin in Geweben und Ausscheidungen. Naunyn-Schmiedeberg's Arch. Exp. Path. Pharmak. *233:* 261–270 (1958).
14. REPKE, K.: Metabolism of cardiac glycosides in: WILBRANDT, W. and LINDGREN, P. Mode of Action of Drugs – Vol. 3, pp. 47–73 (Pergamon, Oxford 1963).
15. REPKE, K. and LAUTERBACH, F.: Cardenolide aglycones as hydrogen carriers between pyridine-nucleotide systems. Naunyn-Schmiedeberg's Arch. Exp. Path. Pharmak. *238:* 46–47 (1960).

16. ROGERS, L. A. and FOUTS, J. R.: Some of the interactions of SKF 525-A with hepatic microsomes. J. Pharmacol. Exp. Ther. *146:* 286–293 (1964).

17. SELYE, H.: Adaptive steroids: retrospect and prospect. Perspect. Biol. Med. *13:* 343–363 (1970).

18. SELYE, H.; JELINEK, J. and KRAJNY, M.: Prevention of digitoxin poisoning by various steroids. J. Pharm. Sci. *58:* 1055–1059 (1969).

19. SELYE, H.; KRAJNY, M. and SAVOIE, L.: Digitoxin poisoning: prevention by spironolactone. Science *164:* 842–843 (1969).

20. SELYE, H.; MECS, I. and TAMURA, T.: Effect of spironolactone and norbolethone on the toxicity of digitalis compounds in the rat. Brit. J. Pharmacol. *37:* 485–488 (1969).

21. SNEDECOR, G. W. and COCHRAN, W. G.: Statistical Methods (6th edition). The Iowa State University Press, Ames, Iowa, U.S.A. (1967).

22. SOLYMOSS, B.; CLASSEN, H. G. and VARGA, S.: Increased hepatic microsomal activity induced by spironolactone and other steroids. Proc. Soc. Exp. Biol. (N.Y.) *132:* 940–942 (1969).

23. SOLYMOSS, B.; VARGA, S. and CLASSEN, H. G.: The effect of various steroids on microsomal aliphatic hydroxylation and N-dealkylation. Europ. J. Pharmacol. *10:* 127–130 (1970).

24. SOLYMOSS, B.; VARGA, S.; KRAJNY, M. and WERRINGLOER, J.: Influence of spironolactone and other steroids on the enzymatic decay and anticoagulant activity of bishydroxycoumarin. Thrombos. Diathes. haemorrh. (Stuttg). *23:* 562–568 (1970).

25. STRAUB, W.: Chemischer Bau und pharmakologische Wirksamkeit in der Digitalisgruppe. Biochem. Z. *75:* 132–144 (1916).

26. WELLS, D.; KATZUNG, B. and MEYERS, F. H.: Spectrofluorometric analysis of cardiotonic steroids. J. Pharm. Pharmacol. *13:* 389–395 (1961).

27. WAELSCH, H. and SELYE, H.: Beitrage zur Entgiftung im tierischen Organismus. III. Bedeutung der Leber bei Avertin- und Magnesium- narkosen. Naunyn-Schmiedeberg's Arch. Exp. Path. Pharmak. *161:* 115–118 (1931).

Mailing address: Dr. B. Solymoss, Institut de Médecine et de Chirurgie expérimentales, Université de Montréal, Faculté de Médecine, Case Postale 6128, Montréal, Que. (Canada).

Electrolytes and Cardiovascular Diseases

Schering Corporation, Bloomfield, N. J.,
New York Medical College, Pharmacology Department,
Maimonides Medical Center, New York, N. Y., USA

MYOCARDIAL LOSS OF FUNCTIONAL MAGNESIUM

I. EFFECT ON MITOCHONDRIAL INTEGRITY AND POTASSIUM RETENTION

M. S. SEELIG

There is growing interest in the importance of magnesium in myocardial structure and function. Most clinicians in the United States still place greatest emphasis on potassium loss in the etiology or intensification of cardiomyopathy. This paper presents an analysis of the data that indicate that the magnesium loss from the heart, as from other tissues, interferes with the cellular machinery without which cardiac potassium cannot be maintained. Thus, magnesium loss predisposes to potassium loss.

The cardiac loss of potassium seen in magnesium deficiency was first attributed by Vitale and his associates (14, 57) to the necessity of magnesium for the normal functioning and structural integrity of heart mitochondria. They correlated the decreased oxidative activity and/or uncoupling of oxidative phosphorylation by cardiac mitochondria from their magnesium-deficient animals (14, 55, 58) with the biochemically demonstrated importance of oxidative phosphorylation in the active transport of potassium (33, 54). Welt and his associates (59–61) have implicated defective functioning of membrane adenosine triphosphatase (ATPase) as a probable cause of the loss of K by Mg-deficient cells. On the basis of the biochemical work on subcellular membrane particles that showed that Na^+-K^+-ATPase is Mg-dependent and is involved in the transport of Na^+ and K^+, they concluded that magnesium is necessary for the action of "pump" ATPase. Without its optimal activity, the cells apparently lose the capacity to maintain an appropriate concentration gradient across the cell membranes (table I).

Skou (49–52), first utilizing microsomes from crab nerves and then from mammalian brain and kidney preparations, showed that Mg^{++} is necessary for the activity of ATPase in active transport of Na^+ and K^+ across an electrochemical gradient. Post *et al.* (38) using erythrocyte membranes showed that both Na^+ and K^+-dependent and independent ATPases are activated by Mg^{++}. Dunham and Glynn (15, 19) have also demonstrated that the ATPases require Mg^{++} for full activity, and that Na^+-K^+-ATPase is directly involved in the ionic pump. The presence of Mg-dependent, Na^+-K^+-ATPase has been demonstrated in cardiac microsomes by Auditore and Murray (1, 2) and by Schwartz and Laseter (42, 44).

Table I

Magnesium-dependent Particulate Adenosine Triphosphatases

Investigators	Source of ATPase	Role of Mg^{++}	Inhibitors
Skou (49–52)	Microsomal particle (Crab nerve, guinea pig, rat, rabbit) Brain Kidney	Mg^{++} requirement for Na^+-K^+ transport across electrochemical gradient	Ca^{++} (of Mg-ATPase) Strophanthin (of Na-K-Mg ATPase)
Post *et al.* (38)	Human red blood cell ghosts Guinea pig kidney	Mg^{++} requirement for Na^+-K^+ dependent / independent ATPase	
Dunham and Glynn (15) Glynn (19)	Human red blood cells Microsomes (electric organ)	Mg^{++} requirement for Na^+-K^+ dependent / independent ATPase	Ca^{++} Cardiac glycosides
Auditore (1) Auditore and Murray (2)	Mitochondria; microsomes Rabbit heart	ATPase role in ionic pump Mg^{++} activates: Na^+-K^+ dependent / independent ATPase	G-Strophanthin
Schwartz (41) Schwartz and Laseter (44)	Microsomes Rat heart	Mg^{++}-Na^+-K^+ ATPase Mg^{++}-ATPase	Ouabain

Because under most circumstances magnesium deficiency is associated with hypercalcemia and because agents that cause hypercalcemia have cardiotoxic potential, the demonstration by some investigators (15, 19, 49–52) of inhibition by Ca^{++} of Mg-activated ATPase is especially interesting. The inhibition by cardiotonic agents of the ATPase that is dependent on Na^{+}, K^{+}, and Mg^{++} (1, 2, 15, 19, 42, 44) is also important, because of the loss of myocardial Mg from hearts of animals with digitalis intoxication (25), the increased sensitivity to glycosides of Mg-deficient animals (31, 45, 56, 57), and the efficiency of magnesium in counteracting digitalis toxicity (45, 53, 62).

Review of some of the basic work on the role of magnesium in mitochondrial structure and in the activity of microsomal ATPase reveals a pivotal position of this cation in normal physiology; the present paper cannot encompass the many additional Mg-dependent enzyme systems. The efficacy of Mg in preventing and reversing early damage to mitochondria caused by a variety of challenges provides a clue to the efficacy of magnesium salts against a number of diverse cardiotoxic agents (for reviews of cardioprotective effect of Mg salts, see 3, 4, 46, 47).

Wherever possible, data have been selected from studies with cardiac mitochondria, because of evidence that mitochondria from the heart differ from the parenchymal mitochondria that have been more commonly used in the basic studies. Green and Fleischer (20) have observed that cardiac mitochondria are jam-packed with cristae, whereas liver mitochondria have relatively few cristae, and they are more widely spaced. The greater the oxidative-phosphorylative activity of the cells, the tighter the arrangement of cristae. DiGiorgio *et al.* (14) observed that heart mitochondria are more susceptible than are liver or kidney mitochondria to the impairment of oxidative phosphorylation that is caused by magnesium deficiency and/or thyroxine. Carafoli *et al.* (10) observed that active accumulation of Mg^{++} and Ca^{++} by the liver mitochondria they employed was not great, whereas the heart mitochondria used by Brierley *et al.* (7, 8) had considerable avidity for Mg^{++} and inorganic P ions, coincident with which H^{+} was released. Ernster (16) has considered the evidence that mitochondrial accumulation of divalent cations with inorganic phosphate functions as a buffering system.

Mitochondrial membranes have been described by Gebicki and Hunter (18) as forming a backbone on which the components of the electron transport system and oxidative phosphorylation system are arranged. Green and Fleischer (20) have described the mitochondrion as the structure that provides utilizable energy for the basic cellular functions (table II). Mitochondria are devices for oxidative phosphorylation, or coupling the synthesis of ATP to the oxidative reactions of the citric acid cycle, for fatty acid oxidation, and for utilization of fatty acids in the formation of acetyl coenzyme A. The electron transport system, which participates in oxidation-reduction reactions and active ion transport, is also a

Table II

Mitochondrial Functions

• Provides utilizable energy for basic cellular functions
• Oxidative-phosphorylation (couples synthesis of ATP to the oxidative steps of the citric acid cycle)
• Fatty acid oxidation (in heart mitochondria)
• Fatty acid conversion to acetyl coenzyme A (coupling fatty acid to citric acid oxidation)
• Electron transport system: respiratory chain oxidation-reduction reactions
• Energy for active ion transport

mitochondrial subunit. The arrangement and organization of enzymes and substrates of the mitochondrion implement its functions in cellular metabolism (20). Physiologic swelling and contraction of the mitochondria lead to altered biochemical activity in response to changes in cellular activity; pathologic swelling and disruption of the mitochondria lead to abnormalities in biochemical activity and ultimately to cell death.

Magnesium is intimately involved in mitochondrial structure and function (fig. 1). It is necessary for formation of the compact Mg-ATP complex (5, 17, 36) at which phase the mitochondrion is "contracted." Calcium, on the other hand, favors the release of ATP from the mitochondria. Ernster and Low (17) have described the mitochondrial arrangement as then being "loose." This is the swollen mitochondrion. As Ca^{++} is taken up by mitochondria (liver), there is an increase in intramitochondrial Na^{+} and a decrease in K^{+}, with reversal of the K : Na ratio (10). It has recently been suggested that, at this phase, there is ATP-induced uptake of Ca^{++} by sarcosomal membranes (11), which is associated with contraction of muscles. (21). Mg^{++}, possibly under the influence of parathyroid hormone, enters the mitochondria in association with phosphate, complexes with ATP in the mitochondria, which then contract. At this phase, K^{+} enters the mitochondria and Na^{+} leaves[1]. Magnesium's complexing with diphosphopyridine nucleotide (DPN) participates as part of the electron-transport respiratory chain in activation of mitochondrial ATPase (37, 48).

Recent findings by Schwartz and his colleagues (28, 29, 43) indicate that K^{+} efflux from mitochondria is an energy-dependent process that is effected by polycationic proteins, including parathyroid hormone and histone. These authors have suggested that histone may enter the mitochondrion with phosphate via an energy-dependent uptake mechanism. Within the mitochondrion, the histone-PO_4 complex would lead to efflux of K^{+}, and possibly of

[1]For more detailed and additional information on the steps involved in the ion transfer reactions, see: Carafoli *et al.* (10), Ernster (16), Rasmussen *et al.* (30, 39), and Sallis *et al.* (41).

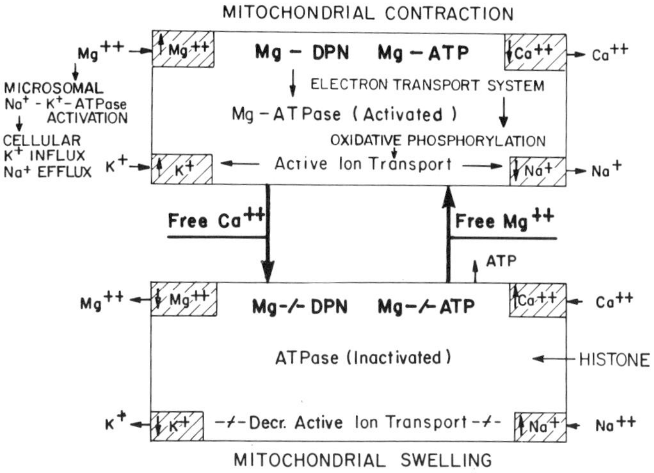

Fig. 1 Physiologic mitochondrial processes.

Mg^{++}, in exchange for the polycationic protein. The mitochondrial swelling and K^+ efflux thus induced is prevented by addition of Mg^{++}.

In their detailed review of the literature on the respiratory chain and oxidative phosphorylation up to 1956, Chance and Williams (12) considered the evidence substantial that Mg acts to preserve the mitochondrial structure. They observed that this function logically explains the prevention by magnesium of the mitochondrial swelling caused either by Ca^{++} or thyroxine. Ernster and his associates (17, 48), Baltscheffsky (5), and Hunter *et al.* (26) have attributed magnesium's stabilization of mitochondria to its complexing with chemical groups on its membrane.

Magnesium deficiency thus causes cellular loss of potassium by several means: a) through decreased microsomal Mg-Na-K-ATPase activity, as a result of which there is decreased ability to maintain an appropriate concentration gradient; b) through basic protein (*i.e.*, histone)-induced K^+ efflux from the mitochondrion; and finally c) through mitochondrial disruption, with destruction of the cellular machinery required for cellular activity (fig. 2).

Lehninger (32) has observed that mitochondrial swelling caused by a wide diversity of chemical agents is preventable or reversible by what appears to be a single, basic mechanism that is ATP-driven. It should be noted, here, that Nanninga (36) has provided evidence that 90% of muscle ATP exists as Mg-ATP, the Mg^{++} probably acting as a bridge, binding substrate to enzyme. In fact, the extent of the mitochondrial swelling has been shown to be inversely proportional to the concentration of Mg^{++} (13).

Brierley (6) has recently suggested that cardiac mitochondrial membrane lesions may favor massive accumulation of magnesium and inorganic phosphate.

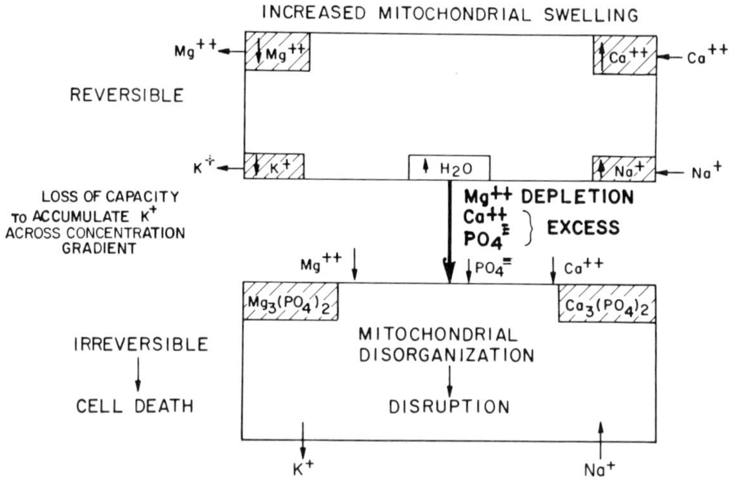

Fig. 2 Mitochondrial changes with magnesium deficiency.

Isolated mitochondria with tightly coupled oxidative phosphorylation accumulated the magnesium more slowly than did those with impaired respiratory control (caused by mechanical damage or by heavy metals). Brierley (6) considers the resultant precipitation of $Mg_3\ (PO_4)_2$ within the damaged mitochondria as a probable pathologic reaction (fig. 2).

This provides the first clear indication that not all of the magnesium in the heart is necessarily functional. Early studies of cardiac mitochondria from magnesium-deficient animals suggested this possibility. For example, Nakamura *et al.* (35) reported that rats on a Mg-deficient diet for 12 days had swollen cardiac mitochondria that contained about the same amount of magnesium (or even slightly more) than did the mitochondria from control rats. That these were not physiologically swollen mitochondria is indicated by the morphologic disorganization of the cristae seen on electron microscopy. Di Giorgio *et al.* (14) proposed that cardiac sarcosomes from Mg-deficient ducks contained Mg which was either insufficient or present in a form unsuitable for coupling of oxidation to phosphorylation.

The particulate electron-dense material seen in the cardiac sarcosomes of Mg-deficient rats was interpreted by Heggtveit *et al.* (22–24) and Mishra and Herman (34) as probable evidence of calcification (fig. 3). That these granules were, indeed, probably calcium is indicated by the observation by Brierley and Slauterback (9) that the fixation procedure for preparation of cardiac mitochondrial pellets, employing osmium tetroxide, removes 80% or more of the inorganic phosphorus associated with magnesium, but only about half that associated with calcium.

Jennings (27) has very recently observed that the development of dense

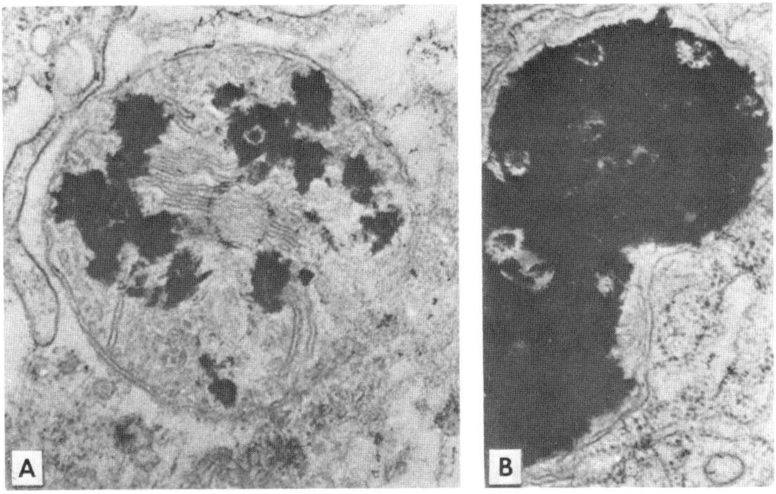

Fig. 3 Degree of filling of mitochondria with electron-dense granules, presumably Ca, in Mg-deficient, stressed rat, day 29. A, Solid electron-dense particles; B, sarcosome filled with tightly packed particles. From Heggtveit *et al.* (24).

granules in the cardiac mitochondria 20 min after coronary occlusion may well be caused by redistribution of Ca^{++} or Mg^{++} within the mitochondria. He has suggested that the crystallization or binding of essential cofactors like inorganic phosphate and Mg^{++} in the granules might contribute to irreversible mitochondrial failure. Demonstration of this shift from functional to nonfunctional Mg may have been provided by Hochrein *et al.* (25) (table III). The progressive drop of myocardial Mg, seen in guinea pigs for the first 4 min of asphyxia, and the

Table III

Myocardial Magnesium in Asphyxia (Guinea Pigs)

Duration of asphyxia (min)	mEq/l	
0	16.6	Control
0.5	14.1	Tachycardia
1.0	11.4	Onset: hypoxic dilatation
1.5	11.4	Complete: hypoxic dilatation
2.0	11.1	AV–block I
2.5	10.9	AV–block II
4.0	10.9	AV–dissociation
8.0	11.7	Cardiac arrest (2 min)
10.0	12.5	Cardiac arrest (4 min)
10.5	15.5	Cardiac arrest (4.5 min)

Translated from Hochrein *et al.* (25).

subsequent rise to almost the control level after 4 1/2 min of cardiac arrest, may have been caused by acute loss of functional Mg, followed by precipitation of the inorganic phosphate salt.

Magnesium normally functions in the high energy phosphate reactions involved in citric acid and fatty acid metabolism, active potassium and sodium transport, in muscle contraction and relaxation, and in maintenance of mitochondrial integrity. When present as an insoluble phosphate salt, magnesium is at best nonfunctional. Whether phosphate precipitates of calcium and magnesium can contribute to cell damage remains to be proved.

SUMMARY

Magnesium is necessary for mitochondrial function and integrity. Cardiac mitochondria are most susceptible to loss of magnesium. Under physiologic conditions, the shift of magnesium out of and of calcium into the mitochondria is associated with impaired mitochondrial structure and function. Inadequate Mg interferes with activation of Na^+-K^+-ATPase, with resultant shifts of K out of and Na into the cell. With pathologically decreased magnesium levels, there is irreversible loss of cellular potassium and, ultimately, mitochondrial disruption and deposition of crystalline phosphates of both calcium and magnesium into mitochondria and microsomes.

REFERENCES

1. AUDITORE, J. V.: Sodium-potassium activated G-strophanthin sensitive ATPase in cardiac muscle. Proc. Soc. Exp. Biol. Med. *110:* 595–597 (1962).
2. AUDITORE, J. V. and MURRAY, L.: The localization of the Mg+Na+K activated G-strophanthin sensitive ATPase in cardiac muscle. Arch. Int. Pharmacodyn. *145:* 137–146 (1963).
3. BAJUSZ, E.: Electrolytes and Cardiovascular Diseases (Karger, Basel 1965).
4. BAJUSZ, E.: Nutritional Aspects of Cardiovascular Diseases. (Lippincott, Philadelphia 1965).
5. BALTSCHEFFSKY, H.: Mitochondrial respiratory control and phosphorylative activities in a Mg-free medium. Biochim. Biophys. Acta. *25:* 382–388 (1957).
6. BRIERLEY, G. P.: Ion transport by heart mitochondria. J. Biol. Chem. *242:* 1115–1122 (1967).
7. BRIERLEY, G. P.; BACHMANN, E. and GREEN, D. E.: Active transport of inorganic phosphate and Mg ions by beef heart mitochondria. Proc. Nat. Acad. Sci. *48:* 1928–1935 (1962).
8. BRIERLEY, G.; MURER, E.; BACHMANN, E. and GREEN, D. E.: Studies on ion transport. II. The accumulation of inorganic phosphate and Mg ions by heart mitochondria. J. Biol. Chem. *238:* 3482–3489 (1963).
9. BRIERLEY, G. P. and SLAUTTERBACK, D. B.: Studies on ion transport. IV. An electron microscope study of the accumulation of Ca^{2+} and inorganic phosphate by heart mitochondria. Biochim. Biophys. Acta. *82:* 183–186 (1964).
10. CARAFOLI, E.; ROSSI, C. S. and LEHNINGER, A. L.: Cation and anion balance during active accumulation of Ca^{++} and Mg^{++} by isolated mitochondria. J. Biol. Chem. *239:* 3055–3061 (1964).
11. CARVALHO, A. P. and LEO, B.: Effects of ATP on the interaction of Ca^{++}, Mg^{++} and K^+ with fragmented sarcoplasmic reticulum isolated from rabbit skeletal muscle. J. Gen. Physiol. *50:* 1327–1352 (1967).

12. CHANCE, B. and WILLIAMS, G. R.: The respiratory chain and oxidative phosphorylation. Adv. Enzymology *17:* 65–134 (1956).

13. CONNELLY, J. L. and LARDY, H. A.: The effect of adenosine triphosphate and substrate on orthophosphate-induced mitochondrial swelling at acid pH. J. Biol. Chem. *239:* 3065–3070 (1964).

14. DI GIORGIO, J.; VITALE, J. J. and HELLERSTEIN, E. E.: Sarcosomes and magnesium deficiency in ducks. Biochem. J. 82: 184–187 (1962).

15. DUNHAM, E. T. and GLYNN, I. M.: Adenosinetriphosphatase activity and the active movements of alkali metal ions. J. Physiol. *156:* 274–293 (1961).

16. ERNSTER, L.: Control of cell metabolism at the mitochondrial level. Fed. Proc. *24:* 1222–1236 (1965).

17. ERNSTER, L. and LOW, H.: Reconstruction of oxidative phosphorylation in aged mitochondrial systems. Exp. Cell Res. (Suppl. *3:*) 133–153 (1955).

18. GEBICKI, J. M. and HUNTER, F. E.: Determination of swelling and disintegration of mitochondria with an electronic particle counter. J. Biol. Chem. *239:* 631–639 (1964).

19. GLYNN, I. M.: "Transport adenosinetriphosphatase" in electric organ. The relation between ion transport and oxidative phosphorylation. J. Physiol. *169:* 452–465 (1963).

20. GREEN, D. E. and FLEISCHER, S.: The Mitochondrial System of Enzymes in Metabolic Pathways. (Academic Press, New York 1960).

21. HAUGAARD, N.; HAUGAARD, E. S.; NAM HEA LEE and HORN, R. S.: Possible role of mitochondria in regulation of cardiac contractility. Fed. Proc. *28:* 1657–1662 (1969).

22. HEGGTVEIT, H. A.: The cardiomyopathy of magnesium deficiency in: Electrolytes and Cardiovascular Diseases pp. 204–220 (Karger, Basel 1965).

23. HEGGTVEIT, H. A.: Myopathy in experimental magnesium deficiency. Ann. N.Y. Acad. Sci. *162:* 758–765 (1969).

24. HEGGTVEIT, H. A.; HERMAN, L. and MISHRA, R. K.: Cardiac necrosis and calcification in experimental magnesium deficiency. Amer. J. Path. *45:* 757–782 (1964).

25. HOCHREIN, H.; KUSCHKE, H. J.; ZAQQA, Q. und FAHL, E.: Das Verhalten der intracellularen Magnesium-Konzentration in Myokard bei Insuffizienz, Hypoxie und Kammerflimmern. Klin. Wschr. *45:* 1093–1096 (1967).

26. HUNTER, F. E.; LEVY, J. F.; FINK, J.; SCHUTZ, B.; GUERRA, F. and HURWITZ, A.: Studies on the mechanism by which anaerobiosis prevents swelling of mitochondria *in vitro:* Effect of electron transport chain inhibitors. J. Biol. Chem. *234:* 2176–2186 (1950).

27. JENNINGS, R. B.: Symposium on the pre-hospital phase of acute myocardial infarction. Part II – Early phase of myocardial ischemic injury and infarction. Amer. J. Card. *24:* 753–765 (1969).

28. JOHNSON, C. J.; MAURITSEN, C. M.; STARBUCK, W. C. and SCHWARTZ, A.: Histones and mitochondrial ion transport. Biochem. *6:* 1121–1127 (1967).

29. JOHNSON, C. L.; ORO, J. and SCHWARTZ, A.: Basic protein induction of low amplitude energy-linked mitochondrial swelling. Arch. Biochem. Biophys. *131:* 310–315 (1969).

30. KIMMICH, G. A. and RASMUSSEN, H.: The effect of parathyroid hormone on mitochondrial ion transport in the terminal portion of the cytochrome chain. Biochim. Biophys. Acta. *113:* 457–466 (1966).

31. KLEIGER, R. E.; SETA, K.; VITALE, J. J. and LOWN, B.: Effects of chronic depletion of potassium and magnesium upon the action of acetylstrophanthidin on the heart. Amer. J. Card. *17:* 520–527 (1966).

32. LEHNINGER, A. L.: Reversal of various types of mitochondrial swelling by adenosine triphosphate. J. Biol. Chem. *234:* 2465–2471 (1959).

33. LEHNINGER, A. L.; WADKINS, C. L.; COOPER, C.; DEVLIN, T. M. and GAMBLE, J. L: Oxidative phosphorylation. Experiments with fragments of mitochondria offer new information about respiratory energy conversion. Science *128:* 450–456 (1958).

34. MISHRA, R. K. and HERMAN, L.: Preliminary observation on the ultrastructure of rat cardiac muscle following a Mg-deficient regime and cold stress in: Proc. European Reg. Conf. on Electron Microscopy, pp. 907–911 (1960).

35. NAKAMURA, M.; NAKATINI, M.; KOIKE, M.; TORII, S. and HIRAMATSU, M.: Swelling of heart and liver mitochondria from magnesium deficient rats and its reversal. Proc. Soc. Exp. Biol. Med. *108:* 315–319 (1961).

36. NANNINGA, L. B.: Calculation of free magnesium, calcium, and potassium in muscle. Biochim. Biophys. Acta. *54:* 338–344 (1961).

37. PACKER, L.: Metabolic and structural states of mitochondria. II Regulation by phosphate. J. Biol. Chem. *236:* 214–220 (1961).

38. POST, R. L.; MERRITT, C. R.; KINSOLVING, C. R. and ALBRIGHT, C. D.: Membrane adenosine triphosphatase as a participant in the active transport of sodium and potassium in the human erythrocyte. J. Biol. Chem. *235:* 1796–1802 (1960).

39. RASMUSSEN, H.; FISCHER, J. and ARNAUD, C.: Parathyroid hormone, ion exchange and mitochondrial swelling. Proc. Nat. Acad. Sci. *52:* 1198–1203 (1964).

40. SALLIS, J. D. and DELUCA, H. F.: Action of parathyroid hormone on mitochondria. Magnesium- and phosphate-independent respiration. J. Biol. Chem. *241:* 1122–1127 (1966).

41. SALLIS, J. D.; DELUCA, H. F. and RASMUSSEN, H.: Parathyroid hormone-dependent uptake of inorganic phosphate by mitochondria. J. Biol. Chem. *238:* 4098–4102 (1963).

42. SCHWARTZ, A.: A sodium and potassium stimulated adenosine triphosphatase from cardiac tissues–I Preparation and properties. Biochem. Biophys. Res. Comm. *9:* 301–306 (1962).

43. SCHWARTZ, A.: The effect of histones and other polycations on cellular energetics I Mitochondrial oxidative phosphorylations. J. Biol. Chem. *240:* 939–943 (1965).

44. SCHWARTZ, A. and LASETER, A. H.: A sodium- and potassium-stimulated adenosine triphosphatase from cardiac tissues–II The effects of ouabain and other agents that modify enzyme activity. Biochem. Pharmacol. *13:* 337–348 (1964).

45. SELLER, R. H.; NEFF, M. H.; MENDELSSOHN, S.; KIM, K. E. and SWARTZ, C. D.: Magnesium sulfate in digitalis toxicity. Amer. J. Card. *25:* 127 (1970).

46. SELYE, H.: The Chemical Prevention of Cardiac Necrosis. (Ronald Press Co., New York, 1958).

47. SELYE, H.: The Pluricausal Cardiopathies. (Charles C Thomas, Springfield, Ill. 1961).

48. SIEKEVITZ, P.; LOW, H.; ERNSTER, L. and LINDBERG, O.: On a possible mechanism of the adenosinetriphosphatase of liver mitochondria. Biochim. Biophys. Acta. *29:* 378–391 (1958).

49. SKOU, J. C.: Enzymatic basis for active transport of Na^+ and K^+ across cell membrane. Physiol. Rev. *45:* 596–617 (1965).

50. SKOU, J. C.: Further investigations on a Mg^{++} + Na^+-activated adenosinetriphosphatase, possibly related to the active, linked transport of Na^+ and K^+ across the nerve membrane. Biochim. Biophys. Acta. *42:* 6–23 (1960).

51. SKOU, J. C.: The influence of some cations on an adenosine triphosphatase from peripheral nerves. Biochim. Biophys. Acta. *23:* 394–401 (1957).

52. SKOU, J. C.: Preparation from mammalian brain and kidney of the enzyme system involved in active transport of Na^+ and K^+. Biochim. Biophys. Acta. *58:* 314–325 (1962).

53. SZEKELY, P. and WYNNE, N. A.: The effects of magnesium on cardiac arrhythmias caused by digitalis. Clin. Sci. *10:* 241–247 (1951).

54. ULRICH, F.: Active transport of potassium by heart mitochondria. Amer. J. Physiol. *198:* 847–854 (1960).

55. VITALE, J. J.; HEGSTED, D. M.; NAKAMURA, M. and CONNORS, P.: The effect of thyroxine on magnesium requirement. J. Biol. Chem. *226:* 597–601 (1957).

56. VITALE, J. J.; HELLERSTEIN, E. E.; NAKAMURA, M. and LOWN, B.: Effects of magnesium-deficient diet upon puppies. Circulat. Res. *9:* 387–394 (1961).

57. VITALE, J. J.; NAKAMURA, M. and HEGSTED, D. M.: The effect of magnesium deficiency on oxidative phosphorylation. J. Biol. Chem. *228:* 573–576 (1957).

58. VITALE, J. J.; VELEZ, H.; GUZMAN, C. and CORREA, P.: Magnesium deficiency in the cebus monkey. Circulat. Res. *12:* 642–650 (1963).

59. WELT, L. G.: Experimental magnesium depletion. Yale J. Biol. Med. *36:* 325–349 (1964).

60. WELT, L. G. and TOSTESON, D. C.: The mode of activation of membrane ATPase by Mg^{++}. South. Soc. Clin. Res. *12:* 73 (1964).

61. WHANG, R. and WELT, L. G.: Observations in experimental magnesium depletion. J. Clin. Invest. *42:* 305–313 (1963).

62. ZWILLINGER, L.: Uber die Magnesiumwirkung auf das Herz. Klin. Wschr. *14:* 1429–1433 (1935).

Mailing address: Mildred S. Seelig, M.D., Department of Medical Statistics, Schering Corporation, Bloomfield, New Jersey 07003 (USA).

Schering Corporation, Bloomfield, N. J.,
New York Medical College, Pharmacology Department,
Maimonides Medical Center, New York, N. Y., USA

MYOCARDIAL LOSS OF FUNCTIONAL MAGNESIUM

II. IN CARDIOMYOPATHIES OF DIVERSE ETIOLOGY

M. S. SEELIG

In his concluding remarks to the conference on the relationship of experimental "metabolic" cardiomyopathies to human-heart diseases (54), Bajusz (3) brought into focus the findings that point towards the probability that underlying almost all types of cardiac necrosis is mitochondrial disorganization. He pointed out that, with the exception of the cardiomyopathy of K deficiency, mitochondrial failure plays an important role in causing the muscle cell to enter a stage of irreversible injury. Bajusz (3) further emphasized that no longer tenable is the earlier belief that losses of K and Mg from the myocardial cells and the uptake of Na and Ca are merely secondary consequences of cellular necrosis.

That both cations are vitally involved in the maintenance of functional and structural integrity of the myocardium is indicated by the cardiac necrosis that develops in animals that are nutritionally depleted of one or both (5). Selye and his co-workers (66) showed, long ago, that deficiency of each of these cations intensifies the necrotizing effect of diverse cardiotoxic agents and that their administration protects against these influences. It should be noted, however, that the first reported application of Mg salts in cardiotoxicity was the use by Hueber and Lehr (28) of each of several Mg preparations to treat and protect against aconitine-poisoning. Mishra (48) showed that Mg-deficient rats had less numerous cardiac mitochondria than did control rats. He postulated that similar mitochondrial abnormalities, possibly caused by loss of myocardial Mg and K, might be responsible for the functional and structural disturbances seen in the electrolyte steroid cardiac necrosis (ESCN). DuRuisseau and Mori (17) found that, in the ESCN model, the loss of Mg from the heart was significant, whereas there was no loss of cardiac K (table I). Associated with the drop to half the level of Mg seen in hearts of control rats, there were significant increases in Na, Ca, and PO_4.

Lehr and Krukowski (38), considering the mechanism by which corticosteroid in combination with NaH_2PO_4 causes myocardial necrosis, proposed that depletion of myocardial Mg might have at least equal, if not greater, pathogenic significance than depletion of myocardial K. Their observation that parathyroidectomy intensified cardiac susceptibility to NaH_2PO_4 loads (38, 40) is of particular interest, in view of Kimmich and Rasmussen's (34) demonstration that

Table I

Cardiac Electrolytes in the ESCN Model (9-Day Study)

	Mg	K	Na	Ca	PO4	Cardiac necrosis
			mEq/kg			
Control	30.4	78.4	55.1	22.5	59.0	0
NaH_2PO_4 (2 mM orally, b.i.d.)	32.5	83.5	59.5	31.2	59.6	0
Me-Cl-Col-AC* (100 μg subcutaneously)	23.2	80.5	58.9	33.2	62.3	0
NaH_2PO_4 (orally) + MeCl-Col-AC* (subcutaneously)	15.0	81.1	69.0	35.1	68.0	Present

* Corticosteroid with mineralo- and glucocorticoid activities.
$p = < 0.05$ developed. Adapted from DuRuisseau and Mori (17).

parathyroid hormone plays a role in the transport of Mg as well as of inorganic PO_4 into the mitochondria. The subsequent studies by Lehr and his associates (36, 37, 39) provided direct evidence that PO_4-loaded, parathyroidectomized rats lost myocardial Mg to a much greater extent than K, and gained Na (table II). Prevention of the parathyroid-depletion effect by administration of parathyroid extract to parathyroidectomized rats resulted in only a slight decrease of cardiac Mg from control levels but in a further rise in PO_4. This observation raises the question as to whether all of that Mg was necessarily functional or whether some may have been in association with the elevated inorganic PO_4.

Table II

Cardiac Electrolytes in the Parathyroidectomized PO_4-Loaded Model of Cardiac Necrosis

	Mg	K	Na	Ca	PO4	Cardiac necrosis
			mEq/kg wet wt.			
Controls	19.1	83.6	44.1	2.8	57.6	0
PT_x* only	19.7	83.8	43.1	2.9	58.4	
NaH_2PO_4 only	19.2	82.1	42.4	3.5	61.1	0
PT_x + NaH_2PO_4	15.7	81.3	49.0	3.2	53.	100%
PT_x + NaH_2PO_4 + PTE†	18.4	82.9	43.8	3.2	61.4	0

* PT_x = parathyroidectomized.
† PTE = parathyroid extract.
Adapted from Lehr (37).

Table III

Cardiac Electrolytes in the
Isoproterenol Model of Cardiac Necrosis

	Time after injection	Mg	K	Na	Ca	PO_4	Cardiac necrosis
	h			*mEq/kg wet wt.*			
Controls		19.2	83.6	44.1	2.8	57.6	0
Isopro-	3	17.0	83.0	42.4	3.8	53.9	0
terenol	7	16.3	83.0	47.1	4.8	53.6	60%
	24	15.6	81.3	53.0	4.3	48.1	100%

Adapted from Lehr (37).

Both in the parathyroidectomized PO_4-loaded model and in the isoproterenol model of cardiac necrosis, Lehr and Krukowski (37, 39) found the K loss to be apparently subordinate to the Mg loss (table III). In the isoproterenol model, a time sequence study revealed that the Mg loss occurred earlier than did the K loss and was greater in degree. The prompt uptake of Ca in the hearts of isoproterenol-treated rats may be explained by the work of Grossman and Furchgott (20), who showed that catecholamine (norepinephrine) causes significant increases in the exchange of Ca by the isolated guinea-pig auricle. This effect was associated with the positive inotrophic effect of norepinephrine, which may be mediated by the effect of intracellular Ca in the mechanism of muscle contraction.

Table IV

Myocardial Electrolytes in Hamsters of the Cardiomyopathic
(BIO 14.6) Strain and an Unrelated (LSH) Strain

		Age (days)		Myocardial electrolytes* (mg/100 g dry wt.)			
Group	Strain	Mean	Range	Ca	Mg	K	Na
1a	Control (LSH)	29	23–33	12.1 ± 0.51	115 ± 2.33	1535 ± 16	421 ± 13
1b	Cardiomyopathic (BIO 14.6)			17.2 ± 0.46	73 ± 0.80	1575 ± 23	419 ± 11
2a	Control (LSH)	62	56–71	14.6 ± 0.44	109 ± 0.98	1496 ± 20	342 ± 2
2b	Cardiomyopathic (BIO 14.6)			215.2 ± 33.90	109 ± 1.25	1422 ±	479 ± 15

* Lower two-thirds of both ventricles. Values represent mean ± standard error. From Bajusz and Lossnitzer (6).

Table V

Serum Electrolytes in Hamsters of the Cardiomyopathic (BIO 14.6) Strain and an Unrelated (LSH) Strain

Group	Strain	Age (days)		Serum electrolytes* (mg %)			
		Mean	Range	Ca	Mg	K	Na
1a	Control (LSH)	29	23–33	9.4 ± 0.40	3.2 ± 0.32	16.3 ± 0.49	309 ± 4.8
1b	Cardiomyopathic (BIO 14.6)			12.0 ± 0.45	4.1 ± 0.19	15.4 ± 0.65	303 ± 2.6
2a	Control (LSH)	62	56–71	10.1 ± 0.37	3.4 ± 0.15	13.8 ± 0.22	316 ± 4.5
2b	Cardiomyopathic (BIO 14.6)			10.2 ± 0.31	3.5 ± 0.15	13.9 ± 0.34	307 ± 6.3

* Mean ± standard error. From Bajusz and Lossnitzer (6).

The cardiomyopathic BIO 14.6 strain of Syrian hamsters has been shown by Bajusz and Lossnitzer (6) also to exhibit decreased myocardial Mg during the prenecrotic phase, at 23–33 days of life (table IV). The fact that the myocardial Mg had returned to normal levels at 56–71 days, at which time the hamsters presented calcifying myocardial lesions, further suggests that not all of the myocardial Mg was functional. The elevation of serum Mg (table V) at the time that myocardial Mg was depressed indicates that serum Mg levels cannot be relied upon as an index of cellular levels. The demonstration by Angelakos (2) that myocardial norepinephrine levels are elevated (table VI) at about the same age as the drop in myocardial Mg was seen, is an indirect indication of the Mg-tissue-depleting effect of catecholamine.

Raab (58) recently evaluated the interrelationships of the increased metabolic

Table VI

Heart Norepinephrine (Means of Determinations in 6 Pools, 5 Hearts/Pool)

Age (days)	NE μg/g		NE μg/heart	
	Controls	Dystrophic	Controls	Dystrophic
35	0.99	1.18*	0.18	0.20
55	1.03	1.28*	0.23	0.27*
85	1.08	1.12	0.30	0.33
105	1.08	0.93	0.33	0.32
120	1.68	0.47*	0.43	0.26*

* Significantly different from the controls ($p < 0.05$). From Angelakos (2).

demands on the heart caused by catecholamine in the light of the myocardial electrolyte derangement seen in what he terms "dysionic cardiopathy" or "pluricausal," so-called coronary heart disease. He pointed out interrelationships of the neurohormonal axis with cardiac metabolic dysfunctions in which a cardiac deficit of Mg, as well as of K, has been shown to play an important role.

Additional to the foregoing examples of drug-induced prenecrotic losses of myocardial Mg, Hochrein *et al.* (26) demonstrated that digitalis or phenylbutazone intoxication of guinea pigs caused substantial drops in intracellular myocardial Mg, as did hypoxia, overload, or hyperkalemia. They concluded that Mg^{++} occupies a central position in the structure and metabolism of myocardial cells.

Myocardial loss of Mg has also been reported in the infarcted area of the heart, subsequent to experimental ligation of a coronary artery (15, 30), and in patients who died suddenly of myocardial infarction (29, 58). Cummings (15) reported that the Mg content of the infarcted left ventricle of dogs, whose left anterior descending coronary artery had been ligated 11 h before sacrifice, was 1.18 mEq/100 g wet weight, as compared with 2.30 for the left ventricle of control, sham-operated dogs. Even the noninfarcted ventricle showed a lower than control level of Mg (2.11 *vs* 2.26), possibly as a consequence of the stress. The drop in myocardial K of the infarcted segment was much more profound (4.5 *vs* 8.5 mEq/100g). The analysis by Iseri *et al.* (29) of infarcted and noninfarcted segments of hearts from patients who had died of myocardial infarction similarly showed lower than normal Mg levels in both segments (table VII). A much greater loss was seen in the infarcted area. Raab (58) recently reported significantly diminished Mg and K and elevated Na contents, even of the structurally noninvolved parts of heart muscle obtained from patients who had died of an infarction within several hours to several days.

Table VII

Cardiac Electrolytes: Clinical Myocardial Infarction

	Mg	K	Na	P
	mEq/100 mg wet tissue (mean values)			
Heart muscle (patients who died of other causes than heart disease)	0.45	8.32	4.96	3.88
Acute myocardial infarction				
Non-infarcted segment	0.30	5.91	6.68	3.23
Infarcted segment	0.22	4.19	8.11	2.32

Adapted from Iseri *et al.* (29).

As indicated by Lehr and his associates (36, 37, 39) in their isoproterenol studies, and by Hochrein *et al.* (26) in their asphyxia study, the loss of Mg from the heart was greatest shortly after the insult. The loss of K from the isoproterenol-damaged heart was delayed. A 6 h delay before 50% of myocardial K was lost from infarcted dog hearts had been reported in 1957 and 1964 by Jennings *et al.* (31, 33) as attributable to the time taken for failure of the energy-producing mechanisms involved in maintaining intracellular K. In 1969 (30, 32), these investigators reaffirmed the observation, supporting it with electron microscopic findings. Their data indicate that the myocardial cellular injury was reversible after 15–18 min of ischemia, but that cells were irreversibly damaged after 40 min (32). The authors commented that the spatial disorganization and/or inactivation of the mitochondrial enzymes were associated with loss of essential cofactors and maldistribution of essential anions and cations within the mitochondrion. These changes were reflected ultramicroscopically by sarcolemmal and mitochondrial disruption.

That the aging process and long-continued stress situations may be associated with decreases in myocardial Mg, and that such losses may predispose to impaired resistance to other cardiopathic agents, is suggested by the work of Mori and DuRuisseau (51) and Raab *et al.* (59). The former authors reported a substantial drop in myocardial Mg and a lesser drop in K as the rats aged (table VIII). They attributed the decrease in Ca seen in the aged animals to the loss of nucleic acids, to which both Ca and Mg are closely linked. The decrease in Na was attributed largely to the shrinking of extracellular phase. Raab *et al.* (59) demonstrated decreases in both myocardial Mg and K with increasing duration of isolation (table IX). They showed that the rats kept in isolation were, in fact, more susceptible to the cardiotoxic effect of epinephrine.

That the K loss by Mg-deficient isolated mitochondria is relevant to the intact animal or man is indicated by Mg-deficiency studies in animals from which tissue analyses were made, and in human subjects undergoing metabolic balance determinations. A decrease in muscle K in Mg-depleted animals was first demonstrated in 1951 by Cotlove *et al.* (12); in that study, the serum K

Table VIII

Cardiac Electrolytes in Aging Rats

Age	Mg	K	Na	Ca	P
mo.	*mEq/kg dry, fat-free tissue*				*mM*
0.9–1.2	86.4	385	230	41.2	335
1.5–2.0	82.7	396	247	30.7	344
10–12	68.8	389	190	25.5	348
34–36	47.4	355	112	21.2	339

Adapted from Mori and DuRuisseau (51).

Table IX

Cardiac Electrolytes in Isolated Rats

	Mg	K	Na	Ca
	mg/100 g wet wt.			
Control rats (in colony)	25.4	342	81	5.0
4 Months isolation	23.9	309	79	5.5
14 Months isolation	22.6	298	102	5.7
2 Weeks in colony (after 14 months isolation)	25.4	314	93	No data

Adapted from Raab *et al.* (59).

remained unaffected. Many subsequent investigations have shown substantial drops of muscle Mg and K, and often also of serum Mg and K (22, 35, 42–45, 68–71, 76, 80, 81, 84–90); in most instances, elevated Na levels were reported (22, 42–45, 68, 84, 88–90). Human metabolic studies with subjects on Mg-deficient diets have provided proof that Mg loss is followed by K loss; in the longer term studies, hypokalemia developed, and this could not be corrected until the Mg was repleted (16, 19, 72–74). In several instances, ECG changes of hypokalemia were recorded as the period of Mg-deficiency lengthened (16, 72, 74).

Despite the substantial laboratory evidence that Mg loss is incompatible with the maintenance of cardiac integrity and function, be it of the isolated cardiac mitochondria, or of the heart in the intact animal, it has been difficult to provide convincing evidence that these findings are relevant to clinical cardiovascular disease. It has been widely assumed that the Mg intake is adequate in the Occident, where cardiovascular disease is a major cause of death. However, a 1964 review (64) of Mg balance data obtained from studies throughout the world indicates that the Occidental diet delivers substantially less Mg than does the Oriental diet (fig. 1). On diets that delivered amounts of Mg found in the typical American diet (daily Mg intake of 4–5.9 mg/kg), young men often showed negative Mg balance; young women tended to stay in balance at such intakes. On diets delivering as much as has been reported from Oriental studies (6–10 mg/kg/d), the test subjects stored Mg.

Perhaps it is an underlying suboptimal intake of Mg that contributes to the greater incidence of cardiac deaths in soft water areas as compared with the incidence in hard water areas (1, 4, 7, 13, 14, 52, 63, 78). Crawford and Crawford (14) found that young accident victims from a soft-water area (Glasgow), where myocardial disease develops early and is more lethal than it is in a hard-water area (London), had much lower concentrations of coronary Mg (fig. 2). The authors attributed the greater incidence of high coronary Mg levels

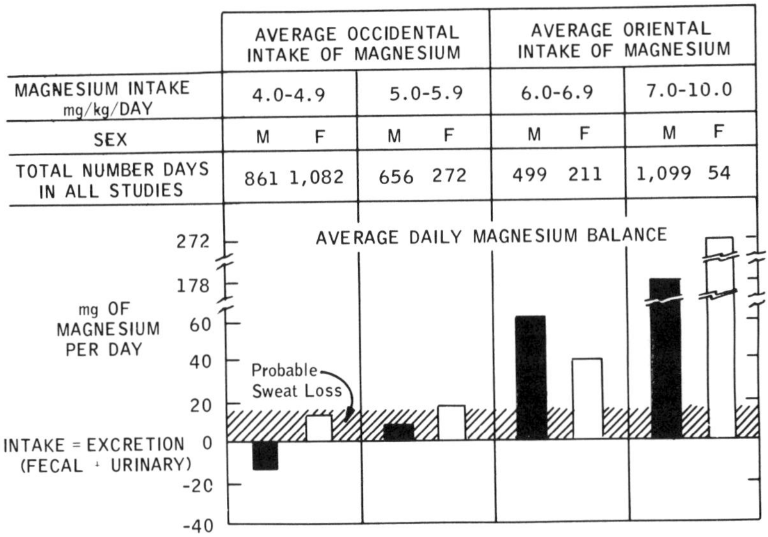

Fig. 1 Influence of sex on magnesium balance at different intakes of magnesium. Adapted from Seelig (64).

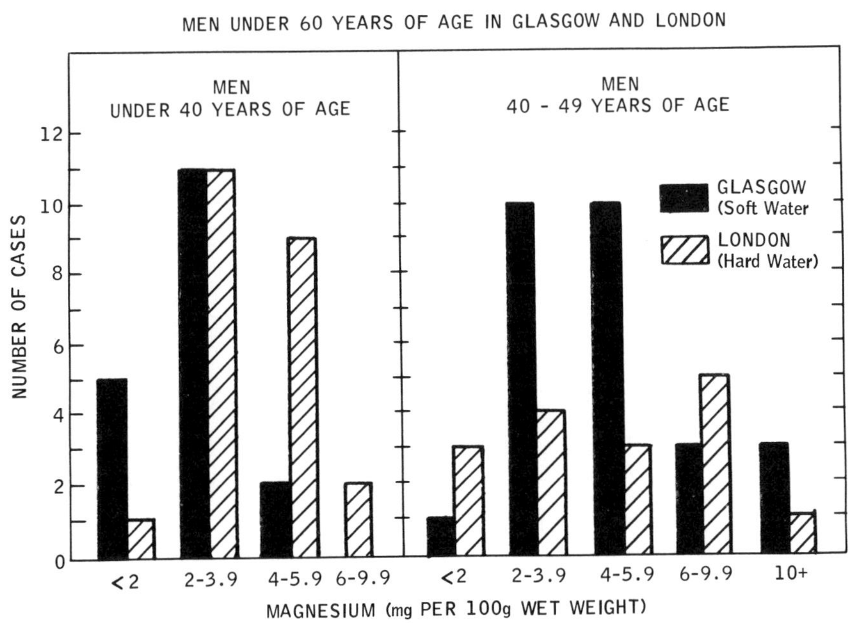

Fig. 2 Coronary magnesium in accident cases. From Crawford and Crawford (14).

in older men from the hard-water group to the deposition of minerals in established lesions.

Manifestly, variations in the amount of Mg in the diet cannot be the sole explanation of clinical heart disease. Quite apart from the hormonal and drug insults that have cardiomyopathic potential, we ingest nutrients that increase Mg requirements. High protein intakes have long been known to increase Mg requirements and to intensify Mg deficiency (10, 11, 18, 47, 82) as have high Ca and/or PO_4 intakes (9, 11, 27, 41, 46, 49, 50, 53, 55–57, 62, 77). High cholesterol intake has also been shown to increase the Mg requirement (24, 25, 49, 79, 82). Vitamin D and its analogue, dihydrotachysterol, have long been known to cause cardiac necrosis in experimental animals (for reviews, see 5, 66, 67). Vitamin D excess, or hyper-reactivity to vitamin D, has more recently been implicated in supravalvular aortic stenosis and myocardial necrosis in infants and children (for reviews, see 8, 65). It is thus provocative: (1) that toxic doses of vitamin D have been shown to cause Mg loss and/or hypomagnesemia in laboratory animals (21, 23, 60, 75, 83, 84, 90); and (2) that 5-fold higher than normal Mg intakes have protected against cardionecrosis caused by feeding rats a diet that was rich in vitamin D, as well as in Ca, PO_4, and protein (61).

SUMMARY

A number of diverse cardiopathic agents, including hormones, ischemia, stress, aging, and nutritional imbalances, have been shown to cause decreased myocardial Mg and mitochondrial disruption. Dietary Mg-deficiency intensifies the cardiotoxicity of many such agents. Mg is necessary for the integrity of the mitochondrial system, which is responsible for the maintenance of cellular K as well as for the metabolic processes. Depletion of cardiac Mg may be contributed to by disease, drugs, or diet. This may be a pivotal factor in the "pluricausal" cardiomyopathies.

REFERENCES

1. ANDERSON, T. W.; RICHE, W. H. and MACKAY, J. S.: Sudden death and ischemic heart disease. New Eng. J. Med. *280:* 805–807 (1969).
2. ANGELAKOS, E. T.: Electrocardiographic (ECG) changes and cardiac catecholamine concentrations in the dystrophic hamster strain (BIO 14.6 line) at various ages. Trans. N.Y. Acad. Sci. Ser. II *30* (7) *:* 955–956 (1968).
3. BAJUSZ, E.: Concluding remarks (re experimental "metabolic" cardiopathies and their relationship to human heart diseases). Ann. N.Y. Acad. Sci. *156:* 620–625 (1969).
4. BAJUSZ, E.: Heart disease and soft and hard water. Lancet *i:* 726 (1967).
5. BAJUSZ, E.: Nutritional Aspects of Cardiovascular Diseases (Lippincott, Philadelphia (1965).
6. BAJUSZ, E. and LOSSNITZER, K.: A new disease model of chronic congestive heart failure: Studies on its pathogenesis. Trans. N.Y. Acad. Sci. II *30* (7) *:* 939–948 (1968).
7. BIORCK, G.; BOSTROM, H. and WIDSTROM, A.: On the relationship between water hardness and death rate in cardiovascular diseases. Acta. Med. Scand. *178:* 239–251 (1965).
8. BLACK, J. A.: Idiopathic hypercalcemia and vitamin D. Germ. Med. Mth. *9:* 290–297 (1964).

9. BUNCE, G. E.; CHIEMCHAISRI, Y. and PHILLIPS, P. H.: The mineral requirements of the dog. IV. Effect of certain dietary and physiologic factors upon the magnesium deficiency syndrome. J. Nutrition *76:* 23–29 (1962).

10. CARRILLO, B. J.; POND, W. G.; KROOK, L.; LOVELACE, F. E. and LOSSLI, J. K.: Response of growing rats to diets varying in magnesium, potassium and protein content. Proc. Soc. Exp. Biol. Med. *107:* 793–796 (1961).

11. COLBY, R. W. and FRYE, C. M.: Effect of feeding high levels of protein and calcium in rat rations on magnesium deficiency syndrome. Amer. J. Physiol. *166:* 408–412 (1951).

12. COTLOVE, E.; HOLLIDAY, M. A.; SCHWARTZ, R. and WALLACE, W. M.: Effects of electrolyte depletion and acid-base disturbance on muscle cations. Amer. J. Physiol. *167:* 665–675 (1951).

13. CRAWFORD, M. D.; GARDNER, M. J. and MORRIS, J. N.: Mortality and hardness of local water-supplies. Lancet *i:* 827–831 (1968).

14. CRAWFORD, T. and CRAWFORD, M. D.: Prevalence and pathological changes of ischaemic heart-disease in a hard water and in a soft water area. Lancet *i:* 229–232 (1960.)

15. CUMMINGS, J. R.: Electrolyte changes in heart tissue and coronary arterial and venous plasma following coronary occlusion. Circulat. Res. *8:* 865–870 (1960).

16. DUNN, M. J. and WALSER, M.: Magnesium depletion in normal man. Metabolism *15:* 884–895 (1966).

17. DURUISSEAU, J. P. and MORI, K.: Biochemical studies on experimental cardiopathy: Electrolytes in rat tissues. Brit. J. Exp. Path. *40:* 250–254 (1959).

18. FEATHERSTON, W. R.; ROGLER, J. C. and PARKER, H. E.: Metabolic inter-relations of magnesium and protein in the chick. Fed. Proc. *23:* 290 (1964).

19. FITZGERALD, M. G. and FOURMAN, P.: An experimental study of magnesium deficiency in man. Clin. Sci. *15:* 635–647 (1956).

20. GROSSMAN, A. and FURCHGOTT, R. F.: The effects of various drugs on calcium exchange in the isolated guinea-pig left auricle. J. Pharmacol. Exp. Therap. *145:* 162–172 (1964).

21. HANNA, S.: Influence of large doses of vitamin D on magnesium metabolism in rats. Metabolism *10:* 735–743 (1961).

22. HANNA, S. and MACINTYRE, I.: The influence of aldosterone on magnesium metabolism. Lancet *ii:* 348–350 (1960).

23. HARRISON, H. E. and HARRISON, H. C.: The interaction of vitamin D and parathyroid hormone on calcium, phosphorus and magnesium homeostasis in the rat. Metabolism *13:* 952–958 (1964).

24. HELLERSTEIN, E. E.; NAKAMURA, M.; HEGSTED, D. M. and VITALE, J. J.: Studies on the interrelationships between dietary magnesium, quality and quantity of fat, hypercholesterolemia and lipidosis. J. Nutrition *71:* 339–346 (1960).

25. HELLERSTEIN, E. E.; VITALE, J. J.; WHITE, P. L.; HEGSTED, D. M.; ZAMCHECK, N. and NAKAMURA, M.: Influence of dietary magnesium on cardiac and renal lesions of young rats fed an atherogenic diet. J. Exp. Med. *106:* 767–775 (1957).

26. HOCHREIN, H.; KUSCHKE, H. J.; ZAQQA, Q. und FAHL, E.: Das Verhalten der intracellularen Magnesium-Konzentration im Myokard bei Insuffizienz, Hypoxie und Kammerflimmern. Klin. Wschr. *45:* 1093–1096 (1967).

27. HOUSE, W. B. and HOGAN, A. G.: Injury to guinea pigs that follows a high intake of phosphates. J. Nutrition *55:* 507–517 (1955).

28. HUEBER, E. F. und LEHR, D.: Wirkung von Magnesium auf die Vergiftung mit Akonitin. Arch. Exp. Path. Pharmakol. *189:* 25–44 (1938).

29. ISERI, L. T.; ALEXANDER, L. C.; MCCAUGHEY, R. S.; BOYLE, A. J. and MYERS, G. B.: Water and electrolyte content of cardiac and skeletal muscle in heart failure and myocardial infarction. Amer. Heart J. *43:* 215–227 (1952).

30. JENNINGS, R. B.: Symposium on the pre-hospital phase of acute myocardial infarction Part II – Early phase of myocardial ischemic injury and infarction. Amer. J. Card. *24:* 753–765 (1969).

31. JENNINGS, R. B.; CROUT, J. R. and SMETTERS, G. W.: Studies on distribution and localization of potassium in early myocardial ischemic injury. Arch. Path. *63:* 586–592 (1957).

32. JENNINGS, R. B.; SOMMERS, H. M.; HERDSON, P. B. and KALTENBACH, J. P.: Ischemic injury of myocardium. Ann. N.Y. Acad. Sci. *156:* 61–78 (1969).

33. JENNINGS, R. B.; SOMMERS, H. M.; KALTENBACH, J. P. and WEST, J. J.: Electrolyte alterations in acute myocardial ischemic injury. Circulat. Res. *14:* 260–269 (1964).

34. KIMMICH, G. A. and RASMUSSEN, H.: The effect of parathyroid hormone on mitochondrial ion transport in the terminal portion of the cytochrome chain. Biochim. Biophys. Acta. *113:* 457–466 (1966).

35. KLEIGER, R.; SETA, K.; HELLERSTEIN, E. E.; VITALE, J. J. and LOWN, B.: Electrolyte and ECG changes in K^+ and/or Mg^{++} deficient dogs infused with acetylstrophanthidin. Fed. Proc. *23:* 158 (1964).

36. LEHR, D.: The role of certain electrolytes and hormones in disseminated myocardial necrosis in: Electrolytes and Cardiovascular Diseases, pp. 248–273 (Karger, Basel 1965).

37. LEHR, D.: Tissue electrolyte alteration in disseminated myocardial necrosis. Ann. N.Y. Acad. Sci. *156:* 344–378 (1969).

38. LEHR, D. and KRUKOWSKI, M.: About the mechanism of myocardial necrosis induced by sodium phosphate and adrenal corticoid overdosage. Ann. N.Y. Acad. Sci. *105:* 135–182 (1963).

39. LEHR, D.; KRUKOWSKI, M. and COLON, R.: Correlation of myocardial and renal necrosis with tissue electrolyte changes. JAMA *197:* 105–112 (1966).

40. LEHR, D.; KRUKOWSKI, M. and COLON, R.: A simple method for the production of acute cardiorenal necrosis. Arch. Int. Pharmacodyn. *168:* 251–277 (1967).

41. MCALEESE, D. M. and FORBES, R. M.: The requirement and tissue distribution of Mg in the rat as influenced by environmental temperature and dietary calcium. J. Nutrition *73:* 94–106 (1961).

42. MACINTYRE, I. and DAVIDSSON, D.: The production of secondary potassium depletion, sodium retention, nephrocalcinosis and hypercalcaemia by magnesium deficiency. Biochem. J. *70:* 456–462 (1958).

43. MANITIUS, A. and EPSTEIN, F. H.: Some observations on the influence of a magnesium-deficient diet on rats, with special reference to renal concentrating ability. J. Clin. Invest. *42:* 208–215 (1963).

44. MARTIN, H. E.. and WILSON, M. L.: Effect of magnesium deficiency on serum and carcass electrolyte levels in the rat. Metabolism *9:* 484–491 (1960).

45. MARTINDALE, L. and HEATON, F. W.: Magnesium deficiency in the adult rat. Biochem. J. *92:* 119–126 (1964).

46. MAYNARD, L. A.; BOGGS, D.; FISK, G. and SEGUIN, D.: Dietary mineral interrelations as a cause of soft tissue calcification in guinea pigs. J. Nutrition *64:* 85–97 (1958).

47. MENAKER, W.: Influence of protein intake on magnesium requirement during protein synthesis. Proc. Soc. Exp. Biol. Med. *85:* 149–151 (1954).

48. MISHRA, R. K.: Studies on experimental magnesium-deficiency in the albino rat. 2. Influence of magnesium-deficiency diet on mitochondrial population of heart, kidney, and liver. Rev. Canad. Biol. *19:* 136–142 (1960).

49. MISHRA, R. K.: Studies on experimental magnesium deficiency in the albino rat 3.

Influence of certain salts and dietary constituents on the electroseizure threshold of rats on Mg-deficient diet. Rev. Canad. Biol. *19:* 143–153 (1960).

50. MISHRA, R. K.: Studies on experimental magnesium deficiency in the albino rat. 6. Effect of phosphate-iontophoresis into the hearts of animals on Mg-deficient diet. Rev. Canad. Biol. *19:* 163–167 (1960).
51. MORI, K. and DURUISSEAU, J.: Water and electrolyte changes in aging process with special reference to calcium and magnesium in cardiac muscle. Can. J. Biochem. Physiol. *38:* 919–928 (1960).
52. MORRIS, J. N.; CRAWFORD, M. D. and HEADY, J. A.: Hardness of local water supplies and mortality from cardiovascular disease in the county boroughs of England and Wales. Lancet *i:* 860–862 (1961).
53. MORRIS, E. R. and O'DELL, B. L.: Relationship of excess calcium and phosphorus to magnesium requirement and toxicity in guinea pigs. J. Nutrition *81:* 175–181 (1963).
54. New York Academy of Sciences Conference: Experimental "Metabolic" cardiopathies and their relationship to human heart diseases. Ann. N.Y. Acad. Sci. *156:* 1–625 (1969).
55. NUGARA, D. and EDWARDS, H. M.: Effect of calcium and phosphorus on magnesium metabolism in chicks. Fed. Proc. *20:* 294 (1961).
56. NUGARA, D. and EDWARDS, H. M.: Influence of dietary Ca and P levels on the Mg requirement of the chick. J. Nutrition *80:* 181–184 (1963).
57. O'DELL, B. L.; MORRIS, E. R. and REGAN, W. O.: Magnesium requirement of guinea pigs and rats. Effects of calcium and phosphorus and symptoms of magnesium deficiency. J. Nutrition *70:* 103–110 (1960).
58. RAAB, W.: Myocardial electrolyte derangement: Crucial feature of pluricausal, so-called coronary, heart disease. Ann. N.Y. Acad. Sci. *147:* 627–686 (1969).
59. RAAB, W.; BAJUSZ, E.; KIMURA, H. and HERRLICH, H. C.: Isolation, stress, myocardial electrolytes, and epinephrine cardiotoxicity in rats. Proc. Soc. Exp. Biol. Med. *127:* 142–147 (1968).
60. RICHARDSON, J. A. and WELT, L. G.: The hypomagnesemia of vitamin D administration. Proc. Soc. Exp. Biol. Med. *118:* 512–514 (1965).
61. RICO, J.; SIMON, G.; HEGYVARI, C. and SOS, J.: Effect of magnesium on dietary infarctoid changes in the heart. Acta. Med. Acad. Sci. Hung. *19:* 231–236 (1963).
62. SCHMIDT, C. L. A. and GREENBERG, D. M.: Occurrence, transport and regulation of calcium, magnesium, and phosphorus in the animal organism. Physiol. Rev. *15:* 297–434 (1935).
63. SCHROEDER, H. A.: Relations between hardness of water and death rates from certain chronic and degenerative diseases in the United States. J. Chron. Dis. *12:* 586–591 (1960).
64. SEELIG, M. S.: The requirement of magnesium by the normal adult. Amer. J. Clin. Nutr. *14:* 342–390 (1964).
65. SEELIG, M. S.: Vitamin D and cardiovascular, renal, and brain damage in infancy and childhood. Ann. N.Y. Acad. Sci. *147:* 537–582 (1969).
66. SELYE, E.: The Chemical Prevention of Cardiac Necrosis (Ronald Press, New York 1958).
67. SELYE, E.: The Pluricausal Cardiopathies (Charles C Thomas, Springfield, Ill. 1961).
68. SETA, K.; HELLERSTEIN, E. E. and VITALE, J. J.: Myocardium and plasma electrolytes in dietary magnesium and potassium deficiency in the rat. J. Nutrition *87:* 179–188 (1965).
69. SETA, K.; KLEIGER, R.; HELLERSTEIN, E. E.; LOWN, B. and VITALE, J. J.: Effect of potassium and magnesium deficiency on the electrocardiogram and plasma electrolytes of pure-bred beagels. Amer. J. Card. *17:* 516–519 (1966).
70. SETA, K.; KLEIGER, R.; HELLERSTEIN, E. E.; LOWN, B. and VITALE, J. J.:

Electrolyte and ECG changes in beagles deficient in K^+ and/or Mg^{++} Fed. Proc. *23:* 158 (1964).

71. SETA, K.; VITALE, J. J. and HELLERSTEIN, E. E.: Interrelationships between low and high dietary magnesium and potassium. Fed. Proc. *22:* 261 (1963).

72. SHILS, M. E.: Experimental human magnesium depletion. Amer. J. Clin. Nutrition *14:* 240 (1964).

73. SHILS, M. E.: Experimental human magnesium depletion. Medicine *48:* 61–85 (1969).

74. SHILS, M. E.: Experimental production of magnesium deficiency in man. Ann. N.Y. Acad. Sci. 162: 847–855 (1969).

75. SMITH, R. H.: Calcium and magnesium metabolism in calves. 2. Effect of dietary vitamin D and ultraviolet irradiation on milk-fed calves. Biochem. J. *70:* 201–205 (1958).

76. SUTER, C.; KLINGMAN, W. O.; BOGGS, D.; MARKS, R. D.; COPLINGER, C. B. and RANDOLPH, V.: Sound-induced seizures in animals: magnesium deficiency and sound-induced seizures in rats. Neurology *8:* 125–128 (1958).

77. TUFTS, E. V. and GREENBERG, D. M.: The biochemistry of magnesium deficiency. II. The minimum magnesium requirement for growth, gestation, and lactation, and the effect of the dietary calcium level thereon. J. Biol. Chem. *122:* 715–726 (1937).

78. UNSIGNED EDITORIAL: Hardness of water and cardiovascular disease. Brit. Med. J. *i* 1429 (1963).

79. VITALE, J. J.; HELLERSTEIN, E. E.; HEGSTED, D. M.; NAKAMURA, M. and FARBMAN, A.: Studies on the interrelationships between dietary magnesium and calcium in atherogenesis and renal lesions. Amer. J. Clin. Nutrition *7:* 13–22 (1959).

80. VITALE, J. J.; HELLERSTEIN, E. E.; NAKAMURA, M. and LOWN, B.: Effects of magnesium-deficient diet upon puppies. Circulat. Res. *9:* 387–394 (1961).

81. VITALE, J. J.; VELEZ, H. M.; GUZMAN, C. and CORREA, P.: Magnesium deficiency in the cebus monkey. Circulat. Res. *12:* 642–650 (1963).

82. VITALE, J. J.; WHITE, P. L.; NAKAMURA, M.; HEGSTED, D. M.; ZAMCHECK, N. and HELLERSTEIN, E. E.: Interrelationships between experimental hypercholesteremia, magnesium requirement, and experimental atherosclerosis. J. Exper. Med. *106:* 757–766 (1957).

83. WALLACH, W.; BELLAVIA, J. V.; SCHORR, J. and GAMPONIA, P. J.: Effect of vitamin D on tissue distribution and transport of electrolytes, Ca and Mg. Endocrinology *79:* 773–782 (1966).

84. WELT, L. G.: Experimental magnesium depletion. Yale J. Biol. Med. *36:* 325–349 (1964).

85. WHANG, R. and MOROSI, H. J.: The influence of continuing Mg deficiency on muscle K repletion. Clin. Res. *13:* 117 (1965).

86. WHANG, R.; MOROSI, H. J.; RODGERS, D. and REYES, R.: The influence of sustained magnesium deficiency on muscle potassium repletion. J. Lab. Clin. Med. *70:* 895–902 (1967).

87. WHANG, R.; OLIVER, J.; MCDOWELL, M. and WELT, L. G.: The renal lesion of magnesium depletion. Clin. Res. *10:* 257 (1962).

88. WHANG, R.; OLIVER, J.; WELT, L. G. and MACDOWELL, M.: Renal lesions and disturbance of renal function in rats with magnesium deficiency Ann. N.Y. Acad. Sci. *162:* 766–774 (1969).

89. WHANG, R. and WELT, L. G.: The *in vitro* effects of Mg^{++} and Ca^{++} on intracellular K^+. Clin. Res. *11:* 48 (1963).

90. WHANG, R. and WELT, L. G.: Observations in experimental magnesium depletion. J. Clin. Invest. *42:* 305–313 (1963).

Mailing address: Mildred S. Seelig, M.D., Department of Medical Statistics, Schering Corporation, Bloomfield, New Jersey 07003 (USA).

Pathology Department, Northwestern University Medical School,
Chicago, Illinois, USA

CALCIUM IN EXPERIMENTAL MYOCARDIAL ISCHEMIA [1]

R. B. JENNINGS and A. C. SHEN

Severely ischemic myocardial cells develop striking changes in mitochondrial structure and function at about the same time that lethal or irreversible injury becomes manifest (8, 10, 13). The mitochondria of the affected cells exhibit increased fragility, a decreased capacity to perform certain integrated functions (10) and, lastly, develop prominent intramitochondrial dense bodies (9). These dense bodies are a characteristic feature of severe ischemic cellular injury in dog myocardium, but also have been reported to occur in myocardial cells after several other types of injury, such as Mg^{2+} deficiency (5), catecholamine excess (17), and electric shock (18). The present paper will describe some of the structural and chemical features of these bodies developing, *in vivo,* in dog heart mitochondria after episodes of permanent or transient ischemia. The dense bodies found in canine myocardial mitochondria after periods of autolysis *in vitro* and after accumulation of Ca^{2+} *in vitro* will also be described briefly.

BIOLOGIC AND STRUCTURAL FEATURES OF MITOCHONDRIAL DENSE BODIES

Control

Mitochondria of control myocardium of dog heart are illustrated in figure 1. These organelles are abundant and comprise approximately 10–20% of the dry weight of the heart. The cristae are tightly packed and contain no dense bodies. Isolated mitochondria (fig. 2) occasionally show tiny intracristal foci that may be dense bodies but never show the large matrix granules found in mitochondria in rat liver or kidney mitochondria (16, 26).

Permanent ischemia

Small intramitochondrial dense bodies are first noted in some mitochondria in the posterior papillary muscle infarct of the dog heart (13, 15) after 40 min of ischemia (7, 9). After 60 min of ischemia, at which time most of the cells in the posterior papillary muscle infarct are lethally or irreversibly injured (13), all mitochondria contain one or more dense bodies, which now appear to be larger than the first detectable forms (fig. 3a). Typically (fig. 3b and c), they exhibit

[1] Supported in part by research grants from the National Heart Institute (HE 08729) and the Chicago and Illinois Heart Associations.

Fig. 1 Control. Portion of a myocardial cell from the posterior papillary muscle of a sham-operated dog. The mitochondria have relatively dense matrices, with closely packed cristae. Tiny, intramitochondrial granules are seen in a few mitochondria. Lead acetate stain, × 13,000. Reprinted from Lab. Invest. (10).

irregular, generally circular profiles, and measure about 80 mμ in diameter. Occasional dense bodies reach 150 mμ in greatest diameter (10). They usually have a fine granular structure, especially on the periphery, but an electron-lucent core is not seen.

Autolysis

Mitochondrial dense bodies first appear after 60 min of incubation (6) of dog myocardium *in vitro* in a moist atmosphere at 37°C. As incubation is prolonged, they increase in number and size, up through 120 and possibly 180 min. These

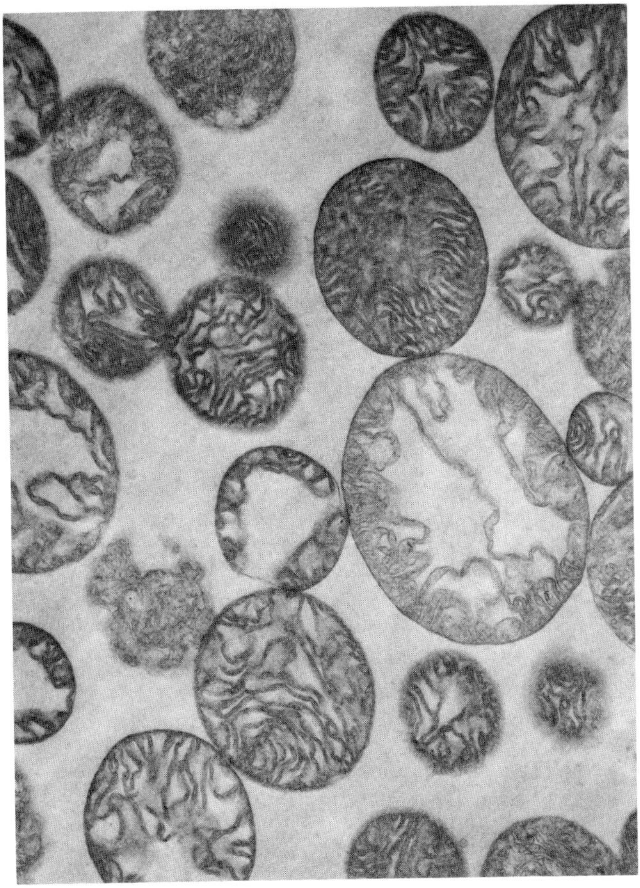

Fig. 2 Control. Appearances of mitochondria isolated in 0.18 M KC1 from the posterior papillary muscle of the left ventricle of a sham-operated dog. In some mitochondria, the cristae are evenly distributed and relatively closely packed; in others, the cristae are arranged peripherally with a central clear area. Occasional, small intramitochondrial granules are seen. Lead acetate stain, ×20,000. Reprinted from Lab. Invest. (10).

dense bodies are similar in structure to those found after equivalent periods of permanent ischemia (fig. 4). Recent experiments (12) on mitochondria isolated in sucrose from dog heart after 60, 120, and 180 min of autolysis have shown that there is an apparent increase in the Ca^{2+} content of these organelles after 1 h of autolysis, and a loss of K^{1+}, Mg^{2+} and nitrogen, which is maximal after 2 h of autolysis. The Ca^{2+} increase coincides with the appearance of the intramitochondrial dense bodies and suggests that these structures may contain this divalent cation (12).

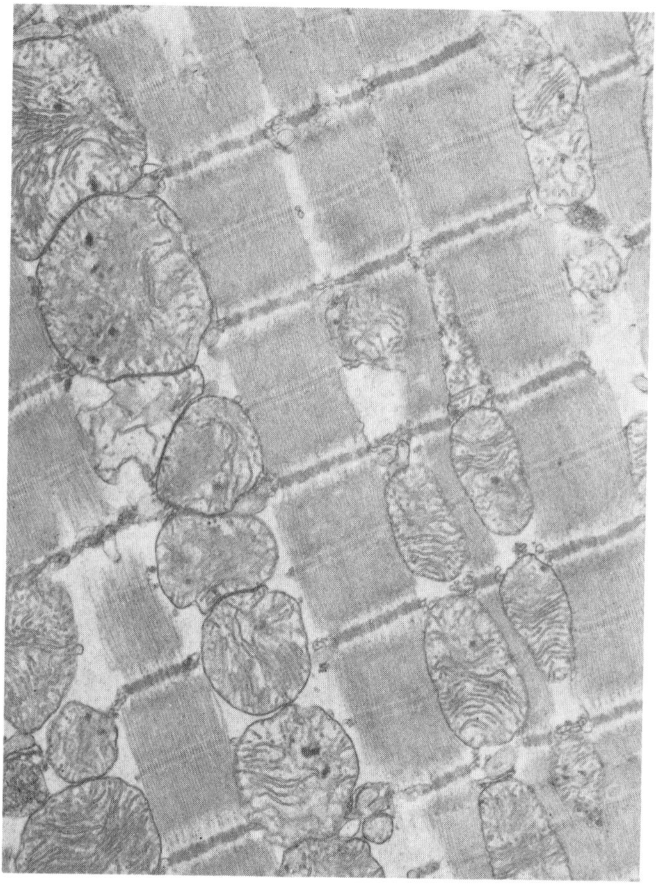

Fig. 3 (a) Ischemia. Portion of a myocardial cell of the posterior papillary muscle of a dog heart after 60 min of ischemia induced by ligation of the circumflex branch of the left coronary. Reprinted from Lab. Invest. (10).

Transient ischemia

Episodes of transient ischemia of 40 min duration are sufficient to irreversibly injure about 55% of the cells in the posterior papillary muscle of the dog heart (13). After 20 min of arterial reflow, the affected cells show striking changes in structure by light and electron microscopy. Interstitial and cellular edema are prominent; contraction bands are noted in all affected cells. Moreover, the mitochondria of each involved cell show prominent intramitochondrial dense bodies (fig. 5). These dense bodies are usually associated with the cristae and exhibit a definite granular substructure, which led us to label them granules in an earlier paper (7). These structures measure up to to 200 mμ

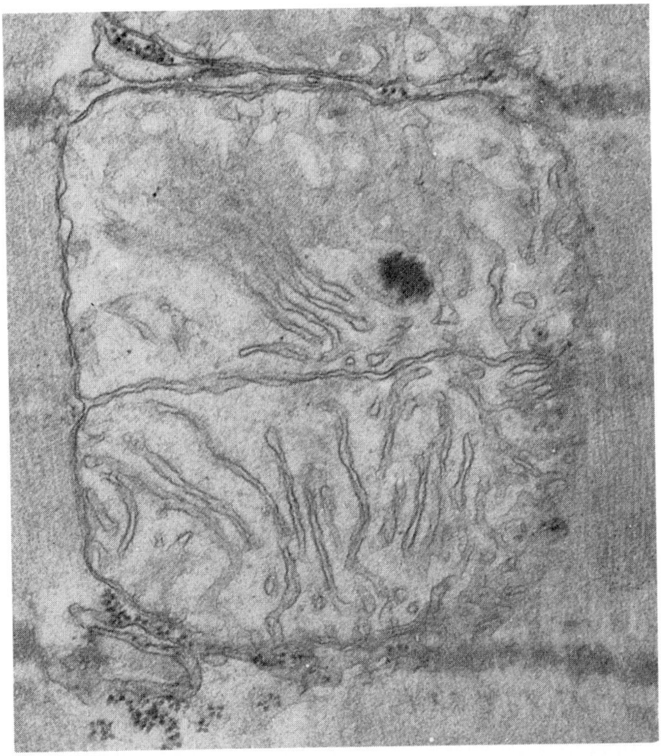

Fig. 3 (b) Ischemia. High power view of a mitochondrial dense body from the posterior papillary muscle of the dog shown in figure 3a. Lead acetate stain, ×45,000. Reprinted from Lab. Invest. (10).

in greatest diameter; a central electron-lucent core is often present. As many as four granules may be present in one section of an individual mitochondrion. Undamaged cells in the focus of ischemic injury show no changes in mitochondrial or cellular morphology.

Periods of ischemia of up to 18 min duration are not associated with cell death (13). Mitochondria of such cells show no change in structure when studied after varying periods of reflow of arterial blood (7).

In vitro Ca^{2+} accumulation

Canine myocardial mitochondria will accumulate Ca^{2+} if incubated with succinate, ATP, phosphate, NaCl, $MgCl_2$, and tris-maleate buffer according to the conditions of Rossi and Lehninger (20). The Ca^{2+} is found in granular dense bodies, which increase in size and number during the first 10 min of incubation. These dense bodies appear to be located in or on the cristae and are not found in

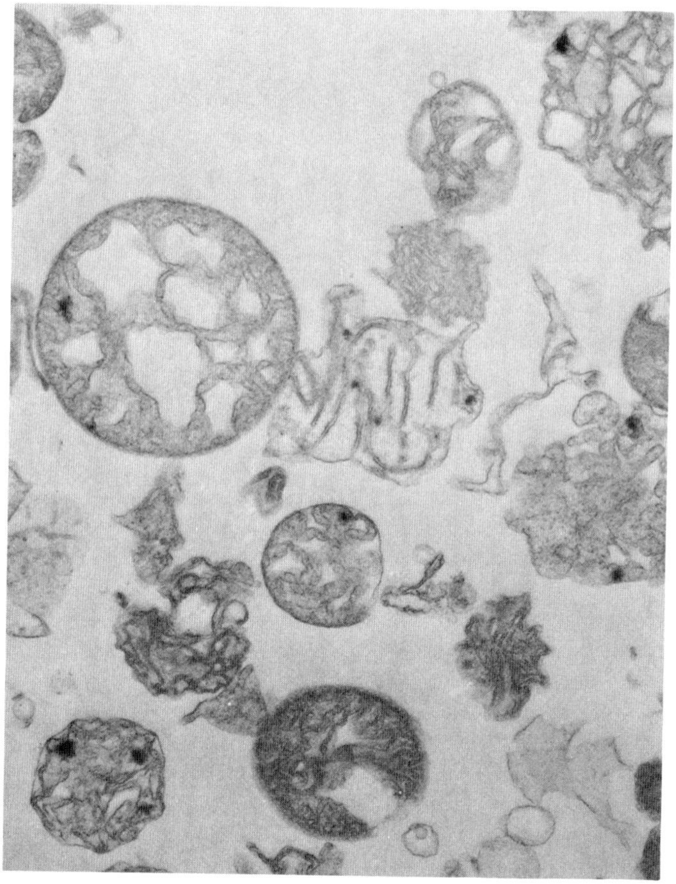

Fig. 3 (c) Ischemia. Appearances of mitochondria isolated in 0.18 M KCl from the posterior papillary muscle after 60 min of ischemia. Many mitochondria are disrupted, and prominent dense bodies are present both in these fragments and the apparently intact mitochondria. Lead acetate stain, × 20,000. Reprinted from Lab. Invest. (10).

the matrix (fig. 6) (11). Greenawalt *et al.* (4) have made similar observations in liver mitochondria. Thomas and Greenawalt (25) and Weinbach and Von Brand (27) have shown that these granules contain calcium phosphate in a noncrystalline form.

ROLE OF Ca^{2+} IN THE FORMATION OF INTRAMITOCHONDRIAL DENSE BODIES

Tissue analysis

We were intrigued by the fact that mitochondrial dense bodies in myocardial cells irreversibly injured by permanent ischemia differed significantly in

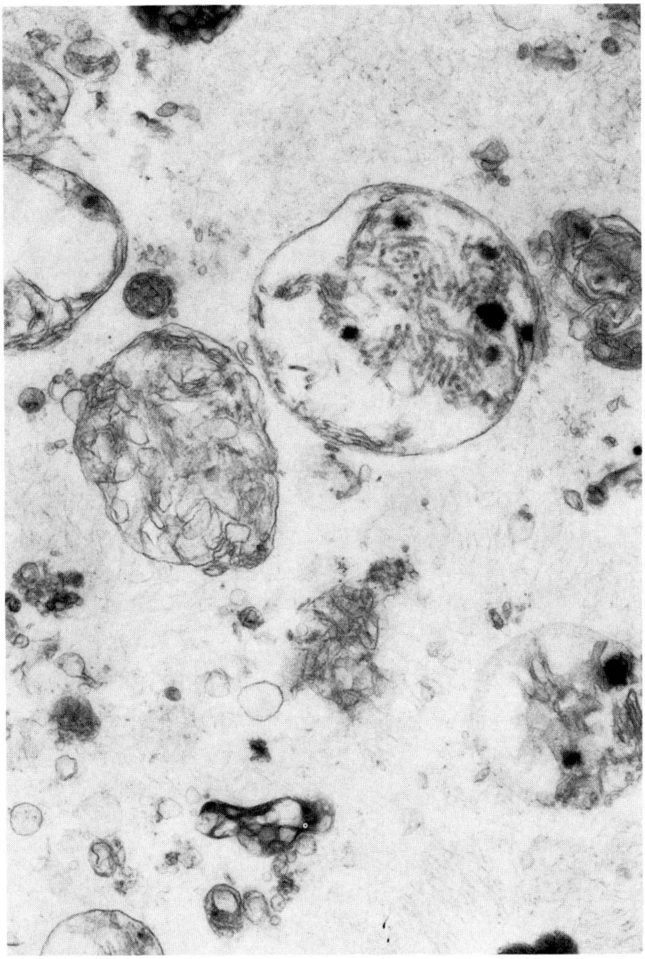

Fig. 4 Autolysis. Appearance of mitochondria isolated in 0.25 M sucrose from dog left ventricular myocardium after 2 h of autolysis at 37°C. Many mitochondrial fragments are present as are numerous intramitochondrial dense bodies. Lead acetate stain, ×20,000. Reprinted from Proc. Soc. Exp. Biol. Med. (12).

structure from those found in transient ischemia. In particular, the transient ischemia granules were larger and more dense (compare figs. 3b and 5). Also, the structure of the transient ischemia granules was very similar to that of the granules found in mitochondria that had accumulated Ca^{2+} in vitro (compare figs. 5 and 6). These observations led us to hypothesize that a significant amount of calcium should be present in the granules of transient ischemia. If this hypothesis is true, the calcium content of the damaged tissue should be

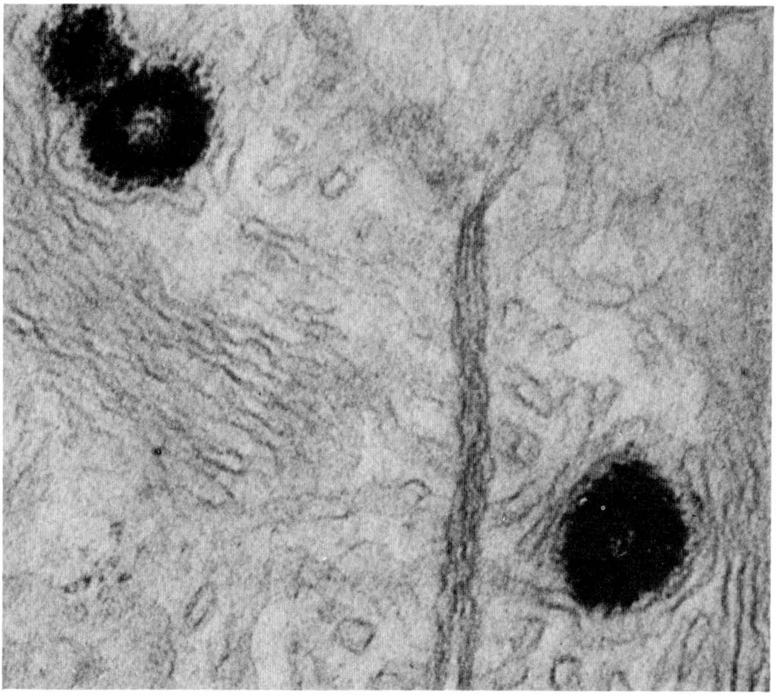

Fig. 5 Transient ischemia. Appearance of mitochondria of posterior papillary muscle of dog left ventricle after exposure to 40 min of ischemia followed by 20 min of reflow of coronary arterial blood. Three granules, two with typical central electron-lucent cores, are present inside the mitochondria. Lead acetate stain, ×117,000. Reprinted from Amer. J. Path. (7).

increased, especially if an adequate sample of damaged tissue containing many mitochondrial granules could be obtained for analysis. The 40 min temporary ischemia model described earlier provides sufficient tissue for direct analysis of Ca^{2+} and was used in the experiment to be described. Although this tissue may often include viable cells that contain no mitochondrial granules (7, 14), it seemed likely that enough Ca^{2+} would be contributed by the granules of the affected cells to yield a detectable increase in total tissue calcium.

Accordingly, we compared the calcium content of posterior papillary muscle with control, nonischemic myocardium from the anterior superior septum (22). The results are summarized in table I. The injured myocardium showed a 10-fold increase in Ca^{2+} per 100 g of dry weight. Sodium and water were considerably increased, and Mg^{2+} showed a marked decrease. The changes in sodium and water have been described in an earlier paper (14). The results of these earlier experiments also showed an increase in tissue chloride and a marked decrease in tissue K^{1+} (14).

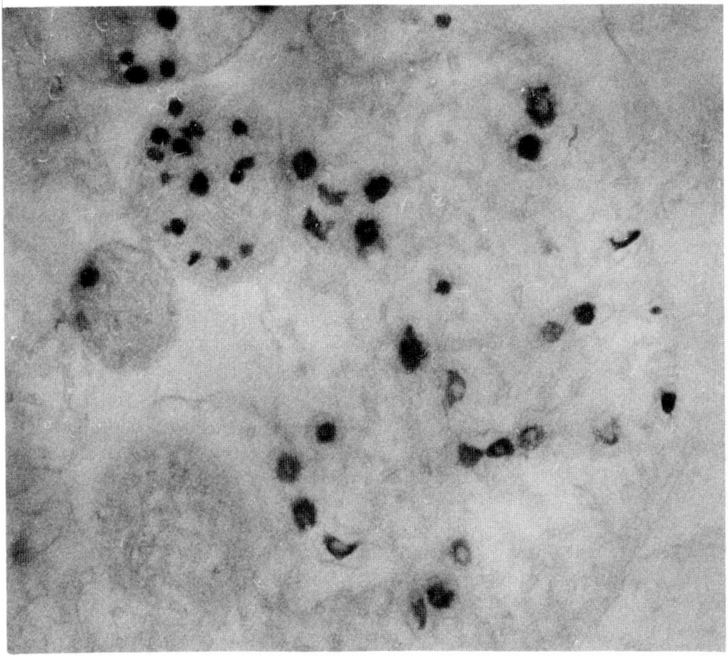

Fig. 6 Ca^{2+} accumulation. Mitochondria of control dog left ventricular myocardium, which have accumulated calcium *in vitro*. Note that many of the intramitochondrial granules exhibit a central electron-lucent core. Osmium fixation; unstained, ×39,000.

Table I

Myocardial Electrolytes of Permanent and Transient Ischemia[a]

	Calcium	Magnesium	Sodium	Water
	(mmole/100 g dry weight)			(g/100 g dry weight)
Control[b]	0.31 ± 0.06	5.06 ± 0.31	16.35 ± 0.63	78.36 ± 0.89
Permanent Ischemia[c]	0.39 ± 0.03	4.99 ± 0.21	17.51 ± 2.25	78.90 ± 1.30
Transient Ischemia[d]	4.59 ± 0.42	3.60 ± 0.35	46.05 ± 5.87	83.02 ± 1.64
	$p < 0.025$	$p < 0.025$	$p < 0.05$	$p < 0.005$

[a] Values shown are means ± standard error of the mean (N = 4). Confidence levels (p), obtained by the paired-t test, refer to the differences between transient ischemia values and the control values.

[b] Nonischemic anterior superior left ventricle N = 8.

[c] 60 min of permanent ischemia of left ventricle.

[d] 40 min of transient ischemia, followed by 20 min of reflow of arterial blood.

The calcium content of the posterior papillary muscle after an equivalent period of permanent ischemia showed no change from control values (table I). Mg^{2+}, Na^{1+}, and H_2O, also, were not altered significantly.

Thus, there is an association between the larger granular dense bodies of transient ischemia and increased tissue Ca^{2+}. The increased Ca^{2+} appears to be intracellular, since interstitial fluid alone cannot account for such a large increase in tissue Ca^{2+}. Although tissue Na^{1+} was also increased, the increase was proportional to the increase in tissue fluid. Moreover, accumulation of this monovalent cation *in vitro* is not associated with development of intramitochondrial granules and it seems unlikely that mitochondria, *in vivo,* would differ significantly in this regard from those *in vitro.* The massive, acute shifts of electrolyte in the transient ischemia model system were shown earlier to be secondary to acute cell death occurring in the presence of a freely flowing arterial blood supply (14). The decrease in Mg^{2+} content has not been shown previously and also appears to be a function of increased diffusion of this ion from the drying cells to the general circulation.

^{45}Ca accumulation

By labelling serum Ca^{2+} with a tracer dose of $^{45}CaCl_2$, Shen and Jennings (22) investigated the role played by the arterial blood reperfusing the ischemic myocardium in providing Ca^{2+} to the damaged cells. Figure 7 shows a typical exponential disappearance curve of ^{45}Ca from the plasma after intravenous injection of this isotope in the dog. The disappearance curve varied slightly from dog to dog, but equilibration of ^{45}Ca was always completed by 30 min. After

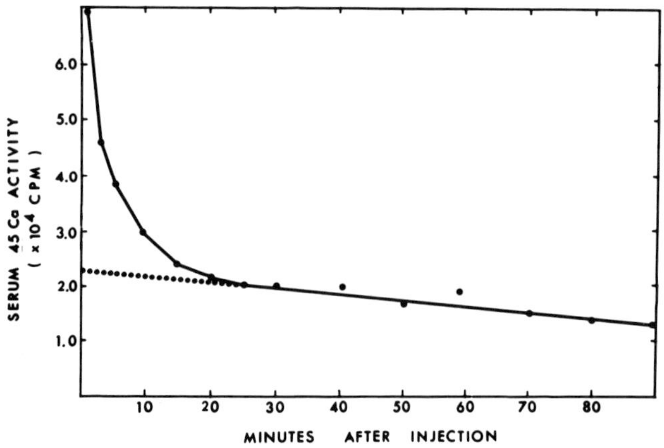

Fig. 7 Disappearance curve of ^{45}Ca from the vascular compartment of the dog; 50 μc of ^{45}Ca (50 μg $CaCl_2$ in 5.0 ml saline) was injected. Serum ^{45}Ca activity was determined at timed intervals after injection.

this time, the serum level of ^{45}Ca was quite stable. Experiments on tissue ^{45}Ca uptake were performed on the flat portion of the ^{45}Ca disappearance curve.

The first experiment on myocardial ^{45}Ca uptake was designed to test the hypothesis that all of the increase in Ca^{2+} shown by chemical means to be in the injured tissue came from the plasma of the blood reperfusing the damaged tissue. Accordingly, ^{45}Ca was injected a few minutes after occlusion of the circumflex artery. The occlusion was maintained for either 60 min without reflow or for 40 min plus 20 min of reflow of arterial blood. The results of such an experiment in permanent ischemia should show little ^{45}Ca activity in an area of maximum ischemia, because of the reduced flow of arterial blood. Any Ca^{2+} changes occurring in the permanently ischemic tissue should be due chiefly to redistribution of endogenous Ca^{2+}. On the other hand, with reflow of blood, not only should there be ^{45}Ca uptake, but the tissue calcium calculated from the tissue ^{45}Ca activity and from the specific activity of the serum calcium should yield a value approximating that found by direct chemical analysis. Our results showed that without reflow of arterial blood, there was practically no ^{45}Ca activity in the posterior papillary muscle (PP). With reflow of arterial blood, the tissue ^{45}Ca activity was markedly increased. Furthermore, the observed tissue ^{45}Ca activity yielded a calculated tissue calcium of 4.31 mmole/100 g dry weight as compared with 4.59 mmole of Ca^{2+} obtained by direct chemical analysis of this tissue in another group of dogs.

The kinetics of the Ca^{2+} accumulation in the transient ischemia model were established by Shen and Jennings (22) by injecting ^{45}Ca and waiting 30 min for it to equilibrate before ligating the circumflex artery for 40 min. Tissue was sampled after periods of reflow of 2, 5, 10, or 20 min, and the ^{45}Ca content of PP and LV samples were subsequently determined. Results were referred to the mean serum ^{45}Ca activity found 45 min after injection of the $^{45}CaCl_2$. This allowed comparison of tissue ^{45}Ca content in different animals.

The results are given in table II and show that the ^{45}Ca content of the PP increased immediately following occlusion and reached a maximum after 10 min of reflow. The ^{45}Ca content either leveled off or decreased after 10 and 20 min of reflow. However, the difference between the means at 10 and 20 min of arterial reflow was not significant. The experimental procedures were associated with no change in the ^{45}Ca of control left ventricle.

Thus, Ca^{2+} accumulation is rapid and appears to be complete by 10 min. The accumulation process appears to be an active one that ceases for unknown reasons. Perhaps the mitochondria are saturated with Ca^{2+} or the accumulation of this quantity of mitochondrial Ca^{2+} *in vivo* inhibits further mitochondrial function. Sommers and his co-workers (13, 23) have shown recently that the first 4 min of reflow of arterial blood are associated with a marked increase in the temperature of the damaged tissue. This is the time during which the rate of mitochondrial Ca^{2+} accumulation is maximal. The temperature decreases to very

Table II

Uptake of $^{45}Ca^{++}$ by Dog Myocardium[a]

Tissue $^{45}Ca^{++}$ activity ($\times 10^2$ cpm/g wet wt./10^4 cpm serum $^{45}Ca^{++}$ activity)				
	Duration of reflow of blood			
	2 min	5 min	10 min	20 min
Posterior papillary muscle[b]	3.02 ± 0.27	4.85 ± 0.39	8.25 ± 1.46	5.03 ± 1.10
	(6)	(4)	(8)	(6)
Interventricular septal tissue[c]	1.94 ± 0.12	2.00 ± 0.05	2.13 ± 0.05	2.26 ± 0.10
	(6)	(4)	(8)	(6)
	$p < 0.001$	$p < 0.001$	$p < 0.025$	$p < 0.05$

[a] Values are means ± standard error of the mean. The number of samples are in parenthesis. Confidence levels (p), obtained by the paired-t test, refer to the differences between transient ischemia values and the control values.

[b] Posterior papillary muscle = transiently ischemic tissue; 40 min of ischemia followed by 20 min of reflow of arterial blood.

[c] Interventricular septal tissue = control tissue.

low levels shortly thereafter, suggesting that metabolism, as measured indirectly by heat production, is diminishing in the dying cells.

Localization of intracellular Ca^{2+}

Preliminary experiments show that isolated mitochondria contain excess calcium phosphate in a molar ratio of 3 Ca to 2 P (21). Direct proof that the intramitochondrial dense bodies contain the calcium phosphate is lacking, but strong suggestive evidence is available. Electron microscopic microincineration studies have shown that the dense bodies of transient ischemia leave an ash (figs. 8 and 9), a finding that supports the idea that mineral, presumably calcium phosphate, is present in the mitochondrial granules. The granules of canine mitochondria that have accumulated Ca^{2+} *in vitro* also leave an ash. Thomas and Greenawalt (25) have observed the same phenomenon in liver mitochondria that have accumulated calcium phosphate *in vitro*. In contrast, the dense bodies of mitochondria of autolysis or permanent ischemia failed to leave a significant amount of ash after microincineration.

DISCUSSION AND SUMMARY

Mitochondria of myocardial cells developed prominent intramitochondrial granules after lethal injury produced by an episode of transient ischemia. These granules were structurally indistinguishable from those found in isolated mitochondria that had accumulated calcium phosphate *in vitro* in a medium containing succinate and ATP. The damaged tissue, after 20

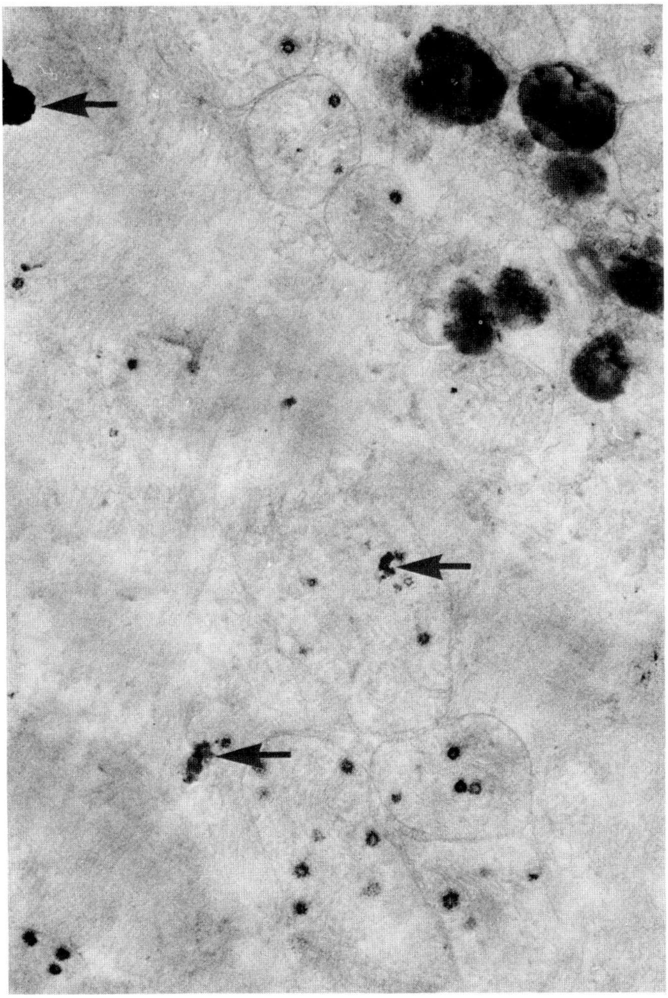

Fig. 8 Transient ischemia. Section of posterior papillary muscle of a dog heart injured by 40 min of transient ischemia followed by 20 min of reflow of arterial blood. Several intramitochondrial granules are shown. Osmium fixation; unstained, ×20,000.

min of reflow, exhibited a 10-fold increase in tissue calcium as well as increased H_2O and Na^{1+}. Tissue Mg^{2+} and K^{1+} were decreased. Mitochondria isolated from the damaged cells contained increased calcium and phosphate, which, according to microincineration studies, were localized in the intramitochondrial granules.

Studies with intravenous ^{45}Ca established that the source of the intracellular calcium was the arterial blood reperfusing the previously ischemic focus. Uptake of ^{45}Ca occurred quickly and was maximal after 10 min of reflow. Intervals of arterial reperfusion longer than 20 min have not been studied.

The intramitochondrial dense bodies of ischemia and autolysis differ structurally from those found in transient ischemia in that they rarely exhibit a central core, are less granular,

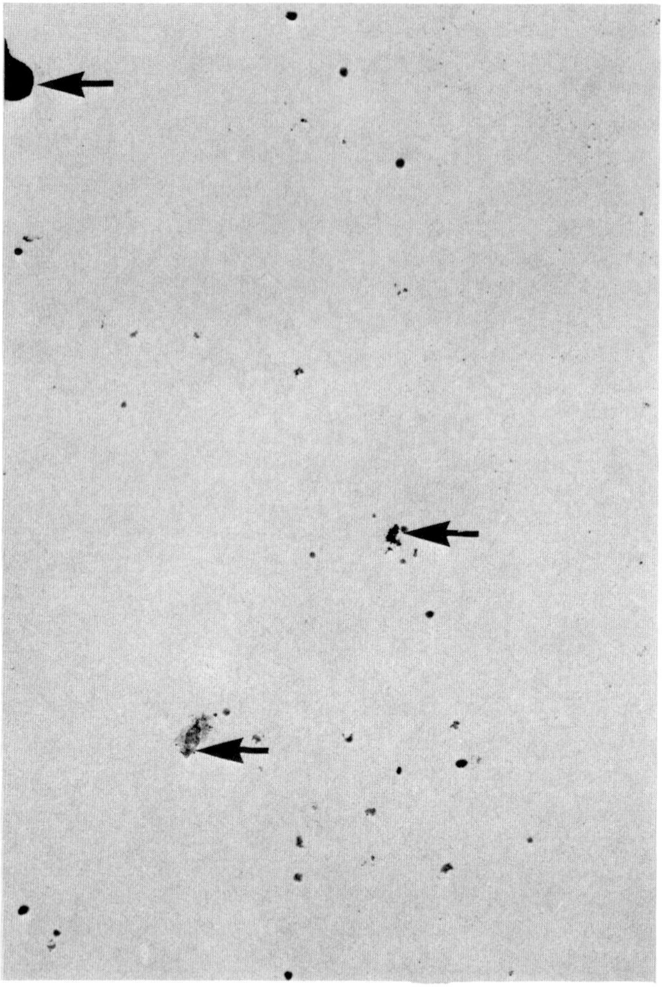

Fig. 9 Microincineration. The same area as that shown in figure 8, after incineration at 550°C for 10 min. Identifiable sites allowing orientation are indicated by arrows in figures 8 and 9. The organic materials are burned away, leaving only ash deposits that correspond exactly to the locations occupied by the intramitochondrial granules prior to incineration. ×20,000.

and leave no ash after microincineration. According to indirect evidence obtained recently (12), they appear to contain small quantities of calcium. The low Ca^{2+} content is not unexpected, since only a finite quantity of Ca^{2+} is available in dog myocardium (0.41 mmole of Ca^{2+} per 100 g of dry weight) to serve as a source for intramitochondrial accumulation of this ion. If more Ca^{2+} was present, we would predict that a greater quantity of this ion would probably accumulate in mitochondria after episodes of either autolysis or permanent ischemia.

The pathogenesis of calcium accumulation in myocardial cells irreversibly injured by 40 min of transient ischemia remains unknown. The experiments summarized in this paper demonstrate that the accumulation of Ca^{2+} is secondary to the increased delivery of this ion to the tissue, via the bloodstream. Further statements about the mechanism through which the mitochondria concentrate this ion *in vivo* are strictly speculative. There are two hypothetical mechanisms that may be operative. The first possibility is that irreversible ischemic injury causes the cell membrane of the affected cells to become more permeable to Ca^{2+}; thus, increased quantities of Ca^{2+} enter the intracellular space where this ion is available to be concentrated by intact mitochondria, which, in health, do not concentrate much Ca^{2+} because they have no access to significant concentrations of it. Alternatively, ischemia may induce a defect in the mitochondria of the irreversibly injured myocardial cells, which leads to excessive ion accumulation, particularly Ca^{2+}. A mitochondrial defect could cause Ca^{2+} accumulation with or without changes in cell membrane permeability, but accumulation probably would be accelerated if cell membrane permeability were increased. The fact that Ca^{2+} does not appear to accumulate in reversibly injured cells after transient episodes of ischemia is consistent with either of the proposed hypothetical mechanisms. We assume that the actual mechanism of transport of Ca^{2+} into the mitochondria is similar to that described by Rossi and Lehninger in isolated mitochondria *in vitro* (20).

The effect of Ca^{2+} accumulation on mitochondrial function *in vivo* remains to be established. However, it seems likely that Ca^{2+} accumulation of the magnitude observed in these experiments would be associated with depressed mitochondrial function. For example, respiration *in vitro* is deficient in mitochondria containing high Ca^{2+} (1). Also, the production of high-energy phosphate in mitochondria would be impaired if the rate of removal of phosphate as calcium phosphate granules sufficiently decreases the inorganic phosphate concentration to depress oxidative phosphorylation. No direct studies of the function of isolated mitochondria in this system have been performed.

It is of interest that similar dense bodies containing Ca^{2+} occur in dying liver cells poisoned with CCl_4 (19, 24), a fact that taken together with the findings in Mg^{2+} deficiency (5) and electric shock (18) suggests that mitochondrial Ca^{2+} accumulation may be a general phenomenon occurring in several different kinds of severe cellular injury, especially acute lethal injury developing in cells perfused by an adequate arterial blood supply. We do not know whether or not massive Ca^{2+} accumulation in mitochondria interferes sufficiently with integrated mitochondrial function to cause cell death. It should be observed, however, that similar dense bodies have been described by Gonzales and Karnovsky (3) in osteoclasts in healing fractures of bone, a biologic phenomenon not known to be associated with either severe or lethal cell injury, but rather a situation in which much Ca^{2+} transport presumably is occurring.

ACKNOWLEDGMENTS

The editors of Laboratory Investigation, the Proceedings of the Society for Experimental Biology and Medicine, and the American Journal of Pathology have granted kind permission to reprint figures 1, 2, 3a, 3b, 3c, figure 4, and figure 5, respectively. Valuable technical assistance during the course of the studies presented here was provided by M. L. Hill, C. M. Moore, B. M. Moulton, and S. Safavi.

REFERENCES

1. CHANCE, B.: The energy-linked reaction of calcium with mitochondria. J. Biol. Chem. *240:* 2729–2748 (1965).
2. D'AGOSTINO, A. N.: An electron microscopic study of cardiac necrosis produced by 9 α-fluorocortisol and sodium phosphate. Amer. J. Path. *45:* 633–644 (1964).
3. GONZALES, F. and KARNOVSKY, M. J.: Electron microscopy of osteoclasts in healing fractures of rat bone. J. Biophys. Biochem. Cytol. *9:* 299–316 (1961).
4. GREENAWALT, J. W.; ROSSI, C. S. and LEHNINGER, A. L.: Effect of active accumulation of calcium and phosphate ions on the structure of rat liver mitochondria. J. Cell Biol. *23:* 21–38 (1964).

5. HEGGTVEIT, H. A.; HERMAN, L. and MISHRA, R. K.: Cardiac necrosis and calcification in experimental magnesium deficiency. Amer. J. Path. *45:* 757–782 (1963).

6. HERDSON, P. B.; KALTENBACH, J. P. and JENNINGS, R. B.: Fine structural and biochemical changes in dog myocardium during autolysis. Amer. J. Path. *57:* 539–557 (1969).

7. HERDSON, P. B.; SOMMERS, H. M. and JENNINGS, R. B.: A comparative study of the fine structure of normal and ischemic dog myocardium with special reference to early changes following temporary occlusion of a coronary artery. Amer. J. Path. *46:* 367–386 (1965).

8. JENNINGS, R. B.: Early phase of myocardial ischemic injury and infarction. Amer. J. Cardiol. *24:* 753–765 (1970).

9. JENNINGS, R. B.; BAUM, J. H. and HERDSON, P. B.: Fine structural changes in myocardial ischemic injury. Arch. Path. *79:* 135–143 (1965).

10. JENNINGS, R. B.; HERDSON, P. B. and SOMMERS, H. M.: Structural and functional abnormalities in mitochondria isolated from ischemic dog myocardium. Lab. Invest. *20:* 548–557 (1969).

11. JENNINGS, R. B. and SHEN, A.: Unpublished observations.

12. JENNINGS, R. B.; SHEN, A. C.; MOORE, C. B. and HERDSON, P. B.: Electrolytes of damaged myocardial mitochondria. Proc. Soc. Exp. Biol. Med. *135:* 515–522 (1970).

13. JENNINGS, R. B.; SOMMERS, H. M.; HERDSON, P. B. and KALTENBACH, J. P.: Ischemic injury of myocardium. Ann. N.Y. Acad. Sci. *156:* 61–78 (1969).

14. JENNINGS, R. B.; SOMMERS, H. M.; KALTENBACH, J. P. and WEST, J.J.: Electrolyte alterations in acute myocardial ischemic injury. Circulation Res. *14:* 260–269 (1964).

15. JENNINGS, R. B.; WARTMAN, W. B. and ZUDYK, Z. E.: Production of an area of homogeneous myocardial infarction in the dog. AMA Arch. Path. *63:* 580-585 (1957).

16. LATTA, H.; OSVALDO, L.; JACKSON, J. D. and COOK, M. L.: Changes in renal cortical tubules during autolysis, Electron microscope observations. Lab. Invest. *14:* 635–657 (1965).

17. MARUFFO, C. A.: Fine structural study of myocardial changes induced by isoproterenol in Rhesus monkeys (Macaca Mulatta). Amer. J. Path. *50:* 27–37 (1967).

18. REICHENBACH, D. and BENDITT, E. P.: Myofibrillar degeneration: A common form of cardiac muscle injury. Ann. N.Y. Acad. Sci. *156:* 164–176 (1969)

19. REYNOLDS, E. S.: Liver parenchymal cell injury. I. Initial alterations of the cell following poisoning with carbon tetrachloride. J. Cell. Biol. *19:* 139–157 (1963).

20. ROSSI, C. S. and LEHNINGER, A. L.: Stoichiometric relationships between accumulation of ions by mitochondria and the energy-coupling sites in the respiratory chain. Biochem. Z. *338:* 698–713 (1963).

21. SHEN, A. and JENNINGS, R. B.: Unpublished observations.

22. SHEN, A. C. and JENNINGS, R. B.: Calcium uptake by ischemic myocardium. Fed. Proc. *29:* 421 (1970).

23. SOMMERS, H. M.: Myocardial temperature changes in reversible ischemic injury. Fed. Proc. *27:* 357 (1968).

24. THIERS, R. B.; REYNOLDS, E. S. and VALLEE, B. L.: The effect of CCl_4 poisoning on subcellular metal distribution in rat liver. J. Biol. Chem. *235:* 2130–2133 (1960).

25. THOMAS, R. S. and GREENAWALT, J. W.: Microincineration, electron microscopy, and electron diffraction of calcium phosphate-loaded mitochondria. J. Cell Biol. *39:* 55–76 (1968).

26. TRUMP, B. F.; GOLDBLATT, P. J. and STOWELL, R. E.: Studies on necrosis of mouse liver *in vitro*. Ultrastructural alterations in the mitochondria of hepatic parenchymal cells. Lab. Invest. *14:* 343–371 (1965).

27. WEINBACH, E. C. and VON BRAND, T.: Formation, isolation and composition of dense granules from mitochondria. Biochem. Biophys. Acta. *148:* 256–266 (1967).

Mailing address: Robert B. Jennings, M.D., Department of Pathology, Northwestern University Medical School, Ward Memorial Building, 303 E. Chicago Ave., Chicago, Illinois 60611 (USA).

New Jersey College of Medicine and Dentistry,
Department of Medicine, Newark, New Jersey, USA

MYOCARDIAL ION AND LIPID CHANGES DURING ISCHEMIA AND CATECHOLAMINE INDUCED NECROSIS: RELATION TO REGIONAL BLOOD FLOW [1]

T. J. REGAN, A. MARKOV, M. I. KHAN, M. U. JESRANI,
H. A. OLDEWURTEL, and P. O. ETTINGER

Ischemia produced by obstruction of a branch of the left coronary artery is associated with myocardial potassium loss (15). Similar leakage of ion from the myocardium has been seen, beginning approximately 90 min after the onset of catecholamine infusion into the left coronary artery (17). In addition, lipid accumulation in the myocardium has been observed in both circumstances. A major contrast exists in the coronary blood increase observed with epinephrine. However, a regional decrease of blood flow and resultant ischemia might explain the potassium loss.

This present study was designed to determine if there was regional ischemia during epinephrine injection and consequent regional electrolyte abnormalities. Experiments have been conducted in a group of animals undergoing ischemia as well as in animals receiving catecholamines. Regional blood flow was evaluated by analysis of the distribution of radioactive krypton in the outer and inter walls of the myocardium and the content of K and Na in the segments of muscle were determined approximately 1 h after onset of injury. In addition, the lipid composition of the ventricle in these states was determined, and compared with that of control animals.

METHODS

Male, mongrel dogs, weighing between 19 and 24 kg, were anesthetized 18 h post-prandially with morphine sulfate 3 mg/kg and pentobarbital (Nembutal) 12 mg/kg and studied without opening the chest. After insertion of an endotracheal tube, respiration was regulated with a Harvard respiratory pump, facilitating the maintenance of arterial oxygen saturation and pH in the normal range. Catheters were placed in the aorta and left ventricle for blood sampling and pressure determinations and in the left coronary artery for myocardial blood flow

[1] Supported by NIH Research Grant HE 09914, Graduate Training Grant HE 05510, and MHI Grant MH 17007, and a grant from the Union County Heart Association of New Jersey.

measurements and infusion. Although the catheters were initially filled with dilute heparin, patency was maintained by slow saline infusions or intermittent flushes during the experiment.

A closed-chest preparation was used in the production of coronary artery obstruction to utilize the greater hemodynamic stability. An electrode catheter was placed in the left anterior descending coronary artery to permit passage of 300 μA to effect thrombus formation in the Group I anesthetized dogs. Measurement of coronary blood flow to the ischemia area was obtained by injection of ^{85}Kr distal to the thrombus site. This inert gas method appears to give valid flow measurements over a wide range of tissue perfusions (8, 19). Approximately 100 μc of ^{85}Kr were injected at 6–10 min intervals before ischemia and at 12–20 min intervals thereafter. The myocardial clearance of isotope was obtained by precordial scintillation counting, and background radioactivity was subtracted in the usual manner. For the purpose of estimating the size of the ischemic site, Evans blue dye was injected via a coronary artery catheter into the area of ischemia at the conclusion of each experiment and the dyed tissue was weighed. Thrombus formation was confirmed at post-mortem, and arterial obstruction in these animals appeared to be complete.

Animals of Group II received epinephrine, 5μg/min of the base being infused into the left anterior descending coronary artery at the rate of 0.1 ml/min for 90 min. When changes in heart rate or rhythm occurred at the onset of infusion, the dosage was reduced for a few minutes and then resumed with maintenance of the control rate. Minimum recirculation of the hormone occurred as judged by the lack of increments in heart rate, arterial pressure, and arterial free fatty acid levels, presumably due to the large myocardial capacity for storage of catecholamines.

Epicardial ECG from saline-filled coronary arterial and venous catheters were monitored for evidence of injury potentials, and left ventricular pressures were examined for contractility change using the relation of maximum rate of left ventricular pressure rise to end-diastolic pressure (14). Ventricular and femoral arterial pressures were measured through 50 cm Goodale Lubin catheters (U.S. Catheter Co., Glen Falls, N.Y.), 8F catheters connected directly to Statham strain gauge transducers (P 28Db, Statham Instruments, Inc., Los Angeles, Calif.). Intermittent photographic recordings were made from a multichannel oscilloscope recorder (Electronics for Medicine, Inc., White Plains, N.Y.). The first derivative of left ventricular pressure pulse was computed by an R-C differentiating circuit converting into mm Hg/sec (14). Statistical variations are expressed as standard error, and Student's t test was paired or nonpaired, as appropriate (3).

Substrate determination on tissue was done in duplicate and included triglyceride (4), phospholipid (2), and free fatty acid (10). The triglyceride method was modified (5), using a Florisil column in place of silicic acid to exclude phospholipid more effectively from the chloroform eluate, as verified by thin layer chromatography. To obtain left ventricular tissue, the chest was quickly opened and the heart arrested with iced sucrose. A transmural sample was rapidly obtained from the center of the perfused area and separated into inner, mid, and outer layers. The latter was carefully trimmed to remove pericardial and adipose tissue and all samples were placed in liquid nitrogen. The tissue was homogenized in phosphate buffer and the lipid extracted in chloroform-methanol. Separate portions of these samples were homogenized in distilled water for analysis of potassium and sodium on the autoanalyzer with flame attachment. Dry weight was obtained by placing in an oven at 100°C for 18 h. Separate samples from the same central area were divided into inner and outer halves and ^{85}Kr counts determined immediately in a Nuclear-Chicago scintillation counter. Isotopic activity was related to wet weight.

RESULTS

Figure 1 illustrates the rate of decay of radioactive ^{85}Kr as counted during precordial counting after injection into the left anterior descending artery distal to the obstruction during ischemia. The lower curve indicates the normal rate of coronary blood flow prior to the production of obstruction. After thrombus is produced, there is a marked decrease in the rate of clearance of ^{85}Kr in the myocardium as illustrated by the prolonged isotope decay slope. Coronary blood flow during epinephrine increased approximately 15%. Since varied rates of blood flow within regions of an organ have been demonstrated by organ sampling at different times during clearance of inert gas (22), we undertook such a sampling procedure of the myocardium. The arrows indicate the temporal points in the clearance curve at which the myocardium was sampled in different groups of animals to determine the content of ^{85}Kr. Myocardium was obtained at about 75% of peak concentration, at 50%, and late in the curve when the

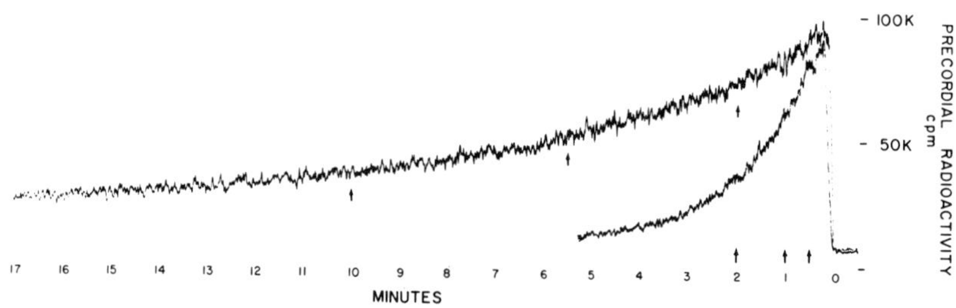

Fig. 1 Zero time indicates point of injection of 0.2 ml of ^{85}Kr in saline into left anterior descending artery. Lower curve of precordial radioactivity represents the normal coronary blood flow prior to obstruction of the coronary artery; upper curve illustrates the prolonged clearance time after arterial obstruction. Arrows indicate the temporal points at which hearts were taken in the different groups of animals to determine the inner/outer wall ratio of ^{85}Kr as a measure of regional blood flow.

concentration was about 30% of the maximum values. In view of the slow clearance of ^{85}Kr in ischemia, the times of tissue sampling were necessarily later than in the control and epinephrine groups.

In the latter two groups, at no time in the course of the clearance of radioisotope was there a significant concentration gradient between the inner and outer wall of the left ventricle near the apex (fig. 2). In contrast, during ischemia, there was a marked gradient in the samplings at 2 min due to a marked reduction in the concentration of ^{85}Kr in the inner wall of the left ventricle. The isotope ratio progressed towards normal as the isotope was cleared from the outer wall.

Figure 3 illustrates the electrolyte concentrations in the groups of normal, ischemic, and epinephrine-treated animals. The tissue samples were taken after 1 h of ischemia or approximately 1 h after the onset of injury in the epinephrine-treated animals. The concentrations of potassium ion in both the ischemic and epinephrine groups showed a graded reduction from outer to inner wall, compared with the control animals (fig. 3). In addition, the K^+ concentration was significantly less in ischemia compared with the epinephrine

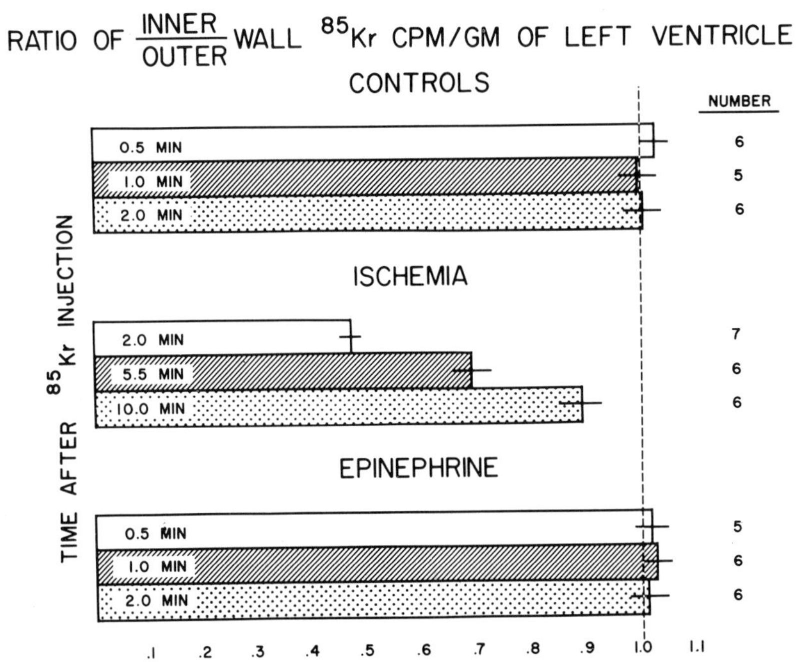

Fig. 2 Time in minutes refers to the period after ^{85}Kr injection into the left anterior descending coronary artery when the heart was arrested for isotope determination. Number refers to the numbers of animals at the indicated times in each group. Only during ischemia was there a marked reduction of isotope in the endocardium.

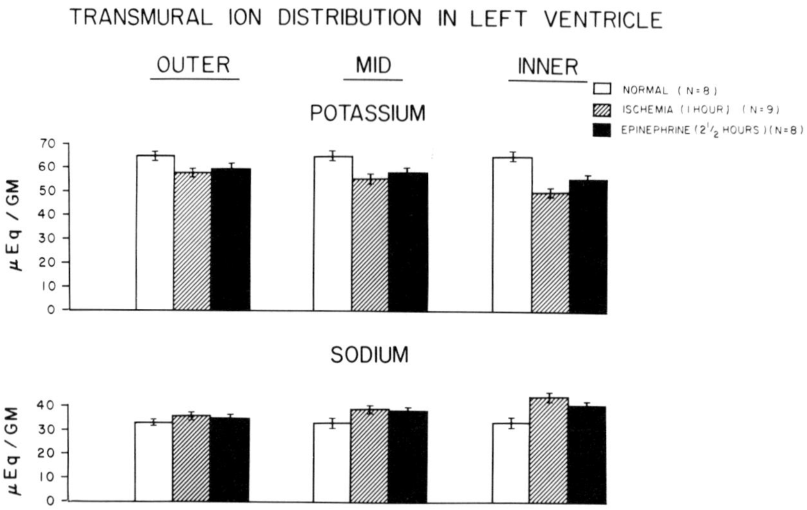

Fig. 3 The myocardial gradient of K^+ loss and Na^+ gain during both forms of injury is illustrated. The endocardial changes were significantly greater during ischemia as compared with epinephrine injury.

group ($p < 0.01$). Graded increments of sodium ion concentrations across the ventricular wall were observed to be less on a milli-equivalent basis than the loss of K^+. The sodium gain in the endocardium was significantly greater after ischemia than after epinephrine.

Analysis of the same tissue at about 1 h after induction of injury due to ischemia or epinephrine origin showed no significant difference in the concentration of free fatty acid or phospholipid in the three layers of myocardium (fig. 4). Triglyceride concentrations were not significantly changed in the ischemic group; however, there was a significant elevation in the epinephrine-treated animals in all three layers, with the greatest increase in the endocardium.

DISCUSSION

An uneven distribution of blood flow during ischemia has been previously observed by Griggs (7), after substantial reduction of total left coronary artery inflow. He observed the reduced concentrations of the ^{131}I labelled iodoantipyrine in the inner as compared with the outer wall. Since obstruction of a left coronary artery branch, rather than of the left main artery itself, may be associated with a greater degree of collateral blood flow to the ischemic site, one could not *a priori* assume that endocardial layers would show a larger reduction of blood flow during obstruction of the branched artery. However, marked reductions of endocardial isotope concentrations were observed in the early

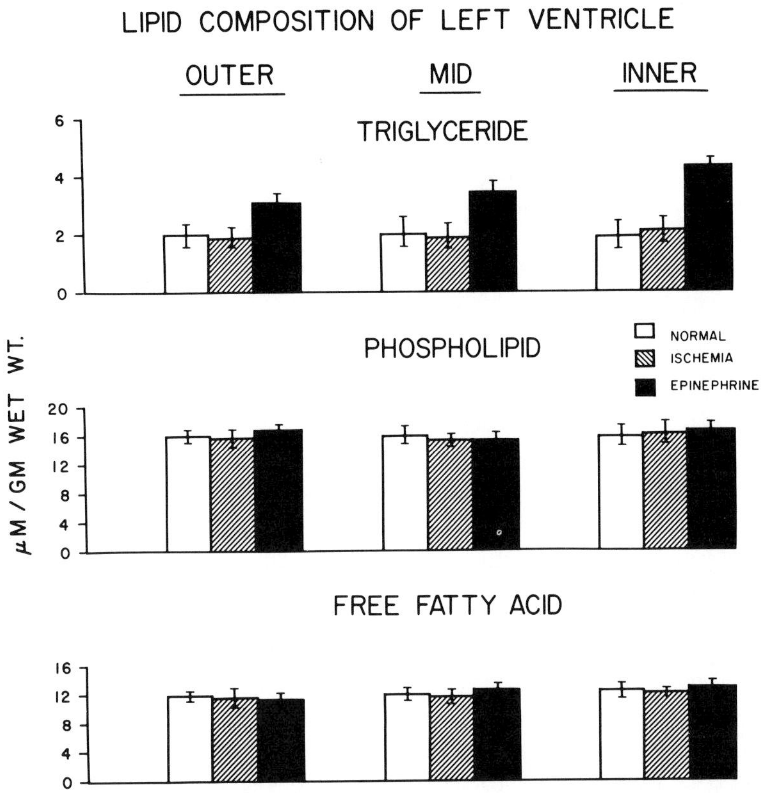

Fig. 4 No significant changes in the transmural concentrations of free fatty acid and phospholipid were seen 1 h after the onset of ischemia or epinephrine injury. Triglyceride content was increased only after epinephrine, and the increment was greatest in the endocardium.

course of clearance of the isotope from myocardium. The mechanism of this differential flow during ischemia remains to be determined, although a larger intramural pressure in the endocardial area has been postulated as contributing to the greater flow reduction in these vessels.

Reduction of myocardial potassium ion has previously been observed during ischemia by Jennings (9) in transmural sections of the myocardium. In our study, the analysis was conducted on the three layers of myocardium at the presumed center of the ischemic area as determined by the distribution of Evans blue dye injected distal to the obstruction site. Analysis of these segments of muscle revealed a significantly greater reduction in the endocardial wall, consistent with the larger reduction of coronary blood flow. Since potassium has been shown to be closely associated with the genesis of ventricular ectopic beats during ischemia (15), it is possible that the gradient of potassium ion is a major factor in the arrhythmia production.

The greater reduction of potassium ion in the endocardium in animals undergoing epinephrine-induced necrosis is consistent with previous observations using histochemical techniques (13). Although total coronary blood flow measurements have shown no decrease in perfusion rates of the myocardium (17), it seemed conceivable that a localized reduction of blood flow might adversely affect the endocardium. Since a reduced content of ^{85}Kr in the inner layers was not present, the hormone may have a more intense metabolic effect on the cells of the endocardium, as has been suggested in a study of turtle heart (12). One cannot exclude the contribution of enhanced contractility that occurs during the initial 60 min of epinephrine infusion in contributing to the localization of potassium depletion. It is noteworthy that the K^+ depletion was less in extent than in the ischemic animals and the epinephrine group also had a lesser incidence of ventricular ectopic beats.

Triglyceride increments in the myocardium were seen only in the epinephrine group with larger concentrations in the endocardium, suggesting a different metabolic reaction in the tissues than during ischemia. While the greater triglyceride increment in the inner wall may have contributed to the leakiness of membranes to potassium ion, preliminary evidence, in which lipid increments were prevented by heparin administration, suggests that one can dissociate the lipid change and the potassium loss.

The failure to observe an increment of triglyceride in the early stages of myocardial ischemia differs from some views of the metabolic response to ischemia. However, there are several histochemical studies indicating that in the central ischemia area, lipid accumulation may not be seen, at least in the early hours after the onset of ischemia (6, 20, 21). The altered lipid concentration during the early stages of epinephrine injury may contribute to the differentiation of ischemia from stress-induced injury in other species, particularly man, if a large species difference in the response to catechols does not exist.

Finally, since we previously observed that the acute ischemic process in the intact dog is associated with increased arterial levels of free fatty acid and a relatively increased uptake of this substrate (16), it is of interest that no accumulation of free fatty acid was observed in the ischemic myocardium in the three layers examined. Hence, the view that enhanced myocardial free fatty acids may contribute to the genesis of arrhythmias during myocardial ischemia (11) was not supported by the studies in these animals who had sporadic ventricular ectopic activity in the ischemic period.

SUMMARY

The metabolic alterations produced by ischemia and excess catecholamines have not been differentiated. Ischemia was produced in intact anesthetized dogs by thrombosis of the left anterior descending artery via catheter electrode. Coronary blood flow was reduced to about 25% of normal as measured by clearance of ^{85}Kr injected distal to obstruction. Ectopic activity occurred, but no antiarrhythmic drugs were given. After 1 h, the heart was cold-arrested. The distribution of ^{85}Kr was diminished in the endocardial layer of ischemic tissue as was the concentration of K^+ *versus* outer-wall. There was no significant change in lipid classes.

Evidence of nonischemic necrosis was obtained 75 min after the start of epinephrine infusion into the left coronary artery; 60 min later, the heart was cold-arrested. There was no transmural gradient of ^{85}Kr. Endocardial K^+ was reduced, but was higher than in ischemic tissue. While phospholipid and free fatty acid concentrations were normal, there was a significant increase in triglyceride. Thus, hormonal-induced necrosis in its early stage is significantly different from ischemia.

REFERENCES

1. BARGER, A. C.; HERD, J. A. and LIEBOWITZ, M. R.: Chronic catheterization of coronary artery: Induction of ECG pattern of myocardial ischemia by intracoronary epinephrine. Proc. Soc. Expt. Biol. Med. *107:* 474–477 (1961).
2. BARTLETT, G.: Phosphorous assay in column chromatography. J. Biol. Chem. *234:* 466–468 (1959).
3. BATSON, H. C.: An Introduction to Statistics in the Medical Sciences (Burgess, Minneapolis, Minn. 1957).
4. CARLSON, L. A. and WADSTRON, L, B.: Determination of glycerides in blood serum. Clin. Chim. Acta. *4:* 197–205 (1959).
5. CARROLL, K. K.: Separation of lipid classes by chromatography on Florisil. J. Lipid Res. *2:* 135–141 (1961).
6. DENKER, M. W.; BERGMAN, R. A. and NACHLAS, M. M.: Ultrastructural changes in myocardium during experimental ischemia. Johns Hopkins Med. J. *124:* 311–330 (1969).
7. GRIGGS, D. M. and NAKAMURA, Y.: Effect of coronary constriction on myocardial distribution of iodoantipyrine–131 I. Amer. J. Physiol. *215:* 1082–1088 (1968).
8. HERD, J. A.; HOLLENBERG, M.; THORBURN, G. D.; KOPALD, H. H. and BARGER, A. C.: Myocardial blood flow determined with Krypton-^{85} in unanesthetized dogs. Amer. J. Physiol. *203:* 122–124 (1962).
9. JENNINGS, R. B.; SOMMERS, H. M.; KALTENBACH, J. P. and WEST, J. J.: Electrolyte alterations in acute myocardial ischemic injury. Circulat. Res. *14:* 260–269 (1964).
10. KELLY, T. F.: Improved method for microdetermination of fatty acids. Analyt. Chem. *37:* 1078–1079 (1965).
11. KURIEN, V. A. and OLIVER, M. F.: A metabolic cause for arrhythmias during acute myocardial hypoxia. Lancet *i:* 813–815 (1970).
12. OSTADAL, B.; RYCHTEROVA, V. and POUPA, O.: Isoproterenol-induced acute experimental cardiac necrosis in the turtle. Amer. Heart J. *76:* 645–649 (1968).
13. RAAB, W.: Prevention of ischemic heart disease: Early metabolic abberrations through which epinephrine elicits myocardial necrosis (Charles C Thomas, Springfield, Ill. 1966).
14. REEVES, T. J.; HEFNER, L. L.; JONES, W. B.; GOGLAN, C.; PRIETO, G. and CARROLL, J.: The hemodynamic determinants of the rate of change in pressure in the left ventricle during isometric contraction. Amer. Heart J. *60:* 745–761 (1960).
15. REGAN, T. J.; HARMAN, M. A.; LEHAN, P. H.; BURKE, W. M. and OLDEWURTEL, H. A.: Ventricular arrhythmias and K^+ transfer during myocardial ischemia and intervention with procaine amide, insulin, or glucose solution. J. Clin. Invest. *46:* 1657–1668 (1967).
16. REGAN, T. J.; MARKOV, A.; OLDEWURTEL, H. A. and BURKE, W. M.: Myocardial metabolism during acute ischemia: Response to norepinephrine. Cardiovasc. Res. *4:* 334-342 (1970).

17. REGAN, T. J.; MOSCHOS, C. B.; LEHAN, P. H.; OLDEWURTEL, H. A. and HELLEMS, H. K.: Lipid and carbohydrate metabolism of myocardium during the biphasic inotropic response to epinephrine. Circulat. Res. *19:* 307–316 (1966).

18. REGAN, T. J.; PASSANNANTE, A. J.; OLDEWURTEL, H. A.; BURKE, W. M. and ETTINGER, P. O.: Metabolism of C^{14} labelled triglyceride by the myocardium during injury induced by norepinephrine (abstract). Circulation *38* (6) : 162 (1968).

19. ROSS, R. S ; and UEDA, K.; LICHTLEN, P. R. and REES, J. R.: Measurement of myocardial blood flow in animals and man by selective injection of radioactive inert gas into the coronary arteries. Circulat. Res. *15:* 28–41 (1964).

20. SOMMERS, H. M. and JENNINGS, R. B.: Experimental acute myocardial infarction. Lab. Invest. *13:* 1491–1503 (1964).

21. TAKAHASHI, A.: Myocardial protein metabolism following coronary occlusion. Jap. Circ. J. *4:* 581–600 (1967).

22. THRONBURN, J. D.; KOPALD, H. H.; HERD, J. A.; HOLLEMBERG, M.; MORCHOE, C. C. and BARGER, A. C.: Intrarenal distribution of nutrient flow determined with Kr^{85} in the unanesthetized dog. Circulat. Res. *13:* 290 (1963).

Mailing address: Timothy J. Regan, M.D., Department of Medicine, Division of Cardiovascular Diseases, New Jersey College of Medicine and Dentistry, 100 Bergen Street, Newark, New Jersey 07103 (USA).

Laboratory of Experimental Cardiology, St. Luke's Hospital Center
New York, N. Y., USA

MYOCARDIAL POTASSIUM LOSS DURING ISCHEMIA [1]

R. B. CASE

INTRODUCTION

Intracellular potassium loss appears to be a general tissue response to a disturbance of oxygen supply (1, 2, 4, 5). This loss becomes critical in the case of cardiac muscle, with subsequent deleterious effects on rhythm and contractility. There is mounting evidence that such ischemic-induced potassium loss may also trigger the onset of myocardial necrosis (12). These considerations demand that the exact relation of potassium loss to oxygen deficiency be defined. Although there is a considerable body of evidence relative to ischemic loss of potassium from the heart, this has not been related to the degree of oxygen deprivation. Moreover, there has been a confusing, interchangeable use of terms such as "ischemia," "hypoxia," "anoxia," and "infarction," most of which have referred essentially to coronary artery occlusion. The pathological process initiated by occlusion has many facets, ranging from the zone of necrosis with no flow to a truly ischemic zone with continued, but inadequate, blood supply. Not only is this process difficult to quantitate, but it may bear little resemblance to the clinical syndrome of myocardial infarction, as has already been pointed out in this symposium. Blood flow to an ischemic area has a profound effect on tissue potassium loss, as Jennings *et al.* have clearly demonstrated (7). This fact is often ignored or cannot be evaluated in ligation experiments. Other questions that need answers are exactly when potassium loss begins in relation to the ischemic episode, how long it continues after the ischemic period, and whether replacement of lost potassium occurs. Previous studies suggested a variable delay in the onset of potassium loss. In order to provide a more quantitative, precise, and complete measure of myocardial potassium balance, a continuous sampling technique was devised. Atrial and coronary sinus blood were withdrawn continuously by a pump and delivered into a fraction collector set to change position each minute. Thus, each sample represented the mean value of all blood passing the catheter tip over the previous minute. Each plasma sample was analyzed for potassium and lactate, the latter being used as an index of adequacy of oxygen supply. Employing an automated flame photometer method, the accuracy of analysis of potassium was unusually high, with the standard

---

[1] Presented at the Conference on Preventive Myocardiology of the American College of Cardiology, June 26–28, 1970, Stowe, Vermont.

deviation being less than 0.03 mEq/l. Using this technique, myocardial ischemia has been studied in the dog and in man in four different situations: 1) ischemia produced in the dog by progressive reduction in flow to the cannulated and perfused left main coronary artery (2); 2) angina produced in patients with coronary artery disease by the use of atrial pacing (11); 3) during cessation of coronary flow for periods of 10–15 min in man and at the time of open heart surgery (9); 4) in the cadaver dog-heart devoid of any flow for periods up to 3 h with subsequent reperfusion (10). The first two methods do not suffer from the problems associated with cessation of flow, and should give a clear picture of the effect of ischemia on potassium loss. The best data are obtained from the dog study, since coronary flow is known, so that a continuous determination of quantitative potassium loss is presented. Nevertheless, considerable information, particularly in regard to potassium uptake after a period without coronary flow, was obtained in the latter two methods. Results from these four methods have been presented elsewhere in greater detail (2, 9–11).

MEASUREMENT OF POTASSIUM LOSS IN EXPERIMENTALLY-INDUCED ISCHEMIA IN THE DOG

This experiment was performed by perfusion of the left main coronary artery with well-oxygenated blood from the femoral artery, using a precisely controlled pump to deliver the blood (2). The circulation was otherwise intact, with a constant, normal aortic pressure. Measurements were made of aortic pressure, left atrial pressure, and coronary artery pressure, and coronary vascular resistance was calculated. During the control period, coronary arterial and coronary sinus potassium values were stable and approximately equal (fig. 1), although on occasion a small, steady arteriovenous difference was present at this time. The sensitivity of the potassium method used was such that deviations greater than 0.06 mEq/l were considered significant changes. Lactate extraction during the control period averaged 27.6%, the average for the individual control periods ranging from 17.5 to 31.8%. In each experiment, the level of lactate extraction was stable during the control period. After graded reduction in coronary flow, a point was reached in all experiments at which a significant rise in left atrial pressure occurred. In each experiment, the level of coronary sinus potassium increased abruptly at this point, considerably exceeding that of arterial potassium and thus showing evidence of myocardial potassium loss. Coronary sinus potassium levels continued to rise with further slight reduction in coronary flow, but reached a steady state of loss if the ischemia was of sufficient duration. The mean rate of potassium loss from the ventricle was 11.8 μEq/min per 100 g of left ventricle at the time a steady rate of loss was present. The total amount of potassium lost by the left ventricle was estimated to be 1.% of its estimated normal content (13) after 7 min of ischemia and 5.4% after 17 min of ischemia. While this rate of potassium loss may seem to be small, the periods of

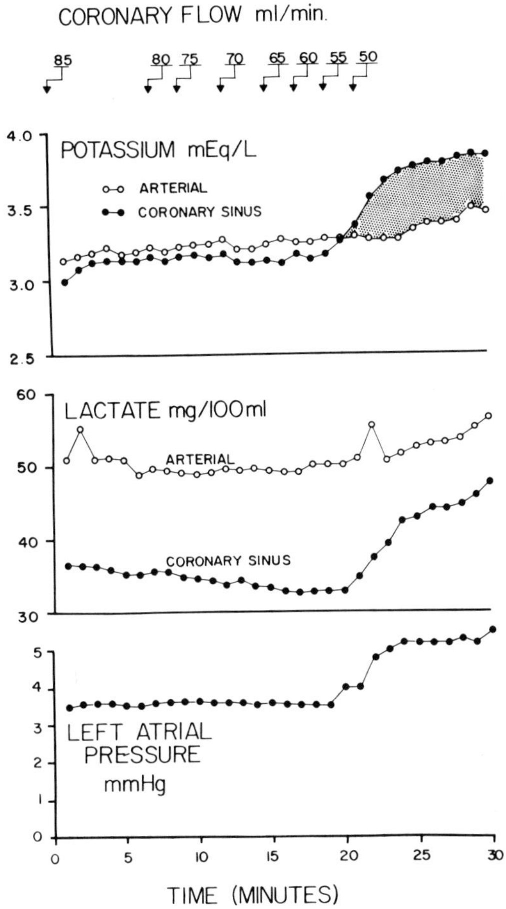

Fig. 1 Coronary flow, potassium and lactate levels, and left atrial pressure in ischemic dog.

ischemia were brief and, moreover, it should be remembered that potassium was being lost from localized areas of the myocardium and, hence, the deficiency was much greater in this area. In this preparation, it has been shown that ischemia is much more severe in the subendocardium (8) and this is presumably the site of major potassium deficiency.

The rate of potassium loss showed a linear correlation ($r = 0.97$) with the rate of lactate production as calculated from increases in coronary sinus concentration during ischemia. On a molar basis, approximately 1 mEq of potassium is lost with each 2 mmole of lactate produced. Since acidosis of coronary venous blood is also a feature of myocardial ischemia (3), it does not seem unreasonable to suggest that each lactate molecule leaving the cell is accompanied by 1 potassium and 1 hydrogen ion.

Abrupt restoration of coronary flow to levels of adequate perfusion resulted in a prompt fall in the elevated level of coronary sinus potassium concentration, reaching the level of arterial potassium in 2–5 min. This decline of coronary sinus potassium continued, falling below that of arterial potassium indicating potassium uptake during the postischemic period. The experiments were not continued to determine whether or not exact replacement occurred, but in each experiment potassium uptake continued to the point of termination of the experiment, at which time approximately 60% of the calculated potassium lost had been replaced. The rate of potassium uptake was approximately half that of its loss, but this could be increased by an increase in the coronary flow at this time. It thus seems quite possible for complete replacement of potassium to occur rather promptly after return of coronary flow and the normal post-ischemic elevation in flow should assist this return. Although coronary sinus potassium concentration fell rapidly after flow restoration, there was a much slower fall in lactate concentration, probably due to the washout of accumulated lactic acid in the myocardium.

The onset of potassium rise in the coronary sinus blood occurred simultaneously with a rise in left atrial pressure, a rise in coronary sinus lactate concentration, and depression of the S-T segment of the electrocardiogram. Coronary vascular resistance was found to fall prior to the onset of these changes and to reach a minimum value at the onset of potassium loss.

POTASSIUM LOSS DURING PACING-INDUCED ANGINA IN MAN

The same sampling system was used in patients undergoing cardiac catheterization for investigation of coronary artery disease (11). Angina was produced in 18 patients by means of atrial pacing from the coronary sinus. The rate was abruptly increased from control values to approximately 140–150 per min and maintained at this value for 8–10 min. Coronary sinus and arterial blood were withdrawn continuously during a control, a pacing, and a recovery period, each lasting 10 min. Right and left heart pressures, cardiac output, arterial and coronary sinus blood gases, and pH were determined. In the patients who developed angina, potassium loss began abruptly at the onset of pacing and continued throughout the pacing period. The arteriovenous potassium difference averaged -0.23 ± 0.02 mEq/l during the pacing period as compared with a control value of -0.05 ± 0.01 mEq/l during the control period. Pacing was stopped abruptly and within 1–2 min coronary sinus potassium had fallen below arterial levels, indicating potassium uptake occurring promptly after cessation of this stress. S-T depression and impaired left ventricular function occurred in all patients who had angina during this pacing period. Since there is evidence that an increased rate will also produce potassium loss (6), this factor was studied in

4 normal and 8 asymptomatic patients during a similar pacing period. Potassium loss was also present in this group, but it occurred primarily in the initial portion of the pacing period, and was much less than that seen during angina. During the first portion of the pacing period, an arteriovenous difference of -0.14 ± 0.01 mEq/l was present as compared with a control of -0.02 ± 0.01 mEq/l. This value was significantly different from the angina group. The arteriovenous potassium difference in this normal group was linearly related to the heart rate increment. When examined at the latter part of the pacing period, at which point potassium loss due to rate alone was minimal, the relation between potassium loss and lactate production showed findings similar to those seen in the dog studies; namely, that 1 mEq of potassium was lost for each 2 mmole of lactate produced. Although flow was not measured in man, these studies indicated that potassium loss begins with the onset of ischemia, is closely correlated to metabolic evidence of ischemia, and potassium replacement begins promptly after cessation of the ischemic episode.

MEASUREMENT OF POTASSIUM LOSS FOLLOWING CESSATION OF CORONARY FLOW IN MAN DURING OPEN HEART SURGERY

Interruption of coronary artery blood flow for variable periods of time is frequently necessary during valve replacement surgery (9). It is difficult to obtain data in man under controlled conditions, relative to the interruption of blood supplied to the myocardium. In order to take advantage of this situation, the standard right heart bypass was altered so that a catheter was introduced into the right atrium, and coronary venous return allowed to flow by gravity drainage from the right atrium to the heart-lung machine. A sterile glass graduated cylinder was interposed in the coronary venous return line to permit intermittent direct measurement of coronary flow. Continued sampling was carried out as described previously, with samples being withdrawn from the coronary venous line and from the reservoir of the heart-lung machine for the arterial line. Coronary flow was interrupted for periods of 12–15 min, followed by a 3–5 min period of direct coronary perfusion. Continuous coronary perfusion in 3 patients was associated with continuous lactate production and potassium loss. During intermittent coronary perfusion, sampling was done during the period of perfusion and following the restitution of normal aortic pressure. In all patients, there was evidence of a marked potassium loss from the heart, becoming apparent immediately after flow was restored. Coronary venous potassium fell as flow continued and, in most patients, rapid potassium uptake occurred while normal coronary perfusion was reestablished while still on bypass. It was concluded from these studies that potassium loss occurred owing to a period of interruption of coronary flow and that this is also followed by a rapid restitution of potassium to the heart after flow is restored. One of the original purposes of evaluating potassium loss during open heart surgery was to try to explain the high incidence of postoperative arrhythmias on this basis; but the rapid restitution of potassium to the heart would rule out this hypothesis.

ANALYSIS OF POTASSIUM IN THE REVIVED DOG-CADAVER HEART

Hearts that have been anoxic at 37°C for more than 1 h can be resuscitated, but mechanical function is poor and transplantation is rarely successful. The ionic and metabolic defects were studied in a series of dog hearts following a variable period of cardiac arrest (10). After control samples were taken, the dogs were rapidly exsanguinated and the hearts left *in situ* at 35–37°C for 30–180 min. The isolated heart was then perfused through the aorta from a support dog at a flow of 80–210 ml per min per 100 g of left ventricle, with return of coronary venous blood to the support dog. Arterial and coronary sinus blood were sampled continuously for the determination of potassium, lactate, glucose, free fatty acids, and sodium. The most striking finding was the preservation of the cation exchange mechanism, following periods of arrest up to 2 h. There was an initial potassium loss at the start of perfusion, followed in 2–6 min by a rapid potassium uptake with a marked A-V difference of 1.2 mEq/l and a maximal potassium uptake rate of 104 μEq/min per 100 g of left ventricle (fig. 2). Following this, potassium and coronary sinus values became approximately equal with no further evidence of alterations in potassium balance. Hearts that were anoxic for more than 2 h were in rigor when perfusion was resumed. Although defibrillation was accomplished, no effective mechanical activity was produced. In these animals, there was no evidence of potassium uptake following restitution of perfusion. Potassium loss was continuous and, in one dog, total potassium loss of 1 h perfusion was 11.1 mEq, or 106% of estimated total

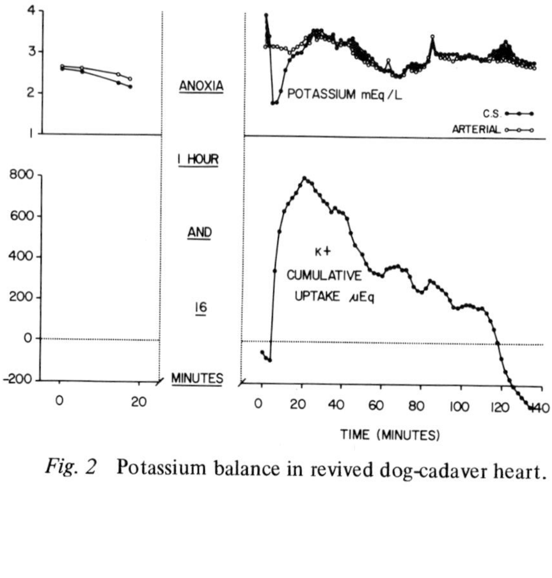

Fig. 2 Potassium balance in revived dog-cadaver heart.

myocardial potassium. Since potassium loss appears to be so sensitive to a deficiency of oxygen, as seen in the first type of experiments (Section 1), it was surprising that potassium uptake returned so quickly following a prolonged period of anoxia. Moreover, the remarkable avidity with which potassium was taken up would suggest that perhaps factors other than energy supply are important in controlling myocardial potassium balance.

SUMMARY

Studies relating potassim loss to myocardial ischemia show that potassium begins to leave the cell at the very onset of ischemia, that it begins to return to the cell almost immediately after the ischemic stress is discontinued, and that prompt replacement of potassium begins in the immediate postischemic period. These studies also indicate that the heart previously made anoxic can be depleted of its entire potassium content within 1 h in the presence of adequate flow, provided there has been a sufficiently long period of anoxia. On the other hand, it is also apparent that the cation exchange mechanism is quite resilient and it is still operative after a prolonged, apparently lethal period of anoxia. The data presented here have supplied additional knowledge regarding the close relationship between ischemia and myocardial potassium loss, and it is hoped that using similar methods we will be able to evaluate the effect of steroids, catecholamines, and other agents.

REFERENCES

1. CALKINS, E.; TAYLOR, I. M. and HASTINGS, A. B.: Potassium exchange in the isolated rat diaphragm; effect of anoxia and cold. Amer. J. Physiol. *177:* 211–218 (1954).
2. CASE, R. B.; NASSER, M. G. and CRAMPTON, R. S.: Biochemical aspects of early myocardial ischemia. Amer. J. Cardiol. *24:* 766–775 (1969).
3. CRAMPTON, R. S.; ROSELLE, H. A. and CASE, R. B.: Coronary sinus blood pH during myocardial ischemia. Fed. Proc. *25:* 402 (1966).
4. DIXON, K. C.: Anaerobic leakage of potassium from brain. Biochem. J. *44:* 187–190 (1949).
5. FLINK, E. B.; HASTINGS, A. B. and LOWRY, J. D.: Changes in potassium and sodium in liver slices accompanying incubation *in vitro*. Amer. J. Physiol. *163:* 598–604 (1950).
6. GILMORE, J. P. and GERLINGS, E. D.: Influence of interstimulus interval on myocardial potassium balance. Amer. J. Physiol. *217:* 136–141 (1969).
7. JENNINGS, R. B.; SOMMERS, H. M.; KALTENBACH, J. P. and WEST, J. J.: Electrolyte alterations in acute myocardial ischemic injury. Circulat. Res. *14:* 260–269 (1964).
8. LEUNISSEN, R. L. A.; PIATNEK-LEUNISSEN, D. A.; NAKAMURA, Y. and GRIGGS, D. M., Jr.: Regional metabolism of the heart during reduced coronary flow. Circulation *34:* (Suppl III): 155–156 (1966).
9. MCCORD, C. W.; CRAMPTON, R. S.; NASSER, M. G. and CASE, R. B.: Myocardial potassium and lactate balance during valve-replacement surgery. Circulation *39:* (Suppl. I) 243–248 (1969).
10. MCCORD, C. W.; NASSER, M. G.; CARTER, P. and CASE, R. B.: Preservation of cation exchange mechanism in the revived dog cadaver heart. Surg. Forum *20:* 143–145 (1969).
11. PARKER, J. O.; CHIONG, M. A.; WEST, R. O. and CASE, R. B.: Myocardial electrolyte balance during pacing induced angina pectoris. Circulation *39:* (Suppl III) 158 (1969).

12. RAAB, W.: The non-vascular metabolic myocardial vulnerability factor in coronary heart disease. Amer. Heart J. *66:* 685–706 (1963).

13. SPURR, G. B. and BARLOW, G.: Influence of prolonged hypothermia and hyperthermia on myocardial sodium, potassium and chloride. Circulat. Res. *7:* 210–218 (1959).

Mailing address: Robert B. Case, M. D., Laboratory of Experimental Cardiology, St. Luke's Hospital Center, Amsterdam Avenue at 114th Street, New York, N. Y. 10025 (USA).

University Health Service, University of Nebraska
Lincoln, Nebraska, USA

EXERCISE AND SERUM POTASSIUM FLUX: MYOCARDIAL METABOLIC IMPLICATIONS [1, 2]

K. D. ROSE, J. A. URSICK, and R. D. MACA

No concensus exists concerning the positive value of exercise programs in the prevention of disease. Opponents call attention to the disturbing fact that the evidence supporting claims for exercise disease prevention is largely circumstantial (5). The frequently-made statement that a physically fit individual feels and lives better, although true, is inadequate to sway the sedentary man ill-disposed to exercise any muscle even to the point of incipient fatigue. Positive relationship between fitness in youth and longevity is still incompletely proven and lifetime fitness studies are not yet available.

The type of controlled study in humans needed to resolve these questions is probably impossible in a society of free people. However, modern biomedical technology is making possible dynamic physiology studies that, hopefully, will improve understanding of the effect of exercise on human cardiovascular physiology (1). The data reported here are from such a study. By comparing serum biochemical transients with concomitant electrocardiographic transients during intensive short-term physically stressful work, certain correlations are suggested.

MATERIALS AND METHODS

Subjects

The experimental sample consisted of 294 separate tests on male subjects, of which 127 were trained sprint, middle distance and distance runners and 152 were untrained students. The latter were about equally divided amongst general students and physical education majors. Ages in both groups ranged from 18 to 25. The trained subjects had had high school and one or more years of varsity track training as well as participation in a continuing running program during nonseasonal months. On that basis, they could be considered in a state of maximum cardiovascular fitness. The untrained cohort was for the most part made up of typical sedentary university students. There were no individuals on a regular cardiovascular fitness program.

---

[1] Presented at the Conference on Preventive Myocardiology of the American College of Cardiology, June 26–28, 1970, Stowe, Vermont.

[2] Supported by grants from the University of Nebraska Research Council and the United States Public Health Service, National Institutes of Health Grant HE-06402.

Tests

Each subject was instrumented prior to warm up and a resting electrocardiogram recorded. The transmitter was then detached during a warm-up period and reattached just prior to the test exercise, the electrodes being left in place. All of the exercises were of sufficiently short duration to demand primarily anaerobic metabolism. Within the first 10–15 sec after cessation of exercise, immediate post-exercise tracings were recorded and repeated after 3 and 10 min rest. Blood samples in special subgroups were drawn from the antecubital vein concomitant with the electrocardiogram.

Telemetry System

The biaxillary, bipolar (CR5-C5) electrocardiographic method used in these studies has been described in detail in a previous publication (12). Custom-built AM-FM telemetry transmitters constructed and tested by the University of Nebraska College of Medicine's Department of Biomedical Instrumentation served as the data link between subject and a receiver-recorder system composed of FM receivers, preamplifiers, Tectronix dual-beam oscilloscope for monitoring, and a Sanborn twin-viso electrocardiograph for strip recording of information. Frequency spectrum of this system (3 db loss) was from 0.04 to 150 cps (Hertz) with a roll off of 6 db at the low end and 10 db at the high end of the spectrum. Decay time constant for the entire system was 1.5 sec. Electrode resistance was kept <4000 ohms by careful skin preparation to accomodate the 130–150 K ohm input impedance of the system. Recording speed was accurately measured and controlled at 9.5 mm/sec for pulse and wave amplitude measurements, and 95 mm/sec for interval measurements. The high speed record was necessary for accurate measurement at the rapid pulse rates of the post-exercise period.

Biochemical analysis

Subgroups totalling 139 subjects (64 trained and 75 untrained) were used for studying serum biochemical changes attending strenuous anaerobic exercise (440 yard run). Venous blood was drawn from the antecubital vein in vacuum tubes, permitted to clot, and the serum removed after centrifugation. Tubes showing hemolysis were discarded and results from such tests do not appear in the data. Serum potassium and sodium were determined by flame photometry. The method of Ferro and Hamm was used for total calcium analysis (6). Ionized calcium was determined by the ultra-filtration technique of Nicholas (10). Magnesium was analyzed by the method of Denis (4). Total serum protein and a/g ratio were determined by methanol precipitation (11), while lactic acid was analyzed by the Sigma technique (14). Heparinized blood collected in 2 ml vacuum tubes was analyzed for pH with an Instrumentation Laboratories pH meter standardized against known buffer systems at pH 6.8 and 7.4. All methods were subjected to rigid quality control and unsatisfactory data were discarded.

Data analysis

Multiple electrocardiographic complexes from each tracing were measured in the standard manner, using the isoelectric phase of the P-R interval as the baseline. The means for all measurements were recorded on IBM punch cards and the total subjected to data analysis for mean, one standard deviation, and level of significance by the Student *t* test. A $p < 0.05$ was considered within significant levels. Data from biochemical analysis were treated in a similar manner.

RESULTS

Electrocardiographic measurements

There is a statistically significant change ($p < 0.001$) in P-R and Q-T intervals in the electrocardiograms of both trained and untrained subjects following exercise. This common finding is rate related, as can be seen by comparing the curves in figure 1B and 1C with the inverse pulse rate curve in figure 1A. The difference between the resting pulse of trained and untrained subjects ($p < 0.01$) also is rate related, and is to be expected from the known bradycardia at rest of the well-trained athlete. At similar peak pulse rates immediately following exercise, trained *vs.* untrained P-R and Q-T interval variables become nonsignificant ($p => 0.05$). Thus, these measurements furnish no essentially new knowledge concerning differences in physiology between the untrained and the well-trained heart.

The data in figure 1D, however, are enlightening. It is quite clear that a minor shortening of QRS interval occurs during peak pulse rates immediately after a run. This is followed by an abrupt lengthening of QRS to the extent of a marked overshoot. The first phase is probably rate related, but the second phase must involve factors other than rate. Lengthening of the QRS is a potassium effect. The interval between the immediate and 3 min analyses is attended by rapid uptake of potassium by the tissues, as will be shown later. A potassium effect appears, therefore, in the post-run electrocardiogram of subjects who have undergone strenuous anaerobic exercise. Both trained and untrained subjects evidence the same significant change ($p < 0.001$), but the difference between the cohorts, although apparent from the curve of the means, is not statistically significant ($p = 0.05$).

Amplitude measurements result in data of a similar nature. Q and S waves increase in negative amplitude, while the R wave decreases in positive amplitude. This is a reflexion of the shift in cardiac axis attending the relatively prolonged

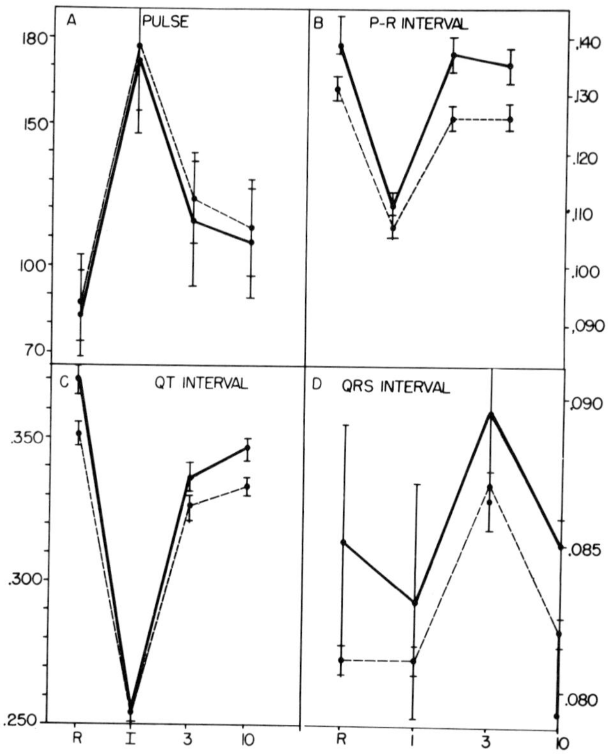

Fig. 1 Pulse and electrocardiogram interval measurements before and after 440 yard test run. •——• = trained track athletes; •-----• = untrained male students. R = at rest before run; I = immediately following run; 3 = 3 min post run; 10 = 10 min post run.

systolic contraction time associated with exercise tachycardia. It is a well-known phenomenon occurring both in trained and sedentary subjects undergoing strenuous physical exercise. The difference between trained and untrained subjects in this study is found to be significant ($p < 0.001$) only in the R and T wave amplitude measurements. These obvious differences, seen in figure 2, are a manifestation of the larger electrical output from a larger heart muscle mass and probably have no other particular physiological significance.

In the interval between the immediate and 3 min rest analyses, amplitude of the Q, R, and S waves returned toward normal, as would be expected. T wave amplitudes continued to rise, however, reaching their peak at 3 min post-exercise. This is a potassium effect, and occurs concomitant with rapid disappearance of potassium from the circulating blood stream. By 10 min, homeostasis has returned and the electrocardiogram, with the exception of the S wave, is back to normal.

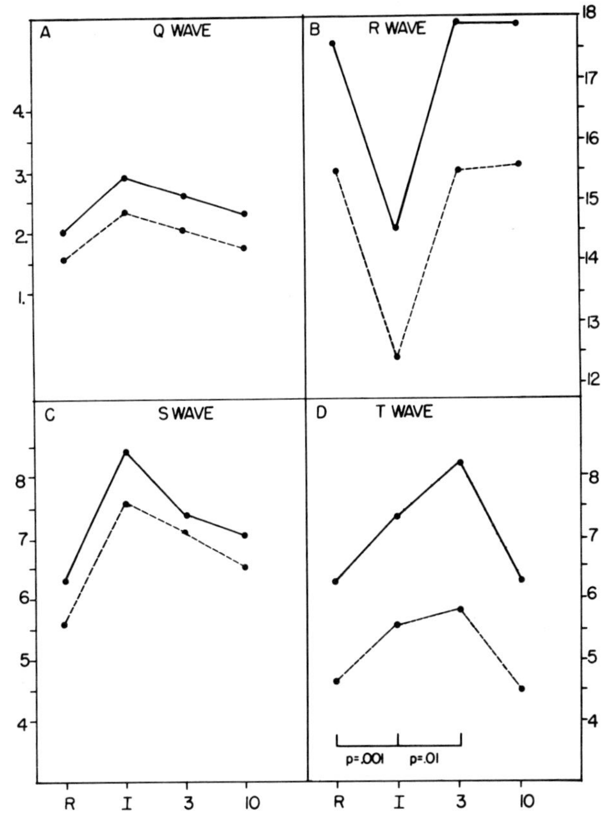

Fig. 2 Electrocardiogram wave amplitudes before and after 440 yard test run (see fig. 1).

Biochemical measurements

A loss in blood volume occurs during the running of a race (15), its magnitude depending on the duration of the run, temperature, humidity, etc. An estimate of this volume loss can be obtained if one is prepared to accept the assumption that protein does not leak from the vascular bed during the brief time it takes to run a sprint race. Albumin would leak first should there be protein loss, and a progressive change in albumin concentration in the blood serum would reflect that loss. Thus, measurement of total protein should reveal a progressive increase in concentration of serum protein if fluid is lost from the vascular bed. If the increase in protein concentrations is due solely to fluid loss, *i.e.*, if no albumin loss is occurring, TSP and albumin measurements should afford a good indirect estimate of hemoconcentration. It is important to know whether hemoconcentration is occurring if one wishes to evaluate concomitant changes in concentrations of other solutes within the serum.

The data on a 44-subject subgroup before and after running a 440 yard race reveal that there was a small increase in serum protein concentration (fig. 3). The magnitude was of the order of 3.7%. It was similar in both trained and untrained subjects ($p > 0.05$) and was attended by no change in the percentage of albumin. Within the limits of this indirect technique, it can be concluded, therefore, that a 3.7% hemoconcentration occurred during the running of a 440 yard race, even as Wit *et al.* (15) have found in their marathon runners. Any concurrent change in serum solute concentration greater than 3.7% is considered a true change in ionic concentration.

Potassium

Serum potassium increases significantly ($p < 0.001$) during the running of a race, but returns to below resting levels at 3 min, and back to near resting levels by 10 min. Trained athletes mobilize more K^+ during muscular work than do

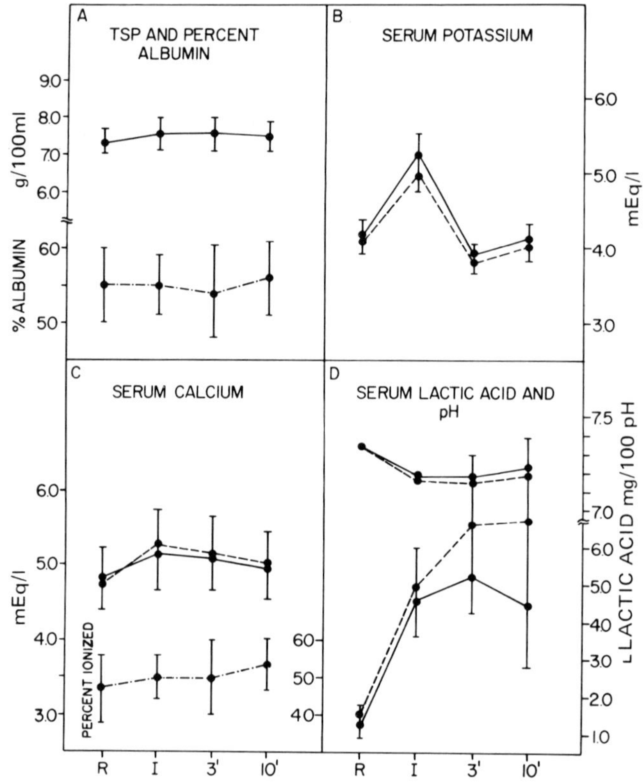

Fig. 3 Certain serum biochemical exercise parameters before and after 440 yard test run (see fig. 1).

their untrained counterparts ($p < 0.01$), but this small difference is rapidly erased as homeostasis is reestablished. The post-exercise 1.09 mEq/1 increase in trained and 0.95 mEq/1 increase in untrained subjects represents 25% and 23% of their respective circulating potassium levels, an increase greater than that due solely to hemoconcentration. Rapid decline in serum potassium after cessation of physiological perfusion represents a shift back into muscle tissue including the heart (8). In the trained subject, this represents a 1.35 mEq/l uptake in 3 min. In the untrained it is 1.2 mEq/l. A 4 month training program decreased post race K^+ increase by 0.12 mEq in trained and 0.33 mEq in untrained subjects.[3]

Calcium

There is a significant ($p < 0.001$) increase in serum calcium following the running of a 440 yard sprint race. In the trained subjects, it represents 10% of the resting level, while that of the untrained control represents 11.2%. These exceed hemoconcentration by 6–7% and represent a transmembrane shift from cellular compartments, most likely muscle, into the circulating blood stream. A definite tendency is observed for the exercise shift to be greater in the untrained controls than in the trained cohort, although the difference is not highly significant ($p < 0.05$). This has been reported by us previously (13), and confirmed by Coester *et al.* (3) It was further observed that the degree of post-exercise calcium mobilization was decreased by a 4 month exercise program. Post training curves were essentially identical to those of trained subjects.

The percentage of ionized calcium increased slowly and became barely significant at the 10 min sample ($p < 0.01$). There was no difference between trained and untrained. The importance of this observation is not readily apparent, but the 9% increase over resting ionized (nonprotein bound) calcium concentration, exceeding hemoconcentration by some 5%, suggests that strenuous physical exertion is attended by a relative loss of this physiologically active ion from muscle tissue.

Sodium and magnesium

A small increase in sodium ion concentration, equivalent to hemoconcentration, occurs during exercise. Magnesium ionic concentration increased erratically and statistically nonsignificantly in a small test subgroup.

Metabolic acidosis

Lactic acid levels increase and pH decreases significantly ($p < 0.001$) in both trained and untrained subjects. Differences between the groups of test subjects

[3] In a 28 subject subgroup.

becomes statistically significant at 10 min post-exercise ($p < 0.05$), reflecting the trained subject's improved recovery from lactacid oxygen debt. These data confirm those of Johnson *et al.* (9), who were able to demonstrate similar differences between trained and untrained subjects not only in carbohydrate energy pathways but also in lipid metabolic pathways.

CONCLUSION

A sudden burst of anaerobic exercise such as occurs during the running of a short sprint race is attended by a significant increase in circulating serum potassium and calcium. The major source of these ions is from muscle, but other tissues undoubtedly also serve. Caswell (2), for instance, has shown that there is a rapid efflux of K^+ and Ca^{++} from liver mitochondria when the medium becomes anaerobic. He attributes this to a natural mechanism involving reduction of the energy level of high energy metabolic intermediates resulting in an increase in membrane permeability to these ions. Similar membrane changes are probably active in muscle during anaerobic exercise.

In the immediate recovery phase, there is a rapid decrease in serum K^+ representing a shift back to muscle, including the heart. In this regard, Grupp (8) has demonstrated a rapid uptake of K^+ by the dog heart during K^+ perfusion of the coronary circulation. Increase in circulating K^+ during anaerobic exercise can be thought of as a physiological perfusion. Uptake following exercise is of the magnitude of 1.2–1.35 mEq/l. Since Fisch *et al.* (7) have shown that short term K^+ infusion resulting in an increase of 1.2 mEq/l/min can cause potassium effects in the electrocardiogram of dogs, and have concluded that this effect can be the result of *rate of perfusion* at physiological serum K^+ levels, it is not surprising that the telemetered exercise ECG in humans reveals such changes after a 50 sec anaerobic run. It has been shown in these experiments that the early K^+ effects of T wave peaking and QRS prolongation occur concomitant with rapid uptake of physiologically perfused K^+ during the first 3 min of post-exercise rest. It is suggested that these observed electrocardiographic changes are the result of that K^+ uptake by the heart.

Findings of this nature are helpful in understanding what the late Joseph Wolffe (16) called "electrical" heart deaths, in which autopsy revealed normal coronary vascular systems. Although obviously physiological in nature and well within tolerance of the normal heart, sudden changes in electrolyte levels may precipitate gross arrhythmias in susceptible individuals. An example will demonstrate this. Figure 4 is the telemetered electrocardiogram of a healthy young male whose K^+ increased 1.2 mEq/l and Ca^{++} 3.2 mg% during the running of a 440 yard race. Figure 4a and b identify sinoatrial blocks and 4c is a nodal premature contraction. The arrhythmia disappeared with 10 min rest. Following 4 months of training, his K^+ increase following a 440 yard race was 0.6 mEq/l

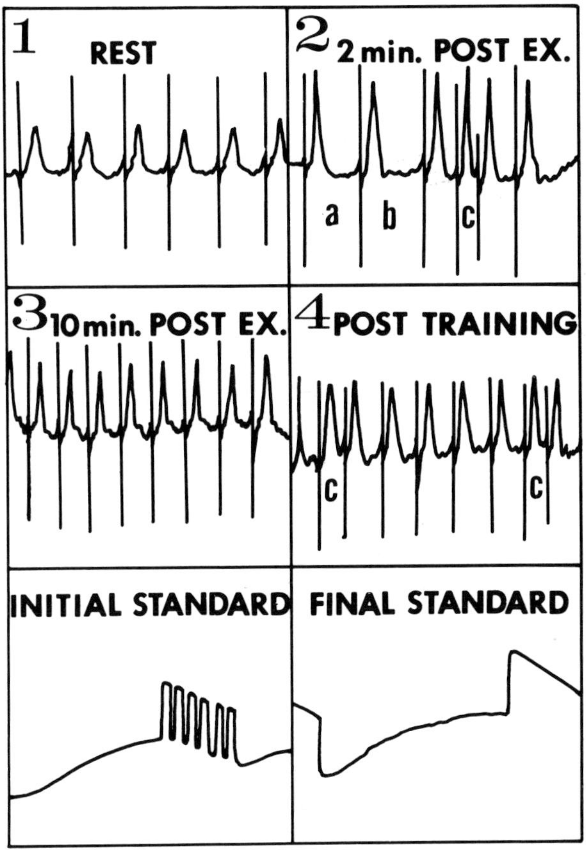

Fig. 4 Cardiac arrhythmia appearing during immediately post-exercise period (see text for explanation).

and his Ca^{++} increase was 1.1 mg%. The telemetered electrocardiogram in figure 4 still showed an occasional nodal extra systole, but his post-exercise rhythm was essentially regular and was normal by 4 min post-exercise. Such electrolyte-induced arrhythmias in an older sedentary individual running to catch a bus might terminate fatally, especially if the patient's cardiac metabolism was compromised by C.H.D., digitalis, catecholamines of stress, metabolic acidosis, and hyperventilation with rapid decrease in CO_2, all of which contribute to lowering the fibrillation potential.

Our data indicate that cardiovascular fitness exercise reduces the magnitude of electrolyte shift with acute anaerobic exercise, supporting the thesis that the philosophy of physical fitness has a demonstrably objective as well as a subjective basis.

SUMMARY

The effect of short term (anaerobic) strenuous exercise on the radiotelemetered electrocardiogram, serum electrolytes, lactic acid, and pH has been studied in 127 trained and 152 untrained subjects. QRS interval lengthened and T wave amplitude increased ($p < 0.001$) in the first 3 min rest. There was an abrupt decline of the physiological hyperpotassemia of exercise in the magnitude of 1.28 mEq/l during this resorptive period. Ca^{++} and lactic acid increased and pH fell during exercise, returning slowly toward basal level with rest. Na was constant. K^{+}/Ca^{++} imbalance existing during the immediate post-run interval, in the presence of metabolic acidosis, may produce a conduction instability leading to arrhythmia in the susceptible heart.

ACKNOWLEDGMENTS

The authors wish to thank Donald Tuch, Dennis Bargen, Gary Bargen, Steven Wilson, Sharon Mathena Berry, Cheryl Marr, and the Biomedical Instrumentation Laboratory of the University of Nebraska College of Medicine for their assistance during the several years of this study.

REFERENCES

1. CACERES, C. A.: Biomedical Telemetry (Academic Press, New York 1965).
2. CASWELL, A. H.: Proton permeability and the regulation of potassium permeability in mitochondria by uncoupling agents. J. Membrane Biol. *1:* 53–78 (1969).
3. COESTER, W.; MCGILL, F.; ELLIOTT, J. C. and LUFT, U. C.: The electrocardiogram during and after exercise in relation to serum electrolytes and pH. Lovelace Foundation; Presented at Amer. College Sports Med. 17th Ann. Meeting, Albuquerque, May 8, 1970.
4. DENIS, W. J.: The determination of magnesium in blood, plasma and serum. J. Biol. Chem. *52:* 411 (1922).
5. EDITORIAL: Physical fitness for what? Massachusetts Physician *28:* 23 (1969).
6. FERRO, P. O. and HAMM, A. B.: A simple spectrophotometric method for the determination of calcium. Amer. J. Clin. Path. *28:* 208 (1957).
7. FISCH, C.; GREENSPAN, K. and EDMANDS, R. E.: Complete atrioventricular block due to potassium. Circulat. Res. *19:* 373–377 (1966).
8. GRUPP, G.: Potassium exchange in the dog heart, *in situ*. Circulat. Res. *13:* 279–289 (1963).
9. JOHNSON, R. H.; WALTON, J. L.; KREBS, H. A. and WILLIAMSON, D. H.: Metabolic fuels during and after severe exercise in athletes and non-athletes. Lancet *ii:* 452–455 (1969).
10. NICHOLAS, H. O.: Diffusable serum calcium by high pressure ultrafiltration. J. Biol. Chem. *97:* 457 (1932).
11. PILLEMER, L. and HUTCHINSON, M. D.: The determination of the human albumin and globulin content of human serum by methanol precipitation. Biol. Chem. *158:* 299 (1945).
12. ROSE, K. D.: Telemetry in the study of the heart in athletes Pts I and II. Amer. Acad. Orth. Surg. Symposium on Sports Med. pp. 37–61 (C. V. Mosby Co., St. Louis 1969).
13. ROSE, K. D.; DUNN, F. L. and BARGEN, D.: Serum electrolyte relationship to electrocardiographic change in exercising athletes. J. Amer. Med. Ass. *195:* 111–114 (1966).
14. TECHNICAL BULL. 825-UV: Determination of L(+) lactic acid in blood, plasma or other fluids at 340 mμ, Sigma Chem. Co., St. Louis (1965).

15. WIT, A. G.; DALDERUP, L. M. and POL, G.: Muscular work and potassium supply. Acta. Physiol. Pharmacol. Neerlandica *11:* 405–410 (1962).

16. WOLFFE, J. B.: Contraindications to additional physical exercise in every day living with special reference to cardiovascular disease. J. Sports Med. Phys. Fit. *2:* 228 (1962).

Mailing address: Kenneth D. Rose, M.D., University of Nebraska Health Service, Lincoln, Nebraska 68508 (USA).

New York Medical College, Flower & Fifth Avenue Hospitals,
New York, N.Y., USA

PREVENTION OF EXPERIMENTAL MYOCARDIAL NECROSIS BY ELECTROLYTE SOLUTIONS [1,2]

D. LEHR, R. CHAU, and J. KAPLAN

In earlier studies from this laboratory, it was shown that exposure of parathyroidectomized (PTx) rats to repeated oral sodium phosphate (NaH_2PO_4) loads resulted, invariably, in the emergence of severe disseminated myocardial necrosis within a period of 54 h. It was shown, further, that evolvement of the cardiac injury was accompanied by pronounced and stereotyped alterations of the myocardial electrolyte content ("dysionic cardiopathy" of Raab (15)) with magnesium (Mg) loss presenting as the earliest and most persistent derangement, followed by reduction of inorganic phosphate (P) and accumulation of sodium (Na), and finally, unexpectedly late, loss of myocardial potassium (K). A closely similar pattern of electrolyte changes was observed by Lehr (6), in myocardial necrosis produced by adrenergic amines in excessive dosage.

It was proposed that, in cardiac cell injury, depletion of the vital intracellular ion, Mg, is one of the events that probably occurs in the living cell. Hence, Mg loss may contribute to disturbances that lead to irreversible failure of cellular function rather than represent simply one of the manifestations of cellular derangement and death. This view was based on our finding that significant reduction of the Mg content represented both a very early and consistent phenomenon in the evolvement of experimental myocardial necrosis, and on the well-known observation, as cited by other workers (2, 12, 13), that dietary Mg deficiency may in itself lead to the development of disseminated myocardial necrosis.

It seemed of interest, therefore, to ascertain whether the administration of Mg salts has a protective effect against myocardial necrosis produced by our method of NaH_2PO_4 loading of PTx rats, and to what extent such protective activity would be reflected in prevention of changes in the total electrolyte content. This paper deals, in the main, with an evaluation of the beneficial effects exerted by $MgCl_2$ and Mg gluconate against myocardial injury.

MATERIALS AND METHODS

A total of 410 adult male albino rats of the Wistar strain, weighing between 300 and 350 g, was employed. Standardized myocardial injury was produced by

[1] Presented at the Seminar on Preventive Myocardiology of the American College of Cardiology, June 26–28, 1970, Stowe, Vermont.

[2] Supported by Grant HE-00890-18,19 from the National Institutes of Health, National Heart Institute, United States Public Health Service.

a simple, short-term procedure developed in this laboratory and described in detail earlier (10). To recapitulate briefly: Parathyroidectomized (PTx) rats are given 4 ml of a 15% aqueous solution of NaH_2PO_4, twice daily, by stomach tube. This procedure results consistently in the development of extensive disseminated myocardial necrosis, readily apparent to the naked eye within 2 to 2½ days. If NaH_2PO_4 administration is continued beyond this period, the rats invariably succumb in a maximum of 4 days, with signs of hypocalcemic tetany.

In order to test for the potential protective activity of the Mg ion against this myocardial injury, $MgCl_2$ and Mg gluconate were administered in the following manner: In experiments utilizing the *oral* route, $MgCl_2$ was given by stomach tube as an aqueous 5.1% solution in a dose of 0.2 g, twice daily, 30 min prior to each intubation of NaH_2PO_4. Control groups received equimolecular and twice equimolecular amounts of NH_4Cl in the same manner to study the influence of the Cl ion.

In experiments employing the *subcutaneous* route, aqueous solutions of either $MgCl_2$ or Mg gluconate were administered in equimolecular amounts, again in an effort to distinguish the protective activity of the Mg ion from that of the Cl ion. Injections of the therapeutic agents were given twice daily, 30 min prior to each NaH_2PO_4 load.

Two groups of intact rats served as over-all controls. One of these received the usual NaH_2PO_4 dosage, while the other was left untreated.

All rats were killed after 2½ days by exsanguination under ether anesthesia. At that time, experimental animals had received a total of five NaH_2PO_4 loads and five treatments with one of the Mg salts or NH_4Cl. Animals that died before completion of the 2½ day experimental period were not included in this study. Following post-mortem examination of each animal, pieces of heart (base and tip), kidney, stomach wall, and diaphragm were weighed and macerated separately in a mechanical blender. Aqueous extracts of the tissue brei of these organs were utilized for automated chemical analysis of individual tissues for calcium (Ca), P, Mg, Na and K. The same electrolytes were also determined in the serum obtained from terminal blood samples.

Adjacent pieces of organs used for chemical analysis were fixed in neutral formaldehyde and sectioned for examination under the microscope, using hematoxylin and eosin, and periodic acid-Schiff stains. Thus, morphological changes of an individual organ could be correlated directly with alterations of its specific electrolyte content. As in earlier studies, myocardial lesions were classified on an arbitrary scale of 1 to 4 plus. For the purpose of tabulation, necrosis in the 1 to 2 plus range (from a single to several small foci) was designated as mild, and in the 3 to 4 plus range (numerous small or large foci), as severe (8).

Information on dosages and other details of experimental design will be found in the descriptions of the particular studies.

RESULTS

In agreement with earlier observations in this and other laboratories on the prophylactic effectiveness of Mg salts against cardiac injury, it was found that the Mg ion had a definite protective effect against the development of disseminated myocardial necrosis produced by NaH_2PO_4 loading of PTx rats. It should be emphasized that the emergence of necrotic foci in the heart was consistently associated with severe renal tubular necrosis and pronounced changes of the renal electrolyte content. On the other hand, no substantial structural changes or deviations of the electrolyte content were noted in the smooth muscle of the gastrointestinal tract (stomach wall) or the skeletal musculature (diaphragm). These findings have been discussed in detail in an earlier publication (9).

Oral administration of $MgCl_2$ and NH_4Cl in NaH_2PO_4 loaded PTx rats

The oral administration of $MgCl_2$ was exceedingly well tolerated and had, moreover, a remarkably curative influence. Hypocalcemic tetany was completely obviated and the premonitory jitteriness, spasticity, muscular rigidity, depression, and respiratory difficulty regularly encountered in unprotected animals failed to appear. Rats that otherwise would have been moribund following completion of the full series of five NaH_2PO_4 intubations looked healthy, alert, and calm at that time and survived even the continuation of phosphate loading for many additional days, without apparent ill effects, as long as $MgCl_2$ treatment was continued. Protected animals manifested substantially lessened hyperphosphatemia and hypocalcemia at the termination of an experiment and, most importantly, showed a sharp drop in incidence and severity of myocardial necrosis. Among a total of 29 rats killed after 2½ days of NaH_2PO_4 administration under $MgCl_2$ protection, only 1 severe, and 2 mild myocardial lesions were encountered (table I, group C). It should be noted that the myocardial electrolyte deviations produced in the unprotected group B were prevented in group C rats. On the other hand, a significant decrease of myocardial K was noticed here for the first time.

In order to distinguish a possible contribution of the Cl ion in the beneficial effect of $MgCl_2$ from that of the Mg ion, NH_4Cl in equimolecular and twice equimolecular amounts was substituted for $MgCl_2$ in additional experiments (table I, groups D and E). This treatment was likewise well tolerated.

It can be seen that rats receiving half the amount of Cl than those in group C, but in the absence of Mg, show a substantial, though less marked reduction of the incidence of myocardial necrosis (group D). Doubling the dose of NH_4Cl (equal in Cl content to that of the $MgCl_2$ dose of group C), caused a significant increase in protection with only 4 mild heart lesions in a total of 18 rats (group E). Prevention of myocardial electrolyte changes was almost equal in both

Table I

Effect of Oral Administration of $MgCl_2$ or NH_4Cl on Myocardial Necrosis and Electrolyte Alterations Induced by NaH_2PO_4 Loading of Parathyroidectomized (PTx) Rats

			Myocardial necrosis			Myocardial electrolytes (Mean mEq/kg ± S.E. wet wt.)					Serum electrolytes (mg%)	
		No. of	No.		Total							
Group	Treatment	rats	Mild	Severe	%	Ca	P	Mg	Na	K	Ca	P
A	Intact + PO_4	12	–	–	–	3.5 ±0.7	61.1 ±4.1	19.2 ±1.0	42.4 ±2.5	82.1 ±3.2	9.1 ±0.2	13.4 ±0.7
B	PTx-PO_4	16	4	12	100	3.4 ±0.4	58.5 ±1.5	17.0 ±0.4	49.3 ±1.5	82.4 ±1.4	3.8 ±0.2	26.3 ±3.1
C	PTx-PO_4 + $MgCl_2$ (1 mM) 0.2 g	29	2	1	10	4.0 ±0.3	62.4 ±1.3	18.9[c] ±0.3	41.5[c] ±0.9	76.7[c] ±1.0	5.0[c] ±0.1	17.8[a] ±0.7
D	PTx-PO_4 + NH_4Cl (1 mM) 0.05 g	18	1	9	55	3.8 ±0.2	62.8[a] ±1.3	19.5[c] ±0.4	44.0[b] ±1.2	79.8 ±1.6	4.6[a] ±0.2	25.8 ±1.9
E	PTx-PO_4 + NH_4Cl (2 mM) 0.1 g	18	4	0	22	3.3 ±0.1	62.3[a] ±1.6	19.5[c] ±0.5	39.3[c] ±0.6	75.7[c] ±1.3		

[a] $p < 0.05$ compared with group B.
[b] $p < 0.01$.
[c] $p < 0.001$.

groups, except for the K reduction, which occurred only with the larger dose and corresponded to the one encountered with $MgCl_2$ (group C). The results achieved with group E underscore the importance of the Cl ion in the protective activity of $MgCl_2$ when administered orally, and do, in fact, cast doubt on any substantial contribution of the Mg ion to this effect, since this bivalent cation is known to be poorly adsorbed from the gastrointestinal tract.

Subcutaneous administration of $MgCl_2$ and Mg gluconate in NaH_2PO_4 loaded PTx rats

Despite the fact that oral administration of protective electrolytes was spaced 30 min apart from NaH_2PO_4 loading, the possibility of direct interchange in the gut, be it chemical or physical in nature, could not be excluded, such as partial prevention of phosphate absorption by the formation of Mg phosphate or by increased cathartic activity and, thus, more rapid elimination of the offending agent. To obviate this possibility, the subcutaneous route was utilized exclusively in additional experiments. Comparison of the effects of equimolecular amounts of $MgCl_2$ and Mg gluconate was designed to distinguish the protective qualities of the Mg ion from that of the Cl ion.

It should be emphasized that parenteral administration of Mg salts was not well tolerated. This is not surprising, in view of the highly efficient absorption of Mg salts from this route and the predisposition to accumulation resulting from the severe embarrassment of renal excretory function that invariably accompanied the cardiac injury. It was noted that each subcutaneous injection of even the smallest dosages of $MgCl_2$ or Mg gluconate employed was followed by a period in which experimental rats exhibited increasing lassitude, often progressing to complete skeletal muscle flaccidity and paralysis. With the largest dosage of $MgCl_2$ used (0.2 g) and the corresponding equimolecular dose of Mg gluconate (0.44 g), death from respiratory paralysis was encountered consistently prior to completion of the series of five injections. The presence of pronounced hypermagnesemia (6 to 19.7 mEq/l) was demonstrated in terminal serum samples in all such instances.

Table II contains the data from a series of experiments carried out with the *subcutaneous* administration of Mg salts. It is apparent that the smaller dose of $MgCl_2$ (only one-fourth the amount used in the *oral* route study) provides considerable protection against myocardial necrosis and electrolyte changes (group C). While the total incidence of cardiac damage remains high, there is a clear-cut shift from severe to mild lesions, which is reflected in a corresponding prevention of deviation in the myocardial electrolyte content, in the face of a moderate reduction of hypocalcemia and hyperphosphatemia. Doubling the $MgCl_2$ dose is considerably more effective if judged by the far smaller incidence and severity of structural myocardial injury (group D). On the basis of this same standard, the equimolecular dose of Mg gluconate appears to be less effective

Table II

Effect of Subcutaneous Administration of Magnesium Chloride and Gluconate on Myocardial Necrosis and Electrolyte Alterations Induced by NaH_2PO_4 Loading of Parathyroidectomized (PTx) Rats

			Myocardial necrosis			Myocardial electrolytes (mean mEq/kg ± S.E. wet wt.)					Serum (mg%)	
			No.		Total							
Group	Treatment	No. of rats	Mild	Severe	Total %	Ca	P	Mg	Na	K	Ca	P
A	Intact controls	26	–	–	–	3.2 ±0.2	60.1 ±1.0	19.8 ±0.3	39.3 ±0.4	71.6 ±1.0	9.8 ±0.1	9.4 ±0.2
B	PTx-PO_4	15	1	14	100	3.0 ±0.2	54.4[b] ±1.8	17.0[c] ±0.5	45.0[c] ±0.9	71.9 ±0.9	3.4[c] ±0.1	27.1[c] ±1.7
C	PTx-PO_4 + $MgCl_2$ 0.05 g	16	11	3	88	3.4 ±0.2	56.4 ±1.8	20.5[f] ±0.5	40.7[f] ±1.0	70.8 ±0.8	4.2[c,e] ±0.2	21.5[c,e] ±1.5
D	PTx-PO_4 + $MgCl_2$ 0.1 g	9	3	0	33	3.5 ±0.2	65.3[a,e] ±2.9	19.9[e] ±0.6	39.3[f] ±1.1	70.1 ±1.2	4.6[c,f] ±0.2	17.5[c,d] ±1.2
E	PTx-PO_4 + Mg Gluconate 0.11 g	11	2	5	64	3.5[d] ±0.2	56.6 ±1.4	19.9[f] ±0.3	39.9[e] ±1.1	69.1 ±1.1	4.5[c,f] ±0.2	18.8[c,e] ±1.6
F	PTx-PO_4 + Mg Gluconate 0.22 g	11	8	0	73	2.8 ±0.2	59.3[d] ±1.0	19.9[d] ±0.2	41.9[a] ±1.5	71.4 ±0.8	4.8[c,f] ±0.2	19.1[c,e] ±1.6
G	PTx-PO_4 + Mg Gluconate 0.33 g	6	3	0	50	2.2[a,d] ±0.2	56.5 ±3.4	18.2[a] ±0.5	40.6[e] ±0.7	66.1[a,d] ±2.8	5.3[c,f] ±0.3	15.7[c,d] ±1.1

[a] $p < 0.05$ compared with group A.
[b] $p < 0.01$.
[c] $p < 0.001$.
[d] $p < 0.05$ compared with group B.
[e] $p < 0.01$.
[f] $p < 0.001$.

(group F). Further increase in the Mg gluconate dose does not seem to provide any dramatic improvement of results, although the number of rats in group G is too small to permit definite conclusions. It is of interest that this group manifests the least deviations from the norm of Ca/P levels in the serum and that it is the only one showing some loss of K from the heart, as was also seen in two orally treated groups (table I). The small number of rats in group G is due to loss of several animals as a result of Mg accumulation.

It is apparent from table II that only a high incidence of severe necrosis will distort the mean values of myocardial electrolytes sufficiently to result in significant deviations of the pattern from the norm. This is not to say, however, that individual animals cannot be readily distinguished as regards the severity of myocardial necrosis by determination of their individual myocardial electrolyte patterns. To illustrate this point, table III exemplifies the individual electrolyte values from rats representing subgroups of groups A, B, C, and D of table II. Note that rat number 12 in group C, which was not protected by $MgCl_2$ treatment, has myocardial electrolyte changes fully comparable to those of the unprotected animals 8 to 11 of group B. On the other hand, the remaining rats in group C, which had mild injury only, show considerable normalization of the myocardial electrolyte content in the face of uncorrected hypocalcemia and hyperphosphatemia. The greater effectiveness of a double dose of $MgCl_2$ (group D) is recognizable here, primarily by the uniformly reduced deviation of the serum Ca/P levels and the complete prevention of myocardial P deviation.

Finally, table III demonstrates that our procedure of classifying myocardial lesions by careful gross inspection with the naked eye did yield data that, on the whole, were well correlated with evaluations based on subsequent microscopic findings. This point deserves emphasis because it does not consistently hold for myocardial necrosis induced by other methods, *e.g.*, by injection of isoproterenol.

Absorption and renal excretion of $MgCl_2$ and Mg gluconate administered orally and subcutaneously

A comparison was made, in normal albino rats, of absorption and renal excretion of the largest oral dose of $MgCl_2$ (0.2 g) and the largest subcutaneous dose of Mg gluconate (0.33 g) tolerated under our experimental conditions. A single dose of either Mg salt was given both orally and subcutaneously to four groups of five rats each. Samples of tail blood were drawn at 1, 3, 5 and 24 h after drug administration and urines were collected without loss at the same time intervals. Figure 1 depicts comparative serum Mg concentrations and cumulative urinary excretion of Mg per animal as percentage of the administered dose. It should be emphasized that 0.2 g $MgCl_2$ is equivalent to 0.44 g of Mg gluconate. Hence, the dose of gluconate used here actually contains less Mg than does the dose of $MgCl_2$.

Table III

Correlation of Degree of Myocardial Necrosis with Alteration of Electrolyte Content in the Heart and Serum of *Individual* Albino Rats

Treatment	Rat no.	Myocardial necrosis: Gross Mild	Gross Severe	Microsc. Mild	Microsc. Severe	Myocardial electrolytes (mean mEq/l ± S.E. wet wt.): Ca	P	Mg	Na	K	Serum electrolytes (mg%): Ca	P
A	1					3.6	60.3	19.6	38.0	66.9	9.3	10.6
Intact	2					2.8	59.1	18.9	36.3	66.0	9.1	10.4
controls	3					3.6	57.0	17.9	37.2	68.1	9.2	10.8
	4					3.9	64.1	22.0	42.5	79.5	10.4	8.6
	5					3.3	61.7	21.0	40.3	74.6	9.7	8.4
	6					4.0	58.3	20.6	44.3	71.0	10.3	8.8
	7					4.4	61.8	21.5	40.2	73.0	10.0	9.4
						3.7	60.3	20.2	39.8	71.3	9.7	9.6
B	8	–	++++	–	+++(+)	1.9	57.3	13.5	42.8	70.6	3.4	19.8
PTx-PO_4	9	–	++++	–	++++	1.9	49.6	15.4	40.9	75.8	3.4	21.3
	10	–	++++	–	++++	3.9	42.9	15.5	55.3	70.4	1.6	22.2
	11	–	++++	–	++++	3.5	50.3	15.9	49.7	77.3	3.4	40.0
						2.8	50.0	15.1	47.2	74.3	3.0	25.8
C	12	–	+++	–	++++	2.3	49.5	15.9	45.5	66.6	3.7	23.2
PTx-PO_4 +	13	+	–	++	–	2.7	41.4	19.5	42.0	65.1	3.5	25.6
$MgCl_2$	14	++	–	++	–	3.0	48.8	19.8	44.9	71.2	4.0	24.6
0.05 g	15	++	–	++	–	2.4	54.2	20.0	40.9	71.0	3.8	35.7
	16	+(+)	–	++	–	3.0	55.9	19.3	40.1	68.2	3.5	31.5
	17	+	–	+(+)	–	2.8	52.8	18.9	38.3	71.4	4.4	20.4
						2.7	50.4	18.9	42.0	68.9	3.8	26.8
D	18	++	–	++	–	3.4	57.2	19.2	43.0	70.5	4.0	18.8
PTx-PO_4 +	19	++	–	++	–	3.2	55.6	19.2	39.8	69.9	4.3	16.8
$MgCl_2$	20	–	–	+	–	3.4	57.4	19.5	38.9	68.1	4.9	17.8
0.1 g	21	+	–	++	–	3.2	65.9	19.2	41.5	75.9	4.6	16.0
	22	–	–	–	–	3.1	62.3	17.6	42.9	70.6	5.4	16.5
						3.3	59.7	18.9	41.2	71.0	4.6	17.2

Interestingly, blood levels with the gluconate salt are somewhat higher and better sustained than with $MgCl_2$, and initial urinary elimination is distinctly slower. Total recovery after 24 h is, however, about equal (90% for gluconate and 82% for $MgCl_2$), indicating very efficient absorption for both salts.

On the other hand, when given by stomach tube in identical dosage, neither salt appears to be absorbed to any significant extent if judged on the basis of urinary excretion. The figures of urinary elimination depicted are well within the

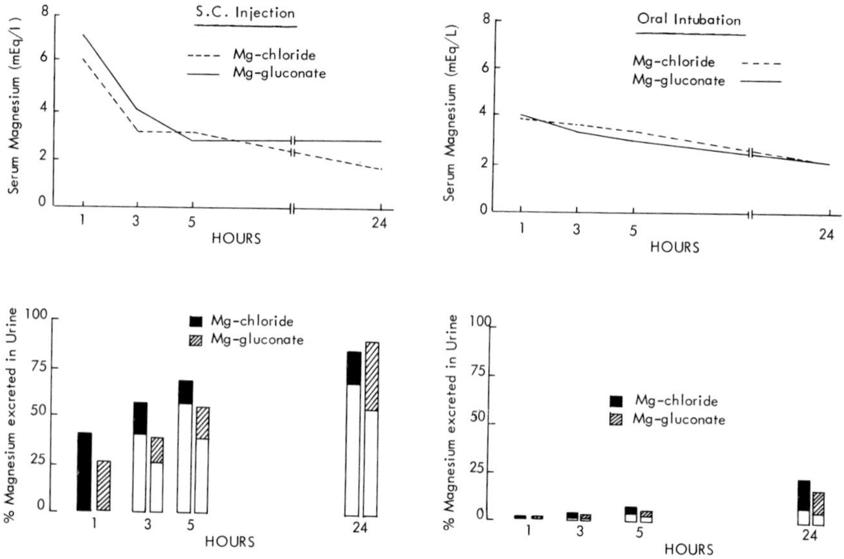

Fig. 1 Comparative absorption and renal excretion of magnesium administered as the chloride or gluconate salt subcutaneously and orally.

range of baseline excretion values of control animals and only the slight initial elevation of blood levels suggests some minimal absorption. Moreover, administration of chloride in the form of NH_4Cl in amounts equal to the Cl content of $MgCl_2$, either subcutaneously or orally, had no effect whatsoever on urinary elimination of dietary Mg in control rats.

In additional experiments, $MgCl_2$ was added to the drinking water of groups of rats in three different concentrations (0.5, 1, and 1.4%) for a period of 3 successive days. Even under these conditions, maximal total recovery of Mg from the urine did not exceed 10% of the intake.

Effect of $MgCl_2$ and NH_4Cl upon isoproterenol-induced myocardial necrosis

Since changes of the total myocardial electrolyte content in isoproterenol-induced myocardial necrosis had been found to be closely similar to those produced by our procedure (6) and to exhibit extrusion of Mg within hours after a single injection of the beta-adrenergic agonist, it appeared of interest to test the preventive effectiveness of the same electrolyte solutions in isoproterenol-induced myocardial injury. In this particular study, myocardial necrosis was produced by a single subcutaneous injection of the beta-adrenergic amine (5.25 mg/kg body weight).

Table IV indicates that $MgCl_2$ given parenterally in a dose and regimen identical to the one used in the experiments just described was only moderately effective. Under its influence, the extent of necrosis is seen to be distinctly

Table IV

Effect of $MgCl_2$ and NH_4Cl on Myocardial Necrosis and Electrolyte Alterations Induced by Isoproterenol

Group	Treatment	No. of rats	Myocardial necrosis (microscop. exam.) No.			Myocardial electrolytes (mean mEq/kg ± S.E. wet wt.)				
			Mild	Severe	Total %	Ca	P	Mg	Na	K
A	Controls	22	–	–	–	2.8 ±0.2	59.0 ±1.0	19.1 ±0.3	39.1 ±1.0	73.5 ±1.0
B	Isoproterenol 5.25 mg/kg	26	2	24	100	4.3^{c} ±0.3	48.4^{c} ±1.3	15.9^{c} ±0.2	50.8^{c} ±1.0	78.9^{c} ±1.2
C	Isoproterenol 5.25 mg/kg + $MgCl_2$ (s.c.) 0.1 g bid (½ mM)	10	7	3	100	3.7^{a} ±0.2	57.0^{e} ±1.4	$17.5^{a,e}$ ±0.4	$43.6^{a,f}$ ±0.8	75.8 ±1.0
D	Isoproterenol 5.25 mg/kg + NH_4Cl (s.c.) 0.05 g bid (1 mM)	8	4	4	100	4.3^{c} ±0.5	$52.1^{b,d}$ ±1.2	15.6^{c} ±0.4	$45.2^{b,d}$ ±1.9	74.5^{d} ±1.0
E	Isoproterenol 5.25 mg/kg + NH_4Cl (oral) 0.05 g bid (1 mM)	10	9	0	90	4.4^{c} ±0.3	$53.0^{b,d}$ ±1.1	16.8^{b} ±0.6	$46.4^{b,d}$ ±1.4	78.3^{a} ±1.3

[a] $p < 0.05$ compared with group A.
[b] $p < 0.01$.
[c] $p < 0.001$.
[d] $p < 0.05$ compared with group B.
[e] $p < 0.01$.
[f] $p < 0.001$.

diminished, with a corresponding clear-cut reduction of electrolyte deviations. NH_4Cl in a dose containing the same amount of Cl appears equally effective when given orally and less effective when injected subcutaneously, if judged on the basis of structural injury. It is, however, inferior to $MgCl_2$ as regards correction of electrolyte disturbances.

Preliminary studies using the alpha-adrenergic agonist, phenylephrine, and epinephrine, which has powerful alpha activity in addition to its beta-adrenergic potency, gave equivocal results.

DISCUSSION

The results obtained in the present study clearly indicate that, in the albino rat, the administration of electrolyte solutions containing either Mg or Cl may offer substantial protection against the development of experimental myocardial necrosis. Since acute absorption of Mg from the gastrointestinal tract was shown to be minimal when Mg was administered as the Cl or gluconate salt, it was concluded that no substantial protective activity could be ascribed to Mg salts administered by stomach tube, and that beneficial effects reported with oral $MgCl_2$ administration may have been due primarily to the action of the Cl ion. In line with this thinking, it is readily understandable why, among all Mg salts administered orally, only $MgCl_2$ is reported to possess substantial protective activity (17). The explanation usually offered or implied for this exceptional situation is that $MgCl_2$ is the only one among the Mg salts that is "adequately absorbed" from the gastrointestinal tract. That this is not necessarily correct was shown in our absorption-excretion studies. It is pertinent, therefore, to emphasize that the important beneficial effect of the Cl ion should be taken into consideration in all instances in which Cl-containing salts are administered for the prevention or treatment of myocardial injury.

Our observations of the excellent protective activity of NH_4Cl strongly support this concept. They are, incidentally, in line with findings published from this laboratory many years ago, when it was shown that NH_4Cl exhibited dramatic protective activity against the emergence of cardiovascular necrosis resulting from obstructive nephropathy (5, 7).

There is no doubt, on the other hand, that Mg can offer considerable protection against the development of cellular injury when Mg salts are administered by parenteral routes. In fact, comparison of the results achieved with the subcutaneous injection of equimolecular amounts of $MgCl_2$ and Mg gluconate does not reveal dramatic superiority of the former over the latter. This suggests that the effectiveness of the Cl ion might be obscured by the simultaneous presence of Mg in high concentration, and perhaps also that the parenteral injection of chlorides is less effective than their oral administration. The beneficial influence of the Cl ion could, nevertheless, be deduced in the studies using the subcutaneous route from the fact that doubling the dose of

$MgCl_2$ seemed to offer a significant increase in protection, whereas doubling and even tripling the dose of Mg gluconate did not achieve such enhancement of protection. Apparently, injured myocardial cells capable of recovery need, or can hold on to, only limited amounts of Mg offered. Any excess above this quantity will, therefore, not cause further improvement, whereas the Na-removing activity of the Cl ion increases within limits in proportion with the dose.

The effectiveness of parenterally administered Mg salts militates against the importance of direct chemical combination or physical effects in the gut when Mg salts are given orally, although the contribution of NaH_2PO_4 removal before absorption by Mg phosphate formation, and increased cathartic activity in using the oral route, cannot be totally ruled out.

It is expected that further light will be shed on these suggested explanations when ongoing studies with the subcutaneous administration of NH_4Cl have been completed.

As regards the NH_4Cl, it is well known that following *oral* administration the ammonium ion, upon reaching the liver, is converted to urea, leading to an excess of Cl, which is retained in the plasma at the expense of bicarbonate. The excess Cl is excreted with an equivalent amount of Na. Of the several mechanisms restricting Na loss, the most important is the increased production of ammonia by renal tubular cells. The excess Cl is then excreted as NH_4Cl. However, in the face of the severe renal tubular injury that is an invariable accompaniment of myocardial necrosis induced by our method, this compensatory mechanism would be expected to be impaired, so that Na loss probably continues. Moreover, lesser effectiveness of NH_4Cl as a Na-removing agent would have to be expected with parenteral administration, since the initial by-pass of the liver delays urea formation. The beneficial effect of NH_4Cl is probably based primarily upon removal of Na. Prevention of intracellular Na accumulation, in turn, may contribute to the reduction of Mg extrusion from the myocardial cell.

The significant drop of myocardial K encountered with the oral administration of $MgCl_2$ and NH_4Cl in dosages containing equal amounts of Cl (table I) might be due to the secondary loss of K known to be following in the wake of massive and continued removal of Na. However, this explanation could not apply to the preliminary observation of a similar K-lowering effect of Mg gluconate in large dosage when given by the subcutaneous route (table II, group C). Here, one might have to consider partial exchange of K for Mg.

Direct prevention of Mg loss by administration of Mg salts should be expected to be successful if this procedure prevents cellular damage or if the injury under treatment is not of such severity that the cell is incapable of replenishing its transiently reduced Mg content. It should be remembered that 45% of body Mg is contained in cells as intracellular cation and only 5% is present as extracellular cation, while the remaining 50% exists in the insoluble state in the bone (14). Mg is an important prosthetic or activator ion, participating in the function of many enzymes involved in phosphate transfer reactions (4), including those requiring ATP or other nucleotide triphosphates as coenzymes (3). The manner in which Mg participates in the activation of the enzyme system is thought by some to relate to the formation of Mg-ATP chelate, which is the favored substrate for the enzyme (11).

Mg chelation may also be a primary mechanism for the maintenance of subcellular structure in general. This is suggested by the growing body of data implicating metal ions in the reversible association of particles (*e.g.,* the ribosomal subunits, whose functional integrity depends upon an adequate concentration of Mg ion) and in the formation of complexes between macromolecules and particles (*e.g.,* the Mg-dependent binding of messenger -RNA to ribosomes) (1, 16).

Thus, depletion of Mg should be expected to lead to severe derangement of cellular function. Replacing this vital cation in the Mg-depleted cell may be as essential as that of supplying oxygen to the hypoxic cell. Our findings appear to bear this out.

The absence of significant changes in myocardial electrolyte content in cardiac necrosis classified as mild by our definition (few small foci of necrosis), indicates that major electrolyte alterations are probably limited to areas of injury and, thus, may become too dilute within the overwhelming mass of healthy muscle tissue to be noticeable by our methodology. Studies of the transmembrane action potential (TAP) of single ventricular cells of the beating heart, *in situ,* appear to confirm this point of view, since even in severely damaged hearts, TAP abnormalities were found to be confined to myocardial cells in areas of injury (Omura and Lehr, unpublished data). On the basis of these findings, it was proposed that significant deviations of the myocardial electrolyte content can be used in place of histological examination as a reliable index of severe degrees of myocardial necrosis and that the effacement of such electrolyte abnormalities can serve as evidence of successful prophylactic or therapeutic intervention (9).

The fact that the development of isoproterenol-induced myocardial necrosis could be partially prevented by administration of $MgCl_2$ and NH_4Cl, whereas closely similar injury caused by adrenergic amines having potent additional or exclusive alpha-adrenergic activity was apparently not as readily affected by these measures, indicates that the direct cardiotoxic, oxygen-wasting effect caused by stimulation of the beta-adrenergic receptors in the myocardium can be successfully counteracted by removal of Na or the supply of Mg (perhaps by a direct decrease of myocardial contractility), whereas the indirect oxygen-wasting effect of increased peripheral resistance caused by alpha-adrenergic stimulation of the arteriolar system is not amenable to the same protection by electrolyte administration.

In deference to Dr. Raab (15) and his emphasis upon the vital importance of intracellular-extracellular K shifts in the damaged myocardium, which would not be detectable by our method of chemical determination of total myocardial electrolyte content, I would like to emphasize that the absence of electrolyte changes in the myocardium detectable by chemical means does not exclude the presence of significant shifts in individual compartments of damaged cells or between extra- and intracellular space. The demonstration of such shifts is simply beyond the reach of the methodology employed.

Finally, it should be stressed that dosage as well as regimen of Mg administration was arbitrarily selected. It can be expected, therefore, that further improvement of results could be achieved by modifications of these parameters.

SUMMARY

It was found that, in the albino rat, the administration of electrolyte solutions containing either the Mg or Cl ion may offer highly significant protection against the development of severe disseminated myocardial necrosis and concomitant disturbances of the total myocardial electrolyte content induced by NaH_2PO_4 loading of parathyroidectomized rats.

The same electrolyte solutions, given under the same dosage regimen, were found less effective against similar myocardial necrosis produced in the albino rat by the beta-adrenergic agonist, isoproterenol. No benefit was achieved, thus far, in cardiac injury caused by amines possessing potent additional or exclusive alpha-adrenergic activity (epinephrine and phenylephrine).

REFERENCES

1. GOLDSTEIN, A.; ARONOW, L. and KALMAN, S. M.: Principles of drug action (Harper & Row, New York 1968).
2. HEGGTVEIT, H. A.; HERMAN, L. and MISHRA, R. K.: Cardiac necrosis and calcification in experimental magnesium deficiency: a light and electron microscopic study. Amer. J. Path. *45:* 757–782 (1964).
3. HILZ, H. and LIPMANN, F.: The enzymatic activation of sulfate. Proc. Nat. Acad. Sci. U.S. *41:* 880–890 (1955).
4. LEHNINGER, A. L.: Role of metal ions in enzymatic systems. Phys. Rev. *30:* 393–429 (1950).
5. LEHR, D.: On methods which inhibit or prevent intrarenal precipitation of compounds of the sulfonamide series. Bull. N.Y. Med. Coll., Flower Fifth Ave. Hosps. *6:* 70–77 (1943).
6. LEHR, D.: Tissue electrolyte alteration in disseminated myocardial necrosis. Ann. N.Y. Acad. Sci. *156:* 344–378 (1969).
7. LEHR, D.; CHURG, J. and MILORA, R.: The production of experimental arteriosclerosis and some factors important to its prevention. Bull. N.Y. Med. Coll., Flower Fifth Ave. Hosps. *15:* 130–152 (1952).
8. LEHR, D. and KRUKOWSKI, M.: About the mechanism of myocardial necrosis induced by sodium phosphate and adrenal corticoid overdosage. Ann. N.Y. Acad. Sci. *105:* 135–182 (1963).
9. LEHR, D.; KRUKOWSKI, M. and CHAU, R.: Acute myocardial injury produced by sympathomimetic amines. Israel J. Med. Sci. *5:* 519–524 (1969).

10. LEHR, D.; KRUKOWSKI, M. and COLON, R.: A simple method for the production of acute cardiorenal necrosis. Arch. Int. Pharmacodyn. Therap. *168:* 251–277 (1967).
11. LIEBECQ, C.: The hexokinase activity of rat muscle extract and the liability of the magnesium ATP complex. Biochem. J. (London) *54:* 22–23 (1953).
12. LOWENHAUPT, E.; SCHULMAN, M. P. and GREENBERG, D. M.: Basic histologic lesions of magnesium deficiency in the rat. Arch. Pathol. *49:* 427–433 (1950).
13. MISHRA, R. K.: Studies on experimental magnesium deficiency in the albino rat. Rev. Canad. Biol. *19:* 122–135 (1960).
14. MORGULIS, S.: Studies on the chemical composition of bone ash. J. Biol. Chem. *93:* 455–466 (1931).
15. RAAB, W.: Myocardial electrolyte derangement: Crucial feature of pluricausal so-called coronary heart disease (dysionic cardiopathy). Ann. N. Y. Acad. Sci. *147:* 627–686 (1969).
16. SEELIG, M. S.: Myocardial loss of functional magnesium. Part I: Effect on mitochondrial integrity and potassium retention. This Book, pp. 615–625.
17. SELYE, H. and GABBIANI, G.: The role of electrolytes in the pathogenesis of experimental cardiopathies without vascular involvement; in: BAJUSZ, E., Electrolytes and Cardiovascular Diseases, Vol. 1, pp. 135–160 (Karger, Basel 1965).

Mailing address: David Lehr, M.D., Department of Pharmacology, New York Medical College, Flower & Fifth Avenue Hospitals, Fifth Avenue at 106th Street, New York, New York 10029 (USA).

Stress and Cardiovascular Diseases

Institut de Médecine et de Chirurgie expérimentales
Université de Montréal, Montréal, Canada

STRESS, HORMONES, AND CARDIOVASCULAR DISEASE [1]

H. SELYE

The extensive literature concerning the role of hormones in the pathogenesis, prophylaxis and therapy of cardiovascular disease has been exhaustively surveyed in a monograph first presented before this meeting (36). There can be no doubt that many hormones, especially catecholamines and steroids, play important roles in the pathophysiology of the cardiovascular system. However, now I should like to limit my remarks to the most recent developments in our understanding of the manner in which steroid hormones and artificial steroids can regulate resistance, in general, and the responses of the cardiovascular system, in particular.

It has long been known that hormones can greatly alter the resistance of the body to various types of injuries. Our own interest in this matter stems from the observation that increased glucocorticoid production is a characteristic feature of the alarm reaction, elicited by any stressor. This endocrine response helps to maintain homeostasis, since resistance to stress is greatly impaired after adrenalectomy but can be restored to near normal levels by exogenous corticoids (12, 16).

Curiously, stress produced by many agents can raise the tolerance to a variety of unrelated toxic compounds. Often, this stress-induced nonspecific tolerance, or "cross-resistance," results from an increased corticoid secretion (21). However, our hopes of raising the stress tolerance of intact animals far above the norm by treatment with glucocorticoids did not materialize, presumably because a near optimal corticoid supply is assured by the physiologic activity of the adrenal cortex (16). Actually, in intact animals, glucocorticoids offer considerable protection against only a few agents, such as inflammatory irritants, bacterial endotoxins (16, 22), lathyrogens (19) and certain ganglioplegics (39). It seems that glucocorticoids act as "syntoxic" hormones, by merely suppressing pathologic tissue reactions; thus, they maintain a balanced "symbiosis," or coexistence, with pathogens (*e.g.*, inflammatory irritants) without destroying the latter.

However, numerous observations that came to light did not fit this simple

[1] Supported by le Ministère de la Santé, Québec, the Medical Research Council of Canada (Block Term Grant MT-1829), and the Quebec Heart Foundation, Montreal; this work was undertaken as a special project of the Council for Tobacco Research, U.S.A., and the Canadian Tobacco Industry.

interpretation of the hormonal regulation of resistance. For example, it was noted that, in rats, various steroids devoid of glucocorticoid potency can prevent the production of anesthesia by the subsequent administration of the same or other steroids (14). Furthermore, female rats (unless treated with androgens) are more resistant to steroid anesthesia than males (63). Pentylenetetrazol convulsions can be prevented by steroids other than glucocorticoids (15). The characteristic mineralocorticoid hypertension is more readily produced in females than in males and treatment with methyltestosterone offers definite protection to females (48). The cardiovascular calcification elicited in rats by dihydrotachysterol (DHT) overdosage is also more readily produced in females or castrates than in intact males. We concluded that, here, some "testicular factor," rather than a corticoid, is responsible for resistance (17). In support of this concept we found that methyltestosterone increased the DHT resistance of female rats, whereas cortisol (like estradiol and desoxycorticosterone) diminished it (18).

However, all these "anomalous" phenomena of hormonal resistance that did not depend upon glucocorticoids could not be ascribed to testicular hormones either. Thus, the "infarctoid cardiac necroses" (produced in rats by fluorocortisol plus sodium salts and/or stress) were inhibited by spironolactone, an antimineralocorticoid devoid of both glucocorticoid and testoid (or androgenic) potency (20). Another antimineralocorticoid, SC-11927, was even more active in this respect and it also protected the rat against the particularly severe cardiovascular calcification elicited by DHT plus Na_2HPO_4 (40). Subsequent systematic studies with a very large number of steroids revealed that the anti-DHT effect is much more closely associated with anabolic than with testoid potency (50, 51).

Considerable progress was made in the elucidation of this "not corticoid dependent type of resistance" recently when it became clear that there exists a special class of "catatoxic steroids," which–unlike the syntoxic glucocorticoids–do not merely normalize the body's response to toxicants but actually attack the latter. Usually, though not always, this is achieved by increasing the rate of metabolic degradation in the liver. The catatoxic action is independent of any other known hormonal effect, although it is often associated with anabolic (ethylestrenol, norbolethone, oxandrolone), antitestoid (cyproterone), or antimineralocorticoid (spironolactone, SC-11927, spiroxasone) properties. Among the chemical characteristics that appear to enhance the catatoxic potency of steroids are the possession of nitrile or lactone groups and a 19-nor configuration (35).

The extraordinarily broad activity spectrum of catatoxic steroids is illustrated by the following partial list of toxicants against which they offer protection: numerous digitalis alkaloids (23, 41–43, 47), indomethacin (25), various anesthetics and hypnotics, including many barbiturates and steroids (26, 43, 45, 49), dimethylbenzanthracene, or DMBA (10, 11), and its highly active metabolite 7-OHM-MBA (60), nicotine (52), mephenesin (28), picrotoxin (29), phenindione (30), bishydroxycoumarin (59), hypervitaminosis A (61), cycloheximide (34), meprobamate (33), colchicine (32), methyprylon (31), and a great variety of pesticides (38). Catatoxic compounds also protect against the diverse forms of infarctoid cardiopathies that are produced by steroids or digitoxin on certain diets (27). Even the fatal renal damage produced by $HgCl_2$ can be prevented by certain catatoxic steroids but, in this respect, only those (spironolactone, isoxasone, emdabol) containing thioacetyl groups are active (37, 46). Evidently, the protection offered by catatoxic steroids extends to compounds of vastly different chemical structure and pharmacologic activity.

The fact that hepatic detoxication plays a role in this type of drug resistance was first called to our attention in 1931, when we found that, in rodents, the anesthetic effect of tribromoethanol is greatly prolonged by removal of about 70% of the liver (62). Subsequently, the same operation was shown to increase the duration of steroid anesthesia (13) and to augment the toxicity of indomethacin (24). Furthermore, the protective effect of catatoxic steroids is associated with a proliferation of the smooth endoplasmic reticulum in hepatocytes (5, 9), an increase in hepatic weight and mitotic index, as well as an accelerated regeneration of the liver after partial hepatectomy (7).

The understanding of the catatoxic effect was greatly promoted by the realization of a relationship between the observations just mentioned and completely independent investigations performed in many other laboratories, which showed that drug-metabolizing hepatic microsomal enzymes can be induced by various chemicals, including steroids (1–4, 8). At first, the induction of drug-metabolizing enzymes by steroids was variously ascribed to their glucocorticoid, testoid, or anabolic potencies. However, it is now clear that a similar mechanism of hepatic enzyme induction is also responsible for many of the protective catatoxic steroid actions, irrespective of all their classic hormonal properties. This has been demonstrated in our Institute by the observation that, parallel with their *in vivo* effects upon resistance, these steroids accelerate the blood clearance of toxic substrates (55, 57–59). Furthermore, increased enzyme activity was noted even upon *in vitro* incubation of substrates with the microsomal + supernatant fractions of the liver of animals that were treated *in vivo* with catatoxic steroids (53, 56).

Finally, the protective potency of catatoxic steroids is found to be impaired in animals pretreated with a variety of compounds known to block the activity or the induction of hepatic microsomal enzymes, namely: ethionine (6), metyrapone (44), SKF 525-A (57–59), cycloheximide, actinomycin D, and puromycin (54).

SUMMARY

The prevention of various experimental cardiovascular diseases by steroids having specific hormonal (especially testoid, anabolic, or antimineralocorticoid) actions has been discussed at length in a detailed monograph (36). However, recent experiments suggest that some of these protective properties are entirely independent of all classic steroid hormonal effects.

This communication represents an attempt to summarize the historic development of our concept concerning the actions of these "adaptive hormones," especially the two major groups of the syntoxic and catatoxic steroids.

In relation to cardiovascular disease, perhaps the most interesting aspect of catatoxic steroid actions is their ability to induce hepatic microsomal enzymes capable of destroying such known potentially cardiotoxic agents as mineralocorticoids, dihydrotachysterol, cholesterol, nicotine, anticoagulants, digitalis alkaloids, and many others.

REFERENCES

1. AXELROD, J.: The enzymatic N-demethylation of narcotic drugs. J. Pharmacol. Exp. Ther. *116:* 322–330 (1956).
2. BOOTH, J. and GILLETTE, J. R.: The effect of anabolic steroids on drug metabolism by microsomal enzymes in rat liver. J. Pharmacol. Exp. Ther. *137:* 374–379 (1962).
3. BRODIE, B. B.: Pathways of drug metabolism. J. Pharm. Pharmacol. *8:* 1–17 (1956).
4. CONNEY, A. H.: Pharmacological implications of microsomal enzyme induction. Pharmacol. Rev. *19:* 317–366 (1967).
5. GARDELL, C., BLASCHECK, J. A. et KOVACS, K.: Action de la norboléthone sur le réticulum endoplasmique hépatique du rat. J. Microscopie *9:* 133–138 (1970).
6. GARDELL, C., SOMOGYI, A. et KOVACS, K.: Influence de la dl-éthionine sur l'action protectrice de la spironolactone lors d'intoxication à la digitoxine chez le rat. J. Europ. Toxicol. *3:* 107–109 (1970).
7. HORVATH, E., SOMOGYI, A. und KOVACS, K.: Einfluss von Spironolacton auf das Regenerationsvermogen der Leber bei Ratten. Klin. Wchnschr. *48:* 385–387 (1970).
8. KATO, R., CHIESARA, E. and FRONTINO, G.: Influence of sex difference on the pharmacological action and metabolism of some drugs. Biochem. Pharmacol. *11:* 221–227 (1962).
9. KOVACS, K., BLASCHECK, J. A. and GARDELL, C.: Spironolactone-induced proliferation of smooth-surfaced endoplasmic reticulum in the liver of rats. Z. Ges. Exp. Med. *152:* 104–110 (1970).
10. KOVACS, K. and SOMOGYI, A.: Prevention by spironolactone of 7,12-dimethylbenz(a)anthracene-induced adrenal necrosis. Proc. Soc. Exp. Biol. (N.Y.) *131:* 1350–1352 (1969).
11. KOVACS, K. and SOMOGYI, A.: Suppression by spironolactone of 7,12-dimethylbenz(a)anthracene-induced mammary tumors. Europ. J. Cancer *6:* 195–201 (1970).
12. SELYE, H.: A syndrome produced by diverse nocuous agents. Nature (London) *138:* 32 (1936).
13. SELYE, H.: On the role of the liver in the detoxification of steroid hormones and artificial estrogens. J. Pharmacol. Exp. Ther. *71:* 236–238 (1941).
14. SELYE, H.: Acquired adaptation to the anesthetic effect of steroid hormones. J. Immunol. *41:* 259–268 (1941).
15. SELYE, H.: The antagonism between anesthetic steroid hormones and pentamethylenetetrazol (Metrazol). J. Lab. Clin. Med. *27:* 1051–1053 (1942).
16. SELYE, H.: Stress. Montreal: Acta Inc., Med. Publ., (1950).
17. SELYE, H.: Effect of castration upon arteriosclerosis produced by dihydrotachysterol (AT-10) in male rats. Lab. Invest. *6:* 301–304 (1957).
18. SELYE, H.: Effect of various hormones upon the syndrome of dihydrotachysterol (AT-10) intoxication. Acta Endocr. (Kbh.), *25:* 83–90 (1957).

19. SELYE, H.: Lathyrism. Rev. Canad. Biol. *16:* 1–82 (1957).
20. SELYE, H.: Protection by a steroid-spirolactone against certain types of cardiac necroses. Proc. Soc. Exp. Biol. (N.Y.) *104:* 212–213 (1960).
21. SELYE, H.: Nonspecific resistance. Ergebn. Allg. Pathol. Pathol. Anat. *41:* 208 (1961).
22. SELYE, H.: Thrombohemorrhagic Phenomena (Charles C Thomas, Springfield, Ill. 1966).
23. SELYE, H.: Spironolactone actions, independent of mineralocorticoid blockade. Steroids *23:* 803–808 (1969).
24. SELYE, H.: Role of the liver in the prevention of indomethacin-induced intestinal ulcers by spironolactone. Acta hepato-splenol. (Stuttg.) *17:* 393–397 (1970).
25. SELYE, H.: Prevention of indomethacin-induced intestinal ulcers by spironolactone and norbolethone. Canad. J. Physiol. Pharmacol. *47:* 981–983 (1969).
26. SELYE, H.: Inhibition of anesthesia by steroids. J. Pharmacol. Exp. Ther. *174:* 478–486 (1970).
27. SELYE, H.: Prevention of various forms of metabolic myocardial necrosis by catatoxic steroids. J. Molec. Cell Cardiol. *1:* 91–99 (1970).
28. SELYE, H.: Prevention of mephenesin intoxication by catatoxic steroids. Acta pharmacol. (Kbh.) *28:* 145–148 (1970).
29. SELYE, H.: Resistance to picrotoxin poisoning induced by catatoxic steroids. Agents and Actions *1:* 133–135 (1970).
30. SELYE, H.: Protection by catatoxic steroids against phenindione overdosage. Thrombos. Diathes. haemorrh. (Stuttg.) *23:* 77–81 (1970).
31. SELYE, H.: Protection against methyprylon overdosage by catatoxic steroids. Canad. Anaesth. Soc. J. *17:* 107–111 (1970).
32. SELYE, H.: Prevention of colchicine intoxication by catatoxic steroids. End. exp. *4:* 71–76 (1970).
33. SELYE, H. and SOLYMOSS, B.: Protection by catatoxic steroids against meprobamate. Neuropharmacology *9:* 327–332 (1970).
34. SELYE, H.: Protection by catatoxic steroids against cycloheximide intoxication. Toxicol. Appl. Pharmacol. *17:* 721–725 (1970).
35. SELYE, H.: Pharmaco-chemical interrelations among catatoxic steroids. Rev. Canad. Biol. *29:* 49–102 (1970).
36. SELYE, H.: Experimental Cardiovascular Diseases, 2 Vols., (Springer-Verlag, Berlin 1970).
37. SELYE, H.: Mercury poisoning: prevention by spironolactone. Science *169:* 775–776 (1970).
38. SELYE, H.: Resistance to various pesticides induced by catatoxic steroids. A.M.A. Arch. Environm. Health *21:* 706–710 (1970).
39. SELYE, H.: Protection by glucocorticoids against ganglioplegics. Res. Commun. Chem. Path. Pharmacol. *1:* 572–579 (1970).
40. SELYE, H., GRASSO, S. and PADMANABHAN, N.: Protection by antimineralocorticoid against an otherwise fatal dihydrotachysterol intoxication. Lancet, Dec. 17, 1350–1351 (1960).
41. SELYE, H., JELINEK, J. and KRAJNY, M.: Prevention of digitoxin poisoning by various steroids. J. Pharm. Sci. *58:* 1055–1059 (1969).
42. SELYE, H., KRAJNY, M. and SAVOIE, L.: Digitoxin poisoning: prevention by spironolactone. Science *164:* 842–843 (1969).
43. SELYE, H., MANDEVILLE, R. and YEGHIAYAN, E.: On the catatoxic effect of antimineralocorticoids. Naunyn-Schmiedeberg, Arch. Pharmak. Exp. Pathol. *266:* 34–42 (1970).

44. SELYE, H. and MECS, I.: Blockade of catatoxic steroid actions by metyrapone. Acta Physiol. Acad. Sci. Hung. *37:* 141–144 (1970).

45. SELYE, H., MECS, I. and SAVOIE, L.: Inhibition of anesthetics and sedative actions by spironolactone. Anesthesiology *31:* 261–264 (1969).

46. SELYE, H., MECS, I. and SZABO, S.: Protection by steroids against acute $HgCl_2$ poisoning. Urol. Nephrol. *2:* 287–301 (1970).

47. SELYE, H., MECS, I. and TAMURA, T.: Effect of spironolactone and norbolethone on the toxicity of digitalis compounds in the rat. Brit. J. Pharmacol. *37:* 485–488 (1969).

48. SELYE, H. and ROWLEY, E. M.: Prevention of experimental nephrosclerosis with methyl-testosterone. J. Urol. *51:* 439–442 (1944).

49. SELYE, H., SAVOIE, L. and SAYEGH, R.: Inhibition of anesthetic and sedative actions by norbolethone. Pharmacology (Basel) *2:* 265–270 (1969).

50. SELYE, H., TUCHWEBER, B. and JACQMIN, M.: Protection by various anabolic steroids against dihydrotachysterol induced calcinosis and catabolism. Acta Endocr. (Kbh.) *49:* 589–602 (1965).

51. SELYE, H., YEGHIAYAN, E. and MANDEVILLE, R.: Protection by catatoxic steroids against dihydrotachysterol intoxication. J. Atheroscler. Res. *11:* 321–331 (1970).

52. SELYE, H., YEGHIAYAN, E. and MECS, I.: Prevention of nicotine intoxication by catatoxic steroids. Arch. Int. Pharmacodyn. *183:* 235–238 (1970).

53. SOLYMOSS, B., CLASSEN, H. G. and VARGA, S.: Increased hepatic microsomal activity induced by spironolactone and other steroids. Proc. Soc. Exp. Biol. (N.Y.) *132:* 940–942 (1969).

54. SOLYMOSS, B., KRAJNY, M., VARGA, S. and WERRINGLOER, J.: Suppression by nucleic acid- and protein-synthesis inhibitors of drug detoxication induced by spironolactone or ethylestrenol. J. Pharmacol. Exp. Ther. *174:* 473–477 (1970).

55. SOLYMOSS, B., TOTH, S., VARGA, S. and SELYE, H.: Protection by spironolactone and oxandrolone against chronic digitoxin or indomethacin intoxication. Toxicol. Appl. Pharmacol. *18:* 586–592 (1971).

56. SOLYMOSS, B., VARGA, S. and CLASSEN, H. G.: Effect of various steroids on microsomal aliphatic hydroxylation and N-dealkylation. Europ. J. Pharmacol. *10:* 127–130 (1970).

57. SOLYMOSS, B., VARGA, S. and KRAJNY, M.: Effect of spironolactone, norbolethone, progesterone, hydroxydione and SKF 525-A on the disappearance rate of indomethacin from blood. Arzneim.-Forschg., in press.

58. SOLYMOSS, B., VARGA, S. and KRAJNY, M.: Increased drug metabolism induced by various steroids and its suppression by SKF 525-A. Acta physiol. Acad. Sci. hung. *37:* 145–149 (1970).

59. SOLYMOSS, B., VARGA, S., KRAJNY, M. and WERRINGLOER, J.: Influence of spironolactone and other steroids on the enzymatic decay and anticoagulant activity of bishydroxycoumarin. Thrombos. Diathes. haemorrh. (Stuttg.) *23:* 562–568 (1970).

60. SOMOGYI, A. and KOVACS, K.: Inhibition by spironolactone of 7-hydroxymethyl-12-methylbenz(a)anthracene-induced adrenal necrosis in rats. Endokrinologie *56:* 245–247 (1970).

61. TUCHWEBER, B. and GARG, B. D.: Protection by spironolactone and various anabolic steroids against vitamin-A overdosage. Proc. Canad. Fed. Biol. Soc. *13:* 10 (1970).

62. WAELSCH, H. und SELYE, H.: Beiträge zur Entgiftung im tierischen Organismus. III. Mitteilg: Bedeutung der Leber bei Avertin- und Magnesiumnarkosen. Naunyn-Schmiedeberg's Arch. exp. Pathol. Pharmak. *161:* 115–118 (1931).

63. WINTER, H. and SELYE, H.: Conditions influencing the course of steroid hormone anesthesia. Amer. J. Physiol. *133:* 495 (1941).

Mailing address: Dr. Hans Selye, Institut de Médecine et de Chirurgie expérimentales, Université de Montréal, Case Postale 6128, Montréal, Que. (Canada).

University of Vermont, Department of Medicine,
Burlington, Vermont, USA

CARDIOTOXIC EFFECTS OF EMOTIONAL, SOCIOECONOMIC, AND ENVIRONMENTAL STRESSES

W. RAAB†

The heart muscle of civilized man is exposed to virtually constant simultaneous noxious interferences on three fronts (fig. 1), all of them converging on their ultimate victim: the vitally important electrolyte equilibrium of the myocardial cells.

It is a well-established fact that maintenance of a certain concentration and mutual proportionality of potassium, magnesium, and sodium ions within the cells of the heart muscle constitutes an indispensable prerequisite for normal electrical stimulus formation and conduction, and for myocardial contractility (29). Major derangements of the myocardial electrolyte balance, especially losses

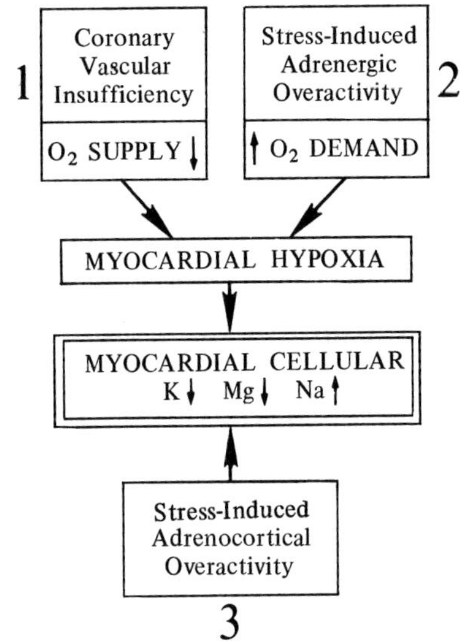

Fig. 1 Tripartite pathogenicity of stress-induced, electrolyte-mediated degenerative myocardial disease: 1. vascular; 2. neurogenic; 3. endocrine.

†*Deceased September 21, 1970*

in K and Mg, are followed by potentially fatal necrotic cell destruction. In man, reduction of the K/Na ratio by more than 35% appeared to be incompatible with life-preserving cardiac contractility (30).

Critical disturbances of the myocardial electrolyte balance are attributable chiefly to two jointly tissue-hypoxia-producing factors, a vascular and a neurogenic one, and, in addition, to a third hormonal element that interferes primarily in carbohydrate metabolism (30).

High degrees of coronary arterial narrowing can, by themselves, suffice to seriously impair coronary flow and tissue oxygenation, but such maximal stenoses are much less common than is generally assumed (30). Furthermore, the belief that massive infarctions are caused primarily by completely occluding arterial thrombi is being increasingly abandoned in favor of the concept that thrombi, whenever found (in approximately 40–50% of infarctions), represent a secondary feature of blood coagulation upstream from metabolically injured myocardial zones (30).

A dominant element in the causation of hypoxia-producing and, consequently, electrolyte-deranging discrepancies between local oxygen supply and consumption consists in the uneconomical, often excessive augmentations of myocardial oxygen consumption that result from the action of adrenosympathetic catecholamines.

Beside these two jointly hypoxiating mechanisms, an overproduction of glucocorticoids (especially 17-hydroxycorticosteroids, cortisol) is apt to disturb the myocardial electrolyte balance in the same direction as the influence of angio- plus neurogenic hypoxia, *i.e.*, loss of K and Mg, and gain in Na.

Presumably due to this equidirectional, additive, and mutually intensifying quality of both catecholamine and corticoid overaction on myocardial electrolyte balance, the combination of both produces a marked aggravation of catecholamine cardiotoxicity.

This widely confirmed phenomenon, which had been first observed by Selye and Bajusz (36), was designated as corticoid-induced "sensitization" of the heart muscle for catecholamine injury. It constitutes one of the most important mechanisms involved in the pathogenesis of degenerative heart disease among those attributable to emotional-environmental stresses.

The primary power stations for the release of both adrenosympathetic catecholamines and adrenocortical steroids are located in intimately connected cerebral cortical and hypothalamic areas, and in the reticular substance of the central nervous system (6, 9, 13, 38). Their direct electrical or mechanical, or stress-induced reflectory stimulation from the periphery and environment is promptly followed by an augmentation of the blood level and urinary excretion of catecholamines as well as glucocorticoids (30). Aside from numerous animal experiments (18, 30) such observations have been made in man in situations of anxiety, tension and excitement (16), *e.g.*, during scholastic examinations (5,

23), on admission to hospital (39), while flying by plane (8) or driving an automobile (3, 35), during embarrassing interrogations (11), viewing of exciting films (15), sexual arousal (15), performance of difficult office work under irritating conditions (10, 15), anticipation of uncomfortable situations (4), and even mild sensory and mental annoyances, such as noise (1, 28), flickering light, and mental arithmetic (28).

Functional responses of cardiometabolically vulnerable hearts to emotional stress are manifold. They range from adrenergic accelerations of the heart beat, shortening of the isometric tension period of the left ventricle, reduction of the coronary oxygen-supply-favoring diastolic "recovery phase," and minor disturbances of rhythm, such as extrasystoles and auricular fibrillation, to atrioventricular block, and to life-threatening ventricular tachycardia and fibrillation (27, 30).

Arrhythmias and electrocardiographic changes are observed also, as a rule, during so-called "sympathetic storms," most of which represent the dramatic clinical manifestation of massive catecholamine discharges into the blood from tumors of the adrenal medulla (pheochromocytomas) or gangliomas.

The long-recognized (26) significance of acutely hypoxiating catecholamine overaction in the origin of anginal chest pain (demonstrated more than 40 years ago in a nearly fatal, extremely distressing self-experiment with injected epinephrine) has been amply confirmed by the association of anginal attacks with acute elevations of the serum catecholamine level (30), *e.g.,* under emotional stimuli (19, 37); the postinjection accumulation of catecholamines in the myocardium (33); the abnormal outflow of both lactic acid (14) and potassium (21) from the heart muscle during anginal pain; the hypoxia-type electrocardiogram; and the therapeutic efficacy of a variety of specifically anti-adrenergic measures (30).

Unexpected sudden death, caused either by apparent vagal stimulation and cardiac standstill or by adrenosympathetically induced ventricular fibrillation (the latter usually associated with pulmonary edema), occurs relatively frequently as the result of an acute emotional upheaval. Sudden fright, in which the central parasympathetic mechanism seems to be prevalent, or emotionally charged arguments, anticipation of an impending crisis, and the like, often precede the fatal event. Instances of unexpected death during sleep are attributed to the effect of emotional dreams in view of the frequent coincidence of recorded ocular REM movements with signs of adrenergic cardiovascular reactions (20). It is also worthy of note that, in its circadian rhythm, the blood level of 17-hydroxycorticosteroids reaches its peak during sleeping hours (40) and may contribute to a fatally cardiotoxic interaction with dream-induced emotional catecholamine discharges.

In a series of unexpected sudden deaths, the *post-mortem* catecholamine concentration of the blood was found to be higher than during pheochromocytoma-induced paroxysms, and as high as in fatal epinephrine poisoning (17).

Although atherosclerotic changes of the coronary arteries are present in the great majority of suddenly deceased humans, instances of cardiac death have been observed where no such changes existed, but where excessively high accumulations of catecholamines were found in the myocardium at autopsy. In both of two young individuals presenting this phenomenon, emotional factors (quarrel with a girlfriend; admission to hospital) seem to have been causally involved in their demise (24, 25).

A cardiometabolic corollary of hypoxic adrenergic structural lesions of the myocardium consists of characteristic hypoxic electrolyte derangements in the affected myocardial cell groups (30). However, not only the areas of established necrotic cell destruction show a severe depletion of K and Mg with augmentation of Na. Significantly, the same pattern appears (although to a less extreme degree) at an earlier, prenecrotic stage (42).

The exact pathogenic mechanism that causes massive myocardial infarctions in man is still under debate, but the concept of confluence of smaller foci to larger necroses, climaxed by secondary thrombus formation near zones of multifocal primary injury, is gaining acceptance (30). Thus, the biochemical cardiotoxic factors that evoke more or less acutely scattered myocardial micro-lesions under the impact of emotional catecholamine and corticoid discharges, especially in vascularly handicapped hearts, can reasonably be assumed to be, at least in part, responsible for the existing rampant cardiac mortality in over-civilized, emotionally, and environmentally denatured societies.

Severe cardiotoxic and, sometimes, ultimately fatal myocardial lesions have been elicited in baboons by prolonged interference in their social and behavioral habits (7), and in rats by periods of induced anxiety and frustration (32), in spite of the absence of coronary vascular abnormalities.

The stress of isolation, which, in rats, provokes an overproduction of glucocorticoids (41) and a correspondingly greatly intensified destructive cardiotoxicity of catecholamines (2, 31), is accompanied by progressive myocardial electrolyte derangements of the hypoxic type (31). After return of the animals to community life, these changes revert toward normal. The markedly increased cardiac mortality of elderly widowers during the first year after the loss of their wives (22) may be due to comparable neuroendocrine stress mechanisms.

Congestive heart failure is regularly associated with a very low K and Mg, and high Na concentration (low K/Na ratio) in the ventricles (12, 30, 34). Presumably, this represents the result of several varying pathogenic factors, including neuroendocrine responses to stresses, mechanical overload, and nicotine-induced overactivity of both the adrenosympathetic and the adrenocortical systems.

SUMMARY

Today's literature suggests that, jointly with vascular oxygen-supply-limiting factors, centrally controlled neuroendocrine mechanisms dominate the myocardial pathogenesis. By interfering with myocardial oxygen economy (catecholamines) and carbohydrate metabolism (glucocorticoids), they derange the vital myocardial electrolyte equilibrium (loss of potassium and magnesium, gain in sodium), thus disturbing stimulus formation and conduction, cell contractility and structure, largely under the influence of civilization-induced emotional and environmental stresses.

REFERENCES

1. ARGUELLES, A. E.; IBEAS, D.; OTTONE, J. P. and CHEKHERDEMIAN, M.: Pituitary-adrenal stimulation by sound of different frequencies, J. Clin. Endocrinol. *22:* 846–852 (1962).
2. BALAZS, T.; MURPHY, J. B. and GRICE, H. C.: The influence of environmental changes on the cardiotoxicity of isoprenaline in rats. J. Pharm. Pharmacol. *14:* 750–755 (1962).
3. BELLET, S.; ROMAN, L. and KOSTIS, J.: Effect of automobile driving on catecholamine and adrenocortical excretion. Amer. J. Cardiol. *24:* 365–368 (1969).
4. BERMAN, L. and GOODALL, McC.: Adrenaline, noradrenaline and 3-methoxy-4-hydroxy-mandelic acid (MOMA) excretion following centrifugation and anticipation of centrifugation. Fed. Proc. *19:* 154 ((1960).
5. BOGDONOFF, M. D.; ESTES, E. H.; HARLAN, W. R.; TROUT, D. L. and KIRSHNER, N.: Metabolic and cardiovascular changes during a state of acute central nervous system arousal. J. Clin. Endocrinol. *20:* 1333–1340 (1960).
6. CHARVÁT, J.; DELL, P. and FOLKOW, B.: Mental factors and cardiovascular disease. Cardiologia *44:* 124–141 (1964).
7. CHERKOVICH, G. M.: Micronecroses in the heart of the monkey, produced by experimental neurosis. Pat. Fiziol. Eksp. Ter. *3:* 22–26 (1959) (Medgiz.) (Russian).
8. EULER, U. S. VON and LUNDBERG, U.: Effect of flying on the epinephrine excretion in Air Force personnel. J. Appl. Physiol. *6:* 551–555 (1954).
9. FOLKOW, B.; HEDNER, P.; LISANDER, P. and RUBINSTEIN, E.: Studies on the release of cortisol as induced by topical stimulation of the hypothalamic "defense area" in cats. Symposium "Emotional Stress: Physiological and Psychological Reactions," Stockholm, 1964. (S. Karger) (Basel 1966).
10. FRANKENHAEUSER, M. and PÁTKAI, P.: Catecholamine excretion and performance during stress. Percept. Motor Skills *19:* 13–14 (1962).
11. HAMBURG, D. A.: Plasma and urinary corticoid levels in naturally occurring psychological stresses, in Ultrastructure and Metabolism of the Nervous System. Res. Publ. Assoc. Res. Nerv. Ment. Dis. *40:* 406–413 (1962).
12. HOCHREIN, H.: Herzinsuffizienz und Myokardstoffwechsel (Aulendorf, Kantor 1965) (14th suppl. Arneimittelforsch.).
13. HOFF, E. C.; KELL, J. F., JR. and CARROLL, M. N.: Effects of cortical stimulation and lesions on cardiovascular function. Physiol. Rev. *43:* 68–114 (1963).
14. KRASNOW, N. and GORLIN, R.: Myocardial lactate metabolism in coronary insufficiency. Ann. Int. Med. *59:* 781–787 (1963).
15. LEVI, L.: Life stress and urinary excretion of adrenaline and noradrenaline, in RAAB, W., Prevention of Ischemic Heart Disease, p. 85–95 (C. C Thomas, Springfield, Ill. 1966).
16. LEVI, L.: Neuroendocrinology of anxiety. Brit. J. Physchiatry (special publication no. 3) (1969).
17. LUND, A.: Adrenaline and noradrenaline in post-mortem blood. Med. Sci. Law *4:* 194–197 (1964).

18. MASON, J. W.; MANGAN, G.; BRADY, J. V.; CONRAD, D. and RIOCH, D. McK.: Concurrent plasma epinephrine, norepinephrine and 17-hydroxycorticosteroid levels during conditioned emotional disturbances in monkeys. Psychosom. Med. *23:* 344–353 (1961).

19. NESTEL, P. J.; VERGHESE, A. and LOVELL, R. R. H.: Catecholamine secretion and sympathetic nervous responses to emotion in man with and without angina pectoris. Amer. Heart J. *73:* 227–234 (1967).

20. NOWLIN, J. B.; TROYER, W. G., JR.; COLLINS, W. S.; SILVERMAN, G.; NICHOLS, C. R.; McINTOSH, H. D., ESTES, E. H., JR. and BOGDONOFF, M.D.: The association of nocturnal angina pectoris with dreaming. Ann. Int. Med. *63:* 1040–1046 (1965).

21. PARKER, J. O.; CHIONG, M. A.; WEST, R. O. and CASE, R. B.: Myocardial electrolyte balance during pacing-induced angina pectoris. Circulation *40*(Suppl. III): 158 (1969) (abstr.).

22. PARKES, C. M.; BENJAMIN, B. and FITZGERALD, R. G.: Broken heart: a statistical study of increased mortality among widowers. Brit. Med. J. *i:* 740–743 (1969).

23. PEKKARINEN, A.; CASTRÉN, O.; JISALO, E.; KOIVUSALO, M.; LAIHINEN, A.; SIMOLA, P. E. and THOMASSON, B.: The emotional effect of matriculation examinations on the excretion of adrenaline, noradrenaline, 17-hydroxy-corticosteroids in the plasma. In Biochemistry, Pharmacology and Physiology pp. 117–137 (Oxford, 1961).

24. RAAB, W.: Sudden death of a young athlete with an excessive concentration of epinephrine-like substances in the heart muscle. Arch. Path. *36:* 388–392 (1943).

25. RAAB, W.: Sudden death with an excessive myocardial concentration of epinephrine-like substances in a case of obesity and cystic thyroid disease. Arch. Path. *38:* 110–111 (1944).

26. RAAB, W.: The sympathetic biochemical trigger mechanism of angina pectoris. Amer. J. Cardiol. *9:* 576–590 (1962).

27. RAAB, W.: Emotional and sensory stress factors in myocardial pathology. Amer. Heart J. *72:* 538–564 (1966).

28. RAAB, W.: Correlated cardiovascular adrenergic and adrenocortical responses to sensory and mental annoyances in man. Psychosom. Med. *30:* 809–818 (1968).

29. RAAB, W.: Myocardial electrolyte derangement: crucial feature of pluricausal, so-called "coronary" heart disease (dysionic cardiopathy). Ann. N. Y. Acad. Sci. *147:* art. 17, 627–686 (1969).

30. RAAB, W.: Preventive Myocardiology (Fundamentals and Targets) (C. C Thomas, Springfield, Ill. 1970).

31. RAAB, W.; BAJUSZ, E.; KIMURA, H. and HERRLICH, H. C.: Isolation-stress, myocardial electrolytes and epinephrine cardiotoxicity in rats. Proc. Soc. Exp. Biol. Med. *127:* 142–147 (1968).

32. RAAB, W.; CHAPLIN, J. P. and BAJUSZ, E.: Myocardial necroses produced in domesticated rats and in wild rats by sensory and emotional stresses. Proc. Soc. Exp. Biol. Med. *116:* 665–669 (1964).

33. RAAB, W. and GIGEE, W.: Specific avidity of the heart muscle to absorb and store epinephrine and norepinephrine. Circulation Res. *3:* 553–558 (1955).

34. RAAB, W. and KIMURA, H.: The myocardial ionogram–A potential diagnostic tool. Amer. J. Cardiol. (in press, 1971).

35. SCHMID, E. und MEYTHALER, CHR.: Untersuchungen über die sympathicoadrenale Reaktion bei Autofahrern mit Hilfe der Vanillinmandelsäure-Bestimmung im Harn. Klin. Wschr. *42:* 139–140 (1964).

36. SELYE, H. and BAJUSZ, E.: Conditioning by corticoids for the production of cardiac lesions with noradrenaline. Acta Endocrinol. *30:* 183–187 (1959).

37. STARCICH, R.: Plasma catecholamines and urinary vanillyl mandelic acid in clinical

ischemic heart disease. In RAAB, W. Prevention of Ischemic Heart Disease, pp. 103–111 (C. C Thomas, Springfield, Ill, 1966).

38. SUZUKI, T.; ROMANOFF, E. B.; KOELLA, W. P. and LEVY, C. K.: Effect of diencephalic stimuli on 17-hydroxycorticosteroid secretion in unanesthetized dogs. Amer. J. Physiol. *198:* 1312–1314 (1960).

39. TOLSON, W. W.; MASON, J. W.; SACHAR, E. J.; HAMBURG, D. A.; HANDLON, J. H. and FISHMAN, J. R.: Urinary catecholamine responses associated with hospital admission in normal human subjects. J. Psychosom. Res. *8:* 365–372 (1965).

40. WEITZMAN, E. D.; SCHAUMBURG, H. and FISHBEIN, W.: Plasma 17-hydroxycorticosteroid levels during sleep in man. J. Clin. Endocrinol. *26:* 121–127 (1966).

41. YEN, H. C. Y.; DAY, C. A. and SIGG, E. B.: Influence of endocrine factors on development of fighting behavior in rodents. Pharmacologist *4:* 173 (1966).

42. ZUGIBE, F. T.; BELL, P.; CONLEY, T. and STANDISH, M. L.: Determination of myocardial alterations at autopsy in the absence of gross and microscopic changes. Arch. Pathol. *81:* 409–411 (1966).

St. Barnabas Hospital, New York, N. Y., and
Hahnemann Medical College and Hospital, Philadelphia, Pa., USA

ETIOLOGIC FACTORS IN ISCHEMIC HEART DISEASE: THE ELUSIVE ROLE OF EMOTIONAL STRESS [1]

H. I. RUSSEK and L. G. RUSSEK

In any disorder of multifactorial origin, such as ischemic heart disease, it is most difficult to rank in the order of etiologic importance the various factors held suspect in the afflicted individual or in the society in which he lives. Nonetheless, in earlier studies (14, 19), we reported that young coronary patients accustomed to a "typical" American diet could be differentiated from healthy controls far more readily by the dimensions of occupational emotional stress than by differences in heredity, diet, obesity, tobacco consumption or physical exercise (table I). In these studies, it was found that, at the time of their attack, 90% of 100 coronary patients (as compared with only 20% of healthy controls) had been holding down two or more jobs, working more than 60 h per week, or experiencing unusual insecurity, discontent or frustration in relation to employment. Subsequently, Buell and Breslow (2) confirmed these findings in their study of the registered mortality of men in California, which indicated that light workers under the age of 45, who are on the job more than 48 h a week, have twice the risk of mortality from coronary heart disease compared with other light workers of similar age.

Recognizing that certain occupational classes are subject to stress more than others, several years ago we compared the prevalence of ischemic heart disease in various occupational categories in which obvious differences exist with respect

Table I

Incidence of Various Factors in Coronary and Control Groups

	No.	Heredity (positive)	High-fat diet	Stress and strain (occupational)	Obesity	Tobacco (30 cigarettes, plus)	Exercise
Coronary group	100	67%	53%	91%	26%	70%	58%
Control group	100	40%	20%	20%	20%	35%	60%
Ratio		1.7:1	2.7:1	4.6:1	1.3:1	2:1	1:1

[1] Presented at Symposium on Preventive Myocardiology, sponsored by American College of Cardiology and the University of Vermont College of Medicine, June 26–28, 1970, Stowe, Vermont.

to "tensions" created by routine demands of the job (15–17). Our questionnaire survey of 25,000 professional men in twenty occupational categories clearly demonstrated a marked gradient in the distribution of coronary disease, which appeared strikingly related to the prejudged stressfulness of occupational activity (fig. 1).

At the present time, there is considerable evidence to suggest that most of the lethality of a high-fat diet in Western society may actually be dependent on the "catalytic" influence of stressful living (14, 17, 19). In cholesterol-fed rabbits, hypothalamic stimulation, like other forms of stress, was shown by Gunn *et al.* (6) to increase the atherogenicity of diet alone. Epidemiologic data obtained in man appear to correlate closely with these recorded observations in experimental animals. Thus, despite high fat intake, the relatively low incidence of ischemic heart disease reported among Benedictine monks (5), inhabitants of Roseto, Pennsylvania (20), Somali camel herdsmen (8), and East African Masai (9) may best be explained by the simplicity of life and relative freedom from serious psychologic stress common to these groups. In sharp contrast, stressful living, when the diet is high in animal fat, not only appears to accelerate atherogenesis but, through its neurogenic and hormonal influences, contributes to myocardial hypoxia and cardiac electrolyte imbalance (11, 12). As a consequence of such myocardial derangement, symptoms frequently surface to clinical view prematurely and may not necessarily mirror the degree or extent of underlying coronary vascular involvement. On the other hand, it is also equally apparent that the atherogenicity of emotional stress itself is markedly diminished, if not nullified, by subsistence on a diet low in fat, and that under such conditions the stress-induced myocardial changes alone are of little clinical import. Consideration of "immune" racial groups in such geographical areas as China, Korea, Japan, and Yemen make such conclusion inescapable (14–17, 19). Obviously,

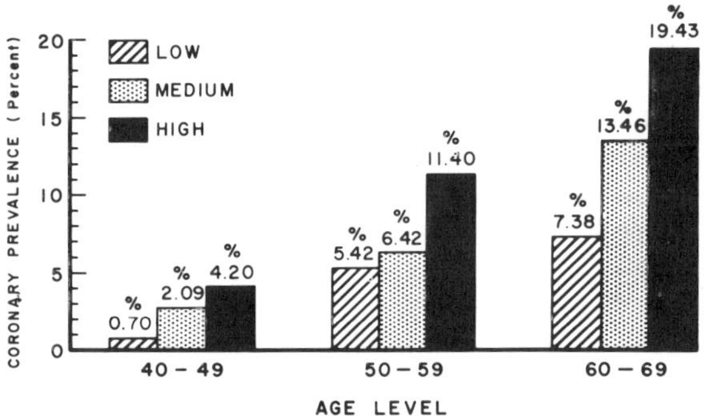

Fig. 1 Coronary prevalence by age and stress group. Classification of occupations according to stress is as follows: *low:* dermatologists, orthodontists, patent lawyers, and periodontists; *medium:* oral surgeons, other lawyers, pathologists, security analysts, and trial lawyers; *high:* anesthesiologists, general practice dentistry, general practice law, general practice medicine, and security traders.

"stressed" humans like "stressed" animals in the laboratory manifest no increased susceptibility to atherosclerosis unless the composition of the diet has been relatively high in animal fat. Thus, while both diet and stressful living appear implicated in the pathogenesis of ischemic heart disease, there is mounting evidence to suggest not only that each is dependent upon the other for pathologic significance but also that together they form a most lethal combination (15–17).

The etiologic importance of emotional stress lends support to the concept of a "coronary prone" personality. Seventy-five years ago, Osler (10) described the typical patient with coronary disease as "a keen and ambitious man, the indicator of whose engines is always set at 'full speed ahead.'" Arlow (1) observed "a compulsive striving for achievement and mastery which never seems to end." Kemple (7) found the coronary patient to be an aggressive, ambitious individual with an intense physical and emotional drive, unable to delegate authority or responsibility with ease, possessing no hobbies and concentrating all his thoughts and energy in the narrow groove of his career. Wolf (21) characterized the coronary-prone individual as one who not only meets a challenge by putting out extra effort, but who takes little satisfaction from his accomplishments. Similarly, Friedman and Rosenman (4) have stressed the prevalence of intense ambition, competitive drive, sense of urgency, and preoccupation with deadlines among persons predisposed to clinical coronary disease. In accord with these observations, we have found the most characteristic trait of the young coronary patient to be his restlessness during leisure hours and his sense of guilt during periods of "relaxation" (19). As a consequence, he rarely took vacations, and such leisure time as he did possess was frequently regimented by obligatory participation in an assortment of social, civic, or educational activities. It can be seen that such behavior characteristics of themselves tend to generate "stress" in a disproportionate fashion. Whether specific personality types contribute in this manner alone to ischemic heart disease etiology or are also associated with an underlying genetic predisposition to the disease requires further study.

The chief risk factors in coronary heart disease are commonly reported as a positive heredity, high-fat diet, hypercholesterolemia, high blood pressure, diabetes, cigarette smoking, obesity and lack of exercise. Although *stressful living* is frequently omitted from authoritative tabulations, every physician knows that nervous strain elevates blood pressure, aggravates diabetes, induces or augments hypercholesterolemia, and increases the tendency to obesity and excessive smoking (17). Psychic stress also frequently contributes to the failure to achieve daily exercise by promoting fatigue, creating a sense of time urgency, and decreasing motivation. Consequently, even such indirect effects of emotional stress, barring all others, should have long ago elevated this factor to a position of considerable significance in the complex etiologic picture of ischemic heart disease (18).

To the physician and patient alike, it is well known that emotional tension results in compulsive smoking in many persons as compensation for anxiety (17). Consequently, even the statistical relationship between smoking and coronary heart disease may reflect the importance of emotional stress rather than cigarettes *per se*. Our previous studies (17) pointed in this direction and new observations seem to lead to similar conclusion. Thus, the report that 100,000 physicians in the United States have given up smoking (13) led us to investigate possible alterations in mortality statistics in this professional group.

A random sampling was made of the age and cause of death among physicians listed in the Journal of the American Medical Association during the months of January through April, inclusive, in the years 1955, 1965, 1967, and 1970. The average age at death from all causes was determined for the respective years, in addition to the average age at death from coronary heart disease alone among those failing to survive to the age of 70 (table II). Classification of the causes and age at death was based solely on the published data. It is evident that there has been no increase in the mean age at death among physicians during the period of the past 16 years (table II). Although the mean life expectancy at birth for all persons in the United States is approximately 70 years, physicians as a group appear to fall short of this mark, particularly since deaths prior to achievement of the M.D. degree do not enter into the present calculation. These findings are consistent with the apparent predisposition of physicians to ischemic heart disease, a proneness attributable to the greater stressfulness of their way of life (15). It is also seen from the data that no favorable change has occurred in recent years in the average age of physicians who succumb to coronary attacks prior to the age of 70.

This is a disappointing observation, if confirmed, since no group in our population has practiced greater abstinence from tobacco in recent years. Moreover, in this span of time we have witnessed a significant change in the saturated-fat content of the American diet along with the introduction of effective hypocholesterolemic agents. We have seen a growing awareness of the need for exercise and normal body weight. Diabetes and hypertension have been more effectively managed with the aid of better understanding and new drugs. Long-term anticoagulant therapy following myocardial infarction has become widely employed. The coronary care unit has made significant impact on early

Table II

Deaths in U.S. Physicians (J.A.M.A.)

	Jan.–April 1955	Jan.–April 1965	Jan.–April 1967	Jan.–April 1970
Aver. age at death (yrs.)	69.6	68.9	69.6	67.2
Aver. age coronary deaths only (> 70 yrs.)	61.0	60.8	60.6	59.4
Total no. of deaths	1148	1091	1251	1013

deaths in patients hospitalized for myocardial infarction. Better diuretics have diminished the incidence of intractable heart failure. New drugs and revascularization procedures have more effectively controlled symptoms and increased productivity in persons with angina pectoris. Despite these seeming gains, there appears to have been no improvement in the mortality risk from coronary heart disease. Are these "advances" therefore inconsequential in dealing with the plague that now confronts us or have the inroads of progress been counterbalanced by the growth of etiologic influences about which we have done or can do little?

In underdeveloped areas of the world, atherosclerosis is seen as an aging process in the latter decades of life; in Western society, it arises almost as a pediatric disorder. No better proof of this need be cited than the results of studies performed on soldiers killed during the Korean War (3). It should be evident, therefore, that if dietary alteration, physical training, and environmental relaxation are effective preventive measures, they must be instituted early in life. Although the dangers inherent in the typical American diet have frequently been emphasized in relation to the growing child, the "stresses" to which he is exposed as student during 16–20 years or more in the educational institutions of this country have not been carefully examined. Quality education administered by dedicated career teachers has become a scarce commodity necessitating more and more self-instruction at home. At the same time, demanding schedules and assignments no longer permit books to be read leisurely for pleasure or profit but only superficially for "grades." As a result, hobbies in the form of art, music and athletics, recreational needs, rest and sleep are being sacrificed to "learning." In the regimented setting at school, prescribed physical training is often neither enjoyable nor relaxing and may actually discourage continuation of athletic pursuits later in life. At the same time, the once popular softball game on city streets after school has been largely eliminated by demanding homework assignments or easy access to television. Nonetheless, educators, through Science Fairs and the like, seek to create "nuclear physicists" out of youngsters who scarcely have time to be "children." Compounding the effects of these pressures, unrelenting competition for high scholastic standing and college placement makes formalized education in the United States a highly traumatic experience for most students.

Who can conceive of a group of candidates more prone to ischemic heart disease than the graduating class of any medical school in the United States–unless it is the same individuals after the completion of their internship or residency? Little wonder that Osler (10), even 75 years ago, called it "the doctor's disease". It should be evident that the mental, emotional, physical, recreational, and spiritual needs of the young are becoming increasingly unfulfilled, both in the academic and home environment, despite unprecedented affluence. The resultant stresses undoubtedly are responsible in large measure for the precocious atherosclerosis observed in our young people. Consequently, preoccupation with primary and secondary preventive measures in persons who have already reached the fourth, fifth or sixth decade of life, commendable as it may be, can have but limited effect upon the results of forces that have already left their mark over scores of years. Hygienic measures and preventive medicine must begin in childhood and continue throughout life if any real impact is to be expected upon the mortality risk from coronary heart disease.

SUMMARY

Despite the application of primary and secondary preventive measures and significant advances in the treatment of angina pectoris and myocardial infarction, it appears that the over-all morbidity and mortality from ischemic heart disease has not been appreciably altered in the past two decades in the United States. It is clearly evident that, to date, efforts have failed dismally in preventing the inexorable progression of coronary atherosclerosis and its disabling and fatal complications. Abundant evidence now exists to indicate not only that *high-fat diet* and *stressful living* form a most lethal combination but also that each is dependent upon the other for pathologic significance. It therefore appears probable that dietary alteration, physical training, and environmental relaxation could prove effective prophylactically, only if instituted early in childhood and maintained as a lifetime practice.

REFERENCES

1. ARLOW, J. A.: Identification mechanisms in coronary occlusion. Psychosomatic Med. *7:* 195 (1945).
2. BUELL, P. and BRESLOW, L.: Mortality from coronary heart disease in California men who work long hours. J. Chron. Dis. *2:* 615 (1960).
3. ENAS, W. F.: HOLMES, R. H. and BEYER, J.: Coronary artery sclerosis in American soldiers killed during the Korean War. J.A.M.A. *152:* 1090 (1953).
4. FRIEDMAN, M. and ROSENMAN, R. H.: Association of specific overt behavior pattern with blood and cardiovascular findings: blood cholesterol level, blood clotting time, incidence of arcus senilis, and clinical coronary artery disease. J.A.M.A. *169:* 1286–1296 (1959).
5. GROEN, J. J.: TIJONG, B. K.: WILLEBRANDT, A. F. and KAMMINGA, C. J.: Influence of nutrition, individuality and different forms of stress on blood cholesterol: results of an experiment of 9 months duration in 60 normal volunteers. Proc. First Int. Congress Dietetics, Voeding *10:* 1959.
6. GUNN, C. G.: FRIEDMAN, M. and BYERS, S. O.: Effect of chronic hypothalmic stimulation upon cholesterol-induced atherosclerosis in the rabbit. J. Clin. Invest. *39:* 1963–1972 (1960).
7. KEMPLE, C.: Rorschach method and psychosomatic diagnosis: personality traits of patients with rheumatic disease, hypertensive cardiovascular disease, coronary occlusion and fracture. Psychosomatic Med. *7:* 85 (1945).
8. LAPICCIRELLA, V.; LAPICCIRELLA, R.; ABBONI, F. et LIOTTA, S.: Enquete clinique, biologique et cardiographique parmi les tribus nomades de la Somalie qui se nourrissent seulement de lait. Bulletin de l'Organisation Mondiale de la Sante *27:* 681 (1962).
9. MANN, G. V.; SHAFFER, R. D. and RICH, A.: Physical fitness and immunity to heart disease in Masai. Lancet *ii:* 1308 (1965).
10. OSLER, W.: Lectures on angina pectoris and allied states. N. Y. Med. J. *4:* 224 (1896).

11. RAAB, W.: Myocardial electrolyte derangement: crucial feature of pluricausal, so-called coronary, heart disease. Ann. N. Y. Acad. Sci. *147:* 627–686 (1969).

12. RAAB, W.; VANLITH, P.; LEPESCHKIN, E., and HERRLICH, H. C.: Catecholamine-induced myocardial hypoxia in the presence of impaired coronary dilatability independent of external cardiac work. Amer. J. Cardiol. *9:* 455 (1962).

13. Report issued by the U.S. Public Health Service, 1968.

14. RUSSEK, H. I.: Role of heredity, diet and emotional stress in coronary heart disease. J.A.M.A. *171:* 503–508 (1959).

15. RUSSEK, H. I.: Emotional stress and coronary heart disease in American physicians. Amer. J. Med. Sci. *240:* 711–721 (1960).

16. RUSSEK, H. I.: Emotional stress and coronary heart disease in American physicians, dentists and lawyers. Amer. J. Med. Sci. *243:* 716–725 (1962).

17. RUSSEK, H. I.: Stress, tobacco and coronary disease in North American professional groups; survey of 12,000 men in 14 occupational groups. J.A.M.A. *192:* 189 (1965).

18. RUSSEK, H. I.: Role of emotional stress in the etiology of clinical coronary heart disease. Dis. Chest *52:* 1 (1967).

19. RUSSEK, H. I. and ZOHMAN, B. L.: Relative significance of heredity, diet and occupational stress in coronary heart disease in young adults. Amer. J. Med. Sci. *235:* 266–275 (1958).

20. STOUT, C.; MORROW, J.; BRANDT, E. N., JR. and WOLF, S.: Unusually low incidence of death from myocardial infarction; study of an Italian-American community in Pennsylvania. J.A.M.A. *188:* 845 (1964).

21. WOLF, S. G.: Cardiovascular reactions to symbolic stimuli. Circulation *18:* 287 (1958).

Mailing address: Henry I. Russek, M.D., 176 Hart Boulevard, Staten Island, New York 10301 (USA).

Division of Cardiology, Department of Medicine,
University of Pittsburgh School of Medicine,
Pittsburgh, Pennsylvania, USA

PATHOGENIC SIGNIFICANCE OF DISPROPORTION BETWEEN MYOCARDIAL OXYGEN SUPPLY AND CONSUMPTION [1,2]

J. SCHEUER[3]

The failure to deliver adequate oxygen supplies to the heart and the manner in which the heart responds to oxygen deprivation are important concerns of cardiologists today. In this brief review, some of the effects of oxygen lack on cardiac metabolism and function will be illustrated. A more extensive treatment of this subject has been published previously (3).

Oxygen is the most important substance to be continuously delivered to the heart. The heart removes approximately 75% of the oxygen that is delivered to it in each ml of arterial blood, so that the coronary sinus contains about 5 ml of oxygen/100 ml of blood. Other tissues, however, generally remove about 20–25% of the oxygen from the arterial blood, so that 15–16 ml/100 ml is the oxygen content of mixed venous blood. Therefore, there is very little capacity for extraction of additional oxygen from blood by the myocardium. The myocardial arteriovenous O_2 difference cannot increase markedly as it can in other tissues. Thus, delivery of oxygen to the heart is determined more precisely by the rate of blood flow through the organ than in other tissues, and changes in oxygen consumption are associated with parallel changes in coronary flow.

This dependence of oxygen delivery upon coronary flow has been well shown by Gregg (1) in exercising dogs. At different levels of exercise, a 4-to 5-fold increase in myocardial oxygen delivery is accomplished almost entirely by a proportionate increase in coronary flow. Oxygen extraction only accounts for about 15% of increased oxygen delivery. Thus, the main reserve capacity for the heart to increase its oxygen delivery is by an increase in coronary flow. It is this reserve that is encroached upon in coronary artery disease. When increased demands are imposed on the heart with sclerosed vessels, oxygen extraction can increase from 75% to about 90%, but this frequently does not make up for the deficit in coronary flow, and the heart becomes hypoxic.

---

[1] Presented at Symposium on Preventive Myocardiology, sponsored by American College of Cardiology and the University of Vermont College of Medicine, June 26–28, 1970, Stowe, Vermont.

[2] Supported by United States Public Health Service Research Grant HE 09727 and Grant-in-Aid 67 744 from the American Heart Association.

[3] Recipient of Research Career Development Award HE 15867 from the United States Public Health Service.

Figure 1 reviews the normal pathways for substrate utilization and energy production in the heart. The major sources of energy are fatty acids, lactate, and glucose; and since these substrates are completely catabolized in the heart to CO_2 and water, an ample supply of oxygen is needed. During the normal resting state in fasted subjects, fatty acids are the predominant source of energy. They are taken up into the myocardium, broken down to two carbon units, and oxidized in the mitochondria in the Kreb's cycle. The reduced flavin and nicotinamide dinucleotides thus formed are oxidized in the electron transport chain, with the efficient production of ATP. Lactate is consistently converted to pyruvate by the normal heart. Glucose is used in modest amounts and there is a relatively steady state of glycogen in the resting condition, although with exercise, glycogen and lipid stores are called upon as extra endogenous energy sources. There is a fine balance between oxygen and substrate supply, ATP production and high energy phosphate utilization, and all these components are in dynamic equilibrium.

Figure 2 shows the effects of hypoxia or ischemia. Oxygen deprivation causes a relative block in the electron transport chain, which causes accumlation of reduced cytochrome components, reduced adenine and flavin dinucleotides, and an inhibition of the Kreb's cycle. Fatty acids continue to be taken up across the cell wall, but instead of being broken down and oxidized they are deposited as tissue lipids. Glucose is taken up in increasing amounts and glycogen is broken down, so that the flux through the glycolytic chain is markedly increased. This increased glycolytic flux results in greater production of pyruvate, which now cannot be oxidized normally in the Kreb's cycle. Reduced NAD also accumulates because of the acceleration of glycolysis and because of shuttle reactions between the mitochondria and cytoplasm. With increased pyruvate and NADH production, the lactate and pyruvate reaction is reversed. The production

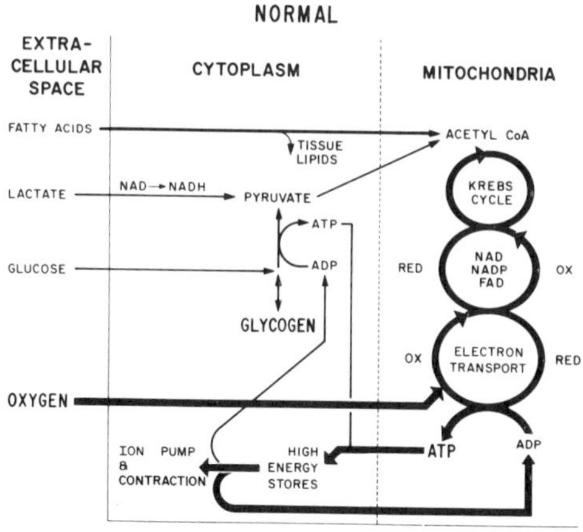

Fig. 1 Energy pathways in normal myocardium (3)

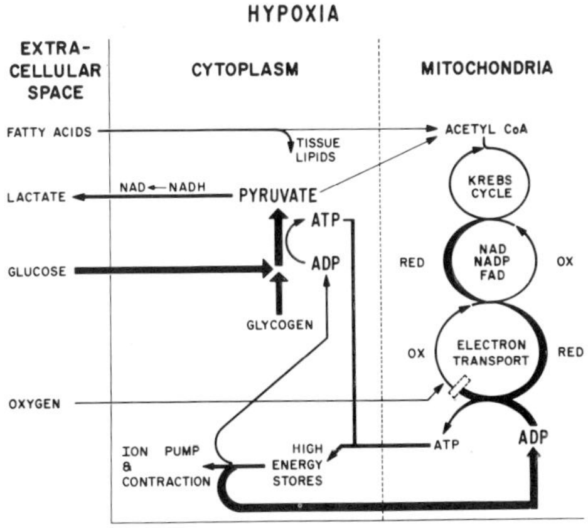

Fig. 2 Energy pathways in the hypoxic myocardium (3).

of lactate has been used clinically as a sign of myocardial hypoxia. Glycolytic production of ATP, which yields only 3 moles of ATP per mole of glycogen or 2 moles per mole of glucose cannot compensate for loss of more efficient oxidative phosphorylating mechanisms. Thus, the high energy phosphate stores fall and electrical and mechanical function become impaired.

Figure 3 shows the effects of acute hypoxia on high energy phosphate stores in paced isolated perfused hearts (5). Anoxia causes an acute decline in creatine phosphate and in ATP, which reach plateaus in about 1 min. These plateaus are maintained for the 5 min during which anoxia is continued and are due to marked compensatory increases in the glycolytic rate. Therefore, glycolytic ATP production can contribute significantly to the anoxic heart. With the abrupt re-introduction of oxygen, there is a rapid rise within seconds of the creatine

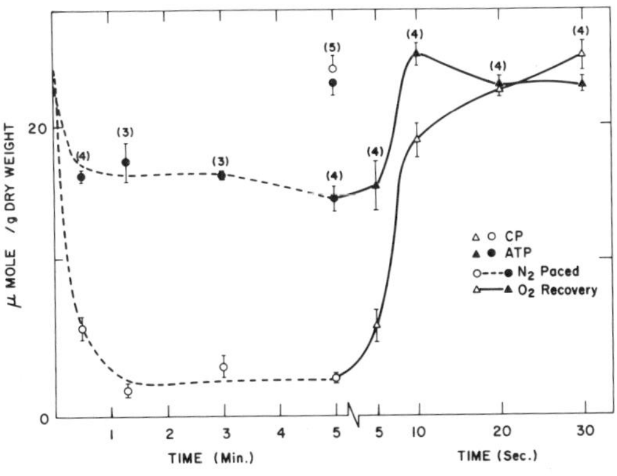

Fig. 3 Changes in myocardial ATP and creatine phosphate with anoxia (N_2 paced) and recovery. Numbers in parentheses indicate the number of hearts (means ± SE) (5).

phosphate and ATP, which are returned to normal levels in about 20–30 sec. This shows the exquisite sensitivity of the metabolism of the heart to oxygen deprivation. Accompanying such an anoxic cycle, there is a rapid rise in lactate efflux and potassium loss from the myocardium and these are maintained over the 5 min period with an abrupt return towards normal when oxygen is re-introduced.

The mechanical changes accompanying such an anoxic cycle are shown in figure 4. There is an abrupt increase in coronary flow at the onset of anoxia. This is the result of coronary arteriolar dilatation, probably through the mechanism of adenosine release. The coronary flow starts to fall during the first

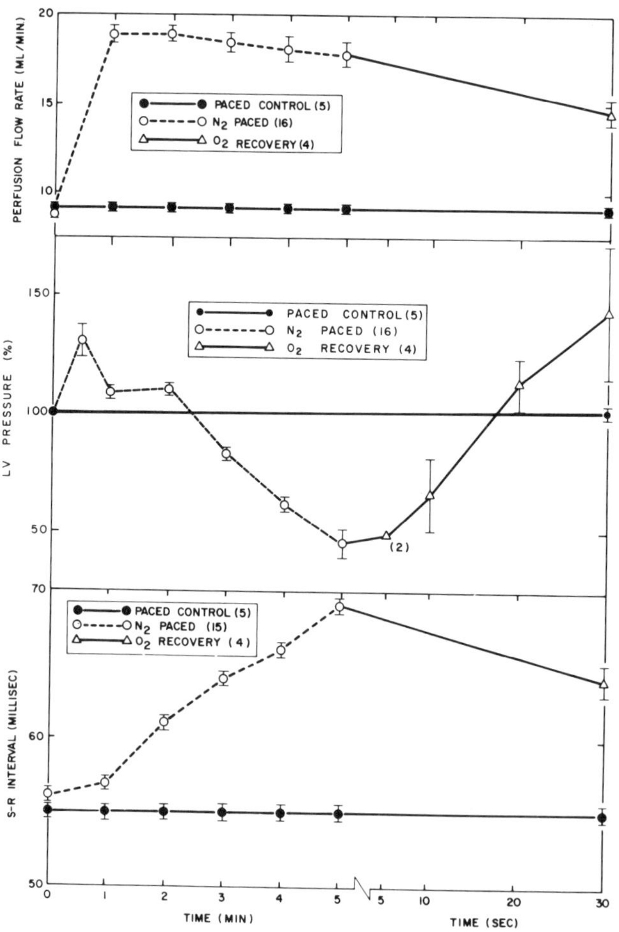

Fig. 4 Dynamic and conduction changes with anoxia (N_2 paced) followed by recovery. S-R interval is the time from the pacing stimulus to the onset of the QRS (5).

30 sec of recovery. The left ventricular pressure initially rises during the first minute and this rise is probably due to a catecholamine release because it is attenuated in reserpinized hearts. The pressure then gradually falls to very low levels, and within seconds of the re-introduction of oxygen there is a recovery of the left ventricular pressure back to the control level. This mechanical change parallels the recovery of high energy phosphates. These mechanical effects of anoxia may not only be due to declines in high energy phosphates, but other factors, such as intracellular acidosis, may be involved. Of particular importance, is the delayed conduction time through the atrium and A-V conduction system with anoxia. This becomes progressively prolonged with anoxia and these hearts actually stop if they are not paced; and if anoxia is maintained, complete atrioventricular block occurs, but the ventricle can still respond to the pacing stimulus. Thus, the cells of the conduction system may be more sensitive to anoxia than contractile cells. These rat heart experiments involve anoxia of the whole heart, whereas in patients, we are dealing with regional ischemia in which the subendocardial layers are most severely affected. We would not expect such profound mechanical and metabolic changes in patients, but the principles we show in these types of isolated heart experiments would still probably obtain.

We have been investigating ways to improve the metabolic state of the heart to better withstand hypoxia. We have raised the level of cardiac glycogen to see if this would furnish more endogenous glycolytic substrate. Figure 5 shows the mechanical effects of anoxia in hearts from reserpinized rats with high glycogen stores and control rats with normal levels of cardiac glycogen. End-diastolic pressure rises only in control hearts; the rate of decline of peak left ventricular systolic pressure and the rate of decline in maximum dp/dt are significantly slower in reserpinized hearts with high glycogen stores than in control hearts.

The reason for this is shown in figure 6. Glycogen levels in reserpinized hearts at onset of anoxia are twice as high as in control hearts. The amount of glycogen that is catabolized during a 5 min period of anoxia is much greater in reserpinized hearts and this is accompanied by a sustained production of lactate in reserpinized hearts but not in control hearts. Thus, glycolysis can continue to compensate better in hearts with high glycogen than with normal glycogen stores. ATP and creatine phosphate levels also are higher in reserpinized hearts during anoxia. We have repeated this kind of experiment by changing the glycogen levels in different ways and have shown that it is the glycogen level and not some unique aspect of reserpine that partially protects these hearts against anoxia. Again, these experiments are conducted with profound anoxia, but we suspect that in a milder hypoxic situation, such as in ischemic heart disease, smaller changes in glycogen levels would be protective.

One of the effects of conditioning on the heart is to reduce the catecholamine content, and the effects of propranolol on the heart have been likened to the

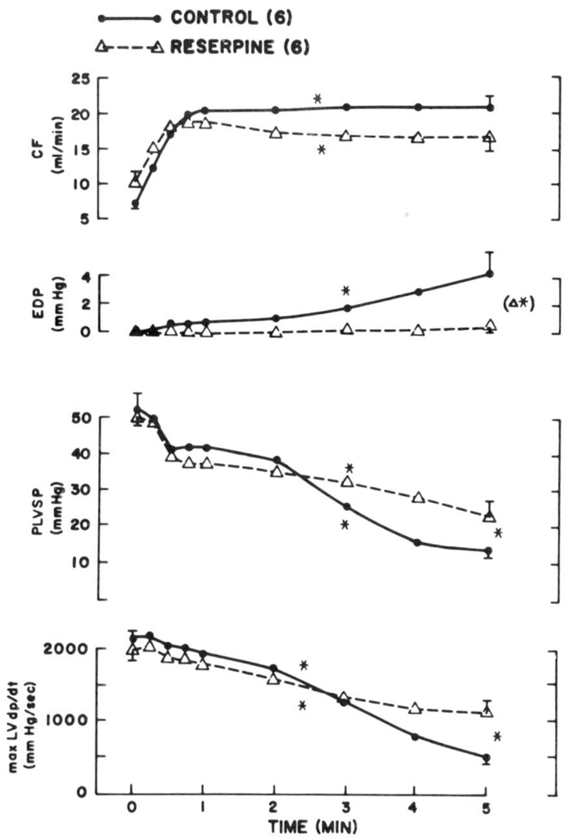

Fig. 5 Dynamic changes during a 5 min anoxic cycle in control and reserpinized hearts. Anoxia begins at time 0. EDP = end-diastolic pressure. PLVSP = peak left ventricular systolic pressure. Asterisks along the lines indicate $p < 0.05$ between the 0 and 5 min value. Asterisks at 5 min indicate $p < 0.05$ between control and reserpine hearts.

effects of conditioning. We have studied glycogen levels in rats undergoing a conditioning program by daily swimming (4). After 5 weeks of conditioning, there are significantly higher levels of glycogen in hearts of conditioned animals than in hearts of control animals. When these hearts are subjected to anoxic perfusion, higher glycogen levels remain in the hearts from conditioned rats than in those of control rats. We have not conducted mechanical experiments with hypoxia with hearts from conditioned animals, although well-oxygenated, conditioned hearts have increased work capacity and have more reserve than well-oxygenated control hearts (2). Thus, one of the beneficial effects of conditioning may be through a modest elevation of glycogen stores, which would protect against mild ischemic insults.

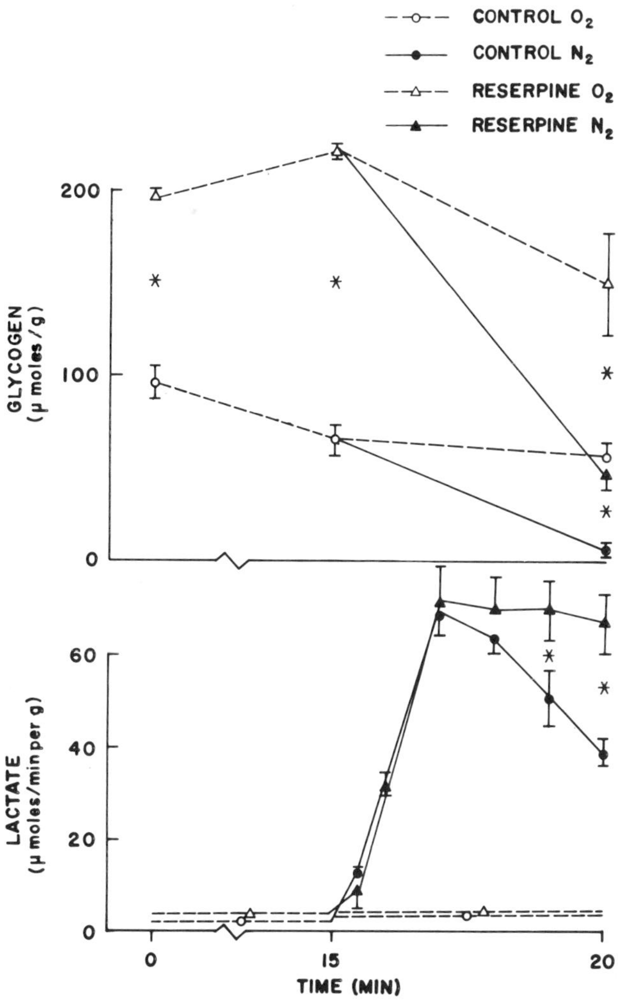

Fig. 6 Glycogen levels and lactate output of control and reserpine hearts during 15 or 20 min of aerobic perfusion (O_2) and a 5 min anoxic cycle (N_2). Results are mean ± SE/g dry weight.

SUMMARY

The changes in substrate utilization and metabolism as they relate to energy production in myocardial ischemia and hypoxia have been briefly outlined. We believe the central component in myocardial failure due to hypoxia is a lower ATP production, although there may be other important factors such as intracellular acidosis. There is a resultant diminution in contractility and, particularly, in passage of the impulse through the conducting system. Raising the glycogen levels of the heart can be protective against hypoxia, and one way to raise glycogen is by reducing the catecholamine content of the heart. Conditioning also raises myocardial glycogen. One of the beneficial effects of conditioning on the heart may be through this mechanism.

REFERENCES

1. KHOURI, E. M.; GREGG, D. E. and RAYFORD, C. R.: Effect of exercise on cardiac output, left coronary flow and myocardial metabolism in the unanesthetized dog. Circulat. Res. *17:* 427–437 (1965).
2. PENPARGKUL, S. and SCHEUER, J.: The effect of physical training upon the mechanical and metabolic performance of the rat heart. J. Clin. Invest. *49:* 18–59 (1970).
3. SCHEUER, J.: Myocardial metabolism in cardiac hypoxia. Amer. J. Cardiol. *19:* 385–392 (1967).
4. SCHEUER, J.; KAPNER, L.; STRINGFELLOW, C. A.; ARMSTRONG, C. L. and PENPARGKUL, S.: Glycogen, lipid, and high energy phosphate stores in hearts from conditioned rats. J. Lab. Clin. Med. *75:* 924–929 (1970).
5. SCHEUER, J. and STEZOSKI, S. W.: Effects of high-energy phosphate depletion and repletion of the dynamics and electrocardiograms of isolated rat hearts. Circulat. Res. *23:* 519–530 (1968).

Mailing address: James Scheuer, M.D., Division of Cardiology, Department of Medicine, University of Pittsburgh School of Medicine, Pittsburgh, Pennsylvania 15213 (USA).

Division of Medical Research, University of
Nebraska Health Service and University of Nebraska
Institute for Cell Biology, Lincoln,
Nebraska, USA

MODIFICATION OF CELLULAR LIPID METABOLISM BY PSYCHOLOGICAL STRESS [1]

K. D. ROSE, J. URSICK, and R. MACA

It was observed in 1960 that male Leghorn chickens exposed to heat and crowding stress manifested a change in their serum, in that NCTC 2445 mouse fibroblasts cultured on such serum produced, within 12–24 h, copious quantities of lipid contained within intracytoplasmic inclusions (4, 5) (fig. 1). Differential staining proved this to be largely cholesterol in nature. The *in vitro* picture so resembled the lipid streaking seen *in vivo* in early, reversible atherosclerosis (2), it was considered possible that this *in vitro* phenomenon might be the metabolic counterpart of the *in vivo* disease atherosclerosis (3). In

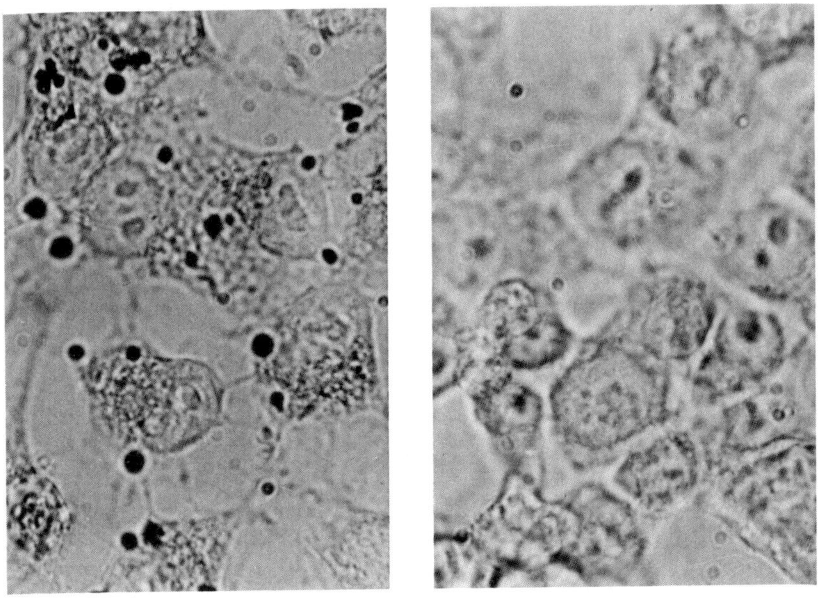

Fig. 1 Sudan III - stained cultures of low-line mouse fibroblasts grown on serum from stressed (left) and non-stressed (right) chickens.

[1] Supported by grants from the University of Nebraska Research Council, Nebraska Heart Association, and United States Public Health Service HE-08769.

view of the importance of this observation in terms of the desirability of having an *in vitro* model with which to investigate some of the basic metabolic factors surrounding the development of this desease, we set about testing the validity of this assumption. We were particularly interested in it because it apparently could be produced by psychological stress alone.

In 1960, it was necessary to show that this phenomenon was not due to cell aging, a common cause of lipid inclusions in old cultures. This was accomplished by demonstrating its occurrence in freshly transferred and logarithmically growing fibroblasts cultured on serum from stressed chickens and/or synthetic media containing unesterified cholesterol (6). These observations were supplemented by time-lapse cinematomicrography, which revealed that cells containing lipid inclusions were fully capable of undergoing the normal mitotic process. (fig. 2)

Lipid accumulation of this nature appeared to be limited to mesenchymal cells of fibroblastic nature. Repeated attempts to demonstrate it in HeLa, lung, liver, skin, and other nonmesenchymal cell strains were without success. L-strain fibroblasts possessed a weaker capability than did NCTC 2445, although both are of mouse origin. NCTC 2445, as well as its parent malignant strain, did not lose this capability in spite of several years of transfer.

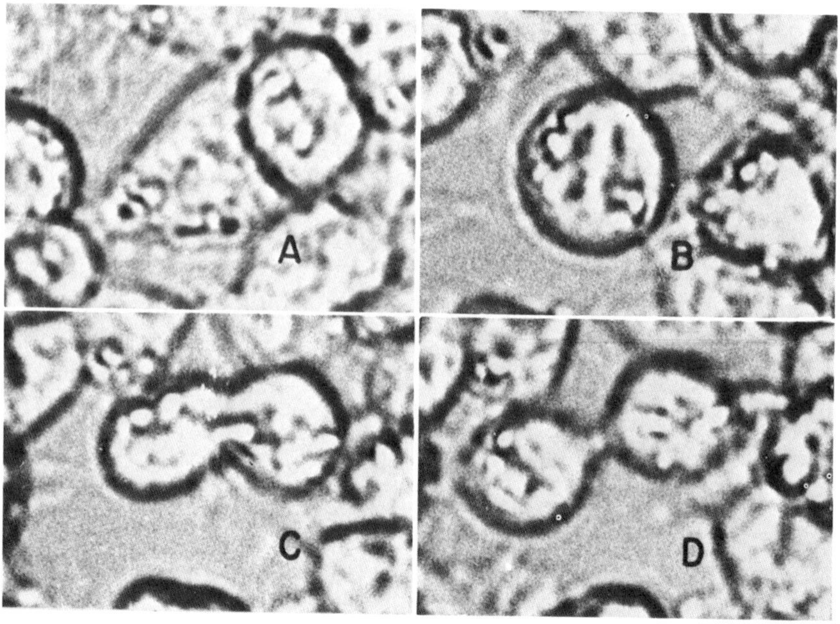

Fig. 2 Normal mitotic development of fibroblast in presence of lipid inclusions (from time-lapse movie). A. Interphase. B. Metaphase. C. Telophase. D. Interphase of daughter cells.

Freshly isolated fibroblastic cell strains were tested. Chicken aortas and chick embryo hearts were grown in organ culture. The principal explants were of fibroblastic nature and possessed strong capability for developing cholesterol-containing lipid inclusions in the presence of serum from a stressed animal (fig. 3). In order to demonstrate that the serum component capable of stimulating this phenomenon was peculiarly related in some manner or another to stress, other sera were tested. Horse serum, one of the standard serum ingredients in our stock culture media (Earle's 109 plus 10% horse serum), did not elicit lipid accumulation, except in 30% or greater concentration. This was found to equate with the relative concentration of low density beta-lipoprotein, which is particularly deficient in horse serum (as will be shown later). Human serum possessed a strong capability as did amniotic fluid. A few attempts to destress humans and reduce the capability for inclusion induction were fruitless. Destressing chickens with reserpine and phenobarbital was unsuccessful, but destressing chickens for 4 days in a specially constructed noise-free, humidity and temperature controlled incubator room was highly successful, not only in reducing the "stress factor", but also in reducing the chickens' resting serum cholesterol, suggesting that the stress factor was in some way related to serum cholesterol levels.

Serum from a stressed chicken developed a strongly positive C-reactive protein (CRP) test. CRP is a lipoprotein containing a multiplicity of enzymes

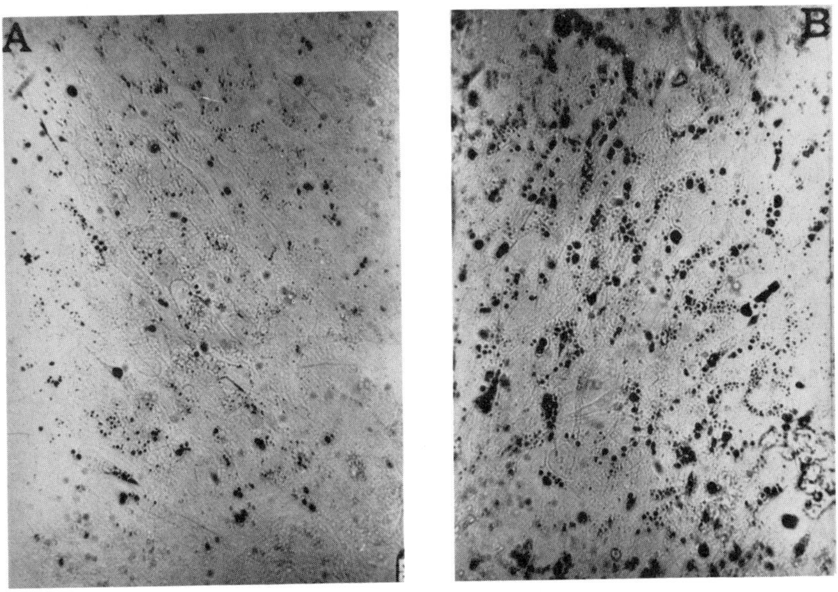

Fig. 3 Chick embryo aortic fibroblastic explants cultured on medium containing serum from (A) destressed chicken and (B) same animal subjected to heat stress.

important in lipid metabolism. Stress factor and CRP possess many properties in common, such as reliance on Ca^{++}, heat lability, lipoprotein nature, deactivation with dextran sulfate precipitation, etc. Stressed chickens' serum has been seen to develop a shift in its beta migrating serum protein, which may be related to the above CRP, which is also stress induced (fig. 4). This observation has not been pursued further. Human serum possessing a strong CRP also possesses a marked capability for inducing lipid droplet accumulation in our cells *in vitro*.

All of the known endocrine glands of the male chicken were excised and immediately extracted at 4°C in isotonic solution (Ringer's), filter-sterilized at 4°C, and applied to fresh 24 h replicate cultures of NCTC 2445 in Earle's 109 plus 10% horse serum. No intracytoplasmic droplets were observed. All of the available hormones, *i.e.*, ACTH, adrenaline, norepinephrine, growth hormone, various adrenal steroids, estrogen, testosterone, plus insulin and thyroxine, were applied to growing cultures in physiological concentrations without eliciting lipid-inclusion formation. It was observed during these experiments that 10 mg% of certain steroids elicited droplet formation. This was considered a substrate phenomenon and not related to the physiological properties of the steroid in question.

Attempts to isolate the stress factor by dialysis and cold-flash evaporation

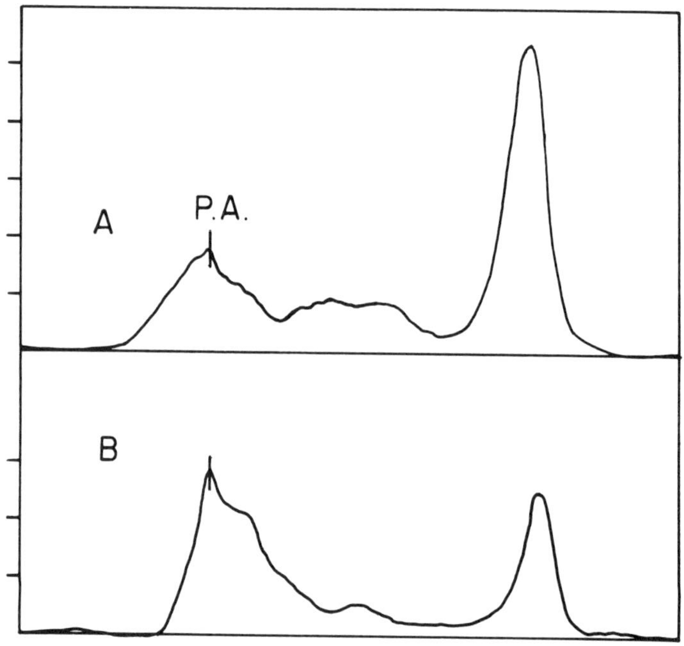

Fig. 4 Electrophoretic pattern of sera from destressed (A) and stressed (B) Leghorn cockerel. P.A. = point of application of serum.

under high vacuum revealed that dialyzed serum lost its capability to induce lipid aggregation in our cell cultures. However, the stress factor could not be recovered from the dialysate concentrated by flash evaporation. This was found to be due to removal of calcium ion by distilled water dialysis. Dialysis against Lincoln city water, which has a high calcium content, failed to remove the stress factor from stressed chicken serum. Since stability of lipoproteins in serum is dependent on ionized calcium, this furnished a clue to the identity of the stress factor.

Removal of beta lipoproteins by precipitation from serum with dextran sulfate successfully eliminated the stress factor without significantly altering the electrophoretic pattern of the serum proteins, other than by removal of all of the lipoproteins except the alpha-migrating component (fig. 5). Serum thus modified lost its lipid-droplet-inducing properties in spite of the continued presence of an abundant supply of cholesterol, presumably serum bound (fig. 6). It appeared that the elusive stress factor was associated with the beta-migrating lipoprotein (7). Beta-lipoprotein, precipitated from either stressed-chicken or human serum can be reconstituted by removal of the dextran sulfate with a slight excess of protamine sulfate. After dialyzing away the excess oxalate used in the process, stabilizing the Ca^{++} content, and millipore filter sterilizing, the beta lipoprotein added to Earle's 109 and 10% horse serum media or the Earle's 109 alone stimulated copious quantities of lipid aggregates. It was thus demonstrated that the stress factor was definitely associated with a beta-lipoprotein component of human and stressed-chicken serum.

A pilot project involving ultracentrifugation of human, horse, and stressed and nonstressed chicken serum was undertaken. The SF 0-10, 10-400, and remainder fractions of these sera were reconstituted, sterilized, and added to

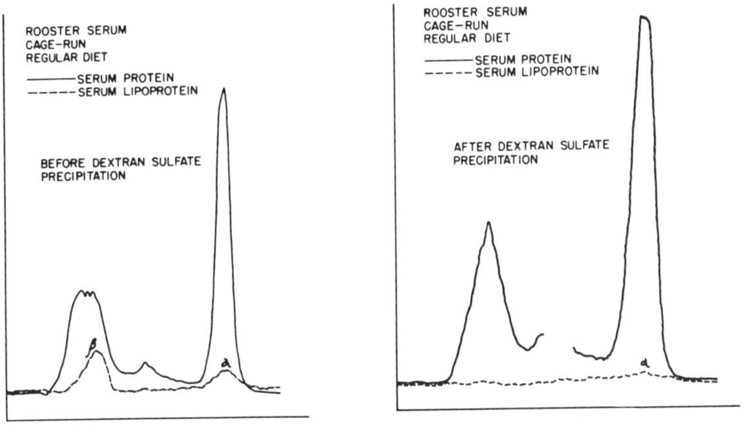

Fig. 5 Protein and lipoprotein electrophoretic patterns of sera before and after dextran sulfate precipitation.

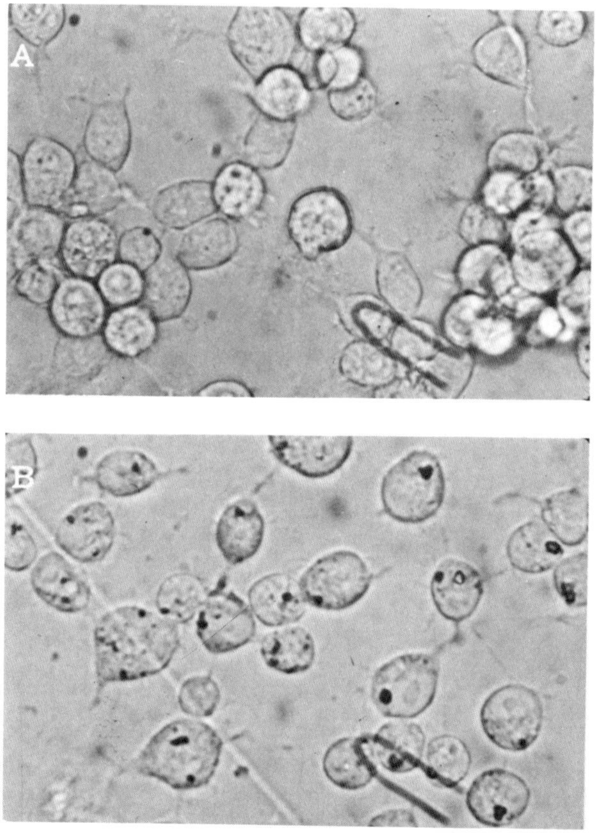

Fig. 6 Fibroblast monolayer cultures grown on media containing stressed chicken serum after (A) and before (B) dextran sulfate precipitation of β-lipoprotein. Sudan IV stain.

culture media. The SF 0-10 in human and stressed-chicken serum contained a high concentration of nonesterified cholesterol and strongly produced inclusions. The same was not true of horse and even less true of nonstressed chicken serum SF 0-10 fractions. Lipid droplet inclusions were also induced by SF 10-400 and the residuals, but in much less quantity. Figure 7 shows the results when cells are cultured for 24 h on stressed chicken SF 0-10 fraction. Thus, the stress factor appeared to reside in the SF 0-10 fraction of the serum lipoproteins, *i.e.*, in the low density fraction.

By serially diluting the various test sera, it was shown that the end-titer for droplet production was directly related to the cholesterol levels, more particularly the nonesterified forms (table I). The ethylester of cholesterol, solubilized in 4 mg% concentration by jet injection of an alcohol solution into serum or serum fractions (globulins), failed to stimulate droplet production,

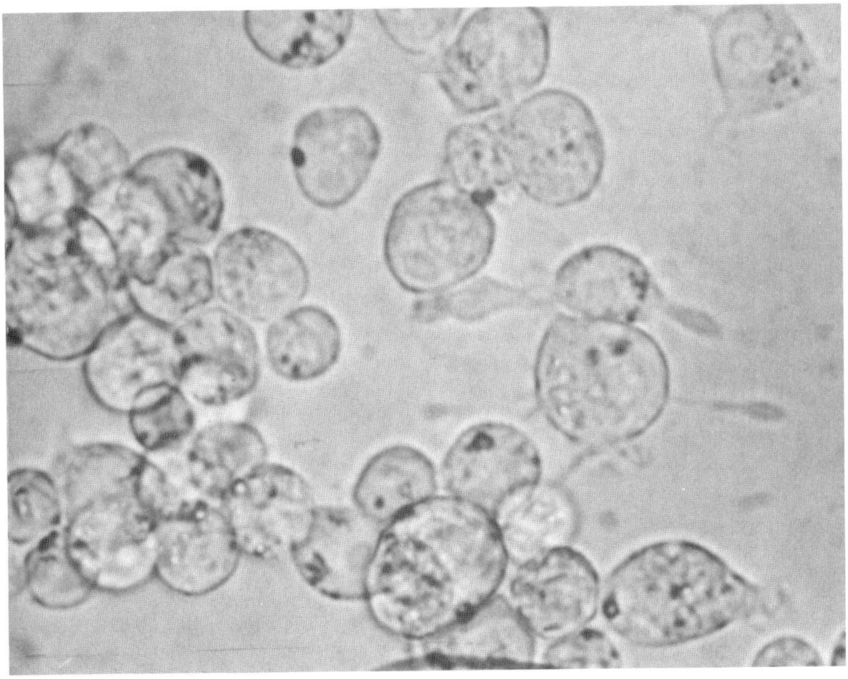

Fig. 7 Twenty-four-hour cultures of fibroblasts grown on medium containing Sf 0-10 low density β-lipoprotein from stressed chicken.

Table I

Effect of Stress and Cholesterol Feeding on Serum Cholesterol Levels and "Stress Factor" Titers

Group and no. of animals	Test condition*	Serum cholesterol: Total mg%	Esterified mg%	Esterified Percent of total	Nonesterified mg%	Nonesterified Percent of total	"Stress factor" titer average
Mature birds (9)	NSR	40.0	34.28	85.6	5.75	14.4	42.5
	SR	46.19	36.14	78.2	10.05	21.8	63.5
	NSHC	108.92	49.85	45.7	59.11	54.3	124.0
Young birds (8)	NSR	95.2	61.25	64.4	33.91	35.6	27.5
	SR	111.9	76.84	69.5	34.12	30.5	47.5
	NSHC	246.2	165.18	67.3	81.07	32.7	93.5

* NSR, Cage-run on regular diet; SR, heat stress on regular diet; NSHC, cage-run with 2% cholesterol added to regular diet.

whereas the unesterfied form did so. From these data, a hypothesis was formed relating the so-called stress factor to levels of serum-beta-lipoprotein-bound unesterified cholesterol.

Numerous experiments wherein lipid droplet aggregation was induced by feeding the cells serum- or serum-component-bound sodium stearate, sodium oleate, and nonesterified cholesterol revealed that the cell cultures consistently were able to clear their cytoplasm of fatty-acid-induced inclusions but not the cholesterol-induced inclusions. Similar inability to clear stress-induced lipid inclusions had been previously observed. This may be of some importance in the argument as to whether atherosclerosis is fatty acid (triglyceride) or cholesterol induced, if the presumption is made that shift from the reversible to the irreversible stages of this disease (Hohman's Stage I and II) (2) is caused by a foreign-body reaction within and without the cells due to the persistence of metabolically inert cholesterol.

At this point in time, there appeared to be some substance to the hypothesis that this *in vitro* observation, similar in nature as it was to the metabolic characteristics of the disease atherosclerosis, might in fact be an *in vitro* model of that disease. Of further interest was its stress induction, the apparent link being stress production of an elevated serum nonesterified, beta-lipoprotein-bound cholesterol, *i.e., chronic stress alone might account for arteriosclerosis dissociated from a high cholesterol diet.* Further studies were undertaken by GLC analysis of serum from stressed and nonstressed chickens and extracts of cells grown on media including these sera. It was found that stress induced an increase in myristic, palmitic, and oleic (with some stearic and linoleic increase) acid fractions in test sera (8) (fig. 8). Hegstead *et al.* (1) have reported that

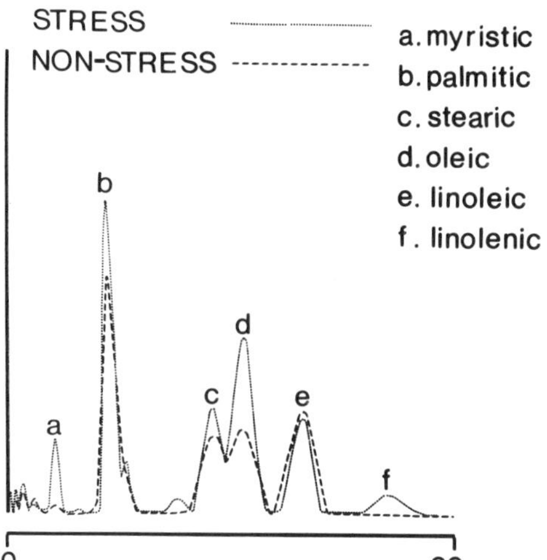

Fig. 8 GLC pattern of methyl esters of fatty acids from destressed and stressed chicken serum.

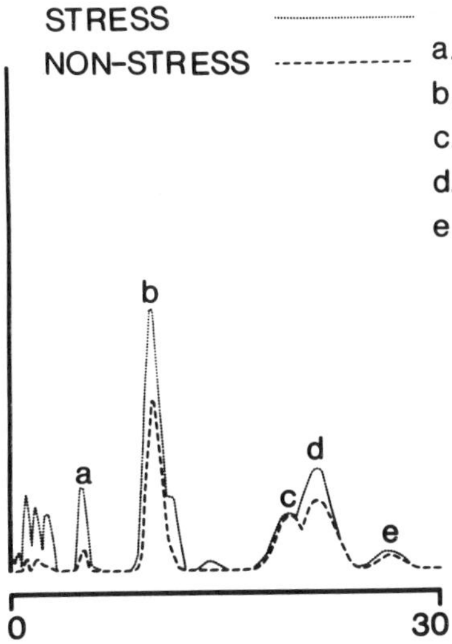

Fig. 9 GLC pattern of methyl esters of fatty acids from fibroblasts grown on serum from destressed and stressed chickens (see fig. 8).

increase in serum cholesterol in humans is significantly related *only* to a concomitant increase in myristic and palmitic acids. Cells cultured on the above sera reflected this increase in myristic, palmitic and oleic, but not stearic acids (fig. 9).

Psychological stress caused a significant increase (approximately 2 times) in serum cholesterol of Leghorn chickens existing on a fat-free diet (synthetic diet), even though their serum cholesterol was considerably lower than when they were kept on a regular diet (table II). This increase is reflected in the concentration of cholesterol per 100,000 cells in NCTC 2445 fibroblasts grown on media containing this serum. These analyses were done by GLC with cholestane as an internal standard for quantitative analysis (9) (table II).

Table II

Effect of Crowding Stress on Serum Cholesterol of Chickens Fed a Fat-Free Synthetic Diet (n = 10) and of Fibroblasts Cultured on these Sera

	Serum cholesterol (mg/%)		Cell total cholesterol (mg/10^5 Cells)*
	Total	Nonesterified	
Nonstressed	70	15	0.072
Stressed	125	35	0.218

* GLC analysis with cholestane as internal standard.

4-C^{14}-tagged nonesterified cholesterol is absorbed by and persists in cell cytoplasm as revealed by serial radioautographic studies (10) (fig. 10). 26-C^{14}-tagged, nonesterified cholesterol-treated cells lose their tag within 24–48 h after removal from the cholesterol media, although they may be stained with lipid stains (fig. 11). The loss is associated with the appearance of $C^{14}O_2$ in the

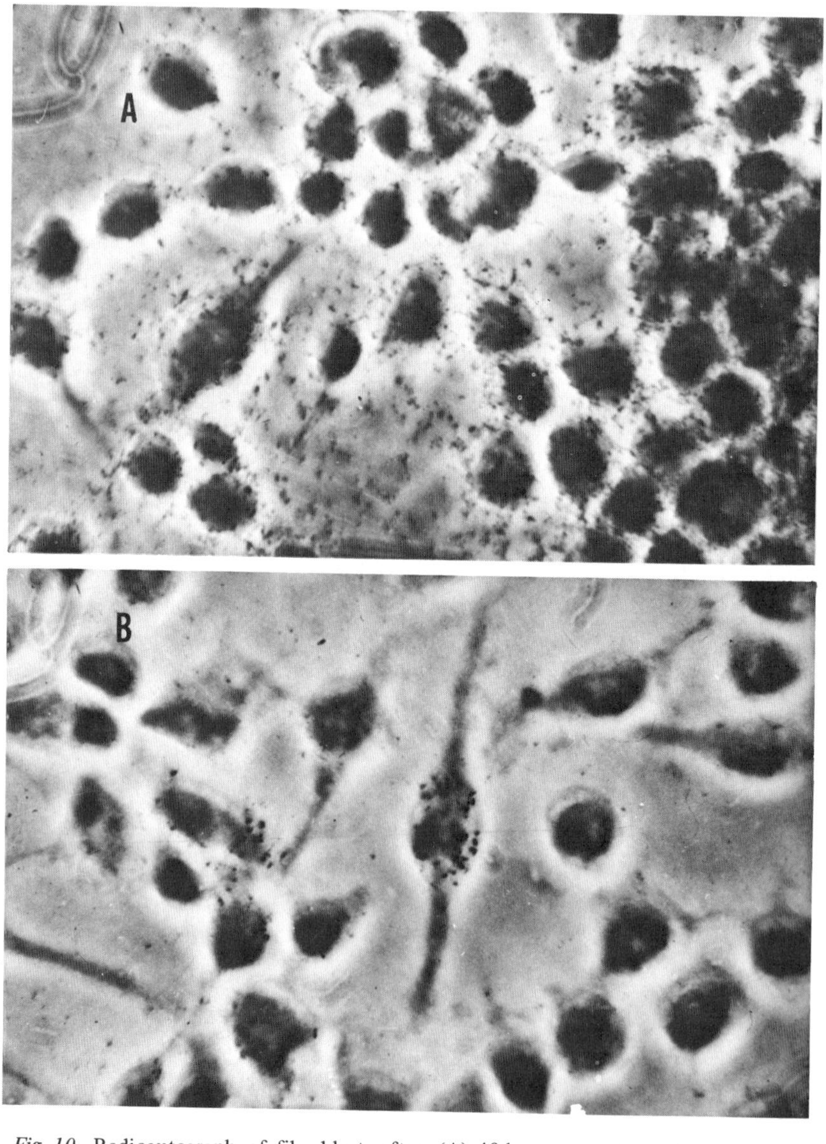

Fig. 10 Radioautograph of fibroblasts after: (A) 48 h exposure to medium containing 4-C^{14} non-esterified cholesterol; (B) after 144 h subsequent growth on standard media replaced every 48 h.

gaseous environment over the culture, an observation not made when 4-C^{14} was the substrate. This suggests that although the steroid nucleus of cholesterol 26-C^{14} persists, the side chain is degraded by the cells. To our knowledge, this was the first time cholesterol side chain degradation by cell cultures had been demonstrated (10).

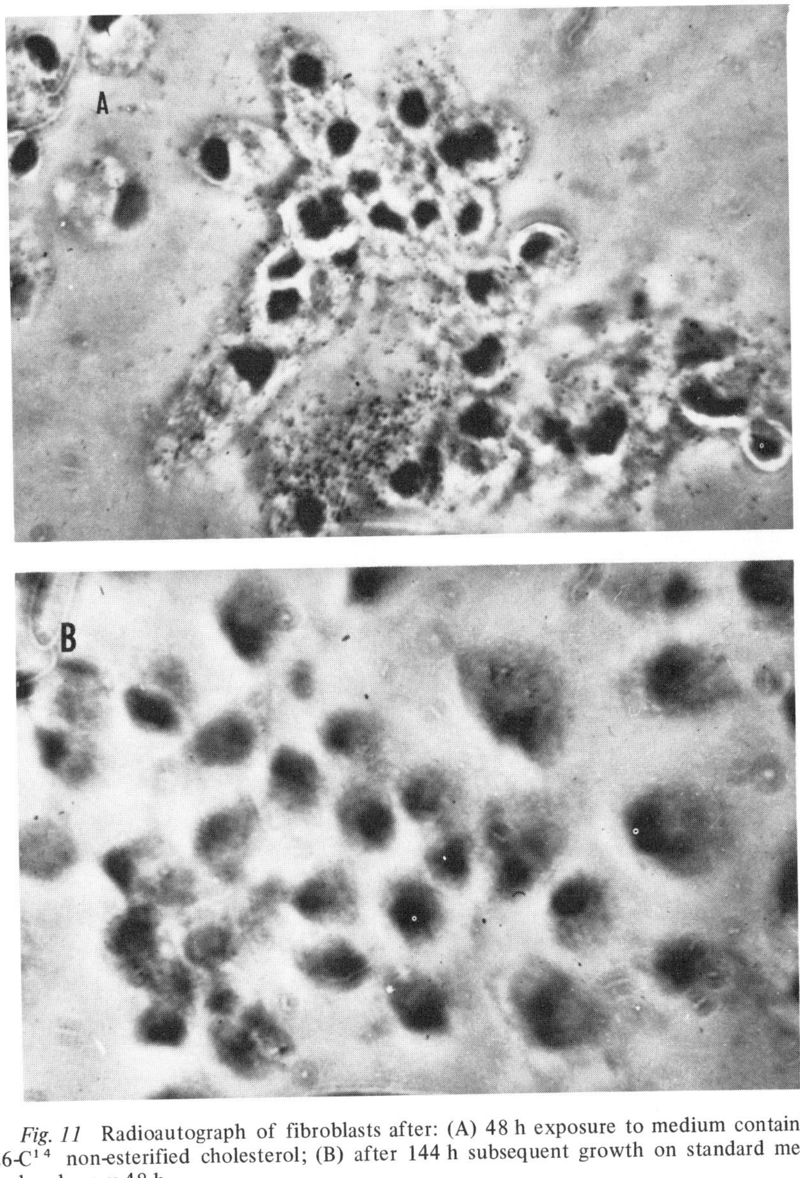

Fig. 11 Radioautograph of fibroblasts after: (A) 48 h exposure to medium containing 26-C^{14} non-esterified cholesterol; (B) after 144 h subsequent growth on standard media replaced every 48 h.

These data were confirmed by liquid scintillation analysis. With the background dpm of cells prior to inoculation taken as zero, dpm rose to 22,842 per ml of cell suspension after exposure for 48 h to cholesterol 4-C^{14}, dropped to 20,005 after 48 h exposure to a cholesterol-free medium, to 13,278 by 72 h, and to 10,622 by 120 h. Thus, a sizable retention of radioactivity existed in spite of considerable loss to the media, which is changed every 48 h, and is presumably due to cell disintegration following natural death (10). DPM of 26-C^{14}-tagged cells dropped to zero by 48 h.

An additional study was accomplished wherein time-lapse phase-contrast observation of NCTC 2445 exposed to stressed serum revealed a marked increase in pinocytotic activity, with accompanying accumulation of cellular lipid aggregates. This led, in several hours, to cell death and expulsion of the inclusions into the surrounding media. We have a time-lapse 35 mm movie demonstrating this, with an accompanying series of the same cell line cultured on nonstressed chicken serum medium, wherein such increased pinocytosis is not evident.

It was necessary, on technical grounds, to determine whether repeated venesection modified chicken serum protein. Blood samples of a like volume used in these studies were drawn at 48h intervals over 6 days and subjected to electrophoretic analysis. There was no significant change in the protein structure of these sera, particularly was there no progressive depletion of any of the major elements. Because a Leghorn chicken is easy to stress, a simple precaution was observed to prevent this during intracardiac bleeding for sample collection. A black hood placed over the chicken's head during the blood letting procedure kept the bird quiet and no significant change in the animal's serum lipids was observed.

SUMMARY

Psychological stress increases the serum cholesterol, particularly the non-esterified fraction, of experimental Leghorn cockerels, regardless of the diet the animals are eating, even a synthetic fat-free diet. Psychological stress causes an increase in the serum myristic, palmitic, and oleic acids of these experimental animals. Such changes are reflected directly in the lipid component of intracytoplasmic inclusions in NCTC 2445 fibroblasts grown in media containing these sera. Fatty acid lipid inclusions are metabolically active and are used as substrate by the cells, disappearing from the cytoplasm within 24–48 h. Cholesterol lipid inclusions are metabolically inert and persist in the cytoplasm, being lost to the cell's cytoplasm only at death, which is attended by cell rupture. We were unable to demonstrate reverse pinocytosis. The presence of some factor, possibly lipoprotein-bound cholesterol, in serum from a stressed animal, markedly enhances pinocytosis, causing the cell to become physiologically overactive, during which process it accumulates an abundance of metabolically inert steroid, leading eventually to cell death and rupture. NCTC 2445 can metabolically degrade the terminal carbon atom of 26-C^{14} tagged cholesterol, releasing the tagged carbon as $C^{14}O_2$.

These data are interpreted as supporting the hypothesis that this *in vitro* phenomenon is metabolically similar to that process in atherosclerosis leading to intracellular accumulation of metabolically inert cholesterol aggregates, which, in turn, through cell rupture and retention of extracellular cholesterol aggregates in the lining of the arterial wall, leads to irreversible changes identified histologically as Holman's Stage II atherosclerosis. This process can be induced by psychological stress.

ACKNOWLEDGMENTS

The authors wish to thank Drs. S. I. Fuenning, Donald R. Pace, and I. W. Brill for their consultation and help in promoting this research project, and members of the Poultry Husbandry Department, University of Nebraska, College of Agriculture, with special reference to Drs. John L. Adams and Thomas W. Sullivan for their advice on poultry endrocrinology and help in animal care.

REFERENCES

1. HEGSTED, D. M.; MCGANDY, R. B.; MYERS, M. L. and STARE, F. J.: Quantitative effects of dietary fat on serum cholesterol in man. Amer. J. Clin. Nutrition *17:* 281–295 (1965).
2. HOLMAN, R. L.; McGILL, H. C.; STRONG, H. P. and GEER, J. C.: The natural history of atherosclerosis. Amer. J. Path. *34:* 209–235 (1958).
3. ROSE, K. D.: The combined effect of stress and dietary cholesterol on sterol accumulation in fibroblasts *in vitro* and its possible relationship to the problem of atherogenesis. J. Amer. Coll. Health Assn. *11:* 76 (1962) (Abs).
4. ROSE, K. D.; FUENNING, S. I. and MACA, R.: Lipid formation by fibroblasts cultured on serum from stressed animals. Proc. Soc. Exp. Biol. Med. *107:* 525–529 (1961).
5. ROSE, K. D.; MACA, R. and PACE, D.: Sterol synthesis by cells cultured on serum from heat stressed chickens. Proc. Soc. Exp. Biol. Med. *108:* 282–285 (1961).
6. ROSE, K. D.; PACE, D. and MACA, R.: Lipid accumulation by actively proliferating fibroblasts cultured *in vitro* . Exerpta Medica *16.2:* #5365, 1072 (1962).
7. ROSE, K. D. and URSICK, J.: Factors affecting lipid storage *in vitro.* In Vitro *II:* 139 (1966) (Abs).
8. ROSE, K. D. and URSICK, J.: Factors affecting lipid storage *in vitro.* Proc. Nebr. Acad. Sci. (April) 1966.
9. URSICK, J.: Lipid accumulation and retention by NCTC 2445 mouse fibroblasts. Ph.D. Thesis, University of Nebraska, Lincoln, 1969.
10. URSICK, J. and Rose, K. D.: Lipid accumulation and retention by NCTC 2445 fibroblasts. In Vitro III: 193 (1968) (Abs).

Mailing address: Dr. Kenneth D. Rose, Division of Medical Research, University of Nebraska Health Service and University of Nebraska Institute for Cell Biology, Lincoln, Nebraska 68508 (USA).

Medical College of Georgia, Augusta; Hames Clinic,
Claxton; and Departments of Biology and Physical Education,
Georgia Southern College, Statesboro, Georgia, USA

PHYSICAL ACTIVITY, CATECHOLAMINES, AND PLATELET STICKINESS [1]

M. T. SIMPSON, C. G. HAMES, D. A. OLEWINE,
F. H. RAMSEY, and R. MEIER

Clinical observations in Evans County, Georgia, suggested ischemic heart disease was much less prevalent among blacks than whites (6). An epidemiological survey for the prevalence of this disease was carried out, which encompassed 92.3% of all inhabitants in the study area aged 40–74, and 50% of those aged 15–40.

Findings from this study confirmed the preponderance of ischemic heart disease in white males; also of interest was a lack of sex differential for the disease in blacks as opposed to the well-known sex differential among whites, in spite of their greater intake of animal fat and the increased hypertension in both black males and females (fig. 1).

However, when comparable groups were classifed by social class, the racial difference disappeared (fig. 2). The data suggested that the prevalence of ischemic heart disease was related to a way of life.

Next, a high coronary-prone group was compared with a low coronary-prone group for a number of different parameters (table I). The less coronary-prone

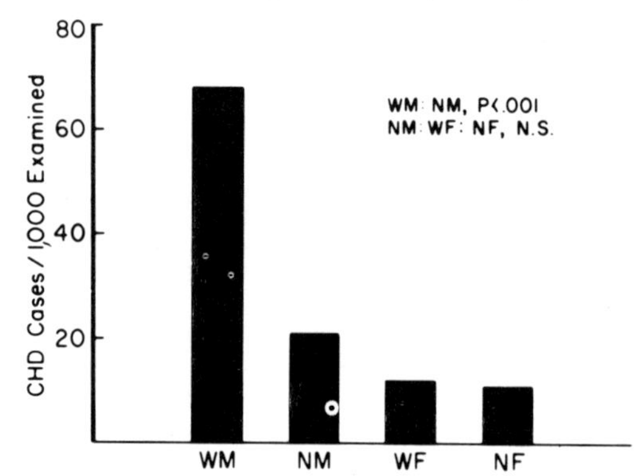

Fig. 1 CHD prevalence among white and Negro males and females aged 40–74 years. Prevalence ratios are age-adjusted.

[1] Presented at the Conference on Preventive Myocardiology of the American College of Cardiology, June 26–28, 1970, Stowe, Vermont.

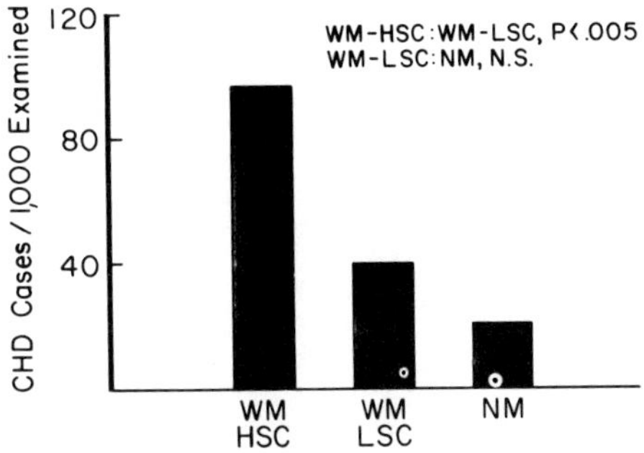

Fig. 2 CHD prevalence among high and low social class white males and Negro males aged 40–74 years. Prevalence ratios are age-adjusted.

group was found to average approximately 20 lb. less in weight, and yet consumed approximately 500 more C. per day.

After metabolic and other factors had been excluded, it appeared this finding could be explained only on the basis of increased physical activity. Prevalence for ischemic heart disease was compared by job classification (fig. 3). Physical activity histories obtained during work and leisure all confirmed that the less coronary-prone group was significantly more physically active.

Zukel *et al.* (11) showed in the North Dakota study that a small amount of physical activity of approximately 1 h per day appeared to make a dramatic difference in ischemic heart disease rates (fig. 4) (5). The data in Evans County suggested this may also be true in this area. Using the Evans County data for diagrammatic purposes, the rates of ischemic heart disease were plotted against the amount of physical activity. Surprisingly, rather than a linear relationship, a curve was produced that suggested there may be a threshold phenomenon in which a certain level of physical fitness could produce dramatic differences in

Table I

Body Size and Daily Calorie Intake among a Random Sample of High and Low Social Class White Males and Negro Males Aged 40–74 Years (Mean and Standard Deviation)

	WM-HSC	WM-LSC	NM	Significance (*p*)		
	N-33	N-34	N-23	HSC:LSC	HSC:NM	LSC:NM
Height (inches)	67.0 (2.67)	66.5 (2.67)	66.3 (2.97)	N.S.	N.S.	N.S.
Weight (pounds)	163 (32.4)	161 (35.3)	144 (26.6)	N.S.	0.02	0.05
Body surface area (sq. meters)	1.85 (0.18)	1.83 (0.21)	1.73 (0.16)	N.S.	0.02	0.05
Calories	3032 (630)	3118 (644)	3535 (803)	N.S.	0.02	0.05
Cal./M^2 body surface area	1639 (300)	1714 (359)	2046 (464)	N.S.	0.001	0.01

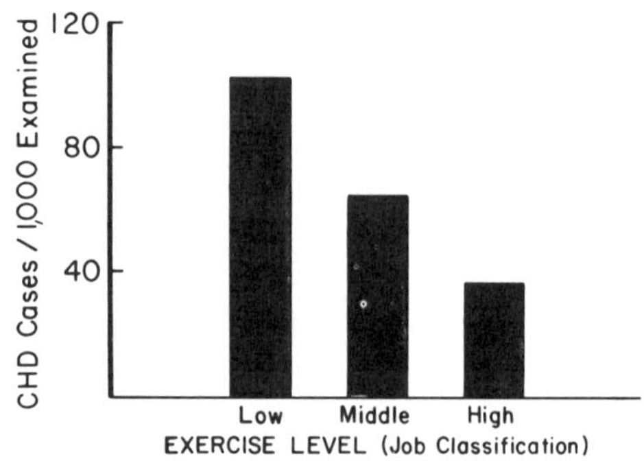

Fig. 3 CHD prevalence among white males aged 40–74 years by exercise level. Prevalence ratios are age-adjusted.

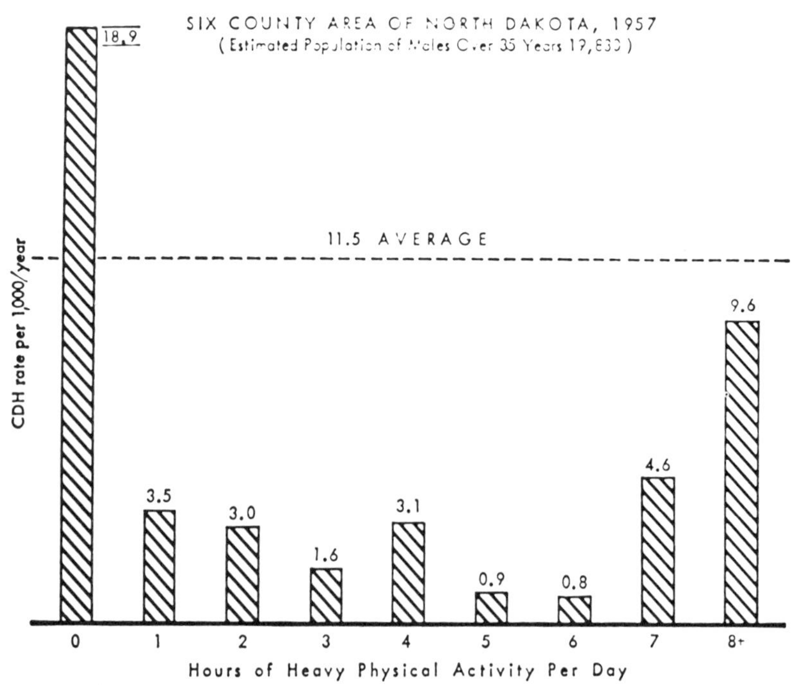

Fig. 4 Incidence of coronary heart disease by hours of heavy physical activity at usual occupation. The study is based on a 10% population sample. It does not include 3% of sample (32% CHD occurrence) on which incomplete occupational physical activity was obtained. Age distribution is considered unlikely to be responsible for lessened incidence in 1–7 h groups.

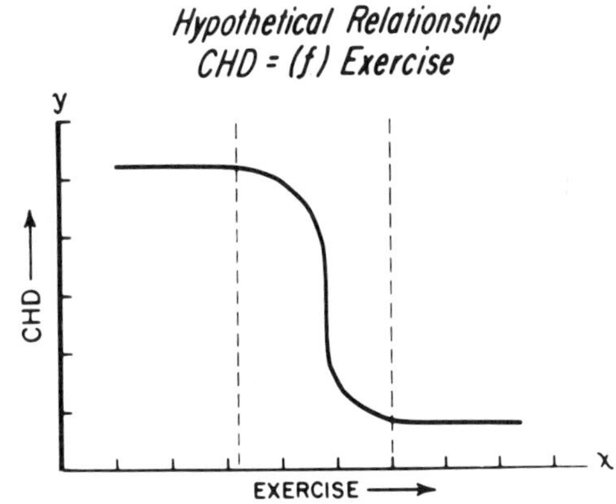

Fig. 5 Hypothetical relationship between coronary heart disease and exercise.

the incidence of ischemic heart disease (fig. 5). The Evans County and North Dakota data are in agreement as can be seen by figure 6, in which the Evans County data are represented by the broken line.

These observations stimulated a search for this critical level of physical fitness if indeed it does exist. Therefore, closely comparable population groups, yet significantly different in terms of physical fitness, would be more appropriate than the two extremes of the fitness range. Two groups of young individuals were chosen: Age: Group I, 24.4 ± 1.5 years; Group II, 24.3 ± 1.4 years. Height: Group I, 71 ± 0.9 inches; Group II, 70 ± 0.8 inches. Weight: Group I, 174 ± 8.5 lb.; Group II, 173 ± 6.8 lb. Those in the relatively fit Group I considered

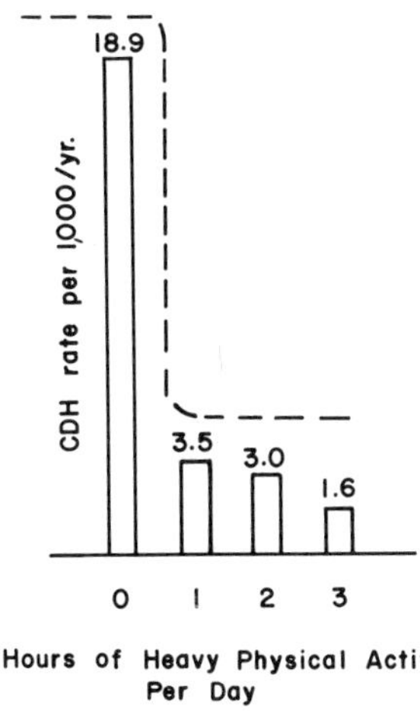

Fig. 6 Six county area of North Dakota, 1957. Studies based on a 10% population sample.

themselves to be physically active and most participated in intramural or varsity sports or some type of fitness program. Those in the relatively unfit Group II considered themselves to be sedentary, although some had participated in athletics in high school.

Studies were designed in an attempt to hopefully illustrate mechanisms that could possibly explain this apparent less ischemic as well as thrombotic vascular disease in the more physically active individuals in this population. One mechanism studied was that of platelet aggregation. The phenomenon of platelet aggregation and the release of platelet contents are generally considered important factors in the genesis of thrombosis and, as suggested by Duguid (4) and others, even in atherogenesis itself through the encrustation hypothesis.

METHODS

Test Procedure

Each subject was given at least one preliminary treadmill test in order to become familiar with the testing procedure. All were asked to conform to the following: 1) no medication of any kind during the 7 days preceding a test; 2) no strenuous activity during the afternoon and evening before a test; 3) no breakfast or coffee until after completion of the test.

All tests were conducted between 6:30 and 10:30 a.m. Following 30 min of quiet rest in a horizontal position, a resting blood sample was obtained. The Balke Test (2) was performed on a Quinton, model 24-72 treadmill at a speed of 3.0 miles/h, beginning at a grade of 2.5% (4 mets). Prior to the test, the subject walked at this grade for 3 min and then rested for 2 min. During the test, the grade was increased 2.5% every 2 min. Blood pressure and heart rate (beats/15 sec) were obtained each minute by auscultation. When the pulse rate reached 180 beats/min or the pulse pressure failed to increase following an increase in grade, the test was terminated. The second blood sample was obtained as quickly as possible after exercise; a third blood sample was taken 15 min later. This procedure was repeated for each subject at least 1 week following his first run. The more fit Group I comprised 7 subjects, to total 14 experiments, while the less fit Group II comprised 8 subjects, to total 16 experiments.

Platelet aggregation

Platelet aggregation was studied by the method of Born (1) and O'Brien (7). Freshly collected citrated platelet-rich plasma containing about 400,000 platelets per mm^3 was warmed at 37°C for 2 min before use; 2 ml were placed in a cuvette and stirred at a constant optimal speed by a magnetic stirrer. The cuvette, kept at 37°C, was placed in an EEL titrometer (a photocolorimeter) and the light transmission through the cloudy suspension was recorded in arbitrary units on a Honeywell recorder. Thus, a second by second record of the degree of dispersion of the platelets was obtained. Normally, the plasma was stirred for a sufficient period to ensure that no spontaneous aggregation would occur.

The degree of aggregation produced by the aggregating agents is influenced by the number of platelets present in the plasma—the higher the platelet count, the greater the reactivity. For each sample of plasma in this study, the platelet count and the observed degree of aggregation were adjusted by the method of O'Brien *et al.* (9) to the degree of aggregation this sample would have produced had its platelet count been exactly 400,000/mm^3. These "adjusted" measurements allow a comparison of the qualitative properties of the average platelet in plasma samples containing different platelet counts.

The line on the left on figure 7 represents the flow of light through the solution of platelets when there is no aggregation or clumping. This example shows a normal response to adrenalin, with increased clumping, increased transmission of light, and, therefore, a downward curve.

Plasma catecholamines

For determinations of plasma catecholamines for the population groups in Evans County, the plasma samples were frozen immediately for analysis by Dr. J. A. Richardson, Medical College of South Carolina, according to his method (10).

Definition of the two groups

The heart rates of the two groups of individuals upon comparison showed significant differences, both at rest and as the exercise load became progressively greater (fig. 8). The less physically fit Group II showed a significantly higher resting heart rate than the more fit Group I. The former group attained the physiological endpoint of 180 beats per min at 8.9 ± 0.3 mets, while the latter group achieved the same physiological endpoint at 11.5 ± 0.4 mets (fig. 8).

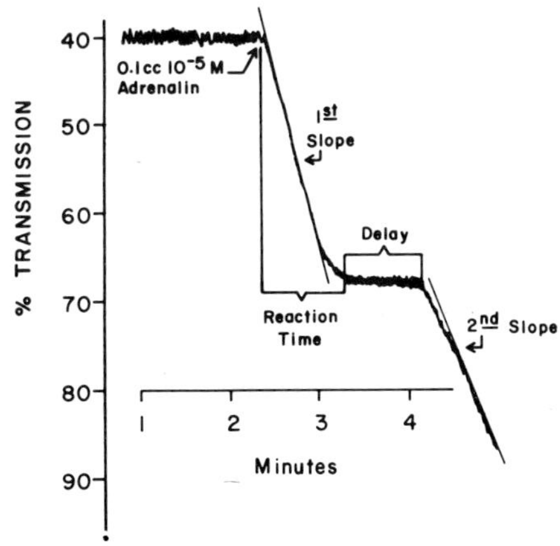

Fig. 7 Typical adrenalin-induced platelet aggregation.

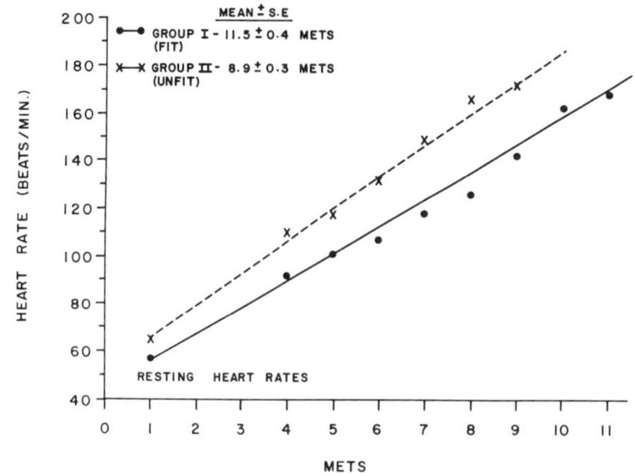

Fig. 8 Heart rates in fit and unfit groups.

The levels of fitness of Groups I and II would be classified as fair and poor, respectively, according to the criteria of Cooper (3), and were significantly different.

Another parameter separating the two groups was that of pulse pressure, the difference between the systolic and diastolic pressures (fig. 9). The more fit

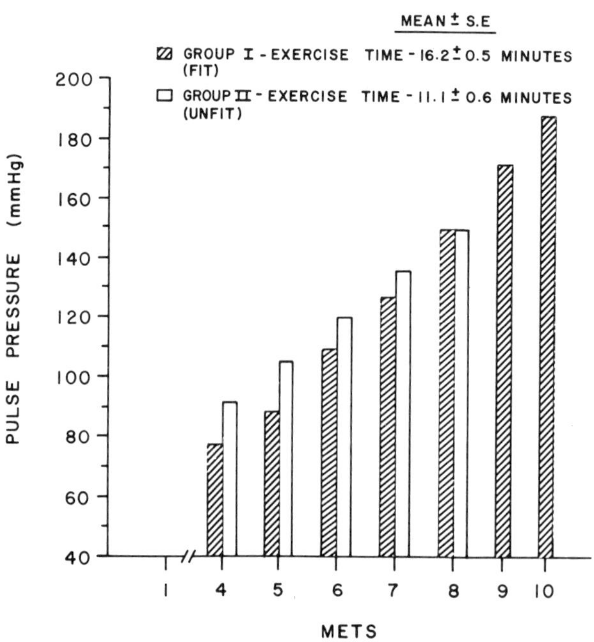

Fig. 9 Pulse pressure in fit and unfit groups.

Group I began with a resting pulse pressure of 76 mm Hg and increased to 188 mm Hg at the physiological endpoint. The less fit Group II began with a higher resting pulse pressure of 96 mm Hg and only increased to 149 mm Hg at the same physiological endpoint.

RESULTS

Platelet aggregation

Two major measurements in platelet aggregation are the reaction time (in sec) and the degree of the slope (fig. 7). Values for the reaction time are related inversely to the aggregation of the platelets, a small value for reaction time indicating a more rapid aggregation. The "adjusted" reaction times for adrenalin 10^{-5} M induced aggregation were compared for the two groups (fig. 10). To illustrate the effect of the exercise on the reaction times for the groups, the resting sample is considered as "1" on the graph and subsequent increases or decreases in reaction times are expressed as ratios of the test sample to the resting sample. As is shown by figure 10, the more fit Group I had an increased reaction time at the peak of exercise, indicating a slower aggregation, while the less fit Group II showed essentially no increase in reaction time.

The degree of the slope of the recorded aggregation is a measure of the aggregation at its maximum rate. It is a direct relationship; thus, a high figure indicates rapid aggregation. The "adjusted" slopes of adrenalin 10^{-5} M induced aggregation were compared in the two groups (fig. 11). The ratios at the peak of exercise show that the less physically fit Group II had an increase amounting to 13% over their slope at rest, while the more physically fit Group I showed a 27% decrease from their slope at rest.

DISCUSSION

Data from this study suggest: 1) Adrenalin-induced platelet aggregation was decreased in a physically fit group of individuals during submaximal physical

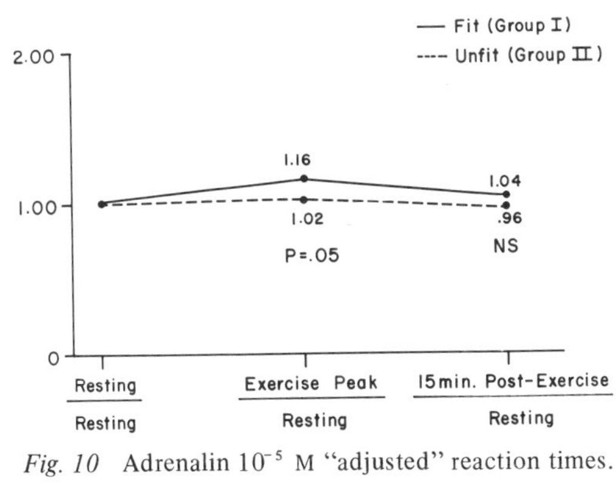

Fig. 10 Adrenalin 10^{-5} M "adjusted" reaction times.

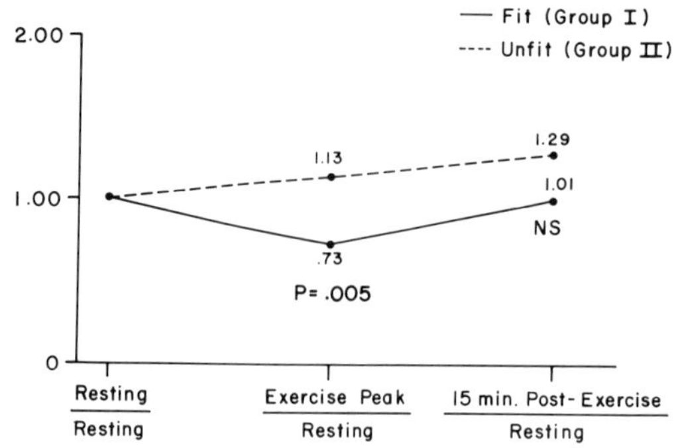

Fig. 11 Adrenalin 10^{-5} M "adjusted" first slopes.

exertion. 2) Adrenalin-induced platelet aggregation in a less physically fit group of individuals was increased as a result of the same exercise.

This was not expected by the authors, since both groups did release increased amounts of catecholamines and it was anticipated that both groups would show a proportionate change in aggregation. Treadmill studies in the Evans County coronary-prone, less physically active population group showed they produced approximately twice as much catecholamines as did the less coronary-prone, physically active group while attaining the same physiological endpoints on a treadmill (table II).

If adrenalin-induced platelet aggregation can be regarded as a valid measure of thrombotic tendency, it would appear that physical activity may mediate less thrombosis through the "fight or flight" mechanism. For example, the "fight or flight" mechanism is called into play as part of the general alarm system. A role of the protective system is to anticipate blood loss and the platelets become more "sticky."

A possible explanation for the more physically active group's platelets

Table II

Post-Exercise Plasma Catecholamine Concentration in a Sample of Adult Male Inhabitants of Evans County, Georgia, by Race and Social Class*

	WM-HSC (N = 14)	WM-LSC (N = 14)	NM (N = 6)
Norepinephrine†	5.33	3.34	2.88
	(0.77)	(0.61)	(0.44)
Epinephrine	1.04	0.68	0.54
	(0.46)	(0.16)	(0.22)

* Values are expressed in μg/l as mean and standard error.

† Difference between means are significant at $p < 0.06$ by the Mann-Whitney U test.

becoming less "sticky" at the peak of exercise was suggested by the work of O'Brien (8). It was shown that platelets incubated *in vitro* with adrenalin became adrenalin-fast, resulting in a reduction in adrenalin-induced aggregation.

With this fact in mind, a possible explanation could be that frequent and sustained exposure to adrenalin, even if in lower quantities in the more physically active individuals, may produce a state of adrenalin-fastness, whereas in the less active group, who may be exposed to the physical-activity-induced adrenalin less often, the body economy views it as a genuine threat and responds by increasing thrombotic tendency in order to conserve blood loss.

SUMMARY

In an epidemiological survey in Evans County, Georgia, the incidence of ischemic heart disease was found to be significantly higher in whites than in blacks. When comparable groups were classified by social class, the racial differences disappeared. After metabolic and other factors had been excluded, it appeared the difference could be explained only on the basis of increased physical activity. The North Dakota study, as well as others, showed that a small amount of physical activity each day appeared to make a dramatic difference in ischemic heart disease rates. Two closely comparable population groups, yet significantly different in terms of physical fitness, were exercised on a treadmill. A comparison of the effect of such exercise on adrenalin-induced platelet aggregation showed significant differences in the two groups. The more physically active group showed a decrease in aggregation at the peak of exercise, while the less physically active group showed an increased aggregation with exercise. It is concluded that physical fitness in the range of 11–12 mets may decrease the tendency to thromboembolic vascular diseases. Further studies are needed to confirm or deny the hypothesis generated by these findings.

REFERENCES

1. BORN, G. V. R.: Quantitative investigations into the aggregation of blood platelets. J. Physiol. *162:* 67P–68P (1962).
2. CONSOLAZIO, C. F.; JOHNSON, R. E. and PECORA, L. J.: Physiological Measurements of Metabolic Functions in Man (McGraw-Hill 1963). pp. 371–374.
3. COOPER, K. H.: Quantifying physical activity–how and why? J. South Carolina Med. Assoc. *65* (Suppl. 1 to 12): 37–40 (1969).
4. DUGUID, J. B.: Thrombosis as a factor in the pathogenesis of coronary atherosclerosis. J. Path. Bact. *58:* 207–212 (1946).
5. FOX, S. M. and HASKELL, W. L.: Physical activity and health maintenance. J. Rehab. *32:* 89–92 (1966).
6. HAMES, C. G.; McDONOUGH, J. R.; STULB, S. C. and GARRISON, G. E.: Physical activity and ischemic heart disease among Negroes and whites in Evans County, Georgia. In: W. RAAB, Prevention of Ischemic Heart Disease: Principles and Practice pp. 244–253 (Charles C Thomas, Springfield, Ill. 1966).
7. O'BRIEN, J. R.: Platelet aggregation: Part II, Some results from a new method of study. J. Clin. Path. *15:* 452–455 (1962).
8. O'BRIEN, J. R.: A comparison of platelet aggregation produced by seven compounds and a comparison of their inhibitors. J. Clin. Path. *17:* 275–281 (1964).
9. O'BRIEN, J. R.; HEYWOOD, J. B. and HEADY, J. A.: The quantitation of platelet aggregation induced by four compounds: A study in relation to myocardial infarction. Throm. Diath. Haem. *16:* 752–767 (1966).

10. RICHARDSON, J. A.; RICHARDSON, A. K. and BRODIE, O. J.: Fluorimetric determination of epinephrine and norepinephrine in plasma. J. of Lab. Clin Med. *47* (5): 832–834 (1956).

11. ZUKEL, W. J.; LEWIS, R. H.; ENTERLINE, P. E.; PAINTER, R. C.; RALSTON, L. S.; FAWCETT, R. M.; MEREDITH, A. P. and PETERSEN, B.: A short term community study of the epidemiology of coronary heart disease. Amer. J. Public Health *49:* 1630–1639 (1959).

Mailing address: Curtis G. Hames, M.D., 2 N. Newton Street, Claxton, Georgia 30417 (USA).

Department of Medicine, Cardiovascular Research Unit and
Department of Community Medicine, University of Vermont
College of Medicine, Burlington, Vermont, USA

EFFECT OF PHYSICAL TRAINING ON THE EXERCISE ELECTROCARDIOGRAM [1,2]

E. LEPESCHKIN and O. A. BRUSIS

During the last twenty years it has been amply demonstrated that a systematic program of physical reconditioning in patients with angina pectoris can lead to a considerable improvement of the anginal symptoms during work and at rest, often allowing resumption of normal activities (8, 19, 21a, 22). We also know that persons who engage in physical work or intensive sports in general show an "ischemic" S-T depression during or after exercise less commonly than sedentary persons of the same age (6, 13, 14). However, studies that would demonstrate an improvement of this S-T depression in a given person during the course of physical training have not been available until recently.

Blomqvist (5) found in 5 normal young men that after a period of bed rest the S-T segment was less elevated at all heart rates during a given work load, while after a period of intensive training it was more elevated and the T wave was taller. Holmgren *et al.* (11) found that the slight S-T depression and inversion of T that was present at rest but accentuated on standing up or after exercise, in 18 persons with vasoregulatory asthenia and apparently normal hearts, disappeared completely in 17 after a physical exercise program lasting 2–13 months. Smith and Kidera (26) observed this also in 49 of 50 persons with "nonspecific" T wave abnormalities but no cardiac complaints. Bruce *et al.* (7) reported on 7 men, 41–69 years old, who were exercised on a treadmill at 64–86% their maximal oxygen intake for 20 min, 3 times weekly, for 2 1/2 months. Among the 7 subjects, 5 showed an increase in their maximal work capacity; the S-T response for a given load remained the same in 4, but was considerably improved in the person showing the greatest work improvement. Decreased work capacity was seen in 1 person, as was an accentuation of S-T depression. In later studies (6), 8 men, 41–66 years old, were subjected to the same kind of training except that the uphill walking period was interrupted every 4 min by 1 min more level walking ("interval training"). At the same submaximal load, the average S-T depression was 0.17 mV at a heart rate of 140 before training and 0.10 mV at a heart rate of 125 after training; the maximal

[1] Presented at the Seminar on Preventive Myocardiology of the American College of Cardiology, June 26–28, 1970, Stowe, Vermont.

[2] Supported by Grant HE-01486 and a Research Career Award 5-K5-HE-440 to Dr. Lepeschkin from the National Institutes of Health, United States Public Health Service.

load increased about 10% after training, but both the heart rate and the S-T depression at this load remained the same (175 and 0.21 mV). Barry *et al.* (2) subjected 6 patients who had suffered an acute myocardial infarction 5–13 months earlier to training on the bicycle ergometer at loads that raised the heart rate to at least 130 but did not cause significant ECG abnormalities; training lasted for 2–13 months. Physical work capacity was increased and the S-T displacement during work was decreased in 5 subjects; but in some cases, this did not occur until after 13 months, while in others there were temporary relapses coincident with episodes of anxiety resulting from environmental or emotional stresses. In 1 subject, there was no improvement in work capacity and no consistent trend in the S-T response. In general, improvement of the S-T response coincided with decrease of oxygen consumption, blood lactate, and pulse rate during exercise as well as of the oxygen debt.

Studies on larger groups are few and partly conflicting. Blohmke *et al.* (4) studied the ECG 2 min after termination of 6 min work on the bicycle ergometer at a load sufficient to elevate the heart rate to 145, in 130 men with a variety of complaints and an average age of 35. After a 1 month period of intensive physical reconditioning consisting of gymnastics, hydrotherapy, running and hiking, a test at the same level of work resulted in an "ischemic" S-T depression (criteria of Lepeschkin (16, 17)) in 16% of the subjects, whereas before reconditioning only 12% showed such a response. At rest, these changes were present in 0.8% before and 8% after reconditioning. The authors conclude that the reconditioning may have been too vigorous for some subjects and advise more individualized supervision during the process. However, in a study of the hemodynamic response to exercise of the same 130 subjects (3), mention is made that in about 1/3 of these subjects the work load after reconditioning was adjusted to give the same heart rate as before reconditioning; since in the other 2/3 the heart rate response to the same load was considerably decreased, it follows that in the first 1/3 the work load after reconditioning was greater than before, and this alone would account for the increased incidence of significant S-T depression during work in the total.

Hellerstein *et al.* (10) found that, in a group of 49 patients (average age 49) with arteriosclerotic heart disease, the incidence of "ischemic" S-T depression during exercise on the bicycle ergometer, which resulted in a heart rate of 150, was 46% before a period of reconditioning. After an average of 17 months of individualized reconditioning, the incidence of this type of S-T depression at the same work load was only 16% in subjects who showed good cooperation with the reconditioning program and 40% in those showing poor cooperation. A later detailed analysis (23, 24) of the same program, which now included 58 normal men and 100 patients (66 of these with healed infarction), with an average duration of reconditioning of 33 months, disclosed that 31% of the patients and 3% of the normals had shown improvement in S-T response to exercise, 41% and

88% respectively showed no change, while in 28% and 15% the response became more abnormal. In subjects showing good cooperation with the reconditioning program, these percentages were (normals in parentheses): 37% (5%), 40% (83%), and 23% (12%); while in those showing poor cooperation they were: 14% (0), 50% (70%), and 36% (30%). Among the subjects with a normal initial ECG response, 39% (11%) of those who showed increased "fitness" (that is, a decrease in heart work, measured as the product of systolic blood pressure and heart rate, for the same amount of exercise) developed an abnormal S-T response and 61% (89%) showed no change. Of the subjects with an abnormal initial response, 79% of those showing an increase in "fitness" developed an improved response and 21% showed no change, while in subjects showing no change or decrease in "fitness," improvement of the S-T response was seen in 31%, no change in 56%, and deterioration in 13%. Smith and Kidera (25) subjected 17 patients with angina to exercise training and saw, in 9 of these, improvement of pain after 6 months but improvement of the S-T response to exercise only after 9–12 months; 6 showed improvement of pain but not of S-T response, and 2 an adverse effect on both.

Figure 1 illustrates the behavior of the exercise ECG during the course of a carefully supervised physical reconditioning program in a 43-year-old professor with typical angina pectoris, who had suffered an anterior myocardial infarction with uneventful recovery several years previously, but still complained of angina after exercise or emotional upsets. The resting ECG had returned to within normal limits except for a slow progression of the R wave amplitude from V_2 to V_3 and a notch in the descending branch of the R wave in lead V_4. During the first double Master exercise test (3 min of exercise), taken 20 days before the start of the reconditioning program, the patient experienced severe pain, accompanied by multiple ventricular extrasystoles, 1 min 34 sec after beginning the exercise test, and had to interrupt the test after 1 min 50 sec. The S-T segment at that time (fig. 1A) showed severe depression (2.0 mm in V_6) immediately after exercise, persisting through the 4th min of rest, while T wave inversion was still present in V_4 after 6 and in V_6 after 8 min.

When the Master test was given after 50 days of jogging and gymnasium exercises, the patient did not start to feel pain until 1 min 45 sec and could carry out the full double Master test. Because of the greater work, the heart rate immediately after exercise was higher (161) and S-T depression in V_6 deeper (3.2 mm); T wave inversion was deeper in V_6 and present also in V_4 after 8 min (fig. 1B). However, only one ventricular extrasystole was registered after the test. After 5 months of almost daily exercise on the bicycle ergometer with a gradually increasing work load, the same amount of exercise now resulted in a slower heart rate immediately after exercise (130) and smaller S-T depression (2.5 mm in lead V_6) (fig. 1C). After 10 months of training (fig. 1D), the heart rate immediately after the same exercise showed a further decrease (127), while

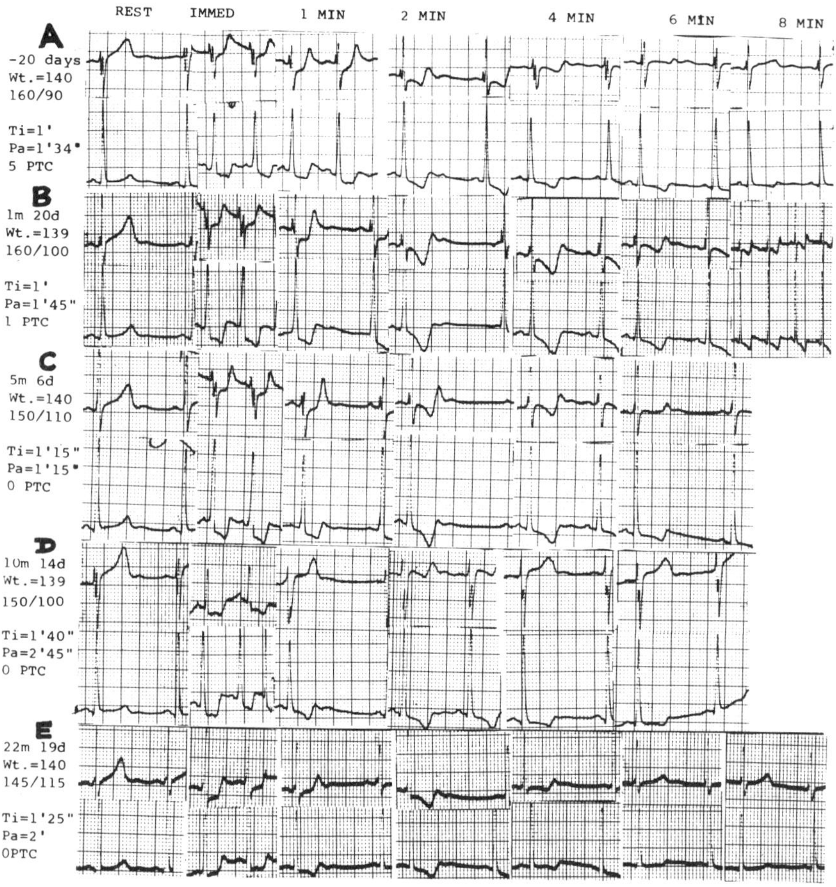

Fig. 1 Leads V_4 and V_6 of the ECG of the patient described in the text, at rest and at specified intervals after the double Master two-step exercise test. *A:* 20 days before start of physical training; because of pain, the test was terminated at 1 min 50 sec. Wt. = weight of patient; 160/90 = resting blood pressure; Ti = time from beginning of exercise at which the patient felt precordial tightness; Pa = time of appearance of precordial pain; PTC = number of premature contractions registered in the post-exercise ECG. *B:* 1 month 20 days after start of training. In this and subsequent tests, exercise was carried out for the full 3 min. Other symbols as in A. The 8 min tracing was registered at 10 mm per sec. *C* and *D:* Symbols as in B.

the S-T depression was even less pronounced (2.0 mm). Furthermore, the time of appearance of precordial tightness as well as of pain was considerably delayed. The T wave became upright in V_4 as early as 4 min after exercise. Finally, after 22 months of training, the heart rate immediately after exercise was slowest of all (120), while S-T depression in V_6 had decreased even further to 1.8 mm and T became normal in V_4 in less than 8 min. During the course of training, there was also a progressive decrease in the intensity and number of spontaneous anginal attacks and a decrease in the nitroglycerine consumption.

DISCUSSION

One possible reason for the favorable effect of physical training on the ECG response to exercise is that it can lead to a decrease of the oxygen consumption needed to carry out a given amount of external work. This can be as much as 20%, but usually does not exceed 10% (12, 18); it can result from an improvement in muscular coordination that leads to decreased use of antagonistic muscle groups and better relaxation of muscles not involved in the particular task, and is greater with the more complicated tasks. In the case of figure 1, this factor could not have played any appreciable role, since training activities were different from those used in the ECG test. A small reduction of oxygen uptake was seen in some studies (2) but it did not parallel the ECG changes, and was completely absent in other studies showing marked improvement of the S-T segment response (6).

The main reason for the improvement of the ECG response after training is definitely the reduction of the elevation of heart rate, blood pressure, and cardiac output for a given amount of external work that accompanies physical training (3, 5, 8, 13, 18, 20–22). Since the S-T segment depression as well as cardiac pain are a sign of a disporportion between the oxygen consumption of the myocardium and the oxygen supply through the coronaries (15, 16, 21, 22), a reduction of cardiac work would result in improvement, even without any change in the coronary circulation. This improvement was actually found related to an improvement of cardiac "fitness," measured as a decrease of the product of systolic blood pressure and heart rate for a given amount of external work (3, 6, 7, 10, 23, 24). The increase of blood pressure and cardiac output that is necessary to supply an increased amount of blood to the working muscles is mediated through stimulation of the central chemoreceptors by an increase in lactic acid and other muscle metabolites in the blood and through direct sensory impulses from the muscles, caused by accumulation of these products in the muscles (12, 18, 22) Both mechanisms lead to a direct nervous sympathetic effect on the heart and circulation as well as to secretion of epinephrine, which reaches the heart through the blood stream. Training leads to an increase in the mass and the capillary supply of the muscles involved in exercise and should decrease the anoxybiotic work component of these muscles for a given amount of external work, and the production of metabolites responsible for the sympathetic effect on the heart and circulation. There is also a reduction of the sensitivity of the sympathetic center (22), possibly because of increased content in enzymes that destroy catecholamines (21). The effect of training is therefore similar to that of beta-adrenolytic drugs (9). In the case of figure 1, the improvement in the ECG response to exercise was accompanied by a definite reduction of the heart rate immediately after exercise and a marked reduction of the systolic blood pressure during exercise.

In the large volume of material of Hellerstein and his co-workers (10, 23, 24), the reduction in the S-T depression during exercise in patients with arteriosclerotic heart disease after physical training showed a much better correlation to the reduction of the heart rate response than to the reduction of the total heart work (increased "fitness"). One of the possible causes for this peculiarity is the fact that the subendocardial layers of the ventricle, which are responsible for the S-T depression, are supplied with blood predominantly in diastole because of the high systolic intramyocardial pressure in these layers (15–17), and reduction of tachycardia would lead to a relative increase in the duration of the diastolic pause (21). Another factor could be that the sympathetic effect on the heart is better related to tachycardia than to the blood pressure, and is responsible for a greater increase of oxygen consumption than would correspond to its effect on the work of the heart (21, 22). One of the causes of this excess effect is the augmentation in the velocity of myocardial contraction and systolic ejection, which increases the energy loss due to blood turbulence; a specific metabolic effect on oxygen consumption is also probable (21, 22).

Another major effect of training on the abnormal S-T displacement during exercise in coronary artery disease may consist of increased formation of collateral circulation to myocardial regions supplied by stenotic blood vessels. This was actually observed in animal experiments (8a) and in some clinical cases by means of coronary arteriography (10). Such a mechanism can be postulated in the 14% of the patients with no change or deterioration of "fitness" who showed an improved S-T segment response (23, 24). The decrease of S-T segment shift caused by hypoxia or pitressin injection in experimental animals after physical training (14) can also be attributed to this effect. Another, at present only hypothetical mechanism for S-T improvement is related to the increase in myocardial potassium after training, which has actually been observed in animals (2a) and may be caused by elevation of serum potassium during exercise, combined with an anabolic effect of small concentrations of epinephrine (22). Since the decrease in membrane potential of subendocardial myocardial cells that is responsible for S-T depression in myocardial ischemia depends on a reduction of the transmembrane potassium gradient (16), the increase in potassium content would require a greater degree of ischemia for production of a given amount of S-T depression.

The effect of training in decreasing the incidence of premature beats after exercise, seen in figure 1, was also observed in other studies (20).

As mentioned earlier, physical training caused an increased S-T segment shift after exercise in 28 of 100 patients with coronary disease (23–24). In 16 of these, this was caused by a decrease of "fitness," that is, by an increase in cardiac work required for a given amount of external work. Such an increase could result from increased sympathetic stimulation, which, in turn, could be caused by fatigue or myositis in muscles unused to physical work, or by an adverse emotional reaction. Such a reaction was observed especially during the initial period of training (2) or after training lasting 4 weeks or less (4). In the remaining 12 of the 28 patients, deterioration of S-T response to exercise appeared in spite of reduction of cardiac work (23, 24). This could have been caused by progression of coronary arteriosclerotic changes, in spite of the previously discussed beneficial effects of training on the heart muscle. A deterioration of the S-T segment response to a standard amount of exercise was found in 8% of middle-aged persons leading their normal life after 1 year (6) and in 46% of the cases after 3–5 years (1). Another possibility could be that, in some cases, the repeated marked myocardial hypoxia during training may cause a sustained decrease in potassium content of subendocardial muscle, so that the same degree of ischemia now causes greater S-T segment shift. This is especially likely when training results in increased "ischemic" T and S-T changes at rest, and is an indication of the need to reduce the intensity of training (4, 25). In many cases, including the one shown in figure 1B, the increase in the intensity of "ischemic" changes during exercise after training was only apparent, since it was caused by increased work performance.

SUMMARY

A survey of the literature on the effect of physical training on the exercise ECG (400 cases) shows that, when the same external physical work as before training was performed after training, the T wave amplitude tended to increase while S-T segment depression decreased in 32% and increased in 11% of the cases. The improvement was greatest in persons with sympathetic predominance, in those showing greatest decrease in the heart rate reached after the same effort, and those showing greatest adherence to the training program. Sometimes, deterioration was observed during the first 4–6 weeks, but an improvement appeared with further training. A case of coronary disease is presented, which showed a progressive decrease in the magnitude of S-T segment depression for the same amount of exercise throughout a period of carefully supervised physical reconditioning. The possible mechanisms responsible for the changes in the ECG response to exercise are discussed.

REFERENCES

1. ASTRAND, I.: Exercise electrocardiograms recorded twice with an 8-year interval in a group of 204 women and men 48–63 years old. Acta Med. Scand. *178:* 27–39 (1965).
2. BARRY, A. J.; DALY, J. W.; PRUETT, E. D. R.; STEINMETZ, J. R.; BIRKHEAD, N. C. and RODAHL, K.: Effects of physical training in patients who have had myocardial infarction. Amer. J. Cardiol. *17:* 1–8 (1966).

2a. BARTA, E.; FIZEL, A. and FIZELOVA, A.: Adrenergic and cholinergic effects on myocardial metabolism (Czech.) Bratislav. Lek. Listy *41:* 321 (1961).

3. BLOHMKE, M.; STELZER, O. and STUMPF, H.: Effect of active physical training on the cardiac and circulatory reactions to exercise of healthy men at rest and during work (German) Z. Kreislaufforschg *55:* 329–341 (1966).
4. BLOHMKE, M.; SCHAEFER, H.; DRESSLER, S.; STELZER, O. and STUMPF, H.: Electrocardiographic findings in active men at rest and after exercise, at the beginning

and end of physical reconditioning or a vacation. (German). Arch. Kreislaufforsch. *52:* 64–78 (1967).

5. BLOMQVIST, C. G.: Variation of the electrocardiographic response to exercise under different experimental conditions: deconditioning, reconditioning and high altitude, In: H. Blackburn, Measurement in Exercise Electrocardiography, p. 323 (Charles C Thomas, Springfield 1969).
6. BRUCE, R. A.; ALEXANDER, E. R.; LI, Y. B.; CHIANG, B. N.; TING, N. and HORNSTEN, T. R.: Electrocardiographic responses to maximal exercise in American and Chinese population samples, In: H. Blackburn, Measurement in Exercise Electrocardiography, p. 413 (Charles C Thomas, Springfield 1969).
7. BRUCE, R. A. and MAZZARELLA, J. A.: Computer quantitation of S-T responses to maximal exercise in middle-aged American and Chinese men, In: Evang, K. and Andersen, K. L. Physical Activity in Health and Disease, p. 90–103 (Williams & Wilkins, Baltimore 1966).
8. CURETON, T. K.: The Physiological Effects of Exercise Programs in Adults (Amer. Lect. Ser. 728) (Charles C Thomas, Springfield 1969).

8a. ECKSTEIN, R. W.: Effect of exercise and coronary narrowing on coronary collateral circulation. Circulat. Res. *5:* 230–235 (1957).

9. FURBERG, C.: Adrenergic beta-blockade and electrocardiographic ST-T changes. Acta. Med. Scand. *181:* 21–32 (1967).
10. HELLERSTEIN, H. K., HORNSTEN, T. R.; GOLDBERG, A. N.; BURLANDO, A. G.; FRIEDMAN, E. H.; HIRSCH, E. Z. and MARIK, S.: The influence of active conditioning upon coronary atherosclerosis, In: Brest, A. N. and Moyer, J. H.: Atherosclerotic Vascular Disease, p. 115–129 (Appleton, New York 1967); also Canad. Med. Assn. J. *96:* 758–760 (1967).
11. HOLMGREN, A.; TONSSON, B.; LEVANDER, M.; LINDERHOLM, H.; SJOSTRAND, T. and STROM, G.: E.C.G. changes in vasoregulatory asthenia and the effect of physical training. Acta Med. Scand. *165:* 259–271 (1959).
12. KARPOVICH, P. V.: Physiology of Muscular Activity. 6th Ed. (W. B. Saunders, Philadelphia 1965).
13. KARVONEN, M. J.: The relationship of habitual physical activity to diseases in the cardiovascular system. In: Physical Activity in Health and Disease, pp. 81–89 (Williams & Wilkins, Baltimore 1966).
14. KOSITSKII, G. I.; ORESHCHUK, F. A. and PRYTKOVA, L. N.: Prophylaxis against coronary insufficiency (Russian). Kardiologiya *7(8):* 58–63 (1967).
15. LEPESCHKIN, E.: Modern Electrocardiography (Williams & Wilkins, Baltimore 1951).
16. LEPESCHKIN, E.: Exercise tests in the diagnosis of coronary heart disease. Circulation *22:* 986–1001 (1960); Physiological factors influencing the electrocardiographic response to exercise. In: Measurement in exercise electrocardiography p. 363 (Charles C Thomas, Springfield 1969).
17. LEPESCHKIN, E. and SURAWICZ, B.: Characteristics of true-positive and false-positive results of electrocardiographic Master two-step exercise tests. New Eng. J. Med. *268:* 511–520 (1958).
18. MOREHOUSE, L. E. and MILLER, A. T.: Physiology of Exercise, 5th Ed. (C. V. Mosby, St. Louis 1967).
19. MURAVOV, I. V., Ed.: Motor activity and Aging (Russ.) Acad. Med. Sci. USSR, Kiev, 1969.
20. NAUGHTON, J.: Cardiopulmonary responses during physical training in patients who have recovered from myocardial infarction. In: W. Raab, Prevention of ischemic heart disease, (Charles C Thomas, Springfield 1966). pp. 332–337
21. RAAB, W.: The adrenergic-cholinergic control of cardiac metabolism and function. Advances in Cardiology *1:* 65–152 (Karger, Basel 1956).

21a. RAAB, W., Ed.: Prevention of Ischemic Heart Disease (Charles C Thomas, Springfield 1966).

22. RAAB, W.: Preventive Myocardiology. (Charles C Thomas, Springfield 1970).

23. RADKE, J. D.; HELLERSTEIN, H. K.; SALZMAN, S. H.; MAISTELMAN, H. M. and RICKLIN, R.: The quantitative effect of physical conditioning on the exercise electrocardiogram of subjects with arteriosclerotic heart disease and normal subjects. In: Brunner, B. and Jokl, E. Physical Activity and Aging, p. 168. (Karger Basel 1970).

24. SALZMAN, S. H.; HELLERSTEIN, H. K.; RADKE, J. D.; MAISTELMAN, H. W. and RICKLIN, R.: Quantitative effects of physical conditioning on the exercise electrocardiogram of middle-aged subjects with arteriosclerotic heart disease, In: Measurement in Exercise Electrocardiography, p. 383 (Charles C Thomas, Springfield 1969).

25. SMITH, J. E. and KIDERA, G. J.: Treatment of angina pectoris with exercise stress. Aerospace Med. *38:* 742–745 (1967).

26. SMITH, J. E. and KIDERA, G. J.: Effect of programmed exercise on nonspecific T wave abnormalities. Aerospace Med. *39:* 528–532 (1968).

Mailing address: Eugene Lepeschkin, M.D., Cardiovascular Research Unit, Department of Medicine, University of Vermont, DeGoesbriand Memorial Hospital, Burlington, Vermont 05401 (USA).

Department of Community Medicine, College of Medicine,
University of Vermont, Burlington,
and
Department of Nutrition, Harvard School of Public Health,
Boston, Massachusetts, USA

CARDIOPROTECTIVE EXERCISE PROGRAMS IN THE UNITED STATES AND CANADA [1]

O. A. BRUSIS

Over the past decade, exercise programs have increased considerably in number in the United States and Canada. Most of these programs are based on one or on a combination of several of the following exercise plans.

CARDIOPROTECTIVE EXERCISE PLANS

The Progressive Rhythmic Continuous Exercise Program of Cureton (7). A conditioning program designed in 21 lessions that last for 6 to 9 months. It consists of vigorous calisthenics, interspersed with frequent walking, jogging, and breathing exercises. There are 5 exercise sessions per week, lasting eventually for 2 h each time. The degree of difficulty of this program, especially at the higher levels, is such that only few exceptional individuals are able and willing to complete it.

The Royal Canadian Air Force Exercises (22), generally known as 5BX and XBX. The XBX stands for 10 basic exercises, a 12 min a day exercise plan for women, and the 5BX for 5 basic exercises, an 11 min plan for men. These two exercise plans are based on calisthenics, with special emphasis on exercising every major muscle group of the body. The exercises also include running in place to enhance cardiovascular fitness. The plan provides an exercise goal for every age group. The program is scheduled according to 4 and 6 achievement charts, respectively. In this way, the trainee is guaranteed a simple and safe, yet progressive increase in physical fitness. The program recommends weight control through a balanced diet.

The Jogging Program of Bowermann and Harris (1). This is a "medically approved program that will reduce the waistline, improve appearance and help prolong life." The program is designed for three different levels of initial physical fitness. It slowly progresses in difficulty at each level over a 12 week period. Weight control and a physical examination for participants older than 30 years are recommended. The program has been based primarily on the experience of a well-known track coach.

---

[1] Supported in part by United States Public Health Traineeship Program (5A03-AH-00502-04).

Aerobics of Cooper (5). This is a system of endurance type or aerobic exercises designed especially to improve cardiovascular and pulmonary capacity. The new revised version (6) is adapted to fit all age groups and fitness levels. This program is unique in that it is based on a point system of accomplishment. One point is equivalent to approximately 7 ml oxygen consumption per kg bodyweight per min. In addition, a simple 12-min field test is described, to test the participant's physical fitness. Oxygen consumption of the field test has been compared with that of the treadmill test and a high correlation ($r = 0.897$) has been found (7). The point system provides the trainee with an easy to understand and simple exercise goal. Since the program considers a number of different exercises, all of which are given point values, an exercise program can be selected to meet the abilities and interests of each individual. This is believed to increase adherence to a greater extent than a "one-exercise-only-program," which can become quite boring. All of these programs stress the importance of weight control and discourage cigarette smoking.

Most exercise programs offered to the public at YMCA's, YWCA's, and other health and fitness centers that use these plans as a basic component for exercise training are not medically supervised or controlled. Although not specifically designed to be cardioprotective, they all are based on exercises that, if performed vigorously, regularly, and long enough, will produce among others the following cardiac and hemodynamic changes: bradycardia during rest and submaximal exercise, an increase in resting and exercise stroke volume, an increase in maximum cardiac output, an increase in velocity of contraction, a decrease in systemic arterial pressure, and a redistribution of blood flow to more active muscles (9). The importance of these changes for the development of arteriosclerotic ischemic heart disease and of its serious consequences consists in the fact that, the lower the oxygen requirement of the myocardium, the lower can be the oxygen supply to suffice for adequate function during rest, exercise, and stress.

Besides these mass-tailored general exercise and fitness programs, there are a great number of programs designed for the specific goal of protecting the participant from arteriosclerotic ischemic heart disease and its most serious consequences, myocardial infarction and death. This is especially true for most community prevention programs, such as that directed by Kasch and Boyer at the San Diego State College in California (15), Juster's program in Glens Falls, N.Y. and vicinity (14), where according to a 1969 survey approximately 1,400 people participate regularly, and the Vermont Coronary Prevention Program (3). All of these programs have in common a multifactorial approach through aerobic exercises, a low calorie-low fat diet, and discouragement of cigarette smoking. Although many of the above-listed programs include in their protocol medical clearance and a baseline evaluation of the participants, consisting of a risk factor profile, physical fitness assessment, and frequently even a medical examination, they all lack control groups. This is understandable, since they are primarily service programs. From this kind of program, one cannot therefore expect an answer to the question whether exercise in mass programs protects from arteriosclerotic ischemic heart disease.

Several investigators (13, 18, 21) were able to demonstrate in their respective prevention studies that adherence to exercise protocol will result in significant reduction of so-called risk factors. However, definite evidence that this reduction means also significant reduction of incidence and mortality is still lacking.

Recent reports from a joint pilot study conducted at the University of Minnesota, The Pennsylvania State University, and the University of Wisconsin, in collaboration with the Heart Disease and Stroke Program of the United States Public Health Service, dealing with the feasibility of mass physical exercise programs for primary prevention, indicate that recruitment and adherence problems are of such a nature that controlled mass field trials that would bring statistically significant results are indeed not feasible, especially not in the present "tight money" situation (21).

A special potentially cardioprotective program, which could be best classified as a "therapeutic" exercise program, is the one directed at patients who have angina pectoris with or without previous myocardial infarction. Evidence from trials conducted by Hellerstein (11), Smith (23), Heller (10), and others (17) indicates that graded exercise in the form of walking, jogging, and cycling can significantly improve objective and subjective well-being, cardiac function, and the exercise electrocardiogram. It should be pointed out, however, that all these programs have in common a lack of control groups. This adds to the dilemma that results are inconclusive as far as impact on endpoints, such as incidence and mortality from acute myocardial infarctions, are concerned.

The outcome of secondary prevention programs emphasizing cardioprotective exercises not only indicates that risk factors after an acute myocardial infarction can be reduced, but it also suggests that recurrence of a second attack is delayed and mortality decreased (12). The lack of matched controls and the small numbers of subjects in these programs do not, however, permit definite scientific conclusions.

Cardiac rehabilitation programs as described by Wenger (26), Torkelson (24), Zohman (29), and Naughton (16, 20) are directed at postmyocardial infarction patients during the period of up to 8 weeks after the event. Usually, they serve a two-fold goal: 1) to maintain and improve cardiovascular reserve and capacity at the pre-infarction level; and 2) to increase myocardial efficiency, thus permitting a more useful and enjoyable life. Some of these programs begin with very specific, passive exercises as soon after the acute attack as the patient's condition allows. Using physical and occupational therapy, they eventually progress to an active fitness program, making it possible for the patient to return to work, in some cases within 6–8 weeks after the attack. At this point, the rehabilitation program usually phases out and is followed by a secondary prevention program to maintain the accomplished state of physical fitness and, hopefully, to slow down the disease process.

BASIC PRINCIPLES OF CARDIOPROTECTIVE PROGRAMS

What are the basic principles for a cardioprotective exercise program, regardless of the target population, and in what aspects do primary and secondary prevention and rehabilitation programs differ from each other?

In all types of program, exercise should be an integral part, but by no means the only one. Weight control, diet adjustment following a diet low in saturated fat and/or cholesterol as described by McGandy (19), discouragement of cigarette smoking, environmental relaxation, and treatment of underlying diseases known to be associated with increased risk, such as diabetes, arterial hypertension, and obesity, should all be included as essential in a cardioprotective program.

Cardioprotective exercises should be of the aerobic or endurance type. It is advisable in designing a program to select exercises that require very little or no equipment or special facilities. Preferred are walking, jogging, hiking, swimming, cycling and, where appropriate, cross-country skiing. Although the major benefit derived from aerobic exercises depends on the amount, duration, and intensity of muscular work, we believe that the emotional component also plays an important role. Therefore, exercising outdoors should be practiced wherever possible and feasible. Competitive exercises should be excluded. Activities that involve breath holding or the Valsalva maneuver *must* be avoided.

Cardioprotective exercise programs should ideally be composed of three parts: a 5 to 10 min warm-up period, which should consist primarily of stretching and relaxation exercises; the actual work phase of aerobic exercises, during which the activity should require an elevated heart rate, placing a certain degree of stress on the cardiovascular and pulmonary systems; and a 5 to 10 min cool-out period, preferably walking. Frequently, the inclusion of a short period of games at the end of the actual exercise session will enhance adherence to the program. The total exercise session should not exceed 1½ h. For deconditioned participants, a 6 to 8 week starter program of slow, continuous, and light exercises is strongly recommended (2, 8).

Primary prevention programs

Categories of individuals suited for participation in a primary prevention program are: 1) "healthy" subjects who are "coronary-prone"; 2) persons in a state of general deconditioning; 3) persons suffering from neurocirculatory asthenia; 4) normal subjects who are neither "coronary-prone" nor afflicted with any conditions that would not permit participation in an active exercise program (2, 8).

In addition to the above indications, chronic pulmonary disease, diabetes mellitus, intermittent claudication, certain emotional disorders, and pre- and postoperative conditions in selected patients are further indications for participation in an exercise program. Ideally, all participants who enter an exercise program should undergo a medical examination with special attention paid to the evaluation of cardiovascular status and function; this should include a multilevel exercise stress test. It is obvious that under the present circumstances (lack of manpower and facilities), exercise stress testing for the population at large is not feasible. Therefore, it should be emphasized again that in deconditioned individuals a starter program is important to avoid undesirable and potentially dangerous complications that might be blamed on exercise programs in general, yet are primarily due to unwise exercise prescription or unreasonable exercise performance.

To measure the intensity of exercises performed during a session, recording of the heart rate seems the easiest and least complicated procedure. A heart rate of 150 beats per min is recommended for age groups lower than 50 years. Once the desired level of fitness and conditioning has been reached, it is generally agreed upon that aerobic exercising 3 times per week for 20 min is sufficient to maintain the acquired level of cardiovascular fitness.

Secondary prevention programs

Indications for a secondary prevention program are: 1) clinically manifest ischemic heart disease, in particular, status after acute myocardial infarction, at least 3 months after the event,and angina pectoris. It should be emphasized that, if performed in groups, exercise programs for patients with these conditions need to be supervised by specially trained group leaders. Such training should include external cardiopulmonary resuscitation procedures. Furthermore, the supervisors should be familiar with signs and symptoms that may occur during exercise and that necessitate a revision of the medical exercise prescription for the individual. The most commonly encountered signs and symptoms are: any chest pain or pain referred to teeth, arm, jaw or ear; syncope, light-headedness, dizziness or irregular heart beat following exercise; persistent fatigue; unusual weight loss; skeletal and muscular problems aggravated by exercise; nausea, vomiting or a feeling of malaise; failure of the pulse rate to return to below the 140 level within 5 min after cessation of exercise (27). For medico-legal reasons, it is also recommended that a physician be always within immediate reach in case of an emergency. Preferably, a defibrillator should be available on the premises.

Participants in secondary prevention programs should not be admitted without an initial exercise stress test. The result of this test will serve as a base-line on which an individually prescribed exercise program can be designed. Heart rate, electrocardiogram, and blood pressure during exercise are good indicators of the individual's response; at least one, preferably two of them should be used wherever possible. Frequent re-evaluations and new prescriptions of exercise, specified with regard to dose, intensity and duration, are necessary. Early in the program, an evaluation every 4–6 weeks is advisable. After 6 months, depending on the individual's condition and his progress, re-evaluation examinations may be scheduled at intervals of 3–6 months. Exercise prescription should aim at a pulse rate that is 5–10 beats per min below the rate recorded in the preliminary exercise tolerance test. In general, therapeutic effects from exercise are observed at rates of 130–140 beats per min, depending on the age of the individual.

Contraindications against participation in secondary prevention programs are: 1) gross cardiomegaly; 2) congestive heart failure; 3) uncontrolled diabetes mellitus; 4) uncontrolled arterial hypertension; 5) severe anemia; 6) certain cardiac arrhythmias, such as second and third degree heart block, fixed rate pacemakers, paroxysmal ventricular tachycardia, untreated atrial fibrillation, frequent ventricular extrasystoles that increase in frequency with exercise; 7) obstructive disease of the ventricular outflow tract; 8) moderate to severe valvular heart disease; 9) impending myocardial infarction in a changing pattern of angina pectoris; 10) myocardiopathies; 11) unstable angina pectoris or angina of less than 2 months' duration. Failure of a favorable response to exercise to appear after 4 weeks requires a thorough medical re-evaluation (28).

Exercise programs for cardiac rehabilitation

As soon as the patient's condition permits release from the intensive care unit, active rehabilitation should commence. Such conditions as congestive heart failure or intractable pain at that time preclude initiation of an active rehabilitation exercise program. Naturally, exercises prescribed at that early stage of the disease have to be simple and to demand only little effort (25). It should be pointed out, however, that conventional medical orders, such as use of a bedside commode, up and about t.i.d., and others, essentially represent prescriptions for exercise; yet the intensity, duration, and frequency of such activities is based entirely on clinical empirical observation. Furthermore, these prescribed activities are frequently monitored by allied health personnel and not by a physician. For these reasons, we believe that early exercise tolerance tests, simple in design and low in demand, depending on the patient's ability and condition, should be more widely explored and employed. Throughout the rehabilitation procedures, the health professionals involved should be familiar with the recognition and management of cardiac emergencies. Appropriate equipment for the control of such incidents ought to be available during the entire program.

SUMMARY

Four major types of exercise programs providing the basis for most mass-tailored exercise and fitness programs in this country have been briefly described. Major research programs dealing with primary and secondary prevention and rehabilitation of ischemic heart disease have been reviewed. All cardioprotective exercise training programs use as their core aerobic exercises. Existing prevention and rehabilitation programs share a lack of well-matched control groups. Some principles and essentials of cardioprotective exercise programs are mentioned and the differences between primary and secondary prevention and rehabilitation programs are discussed, as regards admission requirements, safety measures, indications, and contraindications.

REFERENCES

1. BOWERMANN, W. J. and HARRIS, W. E.: Jogging (Grosset and Dunlap, New York 1967).
2. BOYER, J. L.: Summation of guidelines on exercise. Append. 2. J. South Carol. Med. Assoc. *65:* No. 12 (Suppl. 1), 99–100 (1969).
3. BRUSIS, O. A.: The Vermont Coronary Prevention Program. Vermont Heart Assoc. Publ. (1968).
4. COOPER, K. H.: A means of assessing maximal oxygen intake. J.A.M.A. *203:* 201–204 (1968).
5. COOPER, K. L.: Aerobics (M. Evans & Co., New York 1968).
6. COOPER, K. L.: The new aerobics (M. Evans & Co. New York 1970).
7. CURETON, T. K., Jr.: Physical Fitness and Dynamic Health. Part II. pp. 97–162. (The Dial Press/New York 1965).
8. GIESE, W. K.: Summation of guidelines on exercise. Append. 3. J. South Carol. Med. Assoc. *65:* No. 12 (Suppl. 1), 100–101 (1969).
9. HANSON, S.; TABAKIN, B. S.; LEVY, A. M. and NEDDE, W.: Long term physical training and cardiovascular dynamics in middle aged men. Circulation *38:* 783–799 (1968).
10. HELLER, M.: Four year experience with a graded exercise program for post coronary patient. Appl. Therap. *11:* 386–391 (1969).
11. HELLERSTEIN, H. K.: Exercise therapy in coronary disease. Bull. N.Y. Acad. Med. *44:* 1028–1047 (1968).
12. HELLERSTEIN, H. K.: The effects of physical activity. Patients and normal coronary prone subjects. Minn. Med. *52:* 127–134 (1969).
13. HELLERSTEIN, H. K.: In: BREST, A. N. and MOYER, J. H. The Influence of Active Conditioning upon Coronary Arteriosclerosis in Arteriosclerotic Vascular Disease, (Meredith Publ. New York 1967).
14. JUSTER, I. R.: A community coronary heart disease preventive program. Paper presented at the 1966 Stowe Vermont Conference on Preventive Cardiology.
15. KASCH, F. W.; PHILLIPS, W. H.; CARTER, J. E.; ROSS, W. D. and BOYER, J. L.: Maximum work capacity in middle aged males by a step test method. J. Sports Med. *5:* 198–202 (1965).
16. LATEGOLA, M. T. and NAUGHTON, J.: Restoration of cardiovascular integrity in post-myocardially infarcted aviation personnel. Aerospace Med. No. 10, *38:* 1067–1070 (1967).
17. LEPESCHKIN, E. and BRUSIS, O. A.: Effect of physical training on the exercise electrocardiogram. This Book, pp. 753–761.
18. MANN, G. V.; GARRETT, H. L.; FARHI, A.; MARRAY, H. and BILLINGS, F. T.: Exercise to prevent coronary heart disease. Amer. J. Med. *46:* 12–27 (1969).

19. McGANDY, R. B. and REMMELL, P. S.: The dietary management of coronary heart disease. Cardiovascular Clinics *1:* 34–45 (1969).

20. NAUGHTON, J.; LATEGOLA, M. T. and SANBOUR, K.: A physical rehabilitation program for cardiac patients: A progress report. Amer. J. Med. Sci. *252:* 545–553 (1966).

21. REMINGTON, R. D.; TAYLOR, H. L.; HAMRELL, B.; BLACKBURN, H.; FETCHER, E. S.; BUSKIRK, E.; SKINNER, J.; BALKE, B.; ROBINHOLD, D.; HEINZELMANN, F.; HASKELL, W.; STAMLER, J.; FOX, S. and EPSTEIN, F.: Studies on physical activity in trials of prevention of coronary heart disease. Presented at the Conference on Cardiovascular Disease Epidemiology in New Orleans, March 1970. Abstract in CVD Epidemiology Newsletter, Amer. Heart Assoc. No. 8: 17–19 (1970).

22. Royal Canadian Air Force: Exercise Plan for Physical Fitness (Simon and Schuster, Inc. Canada 1962).

23. SMITH, E. J.: Treatment of angina pectoris with exercise stress. Aerospace Med. *38:* 742–745 (1967).

24. TORKELSON, L. O.: Rehabilitation of the patient with acute myocardial infarction. J. Chron. Dis. *17:* 605–704 (1964).

25. WENGER, N. K.: Summation of guidelines on exercise. Append. 4. J. South Carol. Med. Assoc. *65:* No. 12 (Suppl. 1), 102–104 (1969).

26. WENGER, N. K.: The use of exercise in the rehabilitation of patients after myocardial infarctions. J. South Carol. Med. Assoc. *65:* No. 12 (Suppl. 1), 66–68 (1969).

27. Workshop Report: Exercise program for the prevention of heart disease. J. South Carol. Med. Assoc. *65:* No. 12 (Suppl. 1), 79–82 (1969).

28. Workshop Report: Exercise and rehabilitation. J. South Carol. Med. Assoc. *65:* No. 12 (Suppl. 1), 86–90 (1969).

29. ZOHMAN, L. R. and TOBIS, J. S.: A rehabilitation program for inpatients with recent myocardial infarction. Arch. Phys. Med. *49:* 443–448 (1968).

Mailing address: Otto A. Brusis, M.D., Department of Nutrition, Harvard School of Public Health, 665 Huntington Avenue, Boston, Massachusetts 02115 (USA).

Smith Kline & French Laboratories,
Philadelphia, Pennsylvania, USA

CARDIOTOXICITY OF ISOPROTERENOL IN EXPERIMENTAL ANIMALS. INFLUENCE OF STRESS, OBESITY, AND REPEATED DOSING

T. BALAZS

Overdosage of epinephrine, norepinephrine, and isoproterenol can cause myocardial necrosis in experimental animals and in humans (21). Evidence is growing in support of the view that catecholamines play a role in the etiology of anoxemic heart diseases of man (1, 14, 19). It is, therefore, of interest to determine those factors that may influence the sensitivity of the heart to the cardiotoxic effects of these agents.

We investigated the effects of stress (2), obesity (4), and repeated dosing (3) on the cardiotoxicity of isoproterenol in a number of studies. The results of this work are reviewed herein. Isoproterenol-induced myocardial necrosis in experimental animals resembles an infarct, and hypoxia is believed to be a factor in its causation.

EFFECT OF STRESS ON ACUTE TOXICITY OF ISOPROTERENOL IN RATS

A subtle stressor that does not have a demonstrable acute effect but is expected to act following a prolonged application was chosen in order to examine the sensitizing potential to the cardiotoxicity of isoproterenol. Prolonged isolation of rats in individual cages—with food and water freely available—effectively served this purpose. Sensitization to isoproterenol, as indicated by a reduction of the acute lethal dose, increased until the end of the third month of isolation (fig. 1). Death is preceded by extrasystoles and fibrillation; thus, the median lethal dose also measures the cardiotoxicity of isoproterenol.

In another experiment, identical doses (20 and 50 mg/kg) were injected into isolated and group-caged rats. The cardiac lesions were more severe in the former animals than in the latter, which were caged individually only for a 2-day period of dosing.

When the rats were isolated for 3 months and then returned to group cages for 3 weeks, the acute mortality values were identical to those found in rats group-caged throughout the 3-month period. Thus, a recovery from the effect of isolation occurred.

While the group-caged rats were docile, the isolated ones were agressive; they were difficult to handle.

Serum corticoids were increased by the end of the 3-month isolation period

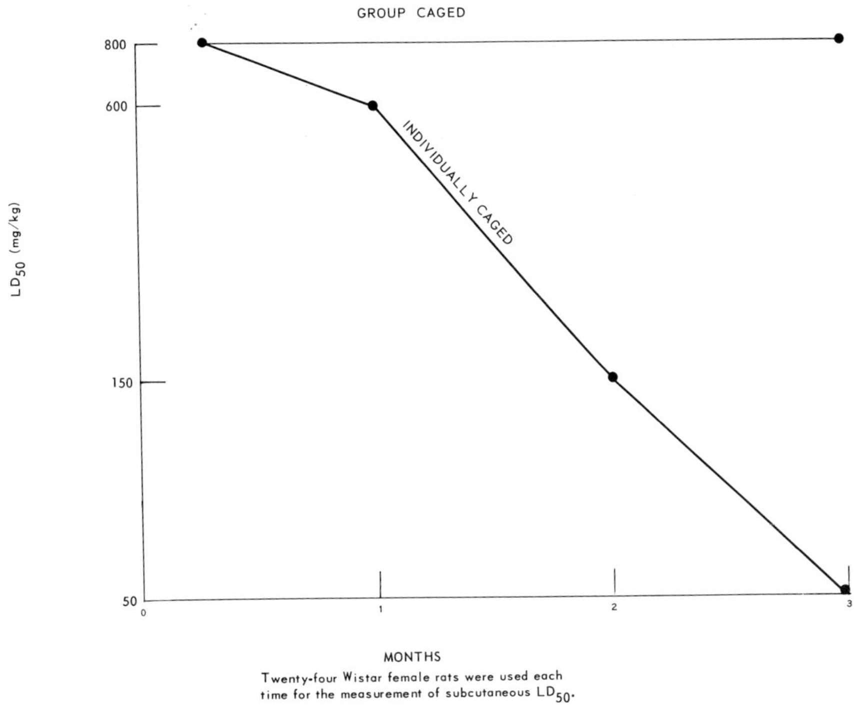

Fig. 1 Isoproterenol toxicity in rats.

as compared with those of group-caged controls. In addition, lymphopenia developed, thymic weight decreased, and adrenal weight increased in the isolated rats (11). No difference was detected, however, in the activity of drug metabolizing enzyme systems between isolated and group-caged rats. Thus, the increased sensitivity of isolated rats to isoproterenol cannot be attributed to an altered rate of drug metabolism (5).

Our finding on the effect of 3-month individual caging has been confirmed by several investigators, using different strains of rats and different routes of administration of isoproterenol (12, 23). Raab *et al.* (15) extended the result of our study and showed an increased sensitivity to the cardiopathic effect of epinephrine in long-term isolated rats. These workers postulated that the decreased intracellular K caused by the sustained hypercorticoid state during isolation is responsible for the augmented vulnerability of the heart. Furthermore, amphetamine, which acts by the release of catecholamine, and tranylcypromine, which in turn acts by the inhibition of catecholamine metabolism, have been found to be more toxic in isolated than in group-caged rats (20, 23). Thus, a sensitization to endogenous catecholamines appeared to develop.

We also examined the effect of an acute stress on the cardiotoxicity of isoproterenol. A group of Wistar rats was exposed to a cold environment ($4° \pm 1°$) for 1 week; they were then returned to room temperature ($29° \pm 1°$) and injected subcutaneously with isoproterenol, simultaneously with their controls. Mortality data (table I) indicated an increased sensitivity of the cold-exposed rats: 1,000-fold in the males and a 10,000-fold in the females. Thus, a potent stressor (cold exposure) that suddenly stimulates adrenocortical function induced an effect of a greater intensity than that of a subtle stressor, *i.e.*, chronic isolation.

EFFECT OF OBESITY ON ACUTE TOXICITY OF ISOPROTERENOL IN RATS

Obesity was produced by high-fat or high-carbohydrate diets in Wistar rats or by *ad libitum* feeding of a normal diet in Sprague-Dawley rats.

Young Wistar rats were placed in 3 groups and fed a high-fat, a high-carbohydrate, or a normal diet freely for 16 weeks. The acute toxicity of isoproterenol markedly increased in the rats kept on the obesogenous diets (table I), which produced a 2-fold increase in body-fat content without increasing the body weight above that of the controls.

In a second experiment, 4 groups of rats received either a normal or a high-fat diet, freely or in restricted amounts (70% of the quantity consumed by the previous group), for 16 weeks. They were then treated with identical doses of isoproterenol on 2 consecutive days. The ensuing heart lesions were found to be the most severe in the animals fed the high-fat diet freely. The proportion of body fat was increased only in this group. There was no significant difference in the severity of cardiac lesions or in the proportion of body fat between the other

Table I

Subcutaneous Median Lethal Doses of Isoproterenol HCl in Rats

Group no.*	Procedure or condition prior to dosings	No. of rats	Average body wt. (g)	LD_{50} (mg/kg)
1	Isolated for 3 months	30	199	<50
2	Exposed to cold for 1 week	24	270	<0.1
3	Group-caged control†	50	203	815
4	Fed a high-fat diet‡	50	216	90
5	Fed a high-carbohydrate diet‡	50	198	67
6	Inherently obese strain, fed *ad libitum*	50	600	0.35
7	Inherently obese strain, restricted in feed	30	400	>900
8	Inherently obese strain, prior to reaching obesity	40	165	1,000

* Groups no. 1–5, Wistar females; groups no. 6–8, Sprague-Dawley derived Charles River males; pair-caged during conditioning period.

† Fed a normal diet.

‡ Group caged.

3 groups. Thus, the increased body fat, and not the presence of fat in the diet *per se,* was responsible for the aggravation of isoproterenol effect.

The acute toxicity of isoproterenol was also measured in inherently obese Sprague-Dawley rats in 2 tests. In the first of these, one group of rats weighing 260 g had access to a normal diet *ad libitum,* while the other group, weighing 360 g, received the same diet in restricted amount. In about 2 weeks, the body weights became comparable (360 g and 330 g, respectively), but the body-fat content differed significantly (19.4 ± 0.4% and 10.9 ± 1.4%). The group that had the greater body-fat content died following a single dose of 10 mg/kg of isoproterenol, whereas the rats of the other group survived.

In a second experiment,[1] rats with an average body weight of 560 g were divided into 2 groups. One group received a normal diet freely, the other was fed the same diet in restricted amounts for one month. The body weights became 603 g (498–785) and 430 g (371–500), respectively. At this time, the subcutaneous acute toxicity of isoproterenol was measured in both groups. The median lethal dose for the group on the restricted diet was found to be higher than 900 mg/kg, while that for the group fed *ad libitum* was 0.35 mg/kg (table I). Thus, an approximately 3,000-fold difference was established. (Restriction of the diet for 2 days had no significant effect.)

The results of these experiments indicated that while obesity is a potent sensitizer to the cardiotoxic effect of isoproterenol in rats, the normal sensitivity to isoproterenol returns if the excess body fat is decreased.

No difference was detected in the chronotropic effect of isoproterenol when it was measured on isolated hearts of obese and lean rats. The activity of the hepatic microsomal drug metabolizing enzyme system decreased in the rats on the restricted diet. On the basis of these findings, it is not likely that altered receptor sensitivity or pharmacokinetic changes (*e.g.,* a decreased biologic half-life of isoproterenol) are responsible for the decreased isoproterenol sensitivity of the rats kept on restricted diet. The resistance appeared to develop to the anoxia-producing but not to the pharmacologic effect of the compound. Metabolic consequences of the weight loss are, therefore, considered to be responsible for this phenomenon.

DEVELOPMENT OF RESISTANCE TO THE CARDIOTOXIC EFFECT OF ISOPROTERENOL IN RATS AND IN DOGS

The cardiac-necrosis-inducing effect of isoproterenol was not detected in the initial multiple-dose toxicity study in experimental animals (8). A compound that produces a myocardial necrosis by its pharmacologic effect is expected to cause such a change at appropriate dose levels also in a multiple-dose study. We

[1] This experiment was performed in the laboratory of Smith Kline & French Laboratories in 1969, while the previous ones were conducted in the Food and Drug Directorate (Ottawa) in 1961.

therefore investigated the cardiotoxicity of isoproterenol in multiple-dosing tests in rats and in dogs.

Thirty Sherman female rats (mean weight 240 g) were injected with 0.5 mg/kg of isoproterenol on 2 consecutive days; dosing continued for 2 weeks in 15 of these animals. Six beagle dogs were injected subcutaneously with 1mg/kg of isoproterenol on 2 consecutive days, dosing being repeated in a number of dogs weekly or biweekly for 2 months. All animals were killed 24 h after the last dosing, and gross and microscopic examinations of the heart were carried out.

While necroses were induced in both species following the initial dosing, no evidence of acute regressive changes was detected in animals that received the dosing for an extended period. Only a scar indicated the site of the initial lesion (table II). While ventricular extrasystoles and paroxysmal ventricular tachycardia developed following the initial isoproterenol dosage in dogs, no electrocardiographic change could be elicited when dosage was repeated at the various intervals. The degree and duration of cardiovascular effects monitored following the initial and terminal dosings in dogs were comparable (table III). Thus, the absence of cardiac necroses is not attributable to an altered pharmacologic effect.

Selye *et al.* (18) and, later, Rona *et al.* (16) demonstrated that isoproterenol failed to induce necrosis in the hearts of rats for about 2 weeks following ligation of a coronary artery. In the hearts of coronary-artery-ligated rats, electron microscopic studies by Rona *et al.* revealed morphologic signs of adaptive response in areas remote from the infarcts.

Poupa *et al.* (13) demonstrated that hearts of isoproterenol-treated rats are less sensitive even *in vitro* to anoxia than their controls, an effect that persists for at least a week. Wenzel and Lyon detected increased formation of lactic dehydrogenase M (skeletal muscle type) in the hearts of rats treated with a single

Table II

Histological Findings following Isoproterenol Treatments

Species*	Treatment	Acute necrosis†	Scar†
Rat	0.5 mg/kg, on 2 consecutive days	+	–
Rat	same for 2 weeks	–	+
Dog	1 mg/kg, on 2 consecutive days	+	–
Dog	same repeated weekly or biweekly for 2 months	–	+

* Animals were killed 24 h after the last dose.
† Present, +; Absent, –.

Table III

Effects of 1 mg/kg Isoproterenol (Subcutaneous) in Dogs

Treatment	Increase in heartbeat/min* following		Decrease in mean arterial pressure* following		Ventricular extrasystoles† following		Myocardial necrosis
	Initial Treatment	Final Treatment	Initial Treatment	Final Treatment	Initial Treatment	Final Treatment	
	%		%				
On 2 consecutive days	110		54		+		++
Same repeated after 1 week	55	62	40	34	+	–	+
Same repeated weekly, several times	48	96	51	46	+	–	–

* Mean of 3 readings (20 min, 1 h, and 4 h postdosing) in 2 dogs each time.
† Recorded 24 h postdosing.

administration of cardiotoxic doses of isoproterenol for about a week after dosing (22). These findings suggest the stimulation of anaerobic metabolic pathways, following cardiotoxic doses of isoproterenol. One can consider these observations as signs of adaptive change. If this concept is proven, it would serve as an example for the development of a resistance via an alternative metabolic pathway.

DISCUSSION

The results of the studies presented herein are an example of the effect of endocrine and metabolic influences on drug toxicity and, in this regard, may have relevance to clinical pharmacology.

The pathogenesis of the isoproterenol-induced heart lesion is explained by a relative coronary insufficiency due to the exaggerated pharmacologic effect of isoproterenol (17). The site of the lesion is the subendocardial area of the left ventricle and, in the dog, primarily the papillary muscles of the left ventricle. This area is a predilectional one for an ischemia due to anatomic as well as to physiologic factors. The same observation is applicable in man, also (6, 7, 9).

Increased emotionality, stress, and obesity have been associated clinically with an increasing incidence of ischemic heart disease in man. Isolation-induced increased emotionality, cold-exposure stress, and obesity increased the myocardial susceptibility to the ischemia-inducing effect of isoproterenol in rats. The protective influence of body weight reduction and repeated anoxia-inducing stimuli was also demonstrated in our experiments.

The basic injury of isoproterenol- or other catecholamine-induced lesions as well as in an ischemic heart disease is a bioenergetic failure (loss of ATP synthesis) (10). Thus, from a molecular biologic point of view, the etiology of the isoproterenol-induced lesion resembles those produced by ischemia. Hence, the factors that have been shown to alter the susceptibility of the heart to isoproterenol may be of a significance that is beyond the scope of pharmacology.

SUMMARY

Isoproterenol-induced heart necrosis in experimental animals has been used as a model to study the significance of certain factors that may influence the susceptibility of the heart to the cardiotoxic effects of catecholamines.

We investigated the effects of stress and obesity on the subcutaneous acute toxicity of isoproterenol in rats. A striking increase of this parameter was found in rats stressed by cold exposure or by long-term isolation prior to dosing, as well as in rats fed obesity-producing diets. Experiments in spontaneously obese animals revealed a 3,000 times increased toxicity of isoproterenol when compared with that observed in rats that received the feed for 1 month in a restricted amount and lost about 20% of their body weight. The LD_{50} for the former = 0.3 mg/kg, while that of the latter = <900 mg/kg. While a single or two subcutaneous doses of 0.5–1 mg/kg of isoproterenol induced myocardial necrosis in young adult rats, the lesion could not be elicited with a similar treatment repeated 2 weeks apart or daily for 2 weeks. This finding implied the development of resistance to isoproterenol.

Subcutaneous injections of 0.5–1 mg/kg of isoproterenol given on 2 consecutive days to dogs caused marked tachycardia, hypotension, and focal necrosis (in the papillary muscles of the left ventricle) accompanied by ventricular extrasystoles, the latter being detectable for 2–3 days. After healing of the lesions, similar doses repeated weekly failed to elicit further lesions. The degree and duration of the cardiovascular effects following repeated injections were similar to those of the initial one; thus, tolerance developed solely to the cardiotoxic effect, which can be explained by a metabolic adaptation of the myocardium in both species.

REFERENCES

1. Annotation: Catecholamines and the heart. Lancet *i:* 1200 (1969).
2. BALAZS, T.; MURPHY, J. B. and GRICE, H. C.: The influence of environmental changes on the cardiotoxicity of isoprenaline in rats. J. Pharm. Pharmacol. *14:* 750–755 (1962).
3. BALAZS, T.; OHTAKE, S. and NOBLE, J. F.: The development of resistance to the ischemic cardiopathic effect of isoproterenol in experimental animals. Tox. Appl. Pharmacol. (in press).
4. BALAZS, T.; SAHASRABUDHE, M. R. and GRICE, H. C.: The influence of excess body fat on the cardiotoxicity of isoproterenol in rats. Tox. Appl. Pharmacol. *4:* 613–620 (1962).
5. DAIRMAN, W. and BALAZS, T.: Comparison of microsomal drug metabolizing enzyme systems in grouped and individually caged rats. Biochem. Pharmacol. *19:* 951–955 (1970).
6. DEBUSK, R. F. and HARRISON, D. C.: The clinical spectrum of papillary-muscle disease. New Engl. J. Med. *281:* 1458–1467 (1969).
7. DEPASQUALE, N. P. and BURCH, G. E.: The necropsy incidence of gross scars or acute infarction of the papillary muscles of the left ventricle. Amer. J. Cardiol. *17:* 169–170 (1966).
8. DERTINGER, B. L.; BEAVER, D. C. and LANDS, A. M.: Toxicity of 1-(3,4-dihydroxyphenyl)-2-isopropylaminoethanol hydrochloride (Isuprel). Proc. Soc. Exp. Biol. Med. *68:* 501–505 (1948).
9. ESTES, H. E.; DALTON, F. M.; ENTMAN, M. L.; DIXON, H. B. and HACKEL, D. B.: The anatomy and blood supply of the papillary muscles of the left ventricle. Amer. Heart J. *71:* 356–362 (1966).
10. FLECKENSTEIN: II. Annual Meeting, International Study Group, Gargnano (Italy), 25–28 September, 1969. Cited by RONA *et al.,* Reference 16.
11. HATCH, A.; BALAZS, T.; WIBERG, G. S. and GRICE, H. C.: Long-term isolation stress in rats. Science *142:* 507 (1963).
12. MOORE, K. E.: Studies with chronically isolated rats: Tissue levels and urinary excretion of catecholamines and plasma levels of corticosterone. Canad. J. Physiol. Pharmacol. *46:* 553–558 (1968).
13. POUPA, O.; TUREK, Z.; PELOUCH, V.; PROCHAZKA, J. and KROFTA, K.: Increased resistance of the myocardium to anoxia in vitro after repeated application of isoprenalin. Physiol. bohemoslov. *14:* 536–541 (1965).
14. RAAB, W.: Key position of catecholamines in functional and degenerative cardiovascular pathology. Amer. J. Cardiol. *5:* 571–577 (1960).
15. RAAB, W.; BAJUSZ, E.; KIMURA, H. and HERRLICH, H. C.: Isolation, stress, myocardial electrolytes, and epinephrine cardiotoxicity in rats. Proc. Soc. Exp. Biol. Med. *127:* 142–147 (1968).
16. RONA, G.; DUSEK, J.; HUTTNER, I. and KAHN, D. S.: Studies on myocardial resistance. Proceedings of II Annual Meeting of the International Study Group for

Research in Cardiac Metabolism. Instituto Lombardo-Fondazione Baselli 173–204 (1970).

17. RONA, G.; KAHN, D. S. and CHAPPEL, C. I.: Studies on infarct-like myocardial necrosis produced by isoproterenol: A review. Rev. Canad. Biol. *22:* 241–255 (1963).

18. SELYE, H.; VEILLEUX, R. and GRASSO, S.: Protection by coronary ligation against isoproterenol-induced myocardial necroses. Proc. Soc. Exp. Biol. Med. *104:* 343 (1960).

19. WALLACE, A. G. and KLEIN, R. F.: Role of catecholamines in acute myocardial infarction. Amer. J. med. Sci. *258:* 139–143 (1969).

20. WELCH, B. L. and WELCH, A. S.: Graded effect of social stimulation upon d-amphetamine toxicity, aggressiveness and heart and adrenal weight. J. Pharmacol. Exp. Ther. *151:* 331–338 (1966).

21. WENZEL, D. G.: Drug-induced cardiomyopathies. J. Pharm. Sci. *56:* 1209–1224 (1967).

22. WENZEL, D. G. and LYON, J. P.: Sympathomimetic amines and heart lactic dehydrogenase isoenzymes. Tox. Appl. Pharmacol. *11:* 215–228 (1967).

23. WIBERG, G. S. and GRICE, H. C.: Effect of prolonged individual caging on toxicity parameters in rats. Fd. Cosmet. Toxicol. *3:* 597–603 (1965).

Mailing address: Tibor Balazs, D.V.M., Smith Kline & French Laboratories, 1500 Spring Garden Street, Philadelphia, Pa. 19101 (USA).

Department of Clinical Physiology, University of Goteborg, Sweden

HEART UNDER UNUSUAL CONDITIONS: EFFECTS OF GRAVITATION AND CHRONIC HYPOXIA [1]

O. POUPA

INTRODUCTION

In the natural history of life, turning points occurred when organisms started to live under new environmental conditions (transition from aquatic to terrestrial mode of life, *etc.*). In the present era, mankind is preparing far-reaching projects: immigration into space and the deep-sea bed. These conquests of men are conditioned by technological support, preserving environmental conditions equivalent to the terrestrial ones. It is not useless, however, to collect data as to how the structure and function of biological systems was rearranged under the impact of unusual environmental factors. Extrapolation of such "biological solutions" verified by survival and further progress of living bodies might be of help in solving pertinent problems of today. In the present paper, data derived from comparative and ecological cardiology are summarized concerning structural, functional and biochemical rearrangements found in cardiac muscle, as well its oxygen supply, in animals living under unusual and different conditions (natural and/or experimental). They concern the influence of: a) the different impact of gravitation; b) chronic hypoxia.

The influence of chronic hypoxia on cardiac muscle was, for many years, the main topic of interest in the Institute of Physiology, Czechoslovak Academy of Sciences, Prague. Comparative material was collected during expeditions of the research ship M/S *Skagerak* and studied in the Department of Clinical Physiology, University of Goteborg, Sweden. Many valuable data are derived from Hesse's monograph (4).

HEART IN ANIMALS PERFORMING DIFFERENT ANTI-GRAVITY EFFORTS

During phylogeny, there occurred two distinct situations where the biological impact of gravitation changed substantially: a) transition from aquatic to terrestrial mode of life; b) transition from walking to flight. The question is raised how far these ecologically conditioned changes in movement habits are reflected in the properties of cardiac muscle.

Figure 1 shows the relations between body weight (gr) and force (p) necessary

[1] Supported by grants from the Swedish Medical Research Council (Nos K 69-14X-2729-01, B 70-14X-2729-02, K 70-14P-3025-01, and B 71-14P-3025-02A) and Ollie and Elof Ericsson's Foundation.

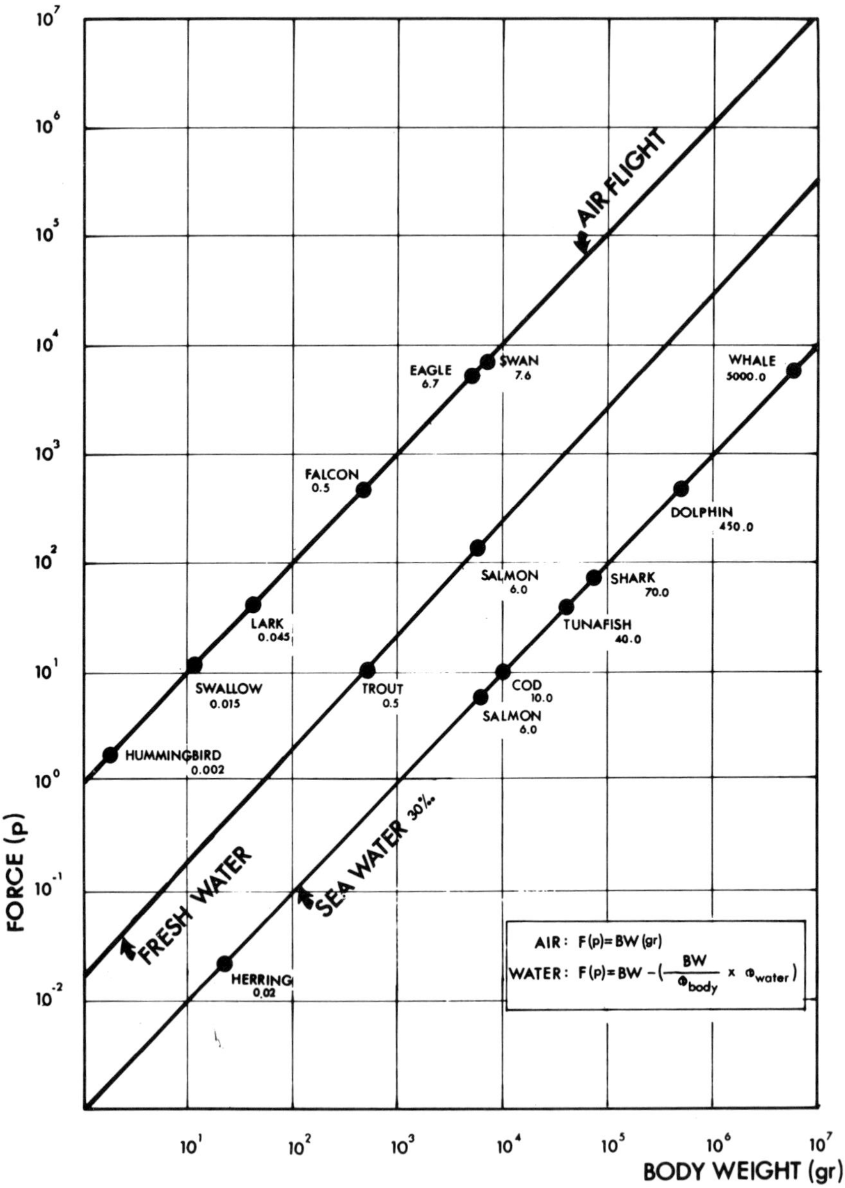

Fig. 1 Body weight and antigravity force relations in different media (air, fresh and salt water). Graph demonstrates force (p) necessary to keep a body of certain weight (gr) floating. Body weights (in kg) of some vertebrates are noted.

to keep the body floating in different media (air, fresh water, salt water of 30% salinity). These data serve to illustrate only the force relations. It is evident that to keep a motionless body of 5,000 kg within sea-water (whale) entails development of a force of the same order required to keep a body of 6–10 kg (swan, eagle) floating in the air. In the light of these relations, it is not surprising that: a) the size of the cardiac "pump" increased during transition from aquatic to terrestrial mode of life; b) among all vertebrates, the highest relative weight of the heart can be found in birds (table I).

Swimmers and nonswimmers; walkers and fliers

Poikilotherm vertebrates were separated into two groups: i) permanent swimmers (aquatic animals); and ii) permanent nonswimmers (terrestrial animals). Homoiotherm vertebrates were grouped into: iii) walkers; and iv) fliers. Relative heart weight was compared between poikilotherm and/or homoiotherm animals, where anti-G effort during movement differs substantially. It was found (fig. 2) that, irrespective of class dependence, the relative heart size in both high-antigravity-effort groups (walking poikilotherms and flying homoiotherms) was greater than in both low-antigravity-effort groups (swimming poikilotherms and walking homoiotherms).

Swim-bladder effect

Transition from aquatic to terrestrial mode of life is connected with considerable structural and functional modifications within the body, terrestrial mode of movement being only one among many operative environmental factors. To demonstrate the influence of increased anti-gravity effort on the heart, it is necessary to select animals that belong to the same vertebrate class and live in an identical environment (water), although antigravity effort performed by them differs substantially. Fishes with and without a swim-bladder fulfill this requirement, since this organ is able to essentially neutralize the

Table I

Heart Ratio in Different Vertebrate Classes*

Classes	n Indiv.	n Spec.	HR	±S.D.
Fishes	171	58	0.081	0.03
Amphibia	138	5	0.299	0.17
Reptilia	50	7	0.188	0.06
Birds	652	129	1.099	0.32
Mammals	370	33	0.643	0.19

* Heart ratio HR = Heart weight × 100/body weight. Total number of individual samples, 1381. Total number of species, 232.

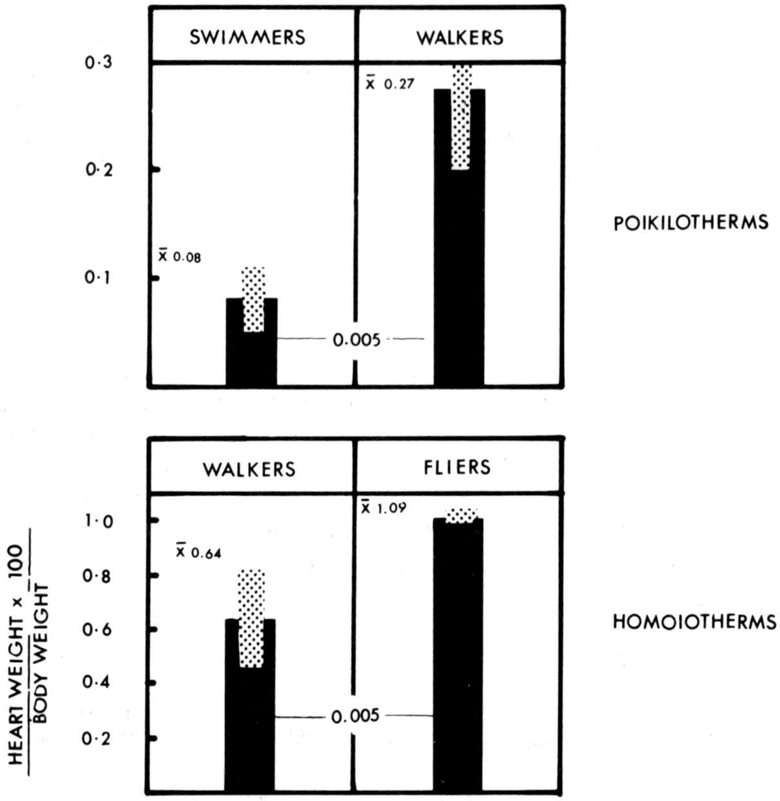

Fig. 2 Differences in relative heart-weight between swimming (upper left) and walking (upper right) poikilotherm and between walking (lower left) and flying (lower right) homoiotherm vertebrates. X and p values. Dotted areas: ± SD.

influence of gravitation. In the majority of Teleosts, the swim-bladder is present whilst, in Selachii, it is generally absent. In accordance with this, a higher relative heart weight was found in Selachii (fig. 3). In Teleosts, the flatfish family (Pleuronectidae) has the lowest relative heart weight among all the fishes studied to date. This finding seems to be paradoxical in view of the fact that they lack a swim-bladder. In species living motionless on the sea-bed, the absence of the swim-bladder does not constitute a handicap, whereas in very active animals, the lack of the swim-bladder represents an extra load during movement. This is the situation in some very active sharks (Selachii), where exceedingly high relative heart weights were found (porbeagle – Lamna cornubica). This tropical shark occasionally visits the North Sea when hunting his favoured migrating prey (mackerel, herrings, and sprats) for long distances.

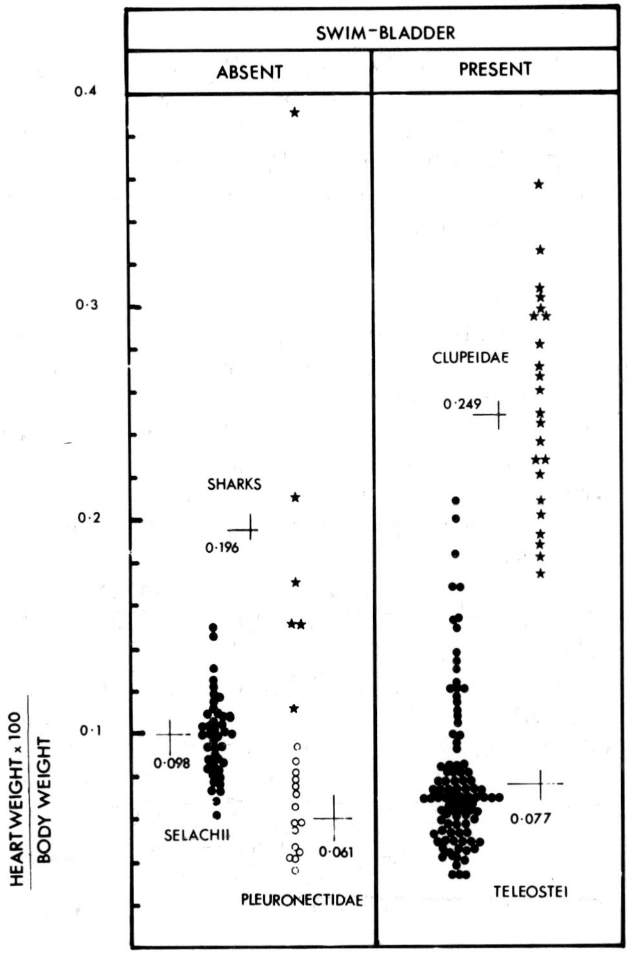

Fig. 3 Relations between relative heart-weight, physical activity, and presence of swim-bladder in salt-water fish. $\overline{X}$ of respective group of fish. 176 samples, 54 species.

Flying fishes

A large effort can be assumed to be made by flying fish (rush movement in the water, anti-gravity effort in crossing water-air barrier and flight). Although information regarding tropical flying fish concerning relative heart weight is lacking, an exceptionally high value was found in the garpike (*Belone belone*). This fish, although it does not fly, belongs to the same suborder as flying fish (Belonidae). It might be of interest to know more about the structure and function of the heart in these peculiar poikilotherms.

Jumping amphibia

In anura, the family of Hyla (tree frogs) have exceptionally large hearts, the relative weight of this organ reaching values that are common in mammals (fig. 4). It has been stated (12) that these animals present an exceptionally high performance as regards jumping: the length of a jump was estimated to be 20–40 times the body length, a value that, in man, means 40–80 m. A jump of such an extent is similar to flight.

Flying mammals

Among all homoiotherm vertebrates, the highest relative heart weight was found in birds, the flight fitness of the respective bird species being highly relevant (24). Some mammals, however, *e.g.*, bats are obligatory fliers, walking as a mode of movement being excluded due to their bodily construction. The heart of these animals has been examined by many authors from other points of view (heterothermy, hibernation). Data concerning the influence of obligatory high anti-gravity effort on the size and other properties of the heart in flying mammals are of interest here.

The relative weight of the bat heart (*Myotis myotis*) is nearly three times that of a walking mammal of the same body size (*Mus musculus;* table II). There are more fibres and more capillaries per mm^2 of cardiac tissue in the bat than in the mouse (fig. 5). The diffusion distance (D/2) is, therefore, shorter in the bat than in the mouse heart. The greater distance between capillaries in the mouse heart is due to the greater size of the cells and not to a greater amount of extracellular water, since the dry to wet weight ratio in both types of myocardium is the same. The diameter of the fibres of the bat heart is about one third smaller than that of the fibres of the mouse heart. The density of the capillary network in the heart of the bat is about one third greater than in the mouse. Apparently, in contrast to the walking mammal, not only the size of the heart increases in flying. The internal structure of the flier's heart is adapted to greater efficiency in at least two ways: a greater supply of oxygen and fuel is ensured by a denser capillary network and, on the other hand, the total mass of the tissue is divided into a higher number of units. In this way, the total surface from which oxygen enters the interior of the cells (total surface of cardiac cells) is much greater in flying mammals. Analogous differences were described between "athletic" and "non-athletic" animals (hare, rabbit (40); wild and laboratory rat (41)), differences found between fliers (bat) and walkers (mouse) being more pronounced.

Growth-pattern of homoiotherm and poikilotherm heart

From the preceding findings, conclusion can be drawn that a high anti-gravity effort of the body correlates with an increased relative weight of the heart in

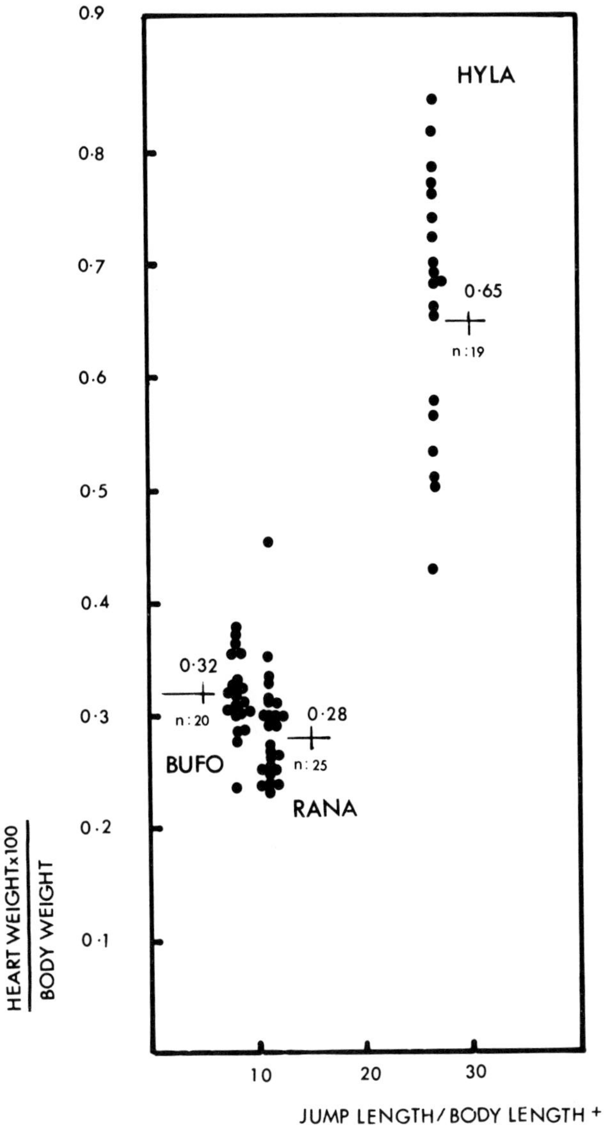

Fig. 4 Relations between relative heart-weight and jumping performance in some anura. n: number of samples within respective frog-family and $\overline{X}$ values; (+ for data see Oliver (12)).

Table II

Heart of Flying (Bat, *Myotis myotis*, Male) and Walking (Mouse, *Musculus musculus* Strain B10 LP, Male) Mammals

Species	n	Body weight g	Heart weight mg	$\frac{\text{Heart weight}}{\text{Body weight}} \times 100$	$\frac{\text{Dry weight}}{\text{Wet weight}} \times 100$	Capillaries/mm^2	Fibres/mm^2
Bat	9	20.0	202.0	1.01	27.3	3717	3328
		±2.0	±20.3	±0.06	±2.7	±310	±272
Mouse	10	28.0	97.9	0.35	28.1	2739	2578
		±1.6	±8.2	±0.02	±1.8	±183	±82
p		0.001	0.001	0.001	–	0.001	0.001

both poikilotherm and homoiotherm animals. The intimate mechanisms of the linkage between both these phenomena are far from being clear. It should be mentioned in this connection that, in poikilotherms and homoiotherms, the growth-pattern of the heart differs substantially. The heart of poikilotherm vertebrates in its most primitive form is a sponge-like system of cells fed by blood flowing through lacunae. In some poikilotherm animals, however, an outer layer of compact musculature can be found. The compact part of the myocardium is supplied by capillaries derived from either coronaries or extracoronary sources. This type of construction of the myocardium exists in species belonging to different animal classes: fishes, amphibia, or reptiles. It cannot, therefore, be interpreted as a systematic class-specific sign. The factors determining the presence or absence of a compact layer within the cardiac wall are unknown. Lacunary and coronary circuits supplying both muscular compartments of the heart ventricle are separated. The sponge-like part of the heart is supplied by pure venous (fishes) or a mixture of venous and arterial blood (amphibia and reptiles), according to whether or not separation of right and left ventricles is anatomically complete, and according to whether or not mixing of arterial and venous blood inside the single ventricle occurs. On the other hand, the outer compact layer receives oxygenated blood only. Questions might be raised concerning the identity (structural, functional, and biochemical) of cardiac cells forming a spongy core, on the one hand, and an outer compact layer, on the other. In this connection, it was found that sensitivity to necrogenic stimuli (isoproterenol) is limited to the sponge-like core of the reptilian heart (13). It is also open to question whether the sensitivity of both these cardiac compartments towards the growth stimuli is of the same order. Growth of capillaries and coronary vessels seems to be directly connected with the evolution of an outer compact muscular layer. It was recently found that, in amphibia (anura) of smaller body weight, coronary vessels do not cross the border between atria and ventricle; in heavier animals of the same species, a rudimentary network of coronary capillaries can be found, but it is limited to the right upper corner of

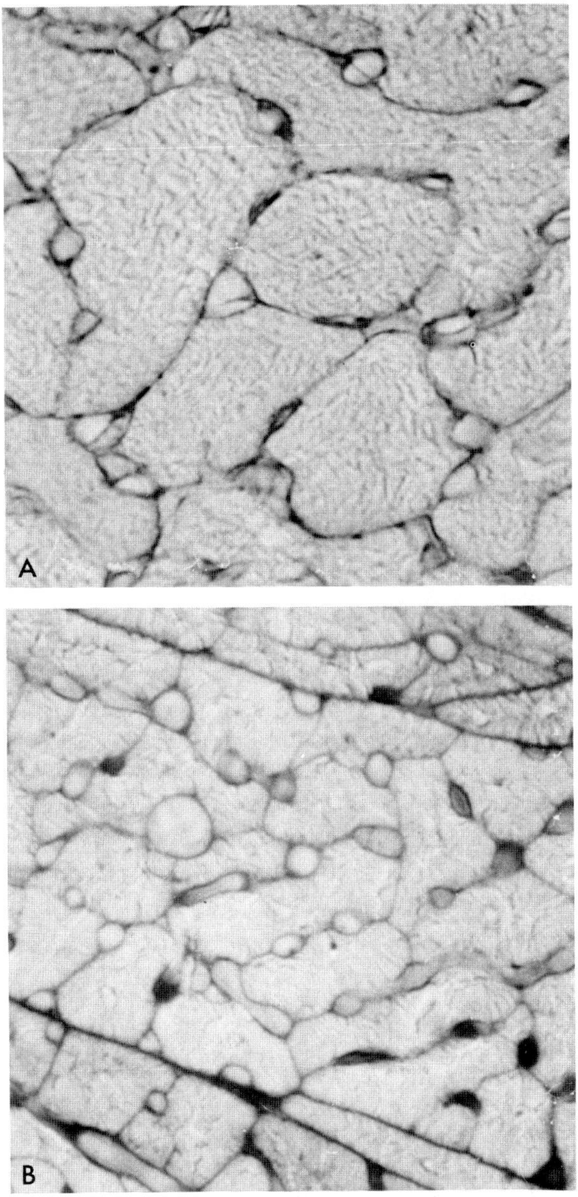

Fig. 5 Cardiac fibers and capillaries in mouse (A) and bat (B). Smaller size of cardiac fibres and higher density of capillary network in bat. PAS stain. ×400 (Wachtlova *et al.* (41a)).

the ventricle (1). It should be stressed in this connection that only this part of the ventricular wall is supplied by purely venous blood. In giant forms of anura (bullfrog), we found a coronary network supplying the whole surface of the cardiac ventricle.

In homoiotherm vertebrates, two types of growth-pattern of the heart can be found: hypertrophic and hyperplastic, the latter being limited to early periods of life. Distinction between both of them is important, since hyperplastic growth only seems to be connected with adequate growth of coronary capillaries (24, 26). Comparative studies concerning the density of the capillary network and the size of cardiac fibres in wild-living animals of highly different body weights (4×10^6 g–2.5×10^5 g) were carried out (29, 40, 41). In all species examined, capillary count and fibre count relation was constant. The size of the cardiac fibres, being proportional to the body size, was inversely proportional to the capillary count. The conclusion can be drawn that, under physiological conditions, there exists a linkage between the number of cardiac fibres and the number of coronary capillaries. On the other hand, such a linkage seems to be lacking between the cardiac fibre size and the number of capillaries.

HEART IN ANIMALS ADAPTED TO HIGH ALTITUDE

Twenty million people live permanently at altitudes above 5,000 m (42). In some instances, the level is in excess of 6,000 m and, exceptionally, 7,000 m is tolerated with no artificial oxygen supply. Interest concerning high-altitude populations was aroused by observations suggesting different patterns of diseases in men living at high altitudes, because such people showed modifications and/or absence of some diseases common in lowland populations (16, 42). This is true of some cardiovascular diseases (16). Among several environmental factors operating at high-altitude, chronic hypoxia seems to be the most relevant. While a large volume of data has been collected concerning human adaptation to high-altitudes, there exists far less information regarding the adaptation of cardiac muscle to chronic oxygen lack.

Resistance of cardiac muscle to acute oxygen lack

Resistance of isolated whole heart or heart muscle tissue to acute anoxia can be determined, in contrast to that of other organs and tissues, since its specific activity, which can be expressed in quantitative terms is: a) maintained *in vitro;* and b) decreased or arrested by acute anoxic load. Two types of anoxic tolerance tests (isolated right ventricle (37); heart-lung preparation (9)) were used to study the question whether cardiac muscle acquires a higher resistance to acute anoxia. In a group of experimental animals previously acclimated to chronic high-altitude hypoxia, it was seen that the anoxic tolerance of the isolated heart muscle was higher than that of the sea-level controls (8, 17, 20,

22) (fig. 6). This finding was confirmed by McGrath and Bullard (11) and by Burlington *et al.* (2). Komives and Bullard (7) observed an analogous effect of high-altitude acclimation in the nerve-muscle preparation of the diaphragm. Recently, Souhrada (35) compared the resistance of the heart of animals acclimated to altitude with that of sea-level controls, using the heart-lung preparation technique; the method employed for testing the anoxic tolerance

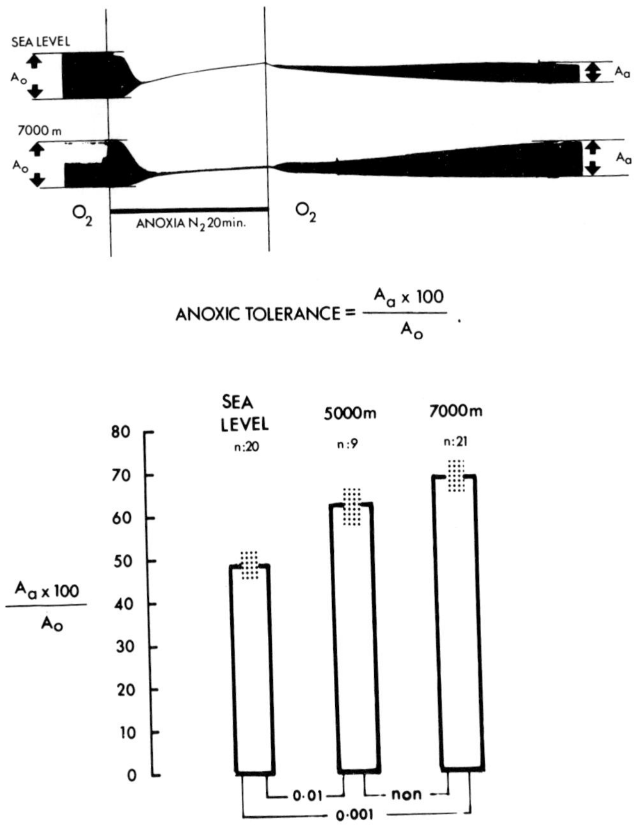

Fig. 6 Difference in resistance to acute anoxia of myocardium isolated from control (sea-level) animals and animals acclimated to high-altitude (7000 m). Upper half of the figure: Isolated right ventricle of the rat (Krebs-Ringer without glucose, 34°C, stimulation-frequency 60/min); medium supplied with 100% O_2. Anoxia: 100% N_2, 34°C, stimulation-frequency 60/min; O_2 supply renewed after 20 min. Anoxic tolerance expressed as $\frac{A_a \times 100}{A_o}$ (height of contractions before (A_o) and after (A_a) anoxic load). Recovery of contractility is better in high-altitude adapted heart (7000 m, lower tracing) than in sea-level heart (upper tracing). Lower half of the figure: Anoxic tolerance of the right ventricle isolated from: sea level controls (n:20) and animals acclimated for 6 weeks to simulated altitudes of 5000 m (n:9) and 7000 m (n:21). $\overline{X}$, ± SD (dotted areas); p values (Poupa *et al.* (20)).

was described in detail elsewhere (37). Anoxic tolerance was expressed: a) as the duration (in min) of anoxia required to stop the ejection of blood into the outflow system of the heart-lung preparation (HR_t); and 2) the degree of recovery of static cardiac work, with the lungs reventilated, after the anoxic period (95% N_2 + 5% CO_2), by oxygen (95% O_2 + 5% CO_2 – Rep%; see table III). Both these values were substantially higher in hearts isolated from altitude acclimated animals. Experimental data on animals living at high-altitude (not artificially acclimated) are not yet available.

The resistance of the myocardium to acute anoxic load can be increased, within several weeks, by exposure to simulated high altitude (5,000–7,000 m). The resistance so achieved is not the result of nonspecific stress: it was increased in adrenalectomized animals, even though adrenalectomy, *per se,* decreases the resistance of the myocardium to acute anoxia (21). A low correlation was observed between an increased concentration of myoglobin in the heart of high-altitude acclimated rats and an increased resistance of cardiac muscle to acute anoxia (23). While this finding does not exclude the possibility that a higher myoglobin content contributes to better survival of the heart in anoxia, it does not seem to be a factor of primary importance.

Metabolic shifts in the cardiac muscle of high-altitude acclimated animals

The observations of Tappan and Reynafarje (38) did not reveal any differences in high-energy phosphate content in the heart of high-altitude and sea-level animals; in the former animals, the activity of ATP-ase was increased. Reynafarje (30) later concluded that, although the concentration of high-energy phosphates is not larger, their turnover in altitude hearts may be faster. Cardiac myosin isolated from hearts adapted to chronic altitude hypoxia shows a certain degree of polymerization (DEAE-cellulose and acrylamide gel separation) and ATP-ase activity is enhanced (15). Komives and Bullard (7) observed that addition of glucose to the medium prolonged survival in anoxia more effectively

Table III

Anoxic Tolerance of the Heart (Heart-Lung Preparation) in Rats (Adult Males) Acclimated to Simulated High Altitude and in Sea-Level Controls (Souhrada (37))*

	n	$R_t$min	Rep%
Sea-level	13	9.5 ± 0.55	37 ± 3
Altitude adapted (9–11 weeks at 7,000 m)	20	16.5 ± 0.65	58 ± 6
Significance		0.001	0.01

* Gas mixtures: 95% O_2 + 5% CO_2 in normoxic test-period and 95% N_2 + 5% CO_2 in anoxic period.

in the organ (phrenic nerve-diaphragm preparation) isolated from altitude-adapted than in that from sea-level animals. Iodoacetate depresses induced resistance of cardiac muscle to anoxic load; an increased capacity for anaerobic glycolysis was therefore assumed by McGrath and Bullard (11). This assumption was recently supported by the findings of Burlington *et al.* (3): the high-altitude adapted heart muscle after initial perfusion by normoxic mixture contains less ATP and CP than the sea-level organ; it produces more lactic acid. After anoxic load (PO_2 = 20 mm Hg), high-energy phosphates and phosphate potential ($ATP/ADP + P_i$) are higher in the high-altitude adapted heart. At the same time, the high-altitude adapted heart produced more lactic acid and the coronary blood-flow was better maintained. Mager *et al.* (10) reported that, within the LDH system, a shift from H to M-type can be found, owing to the altitude adaptation, provided the acclimation was started in young animals. An analogous shift in the LDH system was found in animals suffering from sideropenic anemia (5), but was not observed in adult animals adapted to high-altitude hypoxia. A survey of the data concerning metabolic shifts in high-altitude-acclimated hearts is given in table III.

Role of carbon dioxide

Two conditions were found in which carbon dioxide plays a paradoxical, till now unexplained role in the course of survival in anoxia. It was observed that divers (birds) are more sensitive to anoxia in the absence of CO_2, while the opposite is true in non-divers (21). Paff and Boucek (14) observed that the heart

Table IV

Metabolic Shifts in the Myocardium during High-Altitude Acclimation

Glycolysis	Heart and diaphragm	↑	Komives and Bullard (7)
	Iodoacetate sensitivity	↑	McGrath and Bullard (11)
	LA production	↑	Burlington *et al.* (3)
High energy phosphates	ATP (before load)	↓	
	CP (before load)	↓	
	ATP (after load)	↑	Burlington *et al.* (3)
	CP (after load)	↑	
	$ATP/ADP + P_i$ (after load)	↑	
ATP-ase	Nonpurified	↑	Tappan and Reynafarje (38)
	Purified	↑	Pelouch and Poupa (15)
Myosin	Polymerization	↑	Pelouch and Poupa (15)
LDH system (H to M shift)	In young animals	↑	Mager *et al.* (10)
Myoglobin		↑	Poupa *et al.* (23)

of the mammalian foetus, which has a high tolerance for anoxia (43), is less sensitive to oxygen lack if carbon dioxide is present. The mammalian foetus is a diver and a high-altitude inhabitant, simultaneously. It was of interest, therefore, to examine the effect of the absence and/or presence of CO_2 during anoxic load on the hearts of altitude and sea-level animals (heart-lung preparation (37)). In normal animals, CO_2 shortens the time in which the heart is still able to perform measurable work during anoxia. On the other hand, the altitude heart is able to maintain the work performance for a longer period during anoxia if the lungs are ventilated by $N_2 + CO_2$ mixture rather than by 100% N_2. As a result, the difference in ability to sustain the work-performance during anoxia between the altitude and sea-level heart is less if CO_2 is absent (35) (fig. 7).

Resistance to acute cardiac necrosis in altitude-acclimated animals

The experimental data cited in the preceding paragraphs allow us to conclude that survival of the myocardium in anoxia is better in tissue taken from

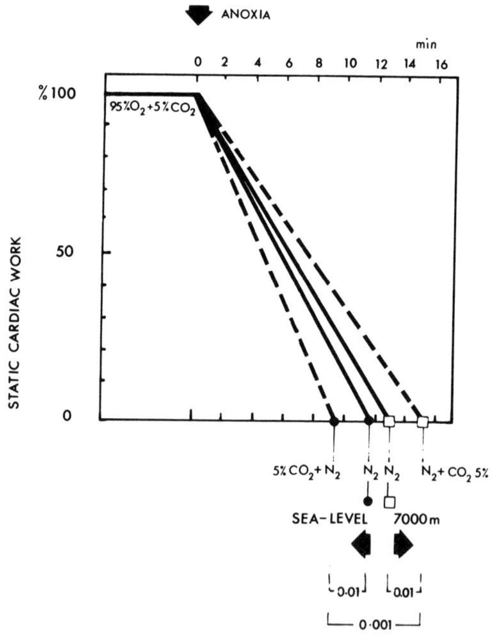

Fig. 7 Effect of carbon dioxide on anoxic tolerance of isolated heart (heart-lung preparation) from sea-level animals and animals acclimated to high-altitude (7000 m). Steady state of static cardiac work (W) in normoxic (95% O_2 + 5% CO_2) period = 100%. Anoxia induced by ventilation of lungs by 100% N_2 or 95% N_2 + 5% CO_2. Period (min) intervening before arrest of static cardiac work by anoxia was shorter in sea-level hearts if CO_2 was present, the reverse being found in high-altitude-adapted hearts (100% N_2 : 95% N_2 + 5% CO_2 ; p: 0.01). (Souhrada *et al.* (37).)

altitude-acclimated than from sea-level animals. Biochemical findings substantiate the view that altitude-acclimated hearts demonstrate higher anaerobic capacity, owing to the metabolic shifts achieved during altitude acclimation. It was therefore worthwhile to examine whether the hearts of altitude animals are more resistant to acute cardiac necrosis (17, 22, 27). Since in such a study the pathogenic impact should be standardized as much as possible, the method used was in accordance with the procedure elaborated by Rona *et al.* (31) in their experiments using isoproterenol (IPR) as the necrogenic agent. The score of cardiac necroses was checked both macroscopically (fig. 8) and microscopically (27). The results were evaluated independently by two observers, neither of whom was aware of the experimental design.

Two groups of rats (fig. 9), one living at 250 m (sea-level group n:33) and one acclimated for 6 weeks at 7,000 m (n:35), received a necrogenic dose of IPR (2 X 40 mg/kg b.w./48 h). All of the animals kept at sea-level showed cardiac

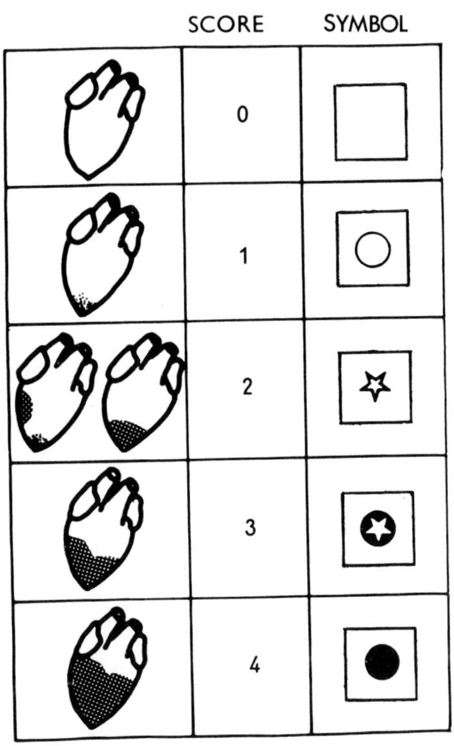

Fig. 8 Score-key to the macroscopical evaluation of IPR-induced cardiac necroses in rats. Score 0 : no pathological change; score 1 : diffuse whitening of the apex of the heart; score 2 : whitening of the apex and well-defined focus in the right ventricle or well-defined focus in the apex; score 3 : well-defined focus in apex and well-defined focus in right ventricle, reaching the left ventricle; score 4 : joined foci in both ventricles. Identical symbols are used in the figure 9. Score-key after Rona *et al.* (31).

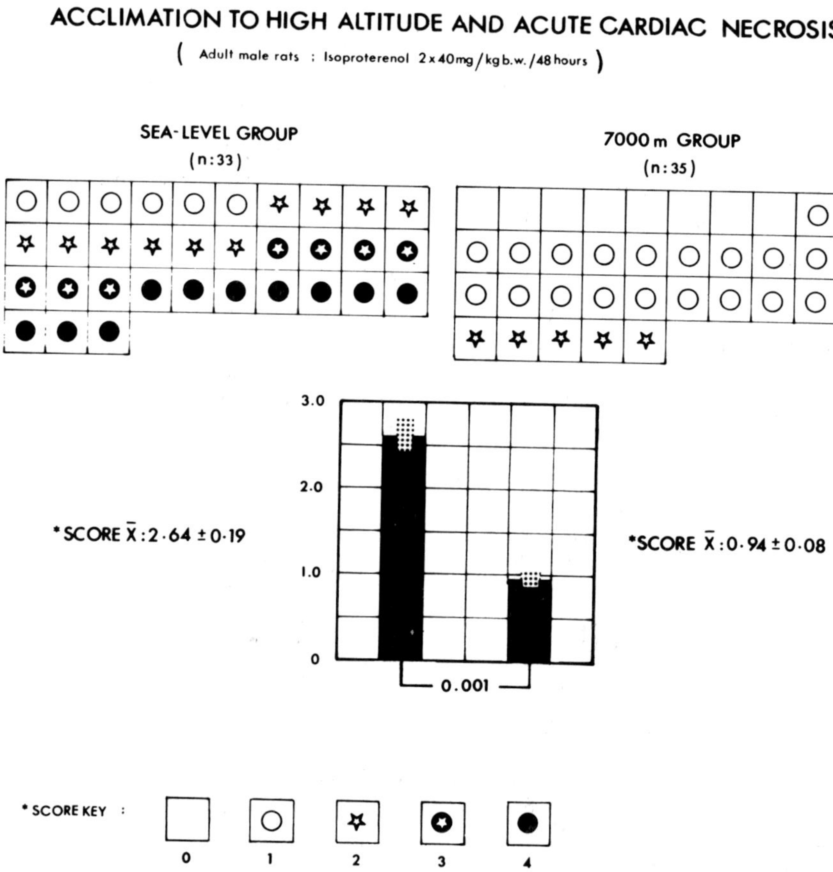

Fig. 9 Resistance of animals acclimated to high-altitude to acute cardiac necrosis. Upper half of the figure: Frequency and extent of cardiac necrosis in individuals of sea-level group (n:33) and altitude-adapted group (7000 m,n:35). Lower half of the figure: Mean score of sea-level (left) and altitude (right) group. Score-key as in figure 8. For explanation, see text. (From Poupa *et al.* (22, 27).)

lesions. In 6 instances, the severity was only slight (whitening of cardiac apex–score 1), while in 10 animals, the degree of necrosis was maximal (massive transmural necrosis and multiple disseminated foci in both ventricles–score 4). Disseminated lesions of the apex and right ventricle (score 2) were observed in 10 animals, and extensive foci in both right and left ventricle (score 3) were found in 7. The total score of lesions in this group was 2.64 ± 0.19. In contrast, out of 35 animals acclimated to 7,000 m, 8 showed no macroscopical lesions. Among the remainder, score 1 (whitening of apex) was observed in most instances (21 animals), and disseminated lesions in the apex and in the right ventricle (score 2) were found in 6. No severe lesions (score 3 and 4) were present in this group. The difference between the mean group-score of the sea-level animals and that of the 7,000 m group (0.94 ± 0.08) was highly significant (0.001). Histological verification of lesions was carried out and the same differences were noted, even though the score of microscopical findings in both groups was somewhat higher than that obtained by macroscopical evaluation. A further analogous experiment was performed, using substantially lower doses of IPR (2 X 0.1 mg/kg b.w./48 h). Macroscopical lesions were not apparent in either group of animals (sea-level and 7,000 m adapted); microscopical lesions were much less frequent in the altitude-acclimated group (27).

It can be concluded from all these experimental series that acclimation to altitude affords a substantial degree of prevention against cardiac necrosis produced by cardiotoxic catecholamines.

Open questions and hypotheses

In interpreting enhanced resistance to IPR-induced cardiac necrosis in animals acclimated to high-altitude, the following questions must be raised: a) the identity (quantitative and qualitative) of catecholamine action in altitude-acclimated and sea-level animals; b) the pathogenetic mechanisms of IPR-induced cardiac necrosis; c) the role of environmental factors that can change resistance and/or susceptibility of the heart to necrogenic stimuli and play a secondary role in the animals during acclimation to altitude; d) whether increased resistance of isolated heart muscle to acute oxygen lack, on the one hand, and increased resistance to necrogens, on the other, are causally connected; and if so, e) whether chronic oxygen lack during altitude acclimation is the stimulus of primary importance.

Not all these questions can be answered at present; most of them deserve further experimental research. Attention might be focused, however, on some recently accumulated experimental data that might help to solve the most pertinent, last-mentioned question. There is no doubt that, during acclimation to altitude, oxygen lack triggers a set of accompanying functional phenomena. Among other factors, carbon dioxide elimination is alleviated owing to the hyperventilation, and acid-base balance must be maintained under new gas-exchange conditions. Attention was recently focused on the relations between the resistance and/or susceptibility of the heart to necrogenic stimuli and acute oxygen lack, on the one hand, and the acid-base balance (local and systemic), on the other. A brief survey of these findings follows:

a) Comparative studies of experimentally induced cardiac necrosis have shown that the resistance to the necrogenic effect of IPR does not correlate with anoxic tolerance of the respective animal species or class (fig. 10). On the other hand, correlation seems to exist with the capacity to eliminate carbon dioxide (18) (fig. 11).

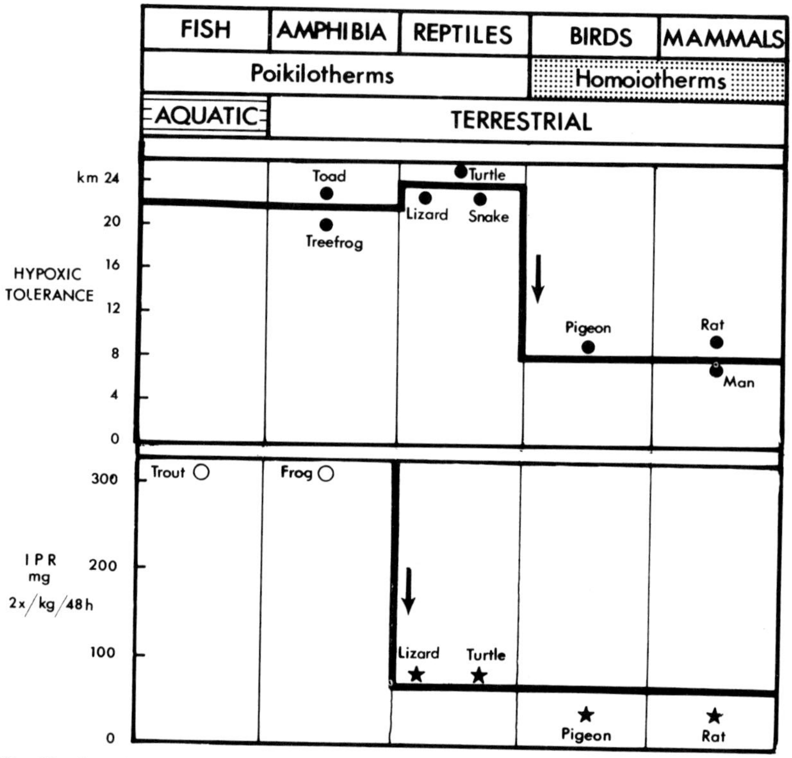

Fig. 10 Anoxic tolerance in vertebrates of different classes (upper curve) and sensitivity to isoproterenol-induced cardiac necrosis (lower curve). Anoxic tolerance expressed as altitude survival (km–data from Hock, Handbook of physiol., sect. 4., p. 841–842 1964). Cardiac-necrosis sensitivity expressed as dose of IPR (2 x mg/kg b.w.) producing lesions in respective animal species (Poupa and Ostadal (24)).

b) In reptiles, cardiac necrosis is localized in the muscular compartment where coronary circulation is lacking (sponge-like core of the heart (13)), the outer compact coronary-supplied layer being left intact. Questions can be raised concerning the differences in the elimination capacity of acid metabolites from capillary-supplied cardiac cortex and by primitive lacunary flow through lacunae from the spongy core of the heart.

c) In amphibia, the cardiac ventricle consists of spongy muscular tissue only, and is supplied by the lacunary blood flow. Cardiac tissue, which proved to be resistant to the necrogenic activity of IPR, was sensitized by a higher environmental temperature. A drop in the blood pH is one of the most prominent phenomena observed in amphibia living in a warm environment (19).

d) Two different procedures were found to induce resistance of mammalian cardiac muscle to necrogenic stimuli: i) ligation of the coronary artery (33, 34); and ii) pretreatment of the animal with small doses of IPR (39). The

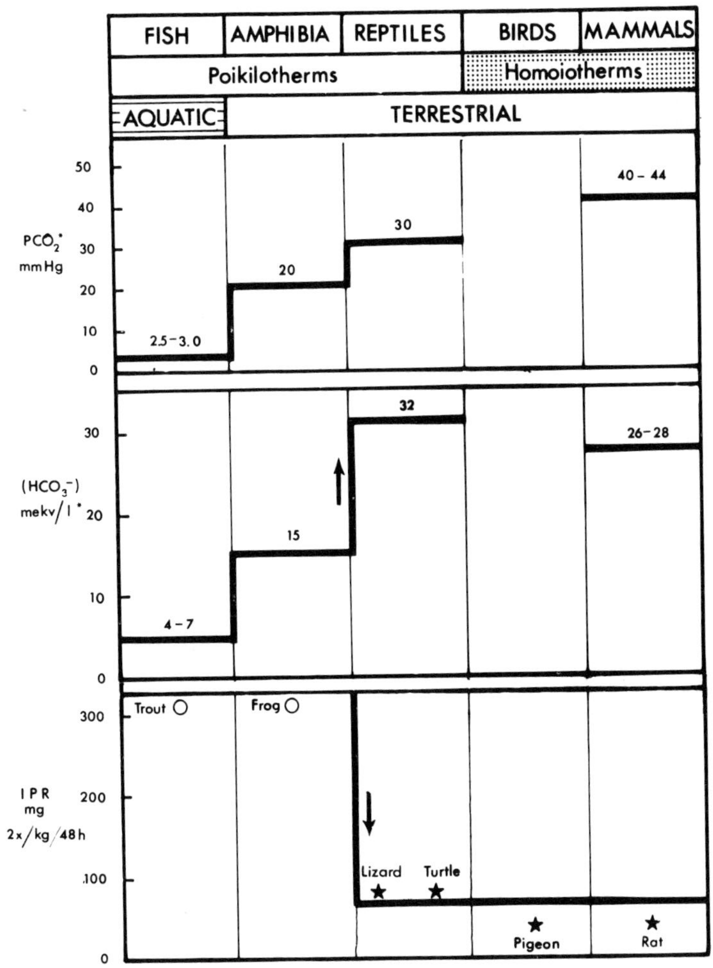

Fig. 11 Blood PCO_2 and (HCO_3^-) changes during phylogeny (upper curves) and sensitivity to isoproterenol-induced cardiac necrosis (lower curve). Due to higher availability of O_2 in the air $\dot{V}$ drops during transition from aquatic to terrestrial environment. CO_2 retention increases in spite of elimination through the skin in amphibia. Skin CO_2 draining capacity being limited in reptiles, blood buffer systems are built up to prevent drop of pH (*for data, see Robin and Murdaugh (30a)). Parallel drop of resistance of cardiac muscle to necrogenic stimuli. Cardiac-necrosis sensitivity expressed as dose of IPR (2 × mg/kg b.w./48 h) producing lesions in respective animal species (for data see Poupa and Ostadal (24)).

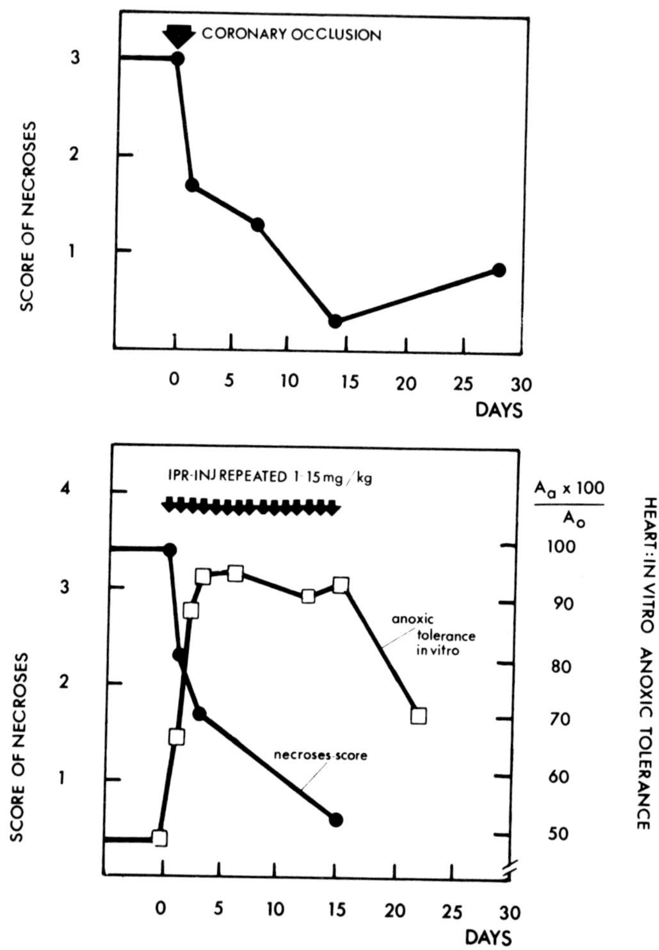

Fig. 12 Protection of cardiac muscle against isoproterenol-induced cardiac necroses by occlusion of left coronary artery or by pretreatment by small doses of isoproterenol. Upper curve: In occluding left coronary artery, rest of myocardium becomes resistant to IPR-induced lesions. Highest drop of sensitivity in 14 days. Test dose of IPR 2 × 50 mg IPR/kg/48 h. (For data see Selye *et al.* (34); Poupa *et al.* (28), and Turek *et al.* (39)). Lower curves: Isoproterenol was applied daily in doses of 1–15 mg/kg during 14 days. Significant drop of sensitivity (●——●) to large necrogenic dose (2 × 40 mg/kg/48 h) was observed 24 h after first injection; peak of protection was achieved within 15 days, the cardiac muscle being practically insensitive to IPR. Right ventricle isolated from IPR-pretreated animals is more resistant to *in vitro* anoxic load (–□–□–), peak being achieved on 3rd day of pretreatment. Induced resistance declines with cessation of IPR treatment (compare 15th and 20th day). For data, see Poupa *et al.* (28) and Turek *et al.* (39).

last-mentioned procedure substantially increases *in vitro* resistance of isolated cardiac muscle to acute oxygen lack (28). The time course and degree of resistance achieved being similar in both experimental procedures (fig. 12), there arises the question concerning a common denominator. It might be of interest to ascertain whether, under both conditions, the buffering capacity of the myocardium increases as a compensatory reaction to the local acidosis induced by coronary ligation or the repeated impact of catecholamine.

e) It was recently found that increased resistance of isolated cardiac muscle to acute anoxia *in vitro* can be achieved, not only by previous altitude adaptation, but also by chronic pretreatment with bicarbonate or tris-buffer, resulting in increased buffering capacity of cardiac muscle (36).

It might be premature to formulate a unifying explanation on the basis of the above-mentioned data. There are, nevertheless, some indications of a convergence to a common focus. Katz and Hecht (6) recently postulated that acidosis is a factor that triggers a chain of derangements leading to interference with the metabolism of Ca-ions (binding with modulatory proteins) and resulting in bulging of the ventricular wall, which seems to be the earliest biochemical injury in initiating acute cardiac necrosis. From all the above-mentioned data concerning conditioned shifts of resistance and/or susceptibility to acute cardiac necrosis, one point seems to be possibly common to all of them: the impact of the acid-base balance of the cardiac cell. This intersecting area appears to be of interest for further research in both environmental and molecular cardiology.

SUMMARY

Data were presented concerning the cardiac muscle of various vertebrates living under conditions where the impact of gravitation differs substantially. A higher heart ratio was found in all vertebrate groups (poikilo- and homoiotherms) where, under natural living conditions, greater antigravity effort must be performed (swimmers *versus* walkers, walkers *versus* fliers, jumpers *versus* nonjumpers, etc.).

The heart in the mammalian flier is relatively bigger, its fibre-size is smaller, and the density of the capillaries higher in comparison with the mammalian walker of the same body size.

Acclimation to altitude substantially decreases sensitivity to experimental acute cardiac necrosis.

Cardiac muscle or perfused heart isolated from altitude acclimated animals is more resistant *in vitro* to acute anoxia. Induced resistance is corticoid-independent. Correlation with cardiac myoglobin content is nonsignificant. Metabolic shifts to higher anaerobic capacity of cardiac tissue were noted.

No parallelism can be found between the resistance of an animal to acute anoxia and the resistance of the myocardium to necrogenic stimuli, when testing different vertebrate species and classes.

The results of experiments concerning shifting sensitivity and/or resistance to acute cardiac necrosis are compared. The role of the acid-base balance of the cardiac cell is stressed.

ACKNOWLEDGMENT

Thanks are due to the Royal Fishery Board of Sweden for permission to participate in the research expeditions of M/S *Skagerak*.

REFERENCES

1. ARNDT, H. und LUBBERS, D.W.: Gefassversorgung und Capillarisierung des Froschherzventrikels. Pflugers Arch. *297:* 115–128 (1967).
2. BURLINGTON, R. F. and MAHER, J. T.: Effect of anoxia on mechanical performance of isolated atria from ground squirrels and rats acclimatized to altitude. Nature *219:* 1370–1371 (1968).
3. BURLINGTON, R. F.; WHITTEN, B. K. and BOWERS, W. D.: Effect of hypoxia on energy metabolism in perfused hearts from altitude-acclimated rats. Fed. Proc. (1970 in press).
4. HESSE, R.: Das Herzgewicht der Wierbeltiere. Zool. Jahrb. Abt. Allg. Zool. Physiol. *38:* 243–364 (1921).
5. JELINKOVA, M.; SOUHRADA, J. and POUPA, O.: The action of sideropenic anaemia on the distribution of lactic dehydrogenase isoenzymes (LDH) in rat heart. Physiol. bohemoslov. *17:* 467 (1968).
6. KATZ, A. M. and HECHT, H. H.: The early "pump" failure of the ischemic heart. Amer. J. Med. *47:* 497–502 (1969).
7. KOMIVES, G. K. and BULLARD, R. W.: Function of the phrenic nerve-diaphragm preparation in acclimation to hypoxia. Amer. J. Physiol. *212:* 788 (1967).
8. KOPECKY, M. and DAUM, S.: Tissue·adaptation to anoxia of rat myocardium. (Tkanova adaptace na anoxii v myokardu krysy). Cs. Fysiol. *7:* 518 (1958).
9. KROFTA, K.; PROCHAZKA, J. and POUPA, O.: The effect of the duration of anoxia, the frequency of stimulation and temperature on the contractility of the rat myocardium injured by anoxia. Physiol. Bohemoslov. *14:* 238–240 (1965).
10. MAGER, M.; BLATT, W. F.; NATALE, P. Y. and BLATTEIS, C. M.: Influence of simulated altitude on the distribution of lactic dehydrogenase (LDH) isoenzymes in the neonatal and adult heart. Fed. Proc. *26:* 665 (1967).
11. McGRATH, J. J. and BULLARD, R. W.: Altered myocardial performance in response to anoxia after high-altitude exposure. J. Appl. Physiol. *25:* 761–764 (1968).
12. OLIVER, J. A.: The natural history of North American amphibians and reptiles. (Van Nostrand, New York 1965).
13. OSTADAL, B.; RYCHTEROVA, V. and POUPA, O.: Isoproterenol-induced acute experimental cardiac necrosis in the turtle (Testudo Horsfieldi). Amer. Heart J. *76:* 645–649 (1968).
14. PAFF, G. H. and BOUCEK, R. J.: Effects of different oxygen and carbon dioxide concentrations on the activity of the embryonic chick heart. Circulation Res. *6:* 88–91 (1958).
15. PELOUCH, V. and POUPA, O.: Changes in the cardiac myosin molecule in animals adapted to high altitude. Physiol. bohemoslov. *18:* 361 (1969).
16. PENALOZA, D.: Physiological patterns: Cardiovascular characteristics of healthy man. In: Life at High Altitudes PAHO/WHO Scientific Publications No 140: p. 27 (1966).
17. POUPA, O.: Resistance of cardiac muscle to acute anoxia in high altitude adaptation. In: Exercise at Altitude (Excerpta medica Monograph 1967).
18. POUPA, O.: Comparative aspects of experimental acute cardiac necrosis with special respect to poikilotherms. In: Drugs and metabolism of myocardium and striated muscle. Symposium International Naucy 1969, pp. 55–75.
19. POUPA, O. and CARLSTEN, A.: Isoproterenol-induced cardiac lesions in frog observed in vivo. Canad. J. Physiol. Pharmacol. *78:* 306–311 (1970)
20. POUPA, O.; KROFTA, K.; PROCHAZKA, J. and CHUAPIL, M.: The resistance of the myocardium to anoxia in animals acclimated to simulated altitude. Physiol. bohemoslov. *14:* 233 (1965).

21. POUPA, O.; KROFTA, K.; PROCHAZKA, J.; RADL, J. and BARBASHOVA, Z. I.: The effect of adrenalectony on changes in resistance of the isolated myocardium to anoxia and on the myoglobin content of the heart during adaptation to high altitude hypoxia. Physiol. bohemoslov. *15:* 447 (1966).

22. POUPA, O.; KROFTA, K.; PROCHAZKA, J. and TUREK, Z.: Acclimation to simulated high altitude and acute cardiac necrosis. Fed. Proc. *25:* 1243–1246 (1966).

23. POUPA, O.; KROFTA, K.; RAKUSAN, K.; RADL, J. and BARBASHOVA, Z. I.: Myoglobin concentration of the heart and resistance of the isolated myocardium to anoxia in vitro during adaptation to high altitude hypoxia. Physiol. bohemoslov. *15:* 450 (1966).

24. POUPA, O. and OSTADAL, B.: Experimental cardiomegalies and "cardiomegalies" in free-living animals. An. N.Y. Acad. Sci. *156:* 445–468 (1969).

25. POUPA, O. and RAKUSAN, K.: The terminal microcirculatory bed in the heart of athletic and non-athletic animals. In: Physical activity in health and disease pp:18–29 (Universitetsforlaget, Oslo 1966).

26. POUPA, O.; RAKUSAN, K. and OSTADAL, B.: The effect of physical activity upon the heart of vertebrates. In: Physical Activity and Aging: Medicine and Sport *4:* 202–233 (Karger, Basel; University Park Press, Baltimore 1970).

27. POUPA, O.; TUREK, Z.; KALUZ, M. and KROFTA, K.: Acute infarct-like necrosis in high-altitude adapted rats. Physiol. bohemoslov. *14:* 542 (1965).

28. POUPA, O.; TUREK, Z.; PELOUCH, V.; PROCHAZKA, J. and KROFTA, K.: Increased resistance of the myocardium to anoxia in vitro after repeated application of isoprenaline. Physiol. bohemoslov. *14:* 536 (1965).

29. POUPA, O.; WACHTLOVA, M.; RAKUSAN, K. and OSTADAL, B.: Capillaries and cells in cardiac muscle of animals of various body size, muscular activity and mode of life. Abstracts of XXIV International Congress of Physiological Sciences, Washington (1968).

30. REYNAFARJE, B.: Physiological patterns: enzymatic changes. In: Life at High Altitudes PAHO/WHO Scientific Publications No 140: p. 401 (1966).

30a. ROBIN, E. D. and MURDAUGH, H. V.: Quantitative aspects of vertebrate gas exchange. In: Development of the lung. Ciba. Found. Symposia (Churchill London 1967).

31. RONA, G.; CHAPPEL, C. I.; BALAZS, I. and GAUDRY, R.: An infarct-like myocardial lesion and other toxic manifestations produced by isoproterenol in the rat. Arch. Pathol. *67:* 443 (1959).

32. SCHOLANDER, P. F.: Experimental investigations on the respiratory function in diving mammals and birds. Hvalradets Skrifter Norske Videnskaps-Akad. Oslo, No 22: p. 131 (1940).

33. SELYE, H.; MENDELL, P. und STREBEL, R. R., Jr.: Verhutung der normalerweise durch Isoproterenol Erfeugten Myokardnekrosen durch einseitige Koronarligatur. Arzneimittelforsch. *11:* 255 (1961).

34. SELYE, H.; VEILLEUX, R. and GRASSO, S.: Protection by coronary ligature against isoproterenol-induced myocardial necroses. Proc. Soc. Exper. Biol. Med. *104:* 343 (1960).

35. SOUHRADA, J.; MRZENA, B.; POUPA, O. and BULLARD, R. W.: Functional changes of cardiac muscle in adaptation to two types of chronic hypoxia. J. Appl. Physiol. *30:* 214–218 (1971).

36. SOUHRADA, J. and BULLARD, R. W.: Tissue buffering capacity and tolerance of the heart to acute anoxia. Fed. Proc. (1970, in press).

37. SOUHRADA, J.; RAKUSAN, K. and POUPA, O.: The heart-lung preparation in the rat: acute anoxia and temperature. Physiol. bohemoslov. *17:* 209–216 (1968).

38. TAPPAN, D. V. and REYNAFARJE, B.: Tissue pigment manifestations of adaptation to high altitude. Amer. J. Physiol. *190:* 99 (1957).

39. TUREK, Z.; KALUS, M. and POUPA, O.: The effect of isoprenaline pretreatment on the size of acute myocardial necrosis induced by the same drug. Physiol. bohemoslov. *15:* 353–356 (1966).

40. WACHTLOVA, M.; RAKUSAN, K. and POUPA, O.: The coronary terminal vascular bed in heart of the hare (Lepus europeus) and the rabbit (Oryctolagus caniculus dom.). Physiol. bohemoslov. *14:* 328 (1965).

41. WACHTLOVA, M.; RAKUSAN, K.; ROTH, Z. and POUPA, O.: The terminal vascular bed of the myocardium in the wild rat (Rattus norvegicus) and the laboratory rat (Rattus norvegicus lab.). Physiol. bohemoslov. *16:* 548–554 (1967).

41a. WACHTLOVA, M.; POUPA O. and RAKUSAN, K.: Quantitative differences in the terminal vascular bed of the myocardium in the brown bat (Myotis myotis) and the laboratory mouse (Mus musculus). Physiol. bohemoslov. *19:* 497–500 (1970).

42. WHO/PAHO/IBP JOINT SESSION: High Altitude Populations, Washington (1967) (not published).

43. WILDENTHAL, K.: Factors promoting the survival and beating of intact foetal mouse hearts in organ culture. J. Mol. Cell Cardiology *1:* 101–104 (1970).

Mailing address: Prof. O. Poupa, Department of Clinical Physiology, University of Goteborg, Goteborg (Sweden).

Park City Hospital, Bridgeport, Connecticut, USA

ROLE OF PLASMALOGEN IN HEART DISEASES

K. A. OSTER

Although a majority of investigators accept the concept of a dietary origin for atherosclerosis and heart disease, there is a wide disagreement as to the nature of the dietary source responsible for the genesis of the atherosclerotic plaque. The controversy between promoters of the saturated fatty acid theory of origin and those who accuse simple carbohydrates is secondary to the almost unanimous indictment of high blood cholesterol levels as a major cause of vascular-induced heart muscle pathology. The recommended dietary changes deriving their rationale from the hypercholesterolemia concept are largely uninspiring, unrealistic, and impractical for the population at large. An increasing need is felt for new ideas to elucidate the origin of the chemical processes leading to both atherosclerosis and heart disease, since present theories are disappointingly unproductive.

This study attempts such a different approach. Generally recognized laws of biology, pharmacology, and epidemiology, together with histochemical observations, have been applied to the metabolism of a phospholipid, plasmalogen, situated strategically in the cell membrane, that critical area which is involved in the electrophysiological phenomena of heart action and is the target of so many drugs used in treating heart disease. The following data are presently available to support the postulated new concept of *plasmalogen disease.*

PLASMALOGENS

Chemistry

Rapport and Franzl (22) described the chemical formula of plasmalogen as a vinyl ether of glycerophosphocholine with one of the hydroxyl groups esterified by an unsaturated fatty acid. Choline plasmalogens are prevalent in heart muscle, comprising 20–30% of all the phospholipids found in that organ. Ethanolamine plasmalogen, on the other hand, is prevalent in the myelin of nerve tissue.

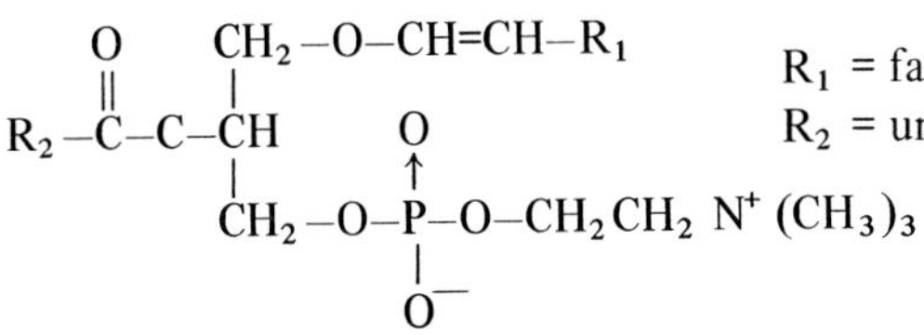

R_1 = fatty aldehyde residue
R_2 = unsaturated fatty acid residue

Choline Plasmalogen

Two enzymes are involved in the metabolic degradation of plasmalogen: phospholipase A, an algesic substance found in heart muscle and also in bee and snake venoms; and vinyl etherase, described in 1965 (31). The pharmacological qualities of the latter enzyme are not yet well known. Magnesium ions enhance the cleavage of ethanolamine plasmalogen (2). Phospholipase A hydrolyzes unsaturated fatty acids, and vinyl etherase hydrolysis results in plasmals, aldehydic compounds, usually stearal or palmital. The metabolism of plasmalogens is described in figure 1.

Histochemistry

Plasmalogens are found in specific locations in many organs. They are abundantly present in muscle tissue and in the myelin sheath of nerve fibers (15), but are not present in the liver or in the mucous membranes of the small intestine.

Pharmacology

Oster (18, 19, 21) has demonstrated that plasmal aldehydes may act as acceptor substances for many active drugs that contain free $-NH_2$ or =S-H

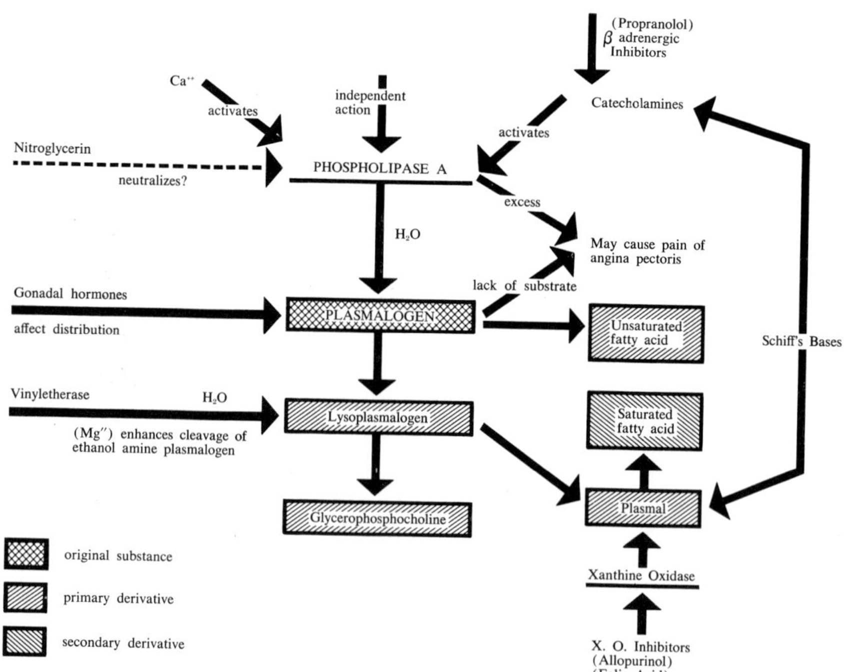

Fig. 1 Proposed pathway of plasmalogen metabolism.

groups. Table I lists some reacting and non-reacting drugs and chemical compounds. Some compounds having cardiac rhythm-controlling effects are found among the combining drugs.

Endocrinology

Oster (13, 14, 20) found that plasmalogen distribution in the intercorticomedullary zone of the rat kidney was influenced by sex hormones, the male hormones effecting a deposition of plasmalogen while estrogens and progesterone exert a plasmalogen-depleting effect. Waxing and waning of plasmalogen in the intercorticomedullary zone of the female rat kidney followed the pattern of the estrous cycle. It was not established which powerful male-like hormone in the rat hormonal equipment possessed the necessary strength to overcome the estrogen and progesterone effect.

Pathology

Miller *et al.* (11) described a diminution of plasmalogen in the atherosclerotic lesion. Oster and Hope-Ross (17) observed the disappearance of plasmalogen from heart muscle without concomitant cell necrosis, in a case of fatal myocardial infarction. On examining the aortas of juvenile drowning victims, Oster (16a) noted that the extensive fatty streaks present did not contain plasmalogen.

XANTHINE OXIDASE

Chemistry

Xanthine oxidase, an enzyme found in certain mammalian tissues, catalyzes the oxidation of purines, aldehydes, and other compounds. Xanthine is oxidized

Table I

Drugs and Chemicals as Reactants with Plasmal Aldehydes

Combining	Noncombining
Barbituric acid	Acetanilide
Thiobarbituric acid	Nitroaniline
β-Phenylethylamine	2-Aminopyridine
Cystein HCl	Benzylamine
Paraaminobenzoic acid	*d,l,α*-Phenylethylamine
Hydralazine HCl	Tyramine HCl
Rhodanine	Amphetamine sulfate
Aminophenol	Ephedrine sulfate
Toluidine	Histamine phosphate
Naphthylamine	Urea
Benzidine	Amino Acids

to uric acid, aldehydes to the corresponding acids. It was found that plasmal aldehydes, when liberated from their plasmalogen bondage by the action of HCl, were oxidized by xanthine oxidase to stearic and palmitic acids (18). Xanthine oxidase has been described as a metallo-flavo-protein, containing molybdenum and iron bound firmly to the apoprotein. Schardinger (27) reported the presence of this enzyme in bovine milk, where it is part of the microsomal particle. It is also found in the milk of sheep and goats, but is absent from the milk of sows and mares (12). Human milk is virtually devoid of the enzyme.

Enzyme absorption

The pharmaceutical probability that orally administered enzymes may be absorbed rather than digested has been shown in the case of chymotrypsin (8). Demonstrable increases in blood levels of chymotrypsin and similar acting enzymes were noted and their clinical efficacy has been utilized in many cases of trauma and blood extravasation.

MICROPARTICLES IN NUTRITION

In his studies on the absorption of microcrystals, Reimann (23) discovered microparticles in heart muscle and photographed them with polarized light. In the generic drug controversy, it was shown conclusively that certain pharmaceutical preparations, though of identical chemical structure, exhibited vastly different levels of biological availability. Most of these well-publicized diversities were caused by variation in particle size and by excipients added to the final pharmaceutical preparation. Micronization of drug particles in the generic preparation as well as in known pharmaceuticals resulted in a higher blood level of the active substance (30).

Principles of pharmaceutical preparation should be extended to foods and their additives. Micronization of the ingredients of a particular food would produce a higher blood level of those ingredients than would be possible with the original particle size. Most foods are chewed, and their particles are digested. There is, however, one almost universally consumed food that has been micronized, especially since 1938 in the U.S.A. That food is bovine milk.

Since bovine milk contains active xanthine oxidase even after pasteurization, homogenization (micronization) of the milk produces at least 3.5 times more particles 1 μ or less in size than are present in unhomogenized milk (4). Although the quantity of xanthine oxidase may be identical in pasteurized milk, pasteurized-homogenized milk, and curdled milk, the enzyme's biological availability would be highest with the microparticle-laden homogenized milk and almost absent from curdled milk with its preponderance of large size particles. On this basis, we would have to discount the belief generally accepted by nutritionists that only the caloric value and chemical content (carbohydrates, proteins, and fats) of foods are important and then consider the preparation and particle size of foods as being important in their biological effects.

Shaper (28) compared blood cholesterol levels of two nomadic tribes in East Africa: the Samburu, who consume bovine milk and have a relatively low serum cholesterol value of approximately 190 mg%, and the Rendille, who drink camel's milk and present a serum cholesterol pattern that closely parallels that of "Western" groups, peaking to 259 mg% at age 55. Here again, particle size may be helpful in explaining the disparity. It is known that camel's milk contains considerable numbers of fat globules of such small size that, despite prolonged churning, it will not yield any great amount of butter.

EPIDEMIOLOGICAL STUDIES

Epidemiological studies concerned with the incidence of atherosclerosis that rely on death rates, number of myocardial infarctions, electrocardiographic findings, and serum cholesterol determinations are statistically unreliable. Undiagnosable atherosclerotic changes and myocardial scarring are present in many more people than the unfortunate number who exhibit often fatal manifest myocardial infarction. Autopsies of young males that have died from war wounds have proven that changes in the arterial wall may be present without causing any circulatory symptoms (5). Among the persons afflicted with atherosclerosis or myocardial scarring, only a limited number suffer actual heart attacks. Many reach a ripe old age despite the presence of extensive changes in the arterial wall or of myocardial involvement, and are not considered in statistical surveys. While bearing in mind these limitations,use was made of published statistics on death rates from myocardial infarction as well as data on the consumption of cows milk, in order to pursue the theory of ectopically deposited xanthine oxidase absorbed in microparticle size from ingested cow's milk. The countries selected for inclusion in the following table were those that occupy extreme positions (low or high) in the scale of reported death rate and extremes in bovine milk consumption. Unlike in other studies, the actual number of deaths is considered arbitrary and without relevance to the statistical accuracy. Therefore, only ranges are utilized.

The Japanese, as well as other oriental populations, do not consume bovine milk in any significant amount. Their rate of death from myocardial infarction is low. The Swiss reportedly drink mostly milk that has been boiled, a process that would inactivate xanthine oxidase; here too, the death rate from myocardial infarction is only moderate. Although milk consumption in the United States is almost identical with that of the Swiss, the death rate from myocardial infarction in the former is on the upper end of the scale, nearly doubled. The United States population consumes milk in homogenized form, and the biological availability of the xanthine oxidase contained in homogenized milk is

Table II

Death Rate from Myocardial Infarction Correlated with Milk Consumption of Various Populations

Country	Death from myocardial infarction	Daily milk consumption (ml)	Particle size milk preparation(μ)	Xanthine oxidase content Lück units*
Japan	Very low	55	1–20 Powdered + homogenized	9 (powdered) 7 (homogenized)
Switzerland	Medium	320	1–20 Raw, boiled milk	0 inactivated
United States	Very high	375	1–5 Homogenized milk	7 (factor 3.5)
Finland	Very high	850	1–20 Pasteurized (+ homogenized)	13
East Africa (Masai)	Almost nil	7000	Macroparticles curdled milk	None biologically available

* 1 Lück unit = $\frac{2500}{t}$, where t = time for decolorization of methylene blue; 2500 = a complex factor related to the number of μmole of xanthine oxidized /min/mg protein at 25°C.

at least 3.5 times greater than that of unprocessed milk. Therefore, the average daily consumption of homogenized milk in the United States is actually the equivalent of approximately 1,300 ml of unhomogenized milk, when the biological availability of milk xanthine oxidase is considered. The Finns report an extremely high milk intake per person, with only some of it homogenized. Their death rate from myocardial infarction is equal to and even surpasses that of the United States.

The Masai of East Africa are the bane of the protagonist of the hyperlipidemia-induced atherosclerosis theory. This well-financed and well-publicized concept is based on data gained from artificially created conditions in animals rather than on conditions found in this tribe of humans who would provide an exceptionally good source material for studies of high saturated fatty acid intake with no resulting atherosclerosis or hypercholesterolemia. These people ingest 7 l or more of curdled milk daily, almost exclusively; thus, saturated fats constitute about 60% of their diet. Curdled milk is characterized by large-sized particles and no biologically available xanthine oxidase. In similar fashion, many high-milk-consuming population groups prefer soured milk or boiled milk to raw or processed milk.

Just as Reimann (23) demonstrated the presence of microparticles in human heart muscle, we can postulate that microparticles of xanthine oxidase in processed milk may be absorbed and deposited ectopically in the heart muscle. Small particles, absorbed from the intestines, are carried via the lymph stream into the subclavian vein, the right heart, the pulmonary circulation, the left heart, and then into the aorta and the generalized circulation. Substances absorbed in this way bypass the liver and enter the arterial system first. Xanthine oxidase finds a ready substrate, plasmal, from plasmalogen *in situ* in the heart muscle and in the arterial wall. Plasmalogen is present in the body in very small amounts during infancy, but increases with the aging process. Therefore, this chemical reaction will not take place in the early development stages when most milk is consumed.

Although normal human blood serum does not contain xanthine oxidase, preliminary findings by Ross *et al.* (26) revealed that the enzyme may be present in humans in the atherosclerotic plaque of the arterial wall and in heart muscle adjacent to the coronary arteries. It is known that human serum contains an inhibitor of the enzyme.

PAIN IN HEART DISEASE

The cause of the intense pain felt in angina pectoris and myocardial infarction has not been explained to anyone's satisfaction by the concept of ischemic heart disease (1). The plasmalogen-based explanation of heart disease may offer a plausible explanation for the origin of the pain. It would be ascribed to the algesic property of the enzyme phospholipase A (6, 10). The pain would be noticeable if the enzyme were to be deprived of one of its substrates, *e. g.*, diminution of plasmalogen in areas of the heart muscle by ectopic xanthine oxidase.

This theoretical consideration was tested by Oster (16), using the xanthine oxidase inhibitor allopurinol in numerous patients suffering from angina pectoris at rest. A quantitative dose response occurred, resulting in a reduction of the severity and frequency of the anginal attacks. The rationale for this treatment arose from the assumption that ectopic xanthine oxidase acts as a constant oxidizer of plasmal liberated from plasmalogen, causing it to disappear completely. Phospholipase A *in situ* in the heart muscle may be activated by calcium ions, catecholamines, or other, as yet not well-known, influences (10). The inhibition of anginal pain by propranolol, a blocker of beta adrenergic receptors, may then be explained by the blocking of the catecholamine influence. These mechanisms are shown schematically in figure 1.

Still unexplained is the mechanism of action of nitroglycerin in the alleviation of anginal pain. Neutralization of phospholipase A, or the inhibition of its effect on plasmalogen, or, lastly, its potential combination as nitrite ion (supplied by nitrate ions) with catecholamines, resulting in the formation of nitrosamines, are chemical possibilities that are under investigation.

INCONSISTENCIES IN THE HYPERLIPIDEMIA RATIONALE

There is a known difference between the incidence of myocardial infarction in women of child-bearing age and that in males of the same age group, the ratio being about 1 : 8. This disparity exists despite consumption of the same diet by males and females, and although both sexes are subject to similar mental stresses. Plasmalogen has been found to be a target substance for sex hormones (13). It is quite possible that the cyclical changes found in such tissues as the rat kidney may also occur in human heart muscle. In the female, therefore, with the changing hormone levels during the menstrual cycle, some of the plasmalogen depleted by the action of ectopic xanthine oxidase may be restored to the heart muscle cells. However, proof of this assumption is almost impossible to obtain. Atherosclerotic lesions found in both sexes are described as almost equal in all age groups (24).

The spotty distribution of so-called ischemic lesions is extremely difficult to explain by the concept of mechanical origin of these lesions. The ectopic xanthine oxidase theory would provide a plausible explanation also for the origin of ectopic foci causing arrhythmia. Forming part of the cell membrane, plasmalogens are in a strategic location to control cell permeability. One might postulate that a disappearance of plasmalogen from the cell membrane may cause a loss of equilibrium between endocellular potassium and ectocellular sodium, resulting in a change in the electrical potentials, which may be recorded electrocardiographically and act as new foci of heart rhythm control. The sarcolemmal membrane is also most probably the site at which many of the drugs that affect the function of the heart exert their influence (20). Proved *in vitro,* plasmal in the cell membrane may also act as receptor for many drugs (19).

APPLICATION AND CONCLUSIONS

There is a basic species difference in the xanthine oxidase content of human and bovine milk. The author has attempted to show that this enzyme, when absorbed in small particle size from the intestinal tract, may react with an essential and strategically situated component of the cell membrane, the plasmal moiety of plasmalogen. Lesions created by this chemical reaction may represent the first damage to the cell integrity that requires repair. Repair mechanisms can vary in different locations: scar formation in the myocardium, and possible cholesterol and fibrin infiltration in the arterial wall. This concept would create a new disease entity, the *Plasmalogen Disease,* resulting from localized phospholipid depletion, as opposed to the concept of tissue damage allegedly caused by a plethora of lipids in the circulating blood.

The augmented biologically available xanthine oxidase in homogenized bovine milk could conceivably be viewed a "food additive" capable of effecting tissue changes that would occur simultaneously in the myocardium and in the arterial wall, both plasmalogen-containing tissues. These pathological events would have their onset in youth, when so much bovine milk is consumed long after weaning, leading to the creation of fatty streaks and culminating in full-fledged atherosclerosis and myocardial scarring in the later decades of life (7).

The proof of this new concept of the genesis of atherosclerosis as an expression of plasmalogen disease is difficult to achieve and does not derive from often misleading results of blood analysis. Significant results may be obtained from tissue examination, pharmacological results of enzyme inhibition studies, and epidemiological investigation of milk consumption and usages of various population groups.

The concept of the mechanical origin of heart disease is under serious challenge (3). Application of this theory to the treatment of myocardial disease has afforded only symptomatic improvement of the victims. Prevention of myocardial infarction by attacking the presence of hyperlipidemia has also encountered adverse criticism. The proponents of this unproven theory, which was promoted chiefly by the dicta of the American Heart Association, advise *ex cathedra* the absolute avoidance of whole milk, butter, and cheese, completely disregarding that the recommended diet is unrealistic and unacceptable to most people. This unfounded attack on the consumption of milk and milk products spawns such wasteful developments as 99% fat-free homogenized milk, which was found to have the same xanthine oxidase content in microparticle size as the normal pasteurized variety (25). The recommended eschewal of dairy fat consumption by population groups is abetted by commercial interests (9).

Application of the expounded theory of ectopic xanthine oxidase affecting the plasmalogen content of myocardial cell membranes would be quite simple to accomplish as a public health measure. Such a step might lower the United States level of myocardial infarction and atherosclerosis to that found, for instance, in Switzerland, representing a reduction of at least 50%. The technological pseudoadvancement of milk homogenization should be replaced by making available to the consumer a biologically active xanthine oxidase-free milk resembling human milk. This proposal would be realistic, even though the results cannot, as yet, be fully guaranteed by the data at hand. In these times of doubt, when the biological effects of food additives, pesticides, and air pollutants are being subjected to question, it would be relatively simple to eliminate a food additive that is produced as a result of various forms of processing.

The proof of the theory would be forthcoming if it could be shown that youth nourished with biologically active xanthine oxidase-free milk exhibit almost no fatty streaks in their arteries and, in later age, display fewer and lesser secondary effects of atherosclerosis and heart disease. At present, the author is attempting to shortcut this rather lengthy process by giving patients with hyperuricemia and angina pectoris at rest another known xanthine oxidase inhibitor, folic acid. Preliminary results have been favorable; elevated uric acid levels were reduced on a dose-response basis, and the frequency of anginal attacks was diminished, although not as spectacularly as with the use of allopurinol. When two different chemicals having one target effect in common, namely inhibition of xanthine oxidase, are shown to also produce the same clinical effect, this should be considered a "giant step" forward in the pharmacological proof of a new concept of the genesis of some heart diseases.

SUMMARY

A biochemically plausible mechanism of the cause of some heart diseases is presented, based on plasmalogen metabolism. The importance of particle size to intestinal absorption is applied to explain the possible dietary origin of both atherosclerosis and heart disease. Such observations as the lower incidence of heart attacks in women of child-bearing age, absence of atherosclerosis in the Masai, cause of anginal pain, absence of coronary artery narrowing in a high percentage of angina pectoris patients, all of which are unexplainable by the hyperlipidemia theories, may well be clarified by the plasmalogen-phospholipase-xanthine oxidase interaction concept of cardiovascular disease. An urgent appeal is made: 1) to immediately discontinue homogenization of milk; and 2) to create a biologically active xanthine oxidase-free milk for the consumer.

REFERENCES

1. AISNER, M.: Angina pectoris 1768–1970. New Engl. J. Med. *282:* 746–748 (1970).
2. ANSELL, G. B. and SPANNER, S.: The magnesium-ion-dependent cleavage of the vinyl ether linkage of brain ethanolamine plasmalogen. Biochem. J. *94:* 252–258 (1965).
3. BAJUSZ, E.: The terminal electrolyte-shift mechanism in heart muscle; its significance in the pathogenesis and prevention of necrotizing cardiomyopthies in: Electrolytes and Cardiovascular Diseases (S. Karger, Basel 1965).
4. DOAN, F. J.: Changes in the physico-chemical characteristics of milk as a result of homogenization. Quart. Rev. Pediat. *8:* 194–200 (1953).
5. ENOS, W. F.; BEYER, J. C. and HOLMES, J. C. and HOLMES, R. H.: Pathogenesis of coronary disease in American soldiers killed in Korea. J.A.M.A. *158:* 912 (1955).
6. GALLAI-HATCHARD, J. J. and THOMPSON, R. H. C.: Phospholipase A activity of mammalian tissues. Biochim. Biophys. Acta. *98:* 128–136 (1965).
7. HOLMAN, R. L.: Atherosclerosis – a pediatric nutrition problem? J. Clin. Nutrit. *9:* 565–569 (1961).
8. KABACOFF, B. L.; WOHLMAN, A., UMBEY, M. and AVAKIAN, S.: Absorption of chymotrypsin from the intestinal tract. Nature *199:* 815 (1963).
9. MANN, G. V.: Nutrition education – U.S.A. Food & Nutrit. News. *41:* 1–4 (1969).
10. MELDRUM, B. S.: The actions of snake venoms on nerve and muscle. The pharmacology of phospholipase A and of polypeptide toxins. Parmacol. Rev. *17:* 393–445 (1965).
11. MILLER, B.; ANDERSON, C. E. and PIANTADOSI, C.: Plasmalogen and glycerol ether concentrations in normal and atherosclerotic tissue. J. Geront. *19:* 430–434 (1964).
12. MODI, V. V.; OWEN, E. C. and PROUDFOOT, R.: Species differences in the occurrence of xanthine oxidase in milk. J. Dairy Sci. *49:* 907 (1959)

13. OSTER, K. A.: On the tissue aldehyde shift in the rat kidney. Endocrinology *36:* 92–97 (1945).

14. OSTER, K. A.: Demonstration of a cyclic change in the histochemistry of the kidney synchronized with the estrous cycle in the rat. Endocrinology *37:* 69–76 (1945).

15. OSTER, K. A.: Tissue aldehydes, a review. Exper. Med. & Surg. *5:* 219–235 (1947).

16. OSTER, K. A.: Treatment of angina pectoris according to a new theory of its origin. Cardiology Digest *3:* 19–34 (1968).

16a. OSTER, K. A.: Plasmalogen diseases: a new concept of the etiology of the atherosclerotic process. Amer. J. Clin. Res. *2:* April, 1971.

17. OSTER, K. A. and HOPE-ROSS, P.: Plasmal reaction in a case of recent myocardial infarction. Amer. J. Cardiol. *17:* 83–85 (1966).

18. OSTER, K. A. and MULINOS, M. G.: Tissue aldehydes and their reaction with amines. J. Pharmacol. & Exper. Therap. *80:* 132–138 (1944).

19. OSTER, K. A. and OSTER, J. G.: Reaction of tissue aldehydes with certain drugs. J. Pharmacol. & Exper. Therap. *85:* 332–335 (1945).

20. OSTER, K. A. and OSTER, J. G.: The specificity of sex hormones on the tissue aldehyde shift in the rat kidney and of fuchsin sulfurous acid reagent on aldehydes. J. Pharmacol. & Exper. Therap. *87:* 306–312 (1946).

21. OSTER, K. A. and PELLETT, C. A.: Interaction of drugs and tissue aldehydes: the histochemical effects of carbonyl-binding substances on four species of laboratory animals. Exper. Med. Surg. *13:* 385–393 (1955).

22. RAPPORT, M. M. and FRANZL, R. E.: The structure of plasmalogen 1. Hydrolysis of phosphatidal choline by lecithinase A. J. Biol. Chem. *225:* 851–857 (1957).

23. REIMANN, H. A.; IMBRIGLIA, J. E. and DUCANES, T.: Crystal-induced myocarditis and pericarditis. Amer. J. Cardiol. *17:* 269–272 (1966).

24. ROBERTS, J. C.; MOSES, C. and WILKINS, R. H.: Autopsy studies in atherosclerosis I. Distribution and severity of atherosclerosis in patients dying without morphology evidence of atherosclerotic catastrophe. Circulation *XX:* 511–519 (1959).

25. ROSS, D.; MALTS, R. and OSTER, K. A.: Xanthine oxidase content of various milk products. (in preparation).

26. ROSS, D.; PTASZYNSKI, M. and OSTER, K. A.: Unpublished data.

27. SCHARDINGER, F.: Über das Verhalten der Kuhmilch gegen Methylen Blau und seine Verwendung zur Unterscheidung von ungekochter und gekochter Milch. Z. Untersuch. Nahr. Genussm. *5:* 1113 (1902).

28. SHAPER, A. G. and JONES, K. W.: Serum cholesterol in camel-herding nomads. Lancet *:* 1305–1307 (1962).

29. TOBIAS, J. M.: A chemically specified molecular mechanism underlying excitation in nerve: a hypothesis. Nature *203:* 13–17 (1964).

30. VARLEY, A. B.: The generic inequivalence of drugs. J.A.M.A. *206:* 1745–1748 (1968).

31. WARNER, H. R. and LANDS, W. E. M.: The metabolism of plasmalogen: Enzymatic hydrolysis of the vinyl ether. J. Biol. Chem. *238:* 2404–2409 (1961).

Mailing address: Kurt A. Oster, M.D., 881 Lafayette Boulevard, Bridgeport, Connecticut 06603 (USA).

Department of Experimental Pathology, School of Medicine,
University of Komensky, Bratislava, Czechoslovakia

PARTICIPATION OF CATECHOLAMINES AND CORTICAL HORMONES IN THE MYOCARDIAL METABOLIC ADAPTATION TO MECHANICAL OVERLOAD OF THE HEART

A. FIZEL and A. FIZELOVA

In previous studies (11, 12), we demonstrated that cardiac overload following aortic insufficiency led to cardiac hypertrophy and to an increase in nucleic acid and protein synthesis. The onset and development of cardiac hypertrophy is accompanied by an augmentation of sympathoadrenal and adrenocortical activity (11). The catecholamine and corticosterone plasma levels are increased, while their concentrations in the adrenals and myocardium are decreased (fig. 1). Since the catecholamines and cortical hormones modify the mechanical activity (7) and metabolic processes of the heart muscle (1, 3, 9, 10, 13, 14, 17–19, 23), we examined the question whether adrenal hormones take part in the development of cardiac hypertrophy and in the associated changes of proteins and nucleic acids.

MATERIALS AND METHODS

In the first part of the experiments, the influence of bilateral adrenalectomy on the onset and development of cardiac hypertrophy following aortic insufficiency

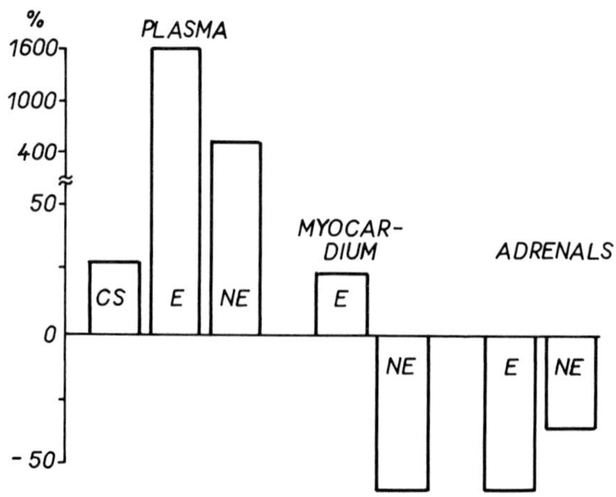

Fig. 1 Changes in the catecholamine and corticosterone concentrations (1% of controls) on the 7th day after cardiac overload following aortic insufficiency. CS: corticosterone; E: epinephrine; NE: norepinephrine.

was examined. The series was performed on 40 rabbits, which were divided into 4 equal groups. Group I served as controls. In group II, bilateral adrenalectomy was performed through a midline abdominal incision. Each animal received hydrocortisone intravenously (15 mg per kg b.w.) prior to and again upon completion of the adrenalectomy. For the next postoperative days, no hydrocortisone was administered and sodium chloride was not added to the drinking water. In group III, aortic insufficiency was performed from the right carotid artery by the method described elsewhere (11). In group IV, aortic insufficiency and bilateral adrenalectomy were performed simultaneously. The size of the heart on X-ray was examined in all animals, daily. The rabbits of group II were killed between the 7th and 12th day, those of groups III and IV between the 15th and 20th day after the interventions. After recording of the heart weight, the nucleic acid and protein fractions were examined in the myocardium of the left ventricle by the methods described in detail elsewhere (12).

In the second part of experiments, changes in the size of the heart on X-ray, in the heart weight, and in nucleic acids and proteins were examined in intact animals, following administration of hormones for 6 days. Group I: untreated controls; group II: adrenaline (350 μg per kg b.w., subcutaneously, twice daily); group III: hydrocortisone (20 mg per kg b.w., intravenously, once daily); group IV: ACTH (4 international units per kg b.w., intramuscularly, once daily); group V: hydrocortisone + ACTH (hydrocortisone 20 mg per kg b.w. and ACTH 4 international units per kg b.w., intramuscularly, once daily). Each group contained 10 rabbits.

RESULTS

Influence of bilateral adrenalectomy

Changes in the size of the heart on X-ray: Following bilateral adrenalectomy (group II), the size of the heart on X-ray gradually diminished, maximum diminution being recorded between the 7th and 12th day (figs. 2 and 3). The animals died 1 to 2 days after the maximum diminution of the heart. Aortic insufficiency led to the opposite changes, in that the size of the heart on X-ray gradually increased (figs. 4 and 5). After aortic insufficiency + bilateral adrenalectomy (group IV), the size of the heart increased initially (figs. 6 and 7); but it began to diminish 2 – 3 days after the operation and, between 15–20 days, was less than before the experiment (fig. 8)

Changes in heart weight and nucleic acids: After bilateral adrenalectomy, the heart weight decreased by 14.6% and the RNA concentration by 19.6%. Aortic insufficiency led to the opposite changes: the heart weight increased by 54.1% and the RNA concentration by 35.2%. Bilateral adrenalectomy prevented the increase in heart weight and in the RNA concentration resulting from aortic

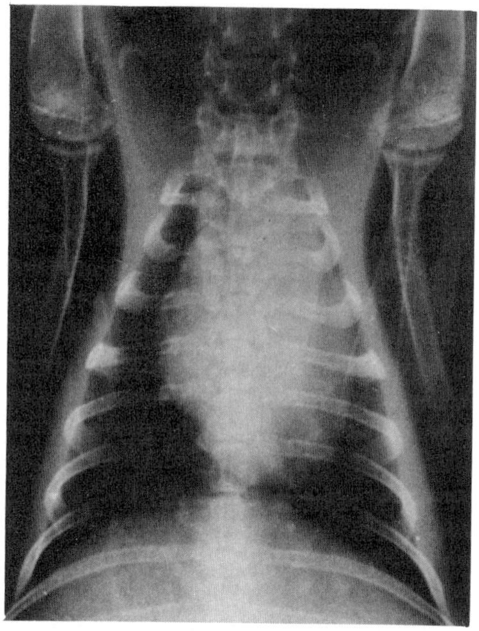

Fig. 2 Heart shadow before experiment.

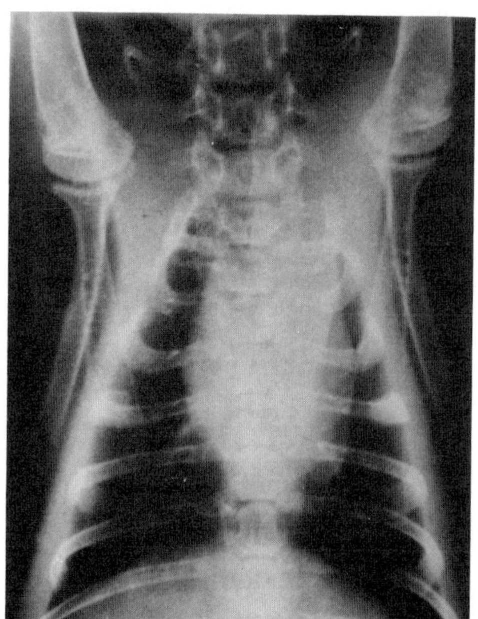

Fig. 3 Diminution of heart shadow on the 8th day after bilateral adrenalectomy.

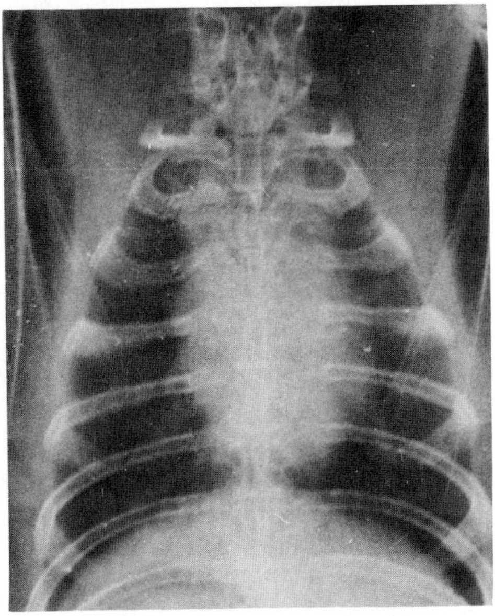

Fig. 4 Heart shadow before experiment.

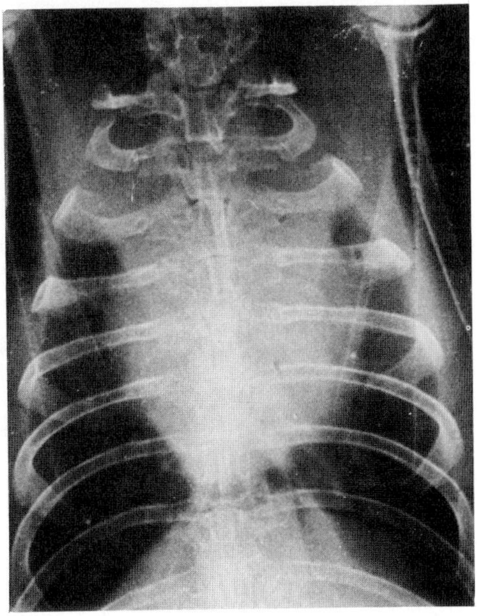

Fig. 5 Enlargement of heart shadow on the 18th day after induction of aortic insufficiency.

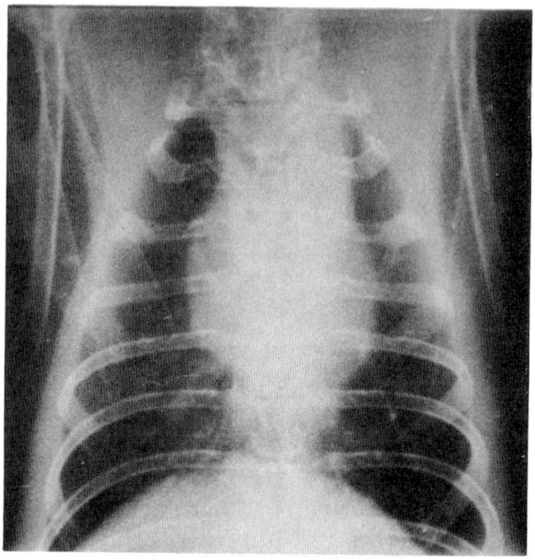

Fig. 6 Heart shadow before experiment.

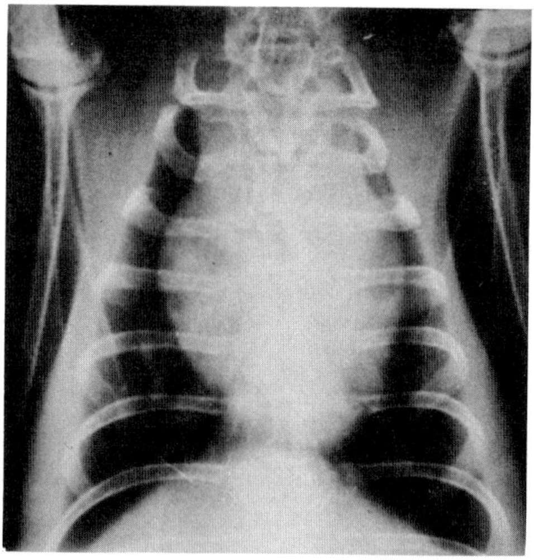

Fig. 7 Enlargement of heart shadow on the 2nd day after induction of aortic insufficiency + bilateral adrenalectomy.

insufficiency (group IV). No significant changes were observed in the DNA concentration of any of the experimental animals (table I).

Changes in the concentration and the total content of proteins: The concentration (mg protein/g wet weight) of mitochondrial proteins decreased by 30.2% after bilateral adrenalectomy, but that of total and sarcoplasmic protein

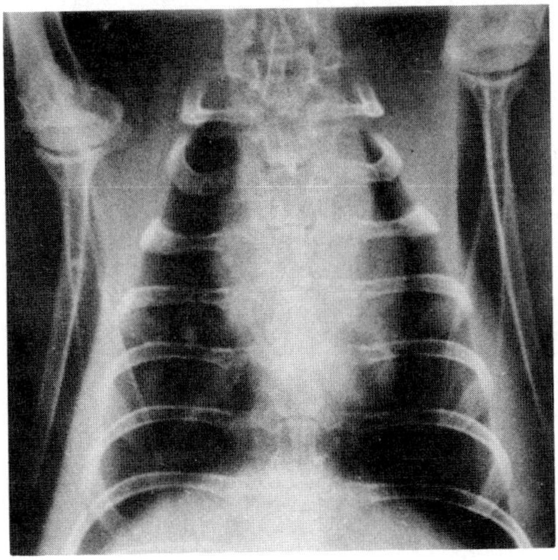

Fig. 8 Heart shadow on the 17th day after induction of aortic insufficiency + bilateral adrenalectomy.

as well as of actomyosin did not change significantly. After aortic insufficiency, the concentrations of total protein and actomyosin were decreased by 7.2% and 22.7%, respectively, while the concentration of mitochondrial proteins was, in contrast, increased by 65.1%. The concentration of sarcoplasmic proteins was not changed. Following aortic insufficiency + bilateral adrenalectomy, no changes in the protein concentrations were observed. The collagen concentration did not change in the experimental animals (table II).

The content (mg protein/myocardium of the left ventricle) of total protein and its fractions were reduced after bilateral adrenalectomy. On the other hand, after aortic insufficiency, the content of all protein fractions (except actomy-

Table I

Changes in the Heart Weight and Nucleic Acids

Group	Heart wt	RNA	DNA
	g/kg b.w.	*mg % P*	
I. Control	1.988 ± 0.212	18.94 ± 2.28	10.28 ± 2.04
II. Bilateral adrenalectomy	1.691 ± 0.120[b]	15.20 ± 2.07[b]	9.15 ± 2.14
III. Aortic insufficiency	3.020 ± 0.326[a]	25.50 ± 2.78[a]	8.78 ± 1.98
IV. Aortic insufficiency + bilateral adrenalectomy	1.880 ± 0.254	17.84 ± 2.14	9.92 ± 1.78

All values expressed as mean ± SD.
[a] $p < 0.001$.
[b] $p < 0.05$.

Table II

Changes in the Concentration of Protein Fractions (mg/g wet weight)

Group	Total protein	Sarcoplasmic proteins	Mitochondrial proteins	Actomyosin	Collagen
I. Control	143.0 ± 12.8	55.0 ± 5.8	4.3 ± 0.9	42.0 ± 4.4	20.4 ± 3.2
II. Bilateral adrenalectomy	139.5 ± 14.2	51.8 ± 4.7	3.0 ± 0.7[b]	38.5 ± 4.0	19.2 ± 2.8
III. Aortic insufficiency	131.2 ± 9.4[c]	53.2 ± 6.4	7.1 ± 1.4[a]	32.4 ± 5.4[b]	17.9 ± 2.2
IV. Aortic insufficiency + bilateral adrenalectomy	143.5 ± 14.4	54.3 ± 6.1	4.9 ± 1.1	40.8 ± 3.8	21.2 ± 3.0

All values expressed as mean ± SD.
[a] $p < 0.001$.
[b] $p < 0.01$.
[c] $p < 0.05$.

osin) was increased, predominantly the content of mitochondrial proteins (138.5%). Bilateral adrenalectomy prevented the changes in the content of proteins resulting from aortic insufficiency (table III).

Administration of hormones to intact animals

Changes in the size of the heart on X-ray: After administration of adrenaline (group II), an increase in the size of the heart on X-ray was observed (figs. 9 and 10). The administration of hydrocortisone + ACTH (group V) led to similar changes. After administration of hydrocortisone (group III), the size of the heart on X-ray gradually diminished, in a manner comparable to that seen after bilateral adrenalectomy (figs. 2 and 3). Administration of ACTH (group IV) did not give rise to any change in the size of the heart.

Changes in heart weight and nucleic acids: Administration of adrenaline led to an increase of 37.1% in the heart weight and of 42.2% in the RNA concentration. Similar changes were observed after administration of hydrocortisone + ACTH. Following hydrocortisone or ACTH alone (groups III and IV), no significant changes were recorded in the heart weight or nucleic acid concentrations (table IV).

Changes in concentration and total content of the proteins: Following administration of adrenaline, the concentrations of total protein and of actomyosin decreased by 7.6% and by 26.9%, respectively, while the mitochondrial proteins and collagen increased by 49.7% and 37.9%. The concentration of sarcoplasmic proteins did not change significantly, and the water concentration increased by 1.4%. After administration of hydrocortisone + ACTH, there was a 31.2% increase in the mitochondrial protein concentrations, but the concentration of the other protein fractions or the water did not change (table V).

The total content of protein fractions, except actomyosin, increased after adrenaline administration. Following hydrocortisone + ACTH, the content of all protein fractions increased (table VI). The concentration as well as the content of protein fractions did not change significantly after administration of hydrocortisone or ACTH alone (tables V and VI).

DISCUSSION

The élimination of cortical and medullary hormones by bilateral adrenalectomy prevented the rise and development of cardiac hypertrophy and the associated changes in nucleic acids and proteins. On the other hand, the administration of adrenaline as well as of hydrocortisone + ACTH led, in the intact animals, to changes in the heart weight, nucleic acids, and proteins similar to those observed after mechanical overload of the heart following aortic insufficiency, when the catecholamine and corticosterone levels were increased in the plasma and decreased in the adrenals.

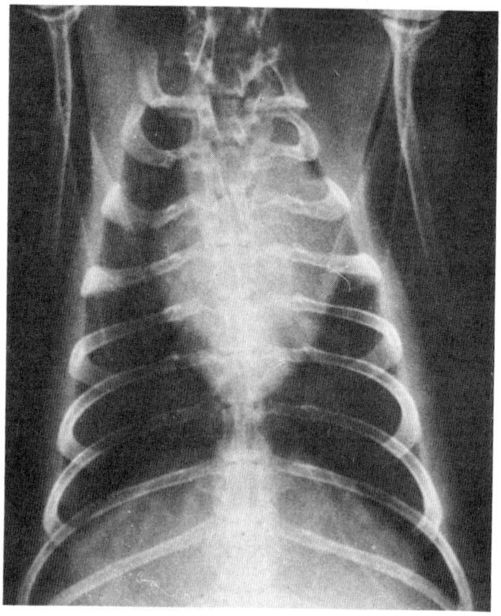

Fig. 9 Heart shadow before experiment.

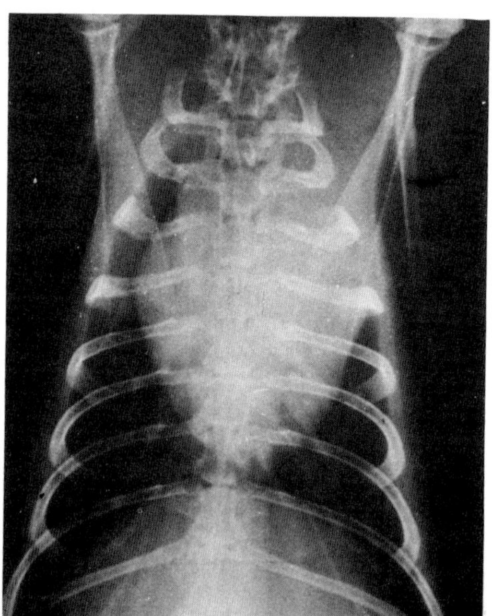

Fig. 10 Enlargement of heart shadow on the 5th day after adrenaline administration.

Table III

Changes in the Total Content of Protein Fractions (mg/myocardium of left ventricle)

Group	Total protein	Sarcoplasmic proteins	Mitochondrial proteins	Actomyosin	Collagen	Weight of left ventricle (g)
I. Control	320.2 ± 30.2	123.1 ± 17.4	9.6 ± 0.9	94.2 ± 4.4	45.7 ± 8.2	2.24 ± 0.19
II. Bilateral adrenalectomy	240.1 ± 44.2[a]	89.2 ± 9.9[b]	5.2 ± 1.1[a]	66.3 ± 8.2[a]	33.1 ± 4.4[b]	1.72 ± 0.36[b]
III. Aortic insufficiency	423.8 ± 79.5[a]	171.5 ± 24.4[a]	22.9 ± 3.6[a]	104.2 ± 11.8[c]	57.6 ± 10.1[c]	3.22 ± 0.42[a]
IV. Aortic insufficiency + bilateral adrenalectomy	298.4 ± 32.8	113.2 ± 19.5	10.2 ± 1.3	85.2 ± 10.9	44.2 ± 8.4	2.08 ± 0.25

All values expressed as mean ± SD.
[a] $p < 0.001$.
[b] $p < 0.01$.
[c] $p < 0.05$.

Our results are in an agreement with the findings of other authors. Rather (20) observed that adrenalectomy interferes to some extent with the hypertrophy of the heart seen in rats with experimental hypertension. It has been reported that administration of DCA to unilaterally nephrectomized rats (22) and to hypophysectomized animals (15, 16), and of ACTH to trained animals (16) elicits cardiac hypertrophy. However, Beznak (4–6) found that adrenalectomy did not abolish the cardiac hypertrophy following aortic constriction; furthermore, cortical extracts and ACTH failed to influence the cardiac atrophy following hypophysectomy and the hypertrophic response to aortic constriction. Beznak (4) therefore concluded that these latter phenomena are not mediated by a secondary adrenal hypofunction due to hypophysectomy, but by an absence of growth hormone, *per se*. Metzler's observations (16) differ from those of Beznak (4), in that they showed that hypophysectomy was not followed by atrophy of the heart and that DCA causes cardiac hypertrophy in hypophysectomized rats.

From the above-mentioned experiments, the conclusion seems to be warranted that the cortical and medullary hormones of the adrenals play an important role in the onset and development of cardiac hypertrophy and in the metabolic adaptation of the heart to mechanical overload. However, it is not possible to say whether these hormones influence the process of metabolic adaptation of the heart directly or indirectly. Since these hormones increase the blood pressure, the blood volume, and the circulation rate (2, 8, 24), and thus alter the cardiac load, they could, in this way, influence the process of cardiac hypertrophy. If cardiac overload is a stimulus to cardiac hypertrophy (21), adrenalectomy could prevent the cardiac hypertrophy following aortic insufficiency by a reduction of blood volume and the circulation rate and thus eliminate the stimulus to cardiac hypertophy. On the other hand, catecholamines and cortical hormones also directly affect the myocardium and modify the metabolic processes taking place in it (1, 3, 9, 10, 13, 14, 17–19, 23); in addition, they

Table IV

Changes in the Heart Weight and Nucleic Acids

Group	Heart wt.	RNA	DNA
	g/kg b.w.	*mg % P*	
I. Control	2.040 ± 0.212	19.20 ± 2.35	10.75 ± 1.95
II. Adrenaline	2.820 ± 0.302[a]	27.31 ± 4.16[a]	10.62 ± 1.77
III. Hydrocortisone	1.902 ± 0.203	17.22 ± 2.40	9.12 ± 1.42
IV. ACTH	2.051 ± 0.196	20.33 ± 2.52	11.32 ± 1.70
V. Hydrocortisone + ACTH	2.614 ± 0.294[a]	25.82 ± 3.62[b]	11.20 ± 2.08

All values expressed as mean ± SD.
[a] $p < 0.001$.
[b] $p < 0.01$.

Table V

Changes in the Concentration of Protein Fractions (mg/g wet weight)

Group	Total protein	Sarcoplasmic proteins	Mitochondrial proteins	Actomyosin	Collagen
I. Control	142.5 ± 12.2	54.8 ± 6.2	4.2 ± 1.0	42.4 ± 4.8	19.5 ± 3.1
II. Adrenaline	131.8 ± 7.5[c]	52.0 ± 4.7	6.3 ± 1.1[b]	31.0 ± 2.1[a]	26.9 ± 4.1[b]
III. Hydrocortisone	145.2 ± 10.3	54.2 ± 4.9	4.4 ± 0.9	44.7 ± 5.2	17.9 ± 3.2
IV. ACTH	140.9 ± 9.7	51.2 ± 5.2	4.9 ± 0.9	43.3 ± 4.7	17.7 ± 3.9
V. Hydrocortisone + ACTH	144.0 ± 10.8	55.2 ± 6.4	5.5 ± 0.8[c]	45.1 ± 4.6	18.9 ± 3.6

All values expressed as mean ± SD.
[a] $p < 0.001$.
[b] $p < 0.01$.
[c] $p < 0.05$.

Table VI

Changes in the Total Content of Protein Fractions (mg/myocardium of left ventricle)

Group	Total protein	Sarcoplasmic proteins	Mitochondrial proteins	Actomyosin	Collagen	Weight of left ventricle (g)
I. Control	316.9 ± 36.4	121.4 ± 16.8	9.3 ± 1.9	94.0 ± 15.1	43.4 ± 10.2	2.22 ± 0.18
II. Adrenaline	399.6 ± 43.8[a]	157.2 ± 26.9[a]	19.2 ± 3.6[a]	93.6 ± 17.4	81.4 ± 19.6[a]	3.02 ± 0.44[a]
III. Hydrocortisone	285.2 ± 36.0	105.5 ± 18.9	8.6 ± 1.2	87.4 ± 9.2	34.9 ± 7.2	1.95 ± 0.31
IV. ACTH	306.2 ± 46.2	111.2 ± 12.4	10.6 ± 2.0	99.8 ± 10.2	38.3 ± 8.3	2.17 ± 0.24
V. Hydrocortisone + ACTH	425.6 ± 74.6[a]	164.2 ± 18.3[a]	16.3 ± 2.9[a]	135.5 ± 25.3[b]	56.2 ± 14.8	2.97 ± 0.40[a]

All values expressed as mean ± SD.
[a] $p < 0.001$.
[b] $p < 0.01$.

could directly influence the process of metabolic adaptation of the heart. This question still remains unanswered, but it is hoped that further studies will serve to elucidate the problem.

SUMMARY

Mechanical overload of the heart, following aortic insufficiency, has been shown to lead to an increase in the heart weight and in nucleic acid and protein synthesis. Bilateral adrenalectomy prevented the rise and development of cardiac hypertrophy so induced, together with the associated changes in the nucleic acids and proteins. On the other hand, following administration of adrenaline as well as of hydrocortisone + ACTH, to intact animals, the usual increase in the heart weight, and nucleic acid and protein synthesis was observed. In intact animals receiving hydrocortisone or ACTH alone, no significant changes were observed in the heart weight or in the nucleic acid and protein synthesis.

REFERENCES

1. BAJUSZ, E. and SELYE, H.: Adaptation to the cardiac necrosis-eliciting effect of stress. Amer. J. Physiol. *199:* 453–459 (1960).
2. BEST, C. H. and TAYLOR, N. B.: The Physiological Basis of Medical Practice (Williams & Wilkins, Baltimore 1955).
3. BEVIZ, A.; MOHME–LUNDHOLM, E. and SVEDMYR, N.: The effect of adrenaline on the carbohydrate metabolism in striated muscle. Acta Physiol. Scand. *69:* 213–217 (1967).
4. BEZNAK, M.: The effect of the adrenals and the pituitary on blood pressure and cardiac hypertrophy of rats. J. Physiol *116:* 219–228 (1952).
5. BEZNAK, M.: The behaviour of the weight of the heart and the blood pressure of albino rats under different conditions. J. Physiol. *124:* 44–63 (1954).
6. BEZNAK, M.: The effect of adrenocortical hormones alone and in combination with growth hormone on cardiac hypertrophy and blood pressure of hypophysectomized rats. J. Physiol. *124:* 75–83 (1954).
7. BRAUNWALD, E.: The control of ventricular function in man. Brit. Heart J. *27:* 1–8 (1965).
8. CHAMBERS, R. and ZWEIFACH, B. W.: Intercelullar cement and capillary permeability. Physiol. Revs. *27:* 436–449 (1947).
9. FIZEL, A. und FIZELOVA, A.: Studium einiger metabolischer Prozesse des versagenden Herzens im Experiment. X. Z. Kreislaufforschg. *55:* 614–627 (1966).
10. FIZEL, A. und FIZELOVA, A.: Studium einiger metabolischer Prozesse des versagenden Herzens im Experiment. XI. Z. Kreislaufforschg. *55:* 804–818 (1966).
11. FIZEL, A. and FIZELOVA, A.: Changes of the sympathoadrenal and adrenocortical activity in the course of cardiac hypertrophy and heart failure. Z. Kreislaufforschg. *58:* 1115–1128 (1969).
12. FIZELOVA, A, and FIZEL, A.: Myocardial metabolic changes in cardiac hypertrophy and heart failure. This Book, pp. 200–212.
13. DEGRANDPRE, R. and RAAB, W.: Interrelated hormonal factors in cardiac hypertrophy. Circulat. Res. *1:* 345–353 (1953).
14. MARCHETTI, G.; MERLO, L. and NOSEDA, V.: Myocardial uptake of free fatty acids and carbohydrates after beta adrenergic blockade. Amer. J. Cardiol. *22:* 370–374 (1968).
15. METZLER, A.: Ueber das Zustandekommen der Herz-und Nebennierenvergrosserung und den Einfluss der Nebenniere auf die Herzgrosse. Arztl. Forsch. *6:* 424–433 (1952).
16. METZLER, A.: Ueber die hormonale Ursache der Herzhypertrophie. Arch. Kreislaufforsch. *18:* 359–368 (1952).

17. RAAB, W.: The adrenergic-cholinergic control of cardiac metabolism and function. Adv. Cardiol. *1:* 65–152 (1956).
18. RAAB, W.: Myocardial electrolyte derangement: crucial feature of pluricausal, so-called coronary, heart disease. Ann. N.Y. Acad. Sci. *147:* 627–686 (1969).
19. RAAB, W.; KIMURA, H.; BAJUSZ, E. and HERRLICH, H. C.: Isolation stress, myocardial electrolytes and epinephrine cardiotoxicity in rats. Proc. Soc. Exp. Biol. Med. *127:* 142–151 (1968).
20. RATHER, L. J.: Experimental cardiac hypertrophy: rate of development and effect of adrenalectomy. Amer. J. Physiol. *159:* 153–161 (1949).
21. SCHREIBER, S. S.; ORATZ, M. and ROTHSCHILD, M. A.: Effect of acute overload on protein synthesis in cardiac muscle microsomes. Amer. J. Physiol. *213:* 1552–1555 (1967).
22. SELYE, H. and HALL, C. E.: Production of nephrosclerosis and cardiac hypertrophy in the rat by DCA overdosage. Amer. Heart J. *27:* 338–347 (1944).
23. SWINGLE, W. W.; DAVANZO, J. P.; CROSSFIELD, H. C.; GLENISTER, D.; OSBORNE, M. and WAGLE, G.: Effect of various adrenal steroids on internal fluid and electrolyte shifts of fasted adrenalectomized dogs. Proc. Soc. Exp. Biol. Med. *99:* 75–78 (1958).
24. WYMAN, L. C.; FULTON, G. P. and SCHULMAN, M. H.: Direct observations on the circulation in the hamster cheek pouch in adrenal insufficiency and experimental hypercorticalism. Ann. N.Y. Acad. Sci. *56:* 643–658 (1953).

Mailing address: A. Fizel, M.D., Department of Experimental Pathology, School of Medicine, Sasinkova 4, Bratislava (Czechoslovakia).

Subject Index

N

O

P

R

S

T

V